Deep Centers
in
Semiconductors

DEEP CENTERS
IN SEMICONDUCTORS

A State of the Art
Approach

Edited by

Sokrates T. Pantelides

IBM Thomas J. Watson Research Center
Yorktown Heights, NY

GORDON AND BREACH SCIENCE PUBLISHERS
New York   London   Paris   Montreux   Tokyo

PHYSICS

Gordon and Breach Science Publishers

P. O. Box 786
Cooper Station
New York, NY 10276
United States of America

P.O. Box 197
London WC2E 9PX
England

58, rue Lhomond
75005 Paris
France

P. O. Box 161
1820 Montreux 2
Switzerland

14-9 Okubo 3-chome,
Shinjuku-ku,
Tokyo 160
Japan

Library of Congress Cataloging in Publication Data
Main entry under title:

Deep centers in semiconductors.

1. Semiconductors—Defects—Addresses, essays,
lectures. 2. Electron donor-acceptor complexes—
Addresses, essays, lectures. 3. Impurity centers—
Addresses, essays, lectures. I. Pantelides, Sokrates T.
QC611.6.D4D418 1985   537.6'22   85-5408

This book is dedicated to the memory of Paul J. Dean, whose untimely death brought an end to a distinguished career in the field of defects and impurities in semiconductors.

"When you set out on the road to Ithaca
pray that the journey is long,
full of adventure, full of knowledge."

C. P. Cavafy

Contents

Preface   ix

Chapter 1   Perspectives in the Past, Present, and Future of Deep Centers in Semiconductors
S. T. Pantelides   1

Chapter 2   Chalcogens in Silicon
H. G. Grimmeiss and E. Janzen   87

Chapter 3   The Lattice Vacancy in Silicon
G. D. Watkins   147

Chapter 4   Oxygen and Oxygen Associates in Gallium Phosphide and Related Semiconductors
P. J. Dean   185

Chapter 5   The Two Dominant Recombination Centers in n-type Gallium Phosphide
A. R. Peaker and B. Hamilton   349

Chapter 6   The Mid-Gap Donor Level EL2 in Gallium Arsenide
G. M. Martin and S. Makram-Ebeid   399

Chapter 7   DX Centers in III-V Alloys
D. V. Lang   489

Chapter 8   Iron Impurity Centers in III-V Semiconductors
S. G. Bishop   541

Chapter 9   Chromium in Gallium Arsenide
J. W. Allen   627

Chapter 10   Chromium in II-VI Compounds
J. Baranowski   691

Chapter 11   Copper in Zinc-cation II-VI Compound Semiconductors
D. J. Robbins, P. J. Dean, P. E. Simmonds, and H. Tews   717

Preface

Semiconductors are the backbone of the electronic industry because their properties can be manipulated over wide ranges through the control of impurities and other imperfections. *Shallow impurities* introduce minor perturbations in the crystal (as manifested by the fact that they give rise to bound states in the fundamental band gap very close to the band edges) and generally contribute extra charge carriers, electrons or holes. Their role, therefore, is primarily to control the type and magnitude of conductivity.

Other impurities and a variety of lattice defects (vacancies, antisite defects, self-interstitials, etc.) constitute a more severe local perturbation, give rise to bound states that are considerably more localized, and often have energies deep in the band gap. We refer to all such impurities, lattice defects, and impurity-defect complexes as *deep centers*. Unlike shallow impurities, deep centers act primarily as carrier traps or recombination centers. Thus, deep centers control the lifetime of charge carriers. As such, they are undesirable in devices where carriers must have long lifetimes, e.g., solar cells. On the other hand, they are useful when the carrier concentration needs to be reduced sharply on a short time scale, as in a fast switch. In addition, when the recombination or carrier-capture energy is released as light, deep centers are used in making light-emitting diodes (LEDs). Finally, many deep centers (vacancies, interstitials, vacancy-impurity complexes, etc.) play a major role in diffusion processes and in reactions that underlie materials modification (e.g. oxidation, recrystallization, etc.)

In addition to shallow impurities and deep centers, a semiconductor may contain *extended defects,* such as dislocations, grain boundaries, stacking faults, or precipitates. Though these defects sometimes affect the electrical properties of the crystal in the same way as deep centers, they generally have a more significant effect on the more macroscopic properties, such as tensile strength and elasticity. In general, in applications, dislocation-free single crystals are desirable, but dislocations and precipitates have been shown to be efficient in gettering impurities from other areas of a crystal (e.g., the active region of a wafer). Thus, all types of imperfections have both useful and detrimental features. It is the expert control of the various types of defects that allows one to manipulate the properties of semiconductors, which form the active part of most electronic devices today.

ix

The microscopic properties and the role of shallow impurities were already quite well understood by the end of the 1950s through a combination of theory (effective-mass theory) and experiments (primarily optical absorption) in silicon and germanium. Deep centers and extended defects, on the other hand, proved far more difficult to investigate. Over the last 25 years, great progress has been achieved. Many experimental and theoretical techniques were developed and applied to particular systems. The choice of systems was dictated sometimes by technological concerns, sometimes by the idiosyncracies of the technique, and sometimes by sheer academic curiosity. In many cases, as new techniques were developed, they were applied to systems that had been extensively studied with earlier techniques. The new information would either confirm, complement, and expand existing knowledge about the center, or contradict it and thus lead to new understanding. Quite often, the desire to understand some of the observed properties of a particular center led to the development of techniques tailor-made for the purpose. Throughout these years, the interplay between theory and experiment has been very strong, even though for many years only simple semiempirical theories were possible.

The field of deep centers is, thus, quite diverse. Many excellent review papers and books have been devoted to the subject over the years, but inevitably they had to focus attention only on a few selected aspects, usually reflecting the authors' specialties and expertise. The idea for the book at hand was to try to accomplish something totally different, cutting across areas of specialty. A number of deep centers that have been extensively studied over the years by a large variety of techniques were chosen. In each case, one of the scientists who played a major role in the developments surrounding the particular center was invited to contribute an article detailing those developments and our current knowledge of that center. Thus, as an alternative to an article or book reviewing, say, experimental or theoretical techniques or a particular topic such as bound excitons or nonradiative transitions, the present book provides a vivid account of the interplay and importance of the various techniques and topics as they actually occur in real systems. The response to the invitations was very positive. Everyone who was invited agreed to contribute. Even though a few of those invited were in the end unable to contribute a paper, despite several deadline extensions, the collection of articles in this volume covers a large variety of topics, techniques, and concepts, and represents a comprehensive picture of the major developments in the field of deep centers.

In addition to the articles containing critical reviews of individual deep centers, the first article attempts to tie the major themes together and draw perspectives in the past and future of the field. Though references to the other articles in this volume are frequently made, this article is not an annotated guide to the rest of the book. Instead, it provides a broad perspective that complements the detailed analyses of individual deep centers contained in the other articles.

Sokrates T. Pantelides

CHAPTER 1

Perspectives in the Past, Present, and Future of Deep Centers in Semiconductors

Sokrates T. Pantelides

*IBM Thomas J. Watson Research Center,
Yorktown Heights, NY 10598*

1. INTRODUCTION
2. THE PAST
  2.1 Experimental Techniques
     2.1.1 The 1960's: Bulk Techniques
     2.1.2 The 1970's: Junction Techniques
     2.1.3 The 1980's: Hybrid Techniques
  2.2 Theoretical Techniques
     2.2.1 The 1950's and 1960's: Formalisms
     2.2.2 The 1970's: Cluster Calculations
     2.2.3 The 1980's: Green's Functions
  2.3 Summary and Discussion of Techniques
  2.4 Objectives and Achievements
     2.4.1 The 1960's and early 1970's: Isolation
        A) Radiation-Induced Defects
        B) Transition-Metal Impurities
        C) Other Impurities
        D) Atomic Diffusion
     2.4.2 The Late 1970's and the 1980's: Seeking the Total Picture
3. MAJOR RECENT DEVELOPMENTS
  3.1 Electronic Structure

    3.1.1   The Single-Particle Picture
            A) Vacancies and sp-Bonded Impurities
            B) Transition-Metal Impurities
    3.1.2   Many-Body Effects
    3.1.3   Excited States
    3.1.4   The Accuracy of Theoretical Calculations
            A) The Choice of Hamiltonian
            B) The Choice of Numerical Technique
  3.2 Local Vibrational Modes
  3.3 Identification of Deep Centers
  3.4 Transition-Metal Impurities
  3.5 Lattice Relaxation
    3.5.1   Causes of Lattice Relaxation
    3.5.2   Consequences of Lattice Relaxation—Negative-U Centers
    3.5.3   Large Lattice Relaxation—Self-Trapping and Persistent
            Photoconductivity
  3.6 Carrier Capture and Recombination
    3.6.1   Mechanisms of Carrier Capture
    3.6.2   Consequences of Carrier Capture and Recombination:
            Enhanced or Athermal Migration and Defect Reactions
            A) Carrier Capture: Charge-State Effects
            B) Recombination
  3.7 High-Temperature Self-Diffusion
4. THE FUTURE

1. Introduction

The systematic investigation of deep centers in semiconductors began in earnest about 25 years ago. Over these years, major changes occurred in the techniques that have been used and in the questions that have been addressed. In this paper, we will look back and try to view the past[1-9] in a structured way and identify some unifying characteristics. In addition we will review the major developments of the last several years in an effort to identify and discuss critically the major topics of current interest. The emphasis will be on concepts and general themes, not on cataloging or tabulating information. There will be no derivations, no equations, no tables. Only general discussion and illustrative figures. Finally, we will take a look at the future and assess some of the potential areas for new developments.

2. The Past

By looking at both the techniques and the primary objectives of investigations, we identify three major periods. The first period begins around 1960 and ends around 1970. We shall, therefore, refer to it as the 1960's. The second period ends in the late 1970's, but we shall, for convenience often refer to it as the 1970's. Finally, the third period brings us to today. In this section, we will identify the major characteristics of these three periods and the major events that justify this particular demarkation of history.

2.1 Experimental Techniques

2.1.1 The 1960's: Bulk Techniques

During the 1960's, deep centers were probed by experimental techniques which exploit phenomena that occur in the *bulk*. Impurities were typically incorporated in bulk samples either during crystal growth or by high-temperature diffusion. Lattice defects were created by irradiation with high-energy electrons. In both cases, the concentrations of deep centers were usually considerably lower than those of shallow dopants.

The following experimental techniques were used most often: Hall-effect and electrical-conductivity measurements, which yield a thermal activation energy corresponding to the ionization of the deep center; optical-absorption, photoconductivity, luninescence, and Raman-scattering measurements, which probe the electronic energy-level structure of a deep center; infrared-absorption measurements which can probe the local vibrational modes; and electron-paramagnetic-resonance (EPR) measurements [also electron-nuclear double resonance (ENDOR) measurements], which probe the local atomic arrangement and the chemical identities of the atoms comprising the deep center.

All the above techniques were developed in earlier years in other contexts. For example, absorption, luminescence, Raman scattering and EPR already had a successful history in color centers in ionic crystals; Hall-effect measurements, conductivity, and absorption evolved in earlier studies of shallow impurities. Applications to deep centers in semiconductors, however, proved quite frustrating for a variety of reasons. One persistent problem was the difficulty of achieving high deep-center concentrations. Thus, signals are weak and difficult to detect, especially if they are superimposed on signals arising from shallow impurities which, typically, have higher concentrations. This problem was exacerbated by the lack of suitable photodetectors in energy regions of interest. In addition, the quality of samples was not particularly good. Stress fields, which are worse when deep centers exist in high concentrations, and unwanted defects, made matters worse. One consequence was too much broadening of signals. For the techniques that depend on excited states, e.g., luminescence, dirty samples affect the lifetime of excitations and can wipe signals out entirely.

Nevertheless, the 1960's are marked by a number of remarkable successes. The most successful technique proved to be EPR, for which, in the case of Si, everything seemed to be just right. EPR lines arising from deep centers are not, in general, superposed on lines arising from shallow impurities. Since the nuclear spin of the most abundant species of Si is zero, EPR lines from deep centers tend to be sharp. On the other hand, the concentration of Si isotopes with non-zero spin is substantial (about 5%), giving rise to detectable satellites. As a result, extensive studies of many deep centers in Si were possible (Refs. 5 and 10; see also Section 2.4.1A below). In compound semiconductors, EPR initially faced broad linewidths, arising in part from the nonzero nuclear spins of the host atoms and from a variety of other effects.[11] These problems were overcome in the 1970's when better samples became available.

From among other techniques, during the 1960's, luminescence had significant success in GaP where everything seemed to work just right (see Section 2.4.1C below). More recently, many of the shortcomings of some of the classic techniques of the 1960's have been overcome, leading to their revival (Section 2.1.3).

2.1.2 The 1970's: Junction Techniques

Beginning at about 1970, a major breakthrough occurred in experimental techniques. For the first time, new techniques were introduced which exploited the unique property of semiconductors to be doped n-type or p-type. The new techniques no longer rely on bulk samples. Instead, measurements are done on *junctions,* such as pn junctions or rectifying semiconductor-metal interfaces. In these junction regions, the Fermi level goes through the midgap region as if the material were intrinsic. Furthermore, by applying appropriate biases, one can let carriers in and out of the junction and move the Fermi level up or down. In particular, in a reversed-biased junction, the junction region is depleted of mobile electrons and holes which allows the linearization of rate equations. As a consequence, perturbations decay exponentially with time and the decay constant is independent of the concentration of deep centers. Typically, one monitors the capacitance or the current across the junction as the charge-state of deep centers in the junction is changed.

Junction techniques were spearheaded around 1970 by Sah and co-workers[12] who described and used several variants. Since then, additional variants have been developed and still newer ones continue to appear occasionally. One particular variant, introduced by Lang[13] and known as Deep Level Transient Spectroscopy (DLTS) is, in fact, a spectroscopic method and allows one to scan a temperature range and observe individual deep levels as peaks in a continuous spectrum. Junction techniques in general and DLTS in particular revolutionized the process of electrical characterization of deep centers.[14,15] For example, DLTS peaks are now routinely recognized as "signatures" of deep centers, which are very useful when one wants to study the same center with other techniques.

The main advantage of the junction techniques is their ability to decouple deep centers from the shallow dopants and focus on signals arising from deep centers. In addition, junction measurements are center specific, in the sense that they can determine the concentration, ionization/capture cross sections, and energy levels of deep centers even when several centers are present in the same sample.

2.1.3 The 1980's: Hybrid Techniques

Since the late 1970's, there has been a new wave of techniques. Some of these are refinements of old techniques triggered by the availability of new tools (e.g., the advent of tunable dye lasers revolutionized the technique of photoluminescence because one can selectively populate excited states). Many new techniques are hybrid, combining EPR with electrical or optical measurements, and thus allowing a simultaneous investigation of the electronic energy levels and the local symmetry and chemical identity. One particular example of these techniques is known as optical detection of magnetic resonance (ODMR). This technique, though not new, was only recently successfully applied to deep centers in semiconductors, beginning with the pioneering work of Cavenett and coworkers.[16]

In recent years, there has also been a revival of the classic techniques of optical absorption and luminescence in bulk samples. A number of the problems faced by these methods in the 1960's and early 1970's were eliminated: New and better detectors allow work in regions of photon energies that were difficult in earlier times, while offering increased resolution; crystal-growth techniques have improved tremendously so that extraneous defects, stress fields and other nuances can be eliminated; Fourier spectroscopy has beat, in some cases, the low-concentration problem faced by optical-absorption measurements by using interference techniques to increase throughput.

One very different technique that appears promising is EXAFS (Extended X-ray Absorption Fine Structure) which provides structural information. Its promise is based on the fact that it probes one particular type of impurity atom at a time and obtains information about the number, distance and chemical identity of its neighbors. The method has proven powerful in the case of molecules, amorphous solids, and high concentrations of shallow impurities[17] but applications to deep centers in semiconductors are hindered by low concentrations. A recent application on Cu-doped ZnSe was successful and encouraging.[18]

Finally, though not a technique in the strict sense of the word, the use of alloys has been a very powerful tool in recent years. By following the states of a given deep center through an alloy series such as $Ga_x As_{1-x} P$, one can study the effect of band-structure changes on the deep levels.[19,20] Similar studies can be made by using pressure, which also alters the band structure. Combining pressure and alloying gives one enormous flexibility and yields very detailed information about the intrinsic properties of deep centers.[19]

2.2 Theoretical Techniques

2.2.1 The 1950's and 1960's: Formalisms

Effective-mass theory[21-23] was developed in the 1950's and was very successful in describing the electronic structure of shallow impurities. This theory made a very simple assumption about the form of the perturbation potential introduced by a shallow impurity (screened Coulombic) and a set of approximations which reduced the resulting Schrödinger equation into a hydrogenic form. Both these approximations were recognized to be unsuitable for deep impurities and other deep centers, which are characterized by strong, short-range potentials. During the 1950's and 1960's, alternative formalisms were developed which were in principle suitable to solve a Schrödinger equation with a potential consisting of a periodic part and a short-range part. In many cases, the formalisms are based on a mathematical construct known as Green's function. The groundwork was laid in 1954 in a classic paper by Koster and Slater[24] and the formalism was expanded upon and further elucidated by Callaway.[25] The corresponding formalism for describing local vibrational modes induced by impurities and defects was described in detail by Maradudin.[26] Another formalism for the electronic structure of deep centers, based on the theory of integral equations with separable kernels, was developed by Bassani, Iadonisi, and Preziosi.[27] Throughout the 1950's and 1960's, these formalisms were used only for simple, square-well-type models. Thus, the question of what type of Hamiltonian is best suited for deep centers was not explored at all.

All the formalisms mentioned above focus on describing the electronic structure or local vibrational modes of deep centers. Another important topic is the interaction of bound electrons with phonons. During the 1950's the basic principles underlying non-radiative transitions via the emission of phonons was also developed. (Huang and Rhys[28], Kubo and Toyozawa[29], Lax[30]).

For practical applications, e.g. as a guide to the analysis of experimental data, theory, in the 1960's, employed only general principles, such as group theory, or simple qualitative models for energy-level structure, vibrational modes, transition rates, etc. One such model, known as the "defect molecule" successfully guided the identification of many lattice defects, especially in the pioneering work of Watkins and Corbett.[31] For example, localized states at a vacancy in Si were viewed to be simply linear combinations of the "dangling bonds" on the four nearest neighbors. One could then predict the number and symmetry of localized states which could be compared with experimental observations.

2.2.2 The 1970's: Cluster Calculations

The main theme of the 1970's was *cluster calculations*. The approach, pioneered by Messmer and Watkins,[32] attempts to simulate the infinite perfect crystal by a small number of atoms, typically 30-60. This cluster of atoms (first a "perfect" cluster, then a cluster containing a deep center) is then treated as a molecule. Initially, calculations were carried out by expanding the wave functions in terms of atomic orbitals and parametrizing the Hamiltonian matrix elements using the prescriptions of extended Hückel theory (EHT) borrowed from molecular chemistry. Later, self consistent calculations were performed using the $X\alpha$ scattered-wave method, again borrowed from molecular theory.[33]

The main advantage of cluster calculations is that, in contrast to Green's-function techniques, are easy to implement. In fact, one can simply use available programs developed for molecules. As a justification for the cluster approximation, one might argue that, by looking at the differences between the "perfect" and the "perturbed" cluster, there may be a certain cancellation of errors. In practice, however, cluster approaches have a number of distinct drawbacks:[22] a) convergence of results with cluster size is usually poor so that the results are sensitive to the cluster boundary conditions; b) it is difficult to determine the precise energy positions of bound states in the gap since the latter, being a bulk property, is not described well by a small cluster; c) it is difficult to get an adequate description of changes occurring in the band continua (resonances and antiresonances) since the continua are replaced by a set of discrete levels.

Cluster calculations using Extended-Hückel Hamiltonians were carried out for many defects in diamond and silicon during the 1970's. The results of these calculations were used extensively as a guide to interpret experimental data. In the last few years, however, it has been recognized that those Hamiltonians had some serious shortcomings, yielding results that are even qualitatively incorrect (see Section 3.1.4A). Similarly, some of the approximations used in the scattered-wave $X\alpha$ scheme, especially the use of muffin-tin spheres, have been found to be inadequate.

The 1970's were also characterized by a persistent search for practical methods that do not employ the cluster approximation. Going back to the mid-1960's, a multiple-scattering technique was implemented by Benneman,[34] but was not carried very far. The classic Green's-function formalism of Koster and Slater[24] and Callaway[25] was implemented by Callaway and Hughes,[35] but was abandoned without producing definitive and reliable results. The biggest hurdle in that work was the construction of perfect-crystal Wannier functions that were to be used as a basis set for the defect calculations. A modified version based on an empirical tight-binding Hamiltonian was used by Lannoo and Lenglart[36] but the approach was also abandoned. Effective-mass theory was generalized by Pantelides and Sah[37] who successfully calculated energy levels for some deep impurities, but the method involved severe approximations[22] and was restricted to cases where the defect potential is dominated by a Coulombic tail. A direct expansion of deep-center wave functions in terms of the perfect-crystal Bloch functions was attempted by Jaros and Ross[38] and then abandoned. The formalism of Bassani, Iadonisi and Preziosi[27] was implemented by Jaros,[39] but, for several years, the results suffered from computational problems and uncertainties in the manner by which the defect potential was constructed.[22] At about the same time, several workers[40] carried out self-consistent pseudopotential calculations using the supercell technique, i.e., using a periodic array of defects in a perfect crystal, preserving periodcity so that energy-band methods are applicable. The supercell method is equivalent to a cluster calculation with periodic boundary conditions. This choice of boundary condition automatically produces the correct band gap for the perfect "cluster," but bound-state energy levels are still difficult to position in the gap because the levels are broadened into bands by interdefect interactions. The method was abandoned after a few applications. Finally, continued-fraction techniques, developed earlier to treat amorphous solids,[41] were used to describe isolated defects in very large clusters (or order 2000 atoms) by Kauffer, Pecheur and Gerl,[42] but the approach is limited to semiempirical tight-binding Hamiltonians.

2.2.3 The 1980's: Green's Functions that Work

In the late 1970's, a breakthrough occurred. Green's-functions, which seemed dead by the end of the 1960's, were resurrected and shown to be very powerful. The first application in this context was by Bernholc and Pantelides[43] who employed tight-binding Hamiltonians. It was soon followed by self-consistent, parameter-free implementations by Bernholc, Lipari and Pantelides,[44] and, independently, by Baraff and Schluter.[45] Since then, a tight-binding formulation for systematic studies of deep impurities was introduced by Hjalmarson, Vogl, Wolford, and Dow,[46] and several variants of the self-consistent formulations were introduced in order to meet different needs (see, e.g., Refs. 47–50). All these formulations have been used for studies of many deep centers. As we did in the case of the experimental techniques, we shall not address here the technical differences between the various Green's-function formulations or discuss their relative advantages and disadvantages. In Section 3, we will have occasions to refer to results obtained with these various techniques.

In contrast to some of the earlier work using Green's functions,[35] the more recent implementations recognized that Wannier functions, though very convenient for formal theory, are not very suitable for practical calculations. Thus, practical basis sets of simple exponential or gaussian orbitals were chosen, following a similar practice in perfect-crystal calculations.

The primary advantage of Green's functions is that they treat an isolated defect in an otherwise infinite perfect crystal with the same accuracy that one chooses to treat the corresponding perfect crystal. The key to this success is the fact that one first treats the infinite perfect crystal, taking advantage of the periodicity and Bloch's theorem. The *changes* induced by a deep center are then calculated by taking advantage of the fact that they are localized in space. In contrast, in a cluster calculation, one does not make this separation. Instead, *all* the eigenstates of the perturbed crystal, i.e. both the bound states and the propagating states, are obtained from a single calculation in which the infinite crystal is truncated. A small cluster clearly cannot treat properly either the propagating states or the interactions between the bound states and the propagating states. These observations are very important because it is often argued that cluster calculations should be adequate because deep centers are localized. A careful scrutiny, however, reveals that it is the *changes* induced by a deep center that are localized and this property is exploited by Green's functions. Cluster calculations, on the other hand, do not exploit this property. Instead, they seek to obtain directly and simultaneously all the states of the perturbed crystal, which is infinite.

Self-consistent Green's-function calculations have demonstrated that "integrated' properties, i.e., those that involve a sum over all occupied one-electron states (e.g., charge density) are considerably more localized in real space than the wavefunctions of individual one-electron states (see, e.g., discussion in Ref. 44. It should also be noted that Green's-function methods are not limited to defect potentials which are localized in space. Potentials with long-range Coulombic tails can be treated[50] by enlarging the basis set to include long-range effective-mass-like orbitals in addition to the localized atomic-like orbitals normally used to describe short-range potentials.

Green's-function methods have the disadvantage that they are more complicated to implement than cluster calculations. Nevertheless, they have caught on because, once the Hamiltonian is chosen, they provide accurate solutions of the defect problem. They are now widely used by many groups with either semiempirical tight-binding or parameter-free self-consistent Hamiltonians. The Bassani Iadonisi-Preziosi method used earlier by Jaros has also been shown to be a Green's-function variant[43] and has since been used in a modified way.[51] Cluster calculations also continue to be used. More recently, the supercell technique has been revived and found to be more powerful when one seeks to compute "integrated" properties such as total energies instead of single-particle eigenvalues.[52] Green's functions themselves are currently being used to compute forces[53] and total energies[54,55] and thus address questions of lattice relaxation, equilibrium sites, formation energies and migration barriers. (For a critique of current theoretical techniques, see Section 3.1.4.)

Green's-function techniques are not limited to calculations of electronic structure. Another natural application is the calculation of vibrational modes associated with impurities or defects. As in the case of electronic structure calculations, the basic principles of the relevant formalism were developed in the 1960's,[26] but useful calculations were carried out only recently.[56,57] In previous years, analysis of vibrational modes was carried out in terms of molecular models.

2.3. Summary and Discussion

In the above account of the historical developments in techniques, certain trends were unmistakably clear. A major break in the development of experimental techniques was the development of junction techniques around 1970. The key feature of these techniques is their ability to focus on the deep centers unhindered by the environment, namely the bulk sample which contains shallow impurities at higher concentrations. A similar major break occurred in theoretical techniques in the late 1970's with the implementation of practical Green's-function techniques. Again, the key feature of these techniques is their ability to focus on the deep centers unhindered by the environment, namely the perfect crystal, which extends effectively to infinity.

It is also worthwhile to explore the forces that have been driving the major developments. Sah[12] has noted that the junction techniques evolved out of a homework problem he assigned to his class, but their evolution was certainly influenced by the background forces that direct attention to important problems. During the 1960's, the electronics industry was beginning to surge with the advent of miniaturization. Semiconductor pn junctions and semiconductor-metal interfaces are the building blocks of electronic devices. Naturally, the properties of junctions were closely studied, and the inevitable occurred. Miniaturization of devices also called for better control of crystal growth, which led to new crystal-growth techniques and better samples for fundamental studies. New technological developments, particularly the invention of lasers and dye lasers, the invention of new detectors, etc., provided powerful new tools that could be put to use in devising new techniques for fundamental studies. In the case of theory, energy-band calculations for perfect crystals went through various stages of development in the 1950's and 1960's and reached a stage of maturity only in the early 1970's. Thus, in the late 1960's and early 1970's, the prospects for comparable calculations for deep centers looked rather bleak, especially after the lack of success in the early attempts with Green's functions.[35] Simple, semiempirical cluster calculations were, therefore a natural choice. The stage was being set for the future developments, however. First, the density-functional theory,[58] developed in the mid-1960's, resolved many uncertainties about effective single-particle potentials. Then, the introduction of pseudopotentials[59] and, later, the construction of accurate and transferable ionic pseudopotentials for self-consistent calculations[60,61] eliminated the need to include unimportant core states. As bulk band-structure calculations matured, it was natural to turn to surfaces and to defects. Surfaces had an early edge[60] because they have two-dimensional periodicity so that standard band-structure programs could be used.[62] An attempt to do the same for defects (the supercell method)[40] had only limited success. The stage was then set to rediscover Green's functions as the mathematical technique that allows computations for defects at the same level of accuracy and sophistication as is possible for bulk crystals and surfaces. In all of these, another important factor should not be overlooked: the rapid improvement of computers made it possible to carry out massive computations that are needed to produce accurate and reliable results.

2.4 Objectives and Achievements

2.4.1 The 1960's and early 1970's: Isolation

During the 1960's and most of the 1970's, work on deep centers was carried out in several areas with very little overlap.

   A) Radiation-Induced Defects. The beginning of the space age in the 1950's led to interest in the effects of radiation on semiconductors. Thus, in the 1960's, extensive studies of radiation-induced defects were carried out. The primary tools were EPR, luminescence, infrared absorption and photoconductivity. Many defects were observed and were labeled by their EPR signature, luminescence lines, etc. The obvious and immediate objective was identification. Noteworthy progress in this quest was made only in the case of several defects detected by EPR because this method could probe the local geometry and symmetry. By combining EPR measurements under various conditions (uniaxial stress, temperature variation, etc.) elementary theoretical models and a few other experimental techniques, e.g., infrared absorption to study vibrational modes, identification was achieved for many radiation-induced defects: the vacancy, divacancy, vacancy-impurity pairs, etc.[31] In a number of cases, it was also possible to obtain some information about the energy-level structure of the defect. Pioneering and very extensive work in this field was done by Watkins, Corbett, Vavilov, Compton, Newman, and many others. One of the articles in this volume is devoted to the vacancy in Si. Most of the work in the field is chronicled in the proceedings of the biannual Conference on Defects and Radiation Effects in Semiconductors which was first held in 1959. We shall reexamine the question of identification in the context of modern developments in section 3.3 below.

   In addition to defect identification, the studies on radiation-induced defects addressed a number of fundamental questions. For example, earlier work on the vacancy in diamond[63], suggested that many-body effects are dominant because of the high degree of localization of the four electrons arising from the four "dangling bonds." Yet, the analysis of detailed experimental data on the vacancy in silicon by Watkins[64] suggested that single-particle theory could account for all observations. The topic was debated for a long time and only recently single-particle theory has been widely accepted as capable of explaining all the observations associated with the vacancy in Si. Just when all the dust settled, however, recent work on radiation-induced defects in GaP indicates that many-body effects dominate in the case of the Ga vacancy (see section 3.3 below).

The process of identification of radiation-induced defects became inexorably connected with the causes and magnitude of lattice relaxation. In particular, it was found that the lattice surrounding the vacancy in Si undergoes reconstruction as predicted by the Jahn-Teller theorem.[65] According to this theorem, any localized system in an electronic state with orbital degeneracy will spontaneously distort and lower its symmetry until a state with no orbital degeneracy is reached. The ground state of the vacancy in various charge states is indeed degenerate so that the observed distortions are consistent with the Jahn-Teller theorem. In another case, however, namely the so-called A-center in Si, after irradiation, oxygen atoms are identified to occupy off-center substitutional positions (hence the center is often described as oxygen-vacancy pair).[66] The question why oxygen moves off center is considerably more difficult. Simple considerations suggest that substitutional oxygen would have a non-degenerate ground state, precluding Jahn-Teller distortions. We will return to this question in section 3.5.1.

Another fundamental question raised by the early studies of radiation-induced defects concerns the mechanisms for defect migration. After irradiation of Si samples, it was possible to isolate and study vacancies.[64] For each vacancy, however, there must be an extra Si atom, a self-interstitial. EPR studies, however, did not produce any signals that could be attributed to self-interstitials. Instead, a particular EPR signal in irradiated Al-doped Si was attributed to interstitial aluminum.[64] Similarly, in B-doped Si, interstitial boron was detected.[64] The message was clear: Si self-interstitials are able to move very efficiently in the Si lattice, find the shallow impurities at substitutional sites, and take their places. The big puzzle is that this efficient migration occurs even at extremely low temperatures, about 4 K. Bourgoin and Corbett[67] in 1972 proposed a mechanism for such athermal migration: Assume that the interstitial in its equilibrium charge state is stable at a certain site A. If by capturing an electron, the interstitial, in its new charge state, is stable at a different site B, the motion from A to B will occur even at 0 K. Subsequent loss of the extra electron will make the interstitial move to another A site and so on. Clearly, for such athermal motion, one needs excess electrons and holes. During radiation experiments, such excess electrons and holes are of course created by the radiation. Thus, energy stored in electron-hole pairs is converted to motional energy: the recombination of one electron-hole pair can move an interstitial from one site A to another site of the same type. In the case of the self-interstitial in Si, cluster calculations[68] using extended Hückel theory led to the suggestion that the most likely path for athermal migration is a winding path through the bonds. As we shall see in Section 3.6.2.B, this path came under scrutiny by recent theoretical calculations which provide a more complete picture of the properties of the self-interstitial.

More direct evidence for the effect of excess electron-hole pairs on defect migration was obtained in the late 1970's. This subject is discussed further in Section 3.6.2 below.

B) Transition-metal impurities. Transition metal-impurities in Si and in compound semiconductors were studied initially almost exclusively by EPR. Pioneering work was done in Si by Ludwig and Woodbury[11] and was aided by theoretical models developed by Ham.[69] The EPR spectra showed clearly that some transition-metal-related centers had tetrahedral symmetry, which implies transition-metal atoms occupying either substitutional or tetrahedral interstitial sites. The distinction between the two was based largely on circumstantial evidence. The strongest piece of evidence was offered in a set of experiments by Woodbury and Ludwig[70] in the case of Mn in Si: Vacancies were introduced in a sample containing Mn centers and the resulting change in the EPR signal was attributed to capture of interstitial Mn atoms by vacancies, i.e. the conversion of interstitial Mn to substitutional Mn. In all cases, the experimental data were found to be consistent with a simple picture. The five orbital states available to d electrons in a free atom, split into a triplet and a doublet when the atom finds itself in a field of tetrahedral symmetry in a tetrahedrally-bonded semiconductor such as Si, GaAs, etc. If the splitting between these two levels is small, then the available electrons are distributed in the different orbital states so as to maximize the net orbital spin while satisfying the Pauli exclusion principle (Hund's rule). By assuming that the doublet state of e symmetry lies higher in energy than the triplet state of t_2 symmetry, the model could account for the EPR spectra of interstitial transition-metal impurities in Si. The magnitudes of the g values and hyperfine interactions could be accounted for by invoking hybridization between the impurity d orbitals and the s-p orbitals of the nearest-neighbor Si atoms.[11] For substitutional transition-metal impurities, the same model was found consistent with EPR data, except that the e state is assumed below the t_2 state

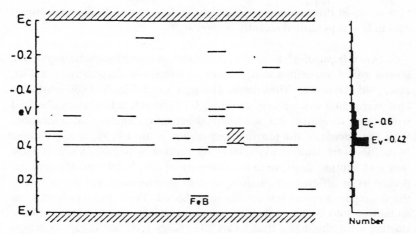

FIGURE 1 Collection of levels reported for iron in Si as given by Graff and Pieper, Ref. 71.

and a number of d electrons are assumed to be transferred to the valence shell in order to complete tetrahedral bonding to the four Si neighbors. (Recent calculations have verified some aspects of the Ludwig-Woodbury model. See Section 3.1.1B).

Following the pioneering work of Ludwig and Woodbury in the early 1960's on transition-metal impurities in Si, EPR studies were also carried out on such impurities in compound semiconductors. In a number of cases, optical-absorption and luminescence measurements provided complementary information. During the 1960's and early 1970's, however, available information was scattered and inconclusive. A large number of papers reported energy levels for many transition-metal impurities in Si, but no systematic efforts were made to determine whether these levels were due to isolated impurities or complexes. An illustration of the rather chaotic proliferation of energy levels is given by Figure 1, taken from a recent review paper by Graff and Pieper[71] on iron in Si. In the case of transition-metal impurities in III-V and II-VI compounds, there were fewer papers, but reflected more systematic efforts to look for evidence suggesting pairs or other complexes.

In recent years, availability of higher resolution absorption, luminescence, and photoconductivity spectra, and the advent of junction techniques and detailed theoretical calculations led to renewed interest in these systems (see Section 3.4). The early history and the more recent advances in our understanding of transition-metal impurities in compound semiconductors are reviewed by Allen, by Baranowski, by Bishop, and by Robbins et al. in this volume. An excellent review of transition-metal impurities in Si was published recently by Weber.[72]

C) Other Impurities. In silicon, in addition to transition-metal impurities, several other impurities were known to constitute deep centers: sulfur, zinc, gold, silver, etc. These impurities were studied in the 1960's with the bulk techniques described in section 2.1.1. The only information obtained from these measurements was thermal activation energies of ionization. With the advent of the junction techniques in the 1970's, many of these impurities were studied extensively by various techniques yielding emission and capture cross sections. Quite often, the results of different experiments or different researchers were in disagreement, but the origins of the discrepancies were not usually investigated. Typically, measured cross sections would be fit to simple model lineshapes for the purpose of extracting a threshold and thus locate the energy level in the gap. This type of work led to a proliferation of "energy levels" quoted in the literature for a given impurity and tables abound in both original and review papers listing such energy levels. There was little effort spent on verifying whether the energy levels thus reported indeed corresponded to simple substitutional impurities or to complexes. This confusion led to the often-mentioned quote: "the deeper the level, the shallower the understanding." This state of affairs lasted until the late 1970's when the entire field of deep centers went into a radically new phase.

In contrast to the rather confusing state of affairs of deep impurities in Si, a number of impurities in GaP proved to be excellent examples for the study of several interesting effects. The major impetus was provided by luminescence experiments in GaP, which detected spectra with many sharp lines. Persistent probing, detailed in this volume by Dean, revealed that the observed spectra were due to oxygen-zinc or oxygen-cadmium pairs. Measurements were possible on this system because a number of factors were particularly favorable. As we noted in section 2.1.1, deep centers usually exist in low concentrations and shallow impurities impede the detection of signals arising from them. Oxygen, however, enters most crystals in large concentrations. In Si, oxygen is electrically inactive (it normally occupies a site between two Si nearest neighbor atoms forming a buckled Si-O-Si chain as in SiO_2 and introduces no levels in the gap.) In contrast, in GaP, oxygen occupies a P site and has a deep level in the gap. It also pairs up with shallow acceptors, forming deep centers which, therefore, exist at high concentrations. Observation of these centers is clearly not impeded by the shallow impurities, since they participate in the formation of the deep center of interest. These oxygen-related centers turned out to be an example for many interesting and new effects: bound excitons, distant-pair luminescence, negatively-charged centers analogous to H⁻ in free space, etc. Though a great deal was learned from these centers over the years, they also spawned many controversies. Some of them have been resolved and some remain. A detailed account can be found in the article by Dean in this volume. Additional discussion can be found in Section 3.1.2 below.

D) Atomic Diffusion. The topic of atomic diffusion is closely related to deep centers. In a perfect crystal, atomic motion can occur in the interstitial channels, in which case the interstitial atoms, whether foreign impurities or host atoms are bona-fide defects. Many impurities in Si, e.g., transition-metal impurities, gold, etc., were recognized as fast interstitial diffusers during the 1950's and 1960's.[72] Atoms that occupy normal atomic sites, however, namely substitutional impurities or host atoms, diffuse by very different mechanisms. In general, they depend on intrinsic defects, such as vacancies, self-interstitials, etc., which are created thermally. For example, an atom (either an impurity or a "marked" host atom) occupying a normal site waits until a vacancy moves to a neighboring site. At that point, the impurity or marked host atom can take one step and the vacancy can go on its way. The impurity or marked atom will then wait until another vacancy comes by. This is the so-called *vacancy mechanism.* When this mechanism is active, the activation energy for self-diffusion (i.e. the diffusion of radioactive tracer host atoms) is simply the sum of the vacancy formation energy (which determines how many vacancies are present at a given temperature) and the vacancy migration energy (which determines the jump rate). Another mechanism for diffusion is the so-called *interstitialcy mechanism.* In this case, the defect mediating diffusion is an interstitial host atom or self-interstitial. When a self-interstitial approaches a marked atom at an atomic site, it may exchange sites with it. In a subsequent step, the marked atom, now occupying an interstitial site, can exchange with another host atom at a normal site. The net result is that the marked atom has gone from one normal atomic site to another. The activation energy for this mechanism is clearly the sum of the formation energy of a self-interstitial (which determines how many self-interstitials exist at a given temperature) and the energy barrier for migration.

The basic principles of the vacancy and interstitialcy mechanisms were worked out in the 1950's and were used to elucidate diffusion in metals. In semiconductors, pioneering experiments on self-diffusion were carried out by Fairfield and Masters[73] in the late 1960's and were followed by

$$\ln D = \ln D_0 - \frac{Q}{kT}$$
$$kT \ln D_0 = kT \ln D_0 - Q$$
$$T \ln D_0 = T \ln D - \frac{Q}{k}$$

others.[74] Measurements of impurity diffusion go back to the 1950's.[74] The primary experimental information on self-diffusion in Si is that the diffusion coefficient obeys an Arrhenius relation $D=D_0 \exp(-Q/kT)$ where the activation energy Q is about 5 eV and the preexponential D_0 is large as compared to values in metals Similarly, impurity diffusion coefficients usually[74] obey Arrhenius relations with Q's typically from 1 to 5 eV.

During the 1960's and 1970's a number of controversies over the mechanisms of self-diffusion in Si dominated the field. The primary controversy was over whether self-diffusion is mediated by vacancies or self-interstitials. Arguments were put forth in favor of one or the other, but the main schools remained entrenched. The problem was that all evidence was indirect, based on other observations, whose interpretations depended to a large extent on assumptions about other, sometimes unrelated phenomena. As a result, in some cases, the same data might be invoked to support opposite points of view by making different assumptions. A critical review of that debate would carry us far afield since we would have to explore the validity of assumptions about other phenomena such as the nature of swirl-type defects, atomic processes during oxidation or stacking-fault growth, etc. A number of reviews, usually advocating one or the other point of view, are available.[74-76] We will return to the question of interstitial vs. vacancy in section 3.7 where we review recent experimental and theoretical advances on the subject. Here, we turn our attention to two other major puzzles that hatched controversies. One of them has its origins in formation/migration *energies (enthalpies)* and the other had its origins in formation/migration *entropies.*

a) The "energy" puzzle: During the 1960's, Watkins[64] carried out extensive EPR studies of irradiated Si. At very low temperatures (less than 50 K) he isolated and identified vacancies. By studying their annealing properties, he concluded that their migration energy is only 0.2–0.3 eV, depending on the charge state. Thus, if the vacancy had the same migration energy at high temperatures where the self-diffusion activation energy is measured to be about 5 eV, the vacancy formation energy would have to be close to 5 eV. In contrast, a variety of quenching data led to vacancy formation energies of order 3 eV[8,64]. In addition, many estimates of vacancy formation energies suggested values of order 2.5-3 eV. It was, therefore, suggested that the high-temperature migration energy of the vacancy is substantially larger than the one measured at low temperatures.[74] A similar problem arose with the self-interstitial. As we already noted, low-temperature irradiation studies led to the conclusion that the self-interstitial migrates athermally, probably by successively capturing electrons and holes (Bourgoin-Corbett mechanism), which implies a small barrier, of order of a band gap. Thus, again, the 5-eV value of the high-temperature self-diffusion activation energy suggests a large formation energy for the self-interstitial, i.e., of order 4-5 eV. In contrast, theoretical estimates of self-interstitial formation energies were of order 1-2 eV.

b) The "entropy" puzzle: Measurements of the self-diffusion coefficient in Si led to values of the preexponential which were several orders of magnitude than corresponding values in many metals.[73] Since the preexponential is related to the entropy of formation and migration of the defect mediating self-diffusion, it was concluded that the relevant defect should have large entropy of formation and/or migration. Monovacancies and simple interstitials did not appear to have such properties.

In 1968, Seeger and Chik[77] made a radical suggestion that has been debated ever since. Motivated by the large preexponential, they suggested that the vacancies and/or self-interstitials that are responsible for self-diffusion at high temperatures are "extended," i.e., some kind of "liquid" bubbles in which all atoms are more or less tetrahedrally coordinated so that the missing or extra atom cannot be singled out. Such extended defects would have high entropy of formation because of the many different ways a liquid bubble can be formed. By the same stroke, one could argue that the migration energy of such complex defects may be larger than the migration energy of the simple defects created by irradiation at low temperatures.

Initially, Seeger and Chik[77] expressed preference for extended interstitials on the grounds that molten Si has higher density than crystalline Si. Other arguments were offered later on[74], but, as we noted already, we will not pursue a critical account of the arguments for one or the other mechanism. We should note that similar arguments were debated with respect to the mechanisms for impurity diffusion. By the late 1970's, no consensus was achieved. We will discuss more recent advances on these problems in Section 3.7.

2.4.2 The Late 1970's and the 1980's: Seeking the Total Picture

Sometime in the late 1970's, the field of deep centers in semiconductors underwent a major change. The isolation of the various communities that was prevalent in the past two decades came to an end. As we have seen, in the past, specific techniques were used to study certain classes of centers (e.g., EPR for radiation-induced defects and transition metal-impurities, electrical measurements for most other impurities, etc.), obtaining only partial information. Since the late 1970's, however, the trend has been to combine techniques and go after the total picture. For example, in recent years, there have been concerted studies of radiation-induced centers and transition-metal centers with EPR, junction techniques, and luminescence measurements. Chalcogens in Si have been studied by a large assortment of techniques, etc. In fact, each article in this volume contains an illustration of how significant progress was made when a number of techniques were combined in recent years. In section 3, we will critically assess the major areas of advances in the last several years. Before we do that, however, it is worthwhile to pause for a moment and ponder over the causes for the major changes that occurred in the field in the last 6-8 years.

The fundamental reason behind the change is the development of a large number of techniques which had a chance to mature. The experimental techniques based on junctions, developed in the early 1970's, had already been established as powerful tools. New techniques were being invented and combined with older techniques, providing the type of information that was not accessible before. For example, recombination-enhanced phenomena, which used to be mere speculations, were observed with junction techniques which allow controlled injection of minority carriers (see Section 3.6.2). At the same time, crystal-growth techniques evolved to the point where samples were of excellent quality and better characterized, making them very suitable for basic research. Finally, theory, with the advent of Green's-function techniques, became a new powerful tool. In the past, theory was primarily used as a guide to interpret experiments. In recent years, however, it became predictive, sometimes leading experiments and sometimes challenging accepted interpretations of experimental observations.

There are many external factors that mark the transition to the new age. During the 1960's, the main preoccupation of the semiconductor physics community was to understand the intrinsic properties of "perfect" crystals (energy band structures, optical spectra, phonon dispersions, etc.) In the early 1970's, clean ideal surfaces came into vogue. Throughout this time, studies of deep centers were motivated largely by applications and focussed on questions of electrical characterization and radiation damage instead of fundamental physics questions. For example, starting with 1959, an international conference on radiation-induced defects in semiconductors was held as a satellite conference to the biennial international conference on the physics of semiconductors. There was little overlap between the two conferences, however. In recent years, on the other hand, this satellite conference has all but abandoned radiation damage, changed its name to "Defects in Semiconductors," and hosts a program that is devoted primarily to the fundamental physics questions underlying studies of defects. In addition, the same topics now take a prominent position in the program of the main conference on the physics of semiconductors.

The recent changes were also catalyzed in great part by a new series of conferences known as "Lund" conferences on deep-level impurities in semiconductors, in honor of the fact that the first of them was organized by a group of researchers at the University of Lund in Sweden. The first of these conferences, held in Sweden in 1977, brought together many prominent scientists who had been or were active in the various fields of deep-center research. In a workshop-style atmosphere, they debated the past and the future of the field. The main theme was the determination to go after the total picture by sharing techniques and samples. The conference has been repeated every two years ever since, providing renewed impetus. Parallel to that, also beginning in 1977, there has been a biennial Gordon conference devoted to point and line defects in semiconductors. All these conferences, perhaps too many, have ended the isolation of the various communities involved in deep-center research, help define the important issues and provide useful forums for the cross fertilization of ideas.

3. Major Recent Developments

In this section we shall attempt to capture some of the spirit of the recent years by reviewing the major topics that have been addressed and the major new developments. There will be no attempt to catalog results or to derive equations. Instead, the emphasis will be on concepts and topics as they evolved in the last few years. The developments will be assessed from a critical point of view and will, therefore, inevitably, reflect the author's opinions, biases, and shortcomings.

3.1 Electronic Structure

A reliable theoretical description of deep centers in semiconductors has been one of the major challenges faced by solid-state theory in the last twenty five years. Green's-function methods have now provided detailed and reliable answers to many of the outstanding questions. In some cases, the results confirmed or complemented earlier pictures obtained with more approximate methods. In other cases, the results were altogether new or even contradicted earlier results. In this section, at first (subsections 3.1.1-3.1.3), we will discuss the current state of *understanding* of the electronic structure of deep centers, without, in general, attempting to track how the various methods, techniques, or individuals contributed to the overall picture. In subsection 3.1.4, we will briefly discuss the quantitative success of electronic-structure calculations.

3.1.1 The Single-particle Picture

The main realization that was brought about by the theoretical activity of recent years is that a deep center is not characterized merely by one or sometimes more bound states in the fundamental energy gap. A deep center introduces a rather severe localized perturbation which modifies significantly the entire energy spectrum. In fact, in order to understand the origin and nature of a bound state in the fundamental gap, one must view it as just one member of a whole series of localized states (bound states or resonances).

A) Vacancies and sp-Bonded Impurities. Vacancies in tetrahedrally-bonded semiconductors were described successfully in the 1960's in terms of a very simple picture:[64] When an atom is removed, four bonds are cut, leaving four "dangling bonds." Since bonds in the perfect crystal are formed by combining sp^3 hybrid orbitals on adjacent atoms,[79] dangling bonds are really dangling hybrids. Localized states in the vicinity of the vacancy are, thus, expected to be composed primarily of these dangling hybrids. Simple symmetry arguments then suggest that there should be a singlet state of A_1 symmetry and a triplet state of T_2 symmetry. Green's-function calculations have confirmed this picture in the case of the vacancy in Si.[44,47] The T_2 state is a bound state in the gap, whereas the A_1 state is a sharp resonance inside the valence bands (Figure 2). In Figure 3, we show the charge density associated with these states, demonstrating that they resemble linear combinations of dangling sp^3 hybrids on the nearest neighbors. In the case of the Ga vacancy in GaP, on the other hand, the T_2 state in the gap, shown in Figure 4, exhibits an essentially pure p-like character. Furthermore, in Figure 2, we see that the vacancy has another prominent resonance at about -8 eV, which is not accounted for by the simple dangling-bond model. In compound semiconductors, this state is a true bound state in the gap that opens up within the valence bands.[80] It turns out that a slightly more general model can account for the results of the numerical calculations. The localized states of the vacancy may be viewed as linear combinations of s-like and p-like orbitals centered on the four nearest neighbors, without imposing an sp^3 admixture on each atom. One can then form two A_1 states (one from a symmetric combination of the s orbitals on the four nearest neighbors and one from a symmetric combination of p orbitals on the nearest neighbors). Thus, one can account for both the A_1 resonances seen in Figure 2. Similarly, T_2 states can be formed separately from the s and from the p orbitals. In both cases, the relative admixture of s and p orbitals is not determined by symmetry, but by the details of the defect potential and the host band structure. It

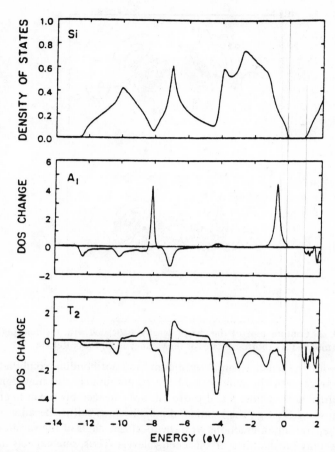

FIGURE 2 Top panel: The density of states of a perfect Si crystal. Middle and bottom panels: The changes in the density of states induced by a vacancy in Si. Only states of A_1 and T_2 symmetry are shown. From Ref. 44.

so happens that the vacancy in Si has an sp^3-like state in the gap, whereas the Ga vacancy in GaP has a mostly p-like state.

The bound states of substitutional impurities can be understood in a number of different, but complementary ways.[80,81] We will discuss one such way when we examine interstitial impurities later on. Here, we start with a crystal containing a vacancy and then consider inserting an impurity at the vacant site. Let us, for a moment, insert a silicon atom at the vacant site. Clearly, its s and p orbitals will interact with the A_1 and T_2 localized states of the vacancy, producing bonding combinations, which

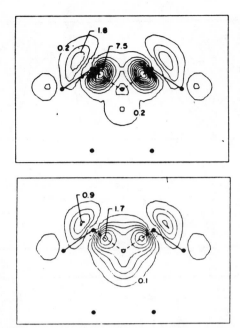

FIGURE 3 Contour plot of the charge density associated with the dangling-bond-like T_2 (top) and A_1 (bottom) states of the vacancy in Si. From Ref. 44.

merge with the valence-band continuum, and antibonding combinations, which merge with the conduction-band continuum. Let us now consider an impurity atom whose s and p orbitals are considerably lower in energy than those of Si, e.g., the nominal donors such as sulfur. Because of the energy separation, interactions with the localized states of the vacancy will produce only small shifts in the energy levels. Thus, one expects an impurity-like A_1-T_2 pair in the lower part of the valence bands and/or below the valence bands and a vacancy-like A_1-T_2 pair slightly than the corresponding vacancy pair. By the same token, for impurities whose s and p energy levels are considerably higher than those of Si, e.g., the nominal acceptors such as zinc, one expects a vacancy-like A_1-T_2 pair slightly below the corresponding vacancy pair and an impurity-like A_1-T_2 pair somewhere in the conduction bands. The intriguing result of this analysis is that, in both cases, the levels in the fundamental gap are expected to be vacancy-like. Indeed, the vacancy states appear to be a limiting case, with nominal donors having their A_1-T_2 states above the corresponding vacancy states in the gap regions and nominal acceptors having their A_1-T_2 states below the corresponding vacancy states. Only nominal donors have bona-fide hyper-deep states.

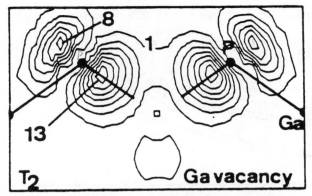

FIGURE 4   Contour plot of the charge density associated with the T_2 state bound state of the Ga vacancy in GaP. From Ref. 80.

We also note that the gap states of donors are antibonding in character, and can, therefore, be viewed as arising primarily from the conduction bands. Indeed, if one slowly changes the donor's s and p energy levels toward those of Si, the gap states slowly move up in energy and merge with the conduction-band continuum. Similarly, the gap states of acceptors are bonding in character, and can, therefore, be viewed as arising primarily from the valence bands. Again, if one slowly changes the acceptor's s and p energy levels toward those of Si, the gap states slowly move down in energy and merge with the valence band continuum. Note that the vacancy levels, being the limiting case of either a deep donor or a deep acceptor, are best viewed as arising equally from the valence and the conduction bands (a dangling bond is the sum of a bonding and an antibonding orbital).

The above qualitative description of sp-bonded substitutional impurities is based on Green's-function calculations using semiempirical tight-binding Hamiltonians[46] and self-consistent local-density theory.[80,81] In fact, in the simplest tight-binding approximation, the vacancy states are a limiting case in a quantitative way: Nominal-donor states lie on a curve which asymptotically goes to the vacancy state from above and nominal-acceptor states lie on a curve that asymptotically goes to the vacancy state from below (Figure 5). This monotonic dependence of impurity levels on atomic energies and the "pinning" to the vacancy states is, however, only approximate. According to the self-consistent calculations of Ref. 82, this monotonicity is drastically violated in the case of substitutional oxygen in Si because of strong non-linear screening effects.

Self-consistent local-density calculations have produced wave functions which are, in general, consistent with the qualitative picture described above: Deep nominal acceptors have a T_2 state in the gap whose wave

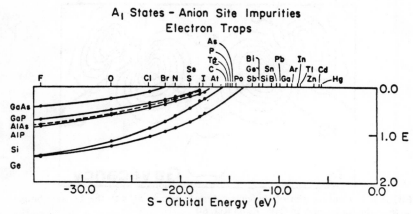

FIGURE 5 Tight-binding energy levels for anion-site impurities in six semiconductors. From Ref. 119.

function is virtually indistinguishable from that of the T_2 vacancy state shown in Figure 3.[81] Similarly, deep nominal donors have an A_1-T_2 pair of states which is vacancy-like. The single exception to the above qualitative picture is the A_1 state of nominal acceptors. Interaction between the vacancy A_1 state and the impurity s state is quite strong so that the net result is a fully bonded state which merges in the valence bands and is no longer distinguishable as a bound state. Additional insights into the structure of these states can be found in Refs. 80 and 81.

In order to complete the picture for substitutional sp-bonded deep impurities, it is desirable to establish a more direct connection with the shallow impurities. For example, we saw that deep nominal acceptors have a T_2 state in the gap whose wave function is virtually identical to that of the T_2 vacancy state. As the acceptor state becomes shallower, we expect the wave-function to become more bonding-like. In the limit of hydrogenic shallow impurities, effective-mass theory actually *assumes* that the wavefunction looks like the Bloch function of the top of the valence bands, which is a bonding function, multiplied by a smooth envelope. When does the transition from a dangling-bond-like wave function to a bonding-like wave function occur? An answer to this question was provided by the calculations of Ref. 50 where the impurity potential was chosen to have a short-range part equal to $\lambda U_{SR}(r)$ and a long-range Coulombic tail of the form e^2/er. The potential U_{SR} was chosen to describe the neutral vacancy. Thus, when $\lambda = 1$, the potential corresponds to the vacancy; when $0 < \lambda < 1$, the potential represents the nominal acceptors; and, when $\lambda = 0$, the potential reduces to the simple hydrogenic potential

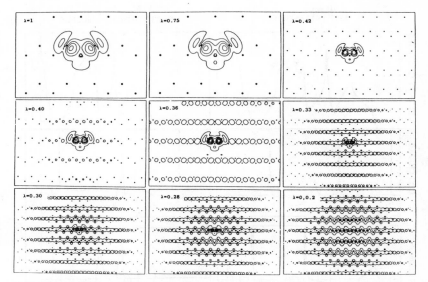

FIGURE 6 Contour plot of the charge density associated with the wavefunction of an acceptor impurity in Si for various strengths of the short-range potential. The maximum strength ($\lambda = 1$) corresponds to the vacancy. From Ref. 50.

of shallow acceptors. By using the mixed-range Green's-function method,[50] solutions for all values of λ were obtained. The results for the bound-state wave function for various values of λ are shown in Figure 6. The transition from the dangling-bond-like wave function of deep states to the effective-mass-like wave wave function of shallow states is apparent.

Interstitial sp-bonded impurities occupying the high-symmetry tetra-hedral site have been found[83] to have a somewhat unexpected energy-level structure: essentially the same as deep substitutional nominal donor im-purities. In Figure 7, we show the state-density changes induced by inter-stitial Si at the tetrahedral site, which is a donor with $\Delta Z = 4$, and compare it with the corresponding state-density changes induced by substitutional sulfur in Si, which is a donor with $\Delta Z = 2$ (there are no $\Delta Z = 4$ donors to compare with). The similarities are astounding and can be understood as follows: At the substitutional site, sulfur has four nearest neighbors. Localized states are thus expected to be composed primarily of the s and p orbitals centered on the central atom and on the four tetrahedrally-arranged nearest neighbors. Similarly, the interstitial has four nearest, neighbors. Again, localized states are expected to be composed primarily of s and p orbitals of the central atom and of the four tetrahedrally-ar-ranged nearest neighbors. Thus, symmetry and the localization of the

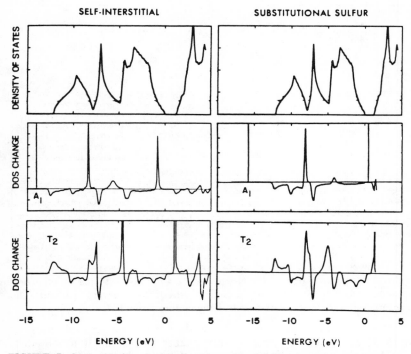

FIGURE 7 State density changes for the sulfur substitutional impurity and the tetrahedral self-interstitial in Si. From Ref. 83.

states require the same overall energy-level structure. Even the total potential is very similar in the local region: in both cases, one has a central atom surrounded by four atoms; important differences occur in more distant shells of neighbors where the bound-state wave functions decay rapidly. Despite these reasonable explanations, one may still be perplexed by the similarity seen in Figure 7 on the grounds that a substitutional impurity interacts with a vacancy and dangling bonds, whereas an interstitial interacts with a fully bonded system. If one wishes to approach the problem from this point of view, then Figure 7 is not relevant. Note that Figure 7 shows the changes to the state density of the *perfect crystal* induced by (a) the replacement of a Si atom with a S atom, or (b) the addition of an interstitial Si atom. If, however, the starting point is a crystal containing a vacancy, the changes brought about by the insertion of a S at the vacant site would be the difference between Figure 7(a) and Figure 2. The resulting figure would indeed be very different from Figure 7 (b) and the inherent similarities of the two systems would be suppressed.

B) Transition-Metal Impurities. The single-particle energy-level structure of transition-metal impurities[84-87] can be understood in general terms along the same lines (for a complete description, one needs to include many-body effects; see section 3.1.2 below). Figure 8 shows calculated energy levels for the series Zn-Ti as substitutional impurities in Si. Zn is really an sp-bonded impurity (i.e., its d orbitals may be viewed as core states) so that the primary localized state is the vacancy-like T_2 state in the lower part of the gap. As we saw above, this state is essentially the vacancy T_2 state which is now pushed from above by the p level of the impurity atom. In addition, Zn has a completely occupied d level at about –12 eV. In the case of Zn, the d level does not play a significant role. For other impurities in the series, however, the d level moves up in energy and delocalizes somewhat so that interactions with the neighbors split it into a triplet T_2 and a singlet E. Thus, if we start with a vacancy, the T_2 state in the gap is now pushed down by the impurity p state and pushed up by the impurity $d(T_2)$ state. The net result is that the vacancy T_2 state slowly moves up in energy as we go through the series from Zn to Ti. It also, to a

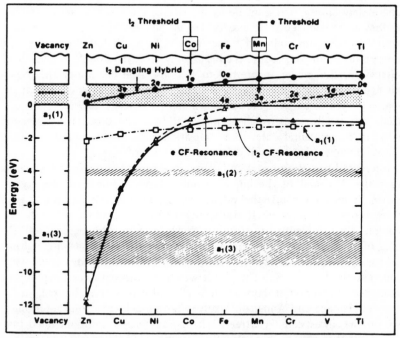

FIGURE 8   Energy levels of substitutional transition-metal impurities in Si as calculated in the single-particle approximation. From Ref. 86.

large extent maintains its vacancy-like or dangling-bond character.[86] The lower T_2 state is primarily impurity-like and is described as a crystal-field resonance. The E state increases monotonically in energy as we go through the series from Zn to Ti. In the region of the gap the rise of the E state is slowed down probably because of strong d-like character in the conduction-band Bloch functions.

The results of the detailed numerical calculations[86] bear an interesting relationship to the energy-level structure deduced from EPR experiments in the early 1960's.[11] As a specific example, let us take neutral Cr in Si which has the free-atom configuration $3d^5 4s^1$. According to the old picture, three of the five d electrons are transferred to the valence sp shell in order to complete tetrahedral covalent bonding and the remaining two electrons occupy the d(E) state, which lies below the d(T_2) state. In contrast, the new calculations suggest that the d(T_2) impurity state hybridizes strongly with the vacancy T_2 state so that the d(e) ends up being sandwiched between *two* T_2 localized states each of which is roughly 50% impurity d(T_2) and 50% "dangling bonds" on the nearest neighbors.[86] The lower one of these T_2 states is completely occupied, while the other is completely empty. Thus, in the final analysis, the resulting states are just as they were deduced from experiment in the early 1960's, i.e., an E state with two electrons followed by an empty T_2 state, but the origins and interpretation of these states is considerably more involved. For additional discussion, the reader is referred to Ref. 86. Another important consequence of this type of electronic structure is that the T_2 states which are strongly hybridized with the vacancy states are considerably more delocalized than the E states which do not hybridize. This observation explains why, in experimental data, some transition-metal states exhibit localized behavior while others exhibit delocalized behavior.[87]

Interstitial transition-metal impurities are simpler to understand. It seems that the d orbital of the impurity interacts very little with the crystal states. It moves through the valence band and into the gap as one goes from Cu to V (Figure 9). It also splits, with the E doublet above the T_2 triplet, in accord with the structure deduced from EPR in the 1960's.[11] One of the most intriguing properties of interstitial transition-metal impurities is that they can have several (up to five!) charge states in the band gap of only 1 eV or so. This result was first observed experimentally in both Si and compound semiconductors[88] and was later accounted for by theory. Haldane and Anderson[89] solved the model Anderson Hamiltonian within the unrestricted Hartree-Fock approximation and concluded that the many charge states are a result of strong rehybridization between the transition-atom d orbitals and the crystalline s-p orbitals which occurs

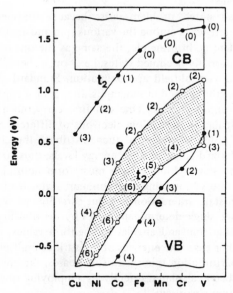

FIGURE 9 Energy levels of interstitial transition-metal impurities in Si as calculated in the single-particle approximation. From Ref. 95.

every time an electron is added to or subtracted from the gap states. The more recent local-density Green's-function calculations of Zunger and Lindefelt[86] confirmed this effect and identified another important factor: dielectric screening is highly nonlinear and reduces electron-electron repulsion by two orders of magnitude instead of one order of magnitude that would be expected on the basis of linear-response theory.

The electronic energy-level structure of more complex deep centers can be understood along the same lines. For example, the localized states of the divacancy and multivacancies are well described in terms of the dangling bonds.[90,90a] The levels of impurity pairs and impurity-vacancy complexes can be understood as linear combinations of the relevant atomic-like orbitals.[90b]

3.1.2 Spin polarization and many-body effects

The above description of the electronic states of deep centers is based on the single-particle approximation according to which all electrons see the same effective potential and the resulting states are occupied in order of increasing energy. It also assumes that electrons with spin $+\frac{1}{2}$ or $-\frac{1}{2}$ see the same effective potential. When this description leads to partially-occupied localized states, problems arise because the electrons can be arranged in different ways among the various spin and angular momentum values. The problem is, in principle, the same as the one occurring in free atoms that have partially-occupied states, i.e., "open shells," when one has to go beyond the central-field approximation. Standard analysis can be found in many textbooks and monographs. In practical applications, there are two possible approaches: a) One can first carry out a spin-polarized single-particle calculation, i.e., allow electrons of different spins to experience a different potential. One can then distribute the electrons in the resulting spin-polarized single-particle energy levels, thus arriving at several single-determinant states. Finally, one must form multiplets, i.e., linear combinations of the various single-determinant states, which are coupled by residual electrostatic interactions. b) One can forgo a spin-polarized calculation, construct single-determinant states by distributing the electrons among the available spin and angular momentum values, and then form multiplets. In either case, the energies of the various multiplets are usually determined by perturbation theory. In some cases, the energies of multiplet states have been calculated directly by employing small clusters and methods of quantum chemistry.[91]

The possible multiplets have been worked out for the various charge states of the vacancy[8,92] and for transition-metal impurities.[93] The prob-

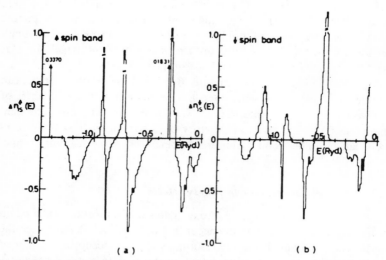

FIGURE 10 State density changes induced by a muonium (or hydrogen) at a tetrahedral interstitial site. From Ref. 98.

lem of determining which is the lowest-energy configuration is difficult and is compounded by the fact that degenerate configurations are unstable and lead to spontaneous, energy-lowering, symmetry-breaking, Jahn-Teller distortions. As a result, the importance of many-body effects is still a matter to be decided by experiment. As an illustration, we consider the vacancy state in any tetrahedral semiconductor when the T_2 state in the gap contains three electrons. The three electrons can be arranged so as to have parallel spins which, according to Hund's rule, should minimize the energy. This state (designated $^4A_2$) has no orbital degeneracy because there is only one way to arrange the three parallel spins, i.e., one each in the x, y, and z orbitals. Alternatively, one can have two parallel spins and one antiparallel. In that case, there are several ways to distribute the three spins among the x, y and z orbitals, leading to three multiplets designated ^2E, $^2T_1$, and $^2T_2$ (see Ref. 8) each of which is degenerate (doublet, triplet, and triplet, respectively). Thus, even though Hund's rule says that these are higher-energy states, their degeneracy would lead to a Jahn-Teller distortion (see sections 2.4.1A and 3.5.1). The big question then is which is going to win? Is the energy gain coming from aligning the spins larger or smaller than the net gain (if any) coming from keeping the spins antiparallel and lowering the symmetry. EPR experiments settled this question for the negatively-charged state of the vacancy in Si, which has three electrons in the T_2 state in the gap, quite some time ago:[64] the Jahn-Teller distortions win, signalling that many-body interactions that would tend to align the spins are weak (see also Ref. 92). Very recently, however, the Ga vacancy in GaP has been identified by EPR and the evidence seems to be that the neutral vacancy with three electrons in the T_2 state exists in the $^4A_2$ singlet state.[94] This state, however, may not be the equilibrium state because the EPR signal attributed to it is seen only after photoexcitation in p-type material. Experiments in n-type material would be informative, but reliable data are not yet available (see discussion in Ref. 80).

Many-body effects play a dominant role in transition-metal impurities. In Si, EPR experiments have established that aligning the spins of individual electrons for maximum total spin is usually energetically favorable even though some of these electrons occupy the higher-energy E level (this is the "weak-field" case of crystal-field theory). EPR experiments do not, usually, measure the value of the E-T_2 splitting. Clearly, the splitting must be small compared with the energy gained from aligning spins. The latter reflects the degree of spatial localization of the wave function since, for localized wave functions, aligning the spins is the only way to keep the electrons apart and thus minimize costly electrostatic repulsions. Theory has been able to calculate the value of the E-T_2 splitting and the degree of

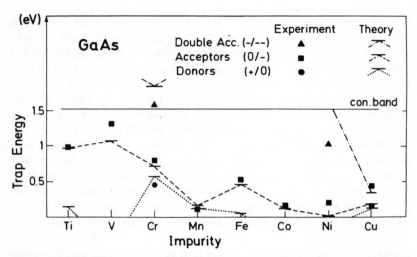

FIGURE 11 Energy levels of transition-metal impurities in GaAs as calculated by Vogl and Baranowski compared with experimental data. From Ref. 87.

localization and successfully account for the observed high-spin states.[95] We will return to this issue when we assess the ability of theory to produce quantitatively reliable results in Section 3.1.4.

In the case of transition-metal impurities in compound semiconductors, a comprehensive picture of the energy-level structure has been developed in terms of a modified tight-binding theory,[87] including a form of self-consistency and spin polarization. The theory contains adjustable parameters, but the results yield a good overall agreement with experimental trends as is illustrated in Figure 11 in the case of GaAs.

During the last few years, the ability of conventional single-particle theory to properly describe first-row impurities such as oxygen and nitrogen has been questioned by Morgan.[96] The challenge is described in this volume by Dean, who opposes it. According to Morgan, the conventional single-particle description of oxygen in GaP (see discussion of sp-bonded impurities earlier in this section) is not capable of explaining all the experimental data. Dean, however, argues that any existing puzzles can be rationalized within single-particle theory and, in any case, are not significant enough to justify throwing out a generally successful theory. Morgan has proposed an alternative description based on many-body states very similar to the multiplets we discussed above in the case of vacancies and transition-metal impurities. He asserts that the ground state is not the multiplet that single-particle theory requires, but another multiplet which is chosen on the grounds that it can explain all the experimental data in a simple way. If one uses conventional single-particle theory as a starting point to form multiplets, as has been successfully done for some transition-metal impurities,[93],[95] Morgan's preferred multiplet would require promoting two electrons to states which lie considerably high in energy with no apparent source for a gain in energy. Morgan's counter argument is that the single-particle picture fails to capture the essential features of the system altogether. At this point, the case that is made on the basis of experimental data alone is not strong enough to justify and uphold such a major breakdown of single-particle theory. We must await more definitive theory or experiments before the issue is resolved to everyone's satisfaction.

We close this section with a brief note on hydrogen, or, more generally, an extra positive charge in the crystal. Both hydrogen and muonium (which differs from a hydrogen primarily in the value of its lifetime) have attracted considerable attention.[97] In a recent self-consistent Green's-function calculation for a muon at a tetrahedral interstitial site in Si, Katayama-Yoshida and Shindo[98] included spin polarization and found a remarkably complex electronic structure, which is significantly different from all previous models that neglected spin polarization. The main result is shown in Figure 10. For one spin species (spin up), a state is pulled below the valence bands (comparable to the hyperdeep state of the Si self-interstitial, Figure 7). It is occupied by one electron which effectively screens the muon. A state of the same spin is also pulled below the conduction-band edge in the fundamental energy gap. The spin down states, on the other hand, experience a repulsive potential, which is a manifestation of exchange forces. The most fascinating consequence of these results is that the value of the electronic spin density at the origin, which is probed by the muon spin resonance (μSR) experiments, is dominated by the hyperdeep state, not the deep state in the gap.[98]

3.1.3 Excited States

Virtually all experiments that probe the electronic structure of deep centers do so by causing an electronic excitation. The excited states are the final states in an optical-absorption experiment and the initial states in a luminescence experiment. Thus, it is worthwhile to consider the various possibilities.

a) Internal transitions: Some deep centers in a certain charge state may have several bona-fide localized states in the gap region. The reconstructed vacancy in Si (Ref. 99) and transition-metal impurities are typical examples of such centers (Figures 8 and 9 and reviews by Allen, by Baranowski, by Bishop, and by Robbins et al., this volume). Thus, internal transitions can occur without changing the charge state of the center. In some cases, states participating in internal transitions may actually be resonances either in the valence or the conduction bands.[100] Thus, a number of excited states of such centers are simply states in which one or more electrons in a localized state are excited into a higher-energy localized state. One such example was discussed recently in Ref. 100 involving a copper-related center in ZnSe (see also discussion by Robbins et al., this volume). Another example is the so-called EL2 center in GaAs, which is discussed in great detail by Martin and Makram-Ebeid in this volume. In both cases, optical-absorption or similar experiments, which do not discriminate between localized or delocalized states, exhibit a broad hump that is not present in the corresponding photo-conductivity spectra, which probe only delocalized states. In the case of EL2, the realization that the hump in the optical-absorption spectrum is due to internal transitions[100a] has been one of the recent major breakthroughs in the struggle to identify the center.

b) Bound excitons: When an electron is excited from the valence into the conduction bands, it may form a bound state with the hole left behind. Such bound states of otherwise free electrons and holes are known as excitons. Excitons themselves are normally free to propagate in the crystal. When an exciton is created in the vicinity of a deep center, however, either the hole or the electron may become bound to it, so that the exciton itself is bound. Such bound exciton states have played an important role in luminescence spectra. They are discussed at length in the articles by Dean and by Robbins et al. in this volume.

c) Rydberg states of electrons or holes: If we start with a neutral deep center which contains one electron in a deep state in the gap, the electron can be excited to the conduction bands leaving behind a positively charged center. The resulting Coulombic potential, screened by dielectric response, produces, in principle,[101] a series of Rydberg-like bound states just below the conduction-band edge. Similarly, if, instead, an electron is excited from the valence bands to the deep state (equivalently, a hole is emitted from the deep state to the valence bands), the center becomes negatively charged and gives rise, in principle, to Rydberg-like states for holes just above the valence bands.

Rydberg series were detected originally in shallow impurities in the 1950's and 1960's.[103] In recent years, refinements of optical-absorption techniques and a number of hybrid techniques led to the detection of Rydberg series with astounding resolution in the deep chalcogen impurities in silicon. A detailed account is given by Grimmeiss and Janzen in this volume.

A novel possibility has been considered recently in Ref. 102: If the deep state is near midgap, the energy to excite an electron to the conduction bands may be comparable or even equal to the energy required to excite a hole to the valence bands. Thus, two very different excited states may be degenerate or nearly degenerate which means that the true excited eigenstates are linear combinations of the two. These type of excitations are called coupled electron-hole excitations and some data suggest that they occur in some centers.[102] The coupling between electron and hole excitations also has significant and observable consequences on capture rates and the interpretation of relevant experiments (see Section 3.6). A detailed theory, which unifies bound excitons, Rydberg states, and coupled electron-hole excitations can be found in Ref. 102.

d) Hybrid excited states: It is possible that internal transitions, as defined in item (i) above, coincide in energy with excitations to Rydberg satates induced by the Coulombic tail as discussed above in item (c) above. The net result is a new set of hybrid excited states that shares both characteristics. The excited statas of ZnSe:Co demonstrate this type of behavior, as discussed in detail by Robbins, Dean, West, and Hayes.[103a]

3.1.4 The Accuracy of Theoretical Calculations

Theoretical calculations have proliferated in the last six or seven years and have been a major topic attracting attention. In closing this section, therefore, we assess briefly the state of the art of theory.

Theoretical calculations produce qualitative understanding of the microscopic properties of deep centers. They also produce quantitative results, e.g., energy levels, which can be compared with experimental data. In section 2.1.2, we reviewed the historical evolution of theoretical methods and discussed briefly their advantages and disadvantages. It is therefore appropriate to ask whether the available theoretical methods can reliably calculate energy levels. Naturally, the ultimate criterion is agreement with experiment. Nevertheless, it is highly desirable that a theory should be able to stand on its own feet by being able to justify all approximations. In addition, one should be careful to determine what the experiment really measures so that a proper comparison with theory can be made. With that, we define a "state" to be a single-particle eigenvalue which may be occupied by one or more electrons. A "level," on the other hand, represents an ionization energy and is characterized by two charge designations. For example a 0/+ level in the band gap, measured from the conduction-band edge, represents the energy needed to transfer an electron from a neutral center to the conduction band, making the center positive. Equivalently, the same level, measured from the valence-band edge, represents the energy needed to transfer a hole from a positively-charged center to the valence bands, making the center neutral.[104,105]

In order to discuss the reliability of theory in determining energy levels quantitatively, we distinguish between A) the choice of Hamiltonian, and B) the choice of technique for solving the relevant Schrödinger equation.

A) Choice of Hamiltonian. Self-consistent Hamiltonians based on the local-density approximation for exchange and correlation[58] have no adjustable parameters. When used for a perfect Si crystal, however, accurate calculations reveal that such Hamiltonians fail to reproduce the fundamental band gap.[106] The theoretical gap is about half of the experimental gap. This is a shortcoming of the local-density approximation. In part, it reflects the approximation made in the exchange-correlation part of the density functional.[58] It also reflects the fact that, for the full density functional, the single-particle eigenvalues do not represent transition energies, as is true when one makes the local-density approximation. The question of how to go beyond the local-density approximation has been the subject of intense study and, very recently, a flurry of activity seems to be leading to some progress.[107] For calculations of defect energy levels in the gap, it is desirable to have a Hamiltonian that produces the correct band gap for the perfect crystal. In most cases, an ad-hoc correction is made, such as, for example, rigidly shifting all conduction bands upwards. Though such as an adjustment can be called an energy-dependent self-energy correction, it introduces an uncertainty whose effect on the final results is difficult to estimate. Comparison with experiment is not possible very often, however, because very few deep centers have been identified unambiguously. In the few cases where both theory and experiment are available, the agreement is quite good.[108] In other cases, questions involving energy levels may be answered even more reliably by theory when systematic cancellations are likely. For example, when one calculates the ordering of the A_1 and T_2 dangling-bond-like states of a vacancy, one can feel fairly confident that local-density uncertainties are likely to affect both states in roughly the same way.[80,109] In the case of transition-metal impurities in Si, Zunger[95] has gone beyond local-density theory by incorporating the so-called self-interaction correction. He found that local density gets the correct ordering of the E and T_2 states, but overestimates their splitting, leading to incorrect predictions about the ground-state multiplet. In the case of interstitial transition-metal impurities inclusion of the self-interaction correction shrinks the value of the splitting and produces the correct ground-state multiplet. The correction has not been included in the case of substitutional transition. Overall, the final word may not yet be in on transition-metal impurities. Spin-polarized calculations by Hemstreet and Dimmock[92] suggest that spin splittings are of the same order of magnitude or even larger than the crystal-field splitting. Thus, inclusion of spin-spin interactions by perturbation theory may be inadequate. Spin-polarized Green's-function calculations would be a more desirable starting point to understand the multiplet structure of transition-metal impurities.

Finally, we note that density-functional theory is truly a ground-state theory. In conjunction with the local-density approximation, it has been a very powerful tool in calculating *total energies* and, hence, ground-state properties of solids[110,111] and solid surfaces.[112] The same approach is currently being used in conjunction with the supercell technique[52] and Green's-function techniques[53-55] to calculate total-energy properties of deep centers.

When *semiempirical Hamiltonians* are used, one makes sure from the start that they yield the correct band gap, but then faces problems in defining the defect Hamiltonian. These problems are minimized by taking some kind of universal prescription for all impurities or classes of impurities, but systemic errors remain. Three types of semiempirical Hamiltonians have been widely used:

a) Hamiltonians based on the Extended-Huckel prescription have recently been found to yield seriously incorrect results.[90,113] It has been shown that the cause of the problem lies with the prescription used to determine Hamiltonian matrix elements.[114]

b) A whole family of semiempirical Hamiltonians which are more sophisticated then Extended Huckel theory (CNDO, MNDO, etc.), also borrowed from molecular chemistry, have been used exclusively with the cluster approximation.[115,116] They have a certain appeal for total-energy calculations,[115,116] but the cluster approximation appears to be a serious limitation. This is illustrated by the results of Ref. 116 where it was found that the addition of an extra shell of atoms altered the results significantly both quantitatively and qualitatively. In most applications of these methods, convergence of the results with cluster size is not pursued (see discussion below under the heading Choices of Numerical Technique). These Hamiltonians have never been used with a Green's-function technique, which would eliminate the uncertainties introduced by the small-cluster approximation.

c) Another type of semiempirical Hamiltonian, usually called "tight binding," is obtained by fitting the known band structure of the perfect crystal and then adopting a prescription for the defect matrix elements.[43,46] Typically, these Hamiltonians are used with the Green's-function technique for solving the Schrödinger equation, so that their accuracy can be judged free of extraneous uncertainties introduced by cluster approximations. In general, tight-binding Hamiltonians are useful for qualitative analysis, but one has to be careful.[117] For quantitative results, the only measure of accuracy is comparison with experiment, when available, or other, more sophisticated calculations. One systematic set of tight-binding Hamiltonians by Vogl, Hjalmarson and Dow[118] have been used extensively with a mixed record. Their first and major success was to reproduce the linear variation of a given deep level as a function of alloy composition (N in GaAsP alloys).[46] The theory's predictions for individual centers or classes of centers are often considerably off, however. For example, these Hamiltonians fail to predict any deep acceptors.[119] Their success in the case of the chalcogen impurities in Si, which are probably the best identified and reliably measured deep centers, is shown in Figure 12 and seen to be rather poor both in actual values and in the overall

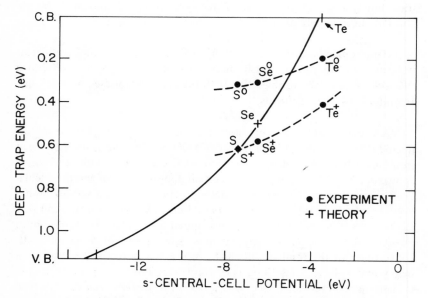

FIGURE 12 Comparison of the energy levels predicted by the tight-binding theory of Refs. 46 and 119 for S, Se, and Te in Si with experimental ionization levels. Note that the theory does not distinguish charge states.

trend. In contrast, the calculated levels of the surface antisite defects in III-V alloys seems to follow the rather complicated pattern formed by the experimental data for Fermi-level pinning at metal contracts of these semiconductors.[120] This agreement between theory and experiment has been interpreted as evidence that Fermi-level pinning is produced by anti-site defects at the relevant interfaces, but such a conclusion is not universally accepted.

Perhaps the most serious shortcoming of tight-binding Hamiltonians is the non-uniqueness of their results. It is generally assumed that it is adequate to fit the perfect-crystal energy bands and that improved fits would yield even better results for defects. It has been pointed out that different parametrizations of the perfect-crystal Hamiltonian yield significantly different vacancy states,[121] but, in that study, none of the Hamiltonians produce accurate condition bands. Thus, it could be argued that the differences in the vacancy states arise simply from the differences in the conduction bands of the various Hamiltonians. In a recent unpublished study, Ivanov and Pantelides included s, p and d basis orbitals and obtained excellent fits to both the valence and conduction bands. By distributing interaction strengths differently between first and second neighbors, it was found that identical fits to the energy bands result in significantly different ideal vacancy states.[122] The conclusion of this study is that the same band structure may have significantly different Bloch functions so that, for a unique tight-binding parametrization, one must also fit the Bloch functions.

B) The Choice of Numerical Technique. Once the Hamiltonian is chosen, one must also choose a technique to solve the Schrödinger equation describing the crystal containing the deep center. Green's-function techniques have been demonstrated to be very accurate because they introduce no new approximations. In some sense, given the Hamiltonian, the defect calculation is carried out as accurately as the corresponding calculation for the perfect crystal. In other words, the Green's-function method for defects is the analog of Bloch's theorem for a perfect crystal. Cluster methods, on the other hand, introduce a drastic approximation in the solution of the Schrödinger equation. Studies have shown that convergence of results is not achieved with the small clusters normally used,[22,116] but this problem is usually ignored. Ironically, the true accuracy of cluster calculations has never been tested properly because of additional approximations. For example, Extended-Hückel Hamiltonians, which have been used extensively for cluster calculations, have been found to have serious parametrization problems.[114] Self-consistent cluster calculations using the so-called scattered-wave $X\alpha$ method have been found to suffer from the fact that they make a spherical approximation to the potential around each atom.[95] It would be desirable to carry out self-consistent cluster calculations that make no other approximations and compare the results with corresponding Green's-function results. By the same token, it would be desirable to test semiempirical Hamiltonians (such as CNDO, MINDO, etc.) using a Green's-function technique without the uncertainties of cluster calculations.

3.2 Local Vibrational Modes

Recent semiempirical Green's-function calculations[56,57] have achieved detailed understanding of the local vibrational modes of deep centers. Some of the new results confirmed existing understanding developed on the basis of molecular (cluster) models, whereas others give totally new insights and quantitative detail. The overall picture is as follows. As in the case of electronic states, deep centers with tetrahedral symmetry exhibit strong A_1 and T_2 modes. The vacancy has A_1 and T_2 modes which correspond to symmetric breathing and vector-like motion of the surrounding lattice, respectively. When an impurity is inserted in the vacancy, the impurity atom cannot move in an A_1 mode so that the A_1 mode of the vacancy is not significantly disturbed. The impurity atoms, on the other hand, introduce a T_2 mode which interacts with the vacancy T_2 mode. As a result, a typical impurity has two prominent T_2 modes, one primarily associated with motion of the impurity atom and one primarily associated with vector-like motion of the surrounding lattice. One of the Green's-function results[57] is that the T_2 mode of oxygen in GaP appears as a two-peaked resonance. Such a structure had been seen, but had not been understood. More complex defects have, of course, more complicated local vibrational structure.

As of now, there exist no parameter-free calculations of local vibrational modes at deep centers. Instead, calculations are parametrized and the objective is primarily to understand experimental data.

3.3 Identification of Deep Centers

During the 1960's, identification of deep centers was a main concern primarily in the field of radiation-induced defects. In contrast, in studies of deep impurities, especially in Si and Ge, the question of identification was virtually ignored. Typically, a deep impurity was diffused in samples, experimental measurements were carried out, and the resulting signals were attributed to that impurity, often implicitly assuming that it occupies a substitutional site. This attitude changed, however, rather drastically in the late 1970's. The risk inherent in the simplistic point of view of the 1960's was highlighted in a series of experiments by Sah and coworkers[123] in the mid 1970's. Since diffusion is usually carried out at high temperatures (typically 1400 K), experiments were carried out in which samples were heat treated under diffusion conditions, but *without* having any impurity to diffuse in. The process introduced deep centers with roughly the same concentration as many known deep impurities, i.e., $10^{13}-10^{14}$ cm^{-3}.

Another illustration of the difficulties involved in identifying deep centers was given in 1977 by Fagelstrom and Grimmeiss.[124] Several authors had previously published absorption cross sections for Cu-doped GaP and the evidence was convincing that the center in question was Cu. The published thresholds, however, showed small but rather disturbing differences. A careful analysis revealed that the different curves correspond to samples with a different shallow-donor dopant. Some experiments were repeated and confirmed that GaP:Cu reproducibly yields a cross section with a threshold that depends on the shallow-donor dopant. Clearly, the shallow dopant is involved, somehow, in the Cu-related defect, but further experiments would be required to establish full identification.

The case of gold in silicon bears a number of similarities to GaP:Cu, in the sense that energy levels measured in gold-doped silicon were, for a long time, attributed to substitutional gold without any supporting evidence. Van Vechten and Thurmond[178] challenged that point of view in terms of thermodynamic arguments, and suggested that the observed energy levels are more likely to be due to a gold-vacancy complex. A recent comprehensive study by Lang, Grimmeiss, Meijer, and Jaros,[125] demonstrated that gold may in fact introduce a whole family of deep centers, some of which may involve shallow dopants or the ubiquitous oxygen. More significantly, Lang et al.,[125] on the basis of concentrations extracted from DLTS measurements, concluded that the familiar donor and acceptor levels associated with gold actually belong to two different centers. More recently, this point of view was challenged by Ledebo and Wang [126] who introduced a general technique designed to test whether two levels belong to the same or different centers. They presented evidence that the donor and acceptor levels in gold-doped Si do in fact belong to the same center, but did not resolve the conflict with the results of Ref. 125. The conflict has since been resolved by Feenstra and Pantelides[102] who introduced the concept of electron-hole exchange transitions. These transitions provide an explanation for the data of Ref. 125 which is consistent with the notion that the donor and acceptor belong to the same center (see section 3.6).

Just about every article in the present volume deals with the question of identification. The main lesson we learn from reading these articles is that no two systems are alike and no general prescription for quick identification is possible. One of the most powerful tools in the case of impurities is use of isotopes which produce small but observable shifts in vibrational spectra. Even when the dominant chemical impurity involved in a center is established, however, the interpretation of all the relevant data and the determination of the detailed atomic and electronic structure proceeds slowly over the years. Oxygen and the oxygen associates in GaP, reviewed in this volume by Dean, the chalcogen impurities in Si, reviewed in this volume by Grimmeiss and Janzen, the chromium impurity in compound semiconductors, reviewed in this volume by Allen and by Baranowski, and the copper impurity in ZnSe, reviewed in this volume by Robbins et al. are such examples. In other cases, deep centers are produced during crystal growth and are known by the particular effect they have on the crystal properties. Major efforts are sometimes undertaken to determine their other properties and to identify their atomic structure. Typical examples are the two "killer" centers in GaP reviewed by Peaker and Hamilton in this volume, and the so-called EL2 center in GaAs, which is primarily responsible for making GaAs semiinsulating. In the case of the killer centers, special experimental techniques had to be designed in order to measure their properties.

In many cases where identification has been pursued, accumulated evidence seems to point in a particular direction until, suddenly, a new observation changes the picture completely. Oxygen in GaAs and EL2 are notable examples. For years, EL2 was attributed to oxygen. Slowly, however, convincing evidence accumulated that oxygen is not involved in EL2. Martin and Makram-Ebeid make that case very strongly in their article in this volume. In recent years, attention has been focusing on the antisite defect or a complex involving the antisite defect. A large assortment of experimental evidence certainly points in that direction, but no definitive and completely specific identification of EL2 can yet be made. At the same time, however, the properties of oxygen in GaAs are slowly being elucidated. Wolford, Modesti, and Streetman[20] tracked the oxygen level from GaP to GaAsP alloys and concluded (Figure 13) that oxygen at As sites induces a deep level with an ionization energy of 0.79 eV, i.e. within the range of ionization energies usually reported for EL2. Since the evidence is quite strong that EL2 does not contain oxygen (EL2 has been found in concentrations that far exceed the net oxygen concentration[127]), Wolford et al.[20] suggested that the apparent equality of the ionization energies of O_{As} and

EL2 is merely accidental. Independent evidence of the existence of such a level associated with oxygen has been provided by capacitance transient measurements[128] and by theoretical calculations.[39,46] All the available evidence, however, still does not preclude that the oxygen-related center in GaAs is not isolated oxygen but some other complex containing oxygen. Additional points for the alloys near the GaAs limit would help elucidate this issue. At the same time, the identification of EL2, remains an open question. Characteristically, within a matter of a few weeks, I recently received two preprints of experimental papers. One of them concludes that EL2 is the isolated As antisite defect and the other concludes that EL2 is not the isolated As antisite defect. Since no definitive conclusion can be reached, we do not discuss any of the recent developments further.

Another example that illustrates many of the points that recur in the discussion of identification of deep centers is the following. Back in 1965,

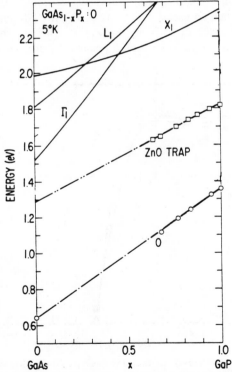

FIGURE 13   The energy level of substitutional oxygen in GaAsP alloys as measured by luminescence. From Ref. 20.

when the first luminescence spectra of irradiated Si were reported, a line at 0.97 eV, known later as the G line, was first attributed to the oxygen-vacancy center known as the A-center simply because of a rough coincidence of their respective energy levels.[129] Correlations with oxygen content quickly ruled that out, however, and all evidence seemed to point toward the divacancy (no dependence on impurity concentration or type, annealing data, symmetry from uniaxial stress measurements).[130] Two key experiments, however, ruled the divacancy out in 1976: Noonan, Kirkpatrick and Streetman[131] carried out low-temperature irradiation and found that the G line appears only after room-temperature annealing, showing that the center responsible for it is not a primary radiation product. Studies on different samples showed that the growth of the line could not be due to pairing of vacancies either. Konoplev, Gippius and Vavilov[132] reexamined the annealing behavior of the G line and the divacancy EPR spectrum and showed that they are significantly different (Figure 14). Having ruled out divacancies, both papers turned to another candidate: Again back in 1965, Watkins[133] reported an EPR spectrum labeled G11, which remained unidentified. The first attempt at identifying this center was made in 1974 by Brower[134] who, on the basis of EPR experiments with ^{13}C-doped samples, proposed a specific model involving

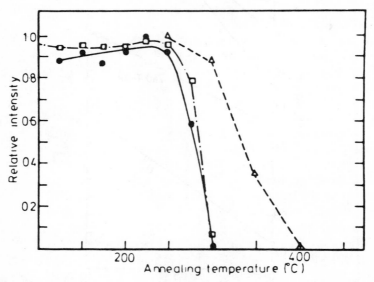

FIGURE 14   Annealing curves for the EPR spectrum of the divacancy and the 0.97-eV luminescence line in irradiated Si. From Ref. 132.

a pair of carbon atoms. In 1976, both Noonan et al.[131] and Konoplev et al.[132] proposed that the G line and the G11 EPR spectrum come from the same center. The primary supporting evidence was the annealing behavior of the two signals (Figure 14). But there was more. Both papers brought into the picture an absorption line that had been observed in the early 1960's and had remained unidentified except for the suggestion that carbon was probably involved.[135] The suggestion that this absorption line, also at 0.97 eV, is due to the same center as the 0.97−eV luminescence G line and the EPR spectrum G11 was subsequently confirmed by various experiments.[136] The detailed nature of the center went through more twists, however. Analysis of the phonon sidebands of the optical spectrum suggested that only one carbon is involved.[136] Subsequent work, however, using the more novel ODMR technique and specially enriched ^{13}C-doped samples led to a more specific model involving two carbon atoms.[137] In more recent work, Davies, Lightowlers, adnd to Carmo[137a] reexamined the vibronic sidebands of this center and confirmed that the local modes involve motion of only one carbon atom, but also suggested that some evidence exists that a second carbon atom may indeed be present. Nevertheless, a number of questions still await definitive answers.[138]

The above historical account illustrates that, until the late 1970's, cross-correlation of data taken by different techniques was limited. In the late 1970's, cross correlation became the primary theme and new techniques, such as ODMR, made the task of cross correlation considerably easier. In other cases, as, for example, the transition-metal complexes with shallow dopants,[139] cross correlations were facilitated by measurements with junction techniques. The chalcogens in Si (Grimmeiss and Janzen, this volume) are one of the best examples of comprehensive cross correlation of results of many techniques and thorough identification.

In all the cases we discussed above, theory played only a supporting role, providing the general principles in terms of which to interpret experimental data. For example, the theoretical description of electrons in the presence of magnetic fields provides a set of equations which are used to fit EPR data. Group theory is usually very important in determining the symmetry of the center responsible for the observed signals. On a number of occasions, however, theory has gone beyond that role. For example, in the early 1970's, Van Vechten[140] developed thermochemical models for predicting equilibrium concentrations of native defects in semiconductors. He predicted that antisite defects and antisite complexes are likely to exist in large concentrations. This prediction catalyzed the subsequent identification of the P antisite defect in GaP,[141] and, more recently, provided added credibility to the suggestion that EL2 in GaAs is an antisite complex (see discussion by Martin and Makram-Ebeid, this volume). The attribution of a luminescence peak to substitutional oxygen in GaAs by Wolford, Modesti and Streetman[20] (Figure 13) was definitely aided by existing theoretical predictions that oxygen induces a deep donor state,[39] and, even more so, by the theoretical prediction of Hjalmarson et al.[46] that the position of the oxygen deep state in the $GaAs_{1-x}P_x$ series varies linearly with x. In the case of the luminescence "killer" centers in GaP, Peaker and Hamilton (this volume) refer to a number of theoretical results as they explore a large number of possible candidates. In the case of the EL2, however, as Martin and Makram-Ebeid unhappily observe in their article, theory has offered very little help.

In 1981, for the first time, an established identification was challenged on the basis of parameter-free Green's-function calculations. Kennedy and Wilsey[142] had earlier attributed an EPR signal in GaP (labeled NRL−1) to the Ga vacancy. The total electronic spin of the center was deduced to be 1/2. In addition, the isotropy of the measured g value implied a center with tetrahedral symmetry. In order to attribute the signal to the Ga vacancy, it was necessary[142] to assume that the ordering of the A_1 and T_2 "dangling-bond" states of the Ga vacancy is inverted by comparison with that occurring in the Si vacancy. Thus, having the A_1 state above the T_2 state would explain the stability of the center against symmetry-lowering Jahn-Teller distortions. The theoretical results of Scheffler, Pantelides, Lipari, and Bernholc,[109] however, predicted normal ordering, even with inward or outward breathing. In contrast, calculations for carbon impurities at Ga sites showed that such centers had properties which were fully consistent with the available experimental data. Subsequent experimental work, aimed at obtaining additional information about NRL-1, revealed that the spin-1/2 assignment was incorrect and that the center actually had spin 3/2.[94] It was then concluded that the only center that could account for the data and also be consistent with the theoretical results is the neutral vacancy with three parallel-spin electrons in the T_2 state in the gap.[94,80] A number of observations still remain puzzling,[80] however, and additional work is needed to clarify the properties of this center.

3.4 Transition-Metal Impurities

We have already discussed a number of recent developments in transition-metal impurities in section 3.1 where we reviewed recent theoretical results. Experimental investigations of transition-metal impurities using a large assortment of experimental techniques (luminescence, optical absorption, photoconductivity, acoustic paramagnetic resonance, Mossbauer spectroscopy, junction techniques) have produced a wealth of information. In many cases, detailed energy-level diagrams for the various charge states and multiplets have been established. Some of these charge states have been found to exhibit Jahn-Teller distortions. Others have been found to bind excitons. Detailed and critical discussions of many transition-metal impurities in compound semiconductors can be found in the articles by Allen, by Baranowski, by Bishop, and by Robbins et al. in this volume. For a critical discussion of transition-metal impurities in Si, the reader is referred to the recent review paper by Weber.[72] Iron and iron complexes are also surveyed by Ammerlaan.[143] All these papers make it very clear that transition-metal impurities have a very complex electronic structure and form a variety of defect complexes. As Allen remarks at the end of his lengthy account of the properties of Cr in GaAs in this volume, "it may be thought that, after so much work, GaAs:Cr is well understood and only details remain to be settled . . . (however) large gaps remain in our knowledge." He concludes that "although some areas are well founded, some others, which are of fundamental importance, have, as yet, hardly been touched."

3.5 Lattice Relaxation

Lattice relaxation in the vicinity of deep centers is quite common. After all, by its very definition, and in contrast to shallow impurities, a deep center introduces a strong and localized perturbation potential. It is, therefore, natural for the host atoms in the immediate vicinity of the deep center to assume new equilibrium conditions. In this section we discuss some of the causes of specific types of lattice relaxation and some of the consequences of lattice relaxation.

3.5.1 Causes of Lattice Relaxation

There are two general types of lattice relaxation. One preserves the symmetry of the deep center (breathing mode), the other reduces the overall symmetry. In the case of symmetry-conserving relaxation, one is usually interested in understanding what causes the sign of the relaxation. For example, one of the age-old questions associated with vacancies is whether the nearest neighbors "breathe" inward or outward. In general, one can identify several effects which induce breathing one way or the other. For example, by breathing toward the vacant site, the nearest neighbors of a vacancy get closer together, leading to energy gain from the formation of stronger bonds among the dangling hybrid orbitals. On the other hand, such motion weakens the back bonds. The final answer can be obtained only from a very careful parameter-free calculation or, perhaps, experiment. Once a calculation is done, one can then look at the pieces and understand the interplay that determines the final result. Such calculations are just now beginning to be possible[52-55] and such analysis is still not available.

Symmetry-breaking lattice relaxation has been probed more extensively. One major cause for such relaxation is the Jahn-Teller instability: a localized system in an orbitally degenerate state distorts spontaneously and lowers its symmetry until the resulting state has no orbital degeneracy. The proof of the Jahn-Teller theorem[65] is quite straightforward and will not be repeated here. In general, the orbital degeneracy is replaced by orientational degeneracy. For example, the ground state of the neutral vacancy in Si, which we discussed in Section 2.4.1, is initially an orbital triplet without any orientational degeneracy. After Jahn-Teller distortions set in, the ground state is an orbital singlet, but there are three different orientations that the distortion can assume.

A number of cases are known where symmetry-breaking distortions occur even though the undistorted configuration has no orbital degeneracy. One such case is substitutional nitrogen in diamond, which was observed by EPR to be distorted.[144] Until recently, the observed distortion was believed to be due to the Jahn-Teller effect on the basis of Extended-Hückel cluster calculations.[32] More recent Green's-function calculations by Bachelet, Baraff, and Schluter,[113] however, found that the ground state of neutral nitrogen in diamond has no orbital degeneracy. This conclusion was supported further by Lannoo[114] who traced the shortcoming of the Extended Hückel Hamiltonian used in Ref. 32, and by calculations on other similar systems.[46,80]

Substitutional oxygen in silicon[66] and substitutional nitrogen in silicon[144] are other examples of centers whose symmetry has been found to be lower than T_d even though the ground state in the T_d configuration is not orbitally degenerate. In particular, as in the case of nitrogen in diamond, the impurity is found to be off-center.

The causes of symmetry-lowering distortions occurring in deep centers without electronic orbital degeneracy are not well understood. One can immediately invoke simple chemical notions that somehow rationalize the observed distortions: Nitrogen and oxygen are first-row elements with small atomic radii. As a consequence, they move off center to attain a more preferred distance to at least some of the neighbors. Clearly, this argument works when the host is Si, but not in the case of nitrogen in diamond. Another argument that is often mentioned[114,116,144] is based on the preferred coordination of elements, in particular the so-called 8–N rule. According to this rule, an atom with N valence electrons prefers to be bonded to 8–N other atoms. Thus, nitrogen (N=5) prefers to be bonded to three atoms (e.g., NH_3), oxygen (N=6) prefers to be bonded to two atoms (e.g., the molecule H_2O and the solid SiO_2), and carbon prefers tetrahedral coordination (e.g., CH_4). This simple observation seems to hold in the first-row impurity cases, since nitrogen moves off-center toward three neighbors,[144,145] oxygen moves toward two neighbors, and carbon in Si seems to stay on-center. There is a definite analogy here with the molecules NH_3, OH_2 (normally written H_2O), and CH_4. Nevertheless, this is merely an analogy, not an explanation. The N–8 rule is not universal (note, for example, the double bonds of organic chemistry and the fact that graphite has larger cohesive energy than diamond). In addition, atomic size and the N–8 rule do not explain why nitrogen in diamond has reduced symmetry whereas phosphorus in silicon has full tetrahedral symmetry.

A more microscopic explanation has been proposed by Lannoo[114] who invoked the so-called pseudo-Jahn-Teller effect. The argument is that a system whose ground state has no orbital degeneracy may distort in a manner analogous to the Jahn-Teller effect if the lowest excited state or states are close by in energy. Then, one can view the ground state as quasi-degenerate and treat the effect of distortion by degenerate perturbation theory as in the case of the Jahn-Teller effect.[65] Energy gain occurs from mixing the excited states with the ground state, but the gain varies quadratically with the amplitude of the distortion. Thus, the pseudo-Jahn-Teller effect does not drive the distortion. It merely stabilizes the distortion, which must be initiated by some other process (e.g., thermal vibrations).

In the one-election approximation, the pseudo-Jahn-Teller effect may arise when the state in the gap containing one or two electrons is not orbitally degenerate (e.g., of A_1 symmetry), while another empty state, say of T_2 symmetry, is slightly higher in energy. Upon distortion, the A_1 and T_2 states interact so that the A_1 state goes down in energy.[114] That is precisely the case treated by Lannoo[114] as an explanation of the observed distortions of first-row substitutional impurities.

Lannoo's suggestion was corroborated by Watkins, DeLeo and Fowler,[116] who carried out cluster calculations for nitrogen and oxygen in Si using the so-called Modified-Neglect-of-Differential-Overlap (MNDO) method, which is a semiempirical method developed originally for molecules. Their total-energy calculations on small clusters reproduced the observed distortions. They then interpreted their results by analysing the wave-function composition of the A_1 and T_2 bound states. It was found that, as a function of distortion, there is strong mixing between the original A_1 state and one of the components of the split T_2 state. It was thus concluded that the relevant mechanism is the pseudo-Jahn-Teller effect.

In more recent work, however, Car and Pantelides[146] raised doubts about the appropriateness of the pseudo-Jahn-Teller effect as the mechanism responsible for the observed distortion of the oxygen and nitrogen centers. As described in Refs. 114 and 116, in the one-electron approximation, the pseudo-Jahn-Teller effect focuses only on interactions between the empty T_2 state and the occupied A_1 state in the gap, ignoring possible mixing between the empty T_2 state and other occupied one-electron states in the valence bands. Such mixing was not looked for in the analysis of Ref. 116. A look at the results reported in Ref. 116, however, reveals that such mixing must indeed occur to a significant extent. For example, the simple pseudo-Jahn-Teller formulas[114,116] require that the total energy gain is smaller than the gain contributed by the shift in the occupied A_1 gap state. Yet, the results of Ref. 116 reveal a total energy gain which is twice as large as the gain arising from the shift of the A_1 gap state.

The above suggestion that mixing between an empty localized state and occupied states in the gap *and* in the valence bands may be viewed as a generalized version of the pseudo-Jahn-Teller effect. The equations given by Lannoo[114] can be generalized to take into account additional occupied states. It is more instructive, however, to analyse the problem in terms of chemical rebonding as follows. As we saw in Section 3.1.1, a deep impurity is characterized by a set of localized states which lie in the fundamental gap, the band continua, and, sometimes, below the valence bands. These states result from interactions between the impurity orbitals and the localized states of the vacancy. In the case of oxygen or nitrogen at a substitutional site in Si, the primary localized states are an A_1-T_2 pair of bonding hyperdeep states in and below the valence bands and an A_1-T_2 pair of antibonding states in the gap region (very similar to the structure of Si:S shown in Figure 7). The bonding pair of hyperdeep states is fully occupied, whereas the antibonding pair is only partially occupied depending on the charge state of the center. According to the conventional pseudo-Jahn-Teller mechanism,[114,116] upon distortion, energy gain arises only from the mixing occurring among the partially occupied states in the A_1-T_2 pair in the gap region. The alternative approach discussed here is to take into consideration the entire manifold of localized states for the following reason: Bonding between the impurity orbitals and the vacancy states determines the actual energy separation between the two A_1-T_2 pairs. Thus, any change in the bonding, such as caused by distortion, will affect this energy separation. If bonding is increased, as one would anticipate for an energy-lowering distortion, the primary effect would be to lower the "center of gravity" of the bonding pair and raise the "center of gravity" of the anti-bonding pair. Thus, energy gain would definitely arise from shifts in localized states below the valence-band top. Note that, even if this shift is small (say of order 0.1 eV), the A_1-T_2 bonding states contain eight electrons so that the net gain is large (of order 1 eV). In addition, the states in each A_1-T_2 pair mix and split, which can contribute to the energy gain depending on the occupation of the levels in the gap region. This gain, however, comes from only one or two electrons so that very large shifts would be needed to produce substantial energy gain.

The above discussion reveals that the causes of lattice distortions are not simple to unravel in systems that are not driven by the classic Jahn-Teller mechanism. The recent advances in total-energy calculations using self-consistent Green's-function methods[52-55] are likely to provide more definitive answers in the future.

3.5.2 Consequences of Lattice Relaxation—Negative-U centers

One of the most fascinating consequences of lattice relaxation around deep centers is the possibility of negative-U centers. The quantity U refers to the energy change that accompanies the addition of an electron to a center in its ground state when the highest-energy occupied orbital is only partially occupied. If no relaxation is involved, the energy change is positive, reflecting the repulsive nature of electron-electron interactions. The energy change is usually called the Hubbard U, after Hubbard[148] who first explored its consequences in a variety of systems. Anderson[149] was the first to note that, if lattice relaxation changes after the capture of the extra electron, the total-energy change would consist of the positive Hubbard term plus a negative term arising from lattice relaxation. There exists, therefore, a possibility that the total energy change, i.e., the effective U, usually called the Anderson U, is negative. The main consequence of a negative U is that a particular charge state is never the equilibrium state no matter where the Fermi level is in the gap. We illustrate this effect in Figures 15 and 16. We choose a defect which is positively charged when the level in the gap contains one electron. On the left of Figure 15, we show the normal ordering of the two different charge states when lattice relaxation is neglected. On the right, we show the negative-U ordering after lattice relaxation. In Figure 16, we show how the total energy of the three charge states of such a defect vary with the Fermi level when U is positive and when U is negative.

Anderson introduced the notion of negative U in an effort to account for the absence of paramagnetism in chalcogenide glasses: if defects in

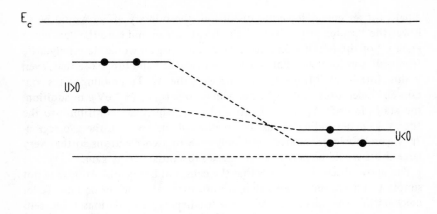

BEFORE RELAXATION   AFTER RELAXATION

FIGURE 15 Schematic levels in the gap of a semiconductor for a deep center with (a) positive Hubbard U, and (b) negative effective U.

those glasses have negative–U character, as in the example above, the ground state will never contain an odd number of spins. Experimental or theoretical confirmation of particular defects having negative–U character in these glasses has not been obtained, however.

The first defect to be recognized as a negative–U center was the vacancy in silicon. On the basis of self-consistent Green's-function results for the energy levels of the vacancy and a phenomenological model for the total energy of three charge states, Baraff, Kane, and Schluter[150] predicted in 1979 that the vacancy has negative–U character. The prediction was consistent with the fact that the EPR-active V^+ state is observed by EPR only after photoexcitation. Detailed quantitative experimental support was lacking, however: The theoretical value for the activation energy for the decay of V^+ was 0.2 eV, which compared only moderately well with the value 0.05 eV extracted from the decay of the EPR signal by Watkins. In addition, the theoretical error bar for the effective negative U spanned negative as well as positive values. Soon afterwards, however, Watkins and Troxell[151] presented a new analysis of the available data in the light of the negative-U idea and presented some new data which supported the negative-U suggestion. In particular, they noted that, in a DLTS experiment, hole emission from V^{++} occurs by the simultaneous emission of two holes. Thus, there should be only one DLTS peak corresponding to the energy needed to release the first hole. By a clever experiment, monitoring the vacancy concentration by the number of vacancy-tin pairs, Watkins

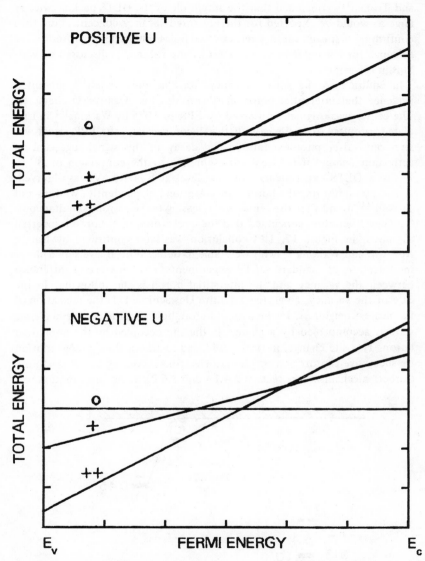

FIGURE 16 Schematic plot of the total energy of various charge states of a deep center as a function of Fermi-level position in the gap of a semiconductor. Top panel: "normal" deep center with positive Hubbard U; bottom panel: "negative-U" deep center. Note that the total energy of the neutral deep center is independent of the Fermi level.

55

and Troxell[151] concluded that the amplitude of the DLTS peak is twice as large as would be expected for the measured vacancy concentration, thus confirming that each vacancy releases two holes simultaneously. Additional evidence for the negative–U character of the vacancy is discussed by Watkins in this volume.

In addition to the silicon vacancy, Watkins and Troxell[151] presented evidence that interstitial boron in silicon also has negative–U character. The center was originally observed by EPR in 1975 by Watkins[152] in irradiated samples. As in the case of the silicon vacancy, the EPR signal was seen only after photoexcitation. The decay of this signal suggested an activation energy of 0.13 eV corresponding to the conversion of B° to B^{+}. In a DLTS experiment, only one level at E_c–0.45 eV was observed. Again, by monitoring the boron concentration independently, Watkins and Troxell[151] found that the center was releasing two electrons simultaneously. They, therefore, concluded that the level structure of boron is inverted as shown in Figure 17. This conclusion was later confirmed by Harris, Newton, and Watkins[153] who were able to detect both B levels by a clever modification of standard DLTS experiments. One interesting difference between the vacancy and the interstitial boron is the following: In the case of the vacancy, a negative effective U is achieved via the relaxation of the nearest neighbors. In the case of boron, however, the change in charge state is accompanied by a change in the site occupied by the interstitial boron. This site change, in turn, could lead to athermal long-range motion by successive capture and emission of electrons according to the Bourgoin-Corbett mechanism (see section 2.4.1A and 3.6.2). Long-range motion was

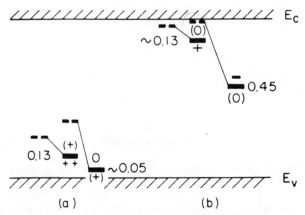

FIGURE 17   The energy levels of the "negative-U" interstitial boron center in Si as deduced from experiments. From Ref. 151.

observed in the case of boron after the injection of minority carriers.[151] However, it was not possible to determine the precise sites and migration path or to demonstrate conclusively that a simple Bourgoin-Corbett mechanism is operative.

More recently, Car, Kelly, Oshiyama, and Pantelides,[55] on the basis of self-consistent Green's-function total-energy calculations, predicted that the silicon self-interstitial is also a negative–U center. Its behavior is predicted to be similar to that of interstitial boron discussed above in the sense that the negative effective U is achieved by a change in equilibrium configuration. More specifically, it is found that I^{++} is stable at the tetrahedral interstitial site, I° is stable at either the hexagonal site or a bond-centered configuration (the two sites are degenerate within the uncertainty of the calculation). I^{+} is predicted to be a metastable state, i.e., it is never the equilibrium state for any Fermi-level position (Figure 18). This prediction may be taken as an explanation of the fact that self-interstitials have not been detected by EPR, but the more likely reason for this state of affairs is the high athermal mobility of interstitials. The predicted change in equilibrium site with charge-state change provides a natural explanation for the athermal migration (see discussion in section 3.6.2).

In addition to inducing negative–U behavior in some centers, lattice relaxation has a variety of other fascinating consequences such as self-trapping, persistent photoconductivity, electron-hole recombination via multiphonon emission, recombination-enhanced defect reactions, etc. We discuss some of these phenomena in other sections in the remainder of this paper.

3.5.3 Large Lattice Relaxation–Self-Trapping and Persistent Photoconductivity

Lattice-relaxation effects were first studied in ionic crystals, particularly alkali halides, where large electron-phonon coupling gives rise to rather dramatic effects. One such effect is self-trapping: in an otherwise perfect crystal, an electron in the conduction band (or hole in the valence bands) can induce a local lattice distortion which sets up a localized potential that traps the electron (hole) in a bound state.

Lattice relaxation of a deep center can be conveniently displayed schematically in terms of configuration-coordinate diagrams. For example, on the left side of Figure 19, the two parabolas may represent the total energy of the crystal as a function of some generalized defect coordinate Q (e.g., in the case of a vacancy in silicon, Q may be the amount of tetragonal distortion). In the example of Figure 19, the lower parabola represents the total energy when the defect is in some particular charge state

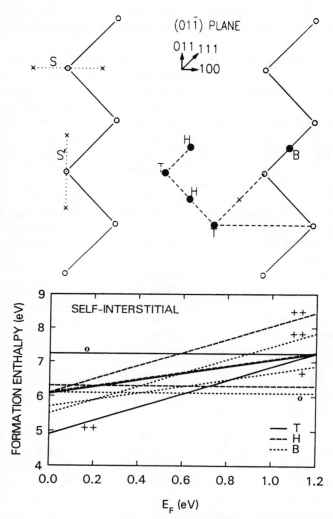

FIGURE 18   (a) Schematic of the Si atoms in a (110) plane indicating three different sites for the self-interstitial in Si: T = site with tetrahedral symmetry, H = site
with hexagonal symmetry, B = bond-centered site, S = split configuration. The figure
does not show the relaxation of the nearest neighbors which is essential in the case of
the B site and S sites. (b) Formation energies of the self-interstitial in Si for three
different sites: T, H, and B. Note the negative-U behavior. From Ref. 55.

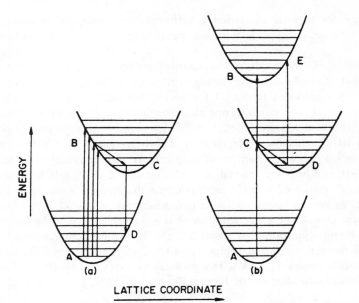

FIGURE 19 Schematic configuration-coordinate diagram for a deep center. Left: Two different charge states. Optical excitation corresponds to one of the transitions labeled AB. The various arrows correspond to phonon sidebands. After the initial optical excitation, the electron will relax to state C and radiative recombination can occur with energy CD. The difference between AB and CD is the Franck-Condon shift. Right: Same as schematic on the left. The new parabola at the top corresponds to the same charge state as the parabola at the bottom but includes an electron-hole excitation across the gap AB. These curves correspond to a "normal" deep center with moderate lattice relaxation.

whereas the upper parabola represents the total energy after an electron has been excited from the valence bands into the gap state, changing the charge state of the center by one unit. The displacement of the parabola to the right suggests that, after the excitation, the defect relaxed to a new value of Q. On the right side of Figure 19, we show a third parabola at higher energies representing the total energy of the system in the original charge state after an electron has been excited from the valence bands to the conduction bands without affecting the charge state of the deep center. This description is a classic description of a "normal" defect that exhibits moderate lattice-relaxation effects. The main characteristic of these defects is the well-known Stokes shift illustrated in Figure 19 on the left: the optical absorption energy (AB) is larger than the corresponding emission (luminescence) energy (CD). In addition, the thermal activation energy for thermal excitation is different from both optical-absorption and emission energies. In our example, it is equal to the energy difference AC.

Negative—U centers do not necessarily have very large lattice relaxations. Instead, enough lattice relaxation is required to overcome the electron-electron repulsion (Hubbard U) when another electron is added to a gap state that already contains one electron. Since, typically, in semiconductors, Hubbard U's are 0.2—0.4 eV[80,81,150] negative—U behavior can occur with lattice-relaxation energies only slightly larger. In the last few years, on the other hand, a number of deep centers with unusually large lattice distortions have been studied. In 1977, Piekara, Langer, and Krukowska-Fulde[154] presented experimental evidence that the In impurity in CdF_2 exists in two different configurations which are separated by an energy barrier. One of these configurations has a shallow level with both optical and thermal ionization of about 0.1 eV. The other configuration has a small thermal ionization energy (about 0.2 eV), but a very large optical ionization energy (1.9 eV). The data can be explained by the configuration-coordinate diagram of Figure 20. Also in 1977, Lang and Logan[155]

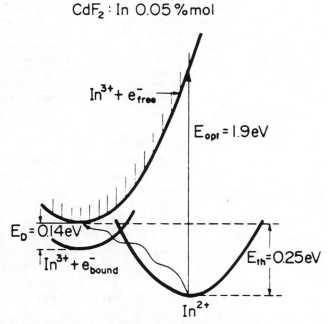

FIGURE 20 Schematic configuration-coordinate diagram for a deep center with large lattice relaxation. From Ref. 154.

proposed a similar configuration-coordinate diagram (Figure 21) to account for persistent photoconductivity in compound semiconductors. Both these diagrams have a unique feature that distinguishes them from the corresponding diagrams for normal defects shown in Figure 19. Note, for example, that the curve labeled D in Figure 21 crosses the curve labeled C to the right of Q=0, whereas, in normal defects, the crossing occurs to the left of Q=0. This feature is the distinguishing mark of deep centers that exhibit "large lattice relaxation." One of the most intriguing consequences of this feature is persistent photoconductivity. For example, in Figure 21, when an electron is bound to the defect, the equilibrium site is at Q_R. Light of frequency 1.1 eV or higher can excite the electron into the conduction band. A number of such electrons will lose energy to lattice vibrations and drop to the bottom of the conduction band, contributing to photoconductivity. At low temperatures, this photoconductivity persists for hours because of the 0.2 eV barrier for recapture. It should be noted, however, that persistent photoconductivity can arise from other mechanisms as well, as described, for example, by Queisser and Theodorou.[156]

In 1978, Toyozawa[157] described the general behavior of defects for a range of strengths of the electron-phonon coupling. He referred to the

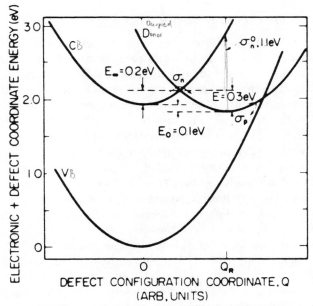

FIGURE 21 Schematic configuration-coordinate diagram for the so-called DX centers in II-V semiconductors exhibiting large lattice relaxation. From Ref. 155.

behavior described by the configuration-coordinate diagram of Figure 21 as "extrinsic self-trapping" by analogy to the self-trapping that occurs in ionic crystals even in the absence of defects.[158] In the latter case, an electron in the conduction bands (or hole in the valence bands) can induce a large local distortion in the lattice which can set up a localized potential strong enough to bind the electron (hole). In the case of extrinsic self-trapping, a defect, in a given configuration (e.g. at Q=0 in Figure 21) does not have a strong enough potential to bind an electron. Similarly, in the absence of a defect, electron-phonon coupling may not be strong enough to induce self-trapping of an electron. In the presence of the defect, however, a combination of the defect potential and electron-phonon coupling may be able to induce binding. The family of deep centers known as DX centers in III–V alloys are examples of self-trapped centers described by the diagram of Figure 21. Their properties are described in detail in the article by Lang in this volume. More recently, a center in InP labeled the M center, has been found to manifest properties that are consistent with extrinsic self-trapping.[159]

3.6 Carrier Capture and Recombination

3.6.1 Mechanisms of Carrier Capture

Deep centers usually trap electrons or holes in localized states in the band gap. Thus, deep centers can be classified as electron traps or hole traps. Recombination centers are those that capture a minority carrier and subsequently capture a majority carrier, the net result being the annihilation of a free electron-hole pair. The energy released during capture can a) be emitted in the form of light (radiative transitions); b) transferred to another electron or hole (Auger process); or c) be transferred to phonon modes (cascade mechanism or multiphonon emission).

Radiative capture is simply photoluminescence and, when it occurs, is a powerful tool for studying the electronic states of deep centers. Quite often, emission of light is accompanied by phonon emission or absorption, resulting in the so-called phonon sidebands, which can provide valuable information about the symmetry of the center. Many examples are described in several of the articles in this volume. Luminescence is also very useful in practical applications, as in LED's and phosphors used in cathode-ray tubes (e.g., ZnS).

Nonradiative Auger transitions are not very common because, in most cases, their likelihood is proportional to the free-carrier concentrations. Thus, in general, they are significant only in heavily-doped crystals. The general theory of Auger processes has been developed,[160] but actual calculations of cross sections are difficult and rather rare.[161] They have not attracted significant attention in recent years. A special kind of Auger transition has been observed in the case of some shallow centers which can bind an exciton.[162] In such a case, there are enough particles bound to the center to allow an Auger process. For example, a bound exciton can recombine, imparting the recombination energy to a trapped electron or hole and ejecting it into the appropriate band.[162] A somewhat analogous process was recently recognized to be playing an important role in the capture of carriers by amphoteric deep centers, namely centers that have both a deep acceptor and a deep donor level in the fundamental energy gap (see below).

Nonradiative capture by phonon emission has, however, received a good deal of attention. Two distinct possibilities were identified back in the 1950's:

Cascade capture[30] occurs when the center has a series of closely spaced energy levels near the relevant band edge. The carrier goes through this ladder of levels by emitting one or a few phonons each time. Eventually, it drops into the ground state, again by emitting a few phonons. The theory of the cascade process[30] reveals that the cross section has a power-law dependence on the temperature of the form T^{-n}. Recent variants of the theory,[163,164] disagree about the value of n, which is usually between 1 and 4 (see discussion by Grimmeiss and Janzen and by Peaker and Hamilton, this volume). Typically, it is expected that the cascade process occurs in shallow centers with a Rydberg series of excited states where the ground state can be reached by emitting only a few phonons.

Multiphonon capture,[128,129] on the other hand, does not rely on closely-spaced Rydberg-like bound states. Instead, it occurs because of the existence of local phonon modes with strong electron-phonon coupling. A configuration-coordinate diagram such as Figure 21 can help illustrate the process. In Figure 21, curve C describes the energy of the system when an electron is in the conduction band in any one of the allowed vibrational states (see Figure 19). Similarly, curve D describes the total energy of the system when the electron is bound at the defect, again in one of the allowed vibrational states. When the two curves cross, the system can pass from the one configuration to the other. In Figure 21, σ_n marks the crossing for electron capture and σ_p marks the crossing for hole capture. Multiphonon capture can occur in both normal defects (Figure 19) and large-lattice-relaxation defects (Figure 21). One of its characteristics is an activated behavior for the cross section because the defect must be excited to the vibrational state at the crossing point.

The formal theory of multiphonon capture is, however, very complex and some problems of principle still remain.[165] The general theory was developed in the 1950's.[28,29] A series of formal papers by Kovarskii and Sinyavskii[166] appeared in the 1960's. Until quite recently, however, multiphonon emission was considered unimportant in semiconductors because theoretical estimates suggested very small capture cross sections. In 1977, Henry and Lang[167] used junction techniques to measure several capture cross sections of deep centers in GaAs and GaP. For most of the measured cross sections, they were able to rule out Auger nonradiative transitions and demonstrated that capture occurs via multiphonon emission. They also developed a simple theory of multiphonon emission. The central contributions of that paper were to include the effects of the breakdown of the adiabatic approximation near the crossing point and to carry out explicit calculations in terms of a simple model demonstrating that multiphonon-emission capture cross sections are large and in overall agreement with observed capture cross sections (see discussion in the article by Lang in this volume). In recent years, attention has focussed on calculating the rate for multiphonon emission in specific models.[167-169] In general, however, it is not straightforward to identify the particular model that applies to a real system.

Carrier capture by an amphoteric deep center is particularly fascinating. Let us consider an amphoteric deep center in its + charge state. Capture of an electron by such a center is believed to occur in two steps: First, the electron is captured into Rydberg-like states by a cascade process, and then drops into the "ground state" by a multiphonon process. It was recently noted by Pantelides and Feenstra,[102] however, that, while the electron is in the Rydberg states in a configuration $(C^+, e^-)^*$, an Auger-like "electron-hole exchange transition" may occur, leaving the center in a configuration $(C^-, h^+)^*$, i.e., a configuration with a hole in a Rydberg state. The hole can then drop into the "ground state" by multiphonon emission. This type of internal transition has important consequences. In particular, it provides a natural explanation for the apparent inequalities in the concentration of the donor and acceptor levels measured in gold-doped Si,[125] eliminating the need to invoke two different centers (see discussion in Section 3.3).

The paper by Peaker and Hamilton in this volume describes two centers in GaP which are quite deep (0.75 and 0.95 eV in a gap of 2.4 eV), but their nonradiative capture cross sections exhibit a T^{-n} dependence on temperature. This observation implies that, after the carrier cascades through the Rydberg series, it drops into the ground state very quickly so that the initial cascade capture is the rate-limiting step. The circumstances under which such behavior occurs are not well understood.

One of the most fascinating consequences of multiphonon capture is its ability to influence atomic motion or enhance defect reactions by supplying energy in excess of what is normally available from lattice vibrations at a given temperature. In some cases, multiphonon capture can provide enough energy for a reaction to occur athermally. We discuss these possibilities in the next section.

3.6.2 Consequences of Carrier Capture and Recombination: Enhanced or Athermal Migration and Defect Reactions

From the point of view of conductivity, the capture or recombination of carriers has one immediate and obvious consequence: the number of carriers and, hence, conductivity is reduced. From the point of view of the defect, capture and/or recombination of carriers can have rather dramatic consequences. These consequences are unique to semiconductors and insulators which have a band gap and, thus, true bound states for carriers.

A) Carrier Capture: Charge-State Effects. Let us first consider carrier capture. Its primary effect is to change the charge state of the defect. One then notes that different charge states of the same deep center may have different barriers for migration or for reacting with other centers. Thus, carrier capture can either enhance or retard defect migration or particular defect reactions. In practice such an effect can be seen by simply changing the doping of the sample. Early examples are the different annealing rates of radiation-induced defects in n-type and p-type Si.[31] More recently, junction techniques, which allow one to inject excess minority carriers, have provided more dramatic illustrations of this effect. Kimerling, de Angelis and Diebold[170] first demonstrated the effect in the case of the phosphorus–vacancy pair (Figure 22). Many such examples have been reported since.

Interstitial aluminum in Si has recently attracted attention as an example of enhanced migration. In irradiated p-type Si, Al was detected at the tetrahedral interstitial site in the + + charge state by EPR in 1965 by Watkins.[133] It was found to anneal thermally with a barrier of 1.2 eV. In 1978, Troxell, Chatterjee, Watkins and Kimerling[171] observed that, under minority-carrier injection conditions, the migration barrier is lowered by 0.93 eV. They interpreted this effect as recombination-enhanced migration (see below). More recently, however, Baraff, Schluter and Allan,[172] on the basis of self-consistent Green's-function calculations interpreted the observed enhancement in terms of a mechanism akin to the charge-state effect we are discussing here. According to Baraff et al., the energy levels of Al interstitials at the tetrahedral and the hexagonal site are as shown in Figure 23. Under injection conditions, Al^{++} first captures one electron and converts to Al^{+} with, presumably, negligible effect on the migration bar-

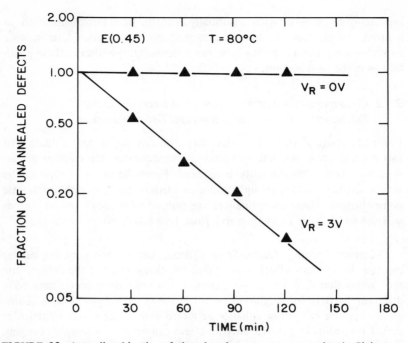

FIGURE 22 Annealing kinetics of the phosphorus-vacancy complex in Si in two different charge states at 80°C. For zero reverse-bias voltage (V_R = 0), most centers are neutral, whereas for V_R = 3 V most centers are ionized and exhibit an enhanced annealing. From A. Chantre, M. Kechouane, and D. Bois, Physica, *116B*, 547 (1983).

bier. It can, however, capture a second electron in a hydrogenic orbit and become neutral. Small vibrations can then move the Al toward the hexagonal site, lowering the energy of the loosely bound electron. Farther down the path, yet another electron is captured. A reduction in the barrier occurs simply because the energy levels containing the extra electrons are lower at the hexagonal site than at the tetrahedral site and, presumably, at sites along the path. Baraff et al. suggested a simple formula for the barrier reduction in terms of the energy levels at the tetrahedral and the hexagonal sites. The calculated barrier reduction was 0.9 eV, in excellent agreement with the experimental value. More recently, Baraff and Schluter[54] confirmed this analysis by carrying out total-energy calculations for the two sites and the various charge states of Al. The essence of the mechanism is a change in the migration barrier resulting from a charge-state change following the capture of one and, subsequently, a second electron.

Even more recently, new insights into the reduction of migration barriers by carrier capture were obtained by Pantelides, Oshiyama, Car, and Kel-

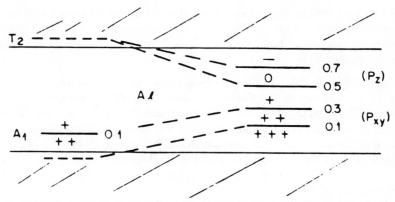

FIGURE 23 Energy levels of interstitial Al at the tetrahedral and hexagonal sites. From Ref. 172.

ly.[191] These authors noted that, contrary to the assumption of Ref. 172, knowledge of the eigenvalues at only the initial equilibrium and saddle points is not, in general, adequate to determine barrier reductions. It may happen that, after the capture of one or two carriers, the new equilibrium and saddle points do not coincide with the old ones. A number of possibilities are illustrated in Figure 24. In the case of interstitial aluminum, the new equilibrium and saddle points were found to remain at the tetrahedral and hexagonal sites, respectively, providing justification for the calculation of the barrier reduction reported by Baraff et al.[172]

In addition, Pantelides et al. noted that, when a center captures two carriers, it becomes unstable against an Auger-like transition: one of the carriers can be ejected into the valence bands while the other is ejected into the conduction bands, with the carrier reverting to its initial charge state. This type of Auger recombination can, therefore, quench migration altogether, if it occurs faster than the time needed to accomplish a jump. In general, however, the Auger recombination merely reduces the jump rate.

B) Recombination. Several mechanisms of recombination-enhanced defect migration and reactions have been described in the literature and have been given different names by different authors.[165,171,173] For example:

a) Energy-release or local-heating mechanism: In this mechanism, the recombination energy is converted into vibrational energy in the mode corresponding to the reaction coordinate for the motion of interest.

b) Local-excitation mechanism: In this mechanism, part of the recombination energy is transmitted to the electronic degrees of freedom. The defect goes into an excited electronic state which may have a lower activation energy for motion. This particular mechanism allows for the possibility of a recombination-retarded process if the activation energy is higher in the excited state.

c) Bourgoin-Corbett or athermal migration: This mechanism, proposed in 1971 by Bourgoin and Corbett,[67] occurs when the ground states of different charge states of a given defect correspond to different sites. Alternate capture of electrons and holes take the defect from one charge state to another and hence from one site to another (see, e.g., Figures 24a and 24b). A quantitative description of this mechanism, including the conditions that must be satisfied for athermal migration, were given recently by Pantelides et al.[191]

Beginning with the initial observation of recombination-enhanced defect annealing by Lang and Kimerling[174] in 1974 (Figure 25), many cases of recombination-enhanced migration and defect reactions have been observed. For example, Feenstra and McGill[175] observed recombination-enhanced dissociation of Zn-O pairs in GaP and discussed various vibrational modes that may be responsible for carrying the energy released by the electron-hole recombination (see discussion by Dean, this volume). More recently, Kimerling and Benton[176] observed recombination-enhanced dissociation of Fe-B pairs in Si.

The recombination-induced athermal migration (Bourgoin-Corbett) mechanism has been proposed by Troxell and Watkins[151] as a possible explanation of the greatly enhanced annealing of boron interstitials in Si. As we saw in section 3.5.2, these centers were found to exhibit negative-U behavior, but the particular configurations of the two stable charge states could not be unambiguously determined.[151]

The athermal migration of Si self-interstitials in Si, which was inferred from EPR data of irradiated Si in 1965 by Watkins (Refs. 64, 133; see also article in this volume), has also attracted renewed interest recently. First, on the basis of self-consistent Green's-function calculations of the energy levels of Si at a tetrahedral site, Pantelides, Ivanov, Scheffler, and Vigneron,[90] showed that early extended-Hückel cluster calculations[68] were incorrect and thus doubted the conclusion that self-interstitials move athermally by winding through the bonds (Path BS in Figure 26). Pantelides et al. also studies the properties of the self-interstitial along the low-density channels (Path TH in Figure 26) using tight-binding Hamiltonians and proposed that the channel path is the more likely path for athermal migration accroding to the Bourgoin-Corbett mechanism. This suggestion was confirmed recently by two independent calculations. Bar-Yam and Joannopoulos[52] used the supercell technique (see section 2.2.2) and carried

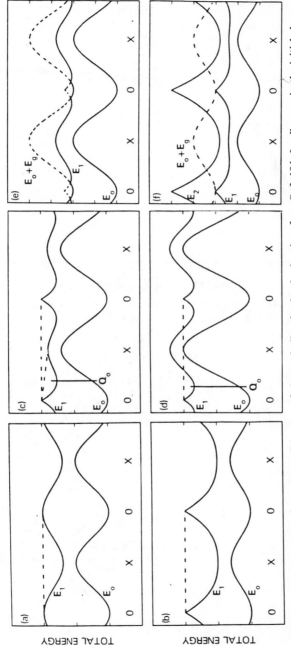

FIGURE 24 Illustration of several possible cases of electronically-stimulated migration, from Ref. 191. In all cases, in the initial charge state, the equilibrium site is 0 and the saddle point is X. After capture of one electron, several possibilities exist (a-e), depending on how the energy level containing the extra electron varies along the path from 0 to X. Enhancement by two-electron capture is illustrated in (f).

69

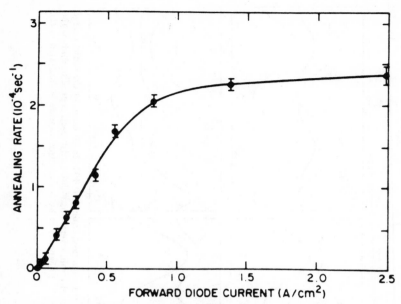

FIGURE 25 The first example of recombination enhanced annealing of a defect in a semiconductor, reported by Lang and Kimerling, Ref. 174.

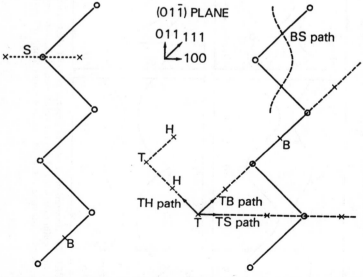

FIGURE 26 Schematic definition of various paths for the migration of interstitials in a Si lattice. After Ref. 55.

out calculations of the barrier for motion from the tetrahedral to the hexa-gonal site fir SI^{++} and Si°. Their results, shown in Figure 27, indicate Bourgoin-Corbett athermal migration via the caputre of two electrons and the subsequent capture of two holes. Independently, Car, Kelly, Oshiyama and Pantelides,[55] used Green's function total-energy calculations and ob-tained a more complete picture of the properties of the self-interstital. As noted earlier, they determined the equilibrium configuration of the self-in-terstitial and found it to have negative-U character (see section 3.5.2 above). In p-type material Si^{++} is the equilibrium charge state. Upon capture of one or two electrons, the Si interstitial was found to move athermally along *several* paths. The total-energy curves are shown along three different paths in Figure 28. In the same figure, the total energy-curves for intrinsic and n-type material are also shown. Athermal migration is possible in all cases. We note that some of these paths involve exchange with atoms at normal atomic sites (termed "interstitialcy paths" by Car et al.; see sec-tion 3.7 below). The migration of the self-interstitial along these other paths and also along a path involving the tetrahedral site and a configura-tion akin to the split (110) configuration has since been independently studied by Bar-Yam and Joannopoulos.[192] They also find athermal migra-tion along the various paths.

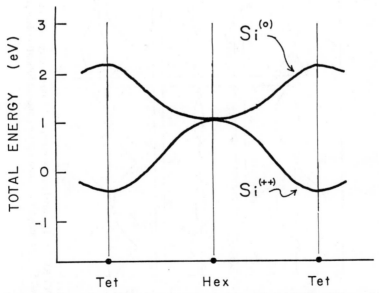

FIGURE 27 The total-energy variation of two charge states of the self-interstitial in Si along the TH path as reported in Ref. 52. The zero of energy is arbitrary.

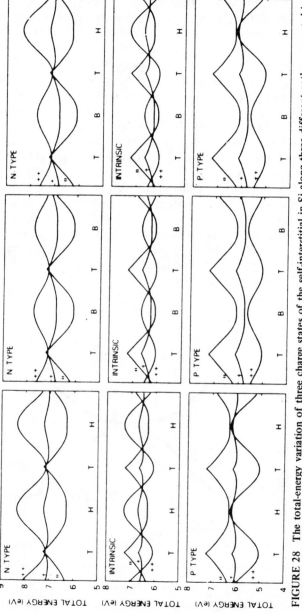

FIGURE 28 The total-energy variation of three charge states of the self-interstitial in Si along three different paths as reported in Ref. 55. The energy scale shown corresponds to formation energies.

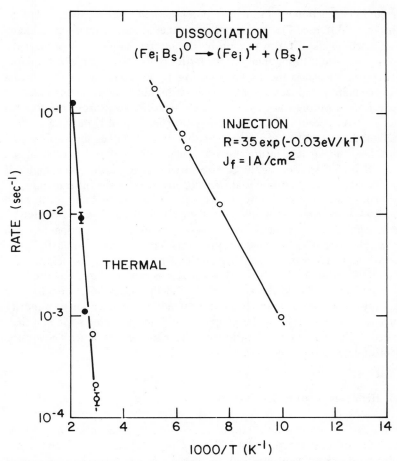

FIGURE 29 Dissociation kinetics of boron-iron pairs in Si, an example of a recombination-enhanced reaction. From Ref. 176.

Both the interstitial Al and the interstitial Si represent a fascinating example of a number of effects. At the tetrahedral site, both have an A_1 state either low in the gap or inside the valence bands and a T_2 resonance near the conduction band edge. In p-type material, the T_2 state is empty and the ground state is non-degenerate and hence stable against Jahn-Teller distortions. As a consequence, both interstitials are stable at the tetrahedral site as long as no excess electrons are around. If excess electrons are around, however, they can be captured in the T_2 resonance in a fashion that is analogous to extrinsic self-trapping: electron capture is the result of combined action of the defect potential *and* coupling to the lattice. The main difference is that coupling to the lattice, in these cases, manifests itself in the form of motion by the interstitial atom away from the tetrahedral site. This coupling to the lattice is also an example of the Jahn-Teller effect because the T_2 resonance is triply degenerate so that its partial occupation induces a symmetry-lowering distortion. In this case, the distortion is simply the motion of the interstitial away from the tetrahedral site. As a result, the gain in energy for motion of I^+ or I° away from the tetrahedral site varies linearly with displacement as shown in Figure 28. Thus, the carrier-capture-enhanced migration of the Al interstitial and the athermal migration of the Si interstitial may be viewed as examples of Jahn-Teller-induced motion that is analogous to extrinsic self-trapping.

Finally, we discuss briefly an example of a recombination-enhanced reaction, in contrast to simple defect migration. Kimerling and Benton[176] used capacitance transient techniques to monitor the pairing reaction between interstitial iron and substitutional boron. They found that association and dissociation can be controlled by electronic mechanisms. In Figure 29, we show the dissociation kinetics as measured by Kimerling and Benton. The figure clearly reveals that both the activation energy and the preexponential factor are reduced dramatically upon injection of minority carriers. The reduction of the preexponential factor reflects the fact that, in recombination-enhanced processes, the jump rate is controlled by the electron-hole recombination rate.

3.7 High-Temperature Self-Diffusion

In the last several years, the controversies over the mechanisms responsible for self-diffusion (section 2.4.1D) took new turns. Lannoo and Bourgoin[8,177] reported semiempirical calculations of the formation entropy of the vacancy and concluded that the entropy can be quite large because of lattice relaxation and modifications of the backbond force constants. This conclusion supported earlier arguments by Van Vechten and Thurmond[178] who attributed large formation entropy to the charged states of the vacancy. Thus, it was argued that there is no need to invoke the rather exotic "extended interstitials" of Seeger and Chik[77] in order to account for large entropies. More recently, Van Vechten[179] also argued that, if one invokes extended defects, extended vacancies would be more likely than extended interstitials. On the other side of the coin, a series of papers analyzing gold-diffusion data presented evidence in support of interstitials. First, Gösele, Frank, and Seeger[180] made a strong case in support of the so-called kick-out mechanism for gold diffusion, ruling out the alternative Frank-Turnbull mechanism, which involves vacancies. The kick-out mechanism involves interstitials. In subsequent papers, Gösele, Morehead, Frank, and Seeger[181] and Stolwijk, Schuster, Hölzl, Mehrer, and Frank,[182] carried the analysis further and actually extracted a value for the self-interstitial diffusivity needed to account for the gold-diffusion data. This value was in good agreement with values extracted from self-diffusion data, providing support for the notion that self-diffusion is also mediated by interstitials. One shortcoming of this analysis is that it assumes that gold occurs in two forms, substitutional and interstitial. There is no evidence, however, indicating that gold occupies a substitutional site. In contrast, since it has only one valence electron, tetrahedral coordination would not appear to be very stable. Furthermore, a variety of experimental data[125,178] suggest that gold forms complexes that may involve vacancies, shallow dopants, or even oxygen.

Additional support for self-interstitials was also recently drawn from observations according to which oxidation-induced stacking faults undergo shrinkage with an activation energy roughly equal to that of self-diffusion.[183] The underlying argument here is that oxidation releases interstitials which, in turn, induce stacking faults. More recently, Gösele and Frank[183] and Gösele and Tan[184] have argued that interstitials are the dominant mechanism at high temperatures whereas vacancies dominate at lower temperatures.

The recent total-energy calculations of Car, Kelly, Oshiyama, and Pantelides[55] have provided the first reliable calculations of formation and migration energies in terms of which to analyze experimental self-diffusion data. The most significant result of this work is that formation energies of vacancies and interstitials are large, of order 5 eV and migration energies are small (Figures 18 and 28). The low- and high-temperature data are therefore reconciled in a simple and elegant way without the need to invoke different configurations or migration energies at the two temperature regimes. In addition, Car et al.[55] found that the self-interstitial has roughly the same formation energy at several different sites which involve no, moderate, or large lattice relaxation. (Figure 18) Thus, the interstitial has large entropy of formation which can account for the observed large preexponential of the self-diffusion coefficient. In addition, large entropies can arise from the low migration barriers (Figure 28) which give rise to low-energy vibrational modes.[185] It should be recognized, however, that not all migration paths of the self-interstitial contribute directly to self-diffusion. Car et al. distinguished between *simple* and *interstitialcy* paths. The latter, which involve exchange between interstitial and substistutional atoms, contribute directly to self-diffusion. Simple paths, which do not involve any exchanges, augment self-diffusion by enhancing the value of the preexponential.

4. The Future

Nobody can prophesy what will be the major developments of the future. One can look in the past, however, and identify gaps in our knowledge which might be eliminated by future work. One can also identify possible future directions by examining the forces that give impetus to certain types of research and drive developments.

In the past, progress in the field of deep centers was driven primarily by technological developments and needs. For example, the extensive work on radiation-induced defects of the 1960's was in direct response to the advent of space exploration; the work on many deep impurities in Si, e.g., gold, was in response to the need to better characterize Si crystals and identify deep centers that can be used to control carrier lifetime; the extensive work on nitrogen and oxygen associates in GaP accompanied the development of commercial LED's using this material; interest in GaAs, especially the ubiquitous EL2 center, was spurred in part by the use of GaAs in optoelectronic devices and its promise for high-speed memory and logic devices; studies of recombination-enhanced processes are intimately connected with efforts to understand degradation mechanisms of solid-state lasers and other optoelectronic devices that operate under injection conditions.

The trend is expected to continue. The new technologies, however, take us away from radiation-induced centers and point into totally different directions.[186] The new technological frontier is very large scale integration (VLSI), which calls for devices to be ever so smaller, junctions ever so shallower, and the number of processing steps larger and larger. The shallower junctions require sharper dopant profiles, which calls for lower processing temperatures in order to avoid unwanted diffusion. In turn, lower processing temperatures imply more residual defects which do not have a chance to anneal out. The new class of defects that are attracting attention are *process-induced defects.* Many of these defects are extended defects (dislocations, stacking faults, precipitates, etc.) which are outside the scope of this volume, but a large number of bona-fide deep centers, in the sense used in this volume, fall in this category. For example, oxygen, carbon, and even nitrogen, which have been ever-present but largely electrically inactive, are increasingly recognized to play some role in device-grade Si. All indications are that these elements form large families of defects, ranging from the very simple to the very complex. One such family is the so-called oxygen thermal donors, namely a number of closely related oxygen centers that appear after a wafer is processed at temperatures of 300–400°C. These centers behave like shallow donors and are believed to contain three or four oxygen atoms.[187] Very recently, a number of studies have been reported combining a variety of techniques such as optical-absorption, EPR, luminescence, etc., aiming at clarifying their atomic structure and other properties.[188] The diffusion of oxygen has attracted renewed attention and led to the discovery of "anomalous diffusion,"[189] which simply means a mode of diffusion that is not understood. At the same time carbon-related centers are being investigated and character-ized.[190] Studies in the future can be expected to focus on the problem of identification of complexes, but also on the dynamics of their formation under different conditions (temperatures, external stimuli such as irra-diation, electron injection, etc.).

Transition-metal-related centers form another class of deep centers that are attracting renewed attention. Large scale integration led to a need to replace aluminum contacts with metals that can form good contacts with sharper interfaces. Transition-metal silicides are quite suitable, because they can be formed by reacting a metallic layer with the Si substrate. Inevitably, such processing raises questions about the silicide growth, the diffusion mechanisms for transition-metal atoms in Si, and the effect of silicide formation on shallow impurities. Interest in these questions is only at a budding stage, but the field appears promising for future work. A number of transition metals, e.g. Cr and Fe play a significant role in con-trolling the properties of compound semiconductors such as GaAs and InP and are very likely to continue attracting attention.

Diffusion processes remain at the heart of many processing steps. Doping by ion implantation is now an established processing tool. After implanta-tion, in most cases, Si turns amorphous and must be recrystallized by some form of annealing. In all cases, the impurities undergo redistribution fol-lowing different diffusion patterns. For example, slow furnace annealing may induce certain diffusion modes, whereas fast annealing by arc lamps or lasers may induce totally different diffusion modes. Impurity diffusion under all kinds of conditions is rather poorly understood, even though much of device design depends on modelling of such processes. These processes are bound to attract considerable attention as the fundamental aspects of atomic migration are established by modern microscopic theor-ies and new experiments.

Finally, VLSI is leading to an area which, for all practical purposes, is virgin territory: deep centers at interfaces and surfaces. In the last few years, Fermi-level pinning at Schottky contacts has attracted considerable attention. A number of models have been proposed that attribute Fermi-level pinning to native defects such as vacancies, antisite defects, etc., *at interfaces*. Evidence in support of such models is largely circumstantial, since most experimental techniques are not suitable for probing defects at interfaces or surfaces. Semiempirical model calculations have been carried out for some defects at surfaces and interfaces, and results have been used to explain Fermi-level pinning.[120] These calculations have not, however, been able to include lattice relaxation, charge redistribution, and charge-state splittings which are so crucial in understanding the properties of deep centers in the bulk. Self-consistent Green's-function techniques are capable of handling these problems, but have not been implemented yet.

Defects at surfaces will inevitably attract attention as technology needs to accurately control epitaxial crystal growth, including the incorporation of dopants, and etching. During epitaxy, impurities tend to segregate at the surface and have a significant effect on the crystal-growth rate. Issues of surface and subsurface diffusion will inevitably need to be addressed, especially when crystal growth is enhanced by laser or ion beams. In the etching process, reactive ions such as fluorine or chlorine stick to or penetrate the surface and form molecules with Si, which are subsequently desorbed. The microscopic processes that underlie etching, therefore, raise the same questions that have been addressed repeatedly in the bulk. For example, what is the equilibrium site for fluorine on a silicon surface as a function of its charge state and the Fermi level? What electronic processes can contribute to enhanced etching by destabilizing local bonding? How do fluorine and silicon atoms diffuse on the silicon surface at various temperatures? Experimental techniques that can shed light on these questions are currently borrowed from studies of free surfaces: photoemission, inelastic electron scattering, etc. In general, these techniques require coverage of the order of a monolayer. On the other hand, techniques that are suitable for point defects in the bulk are not, in general, suitable for point defects on surfaces. Thus, new techniques would have to be invented that combine features of surface and point-defect techniques so that one can study point defects on surfaces.

In a more general sense, the emphasis in the future is likely to be on a large assortment of phenomena that involve *localized interactions* in the bulk, on surfaces, in interfaces. Reactions, migration, diffusion under thermal conditions, under photoexcitation or electronic stimulation are very likely to be the primary areas of interest. Overall, the future looks promising as many new challenges are evolving and new experimental and theoretical techniques can deal with new problems and more complex systems.

Acknowledgment. I would like to thank J. W. Allen, R. Car, R. M. Feenstra, S. Makram-Ebeid, D. J. Robbins, G. D. Watkins, A. R. Williams, and A. Zunger for valuable comments on this manuscript. This work was supported in part by the Office of Naval Research under contract No. N00014-80-C-0679.

References

1. Many good books and review papers are available which contain detailed accounts of special areas. A number of such books are listed in Refs. 2–9. Review papers and Conference Proceedings are referenced throughout this paper where appropriate.
2. C. P. Flynn, *Point Defects and Diffusion* (Clarendon, Oxford, 1972).
3. D. Shaw, ed., *Atomic Diffusion in Semiconductors* (Plenum, New York, 1973).
4. A. G. Milnes, *Deep Impurities in Semiconductors* (Wiley, New York, 1973).
5. J. H. Crawford and L. M. Slifkin, Jr., eds., *Point Defects in Solids* (Plenum, New York, 1975).
6. A. M. Stoneham, *Theory of Defects in Solids* (Clarendon, Oxford, 1975).
7. R. K. Watts, *Point Defects in Crystals* (Wiley, New York, 1977).
8. M. Lannoo and J. Bourgoin, *Point Defects in Semiconductors I, Theoretical Aspects* (Springer-Verlag, Berlin, 1981).
9. M. Jaros, *Deep Levels in Semiconductors,* (Adam Hilger, Bristol, 1982).
10. J. W. Corbett, Solid State Physics, Suppl. 7 (1966).
11. G. W. Ludwig and H. H. Woodbury, Solid State Phys. *13,* 223 (1962).
12. For a review and references to the original papers, see C. T. Sah, Solid State Electron. *19,* 975 (1976). See also Refs. 13–15.
13. D. V. Lang, J. Appl. Phys. *45,* 3023 (1974).
14. H. G. Grimmeiss, Ann. Rev. Mater. Sci. *7,* 341 (1977).
15. G. L. Miller, D. V. Lang, and L. C. Kimerling, Ann. Rev. Mater. Sci. *7,* 377 (1977).
16. For a review and references to the original papers, see B. C. Cavenett, Adv. in Phys. *30,* 475 (1981).
17. For a review, see T. M. Hayes and J. B. Boyce, Solid State Phys. *37,* 173 (1982).
18. A. I. Goldman, E. Canova, Y. H. Kao, B. J. Fitzpatrick, R. N. Bhargava, and J. C. Phillips, Appl. Phys. Lett. *43,* 836 (1983).
19. D. J. Wolford, J. A. Bradley, K. Fry, J. Thompson, and H. E. King, Inst. Phys. Conf. Ser. No. 65, 477 (1982); D. J. Wolford, S. Modesti, and B. G. Streetman, Inst. Phys. Conf. Ser. No. 65, 501 (1982).
20. P. Omling, L. Samuelson, H. Titze, and H. G. Grimmeiss, Il Nuov. Cim. *2D,* 1742 (1983).
21. For a review of the early work, see W. Kohn, Solid State Phys. *5,* 257 (1957). For more recent reviews, see Refs. 22 and 23.
22. S. T. Pantelides, Rev. Mod. Phys. *50,* 797 (1978).
23. M. Altarelli and F. Bassani, in Handbook of Semiconductors, edited by S. Keller, (1982).
24. G. F. Koster and J. C. Slater, Phys. Rev. *95,* 1167 (1954); *ibid. 96,* 1208 (1954).

25. J. Callaway, J. Math. Phys. *5*, 783 (1964); Phys. Rev. *154*, 515 (1967); Phys. Rev. B *3*, 2556 (1971).
26. For a review, see A. A. Maradudin, Solid State Phys. *18*, 331 (1966).
27. F. Bassani, G. Iadonisi, and B. Preziosi, Phys. Rev. *186*, 735 (1969).
28. K. Huang and A. Rhys, Proc. Roy. Soc. *204*, 406 (1950).
29. R. Kubo and Y. Toyozawa, Progr. Theor. Phys. *13*, 160 (1955).
30. M. Lax, J. Chem. Phys. *20*, 1752 (1952).
31. See, e.g., Refs. 5 and 10 for references to the original papers on the vacancy, vacancy-oxygen complex, etc. in Si. See also the article by Watkins, this volume.
32. R. P. Messmer and G. D. Watkins, Phys. Rev. Lett. *25*, 656 (1970); Phys. Rev. B *7*, 2568 (1973).
33. K. H. Johnson and F. C. Smith, Phys. Rev. B *5*, 831 (1972); J. C. Slater and K. H. Johnson, Phys. Rev. B *5*, 844 (1972).
34. K. H. Bennemann, Phys. Rev. *137*, A1497 (1965).
35. J. Callaway and A. J. Hughes, Phys. Rev. *156*, 860 (1967); *ibid. 164*, 1043 (1967).
36. M. Lannoo and P. Lenglart, J. Phys. Chem. Solids *30*, 2409 (1969).
37. S. T. Pantelides and C. T. Sah, Phys. Rev. B. *10*, 621 (1974); *ibid.* p. 638. See also Ref. 22 for subsequent developments.
38. M. Jaros and S. F. Ross, J. Phys. C *6*, 3451 (1973).
39. M. Jaros, J. Phys. C *8*, 2455 (1975); M. Jaros and S. Brand, Phys. Rev. B *14*, 4494 (1976). For a recent review and references to other papers, see M. Jaros, Adv. in Phys. *59*, 409 (1980).
40. A. Zunger, J. Chem. Phys. *62*, 1861 (1975); A. Zunger and A. Katzir, Phys. Rev. B. *11*, 2378 (1975); S. G. Louie, M. Schluter, J. R. Cheilikowshy, and M. L. Cohen, Phys. Rev. B *13*, 1654 (1976).
41. R. Haydock, V. Heine, and M. J. Kelly, J. Phys. C *5*, 2845 (1972); *ibid. 8*, 2591 (1975).
42. E. Kauffer, P. Pecheur, and M. Gerl, J. Phys. C *9*, 2319 (1976); Phys. Rev. B *15*, 4107 (1977).
43. J. Bernholc and S. T. Pantelides, Phys. Rev. B *18*, 1780 (1978).
44. J. Bernholc, N. O. Lipari, and S. T. Pantelides, Phys. Rev. Lett. *41*, 895 (1978); Phys. Rev. B *21*, 3545 (1980).
45. G. A. Baraff and M. Schluter, Phys. Rev. Lett. *41*, 892 (1978); Phys. Rev. B *19*, 4965 (1980).
46. H. P. Hjalmarson, P. Vogl, D. J. Wolford, and J. D. Dow, Phys. Rev. Lett. *44*, 810 (1980).
47. U. Lindefelt and A. Zunger, Phys. Rev. B *24*, 5913 (1981); *ibid. 26*, 846, 5989 (1982).
48. A. R. Williams, P. J. Feibelman, and N. D. Lang, Phys. Rev. B *26*, 5433 (1982).
49. G. A. Baraff and M. Schluter Phys. Rev. B *27*, 1010 (1983).
50. J. P. Vigneron, S. T. Pantelides, and N. O. Lipari, Phys. Rev. B, to be published.
51. S. Brand, M. Jaros, and C. O. Rodriguez, J. Phys. C *14*, 1243 (1981); S. Brand and M. Jaros, Phil. Mag. *47*, 199 (1983).
52. Y. Bar-Yam and J. D. Joannopoulos, Phys. Rev. Lett., *52*, 1129 (1984).
53. M. Scheffler, J. P. Vigneron, and G. B. Bachelet, Phys. Rev. Lett. *49*, 1756 (1982). U. Lindefelt, Phys. Rev. B *28*, 4510 (1983); U. Lindefelt and A. Zunger, Phys. Rev. B *30*, 1102 (1984).
54. G. A. Baraff and M. Schluter, Phys. Rev. B *28*, 2296 (1983); see also Bull. Am. Phys. Soc. *29*, 250 (1984).

55. R. Car, P. J. Kelly, A. Oshiyama, and S. T. Pantelides, Phys. Rev. Lett., *52*, 1814 (1984); see also Bull. Am. Phys. Soc. *29*, 250 (1984).
56. D. N. Talwar, M. Vandevyver, and M. Zigone, J. Phys. C *13*, 3775 (1980).
57. R. M. Feenstra, R. J. Hauenstein, and T. C. McGill, Phys. Rev. B *28*, 5793 (1983).
58. P. Hohenberg and W. Kohn, Phys. Rev. *136*, 864 (1964); W. Kohn and L. J. Sham, Phys. Rev. *140*, A1133 (1965). For a number of review articles, see *The Inhomogeneous Electron Gas*, edited by N. H. March and S. Lundqvist (Plenum, New York, 1984). For applications of the local-density formalism, see especially the article by A. R. Williams and U. von Barth.
59. J. C. Phillips and L. Kleinman, Phys. Rev. *116*, 287 (1959); M. H. Cohen and V. Heine, Phys. Rev. *122*, 1821 (1961).
60. J. A. Appelbaum and D. R. Hamann, Phys. Rev. Lett. *31*, 106 (1973).
61. M. L. Cohen, M. Schluter, J. R. Chelikowsky, and S. G. Louie, Phys. Rev. B *12*, 5575 (1975).
62. M. Schluter, J. R. Chelikowsky, S. G. Louie, and M. L. Cohen, Phys. Rev. Lett. *34*, 1385 (1975).
63. C. A. Coulson and M. J. Kearsley, Proc. Roy. Soc. A *241*, 433 (1957).
64. See G. D. Watkins, this volume, and references therein to original papers. See also G. D. Watkins, J. R. Troxell, and A. P. Chatterjee, Inst. Phys. Conf. Ser. No. 59, 16 (1981).
65. H. A. Jahn and E. Teller, Proc. Roy. Soc. London *161*, 220 (1937); for an interesting historical note by E. Teller on the discovery of the Jahn-Teller theorem, see R. Englman, *The Jahn-Teller Effect in Molecules and Crystals*, (Wiley, London, 1972).
66. G. D. Watkins and J. W. Corbett, Phys. Rev. *121*, 1001, 1015 (1961).
67. J. Bourgoin and J. W. Corbett, Phys. Lett. *38A*, 135 (1972).
68. G. D. Watkins, R. P. Messmer, C. Weigel, D. Peak, and J. W. Corbett, Phys. Rev. Lett. *27*, 1573 (1971); C. Weigel, D. Peak, J. W. Corbett, R. P. Messmer and G. D. Watkins, Phys. Rev. B *8*, 2906 (1973).
69. F. Ham, quoted in Ref. 11
70. G. W. Ludwig and H. H. Woodbury, Phys. Rev. Lett. *5*, 96 (1960).
71. K. Graff and H. Pieper, J. Electrochem. Soc. *128*, 669 (1981).
72. See, e.g., discussion in the review paper by E. R. Weber, Appl. Phys. A *30*, 1 (1983).
73. J. M. Fairfield and B.J. Masters, Appl. Phys. Lett. *8*, 280 (1966); J. Appl. Phys. *38*, 3148 (1967).
74. For a series of review articles on diffusion, see Ref. 3. For more recent reviews, see W. Frank, Festkorperplobleme *21*, 221 (1981) and W. Frank, U. Gösele, H. Mehrer, and A. Seeger, in *Diffusion in Solids II*, edited by A. S. Nowick and G. Murch, (Academic, New York, in press). See also a comprehensive discussion in Ref. 8.
75. The case for the vacancy mechanism is discussed in Ref. 8
76. The case for the interstitial mechanism is discussed by Frank and by Frank et al. Ref. 74.
77. A. Seeger and K. P. Chik, Phys. Stat. Solidi *29*, 455 (1968).
78. In an activated process, such as self-diffusion, the mechanism with the lowest activation energy usually dominates and determines the value of the effective Q. Thus, calculations giving self-diffusion activation energies of order 1–3 eV must be discarded as unreliable. See, e.g., S. M. Hu, in Ref. 3.

79. See, e.g. S. T. Pantelides and W. A. Harrison, Phys. Rev. B *11*, 3006 (1974) and W. A. Harrison, *Electronic Structure and the Properties of Solids* (Freeman, San Francisco, 1980).
80. M. Scheffler, J. Bernholc, N. O. Lipari, and S. T. Pantelides, Phys. Rev. B, *29*, 3269, (1984).
81. J. Bernholc, S. T. Pantelides, N. O. Lipari, and A. Baldereschi, Solid State Comm. *37*, 705 (1981); J. Bernholc, N. O. Lipari, S. T. Pantelides, and M. Scheffler, Phys. Rev. B *26*, 5706 (1982).
82. V. A. Singh, A. Zunger, and U. Lindefelt, Phys. Rev. B *27*, 1420 (1983).
83. J. P. Vigneron, M. Scheffler and S. T. Pantelides, Physica *117B*, 137 (1982).
84. L. A. Hemstreet, Phys. Rev. B *15*, 834 (1977); *ibid.* 22, 4590 (1980).
85. G. G. DeLeo, G. D. Watkins and W. B. Fowler, Phys. Rev. B *23*, 1851 (1981); *ibid.* 25, 4962, 4972 (1982).
86. A. Zunger and U. Lindefelt, Phys. Rev. B *27*, 1191 (1983).
87. P. Vogl and J. Baranowski, Phys. Rev. B, to be published.
88. See, e.g., Ref. 11 and the articles by Allen, by Baranowski, and by Bishop, this volume.
89. F. D. M. Haldane and P. W. Anderson, Phys. Rev. B *13*, 2553 (1976).
90. S. T. Pantelides, I. Ivanov, M. Scheffler, and J. P. Vigneron, Physica *116B*, 18 (1983); see also S. T. Pantelides, in *Methods and Materials for Microelectronic Technology,* edited by J. Bargon (Plenum, New York, 1984).
90a. S. T. Pantelides, in *Defect Complexes in Semiconductor Structures,* edited by J. Giber, F. Beleznay, I. C. Szep, and J. Laszlo, (Springer-Verlag, Berlin, Heidelberg, 1983), p. 75.
90b. O. F. Sankey, H. P. Hjalmarson, J. D. Dow, D. J. Wolford, and B. G. Streetman, Phys. Rev. Lett. 45, 1656 (1980).
91. G. T. Surratt and W. A. Goddard, III, Phys. Rev. B *18*, 1831 (1978).
92. M. Lannoo, G. A. Baraff, and M. Schluter, Phys. Rev. B *24*, 943, 955 (1981).
93. L. A. Hemstreet and J. O. Dimmock, Phys. Rev. B *20*, 1527 (1979).
94. T. A. Kennedy, N. D. Wilsey, J. J. Krebs, and G. H. Stauss, Phys. Rev. Lett. *50*, 1281 (1983).
95. A. Zunger, Phys. Rev. B *28*, 3628 (1983).
96. T. N. Morgan, Phys. Rev. Lett. *49*, 173 (1982); Physica *116B*, 131 (1983).
97. See, e.g., articles in *Muon Spin Rotation,* edited by F. N. Gygax, W. Kundig, and P. F. Meier, (North Holland, Amsterdam, 1979). See also A. Mainwood and A. M. Stoneham, Physica *116B*, 101 (1983) and references therein.
98. H. Katayama-Yoshida and K. Shindo, Phys. Rev. Lett. *51*, 207 (1983).
99. N. O. Lipari, J. Bernholc, and S. T. Pantelides, Phys. Rev. Lett. *43*, 1354 (1979).
100. S. T. Pantelides and H. G. Grimmeiss, Solid State Commun. *35xx*, 653 (1980).
100a. M. Kaminska, M. Skowronski, J. Lagowski, J. M. Parsey, and H. C. Gatos, Appl. Phys. Lett. *43*, 302 (1983).
101. N. F. Mott and F. W. Gurney, *Electronic Processes in Ionic Crystals* (Clarendon, Oxford, 1940).
102. R. M. Feenstra and S. T. Pantelides, Bull. Am. Phys. Soc.. *29*. 207 (1984); phys. Rev. B *31*, 4083 (1985); and to be published.
103. See, e.g., E. Burstein, G. S. Picus, B. Henvis, and R. Wallis, J. Phys. Chem. Solids *1*, 65 (1956); R. L. Aggarwal and A. K. Ramdas, Phys. Rev. *140*, A1246 (1965).
103a. D. J. Robbins, P. J. Dean, C. L. West, and W. Hayes, Phil. Trans. R. Soc. Lond. A *304*, 499 (1982).
104. For informative discussions of various definitions of "levels" see Ref. 105 and the article by Allen, this volume.

105. G. A. Baraff, E. O. Kane, and M. Schluter, Phys. Rev. B *21*, 5662 (1980).
106. See, e.g., D. R. Hamann, Phys. Rev. Lett. *42*, 662 (1979).
107. See, e.g., Z. H. Levine and S. G. Louis, Phys. Rev. B *25*, 6310 (1982); C. S. Wang and W. E. Pickett, Phys. Rev. Lett. *51*, 597 (1983); and Bull. Amer. Phys. Soc. *29*, (1984).
108. See, e.g., the case of the P antisite defect in GaP (Refs. 80 and 109), and the chalcogen impurities in Si [Ref. 81 and V. A. Singh, U. Lindefelt, and A. Zunger, Phys. Rev. B *27*, 4909 (1983)]
109. M. Scheffler, S. T. Pantelides, N. O. Lipari, and J. Bernholc, Phys. Rev. Lett. *47*, 413 (1981).
110. See, e.g., V. L. Moruzzi, J. F. Janak, and A. R. Williams, *Calculated Electronic Properties of Metals,* (Pergamon, New York, 1978).
111. M. T. Yin and M. L. Cohen, Phys. Rev. Lett. *45*, 1004 (1981).
112. J. E. Northrup, J. Ihm, and M. L. Cohen, Phys. Rev. Lett. *47*, 1910 (1981); K. C. Pandey, Phys. Rev. Lett. *49*, 223 (1982).
113. G. B. Bachelet, G. A. Baraff, and M. Schluter, Phys. Rev. B *24*, 4736 (1981).
114. M. Lannoo, Phys. Rev. B *25*, 2987 (1982).
115. Mainwood and Stoneham, Ref. 97; P. Masri, A. H. Harker, and A. M. Stoneham, J. Phys. C *16*, L613 (1983).
116. G. D. Watkins, G. G. DeLeo, and W. B. Fowler, Physica *166B*, 28 (1983).
117. See, e.g., Ref. 82, where the results of tight-binding calculations are at variance with those of self-consistent local-density calculations.
118. P. Vogl, H. P. Hjalmarson, and J. D. Dow, J. Phys. Chem. Solids, *44*, 365 (1983).
119. H. P. Hjalmarson, PhD Thesis, University of Illinois (1980), unpublished.
120. R. Allen and J. D. Dow, Phys. Rev. B *25*, 1423 (1982).
121. S. Das Sarma and A. Madhukar, Solid State Commun. *38*, 183 (1981). For further criticism of tight-binding Hamiltonians, see J. B. Krieger and P. M. Laufer, Phys. Rev. B *23*, 4063 (1981), and V. A. Singh, U. Lindefelt, and A. Zunger, Phys. Rev. B *25*, 2781 (1982). Some of the criticism in these papers is not appropriate, but discussion of such technical issues is outside the scope of this article.
122. For example, an excellent fit to the energy bands can be obtained with only nearest-neighbor interactions. The resulting A_1-T_2 splitting of the ideal vacancy states is essentially zero because the dangling hybrids are centered on second-neighbor atoms and thus do not interact directly. If the perfect-crystal Hamiltonian is refitted with first and second-neighbor parameters, the A_1-T_2 splitting can be varied by changing the relative strength of first- and second-neighbor interactions (I. Ivanov and S. T. Pantelides, unpublished). In the case of the Hamiltonians of Vogl et al. (Ref. 118), only nearest-neighbor interactions are retained so that the A_1 and T_2 vacancy states in the gap region should be very nearly degenerate. Indeed they are, if the ideal vacancy is calculated according to the standard definition given, for example, in Ref. 43, i.e. by removing from the basis set of the crystal all the atomic orbitals centered on the vacant site (I. Ivanov and S. T. Pantelides, unpublished). The vacancy levels usually quoted in conjunction with the Hamiltonians of Ref. 118 (e.g., Refs. 46, 119, 120) are obtained by retaining the s* orbital at the vacant site. No justification for this procedure has been given.
123. L. D. Yau and C. T. Sah, Solid State Electron. *17*, 193 (1974); C. T. Sah and C. T. Wang, J. Appl. Phys. *46*, 1767 (1975).
124. P. O. Fagelstrom and H. G. Grimmeiss, unpublished.

125. D. V. Lang, H. G. Grimmeiss, E. Meijer, and M. Jaros, Phys. Rev. B 22, 3917 (1980).
126. L.-A. Ledebo and Z.-G. Wang, Appl. Phys. Lett. 42, 680 (1983).
127. See discussion by Martin and Makram-Ebeid, this volume.
128. J. Lagowski, D. G. Lin, T. Aoyama, and H. C. Gatos, Appl. Phys. Lett., to be published.
129. A. V. Yukhnevich, Fiz. Tverd. Tela (Leningrad) 7, 322 (1965) [Sov. Phys.-Solid State 7, 259 (1965).]
130. R. J. Spry and W. D. Compton, Phys. Rev. 175, 1010 (1968); C. E. Jones, E. S. Johnson, W. D. Compton, J. R. Noonan, and B. G. Streetman, J. Appl. Phys. 44, 5402 (1973); A. Yukhenvich and A. V. Mudryi, Fiz. Tekh. Poluprov. 7, 1215 (1973) [Sov. Phys. Semicond. 7, 815 (1973)].
131. J. R. Noonan, C. G. Kirkpatrick, and B. G. Streetman, J. Appl. Phys. 47, 3010 (1976).
132. V. S. Konoplev, A. A. Gippius, and V. S. Vavilov, Inst. Phys. Conf. No. 31, 244 (1977).
133. G. D. Watkins, in Radiation Damage in Semiconductors, (Dunod, Paris, 1965), p. 97.
134. K. L. Brower, Phys. Rev. B 9, 2607 (1974).
135. A. R. Bean, R. C. Newman, and R. Smith, J. Phys. Chem. Solids 31, 739 (1970); R. C. Newman and A. R. Bean, Radiat. Effects 8, 189 (1971).
136. K. Thonke, H. Klemisch, J. Weber, and R. Sauer, Phys. Rev. B 24, 5874 (1981).
137. K. M. Lee, K. P. O'Donnell, J. Weber, B. C. Cavenett, and G. D. Watkins, Phys. Rev. Lett. 48, 37 (1982); K. P. O'Donnell, K. M. Lee, and G. D. Watkins, Physica 116B, 258 (1983).
137a. G. Davies, E. C. Lightowlers, and M. do Carmo, J. Phys. C 16, 5503 (1983).
138. In 1974, J. R. Noonan, C. G. Kirkpatrick, and B. G. Streetman [Solid State Commun. 15, 1055 (1974)] reported that the 0.97-eV luminescence line was virtually quenched in Ga-doped Si. No explanation has been advanced. Also, a more definitive explanation is needed for the results of Ref. 136 which suggest that only one carbon atom is involved in the center. A brief discussion of this point is given in Ref. 137.
139. See, e.g., R. Sauer and J. Weber, Physica 116B, 195 (1983).
140. J. A. Van Vechten, J. Electrochem. Soc. 122, 423 (1975).
141. U. Kaufmann, J. Schneider, and A. Rauber, Appl. Phys. Lett. 29, 312 (1976).
142. T. Kennedy and N. D. Wilsey, Phys. Rev. Lett. 41, 977 (1978).
143. C. A. J. Ammerlaan, in Defect Complexes in Semiconductor Structures, edited by J. Giber, F. Beleznay, I. C. Szep, and J. Laszlo, (Springer-Verlag, Berlin, Heidelberg, 1983), p. 111.
144. W. V. Smith, P. P. Sorokin, I. Gelles, and G. J. Lasher, Phys. Rev. 115, 1546 (1959).
145. K. L. Brower, Phys. Rev. Lett. 44, 1627 (1980).
146. R. Car and S. T. Panelides, Bull. Am. Phys. Soc. 28, 288 (1983); also reported at the 4th "Lund" International Conference on Deep-Level Impurities in Semiconductors, Eger, Hungary, 1983; and to be published.
147. R. C. Newman and J. Wakefield, J. Phys. Chem. Solids 19, 230 (1961); R. C. Newman and J. B. Willis, J. Phys. Chem. Solids 26, 373 (1965).
148. J. Hubbard, Proc. Roy. Soc. London Ser. A 276, 238 (1963).

149. P. W. Anderson, Phys. Rev. Lett. *34*, 953 (1975).
150. G. A. Baraff, E. O. Kane, and M. Schluter, Phys. Rev. Lett. *43*, 956 (1979); see also Ref. 105.
151. G. D. Watkins and J. R. Troxell, Phys. Rev. Lett. *44*, 593 (1980); J. R. Troxell and G. D. Watkins, Phys. Rev. B *22*, 921 (1980).
152. G. D. Watkins, Phys. Rev. B *12*, 5824 (1975).
153. R. D. Harris, J. L. Newton, and G. D. Watkins, Phys. Rev. Lett. *48*, 1271 (1982); Phys. Rev. Lett. *51*, 1722 (1983).
154. U. Piekara, J. M. Langer, and B. Krukowska-Fulde, Solid State Commun. *23*, 583 (1977).
155. D. V. Lang and R. A. Logan, Phys. Rev. Lett. *39*, 635 (1977); D. V. Lang, R. A. Logan, and M. Jaros, Phys. Rev. B *19*, 1015 (1979).
156. H. J. Queisser and D. E. Theodorou, Phys. Rev. Lett. *43*, 401 (1979).
157. Y. Toyozawa, Solid State Electron. *21*, 1313 (1978).
158. L. Landau, Phys. Zeits. d. Sowjetunion *3*, 664 (1933); see also Ref. 101.
159. M. Levinson, M. Stavola, J. L. Benton, and L. C. Kimerling, Phys. Rev. B *28*, 5848 (1983); M. Stavola, M. Levinson, J. L. Benton, and L. C. Kimerling, Phys. Rev. B, to be published.
160. P. T. Landsberg and D. J. Robbins, Solid State Electron. *21*, 1289 (1978).
161. M. Jaros, Solid State Commun. *25*, 1071 (1978); N. Itoh, A. M. Stoneham, and A. H. Harker, J. Phys. Soc. Jpn. *49*, 1364 (1980).
162. P. J. Dean, R. A. Faukner, S. Kimura and M. Ilegems, Phys. Rev. B *4*, 1926 (1971). See also H. Queisser, Inst. Phys. Conf. Ser. No. 43, 1255 (1979); P. J. Dean, *ibid.* p. 1259.
163. V. N. Abakumov and I. N. Yassievich, Sov. Phys. JETP *44*, 345 (1976).
164. R. M. Gibb, G. J. Rees, B. W. Thomas, B. L. H. Wilson, B. Hamilton, D. W. Wight, and N. F. Mott, Phil. Mag. *36*, 1021 (1977).
165. See, e.g., the review by A. M. Stoneham, Rep. Progr. Phys. *44*, 251 (1981) and references therein.
166. T. A. Kovarskii, Sov. Phys.–Solid State *4*, 1200 (1962); T. A. Kovarskii and E. P. Sinyarkii, *ibid.* p. 2345.
167. C. H. Henry and D. V. Lang, Phys. Rev. B *15*, 989 (1977).
168. B. K. Ridley, Solid State Electron. *21*, 1319 (1978); H. Sumi, Physica B+C *117+118*, 197 (1983).
169. T. N. Morgan, Phys. B *28*, 7141 (1983).
170. L. C. Kimerling, H. M. DeAngelis, and J. W. Diebold, Solid State Commun. *16*, 171 (1975).
171. J. R. Troxell, A. P. Chatterjee and G. D. Watkins, and L. C. Kimerling, Phys. Rev. B *19*, 5336 (1979).
172. G. A. Baraff, M. Schluter, and G. Allan, Phys. Rev. Lett. *50*, 739 (1983).
173. L. C. Kimerling, Solid State Electron. *21*, 1391 (1978).
174. D. V. Lang and L. C. Kimerling, Phys. Rev. Lett. *33*, 489 (1974).
175. R. M. Feenstra and T. C. McGill, Phys. Rev. B *25*, 6329 (1982).
176. L. C. Kimerling and J. L. Benton, Physica *116B*, 297 (1983).
177. M. Lannoo and G. Allan, Phys. Rev. B *25*, 4089 (1982).
178. J. A. Van Vechten and C. D. Thurmond, Phys. Rev. B *14*, 3539 (1976); *ibid.*, p. 3551.
179. J. A. Van Vechten, Bull. Amer. Phys. Soc. *28*, 329 (1983).

180. U. Gösele, W. Frank, and A. Seeger, Appl. Phys. *23*, 361 (1980).
181. U. Gösele, F. Morehead, W. Frank, and A. Seeger, Appl. Phys. Lett. *38*, 157 (1981).
182. N. A. Stolwijk, B. Schuster, J. Hölzl, H. Mehrer, and W. Frank, Physica *116B*, 335 (1983).
183. U. Gösele and W. Frank, in *Defects in Semiconductors,* edited by Narayan and Tan, (North Holland, Amsterdam, 1981), p. 55.
184. U. Gösele and T. H. Tan, in *Defects in Semiconductors,* edited by S. Mahajan and J. W. Corbett, (Elsevier, New York, 1983).
185. J. R. Leite, R. M. Feenstra, and S. T. Pantelides, to be published.
186. For some thoughts on this point, see G. D. Watkins, Inst. Phys. Conf. Ser. No. 59, 139 (1981).
187. See, e.g., G. S. Oehrlein, J. Appl. Phys. *54*, 5453 (1983), and references therein.
188. R. Oeder and P. Wagner, in *Defects in Semiconductors,* edited by S. Mahajan and J. W. Corbett, (Elsevier, New York, 1983).
189. M. Stavola, J. R. Patel, L. C. Kimerling, and P. E. Freeland, J. Appl. Phys. *42*, 73 (1983).
190. See, e.g., S. Kishino, Y. Matsushita, and M. Kanamori, Appl. Phys. Lett. *35*, 213 (1979); G. S. Oehrlein, J. L. Lindström, and J. W. Corbett, Appl. Phys. Lett. *40*, 241 (1982).
191. S. T. Pantelides, A. Oshiyama, R. Car, and P. J. Kelly, Phys. Rev. B *30*, 2260 (1984).
192. Y. Bar-Yam and J. D. Joannopoulos, Phys. Rev. B *30*, 2216 (1984).

CHAPTER 2

Chalcogen Impurities in Silicon

H. G. Grimmeiss and E. Janzén*

*Department of Solid State Physics, University of Lund,
Box 725, S-22007 Lund, Sweden*

1. INTRODUCTION
2. ENERGY SPECTRUM OF CHALCOGEN ATOMS IN SILICON
3. CAPTURE AND EMISSION PROCESSES
4. SAMPLE PREPARATION AND DEFECT IDENTIFICATION
5. EXPERIMENTAL RESULTS
   5.1 Tellurium
   5.2 Selenium
   5.3 Sulfur
   5.4 Oxygen
6. DISCUSSION
   6.1 Spin-Valley and Multi-Valley Interactions in Si:Te, Si:Se, and Si:S
   6.2 Electron Spin Resonance Data
   6.3 Chemical Trends

1. Introduction

If an atom of the host lattice in silicon is replaced by an atom belonging to the fifth group in the periodic table, the potential binding the extra

*Present Address: ASEA Research and Innovation, Opitcal Sensors Innovation Department, S-721 83 VÄSTERÅS, Sweden

electron to the impurity atom can, in most cases, be approximated by a hydrogen-like potential.[1] This gives rise not only to an energy level in the band gap for the impurity ground state, but, in addition, also to a series of excited states. In silicon, the ground-state energies of these impurities are of the order of 50 meV. Such centers are therefore called "shallow" impurities, and are widely used in semiconductor technology for modifying the type and degree of electrical conductivity. The energy levels of the excited states are almost independent of the ground-state energies and are well described by effective mass theory (EMT).[2,3,4,71] That these excited states are indeed so well descirbed by EMT is one of the reasons for our good understanding of shallow centers. The assignment of these levels and of the ground states has been considerably facilitated by the fact that optical absorption spectra generally exhibit a number of sharp lines.[5]

In, on the other hand, a host atom in silicon is replaced by an atom from the sixth group in the periodic table, two extra electrons are available which may give rise to double donors. The potential binding these two electrons of the impurity atoms is frequently compared to that for the two electrons in a helium atom.[6] Double donors have been treated theoretically quite extensively in the literature.[6-9] In addition, the ground state of chalcogen atoms in silicon, which have been investigated so far, lie at a much greater distance from the conduction band than impurity atoms from the fifth group in the periodic table. They are thus creating so-called "deep" impurity levels. Apart from some transition metals, published optical spectra of crystal defects with large binding energies often show a smooth energy dependence without any of the detailed structure which is otherwise characteristic of shallow energy levels. This has often been considered as an indication for the nonexistence of excited states of deep centers. If excited states are not observed, and, hence, experimental line spectra not available, the energy assignment and the chemical identification of crystal defects with large ground-state energies becomes much more difficult than for shallow centers for which detailed spectra are readily available. Detailed studies of deep energy states have shown that both the measurements and the interpretation are often far more difficult in compound materials than in elementary semiconductors due to residual impurities, nonstoichiometry and the formation of complexes. The reason for this is, of course, the fact that the basic properties of elementary semiconductors such as silicon are, in many respects, better known than those of compounds. Elementary semiconductors such as silicon are therefore particularly suited for the study of deep energy states. Of the possible deep impurities in silicon, elements from the same row of the periodic table as silicon are of particular interest. All of them have the same number of core electrons (isocoric impurities) and, hence, the closest resemblance to the host atom.[3] Sulfur is one of the isocoric atoms in silicon which causes deep energy states. Since the behaviour of Sb, As and P dopants in silicon are very similar,[5] it is no surprise that the electronic properties of tellurium selenium and sulfur are also similar. In addition, the electron photo-ionization cross section spectra clearly demonstrate the existence of excited states for both tellurium, selenium and sulfur. Two types of excited states have been observed: those which are generally referred to as the Rydberg series,[1] and those which originate from the multi-valley nature of the conduction band. It has already been pointed out that the existence of excited states at deep energy levels is of vital importance for a better understanding of deep centers, not only of the capture processes[10],[11], but also of the electronic structure in general. Chalcogen atoms and, in particular, tellurium, selenium and sulfur in silicon are therefore especially suited to improving our knowledge about the electronic properties of deep energy states. However, chalcogen atoms are not only interesting for academic reasons. Several of them are already widely used in many technical applications such as infrared detectors.[12-17] Extrinsic-silicon infrared detectors for wavelengths corresponding to the atmospheric transmission windows (3–5 μm, 8–14 μm) are of great importance because of their possible use in charge-coupled devices in monolithic thermal imaging systems. The limiting sensitivity of an infrared detector is controlled by the noise due to the generation and recombination of charge carriers influenced by the background radiation. To reach this limit, the operating temperature of the device has to be sufficiently low, so that the thermally generated charge carriers are much fewer than those generated by the background radiation. The temperature at which the detector can be operated depends, therefore, strongly on the binding energy of this level giving rise to the photoresponse. At present In is used for this purpose in the 3–5 μm range. Due to its binding energy of 0.16 eV, an operating temperature of less than 60 K is needed. It would be much more convenient if liquid nitrogen at normal pressure could be used as a coolant. This, however, demands a dopant with slightly deeper energy levels. Several chalcogens in silicon have levels between 0.2 and 0.3 eV, which is one of the reasons for the growing interest in these dopants. The possibility of growing epitaxial layers, using vapor transport techniques with chalcogens as carriers, is also important in this context.[18]

This chapter summarizes some aspects of present-day thinking about phenomena particular to chalcogen atoms in silicon. Emphasis is placed on measuring techniques and new experimental results rather than on theoretical considerations. We have chosen not to outline all the different features proposed for the chalcogens in silicon, but rather to concentrate on their electrical and optical properties and, hence, on a discussion of thermal and optical capture and emission rates and energy positions.

In Section 2, the energy spectrum of chalcogen atoms in silicon is discussed, and in Section 3, a brief survey of capture and emission processes is given. Different techniques of sample preparation and defect identification are described in Section 4, and more detailed experimental results are presented in Section 5. In Section 6, a short general discussion of the spin-valley and multi-valley interaction in Si:Te, Si:Se and Si:S is given, followed by some comments on the electron spin resonance data. The section is concluded with a few remarks on chemical trends.

2. Energy Spectrum of Chalcogen Atoms in Silicon

Silicon belongs to group IV of the periodic table. It crystallizes in a diamond structure with a lattice constant a = 5.43 Å. The electron in the outer shell forms four sp^3 hybrid orbitals, directed tetrahedrally towards the nearest neighbours 2.35 Å away.

The band structure of silicon is shown in Figure 1. The band gap is indirect, and its temperature dependence is given in Figure 2. The minimum of the conduction band is in the [100] direction $0.85 \times \frac{2\pi}{a}$ away

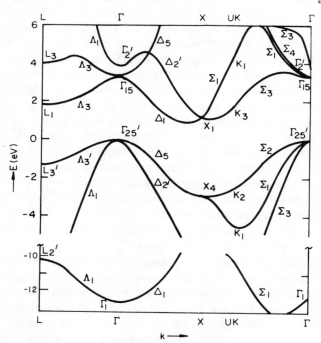

FIGURE 1   The energy band structure of silicon (from Ref. 8).

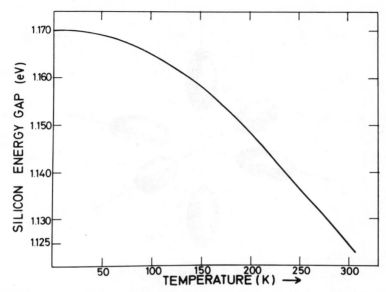

FIGURE 2 Temperature dependence of the intrinsic energy gap of silicon (from Ref. 19). For T < 190K, $E_g(T) = 1.17 + 1.059 \times 10^{-5} T - 6.05 \times 10^{-7} T^2$ eV and for $150 < T < 300K$, $E_g(T) = 1.1785 - 9.025 \times 10^{-5} T - 3.05 \times 10^{-7} T^2$ eV.

from the center of the Brillouin zone, whereas the maximum of the valence band is at the zone center. There are six [100] directions implying six equivalent band minima. Surfaces of constant energy near the conduction band minima are shown as ellipsoids in Figure 3. The corresponding effective masses at low temperatures are m_t (transverse) = 0.19 m_o, and m_l (longitudinal) = 0.92 m_o, where m_o is the free-electron mass.

If a chalcogen atom is incorporated into the silicon crystal lattice, the energy spectrum of the impurity strongly depends on its physical location in the lattice. A descriptive model can easily be constructed if the chalcogen atom simply replaces a silicon atom.[1-4] Such an impurity is known as substitutional.

Chalcogens have six electrons in the outer shell. If the chalgocen atom is substitutional, four of these electrons may be assumed to form sp^3 hybrid orbitals to match those of the silicon host lattice. Let us for a moment assume that the remaining two electrons are taken away from the crystal, which is otherwise perfect, except for the one substitutional chalcogen atom. Originating from the chalcogen core, a screened Coulomb potential

$$U(r) = -\frac{Ze}{4\pi\epsilon\epsilon_o r} \qquad (1)$$

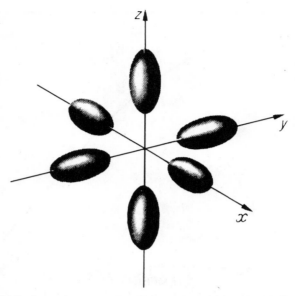

FIGURE 3   Surfaces of constant energy near the conduction band minima of Si.

will exist in the crystal, where Z is the charge of the core (here Z = 2), and
ϵ is the dielectric constant of silicon. [Equation (1) is not valid for values
of r in the vicinity of the impurity cell (the so-called central cell), but
corresponds to a first approximation of the true potential.] If now one
of the two electrons is brought back into the crystal, it will move with an
effective mass m_e^* in the hydrogen-like potential U(r). In analogy with the
hydrogen atom, there will be bound states, usually called the Rydberg
series, below the conduction band, with binding energies E_B

$$E_B = \frac{Z^2 e^4 m_e^*}{(4\pi\epsilon\epsilon_o)^2} \frac{1}{2\hbar^2} \frac{1}{n^2} \tag{2}$$

where n = 1, 2, ... If the first electron is in its lowest bound state, it will
partly screen the core charge. Hence, taking $Z \approx 1$, the introduction of the
second electron into the crystal can, at least for the excited states, be
described in a similar way. In a more rigorous derivation of this "hydrogenic effective mass theory", the electron wavefunction Ψ is expanded in
terms of all the Bloch functions $\Psi_{n\vec{k}}$ of the perfect crystal:

$$\Psi(\vec{r}) = \sum_{n\vec{k}} F_{n\vec{k}} \Psi^o_{n\vec{k}}(\vec{r}). \tag{3}$$

Here, n is the band index and $\vec{K}$ the wave vector. By making some approximations, for example, that only the bottom of the conduction band (n = 0) contributes appreciably to the wavefunction Ψ, the original eigenvalue problem for the imperfect crystal $H\Psi = E_B \Psi$ is reduced to the following equation:

$$\left[-\frac{\hbar^2}{2m_e^*} \left(\frac{\partial^2}{\partial x^2} + \frac{\partial^2}{\partial y^2} + \frac{\partial^2}{\partial z^2} \right) + U(r) \right] F(\vec{r}) = E_B \, F(\vec{r}). \qquad (4)$$

This equation is similar to the Schrödinger equation for the hydrogen atom if modifications due to the dielectric screening, the charge and the effective mass are made. $F(\vec{r})$ is the inverse Fourier transform of $F_{ok}\vec{r}$. Thus the energy solutions of the Rydberg series are given by (2) and the corresponding $F(\vec{r})$ are modified hydrogenic wavefunctions. The complete wavefunction $\Psi(\vec{r})$ is given by

$$\Psi(\vec{r}) = F(\vec{r}) \, \Psi_{k_o}^{0} (\vec{r}) \qquad (5)$$

where $\Psi_{k_o}^{0} (\vec{r})$ is the Bloch function close to the bottom of the conduction band. $F(\vec{r})$ differs for every state in the Rydberg series, whereas $\Psi_{k_o}^{0}$ is the same. The different states are, as usual, labelled 1s, 2s, 2p etc. (see Figure 4a). Equation 5 may be interpreted in terms of a slow modulation of the Bloch state $\Psi_{k_o}^{0}$ by the function $F(\vec{r})$ as a result of the weak Coulomb attraction of the donor ion.

Equation (4) is obtained when the conduction band minimum is isotropic. If a minimum (j) lies at $k_j \neq 0$, the effective mass is generally different in the transverse and longitudinal directions and Equation 4 has to be modified to:

$$\left[-\frac{\hbar^2}{2m_t^*} \left(\frac{\partial^2}{\partial x^2} + \frac{\partial^2}{\partial y^2} \right) - \frac{\hbar^2}{2m_\ell^*} \frac{\partial^2}{\partial z^2} + U(r) \right] F_j(\vec{r}) = E \, F_j(\vec{r}). \qquad (6)$$

The symmetry of the Hamiltonian is now no longer spherical but cylindrical. The non-spherical part of the Hamiltonian will mix states with the same parity (1 odd or even) and the same projection (m) of angular momentum (ℓ) along the z-axis. The result is that the accidental degeneracy of states with the same n but different ℓ no longer holds, and that states with the same ℓ but with different values of $|m|$ will split (see Figure 4b).

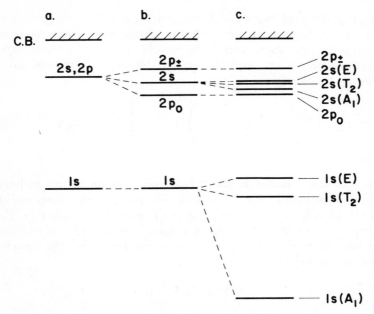

FIGURE 4   The lowest energy states of an electron bound to a donor in a Coulumb potential:

a) if the effective mass is isotropic (e.g., GaAs), the ns and np states will be degenerate.

b) if the effective mass is anisotropic (e.g., Si), this degeneracy will disappear.

c) in addition the 1s and 2s states will split due to multi-valley interactions in silicon.

For convenience, the hydrogenic notation is still used although each level now represents a mixture of all hydrogenic states with the appropriate parity and projection of angular momentum. The notation therefore indicates the hydrogenic state into which each level would reduce in the limit $\dfrac{m_t^*}{m_\ell^*} \to 1$.

Energy level calculations of this kind with $Z = 1$ have been carried out by Faulkner[4] and more recently in Ref. 71. Agreement with experimental results from group-V donors in silicon is extremely good except for the 1s and 2s states. It is straightforward to extend Faulkner's calculation to $Z = 2$. Each energy level will simply be four times deeper. Faulkner's calculation should therefore give information on the energy positions of excited states (except 2s) of double donors in silicon caused by chalcogens (i.e., both $Z = 1$ and $Z = 2$).

A different approach is to use the Green's-function method.[129] This method should be considered to be complementary to the effective mass theory and is appropriate for the ground state. According to Green's-function calculations, the deep state in the gap is only one of several localized states, the rest of which occur within or below the valence bands. The origins and wavefunction character of all these localized states can be understood in terms of s and p atomic orbitals on the impurity atom and on the four nearest neighbors. For example, a bonding combination between the s impurity orbitals and orbitals on the neighbors lies below the valence bands. The corresponding antibonding combination is the deep state in the fundamental gap. More detailed discussion and plots of the localized wavefunctions can be found in Ref. 129.

There are two main reasons why the effective mass theory fails to predict the correct energy positions of the 1s and 2s states. First, an electron in an s-state, especially if it is 1s, will spend an appreciable amount of time close to the donor ion, a region which is sometimes referred to as the central cell. In this region, the attractive potential is stronger than the screened Coulomb potential, thus making the energy level deeper. Second, since there are six equivalent conduction-band minima in silicon, the electron wavefunction of every state should be written as a linear combination of wavefunctions belonging to the different minima[2] (cf. Equation 5):

$$\Psi(\vec{r}) = \sum_{j=1}^{6} \alpha_j \, F_j(\vec{r}) \, \Psi_{\vec{k}}^{o}(\vec{r}).$$

(7)

Here, $F_j(\vec{r})$ is the hydrogen-like envelope function belonging to the jth minimum, and $\Psi_{\vec{k}}^{o}(\vec{r})$ the corresponding Bloch function of the perfect crystal. This makes every s, p_o, $p_\pm$ state six-, six-, and twelvefold degenerate, respectively, excluding spin. If, however, the potential acting upon the electron is strong, $F_j(\vec{r})$ will be localized close to the donor ion. While still *mainly* centered around the jth minimum, $F_j(\vec{k})$, the Fourier transform of $F_j(\vec{r})$ will be spread out into the Brillouin zone. If the $F_j(\vec{k})$ from two different minima (valleys) overlap, the approximations that led to Equation 6 are no longer valid. New terms called inter-valley terms will appear in the Hamiltonian.[7,8,20,21] Only s-states have non-negligible inter-valley terms, since the p, d, ... states all have vanishing amplitudes in the central cell region where the potential is strongest. These latter states are well described by the "one-valley" effective mass theory (Equation 6).

If the chalcogen atom occupies a substitutional site, the overall symmetry is tetrahedral. The central cell potential will therefore split each six-fold degenerate s-state into a singlet (A_1), a triplet (T_2) and a doublet (E) (see Figure 4c). The splitting is, of course, largest for 1s. Sometimes it is referred to as valley-orbit splitting. The ground state should be $1s(A_1)$ since, in contrast to $1s(A_1)$, $1s(T_2)$ and $1s(E)$ have a node at the chalcogen atom and are therefore less affected by the central cell potential.[2] Due to insufficient screening one may expect that the effective charge seen by the outer electron of a donor is larger than one. However, it is found experimentally[22,23] that for neutral chalcogen centers $(Z = 1)$ this effect is significant only for the ground state $1s(A_1)$ for which the two electrons are equivalent.

An additional splitting of the $1s(T_2)$ state due to spin-orbit effects may occur.[5,23-25] If spin is included, the symmetry representations (as shown in Figure 4) are changed as follows: $A_1 \rightarrow \Gamma_6$, $E \rightarrow \Gamma_8$ and $T_2 \rightarrow \Gamma_7 + \Gamma_8$. The splitting of the T_2 level is due to its "valley-induced" p-character which makes an interaction with the spin possible. This spin-"pseudo-orbit" splitting is sometimes referred to as spin-valley splitting[26,27] (see Figure 5). Due to their different degeneracies, the Γ_8 state is shifted less than the Γ_7 state. If spin-valley interactions were absent, the unsplit $1s(T_2)$ state would lie at an energy of $E(T_2) = \frac{1}{3} E(\Gamma_7) + \frac{2}{3} E(\Gamma_8)$.[28] Although the $1s(E)$ state cannot split, the $1s\Gamma_8(E)$ state may contain admixtures of T_2 states. To a first approximation, however, spin-valley interaction does not mix $1s\Gamma_8(E)$ states with $1s\Gamma_8(T_2)$ states, especially when the energy separation $1s(T_2) - 1s(E)$ is large.[25]

It is not clear whether $1s\Gamma_8(T_2)$ or $1s\Gamma_7(T_2)$ should have the larger

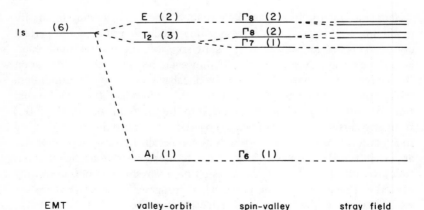

FIGURE 5 Multi-valley splitting of 1s donor states in silicon. The numbers in brackets are the degeneracies (excluding spin) of the states. For details, see text.

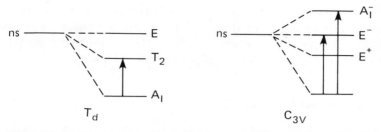

FIGURE 6 The behavior of multi-valley split s-states when lowering the symmetry from T_d to $C_{3v}(D_{3d})$. Symmetry-allowed transitions are marked with arrows.

binding energy. Γ_8 corresponds to a "pseudo" $p_{3/2}$ state and Γ_7 to a "pseudo" $p_{1/2}$. Hund's rules tell us that for a single electron the energy is lowest when the spin is antiparallel to the orbital angular momentum. If Hund rules are applicable for this kind of "pseudo" state, Γ_8 should then lie above Γ_7.

Electric dipole optical transitions from the ground state to excited states are governed by selection rules. For instance, the initial and final states must have different parties, i.e., transitions $s \rightarrow p, f, h$ etc., are allowed whereas transitions $s \rightarrow s, d$ etc., are forbidden. Transitions among multi-valley split s-states are obviously parity-forbidden. They have therefore not been observed for the shallow donors P, As and Sb, since even the ground states of these donors are nearly effective mass like. To be able to see transitions between s-states, at least one of the states has to be mixed with a state of different parity (p, f etc.). The same effect would be obtained if at least one of the states consisted partly of wavefunctions from other bands. Transitions $1s(A_1) \rightarrow 1s(T_2)$ have been observed for Bi and for all chalcogen centers, but not $1s(A_1) \rightarrow 1s(E)$. The reason is that, for T_d symmetry, $1s(A_1) \rightarrow 1s(T_2)$ is symmetry-allowed, whereas $1s(A_1) \rightarrow 1s(E)$ is symmetry-forbidden. In the same way, $1s(A_1) \rightarrow 2s(T_2)$ may be seen, but not $1s(A_1) \rightarrow 2s(A_1)$ or $1s(A_1) \rightarrow 2s(E)$.

Although most of the published data indicate that substitutional chalcogen atoms can be introduced into silicon, one cannot exclude the possibility of any interstitial site or the formation of complexes such as chalcogen pairs, chalcogen vacancies or chalcogen complexes with any other impurities present in the crystal. If the defect consists of a chalcogen pair instead of an isolated substitutional chalcogen atom, the local symmetry will be lowered from tetrahedral to trigonal. An electron far away from the defect will only feel the isotropic coulomb potential.

Excited states like p-states are therefore essentially unaffected by the lowering of symmetry, but it has a dramatic effect on multi-valley split s-states, see Figure 6. The construction of models for the energy spectrum of more complicated complexes will not be further discussed.

3. Capture and Emission Processes

Capture of a free-charge carrier results in a bound-charge carrier while a considerable amount of energy (several tenths of an eV in silicon) has to be carried away. This energy is accounted for by photons, phonons or an increase in the kinetic energy of free charge carriers. The relative importance of the various forms of energy release depends on factors such as temperature, and the amount of energy that has to be carried away etc. Experimentally, it is found that the dominating process for the capture of free charge carriers in silicon are non-radiative capture processes. This implies that in samples with relatively low carrier concentrations the recombination energy has to be carried away by either a cascade or a multi-phonon process. Evidence has been given by Grimmeiss et al.[29,30] that, in sulfur- and selenium-doped silicon, the recombination energy is partly carried away in the form of a phonon cascade process. The electron is first captured in a highly excited state. It then diffuses down the ladder of excited states emitting one or very few phonons at every step. The probability for re-emission back to the conduction band decreases with increasing binding energy. Since excited states are less closely spaced further down the ladder, the electron will finally reach a state where the next step will require a multi-phonon transition or the emission of a photon, both of which are slow processes. If the temperature is low enough, re-emission from this state will be negligible. Hence, the electron capture cross section, as measured by the rate at which electrons leave the conduction band, is not affected by the subsequent transitions.

The cascade capture model was first proposed by M. Lax[10] and could successfully explain the very large capture cross sections observed for shallow donors in silicon and germanium. The calculated temperature dependence varies from T^{-4} to T^{-1}, depending both on the temperature and the type of phonons involved.

In a recent paper, Abakumov et al.[11] claim that the probability for re-emission from highly excited states is smaller than shown by Lax. Hence, their capture cross sections are larger than Lax's and the temperature dependence is T^{-3} or T^{-1}, depending on whether acoustical or optical phonons are involved. Both the Abakumov and the Lax model are only valid at relatively low temperatures.

At high temperatures the situation is different. The electron will still diffuse down to the lowest state below the conduction band accessible by a phonon cascade process (see Figure 7). However, the subsequent slow transition will now have to compete with thermal re-emission back to the conduction band and, hence, the capture rate will decrease. If the lowest state, accessible by a cascade process, lies at an energy E_1 below the conduction band, it has been shown[31,32] that in the high temperature limit the electron capture cross section σ_n^t is given by the expression

$$\sigma_n^t = \frac{\nu_2 \times g_x}{V_{th} \, N_c} \, e^{E_1/kT}, \tag{8}$$

where ν_2 is the rate of the slow transition after the cascade process and g_x the degeneracy of the lowest "cascade" state.

It is not immediately obvious to what energy state in the spectrum of chalcogen atoms in silicon E_1 should correspond. If all kinds of phonons are involved, the lowest "cascade" state for neutral donors would be $1s(T_2)$ and for ionized donors $2p_0$ or perhaps $2s(A_1)$. The energy position of $2s(A_1)$ is unknown for all ionized chalcogen centers. For the corresponding neutral centers, it has been found that $2s(A_1)$ is deeper than $2p_0$.[32a,b] So it is not unreasonable to believe that $2s(A_1)$ is below $2p_0$ also for the ionized centers. In fact, recent calculations by Altarelli[33] show that the $2p_0$ and $2s(A_1)$ states of ionized donors in silicon should lie 46 and 73 meV, respectively, below the conduction band. If only low energy acoustical phonons are involved, the lowest "cascade" state will be $2p_0$ or $2s(A_1)$ for neutral donors. The situation is more complex for singly-ionized donors. Since the excited states are less closely spaced for these

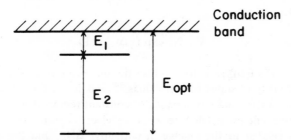

FIGURE 7 Energy level scheme of a center whose dynamical behavior can be described in terms of two-stage capture and emission processes. E_1 is the energy distance between the lowest cascade state and the conduction band, E_2 the energy separation between the ground state and the lowest cascade state, and E_{opt} the ground-state binding energy obtained from absorption measurements.

donors than for neutral donors, the lowest "cascade" state may vary from $2s(A_1)$ to $2p_\pm$, depending on the energy positions of the multi-valley split 2s-states.

The excited states of chalcogen atoms in silicon influence not only capture processes but also emission processes. Gibb et al.[31] and Rees et al.[32] calculated the emission rate of a two-stage emission process. In this model, the transition (energy E_2) from the ground state to the excited state is the rate-limiting process at high temperature and the emission is given by the expression (see Figure 7).

$$e_n^t = \frac{g_x}{g_g} \nu_2 \, e^{-E_2/kT}, \tag{9}$$

where g_g is the degeneracy of the ground state. For low temperatures, they obtained a temperature dependence which is proportional to exp $[-(E_1 + E_2)/kT]$. Since chalcogen atoms in silicon have more than one excited state, it is reasonable to assume that more than one of them is involved in the capture and emission process. Due to the decreasing energy spacing of these excited states, it is sometimes possible to separate them into two groups. Those which have small binding energies and quickly equilibrate with the free-carrier concentration, and those which have larger binding energies and for which the capture processes are much faster than the emission processes. If such a separation is possible, the two-stage model is valid. However, the assignment of the experimentally obtained values for E_1 and E_2 is still not straightforward, since any temperature dependence of ν_2 will alter the energy distribution.

Independent of the emission and capture process involved, the thermal emission rate e_n^t and the capture cross section σ_n^t are always related to the change in Gibb's free energy, ΔG_n, given by the complex balance relationship

$$e_n^t = \sigma_n^t \, V_{th} \, N_c \, \exp(-\Delta G_n/kT). \tag{10}$$

Here, $V_{th} = (3kT/m_n^*)^{\frac{1}{2}}$ is the average thermal velocity of the electrons, m_n^* is the density-of-states effective mass[34] and N_c is the effective density of states in the conduction band. At constant temperature, the change in Gibb's free energy needed to emit an electron into the conduction band[35,36] is related to the change in enthalpy, ΔH_n, and the change in entropy, ΔS_n, by the thermodynamic relationship

$$\Delta G_n = \Delta H_n - T \, \Delta S_n. \tag{11}$$

The total change of entropy, ΔS_n, is often divided into two parts

$$\Delta S_n = \Delta S_{ne} + \Delta S_{na} \tag{12}$$

where

$$\Delta S_{ne} = k \ln g \tag{13}$$

and ΔS_{na} is the sum of all other entropy contributions. It follows from Equations 10 and 11 that the enthalpy, ΔH_n, is obtained from an Arrhenius plot if proper corrections are made for the temperature dependence of σ_n^t. By measuring e_n^t and σ_n^t at different temperatures, the temperature dependence of ΔG_n can be calculated using Equation 10. Knowing the electronic degeneracy factor g, the temperature dependence of the energy position ΔG_n^o of the center is then obtained from the relation[37]

$$\Delta G_n^o = \Delta G_n + kT \ln g \tag{14}$$

which is similar to the optical binding energy obtained from an evaluation of the Rydberg-type excited states if ΔS_{na} is small.

4. Sample Preparation and Defect Identification

Several methods have been used to incorporate sulfur, selenium and tellurium into silicon. Since oxygen is often present in silicon as a residual impurity in rather high concentrations (depending on the crystal growing process used) no extra effort has to be applied to obtaining oxygen doped silicon. One of the most often used techniques for the incorporation of the other chalcogens is diffusion. This method, however, is complicated by the fact that sulfur, selenium and tellurium in the vapour phase erode the silicon surface. With increasing mass of the chalcogen atom, the erosion increases whereas the diffusion coefficient decreases. The incorporation of tellurium into silicon by diffusion is therefore rather difficult. Another method—a modified vapour transport technique—for doping silicon with chalcogen atoms has therefore recently been suggested, making use of the eroding capabilility of these elements. Using the chalcogen atom as a carrier in this method homogenously doped silicon with relatively high concentrations of chalcogens has been fabricated.[18] Volatile silicon chalcogenides are formed at high temperatures in contact with the silicon seed and the source material. Applying a shallow thermal gradient, epitaxial silicon layers are deposited under quasi-equilibrium conditions during which the chalcogen atoms are incorporated at their maximum solubility value. A comparison between selenium-mass transported and selenium-diffused silicon shows[38] that similar electronic properties are observed in both cases for the selenium-induced levels.

Other methods which have been used for the incorporation of chalcogens into silicon are float zone techniques and ion implantation.[39,40,41]

Any progress in our understanding of deep level impurities will depend on the possibility to reveal the chemical identity of the particular crystal defect investigated. It has been shown[22,42] that ESR or SIMS combined with space charge junction techniques[43-45] are very powerful tools. In a SIMS[46] experiment, the sample surface is bombarded with primary ions and the secondary ions sputtering off the surface are analyzed in a mass spectrometer. By "peeling off" successive layers of atoms, the sputtering beam eats its way towards the interior of the sample. In this way, the concentration profile of an impurity can be obtained. It is also possible to use junction techniques in determining concentration profiles. If the profile measured with the electrical method (junction technique) coincides with that obtained from the "atom-identifying" method (SIMS), the element related to the defect studied is revealed.

Information about the local defect structure cannot be gained from a SIMS experiment. This important information—which is essential to the understanding of the electronic behavior of the defect—can instead be obtained from another identification method—ESR.[47,48] The interpretation of the ESR data is particularly simple in the case of S, Se and Te in silicon, since it turns out that they are spin-only centers with $S = \frac{1}{2}$. In such cases, the number of peaks of equal height in an ESR-spectrum is directly related to the nuclear spin of the isotope giving rise to the ESR-signal. The relative magnitude of peaks with different heights corresponds to the natural abundances of the isotopes of a specific element. All isotopes with zero nuclear spin will show up as an unsplit central line. Thus, in most cases, the element giving rise to the ESR-signal may be unambiguously identified. Illuminating the sample with light of different wavelengths and measuring the spectral dependence of the ESR-signal, the spectrum of the photo-ionization cross section of the defect can be obtained. If this spectrum coincides with that obtained from junction techniques, not only the element causing the defect but also the local defect structure is revealed, since interactions with neighbouring nuclei give rise to additional hyperfine structure of every ESR-line.

5. Experimental Results

5.1 Tellurium

Since the interpretation of the experimental data is rather similar for sulfur, selenium and tellurium, a more comprehensive presentation of experimental results is given only for the tellurium case. The data for sulfur and selenium are presented in the next two sections without any further detailed discussion.

Hitherto, no information on polonium-doped silicon is available. Tellurium has therefore to be considered as the largest and heaviest element among all those chalcogens in silicon which have been investigated in more detail. From a naive point of view, one might expect tellurium to create a larger disturbance in the silicon crystal lattice than any of the other chalcogen atoms and, hence, that the ground state of tellurium crystal defects

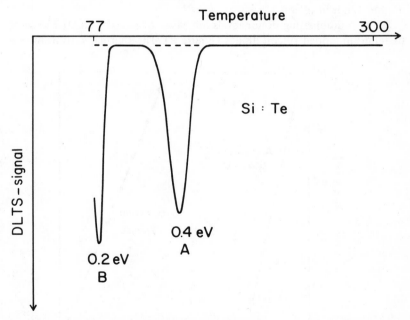

FIGURE 8   DLTS (Deep Level Transient Spectroscopy, see Ref. 45) spectra of tellurium-doped Schottky-diodes (solid curve) and of reference diodes not deliberately doped with tellurium (dashed curve), from Ref. 22.

might be deeper than those of sulfur and selenium. However, the opposite is observed. Chalcogens, therefore, behave similarly to elements in the fifth group of the periodic table. Antimony having the same number of core electrons as tellurium has in fact the shallowest ground state of all the group V donor levels in silicon.[49]

Silicon samples which have been prepared either by the tellurium vapour transport method or by tellurium diffusion at 1200°C for several hours in a sealed evacuated quartz ampoule, always show two dominant donor levels with similar concentrations. These results are in agreement with the model of substitutional double donor levels outlined earlier in the paper. Since these data, however, are only necessary and not sufficient to prove that tellurium forms a double donor in silicon, the two donor levels are labelled A- and B-center. However, considering that most of the data obtained in silicon doped with tellurium are best understood if the deep A-center is identified with Te^+ and the shallow B-center with Te^o, the further discussion in this chapter will be performed as if tellurium formed a double donor in silicon.

The two dominating donor levels are easily observed[22] in DLTS-spectra of tellurium-doped silicon diodes (Figure 8). The thermal activation ener-

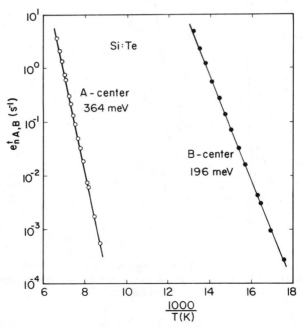

FIGURE 9 Arrhenius plots of the thermal emission rates of electrons for the two dominating donor levels in tellurium-doped silicon (from Ref. 22).

gies, E_n^t, (not enthalpies, cf. Equation 10) of the two centers are obtained from Arrhenius plots of the thermal emission rates giving E_{nA}^t = 365 meV and E_{nB}^t = 196 meV (Figure 9). These data were obtained by employing several different junction space charge techniques.[22] The thermal activation energy of the B-center was also investigated using the Hall effect. The results obtained are E_{nB}^t = 0.14 eV in Ref. 50, 0.185 eV in Ref. 18 and 0.202 eV in Ref. 39 (Figure 10).

The most accurate data on the energy position of the two tellurium centers are obtained from low temperature absorption measurements.[22] Overall transmission spectra, covering the spectral range from 4000 to 1000 cm^{-1}, are shown in Figure 11 for n-type and weakly n-type Si.Te samples. For both samples, traces of interstitial oxygen could be detected by the characteristic local vibrational modes at 1205 cm^{-1} (149.4 meV) and 1136 cm^{-1} (140.8 meV).[51]

The lower spectrum of Figure 11 was recorded from an n-type 0.19 Ωcm tellurium-doped silicon sample. The occupancy of the B-center, i.e., the neutral charge state of the tellurium impurity, Te^o, is stabilized in this sample by the Fermi level. Starting at low energies, the first line due to

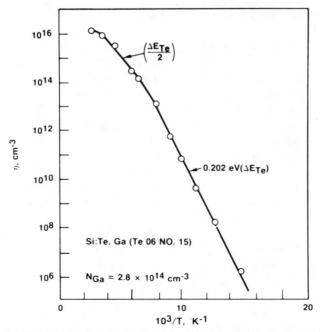

FIGURE 10 Free-electron concentration as a function of inverse temperature for Te-doped Si grown by the float zone technique with Ga compensation (from Ref. 39).

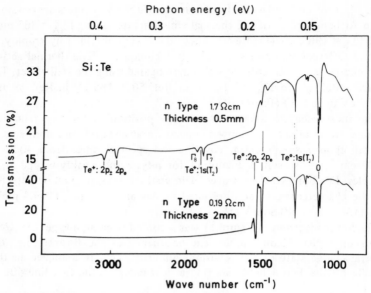

FIGURE 11 Overall transmission spectra of n-type (0.19 Ωcm, lower trace) and weakly n-type Si:Te samples (upper trace) at T ≈ 20K (from Ref. 22).

excitation of an electron from the B-center is seen at 1288 cm^{-1} (159.7 meV) arising from the electric dipole (E1) allowed transition 1s(A$_1$) → 1s(T$_2$). The line at 1217 cm^{-1} (151 meV) is probably due to oxygen, although the relative strength of absorption for the two lines 1217 and 1205 cm^{-1} should be different at 5K.[52]

Shortly before the photo-ionization of electrons directly into the conduction band at photon energies of about 0.2 ev, two sharp absorption lines are observed at 187.3 meV and 192.4 meV (Figure 12). These are due to transitions from the deep 1s(A$_1$) ground state into the shallow Rydberg levels 2p$_0$ and 2p$_\pm$, respectively. The energy separation for these two lines of 5.1 meV is in agreement with the value of 5.11 meV calculated from effective mass theory (EMT).[4] The EMT value for the energy separation of the 2p$_\pm$ level from the conduction band is 6.40 meV. Hence, the binding energy of the B-center [E$_c$ − E$_T$ (Te^o)] is 198.8 meV.

Since the position of the Fermi level of the weakly n-type Si:Te sample shown in Figure 11 was relatively deep, some of the B-centers were already empty at thermal equilibrium. Assuming that the A- and B-centers are different charge states of a double donor, the empty B-centers represent singly-ionized tellurium centers, Te$^+$, and, hence, filled A-centers. It is therefore expected that the excitation of electrons from the A-centers occurs at higher energies than for the B-centers. At 239.6 and 234.2 meV a doublet is observed, which is due to transitions from the ground state $1s(A_1)$ of the Te^+ center into the spin-valley split $1s(T_2)$ states Γ_7 and Γ_8. Prior to the photo-ionization threshold, sharp absorption lines are observed at 364.4 meV and 385.2 meV due to the excitation of electrons from the ground state into the Rydberg states $2p_0$ and $2p_\pm$, respectively (Figure 13). The energy spacing of 20.8 meV between these two lines agrees with the EMT value for an ionized double donor which should be four times 5.11 meV, i.e., 20.44 meV. The binding energy of the A-center at 5 K is obtained by adding the EMT value of $4 \times 6.40 = 25.6$ meV to the energy of the $2p_\pm$ state giving a value of 410.8 meV.

Using junction space charge techniques,[43,44] information about these excitation energies can be obtained at much higher temperatures by mea-

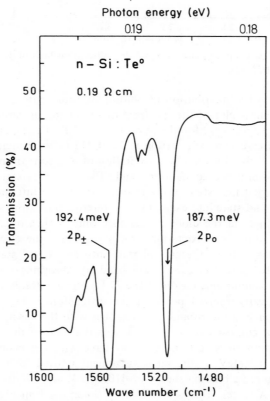

FIGURE 12 $1s(A_1) \rightarrow 2p_0$ and $1s(A_1) \rightarrow 2p_\pm$ absorption lines of Te^0 in silicon at $T \leqslant 5K$ (from Ref. 22).

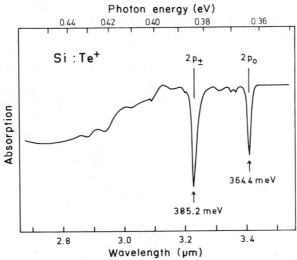

FIGURE 13 $1s(A_1) \to 2p_0$ and $1s(A_1) \to 2p_\pm$ absorption lines of Te$^+$ in silicon at T = 5K (from Ref. 22).

suring the spectral distribution of photo-ionization cross-sections. Figure 14 shows a linear plot of the spectrum of the photo-ionization cross section for electrons, σ_{nA}^o, for the A-center at 77 K. The rapid increase of σ_{nA}^o with increasing energy, starting at about 411 meV (as indicated by the arrow F), is caused by the photo-excitation of electrons from the ground state directly into the conduction band. The structures seen at energies smaller than that indicated by the arrow F are due to internal transitions from the ground state into excited states. Internal transitions are expected to cause changes in the diode capacitance only if they originate from a two-step photothermal excitation process in which the electron is first excited optically from the ground state into an excited state and then further excited into the conduction band by absorbing one or several phonons. Expanding this region of the 77 K spectrum (cf. insert of Figure 15), three clearly-resolved peaks can be distinguished. Figure 15 shows the same data in a logarithmic plot together with the absorption data of Figure 13 for comparison. It is readily seen that two of the peaks of the photo-ionization cross section spectrum agree with the main absorption lines within 0.2 meV. If the same assignment is tried for the spectrum, this would imply a binding energy of 410.6 meV at 77 K for the ground state of the A-center. This is indicated by the arrow F in Figures 14 and 15. The good agreement between the binding energies obtained from

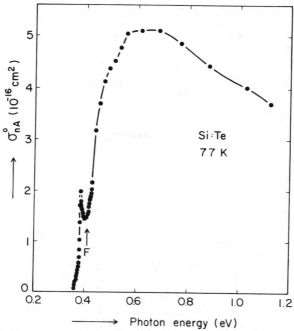

FIGURE 14 The spectral dependence of the photo-ionization cross section of electrons for the A-center plotted on a linear scale. The onset of photo-excitation of electrons from the ground state directly into the conduction band is marked with an arrow (F). The structure at lower energies is due to photothermal ionization processes (from Ref. 22).

absorption measurements at 5 K and from the spectrum of σ^o_{nA} at 77 K may suggest that the A-center is pinned to the conduction band.

From EMT, it is known that the 2s state lies 35.3 meV[4] below the conduction band. The corresponding peak in the spectrum of the photoionization cross section should therefore be seen at 375.3 meV. From investigations of shallow donors in silicon it is known that the $2s(T_2)$ state is somewhat deeper than predicted by EMT.[53] It is therefore not unreasonable to believe that the peak at 374.3 meV in Figure 15 is caused by the absorption of electrons from the ground state into the $2s(T_2)$ state. A closer inspection of the absorption data shows that a weak line at the same energy is also observed in these measurements (Figure 13).

The conclusion which can be drawn from these data is that both the A- and the B-centers have not only multi-valley split-off, but also Rydberg-

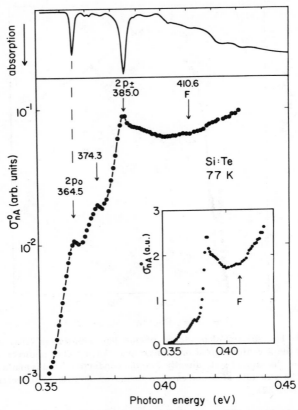

FIGURE 15 Expanded version of the low energy part of Figure 14, plotted logarithmically in the main figure and linearly in the inset. The absorption data above the main figure are the same as in Figure 13 (from Ref. 22).

type excited states. The existence of the latter closely spaced excited states strongly suggests that the electron capture is governed by a cascade process, which is a very fast process. This interpretation is in agreement with the observation that the electron capture cross section of both the A- and the B-centers is too large to be measured yet.[22]

From thermal and optical measurements it has to be concluded that the A- and B-centers always have similar concentrations, and that they are doubly and singly charged, respectively, when empty. These data support the model of tellurium forming a double donor in silicon. Further information on the electronic properties of these centers is obtained from ESR-studies.

Figure 16 shows the ESR[22] spectrum of a tellurium-transported silicon crystal. The pattern is independent of the orientation of the magnetic field and is described by the spin Hamiltonian

$$\mathcal{H} = g_e\, \beta_e\, \vec{H}\vec{S} + a\,\vec{I}\,\vec{S} - g_n\, \beta_n\, \vec{H}\,\vec{I}, \quad S = 1/2, \qquad (2)$$

where the first and last terms represent the electron ($g_e = 2.0023$) and nuclear Zeeman interactions, respectively, and the second term represents the isotropic hyperfine (hf) coupling. The doublet splittings seen in Figure 16 are consistent with the hf interaction of the $I = 1/2$ nuclei of ^{123}Te and ^{125}Te. The hf couplings ($|a| = 966.1 \times 10^{-4}$ and 1164.7×10^{-4} cm^{-1}, respectively) scale as their respective nuclear moments and the relative intensities of the lines correspond to the natural abundances of the Te isotopes. The strong central line arises from the even Te isotopes with zero nuclear spin. The centroids of the Te hf patterns are shifted from the position of this line by terms which are proportional to a^2/H.

Further resolution of the ESR lines of Si:Te$^+$ (Figure 17) reveals a structure which can be attributed to the hf interaction of ^{29}Si ($I = 1/2$ and 4.7% natural abundance) in the various shells surrounding Te.[22] These can be described by the additional terms in the spin Hamiltonian:

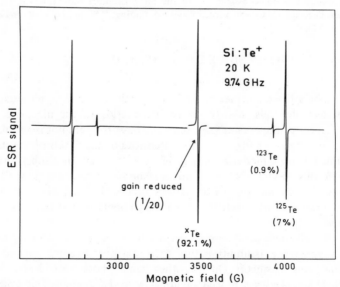

FIGURE 16 ESR of the ground state of the Te$^+$ donor in silicon. The doublets arise from the Te isotopes indicated. Their natural abundancies are also given. ^{x}Te denotes even Te isotopes with zero nuclear spin (from Ref. 22).

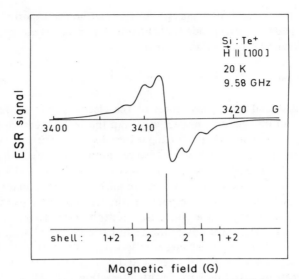

FIGURE 17 High resolution ESR spectrum of the central line of Figure 16. The structure lying symmetrically with respect to the main line is due to hyperfine interactions with ^{29}Si nuclei surrounding Te$^+$. The diagram in the lower part of the figure presents a possible assignment of the lines assuming substitutional Te. The intensities correspond to the probabilities of finding ^{29}Si in the shells indicated (from Ref. 22).

$$\mathcal{H}(^{29}\text{Si}) = \sum_i \vec{S} \; T_i \; \vec{I_i} - g_n \, \beta_n \, \vec{H} \, \vec{I_i}, \qquad (3)$$

where i labels the quantities related to the i-th surrounding nucleus. The Si hf-pattern depends slightly on the orientation of the magnetic field $\vec{H}$ with respect to the crystal axes, since the tensors T_i are not isotropic.

In the lower part of Figure 17, the intensities of the ^{29}Si lines are shown as obtained from a model assuming Te$^+$ to be located on a substitutional site.[22] In this model, Te$^+$ has four neighbours in the first Si shell and twelve in the second one. The pattern obtained clearly resembles the structure of the ESR spectrum if the assignments of the lines are made as shown.

Even though these data cannot be considered as final proof, they nevertheless support the assumption that tellurium may occupy isolated substitutional sites in silicon. On the other hand, photo ESR clearly shows that filled A-centers are indeed identical with Te$^+$-centers. This evidence was obtained by photon-induced valency changes of Te$^+$ and Te^{2+}. It turned out that the Te$^+$ signal could be reduced to zero by illuminating the sample with photons of energy less than 0.7 eV but larger than 0.4 eV.

Figure 18 shows schematically the time dependence of the Te^+ signal, i.e., the concentration of filled A-centers in a p-type sample due to illumination.[22] When the sample was illuminated with photons of a particular energy, specified below, Te^+ was created (I). It should be noted that the initial conditions were chosen to be such that all A-centers were empty and, hence, the initial Te^+ concentration, $n_T(o)$, was zero. After removing the light source, the Te^+ intensity first increased slightly and then decreased slowly (II). Illumination of the sample with photons of another energy during this stage led to a strong decrease in the Te^+ signal (III). The slopes defined in Figure 18 were taken as a measure of the sensitivity of the system to light irradiation.

The increase of the Te^+ signal after removing the light source (II in Figure 18) can be understood as follows:[22] the generation of Te^+ and, hence, the filling of empty A-centers in steady-state is stabilized by the recombination of electrons from the A-centers with holes in the valence band and further optical excitation of electrons from the A-centers into the conduction band. Electrons excited into the conduction band may be recaptured by trapping centers, possibly empty B-centers. The increase in the Te^+ intensity may therefore be caused by the reverse process. After the slight initial increase, the decrease of the Te^+ signal in darkness (II in Figure 18) is known, from photocapacitance measurements,[22] to be due to

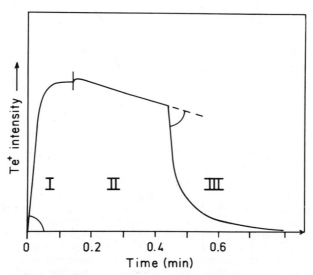

FIGURE 18 Schematic time-dependence of the photoresponse of Te^+ in p-type Si. I: excitation with $h_\upsilon \geqslant E_C - E_T$; II: light off; III: quenching with $h_\upsilon \geqslant E_T$ (from Ref. 22).

room temperature radiation causing photothermal excitation of electrons from the ground state into the conduction band via the Rydberg series.

Figure 19 shows the generation and annihilation of Te$^+$ centers (filled A-centers) in silicon as a function of photon energy.[22] It is easily seen that the A-centers start to fill at about 0.73 eV, in good agreement with the energy position of the centers obtained from junction space charge techniques and absorption measurements (Figures 14 and 15). The rate of filling is given by $\sigma_{pA}^o \phi N_{TT}$, where N_{TT} is the total concentration of A-centers.

The light-induced quenching of the Te$^+$ signal is obviously caused by the optical excitation of electrons from filled A-centers into the conduction band. The rate of the quenching process is given by $\sigma_{nA}^o \phi n_T(o)$. The spec-

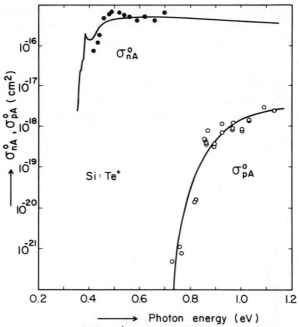

FIGURE 19 Energy dependence of the photo-ionization cross section of Si:Te$^+$ for holes, σ_{pA}^o (○), and electrons σ_{nA}^o (●), as determined by ESR. The vertical scales for both quantities are based on different calibrations. The solid curve in the σ_{nA}^o spectrum is a plot of data obtained from photocapacitance measurements. The absolute values of the σ_{nA}^o data obtained by ESR are adjusted to give optimal agreement with photocapacitance data. The adjustment is within the experimental error. It should be noted that the photocapacitance data for energies smaller than about 0.4 eV arise from photothermal excitation processes and are therefore thermally activated. At low temperatures, these excitation processes are thus not observed in agreement with ESR data shown (from Ref. 22).

tral distribution of the photoionization cross section of electrons, σ_{nA}^{o}. is shown in Figure 19 together with the data obtained from photocapacitance measurements at 77K.[22]

The results so far described were obtained in diffused or tellurium transported samples.[22,39,54,55] Quite often tellurium has been used for ion implantation[41,56-61] in silicon. Rather peculiar properties of the energy levels observed in such samples have been reported.[41] These may result from residual implantation damage effects. On the other hand, several authors claim that most of the implanted tellurium atoms are incorporated on substitutional sites[41,56-61] in agreement with the results obtained in diffused or transported samples.

Even though most of our knowledge about tellurium in silicon refers to centers which are assumed to be single substitutional impurities, there are nevertheless reasons to believe that there may also exist other types of tellurium-related crystal defects in silicon depending on the doping procedure. Such different types of centers have been observed in sulfur and selenium-doped silicon.

5.2 Selenium

If selenium is diffused into silicon at low temperatures or if the cooling procedure is slow, centers with properties similar to the A- and B-centers

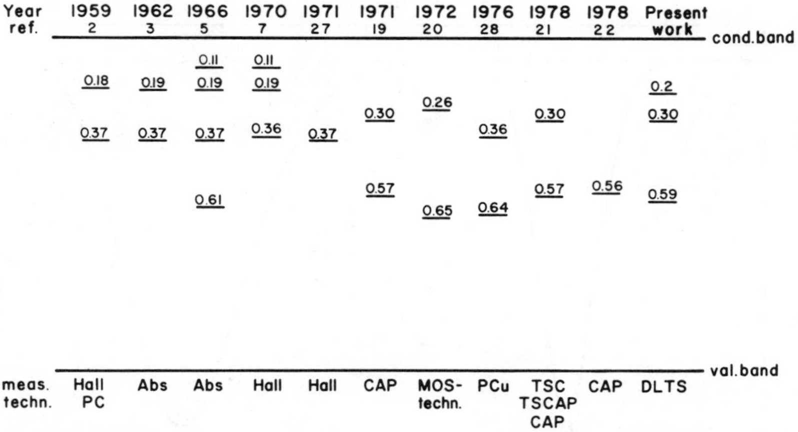

Year ref.	1959 2	1962 3	1966 5	1970 7	1971 27	1971 19	1972 20	1976 28	1978 21	1978 22	Present work
cond. band											
			0.11	0.11							
	0.18	0.19	0.19	0.19							0.2
						0.30	0.26		0.30		0.30
	0.37	0.37	0.37	0.36	0.37			0.36			
						0.57			0.57	0.56	0.59
			0.61				0.65	0.64			
val. band											
meas. techn.	Hall PC	Abs	Abs	Hall	Hall	CAP	MOS-techn.	PCu	TSC TSCAP CAP	CAP	DLTS

FIGURE 20 Survey of published data for the energy positions of selenium-related centers in silicon. PC stands for photoconductivity, DaC for dark conductivity, and Abs for absorption. The DLTS data are taken from a plot of log e_n^t versus $1/T$ [from Ref. 29, Ref. 6 in the figure corresponds to Ref. 63 in this paper (6-63), 7-64, 8-65, 9-55, 10-66, 11-67 and 13-40].

in tellurium doped silicon are observed.[29] Additional centers, however, occur when selenium is diffused at high temperatures.[69,71] As with sulfur-doped silicon[62] some of these centers may originate from selenium pairs. A survey of published data [from Ref. 29 (valid 1979)] on the energy positions of selenium-related centers is shown in Figure 20. In spite of the disagreement in the literature about the values of the energy positions of centers originating from selenium, most data can be classified into three groups with energies 0.2 (C-center), 0.3 (B-center) and 0.5–0.6 eV (A-center). Experimental data suggest that the A- and B-centers are the ionized and neutral versions,[22,29,69,70,71] respectively, of an isolated, substitutional selenium atom, whereas the C-center is the neutral version of a selenium pair.[69,71]

A DLTS spectrum of selenium doped silicon featuring the A- and B-centers is presented[29] in Figure 21. It is easily shown that the two centers have similar concentrations. From Arrhenius plots of the thermal emission rates for electrons, thermal activation energies of $E_{nA}^t = 0.524$ eV and $E_{nB}^t = 0.286$ eV are obtained[29] (Figure 22). These data clearly show that selenium related A- and B-centers in silicon are considerably deeper than the tellurium ones. Since the electron capture into the B-center is slower for selenium than for tellurium, rather detailed information on the capture process is available.[29]

Figure 23 shows the temperature dependence of the electron capture cross section σ_{nB}^t (Ref. 29). The measured dependence of $T^{-3.2}$ is in

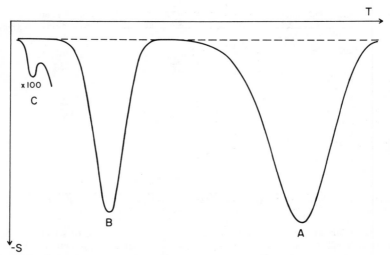

FIGURE 21   DLTS spectra of selenium-doped p⁺n diodes (solid curve) and of reference diodes deliberately not doped with selenium (dashed curve) (from Ref. 29).

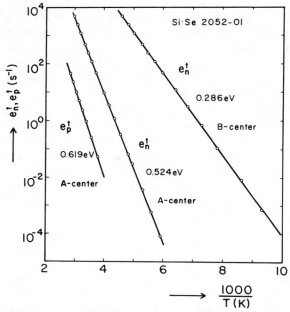

FIGURE 22 Electron and hole thermal emission rates versus inverse temperature for the two dominant centers in selenium-doped silicon (from Ref. 29).

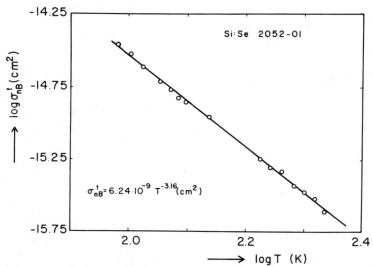

FIGURE 23 Temperature dependence of the electron capture cross section at the B-center, assuming cascade capture (from Ref. 29).

fair agreement with Abakumov's calculations[11] which suggest a T^{-3} dependence. If the experimental data are interpreted in terms of a two stage capture model[31,32] (Figure 24) a binding energy (E_1) of 14 meV is obtained for the intermediate capture state.[29] Using Equation (10) and taking the temperature dependence of the electron capture cross section into account, a temperature independent enthalpy ΔH_{nB} of 0.301 eV is obtained from the Arrhenius plot of the thermal emission rate e^t_{nB} for electrons. Inserting absolute values of the thermal emission rate and the electron capture cross section for different temperatures into Equation (10), it is readily seen that the Gibb's free energy, ΔG_{nB}, is independent of temperature and equal to ΔH_{nB} (Ref. 29). This means that the total change in entropy, ΔS_n, is equal to zero (cf. Equation (11)) and, hence, either the electronic degeneracy factor g is equal to 1 or $\Delta S_{ne} = -\Delta S_{na}$ (cf. Equation (12)). For the neutral charge state of a double donor one would expect g to be 2.

Absorption measurements on the selenium A- and B-centers have been performed recently, revealing well resolved line spectra.[38,69] Figure 25 shows a typical example for the B-center at 10K. Excitation energies in meV relative to the ground state obtained from absorption measurements for both centers are summarized in Table 1.[71] From these data, ground-state binding energies of 593.2 meV (589.4 meV in Ref. 69) and 306.6 meV are obtained for the A- and B-centers respectively.

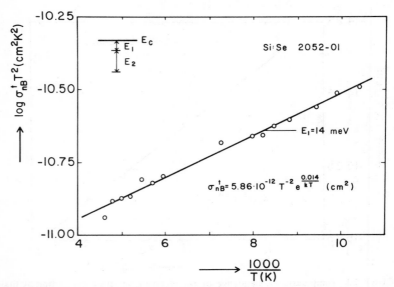

FIGURE 24 Temperature dependence of the electron capture cross section at the B-center, assuming a two-stage capture model (see Equation (8)) (from Ref. 29).

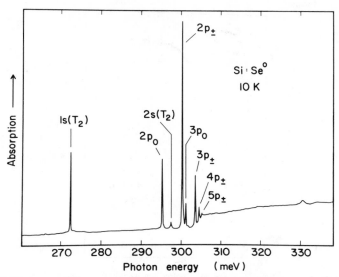

FIGURE 25 Absorption spectrum for the B-center in diffused Si:Se (from Ref. 71).

Structured spectra of photo-ionization cross sections for electrons in selenium doped silicon have been observed by Grimmeiss et al.[70] The spectra were obtained from high resolution photocapacitance measurements. A survey of the absolute values of the photo-ionization cross section σ_{nA}^o, σ_{pA}^o and σ_{nB}^o is shown in Figure 26. The spectral dependence of σ_{nB}^o and σ_{nA}^o are replotted in greater detail in Figures 27, 28 and 29. Considering the data presented in Table 1 with the data in Figure 29, it is quite clear that the structure observed in the spectrum of the photo-ionization cross section is due to excited states. The assignment of the peaks has been obtained by comparison with absorption measurements and differs therefore from the previous assignment in Ref. 70. Note also the similarity between the spectra of σ_{nA}^o in Si:Te (Figure 15), Si:Se (Figure 29) and Si:S (Figure 43).

An ESR measurement of the ground state of the Se^+ donor level in silicon is shown in Figure 30.[22] The spectrum is isotropic and identified as

TABLE 1 Excitation energies (meV) of the A- and B-centers in Si:Se with respect to the ground-state energy (Ref. 71).

	$1s(T_2)$	$2p_o$	$2s(T_2)$	$2p_\pm$	$3p_o$	$3p_\pm$	C.B
Se° B-center	272.2	295.1	297.4	300.2	301.2	303.5	306.6
Se⁺ A-center	427.3 429.5	547.2	553.9	567.6	571.6	581.0	593.2

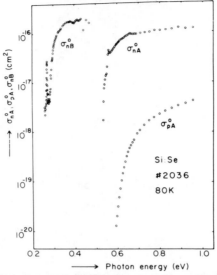

FIGURE 26 A survey of the absolute values of the photo-ionization cross section σ_{nA}^0, σ_{pA}^0 and σ_{nB}^0 versus photon energy in selenium-doped silicon at 80K (from Ref. 70).

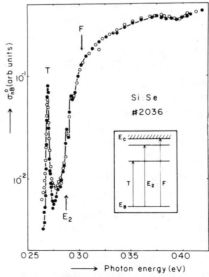

FIGURE 27 Spectra of the photo-ionization cross section of electrons, σ_{nB}^0, for the shallower donor level in Si:Se at two different temperatures ($\bullet$ = 40K, $\circ$ = 80K) (from Ref. 70).

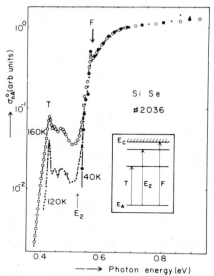

FIGURE 28 Spectra of the photo-ionization cross section of electrons, σ_{nA}^0, for a deeper donor level in Si:Se at different temperatures ($\bullet$ = 40K, + = 120 K, $\circ$ = 160K) (from Ref. 70).

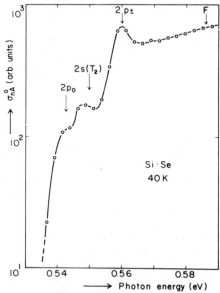

FIGURE 29 An enlarged part of the continuum for σ_{nA}^0 in Si:Se, showing the influence and assignment of the excited states (from Ref. 70).

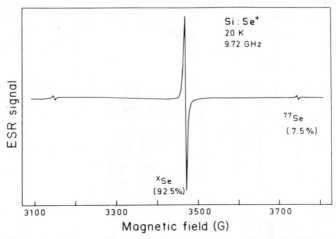

FIGURE 30   ESR of the ground state of the Se$^+$ donor in silicon (from Ref. 22).

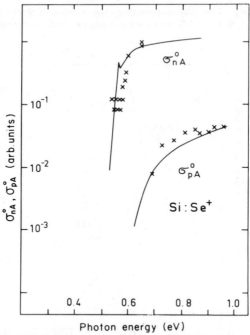

FIGURE 31   Energy dependence of the photo-ionization cross sections of Si:Se, as determined by ESR. The vertical scales of the experimental data have been adjusted to give optimal coincidence with the photocapacitance data (full lines) reported in Ref. 70 (from Ref. 22).

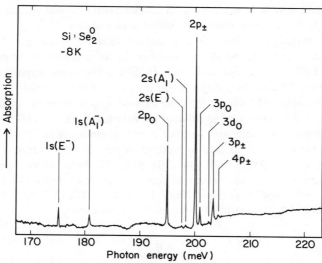

FIGURE 32 Absorption spectrum of the C-center in Si:Se (Se_2^0) (from Ref. 71).

being due to Se^+ on tetrahedral sites (either substitutional or interstitial). From these measurements values of 2.0057 for the g factor, 1/2 for S and $553.2 \times 10^{-4} cm^{-1}$ for the hf coupling $|a|$ for $^{77}Se^+$ are obtained. Furthermore, the identity of the A-center with Se^+ on a tetrahedral site in silicon could unambiguously be proved by photo-ESR spectra[22] (Figure 31). Measurements of the diffusion profiles for the A- and B-centers obtained with junction space charge techniques show[42] that identical profiles are observed for selenium atoms studied using SIMS.

An absorption spectrum of the C-center in Si:Se is shown[71] in Figure 32. From these data, see Table 2, a ground state binding energy of 206.4 meV is obtained. Similar spectra have been measured by Swartz et al.[69] The pattern of the excited states (see Table 2) is very close to that which is expected for the neutral version of a selenium pair. As pointed out earlier, a $1s(T_2)$ state is split into a singlet $1s(A_1)$ and a doublet $1s(E)$ when the symmetry is changed from T_d to $C_{3v}(D_{3d})$.[71] The assignment of the absorption lines shown in Figure 32 is supported by the fact that the peak at 175.1 meV is broadened in the presence of a stray field, whereas the

TABLE 2 Excitation energies (meV) of the C-center in Si:Se with respect to the ground-state energy (from Ref. 71).

$1s(E^-)$	$1s(A_1^-)$	$2p_o$	$2s(E^-)$	$2s(A_1^-)$	$2p_\pm$	$3p_o$	$3p_\pm$	C.B
175.1	180.7	194.9	197.6	198.3	200.1	200.9	203.3	206.4

peak at 180.7 meV is unchanged. A similar center corresponding the C-center in Si:Se has been observed in sulfur-doped silicon.[71,72]

Sclar[15] investigated the effect of dopant diffusion vapor pressure on selenium and sulfur-doped silicon, keeping the diffusion temperature constant at 1200°C (see also Ref. 71). In Si:Se, the main impurity level shifted from 0.30 eV to 0.20 eV when the pressure was increased. In view of the preceding discussion, this can be understood in terms of a shift from the neutral version of an isolated substitutional selenium atom to the neutral version of a selenium pair (or impurity molecule in Sclar's notation). A further level in selenium-doped silicon has been observed by Swartz et al.[69] at 387.9 meV (389.5 in Ref. 71). As in sulfur-doped silicon, this center might be the ionized version of a selenium pair. Two other centers were found[69] at 116.0 meV (neutral) and 213.7 meV (singly ionized) in Si:Se. The origin of these centers is unknown.

5.3 Sulfur

Apart from oxygen sulfur was the first chalcogen in silicon that was investigated thoroughly. As early as 1959, Carlson et al.[73] published Hall effect measurements on sulfur doped silicon (Figure 33). The energy levels found at 0.18 eV and 0.37 eV were assumed to be due to the two charge states of an isolated substitutional sulfur atom. Later this assignment was changed, since it became highly probable that these energy levels originated from a sulfur pair. The energy levels (A and B) which are now ascribed to the isolated substitutional sulfur double donor are very similar to those in selenium-doped silicon. According to Brotherton et al.[62] high diffusion temperatures favor the generation of pairs whereas low temperatures favor the generation of isolated centers (Figures 34 and 35). A survey [from Ref. 30] of published data up until 1979 on the energy positions of sulfur-related centers is shown in Figure 36.

The concentrations of the A- and B-centers in sulfur doped silicon are always found to be very similar.[30,62] This is easily shown by DLTS measurements (Figure 34, 35). In addition to the A- and B-centers, a third center (C-center) is usually observed which is probably the neutral version of a sulfur pair. Arrhenius plots of the thermal emission rates for electrons (Figure 37) give the following activation energies for the A- and B-centers: $\Delta E_{nA} = 0.587$ eV and $\Delta E_{nB} = 0.302$ eV.[30] The capture process of electrons into these centers is very fast, and the temperature dependence of the electron capture cross section has so far been measured[30] only for the B-center (Figures 38 and 39). The enthalpy of the B-center, ΔH_{nB}, obtained from these data[30] is 0.320 eV. Using Equation 10 for calculating the

Gibb's free energy, ΔG_{nB}, it turns out that in contrast to Si:Se the Gibb's free energy is slightly temperature dependent giving values between 0.30 and 0.31 eV. Taking g = 2 a temperature independent value of 0.319 eV for the energy position ΔG^o_{nB} (see Equation 14) of the B-center is obtained in agreement with optical data and supporting the assumption that the A- and B-centers are caused by single substitution impurity atoms. Similar thermal data have been published in Ref. 62.

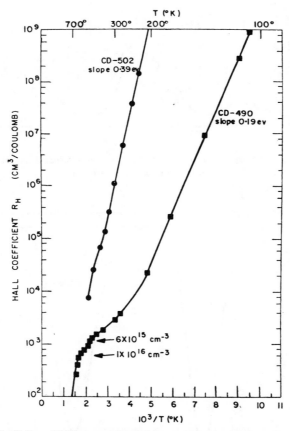

FIGURE 33   Hall coefficient vs. reciprocal temperature for typical samples of sulfur-doped silicon, showing the two observed levels. Samples were p-type before diffusion of sulfur. The slopes of the Hall coefficient curves must be corrected for the $T^{3/2}$ dependence of the density of states near the bottom of the conduction band. This correction places the levels 0.18 and 0.37 eV from the conduction band edge (from Ref. 73).

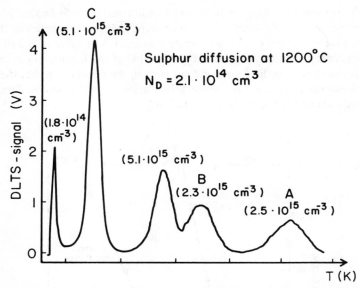

FIGURE 34 DLTS spectrum obtained from a sulfur-diffused sample (1200°C) (from Ref. 62, but the designation of the peaks is different here).

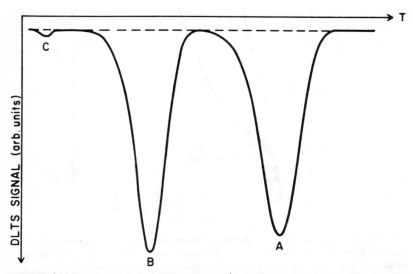

FIGURE 35 DLTS spectra of sulfur-doped p⁺n diodes (solid curve) and reference diodes not deliberately doped with sulfur (dashed curve). Diffusion temperature 825°C (from Ref. 30).

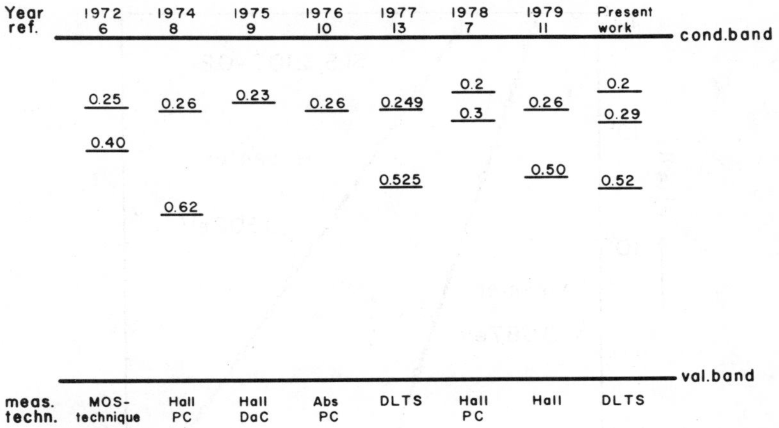

Year ref.	1972 6	1974 8	1975 9	1976 10	1977 13	1978 7	1979 11	Present work	
									cond.band
	0.25	0.26	0.23	0.26	0.249	0.2 / 0.3	0.26	0.2 / 0.29	
	0.40								
					0.525		0.50	0.52	
			0.62						
									val.band
meas. techn.	MOS- technique	Hall PC	Hall DaC	Abs PC	DLTS	Hall PC	Hall	DLTS	

FIGURE 36 Survey of published data on the energy positions of sulfur-related centers in silicon. PC stands for photoconductivity, Abs for absorption, CAP for isothermal capacitance transient, PCu for Photocurrent, TSC for thermally stimulated current, and TSCAP for thermally stimulated capacitance. All CAP and DLTS data are taken from a plot of log e_n^t vs. $1/T$. Some optical data have been excluded [from Ref. 30. Ref. 2 in the figure corresponds to Ref. 73 in this paper (2-73), 3-74, 5-72, 7-75, 19-76, 20-63, 21-78, 22-79, 27-80 and 28-81].

Absorption measurements[71] of the B-center in Si:S were recently reported (see Figure 40) as well as detailed photoconductivity studies.[71,82] The ground state binding energy obtained from these measurements is 318.3 meV, which is in excellent agreement with $\Delta G_{nB}^o = 0.319$ eV from Ref. 30. Detailed absorption measurements of the A-center have been performed by Kleiner et al.[72] and Grimmeiss et al.,[23] giving a binding energy of 613.5 meV.

Several papers on the photocapacitance of the A- and B-centers have been published[70,84-86] (Figures 41, 42, 43 and 44). The spectra obtained for the photo-ionization cross sections are very similar for selenium and tellurium in silicon, and in particular, the same types of excited states are observed.

The first ESR measurement of an S^+ center was performed by Ludwig.[87] The ESR spectrum of a sample enriched to contain 26% sulfur is shown in Figure 45. From this isotropic spectrum with a g-factor of 2.0054, $S = 1/2$ and the hyperfine coupling $|a| = 104.2 \times 10^{-4}$ cm^{-1}, it is concluded that S^+ occupies a tetrahedral site in the silicon lattice. From unpublished photo-ESR results by Kravitz et al.,[75,87] a binding energy of 0.59 eV is suggested for the S^+ center which supports the assumption that the A-center is probably the ionized version of an isolated substitutional S atom.

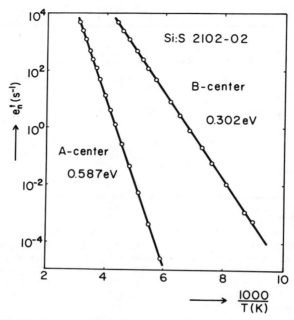

FIGURE 37 Electron thermal emission rates versus inverse temperature for two centers in sulfur-doped silicon (from Ref. 30).

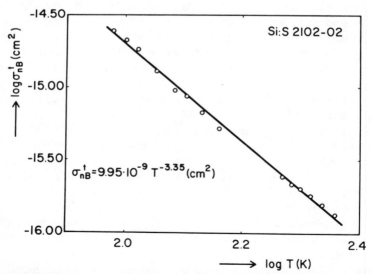

FIGURE 38 Temperature dependence of the electron capture cross section of the B-center assuming cascade capture (from Ref. 30).

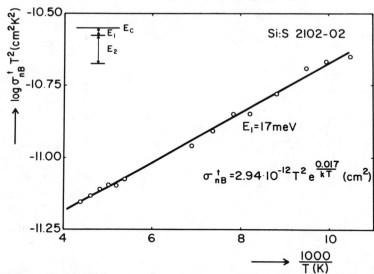

FIGURE 39   Temperature dependence of the electron capture cross section of the B-center assuming a two-stage capture model (from Ref. 30).

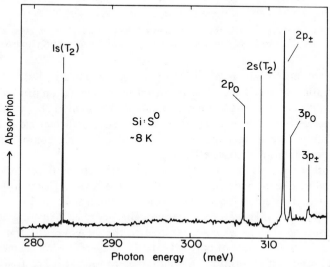

FIGURE 40   Absorption spectrum of sulfur doped silicon showing the B-level (from Ref. 71, Si:S⁰).

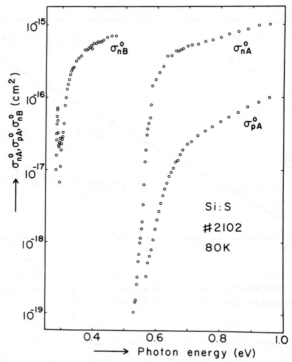

FIGURE 41   A survey of the absolute values of the photo-ionization cross sections σ^0_{nA}, σ^0_{pA} and σ^0_{nB} versus photon energy in sulfur-doped silicon at 80K (from Ref. 70).

In recent papers Myers et al.[88] claim that in implanted samples the concentration of the B-centers depends on the implantation dose whereas the A-center concentration remains essentially constant. These results are not in agreement with the double donor model, and further experiments are probably needed since measurements of this type are complicated by residual implantation damage effects.

In another paper Myers et al.[79] reported an isotope shift for the sulfur deep level in silicon, which is in our notation the A-level. Using isothermal capacitance transient measurements, they found a shift of 14 meV between ^{34}S and ^{32}S. However, in a recent paper, Forman[89] investigated the absorption line that is generally referred to the $1s(A_1) \rightarrow 2p_\pm$ transition. He found an isotope shift of only 0.14 meV. Furthermore, he claimed that the three strongest absorption lines all represent $1s(A_1) \rightarrow 2p_\pm$ transitions of different centers with slightly different binding energies of their $1s(A_1)$ states instead of the $1s(A_1) \rightarrow 2p_0$, $2s(T_2)$ and $2p_\pm$ transitions of a single

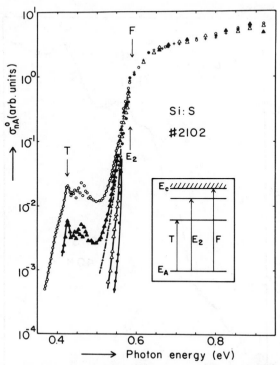

FIGURE 42 Spectra of the photo-ionization cross section of electrons, σ_{nA}^o, for the deeper donor level in Si:S at different temperatures ($\bullet$ = 40K, $\triangle$ = 80K, + = 120K, $\blacktriangle$ = 140K, $\circ$ = 160K) (from Ref. 70).

center. By assuming that "through a presently unknown mechanism" different mixtures of the slightly different centers are present in [34]S and [32]S samples, he could account for the results of Myers et al.,[79] since only an average energy is measured in thermal measurements.

Forman's assumption is supported by the fact that the lines do not display the same annealing behaviour.[89,90] Considering, however, that there are 10^{16} negatively charged boron atoms/cm^3 in his samples, a strong electric field and perhaps even stress may be present in the crystal. During annealing the situation alters. The number of charged impurities may charge and stress disappear. Furthermore, different lines are not eqully sensitive to the electric field as can easily be seen from a comparison of Figures 10 and 12 in Ref. 91.

The C-center (neutral sulfur pair) has been investigated by Brotherton et al.[62] using DLTS and Hall effect. The electron capture into this center

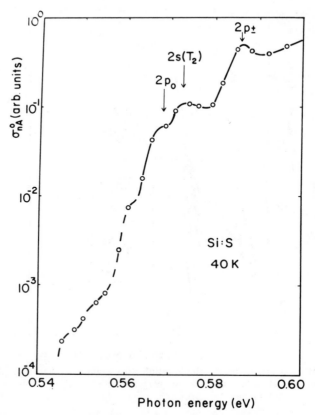

FIGURE 43 An enlarged part of the continuum region for σ_{nA}^0 in Si:S, showing the influence and assignment of the excited states (from Ref. 70).

is independent of temperature. In the same samples the authors found a further center of similar concentration. The thermal emission rates of both centers are non-exponential and the activation energies field dependent. Using the Hall effect and low field thermal emission rate measurements binding energies of 0.18 and 0.38 eV are obtained for the neutral and ionized version of a sulfur pair respectively.

Further evidence for the assumption that these centers originate from a sulfur pair is given by Camphausen et al.[75] using hydrostatic pressure. Krag et al.[72] performed absorption measurements and obtained binding energies of 187.7 and 370.5 meV for the neutral and ionized sulfur pairs, respectively, see also Ref. 71. Part of the absorption spectrum is shown in

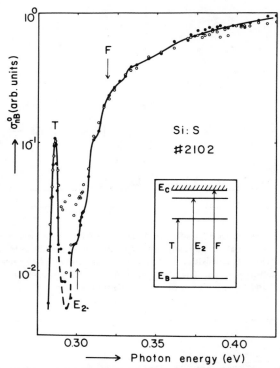

FIGURE 44 Spectra of the photo-ionization cross section of electrons, σ_{nB}^{0}, for the shallower donor level in Si:S at two different temperatures ($\bullet$ = 40K, $\circ$ = 80K) (from Ref. 70).

Figure 46 (cf. also Figure 32). As previously demonstrated for selenium, the $1s(T_2)$ state splits into a $1s(A_1)$ and a $1s(E)$ state due to the fact that the symmetry is changed from T_d to C_{3V} (D_{3d}). Furthermore, it is clearly seen that uniaxial stress splits only the $1s(E)$ state, whereas the $1s(A_1)$ state is unsplit in agreement with the model. The absorption peaks originating from transitions from the ground state into the p-states split even for the stress in the [111] direction.[72] Since the sulfur-and selenium-related C-centers in silicon have very similar properties, there is reason to believe that both centers are caused by pairs. An ESR signal originating from a sulfur pair where the two sulfur atoms occupy equivalent positions has been observed by Ludwig[87] in the same samples that have been studied by Krag et al.[74] and Carlson et al.[73] This signal could be quenched with light of energy greater than 0.37 eV. Due to the striking similarity, it is

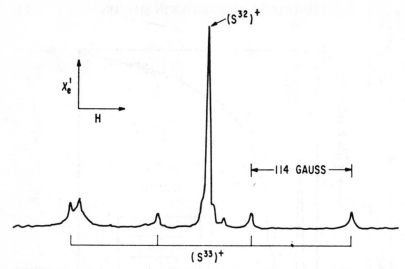

FIGURE 45 Spectrum of S^+ under rapid passage conditions in a sample enriched to contain 26% ^{33}S. The almost equally spaced satellite lines are about 9% as intense as the main line, which is consistent with the above enrichment if the resonant center contains one sulfur nucleus. The two other weak lines are associated with small amounts of Fe^0 and Cr^+ present in the sample (from Ref. 87).

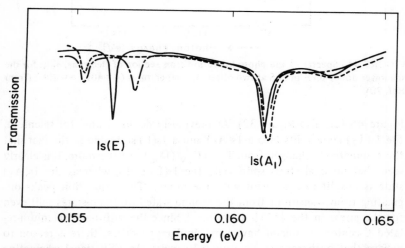

FIGURE 46 Effect of stress on the neutral sulfur pair (C-center) spectrum in silicon.

E   unpolarized, T = o

E ∥ T

E ⊥ T $T \parallel [001] = 2.4 \times 10^8$ dynes/cm^3

(from Ref. 72).

TABLE 3 Ground-state binding energies in eV of known chalcogen donors in silicon in eV. D refers to isolated substitutional chalcogen atoms, D_2 to chalcogen pairs and D_c to some chalcogen related defects, see also Ref. 71.

	D^o	D^+	D_2^o	D_2^+	D_c^o	D_c^+
S	0.32	0.61	0.19	0.37	0.11 0.16	
Se	0.31	0.59	0.21	0.39	0.12	0.21
Te	0.20	0.41				

therefore highly probable that the centers at 0.19 and 0.37 eV in Si:S and at 0.21 and 0.39 eV in Si:Se originate from a pair of sulfur or selenium atoms, respectively.

A few other sulfur related centers have been observed in Si:S, but their origins are as yet unknown (see Table 3).

5.4 Oxygen

Oxygen is considered to be the dominant impurity in device-grade silicon and is typically present in concentrations of 10^{18} cm^{-3}. Oxygen enters the silicon lattice during crystal growth from a melt contained in a quartz crucible. The oxygen atom does not normally occupy a substitutional site. According to Hrostowski and Adler,[92] the oxygen interrupts a normal Si–Si valence bond, forming a non-linear Si–O–Si molecule imbedded in the silicon lattice. The oxygen atom in this "interstitial" position puts the surrounding lattice under a compressive strain, but does not produce electrical activity.[93,94] During annealing, mobile oxygen, which is present in supersaturation concentrations, will cluster and precipitate, giving rise to a variety of electrically active structures.[95] Some of these centers, in particular the oxygen-related donor states, which are produced by neat treatment of the as-grown silicon in the temperature range of 300–500°C, have been the subject of study for more than twenty-five years. They have not been structurally identified or electrically understood, although considerable effort has gone into the study of these defects.[96-108] Most of these results have been summarized in two recent review papers by Patel[109] and Kimerling,[110] and will therefore not be further discussed in this paper.

When a mobile vacancy in silicon is trapped by an interstitial oxygen atom, the so-called Si A-center is formed. This model was first suggested by Bemski[111] and Watkins et al.[112] The A-center with oxygen in the "substitutional" site is comprised of a single oxygen atom in a lattice vacancy. One for each of the four silicon atoms around a vacancy contributes with a broken bond, resulting in four broken bonds around a vacancy. The oxygen bridges two of these broken bonds, forming a Si–O–Si "molecule".[113] The remaining two silicon atoms pull together to form a Si–Si molecular bond (Figure 11, Ref. 114). Evidence for this model was obtained from spin resonance studies,[113,115,116] infrared absorption measurements[114,117] and optical excitation studies.[118] Several important features of the model could be obtained from ESR measurements in which uniaxial stress is applied simultaneously.[113]

The A-center is an electron trap 0.18 eV below the conduction band. It is easily generated by electron irradiation at room temperature. Since no A-centers are observed after a 20°K irradiation, it is safe to say that the A-centers are not a primary defect (i.e., interstitial, vacancy, etc). From capacitance transient spectroscopy an electron capture cross section of about 10^{-14} cm^2 is obtained[119] for this center.

6. Discussion

6.1 Spin-valley and Multi-valley Interactions in Si:Te, Si:Se and Si:S

It is readily seen from Figures 11 and 47 that the absorption peaks corresponding to transitions from the ground state $1s(A_1)$ to the excited state $1s(T_2)$ are split into two peaks for the Te$^+$, Se$^+$ and S$^+$ centers in silicon.[23] As already pointed out in Section 2, there are strong reasons to believe that the splitting is due to a spin-valley interaction. However, it is not clear whether $1s\Gamma_7(T_2)$ or $1s\Gamma_8(T_2)$ should have the larger binding energy. The integrated absorption of the upper energy component is larger than the lower one as shown in Figure 47, suggesting that $1s\Gamma_8(T_2)$ with degeneracy 2 (excluding spin) has a smaller binding energy than $1s\Gamma_7(T_2)$ with degeneracy 1. If any random stress or electric field is present in the crystal, $1s\Gamma_8(T_2)$ will split (see Figure 5) and the corresponding peak become broader. This effect is clearly seen in Figure 47 for Se$^+$ and Te$^+$.[23] The upper component is considerably broader than the lower one, which indicates that $1s\Gamma_8(T_2)$ does lie above $1s\Gamma_7(T_2)$. The broadening, however, is large enough to imply some uncertainty in the magnitude of the splitting, since it cannot be taken for granted that the peak positions are not changed due to the presence of random stress or fields.[23] By using uniaxial stress which splits the upper component into two peaks but

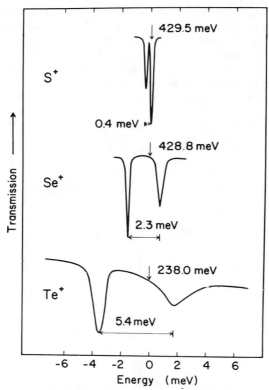

FIGURE 47 Excitation spectra due to transitions from the ground states $1s(A_1)$ to the spin-valley split $1s(T_2)$ states for S^+, Se^+ and Te^+ donors in silicon. The component to the right corresponds to $1s\Gamma_8(T_2)$ and that to the left $1s\Gamma_7(T_2)$. The origin of the energy scale is the transition energy to the unsplit $1s(T_2)$ state into which the two components of the doublet would reduce if the spin-valley interaction were absent (from Ref. 23).

leaves the lower component unsplit (see Figure 5), Krag et al.[25] have shown that the ordering mentioned above is correct for Bi in Si. Although the data[5] for Sb can be interpreted in the same way, the splitting was not discussed in these terms in Ref. 5. All available data on 1s states for group V and VI impurities in silicon are summarized in Table 4.[23] It can be seen that although the ground state $1s(A_1)$ binding energies vary widely, the energy positions of the $1s(T_2)$ states are very similar and close to the EMT value for all D° centers. Only for neutral tellurium is the $1s(T_2)$ level slightly deeper. The deviations are larger for the ionized (D^+) chal-

TABLE 4 Available data on the binding energies (in meV) of 1s donor states for isolated tetrahedral Group V and VI impurities in silicon. The following data are included: a from absorption measurements at 10K in Ref.49, b from electronic Raman scattering at about 20K in Ref 127, c and d from absorption measurements at 30K and 59K respectively in Ref. 5, f from absorption measurements at 10K in Ref. 25, and g and k from photoconductivity and absorption measurements in Refs. 32b and 71, respectively. Data labelled p, q and r are calculated values from Refs. 4, 22 and 21, respectively. All unlabelled data are from Ref. 23.

	$1s(A_1)$	$1s(T_2)$	$1s_7(T_2)-1s_8(T_2)$ exp.	calc.	$1s(E)$	$1s(T_2)-1s(E)$
P^0	45.58^a	33.90^c		0.015	$32.6^b/32.57^c$	1.33^c
As^0	53.77^a	43.65^d		0.086	$31.5^b/31.24^d$	1.41^d
Sb^0	42.77^a	32.88^c 33.17^c	0.29^c	0.34	$30.6^b/30.58^c$	2.40^c
Bi^0	71.00^a	31.92^f 32.92^f	1.00^f	1.03		
S^0	318.32^k	34.62^k		0.017	31.6^g	3.0
Se^0	306.63^k	34.44^k		0.13	31.2^g	3.2
Te^0	198.8	39.1		0.55	31.6^g	7.6
D^0	31.27^p 47.5^r	31.27^p 31.4^r			31.27^p 30.6^r	0.84
S^+	613.5	183.9 184.3	0.4	0.36		
Se^+	593.3	163.7 166.0	2.3	1.6		
Te^+	410.8	171.0 176.4	5.4	6.9		
D^+	125.08^p	125.08^p 155^q			125.08^p 130^q	25^q

cogens. This is not surprising since the Bohr radii for D^+ states are only about 7 Å, which should be compared with 20 Å for D^0 ones. Thus, the $1s(T_2)$ states for D^+ centers are more affected by the central cell potential than those for D^0 centers.[23]

Altarelli[33] has recently carried out more detailed calculations for the binding energies of the excited states of an electron bound to a substitutional +2e point charge in Si. The coulomb potential used was screened by a space dependent dielectric function $\epsilon(r)$ [$\epsilon(r) \to 11.4$ when $r \to \infty$ and

$\epsilon(r) \to 1$ when $r \to 0$]. Only the six lowest equivalent conduction band minima were included, but both direct and umklapp inter-valley interactions were taken into account. The binding energies obtained for the $1s(T_2)$ and $1s(E)$ states were 155 meV and 130 meV respectively (see Table 4). It is interesting to compare these calculations with earlier calculations by Pantelides,[8] from which binding energies of 127 meV [$1s(T_2)$] and 117 meV [$1s(E)$], were obtained when umklapp interactions were neglected. The deviation between the calculations may, however, partly be due to differing treatments of the inter-valley kinetic energy.[120]

A valley-orbit splitting, $1s(T_2)-1s(E)$, of about 25 meV is expected from Altarelli's calculations.[33] This is much larger than the doublet splitting discussed above, in agreement with the assignment suggested. Altarelli's calculated $1s(T_2)$ value is in fair agreement with experimental data,[23] Table 4. If different impurity pseudo-potentials had been used, even the chemical shifts could perhaps have been reproduced. Furthermore, considering the depth of the energy level, it is not unreasonable to assume that wavefunctions originating from other bands should be taken into account. If these bands are conduction bands, the binding energies would increase.

Altarelli et al.[21] have recently performed similar calculations on a substitutional +1e point charge in Si. Although due to insufficient screening of the extra proton these results cannot be directly related to those for neutral chalcogen donors, the data are nevertheless presented in Table 4 for comparison.[23]

Recently, the $1s(E)$ states for S^o,[32a,b,82] Se^o,[32a,b] and Te^o have been observed in photoconductivity and absorption measurements as Fano resonances above the ionization edge. The binding energies inferred were 31.6 meV for S^o, 31.2 meV for Se^o and 31.6 meV for Te^o (Table 4).[32b]

The energy positions of the $1s(E)$ states for Bi^o and the ionized chalcogen donors have not as yet been reported. Since the corresponding absorption lines should be seen if the admixture of T_2 states due to spin-valley interactions is sufficiently strong, this interaction is probably not very strong.[23] Using the data presented in Table 4, one may anticipate that the $1s(E)$ state for Bi^o probably lies 2–5 meV above the $1s(T_2)$ state whereas, in the case of ionized chalcogen donors, the energy difference, according to Altarelli's calculations, should be at least 25 meV.

The experimentally determined spin-valley splittings vary from 0.3 meV (Sb^o) to 5.4 meV (Te^+). The magnitude of this splitting should be approximately equal to the impurity atomic spin-orbit splitting reduced by the fraction of the donor envelope wavefunction in the central core region.[121] By using spin-orbit parameters and atomic radii from Ref. 122 together with Bohr radii deduced from experimentally determined $1s(T_2)$ energy positions, the spin-valley splittings could be calculated.[23] It turned out that the calculated values were 2–3 times smaller than those found experimentally. The best fit between the calculated and measured results was obtained by multiplying the calculated values by 2.4 (see Table 4).[23] Although the calculations are very crude, they probably predict hitherto unobserved splittings within a factor of two. In this context it is interesting to note that unpublished calculations by L. M. Roth suggest a splitting of 0.9 meV for Bi.[24]

It is easily seen (Table 4 and Figure 48) that the $1s(A_1)$ ground states of the neutral chalcogen B-centers are, in all cases, much deepr than the corresponding ground states of the singly-charged A-centers after correction for effective charge. A similar behaviour is found for the He-system. If the effective charge, Z, of the cores of the neutral donors were exactly equal to 1 in all cases, their $1s(A_1)$ ground states would have the same binding energy as the singly-ionized donors in the scheme used in Figure 48, where corrections are made for different charge states. In this discussion the non-coulombic central cell part of the potential is not considered. As for He^o, the deviation from $Z = 1$ is obviously caused by electron-electron correlations which prevent spherically-symmetric screening of the twofold core charge by electron 1. Electron 2 thus experiences an effective charge $1 < Z < 2$. As can be seen from Figure 48, an appreciable part of the ground state energy of the neutral donors is due to this mechanism. This can at least partly explain why, compared with the shallow donors

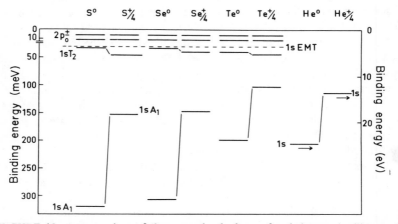

FIGURE 48   A comparison of the energy level schemes for chalcogens in silicon and the helium atom. For energy values and references, see Table 3 (from Ref. 22).

P^o, As^o and Sb^o, these neutral donors are so deep. It should be noted that such correlation effects are much weaker for the excited states, since the electron-electron interaction is strongly reduced, because of the larger orbital radii.[22]

6.2 Electron Spin Resonance Data

The ESR observations (large hf interactions with nuclei of S, Se and Te, isotropy of the resonances, small g-shifts, $S = 1/2$) are consistent with single unpaired electrons in s-type donor wavefunctions of S^+, Se^+ and Te^+.[22] For the chalcogen donor wavefunctions, the value of g is expected to be close to that of a free electron, as has been observed for Te^+. Small positive deviations have, however, been found for Se^+ and S^+ (see Ref. 22). These deviations can be attributed to the admixture of states corresponding to holes in the valence band with the donor wavefunction induced by the local Si spin-orbit coupling.[22,123] Such contributions are expected to become smaller with increasing distance of the donor level from the valence band. The g-shift is therefore smaller for Te^+ than for S^+ and Se^+. There is apparently no influence from the spin-orbit coupling of the donor ions themselves, as indicated by the similarity of the g values of Se^+ and S^+ in spite of the larger difference in spin-orbit coupling.[22]

The hf interactions of the chalcogen nuclei lead to estimates of the probability of finding the unpaired electrons near the donor ions, showing that the density of the unpaired electrons is about 10% at the central ions of the donor centers in all three cases.[22,128]

Turning to the ^{29}Si hf interaction,[22] it should be noted that the couplings with the Si shell closest to the donor are larger in the case of S^+ than for Te^+. One has to conclude that the donor wavefunction for $Si:Te^+$ is less compact and approaches more a conduction band state. This is consistent with the smaller energy distance from the conduction band edge of this donor.

6.3 Chemical Trends

A survey of known chalcogen donors in silicon is given in Table 3. Recently a model for predicting chemical trends of deep impurity levels in covalent semiconductors was developed.[124] By considering the difference in atomic orbital energies between the impurity and the host atom, it was claimed that the isolated substitutional donors Te, Se, S and O are likely to have increasingly larger ionization energies in silicon. In a refined version of the model,[124] an impurity is defined as *deep* if its *central cell po-*

tential alone is sufficiently strong to bind a state within the band gap. In contrast to a shallow energy level, such a deep energy level is not exclusively derived from the nearest band edge. Experimentally this means that the binding energy of the energy level does not follow a nearby band edge when that edge is perturbed by alloying or pressure. The model predicts that S and Se are deep, whereas Te should be shallow (Figure 49). Oxygen is predicted to produce a resonant state slightly below the valence band edge. The levels for both S^+ and Se^+ are in good agreement with the model, whereas Te^+ level is considerably deeper than predicted. However, the model does not take into account the charge of the center. So perhaps it would have been more proper to compare the model with the experimental values of the corresponding neutral centers instead. Hydrostatic pressure DLTS measurements[125] have shown that Te is less sensitive to pressure than S and Se. Furthermore ENDOR measurements at the S^+ site discussed in Ref. 22 reveal that the electron density is larger at the four first neighbors than at the central core which is consistent with the anti-

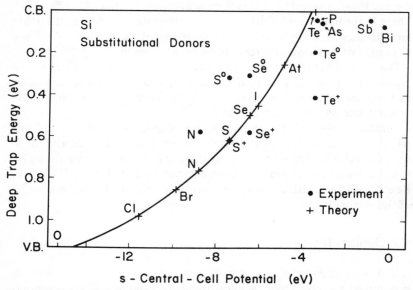

FIGURE 49 Predicted energies (+) relative to the conduction band (C.B.) edge of A_1 symmetric deep impurity levels in Si, compared with experimental data (•). The theory omits the long-range Coulomb potential; hence, all shallow impurities have zero binding energies. The s-central-cell potential is given in terms of the s-orbital energies (from Ref. 124). The experimental data for S, Se and Te are taken from Table 3 and differ from those used in the original figure).

bonding orbital predicted by the model. Te^+, on the other hand, has its highest electron density at the donor core,[22] which is also the case for the shallow donors. This implies a larger influence of the coulomb part of the potential for this element in accordance with the deviation from the prediction by the model.

References

1. W. Kohn, Solid State Physics *5*, 257 (1957)
2. W. Kohn, Phys. Rev. *98*, 915 (1955)
3. S.T. Pantelides, Rev. Mod. Phys. *50*, 797 (1978)
4. R.A. Faulkner, Phys. Rev. *184*, 713 (1969)
5. R.L. Aggarwal and A.K. Ramdas, Phys. Rev. *140*, A1246 (1965)
6. A. Glodeanu, Phys. Status Solidi *19*, K 43 (1967)
7. T.H. Ning and C.T. Sah, Phys. Rev. B *4*, 3468 and 3482 (1971)
8. S.T. Pantelides and C.T. Sah, Phys. Rev. B *10*, 621 and 638 (1974)
9. A.A. Grinberg and E.D. Belorusets, Sov. Phys. Semicond. *12*, 1171 (1978)
10. M. Lax, Phys. Rev. *119*, 1502 (1960)
11. V.N. Abakumov, V.I. Perel' and I.N. Yassievich, Sov. Phys. Semicond. *12*, 1 (1978)
12. P. Migliorato, A.W. Vere and C.T. Elliott, Appl. Phys. *11*, 295 (1976)
13. P. Migliorato and C.T. Elliott, Solid State Electronics *21*, 443 (1978)
14. N. Sclar, Infrared Phys. *16*, 435 (1976)
15. N. Sclar, J. Appl. Phys. *52*, 5207 (1981)
16. H.R. Vydyanath, W.J. Helm, J.S. Lorenzo and S.T. Hoelke, Infrared Phys. *19*, 93 (1979)
17. A.L. Lin, Proceedings IRIS Detector Speciality Group Meeting, 1980 (unpublished)
18. C. Holm and E. Sirtl, J. Crystal Growth *54*, 253 (1981)
19. W. Bludau, A. Onton and W. Heinke, J. Appl. Phys. *45*, 1846 (1974)
20. A. Baldereschi, Phys. Rev. B *1*, 4673 (1970)
21. M. Altarelli, W.Y. Hsu and R.A. Sabatini, J. Phys. C., L605 (1977)
22. H.G. Grimmeiss, E. Janzén, H. Ennen, O. Schirmer, J. Schneider, R. Wörner, C. Holm, E. Sirtl and P. Wagner, Phys. Rev. B *24*, 4571 (1981)
23. H.G. Grimmeiss, E. Janzén and K. Larsson, Phys. Rev. B *25*, 2627 (1982)
24. L.M. Roth, MIT Lincoln Laboratory, Solid State Research Report, No. 3, p 23 (1962) (unpublished)
25. W.E. Krag, W.H. Kleiner and H.J. Zeiger, Proc. 10th Int. Conf. on the Physics of Semicond., Cambridge, Massachusetts, 271 (1970)
26. P.J. Dean, R.A. Faulkner and S. Kimura, Phys. Rev. B *2*, 4062 (1970)
27. P.J. Dean, W. Schairer, M. Lorenz and T.N. Morgan, J. Lumin. *9*, 343 (1974)
28. T.G. Castner, Phys. Rev. *155*, 816 (1967)
29. H.G. Grimmeiss, E. Janzén and B. Skarstam, J. Appl. Phys. *51*, 3740 (1980)
30. H.G. Grimmeiss, E. Janzén and B. Skarstam, J. Appl. Phys. *51*, 4212 (1980)
31. R.M. Gibb, G.J. Rees, B.W. Thomas, B.L.H. Wilson, B. Hamilton, D.R. Wight and N.F. Mott, Philos. Mag. *36*, 1021 (1977)

32. G.J. Rees, H.G. Grimmeiss, E. Janzén and B. Skarstam, J. Phys. C *13*, 6157 (1980)

32a. E. Janzén, R. Stedman and G.H. Grimmeiss, Proc. 16th Int. Conf. Phys. Semic., Montpellier (1982) p 125 (Ed. M. Averous, North-Holland, Amsterdam 1983)

32b. E. Janzen, G. Grossman, R. Stedman and H. G. Grimmeiss Phys. Rev. (1985) in print.

33. M. Altarelli, Proc. 16th Int. Conf. Phys. Semic., Montpellier (1981) p 122 (Ed. M. Averous, North-Holland, Amsterdam 1983)

34. H.D. Barber, Solid State Electron *10*, 1039 (1967)

35. O. Engström and A. Alm, Solid State Electron *21*, 1571 (1978)

36. J.A. van Vechten and C.D. Thurmond, Phys. Rev. B *14*, 3539 (1976)

37. C.O. Almbladh and G.J. Rees, J. Phys. C *14*, 4575 (1981) and Solid State Commun. *41*, 173 (1982)

38. B. Skarstam and L. Lindström, Appl. Phys. Lett. *39*, 488 (1981)

39. A.L. Lin, A.G. Crouse, J. Wendt, A.G. Campbell and R. Newman, Appl. Phys. Lett. *38*, 683 (1981)

40. F. Richou, G. Pelous and D. Lecrosnier, Appl. Phys. Lett. *31*, 525 (1977)

41. T.F. Lee, R.D. Pashley, T.C. McGill and J.W. Mayer, J. Appl. Phys. *46*, 381 (1975)

42. H.G. Grimmeiss, E. Janzén, B. Skarstam and A. Lodding, J. Appl. Phys. *51*, 6238 (1980)

43. H.G. Grimmeiss, Ann. Rev. Mater. Sci. *7*, 341 (1977)

44. H.G. Grimmeiss and C. Ovrén, J. Phys. E. *14*, 1032 (1981)

45. G.L. Miller, D.V. Lang and L.C. Kimerling, Ann. Rev. Mater. Sci. *7*, 377 (1977)

46. A. Lodding, Adv. Mass. Spectrom. *8*, 471 (1980)

47. G.D. Watkins in "Point defects in solids" *2*, 333 (1975), J.H. Crawford, L.M. Slifkin eds, Plenum Press

48. K.W. Blazey, O.F. Schirmer, W. Berlinger and K.A. Müller, Solid State Commun. *16*, 589 (1975)

49. B. Pajot, J. Kauppinen and R. Anttila, Solid State Commun. *31*, 759 (1979)

50. S. Fischler, Metallurgy of Advanced Electronics Materials *19*, 273 (Interscience, New York 1963)

51. D.R. Bosomworth, W. Hayes, A.R.L. Spray and G.D. Watkins, Proc. Roy. Soc. Lond. A*317*, 133 (1970)

52. K. Krishnan, (Digilab), unpublished

53. H.J. Hrostowski and R.H. Kaiser, J. Phys. Chem. Solids *7*, 286 (1958)

54. A.S. Lyutovich, V.P. Prutkin, V.M. Mikhaelyan, D.S. Gafitulina and Zh.V. Abramova, Sov. Phys. Semicond. *2*, 728 (1967)

55. N.S. Zhdanovich and Yu.I. Kozlov, Sov. Phys. Semicond. *9*, 1049 (1976)

56. S.T. Picraux, N.G.E. Johansson and J.W. Mayer, Semiconductor Silicon, 422 (eds. R.R. Haberecht and E.L. Kern), Electrochemical Society (1969)

57. O. Meyer, N.G.E. Johansson, S.T. Picraux and J.W. Mayer, Solid State Commun. *8*, 529 (1970)

58. J. Gyulai, O. Meyer, R.D. Pashley and J.W. Mayer, Rad. Effects *7*, 17 (1971)

59. G. Foti, S.U. Campisano, E. Rimini and G. Vitali, J. Appl. Phys. *49*, 2569 (1978)

60. A. Nylandsted Larsen, G. Weyer and L. Nanver, Phys. Rev. B *21*, 4951 (1980)

61. G.J. Kemerink, Thesis (University of Groningen, the Netherlands, 1981)

62. S.D. Brotherton, M.J. King and G.J. Parker, J. Appl. Phys. *52*, 4649 (1981)

63. W. Fahrner and A. Goetzberger, Appl. Phys. Lett. *21*, 329 (1972)

64. H.R. Vydyanath, J.S. Lorenzo and F.A. Kröger, J. Appl. Phys. *49*, 5928 (1979)
65. N.A. Sultanov, Sov. Phys. Semicond. *8*, 1148 (1975)
66. N.S. Zhadanovich and Yu I. Kozlov, Sov. Phys. Semicond. *10*, 1102 (1976)
67. C.S. Kim, E. Ohta and M. Sakata, Jpn. J. Appl. Phys. *18*, 909 (1979)
69. J.C. Swartz, D.H. Lemmon and R.N. Thomas, Solid State Commun. *36*, 331 (1980)
70. H.G. Grimmeiss and B. Skarstam, Phys. Rev. B *23*, 1947 (1981)
71. E. Janzén, R. Stedman, G. Grossman and H.G. Grimmeiss, Phys. Rev. B (1984)
72. W.E. Krag, W.H. Kleiner, H.J. Zeiger and S. Fischler, Proc. 8th Int. Conf. on the Physics of Semicond., Kyoto, 230 (1966)
73. R.O. Carlson, R.N. Hall and E.M. Pell, J. Phys. Chem. Solids *8*, 81 (1959)
74. W.E. Krag and H.J. Zeiger, Phys. Rev. Lett. *8*, 485 (1962)
75. D.L. Camphausen, H.M. James and R.J. Sladek, Phys. Rev. B *2*, 1899 (1970)
76. L.L. Rosier and C.T. Sah, Solid State Electron. *14*, 41 (1971)
78. R.Y. Koyama, W.E. Philips, D.R. Myers, Y.M. Liu and H.B. Dietrich, Solid State Electron *21*, 953 (1978)
79. D.R. Myers and W.E. Philips, Appl. Phys. Lett. *32*, 756 (1978)
80. A.A. Lebedev, A.T. Mamadalimov and N.A. Sultanov, Sov. Phys. Semicond. *5*, 17 (1971)
81. O. Engström and H.G. Grimmeiss, J. Appl. Phys. *48*, 4090 (1976)
82. R.G. Humphreys, P. Migliorato and G. Fortunato, Solid State Commun. *40*, 819 (1981)
83. G.W. Ludwig, Phys. Rev. *137*, A1520 (1965)
84. T.H. Ning and C.T. Sah, Phys. Rev. B *14*, 2528 (1976)
85. C.T. Sah, T.H. Ning, L.L. Rosier and L. Forbes, Solid State Commun. *9*, 917 (1971)
86. L.L. Rosier and C.T. Sah, J. Appl. Phys. *42*, 4000 (1917)
87. G.W. Ludwig, Phys. Rev *137*, A1520 (1965)
88. D.R. Myers and W.E. Philips, J. Electron. Mater. *8*, 781 (1979)
89. R.A. Forman, Appl. Phys. Lett. *37*, 776 (1980)
90. R.A. Forman, R.D. Larrabee, D.R. Myers, W.E. Phillips and W.R. Thurber, Defects in Semiconductors, (1981), Narayan and Tan eds., North-Holland, Inc.
91. C. Jagannath, Z.W. Grabowski and A.K. Ramdas, Phys. Rev. B *23*, 2082 (1981)
92. H.J. Hrostowski and B.J. Adler, J. Chem. Phys. *33*, 980 (1960)
93. W. Kaiser, P.H. Keck and C.F. Lange, Phys. Rev. *101*, 1264 (1956)
94. J.W. Corbett, R.S. McDonald and G.D. Watkins, J. Phys. Chem. Solids *25*, 873 (1964)
95. L.C. Kimerling and J.L. Benton, Appl. Phys. Lett., Aug. 15, (1981)
96. C.S. Fuller, J.W. Dietzenberger, N.B. Hannay and E. Buehler, Phys. Rev. *96*, 833 (1954)
97. W. Kaiser, H.L. Frisch and H. Reiss, Phys. Rev. *112*, 1546 (1958)
98. G.Feher, Phys. Rev. *114*, 1219 (1959)
99. V.N. Mordkovich, Sov. Phys.-Solid State *6*, 654 (1964)
100. A.R. Bean and R.C. Newman, J. Phys. Chem. Solids *33*, 255 (1972)
101. K. Graff and H. Peiper, J. Electron. Mater. *4*, 281 (1975)
102. O. Helmreich and E. Sirtl in: Semiconductor Silicon 1977 (Ed. H.R. Huff and E. Sirtl, Electrochem. Soc. Ser #PV 77-2, Princeton, N.J., 1977) p 626
103. S.H. Muller, M. Sprenger, E.G. Sieverts and C.A.J. Ammerlaan, Solid State Comm. *25*, 987 (1978)
104. D. Wruck and P. Gaworzewski, Phys. Stat. Sol. (a) *56*, 577 (1979)

105. A. Kanamori and M. Kanomori, J. Appl. Phys. *50*, 8095 (1980)
106. M. Tajima, S. Kishino, M. Kanamori and T. Iizuka, J. Appl. Phys. *51* 4206 (1980)
107. V. Cazcarra and P. Zunino, P. Appl. Phys. Lett. *38*, 274 (1981)
108. D. Rava, H.C. Gatos and J. Lagowski, Appl. Phys. Lett. *38*, 274 (1981)
108a. P. Wagner and Oeder, Proc. MRS, Boston (1981)
108b. B. Pajot, H. Compain, J. Lerouille and B. Clerjaud, Proc. 16th Int. Conf. Phys. Semic., Montpellier (1982) p 110 (Ed. M. Averous, North-Holland, Amsterdam 1983)
109. J.R. Patel, "Semiconductor Silicon 1977", Ed. H.R. Huff and E. Sirtl, (Electrochem. Soc. Princeton, N.J. 1977) p 521
110. L.C. Kimerling in "Defects in Semiconductors" (Elsevier North Holland, New York, 1981)
111. F. Bemski, J. Appl. Phys. *30*, 1195 (1959)
112. G.D. Watkins, J.W. Corbett and R.M. Walker, Bull. Am. Phys. Soc. *4*, 159 (1959)
113. G.D. Watkins and J.W. Corbett, Phys. Rev. *121*, 1001 (1961)
114. J.W. Corbett, G.D. Watkins, R.M. Chrenko and R.S. McDonald, Phys. Rev. *121*, 1015 (1961)
115. K.L. Brower, Phys. Rev. *B4*, 1968 (1971)
116. K.L. Brower, Phys. Rev. *B5*, 4274 (1972)
117. R.C. Newman, "Infra-red Studies of Crystal Defects", Taylor and Francis Ltd; London 1973
118. Y.H. Lee, J.C. Corelli and J.W. Corbett, Physics Letters *59A*, 238 (1967)
119. L.C. Kimerling, Radiation Effects in Semiconductors 1967 (Inst. Phys. Conf. Ser. No. 31), p 221
120 D.C. Herbert and J. Inkson, J. Phys. C *10*, L695 (1977)
121. P.J. Dean, R.A. Faulkner and S. Kimura, Phys. Rev. B *2*, 4262 (1970)
122. S. Fraga and J. Karwowski, K.M.S. Saxena, "Handbook of Atomic Data", Elsevier, Amsterdam, 1976
123. G.D. Watkins and J.W. Corbett, Phys. Rev. *134*, A1359 (1964)
124. P. Vogl, Festkörperprobleme XXI, 191 (1981)
125. W. Jantsch, K. Wünstel, O. Kumagai and P. Vogl, Phys. Rev. (1982)
127. K.L. Jain, S. Lai and M.V. Klein, Phys. Rev. B *13*, 5448 (1976)
128. E. Janzén and H.G. Grimmeiss, J. Phys. C *15*, 5791 (1982)
129. J. Bernholc, N.O. Lipari, S.T. Pantelides and M. Scheffler, Phys. Rev. B *26*, 5706 (1982)

CHAPTER 3

The Lattice Vacancy in Silicon

George D. Watkins

Department of Physics and Sherman Fairchild Laboratory
Lehigh University, Bethlehem, PA 18015

1. INTRODUCTION
2. PRODUCTION OF VACANCIES
3. ELECTRONIC STRUCTURE OF THE VACANCY
   3.1 EPR Spectra and the Identification of the Vacancy
   3.2 Simple LCAO Model for the Vacancy
   3.3 Jahn-Teller Distortions
4. ELECTRICAL LEVEL STRUCTURE
5. VACANCY INTERACTIONS WITH OTHER DEFECTS
6. THERMALLY ACTIVATED MIGRATION OF THE VACANCY
7. VACANCY MIGRATION UNDER ELECTRONIC EXCITATION
   7.1 Recombination-Enhanced Migration
   7.2 Photo-enhanced Migration
   7.3 Mechanism for Enhanced Migration
8. THEORY OF THE ELECTRONIC STRUCTURE OF THE VACANCY
   8.1 Many-Electron Effects
   8.2 Jahn-Teller Effects
9. SUMMARY

1. Introduction

There are two fundamental lattice defects in an elemental semiconductor such as silicon. One is the *lattice vacancy,* which is a missing atom in an otherwise perfect crystalline array. The other is the *self-interstitial,* which is an extra host atom squeezed into the lattice.

Together these simple point defects form the fundamental building blocks from which all other possible intrinsic defects can be considered to be constructed: divacancies, small voids, aggregates, dislocations, grain boundaries, and even surfaces (which are, in effect, an infinite array of vacancies). They are also believed to supply the mechanism by which high temperature diffusion and mass transport processes occur, as they, in turn, migrate and stir up the lattice. Therefore the study of these simple defects in any material represents the first logical step in unraveling all of the complex defect process that can occur.

In this chapter, we consider the lattice vacancy in silicon. In addition to the "structural" aspects as a lattice defect, mentioned above, the vacancy also introduces electrical levels into the forbidden gap. It is therefore an interesting "deep level" in itself and we shall see that it has some remarkable properties.

Finally, understanding the lattice vacancy can be considered a first logical step toward understanding all substitutional impurity deep levels. The lattice vacancy is, after all, the "hole" into which the impurity atom is placed. As such, it provides the electronic and elastic environment to which the impurity atom must couple, and through which it must "sense" the infinite crystal.

2. Production of Vacancies

The conventional method, successful for many materials, of quenching in vacancies for study by rapid cooling from high temperatures does not work for silicon. The lattice vacancy is too mobile. The material simply cannot be quenched rapidly enough to freeze in a measurable concentration of vacancies.

The only successful method so far for isolated vacancy production has been by radiation damage at cryogenic temperatures. Here simple "knock-on" scattering of the incoming particle by a host atom nucleus transmits enough recoil energy to the atom to displace it from its normal lattice site. Cryogenic temperatures are required to prevent the vacancy so produced from diffusing through the lattice and pairing off with other defects.

In Figure 1 we show the damage process deduced from experimental studies[1,2] using high energy electrons (1-3 MeV) at 4.2 and 20.4K. The primary process is the formation of a Frenkel pair (vacancy-interstitial) as the recoiling silicon atom is dislodged into a nearby interstitial site. One of the early surprises of these studies was that the interstitial silicon atom is apparently mobile even at 4.2K! As a result, the Frenkel pair either annihilates itself or the interstitial escapes and is trapped at some impurity, leaving an isolated vacancy. In p-type silicon, isolated vacancy production is a relatively efficient process, perhaps reflecting long range Coulomb attraction to the dominant interstitial trap which has been identified directly in EPR studies to be the group III atom used in the p-type doping. In n-type silicon, isolated vacancy production at these temperatures is very inefficient (a factor of 100 lower), with Frenkel pair annihilation apparently dominating.

The role played by the group III atom in trapping the interstitial and thereby stabilizing the vacancy is demonstrated in Figure 2. Here the relative vacancy production rate, monitored by EPR studies for 1.5 MeV electron irradiation at 20.4K, has been measured for vacuum floating zone p-type material of various boron doping concentrations. The nearly linear dependence on boron concentration ($\sim B^{1.1}$) confirms its role in trapping interstitials in competition with vacancy-interstitial annihilation.

Consistent also with this simple model for the damage process is the observation in n-type material that low temperature vacancy production efficiencies comparable to that found in p-type material can be achieved at higher electron bombardment energies (5,50 MeV).[2,4,5] Here the re-

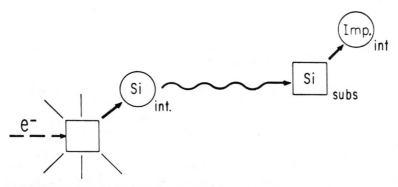

FIGURE 1. Vacancy production at cryogenic temperatures (4.2K, 20.4K) in silicon by 1-3 MeV electrons. The mobile interstitial atoms that escape are trapped by impurities, leaving isolated vacancies.

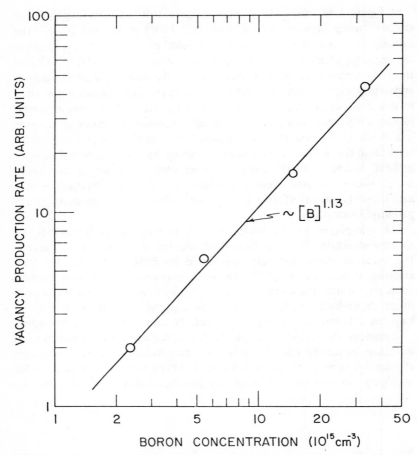

FIGURE 2. Vacancy production rate in p-type silicon for 1.5 MeV electrons at 20.4K vs. concentration of the group III boron dopant.[2,3]

coiling interstitial is displaced farther from the vacancy, reducing correlated vacancy-interstitial annihilation. Consistent also is the observation that the low temperature n-type vacancy production efficiency can be increased by counterdoping with group III atoms,[2] thus providing effective interstitial traps.

The mechanism for the high mobility of the interstitial silicon atom is not understood, although it is generally believed to be driven by the energy released when electrons and holes, present during the irradiation, recombine at the defect.[6-10] This is a subject of current experimental and theoretical interest but will not be treated in this chapter. Instead, let us

accept this remarkable result, for it provides us a way of producing isolated vacancies for studies.

3. Electronic Structure of the Vacancy

3.1 EPR Spectra and the Identification of the Vacancy[1,2,11,12]

The principal experimental tool for the *identification* and the study of
the vacancy has been electron paramagnetic resonance (EPR). EPR has
been observed for two different charge states of the vacancy (V^+ and V^-)
The two $S = \frac{1}{2}$ spectra seen in a p-type sample (irradiated *in situ* at T $\leqslant$
20.4K by 1.5 MeV electrons) are show in Figure 3. In order to reveal
these spectra it is necessary to illuminate the crystal with infrared light.

The spectrum for V^+ is shown in Figure 3a, which is optimized by long
wavelength light, $h\nu < 0.35$ eV (globar source through an InAs filter).
The model for its structure, deduced from the EPR spectrum, is shown
in the inset, the unpaired electron spread equally over the four silicon
neighbors surrounding the vacancy. The identification of this spectrum
as arising from the isolated lattice vacancy comes primarily from the
^{29}Si hyperfine satellites (a' and a'' for component a; b' and b'' for b)
which, in their angular variation reflect a "dangling-bond" character
(15% 3s, 85% 3p, as determined from the isotropic and anisotropic parts
of the hyperfine interaction, respectively) on each of the four neighboring
silicon atoms. Approximately 65% of the wavefunction is accounted for
on these four atoms, the remainder of the wavefunction being more diffuse. A tetragonal distortion, illustrated as a pairing by twos in the figure,
is detected as a small anistropy of the g-tensor ($\sim\frac{1}{2}\%$) as well as in a slight
tilting of the ^{29}Si hyperfine axes (7.2° away from the <111> "dangling
bond" axes toward the <100> tetragonal distortion axis).

Near band gap light, on the other hand, can produce free electrons
which, when trapped at the vacancy give V^-. The resulting spectrum is
shown in Figure 3b. The identification as V^- again stems primarily from
the ^{29}Si satellites. In this case, the vacancy takes on the interesting configuration as shown, in which the unpaired electron sloshes to one side
with $\sim$60% of the wavefunction spread between only two neighboring
silicon atoms. Hyperfine analysis indicates 29% 3s, 71% 3p character on
each of the two silicon sites and with the hyperfine axes now accurately
along the <111> "dangling bond" directions. Recent electron-nuclear-
double resonance (ENDOR) studies[13] have resolved weaker ^{29}Si hyperfine interactions with an additional 26 inequivalent shells of neighboring
silicon atom sites. These studies confirm the symmetry deduced from the

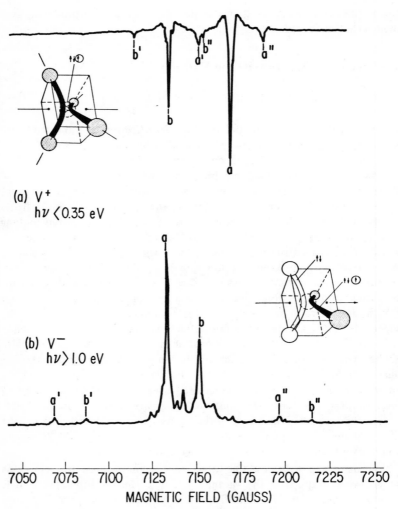

FIGURE 3. Vacancy EPR spectra,[12] H||<100>, ν_o = 20 GHz: (a) V^+ at 4.2K; (b) V^- at 20.4K.

EPR studies (C_{2V}, Figure 4a) and most of the remaining wavefunction for V can be accounted for at these more distant sites.

3.2 Simple LCAO Model for the Vacancy

The EPR results suggest a simple linear combination of atomic orbital-molecular orbital (LCAO-MO) model for the various charge states of the vacancy. This is illustrated in Figure 4. Here the atomic orbitals are the "broken bonds" (a, b, c, d) of the four atoms surrounding the vacancy. In the symmetry T_d of the undistorted vacancy a singlet (a_1) and triplet (t_2) set of molecular orbitals are formed. For V^{++}, Figure 4a, two electrons go into the a_1 orbital paired off, the defect is diamagnetic, and no EPR is observed. For V^+, the third electron goes into the t_2 orbital. Because of the degeneracy associated with the orbital, a tetragonal Jahn-Teller distortion results, as shown, lowering the symmetry to D_{2d}. The

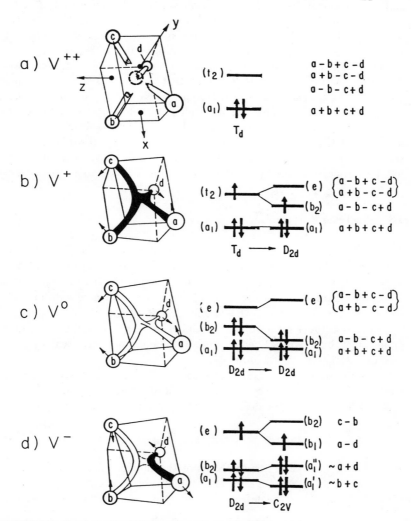

FIGURE 4. Simple LCAO-MO treatment for the lattice vacancy.

resulting orbital (b_2) containing the unpaired electron is spread equally over the four atoms as seen in the ^{29}Si hyperfine interactions for the V^+ spectrum. The axial $<100>$ anisotropy of the g-tensor reveals the tetragonal distortion.

In forming $V°$, Figure 4c, the next electron also goes into b_2, paired off, further enhancing the tetragonal Jahn-Teller distortion. In this state the defect is again diamagnetic and no EPR is observed.

For V^-, Figure 4d, the fifth electron goes into the degenerate e orbital and an additional Jahn-Teller distortion occurs. This distortion is of b_2 symmetry (in D_{2d}) with atoms b and c pulling together and a and d separating slightly. The unpaired electron is now localized on only two of the four atoms, consistent with the observation in the V^- spectrum.

The remarkable success of these simple one-electron models suggests that the electron-electron interactions that tend to favor parallel spin coupling (Hund's rules for atoms, etc.) are small. We are, in effect, in the strong crystal field regime. We fill each level before going to the next. When degeneracy occurs, a Jahn-Teller distortion results which imposes a new crystal field, decoupling the electrons again. We will see later in the section on theory that the relative unimportance of many-electron effects has not been easy to justify theoretically causing considerable controversy.

3.3 Jahn-Teller distortions[2]

When a defect undergoes a static Jahn-Teller distortion, electronic degeneracy is traded for orientational degeneracy, as the defect distorts into one of several possible distortions equivalent by symmetry. The EPR spectrum reflects a superposition of the anisotropic spectra associated with each of the equivalent distortions. The amplitudes of each spectral component therefore is a direct measure of the number of defects with that particular orientation. In an unstrained crystal, the amplitudes are equal because each orientation is equally probable.

The application of uniaxial stress to the crystal provides a powerful method for studying these distortions. Following Kaplyanskii,[14] the coupling of an anisotropic defect to applied strain can be written

$$\Delta E_i = \sum_{m,n} B^i_{mn} \, \epsilon_{mn} \qquad (1)$$

where ϵ_{mn} are the tensor components of the applied strain and B^i_{mn} form the components of a symmetric second rank "piezospectroscopic" tensor $\mathbf{B}$ for the coupling of the defect of orientation i to the strain. If the defect can reorient at the temperature of the applied strain, alignment of the defects will occur. By equating ratios of the observed intensities, n_i and n_j, of the corresponding spectral components to a Boltzmann distribution

$$\frac{n_i}{n_j} = \exp\left(-\frac{\Delta E_i - \Delta E_j}{kT}\right), \tag{2}$$

the ΔE_i can be determined leading to a complete determination of B, with the exception of its trace. (The trace reflects the coupling to the hydrostatic "breathing mode" which affects all defect orientations equally). Here T is the temperature at which equilibrium alignment has been established. Once alignment has been achieved, the kinetics of the reorientation process can also be studied.

For V^+, alignment can occur even at 2K revealing a very low barrier for thermally activated reorientation. In the temperature region 14-21K, the reorientation rate is so rapid as to cause broadening effects in the EPR lines. Study in this temperature region gives for the characteristic lifetime between reorientations[2]

$$\tau^{-1} = 5.6 \times 10^{11} \exp(-0.013 \text{ eV}/kT) \sec^{-1} \tag{3}$$

The components of B^+ deduced from the observed magnitude of alignment for V^+ vs. stress and temperature are given in Table 1.

For V^-, the two distortions (Figure 4d) can be studied separately. At 20.4K, the tetragonal distortion is completely frozen in and only the flipping of the unpaired electron from one side to the other (the b_2 distortion) can occur under stress. At 20.4K the characteristic time constant for the process is $\sim$0.3s and studies vs. temperature indicate[2]

$$\tau^{-1} = 350 \exp(-0.0081 \text{eV}/kT) \sec^{-1}. \tag{4}$$

TABLE 1 Strain tensor coupling coefficients (B_{ij}) and reorientational activation energues (U_Γ) for the Jahn-Teller distortion modes of V^+, V°, and V^-. (The defect axes are defined in Fig. 4a).

	B_{zz} ($=-2B_{xx}=-2B_{yy}$) (eV/Å)	$B_{xy}=B_{yx}$ (eV/Å)	U_e (eV)	U_{b2} (eV)
V^+	−6.5	−	0.013	−
V°	−13.3	−	$\sim$0.23	−
V^-	−9.9	−7.2	$\sim$0.072	0.008

At somewhat elevated temperatures, reorientation from one tetragonal distortion to the other occurs for V^-. Detailed kinetic studies have not been performed for this motion but at 26.8K the time constant was measured to be 30s. Assuming a typical preexponential frequency factor of $\sim kT/h$, this indicates an activation barrier of ~ 0.072 eV. These results are included in Table 1 along with the components of B^- determined from the degree of alignment vs. stress.

Although V° cannot be observed by EPR, stress-induced alignment has also been studied indirectly for it,[2] as follows: In heavily irradiated p-type material with the Fermi level no longer locked to the shallow acceptor, the equilibrium charge state of the vacancy should be V°. Therefore, we apply stress at elevated temperature in the dark, cool to 20.4K and remove the stress. We then illuminate with near band gap light to generate free electrons which are trapped at the neutral vacancies to produce V^-. The alignment observed in V^- is a direct monitor of the alignment initially induced in V°. Isochronal anneals, each time regenerating V^- to monitor the alignment, reveal reorientation of V° with a 15 min time constant at ~ 74K. Again, with a preexponential factor of $\sim kT/h$ this indicates a reorientation barrier of ~ 0.23 eV. The components of B° in Table 1 were calculated from the measured V^- alignment assuming an equilibrium alignment for V° frozen in at $T \sim 74$K.

The fact that reorientation can occur with these low activation barriers provides strong confirmation that the distortions observed in the spectra are indeed of Jahn-Teller origin. Nearly 100% alignment can be achieved with modest stresses confirming that the observed distortion is not due to the presence of a nearby defect but rather is an intrinsic property of each vacancy which has full access to all possible distortions from its otherwise high symmetry T_d undistorted site.

The energy for a defect undergoing a Jahn-Teller distortion can be written

$$E_i = -V_\Gamma Q_{\Gamma_i} + \tfrac{1}{2}k_\Gamma Q_{\Gamma_i}{}^2 \quad . \qquad (5)$$

Here V_Γ is the linear Jahn-Teller coupling coefficient for the local distortion mode of Γ symmetry, Q_{Γ_i} is the amplitude of the i-th normalized component of this mode, and k_Γ is the force constant for this mode, expressing the elastic restoring forces on the atoms. Minimizing Equation (5) with respect to Q_{Γ_i}, the Jahn-Teller stabilization energy becomes

$$E_{JT} = -V_\Gamma{}^2/2k_\Gamma \quad . \qquad (6)$$

The stress alignment studies have been used to provide a rough estimate of V_Γ, which we now summarize. This in turn provides through Eq. (6) a means of estimating the Jahn-Teller engines.

The **B** tensor components given in Table 1 are actually what is measured in the experiment and no approximations are involved in their analyses. They involve the coupling of the defect to long range uniform strain. The V_Γ, on the other hand, are the coupling coefficients to the *local* Jahn-Teller distortion modes Q_Γ. In the limit of a highly localized defect where the coupling arises primarily from the nearest neighbor atom motions only, then Equation (5) gives for the stress-induced energy of the i-th oriented defect,

$$\Delta E_i = - V_\Gamma Q'_{\Gamma_i} \ , \tag{7}$$

where Q_{Γ_i} is the amplitude of the nearest neighbor local mode distortion produced by the applied stress. If it is further assumed that the neighboring atoms move under the applied stress identically to atoms far removed from the defect (so that bulk elastic constants can be used), one obtains

$$V_e = \frac{\sqrt{6}}{3a}(2B_{zz} - B_{xx} - B_{yy}) \ , \quad V_{b2} = -\frac{2\sqrt{2}}{a} B_{xy} \ , \tag{8}$$

where a = 5.43 Å, the lattice constant of silicon. The estimates obtained in this manner are given in Table 2.

Clearly these approximations are difficult to justify for the vacancy. Only 60–65% of the unpaired electron wavefunction is localized on the

TABLE 2   Jahn-Teller coupling coefficients (V_Γ) and Jahn-Teller energies (E_{JT}) estimated from B, Table I, assuming (i) coupling only to nearest neighbor atom displacements, (ii) nearest neighbor atom displacements identical to bulk atom displacements, and (iii) $k_E = 2k_T = 11.9$ eV/Å, see text.

	V_e (eV/Å)	E_{JT}^e (eV)	V_{b2} (eV/Å)	E_{JT}^{b2} (eV)	E_{JT}^{total} (eV)
V^+	2.9	0.4	–	–	0.4
V°	6.0	1.5	–	–	1.5
V^-	4.5	0.9	3.8	1.2	2.1

nearest neighbors and the extended parts of the wavefunction could couple significantly to the applied strain. This would probably cause our stress alignment experiments to overestimate V_Γ because the true Jahn-Teller mode of distortion must be localized in order that small elastic energy (k_Γ) be stored in the lattice. Also, evidence has been cited that the atoms neighboring a vacancy respond differently under stress than host atoms.[15] Still this is the best that can be done experimentally at this stage and the results at least serve as a guide to the magnitudes of the couplings.

The Jahn-Teller energies in turn have been estimated by Equation (6) using these coupling coefficients. Again there is a large uncertainty in the choice of k_Γ, with estimates ranging from

$$k_E = 2k_{T_2} = 11.9 \text{ eV/Å}^2 \ , \tag{9}$$

derived from a simple nearest neighbor central force model,[2,16] to

$$k_E = 1.30 \text{ eV/Å}^2 \ ; k_{T2} = 0.97 \text{ eV/Å}^2 \ ,$$

calculated from a large cluster valence force model.[2,15] A more recent estimate using a valence force model with a different parameterization has given[17-20]

$$k_E = 3.7 \text{ eV/Å}^2 \ ; k_{T2} = 1.82 \text{ and } 3.62 \text{ eV/Å}^2 \ .$$

In Table 2, we present the results using the stiffest force constants Equation (9). We see that even with these force constants the predicted Jahn-Teller energies are large. The use of the softer force constants predict even larger Jahn-Teller energies. These estimates therefore appear to confirm that the Jahn-Teller energies are indeed large enough to overcome the electron-electron couplings, as deduced from the one electron models. We will return to reconsider these estimates briefly in the section on theory.

An interesting feature to note in the tables is that the ratio of the tetragonal coupling coefficients, V_e, is very close to 1:2:1.5 for $V^+:V^\circ:V^-$. This is precisely what would be predicted in a very simple one-electron molecular orbital treatment where the one-electron coupling coefficients are unchanged vs. charge state. In Figure 4 we see that this follows immediately because there is one electron in the lowered b_2 orbital for V^+, two for V°, and for V^- there are two in b_2 but one in the e orbital that is raised by half the amount by which b_2 is lowered. This result must again be considered strong confirmation of the one-electron character of these defects.

Another interesting feature in the table is the large V_{b2} coupling co-efficient for V^-. In a simple linear model this would be equal also to the V_{T2} coupling coefficient before distortion (T_d). Using this value, with the indicated force constants, the principal distortion would have been pre-dicted to be a *trigonal* one rather than the observed tetragonal one. Simi-larly, if this coupling coefficient were independent of charge state, a trigonal distortion would have been predicted for all three charge states.

Experimentally we know that this is not correct. The tetragonal dis-tortion dominates. Again this demonstrates the uncertainties in such a simple analysis. Experiment does confirm, however, that the competition between tetragonal and trigonal distortion is very close: Opik and Pryce,[21] using an approximation equivalent to that implied by Equation (5), have calculated that for a T_2 state coupled to both E and T_2 vibrational modes, the barrier for reorientation from one tetragonal distortion to another is

$$U_{reorient} = \frac{3}{4} [(E_{JT})_{tetrag} - (E_{JT})_{trig}] . \qquad (10)$$

The low barrier implied by the easy reorientation at low temperature ob-served for all three charge states of the vacancy, see Table I, indicates that the two energies are indeed comparable.

4. Electrical Level Structure

In Figure 5b we show a tentative electrical level structure deduced for the isolated vacancy. Shown are four energy levels in the gap, the notation (i/j) meaning that the vacancy has charge state i when the level is occupied by an electron, and charge j when empty. This structure implies therefore that there are five different charge states available to the vacancy which are energetically stable (or metastable) within the bandgap of silicon. EPR has been observed directly for V^+ and V^-. The evidence for $V^°$ has already been described in the stress-induced alignment studies. The existence of the $V^=$ charge state has been inferred from the observation that in low resistivity n-type material V^- is not seen unless generated by light.[2]

The existence of the remaining charge state V^{++} has only recently been postulated. Its existence, and its relative location as shown in Figure 5b with the corresponding second donor state (+/++) above the first donor state (0/+) was first predicted by Baraff, Schluter and Kane.[17-19] We will present experimental evidence that confirms this ordering. But first let us make some general observations.

First let us carefully define what is meant by the electrical level positions in Figure 5. The position of level (o/+) measured with respect to the con-

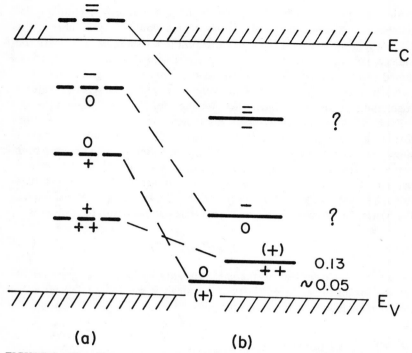

FIGURE 5. Electrical level positions in the gap for the vacancy, (a) before, and (b) after, Jahn-Teller relaxation.

duction band edge is the total energy difference between V° in its relaxed state and the ionized state V^+, again in its fully relaxed state, with the electron in the conduction band. It is therefore the electron ionization energy of V° between relaxed ground and excited states. It is the equivalent of the *zero phonon* optical transition. The level (+/++) is defined in a similar way as the second ionization energy, $V^+ \rightarrow V^{++} + e^-$. The location with respect to the valence band edge can be defined in an analogous way with (+/++) defined as $V^{++} \rightarrow V^+ + h^+$, again between fully relaxed ground and excited states, etc. This definition is the only correct definition of a level position consistent with the conventional level location by a Hall measurement which is one of thermodynamic equilibrium between the Fermi level and the charge state of a defect.

Since *total* energies are involved it means that a change in the Jahn-Teller energy between the two charge states will contribute to the level position. Shown schematically in Figure 5a are possible level positions without Jahn-Teller distortions, each level rising in the gap by U (the

Hubbard correlation energy)[22] as an electron is added reflecting the Coulomb repulsion between the electrons. As can be seen in Table 1, the Jahn-Teller relaxation energy increases with increased electronic charge serving to collapse the energy levels, as illustrated schematically in Figure 5. This provides a partial explanation for why so many charge states can be crowded into the narrow gap of silicon ($\sim$1.2 eV). In effect, the lattice vacancy is a "soft" polarizeable defect which, each time you add an electron, can relax in such a way as to keep the electrons apart and minimize their Coulombic repulsion, while at the same time optimizing their attractive interactions with the nuclear cores. This is the essence of the Jahn-Teller effect.

What Baraff et al. noted, on the basis of their theoretical calculations, was that the gain in Jahn-Teller energy in going from V^+ to V° could be large enough to overcome the Coulomb repulsion energy U, which they estimated to be only $\sim$0.3 eV. They therefore predicted an inverted "negative effective-U"[23,24] ordering for the single and double donor levels of the vacancy as illustrated in Figure 5b.

The experimental evidence in support of this ordering is as follows: In Figure 6, we show the kinetics of the loss of the V^+ EPR spectrum in the dark after photogeneration in indium-doped p-type material.[2] This has been interpreted as hole release to the valence band

$$V^+ \rightarrow V^\circ + h^+ \tag{11}$$

leading to the estimate for the single donor state (o/+) location as $\sim E_V + 0.05$ eV, as indicated in Figure 5b. A different hole release process is monitored in DLTS studies,[10,25-27] as shown in Figure 6b. There a single dominant hole emission peak at $E_V + 0.13$ eV is observed in p-type material irradiated by electrons at 4.2K which in its production and annealing properties can be correlated directly with the vacancy. This is clearly a different emission process from that monitored by EPR, Equation (11), and has therefore been tentatively assigned to emission from the second donor state,[19,28-30] as indicated in Figure 5b.

A critical argument in this identification was the recognition that for negative-U ordering, hole emission from V^{++} is a *two hole* emission process,[28]

$$V^{++} \underset{0.13}{\longrightarrow} V^+ + h^+ \underset{0.05}{\longrightarrow} V^\circ + 2h^+ \ . \tag{12}$$

Here we have indicated the thermal activation barrier associated with the level position (in eV) for each emission process. The limiting process is the first hole release. At temperatures where this is being observed as a

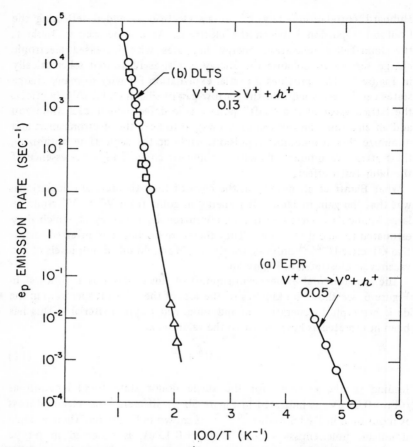

FIGURE 6. (a) Kinetics of the EPR V^+ spectrum decay after photo-generation.[2] (b) Hole emission rate observed by DLTS[10,25-27] for a level at $E_V + 0.13$ eV identified with the vacancy.

thermally activated emission, the second hole release will follow immediately. Therefore, a single peak will be observed in DLTS associated with the first hole level position but its amplitude will be twice its normal size. The evidence cited for this[28] is summarized in Figure 7. Shown is the DLTS spectrum of p-type floating zone silicon containing 10^{18} tin atoms per cm^3, which has been irradiated at 4.2K by 1.5 MeV electrons. From previous EPR studies[31] we anticipate $\cong 100\%$ conversion of vacancies to vacancy-tin pairs $(V \cdot Sn)$ upon thermal annealing as the vacancies diffuse through the lattice and are trapped by tin, the dominant impurity. Instead, we observe that the amplitude of the vacancy level is $\cong twice$ the intensity of the resulting vacancy-tin peaks at $E_V + 0.07$ eV and $E_V + 0.32$ eV.

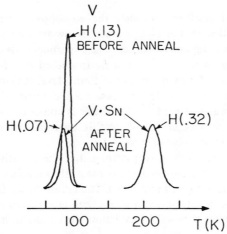

FIGURE 7. Conversion of vacancies to tin-vacancy pairs as monitored by DLTS[29] in electron-irradiated p-type silicon containing 10^{18} Sn/cm³.

Assuming that the V·Sn pair is a normal defect with one-hole emission for each of its two levels (from EPR studies it is confirmed that the neutral state has no Jahn-Teller distortion[31]), this was taken as evidence that the vacancy emits two holes as predicted for negative-U ordering.

The level ordering of Figure 5 implies that the stable charge state in low resistivity p-type material is V^{++}. If this is correct, photogeneration of the EPR V^+ state with less than band gap light at temperatures below carrier freeze-out should release holes

$$V^{++} \overset{h\nu}{\rightarrow} V^+ + h^+ \tag{13}$$

which will be trapped at the compensated shallow acceptors (A_S^-)

$$A_S^- + h^+ \rightarrow A_S^\circ . \tag{14}$$

This has been tested in high quality WASO-S p-type material grown by Wacker Chemitronics where the internal strains are low enough to allow EPR observation of the neutral group III acceptor A_S° as well as V^+. Upon photogeneration of V^+ at 4.2K with $h\nu < 0.35$ eV, a corresponding *increase* in the A_S° resonance was observed as expected.[30-32] In addition, subsequent decay of the photogenerated V^+ signal upon thermal annealing produced an equivalent additional increase in the A_S° resonance.[32] This confirms the mechanism for the decay as hole release according to Equation (11) and confirms the single donor identification of the $E_V + 0.05$ eV level.

(An additional confirmation that the equilibrium charge state of the vacancy is not the neutral charge state in low resistivity p-type material comes from stress-alignment studies. The experiments described in section 3.3 on stress alignment of V^o in heavily irradiated material were repeated on lightly irradiated material, with the Fermi level still locked to the substitutional boron acceptor. In this case no alignment of the photogenerated V^- signal was observed.[29] Again this is consistent with V^{++} as the equilibrium charge state at $\sim$100K in this low resistivity material since no Jahn-Teller distortion occurs for V^{++}.)

The evidence therefore is very strong that the negative-U ordering of Figure 5 is correct. No information is presently available on the positions of the acceptor levels (−/o) and (=/−). The failure to see DLTS levels associated with them in n-type material has been interpreted as indicating that they must be greater than $\cong 0.17$ eV below the conduction band edge.[33,34]

5. Vacancy Interactions with Other Defects

As the temperature is raised, the vacancy EPR spectra disappear and a large assortment of new spectra emerge which have been identified as vacancies paired off with other defects. The annealing therefore is the result of long range migration of the vacancy. We will return to the kinetics of the vacancy migration in the next section but first we identify some of the vacancy-defect pairs.

Some of the ones that have been identified are illustrated in Figure 8. In Figures 8a and 8b the vacancy is trapped next to substitutional germanium. Both the $(V \cdot Ge)^+$ and $(V \cdot Ge)^-$ charge states have been studied.[2,35] The distribution of the wavefunction is altered slightly between the four neighbors from that for V^+ and V^-, Figure 4, but otherwise the electronic structure, Jahn-Teller distortions, level positions (including possible negative-U), etc., appear very similar. The presence of germanium is evidenced by additional $(2I + 1) = 10$ hyperfine lines for ^{73}Ge($I = 9/2$, 7.6% abundant).

The vacancy-oxygen pair, Figure 8c, results from the trapping of a vacancy by interstitial oxygen, a common impurity in silicon. This has been seen both as $(V \cdot 0)^-$ (Refs. 36 and 37) and in a photo-excited $S = 1$ state of $(V \cdot 0)^o$ (Refs. 38 and 39) Two metastable $S = 1/2$ configurations of the $(V \cdot 0)^-$ pair have also been observed which are low temperature precursors to the formation of the final configuration shown in the figure. Vacancies trapped by group V atoms (P,[40] As,[41] Sb[41]) take on the configuration shown in Figure 8d. A photo-excited $S = 1$ state of the aluminum-vacancy pair has been observed,[42] Figure 8e, as has the ground

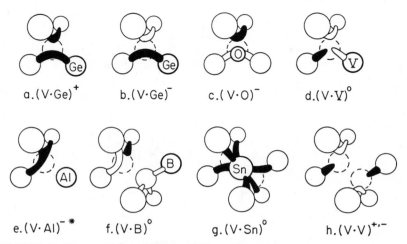

FIGURE 8. Vacancy-defect pairs identified in silicon.

state of a boron-vacancy pair, where the boron atom appears to be in a next-nearest-neighbor position to the vacancy, Figure 8f. The vacancy-tin pair,[44] Figure 8g, takes on the interesting configuration in which the tin atom resides halfway between two atom sites and bonds partially with all six resulting dangling bonds. The divacancy is also produced as two vacancies get together during vacancy anneal. The divacancy[45-51] takes on four charge states (VV^+, VV^0, VV^-, and $VV^=$) and has been observed by EPR in the single plus and minus states with the configuration shown in Figure 8h.

For most of these defects simple one-electron molecular orbital models similar to those of Figure 4 for the vacancy (but with reduced symmetry due to the nearby defect) again provide an adequate description of their structure. For the ground states the levels are populated according to the large crystal field regime and Jahn-Teller distortions occur when degeneracy remains, producing minimum multiplicity for the ground state ($S = 1/2$ or 0) as for the isolated vacancy. The only exception is the vacancy-tin pair which has an $S = 1$ ground state. In the figure, Jahn-Teller distortions are evidenced by the bond reformation by pairs. The localization of the unpaired spin is indicated in black. As for the isolated vacancy, hyperfine interactions and g-shifts are consistent with dangling bond character for the wavefunction, with ∿65% localization on the near atoms of the vacancy.

In Figure 9, the stability of the various vacancy and vacancy-defect pairs is schematically summarized as would be observed in ∿15 min isothermal annealing studies.

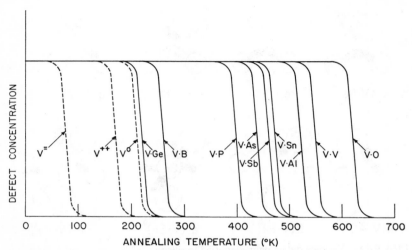

FIGURE 9. Schematic of vacancy and vacancy-defect pair annealing stages (~ 15 min isochronal).

6. Thermally Activated Migration of the Vacancy

The kinetics of the vacancy disappearance upon annealing have been studied both in EPR and in DLTS studies. The activation energy for migration has been found to depend strong upon the charge state of the vacancy. The results are summarized in Figure 10.

In low resistivity n-type silicon, where the vacancy is in the $V^=$ charge state (Figure 5) the migration and trapping by oxygen was studied by EPR giving an activation energy of 0.18 ± 0.02 eV.[2] In low resistivity p-type silicon, DLTS studies under zero bias (V^{++} charge state) revealed an activation energy of 0.32 ± 0.02 eV[10] in good agreement with the result from EPR studies of 0.33 ± 0.03 eV.[11] Under reverse bias, the DLTS studies demonstrated still a third distinct activation energy for the annealing process of 0.45 ± 0.04 eV.[10] From correlative EPR studies a higher stability for $V°$ has been confirmed,[52] suggesting that this higher activation energy should be assigned to $V°$.

These assignments are summarized in Table 3. No charge dependent capture cross section is anticipated for the defect which is trapping the vacancy because in these studies the dominant trap was neutral (interstitial oxygen or substitutional tin). Therefore the measured activation energies have been assigned directly to the barriers for migration of the vacancy.

By themselves, these annealing studies provide independent direct evidence that at least three different stable charge states exist for the

TABLE 3 Activation Energy for Vacancy Migration

Conductivity Type	Charge State	U(eV)
low ρ n-type	$V^=$	0.18 ± 0.02
high ρ p-type (reverse bias)	V°	0.45 ± 0.04
low ρ p-type	V^{++}	0.32 ± 0.02

vacancy. They therefore serve as additional confirmation of the general level structure given in Figure 5.

Such low activation energies for vacancy migration are indeed surprising for a material which doesn't melt until 1450°C, and for which activation energies of self diffusion are ~5 eV.[53] Much controversy and confusion has centered around these results and how they are to be understood in terms of high temperature transport processes in silicon.

7. Vacancy Migration Under Electronic Excitation

7.1 Recombination-enhanced Migration

Vacancies can also be made to migrate through the lattice at temperatures well below those required for thermally activated diffusion, under conditions of minority carrier injection. This was first noted by Gregory[54] in electrical property measurements of n^+/p diodes where an annealing stage at ~160°K attributed to long range migration of the vacancy in the p-type material could be induced by forward bias electrical injection of minority carriers at 77K. This was initially thought to reflect simply a change of the vacancy charge state, the faster diffusing charge state in n-type silicon presumably being generated in the p-type material under injection conditions. Subsequent similar studies in DLTS at lower temperature demonstrated, however, that the injection-enhanced rate is essentially temperature independent and can greatly exceed the fastest thermally activated rate in n-type material.[34,55] These results are also shown in Figure 10. Therefore the enhancement cannot be explained simply by the change in charge state but rather must be the direct result of the electronic excitation as electron-hole pairs recombine at the vacancy.

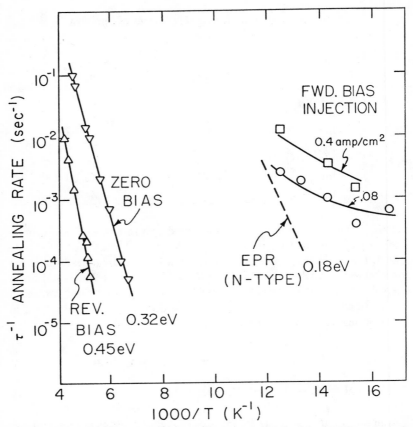

FIGURE 10. Annealing kinetics for the vacancy determined by DLTS studies of the $E_V + 0.13$ eV level in p-type silicon.[10,55] Shown also are the kinetics for vacancy anneal in low resistivity n-type silicon from EPR studies.[2]

This recombination-enhanced migration has been studied directly by EPR.[10,55-60] In a floating zone p-type sample (Al, 3×10^{16} cm^{-3}) which had been doped with 5×10^{18} cm^{-3} germanium, the intensity of the isolated vacancy and vacancy-germanium pair EPR spectra were monitored vs. 2.4 MeV electron fluence at 20.4K. It was found that isolated vacancies were produced linearly with fluence as expected by that V·Ge pairs grew in quadratically, reflecting the reaction

$$V + Ge \underset{K_2}{\overset{K_1}{\rightleftarrows}} V \cdot Ge \qquad (15)$$

where vacancy migration and trapping by Ge must therefore have been occurring at 20.4K during the irradiation. Analysis of the results gave for the rate constant

$$K_{le} = 2.5(10^{-5}) \sec^{-1} watt^{-1} cm^3 \qquad (16)$$

which is the rate per second normalized to an electron energy ionization loss deposited in the sample of 1 watt/cm^3.

7.2 Photo-enhanced Migration

The enhanced migration process has also been monitored in EPR studies by photoexcitation.[10,55-60] A typical result is shown in Figure 11, where bulk excitation of the sample is achieved by illumination with 1.064μ light from a NdYAG laser at 4.2-20.4K. Initially there are some V∘Ge pairs reflecting the formation during the initial electron irradiation. Under photo excitation, the vacancies convert quickly to V·Ge pairs approaching a constant ratio of ∿5:1 pairs to isolated vacancies. This demonstrates

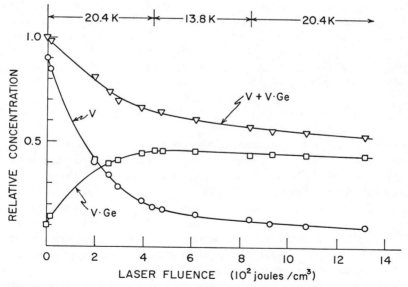

FIGURE 11. Annealing of vacancies and vacancy-germanium pairs under photo-excitation with Nd-YAG 1.064 μ laser light at cryogenic temperatures. The defects are monitored by their sEPR spectra in p-type silicon (Al, 3 × 10^{16} cm^{-3}; Ge, 5 × 10^{18} cm^{-3}).

that in addition to the photo-induced long range migration of the vacancy to produce the pairs (K_1), the reverse reaction to regenerate the vacancies from the breakup of the pairs (K_2) is also being stimulated as indicated in Equation (15). (The break-up has also been convincingly demonstrated by a 180K anneal which converts all to $V \cdot Ge$ pairs, with subsequent regeneration of vacancies by the laser illumination.[10]) Superimposed on this reaction is a loss of the total $V + V \cdot Ge$ pairs which must reflect trapping at a competing trap, not being observed in EPR.

$$V + X \xrightarrow{K_3} V \cdot X , \qquad (17)$$

In these experiments the internal temperature of the sample could be monitored by studying the spin lattice relaxation rate under excitation. Negligible heating ($\sim 1K$) occurred at the low excitation levels used in the study.

The photo-induced reaction rate constants ($K_{i\varrho}$) were found to vary linearly with laser intensity but to be roughly independent of the electron damage fluence (and therefore the vacancy concentration) up to $\sim 4 \cdot 10^{16}$ el/cm^2 (an estimated vacancy concentration of $\sim 6 \cdot 10^{15}$ cm^{-3}). The temperature dependence of the rate constants, determined for damage fluences below this limit are shown in Figure 12. The relative insensitivity of these constants to temperature confirms that the process is essentially *athermal*. That is, the electronic excitation supplies enough energy to surmount the migrational barrier and no additional thermal energy is required.

The solution of the diffusion equation gives

$$K_1 = 4\pi N_{Ge} R_{Ge} D \cong 4\pi N_{Ge} \nu d^3 / 6, \qquad (18)$$

where N_{Ge} is the concentration of traps (Ge), R_{Ge} is the capture radius (set equal to d, the nearest neighbor distance since substitutional Ge is a *neutral* defect), D is the vacancy diffusion constant ($= \nu d^2 / 6$), and ν is the single jump frequency for the vacancy. With $N = 5 \circ 10^{18}$ cm^{-3}, and $k_{1\varrho} = 2.5 \times 10^{-3}$ sec^{-1} watt^{-1} cm^3 from Figure 12, Equation (18) implies a jump frequency of ~ 20 sec^{-1} at an excitation level of 1 W cm^{-3} (5.4×10^{18} photons/sec). Estimating a vacancy concentration of 6×10^{15} cm^{-3}, this means that a vacancy makes one jump for every ~ 50 photons incident on the sample.

Enhanced migration has also been studied by EPR in n-type silicon.[55,56,58-60] Counterdoped n-type pulled silicon P, 3×10^{16} cm^{-3}; B, 1×10^{16} cm^{-3}; O, 8×10^{17} cm^{-3}) was first irradiated with 1.5 MeV electrons at 20.4K, to a fluence of 1.1×10^{17} eϱ cm^{-2}. The migration

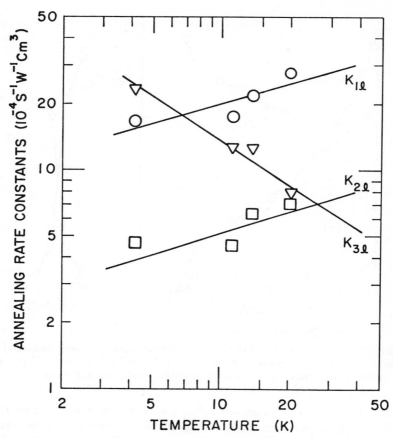

FIGURE 12. Rate constants for photo-enhanced vacancy annealing reactions vs. temperature (1.5 MeV electron damage fluence $\leqslant 4 \times 10^{16}$ eℓ/cm^2).

of vacancies and trapping by oxygen was then monitored vs. laser fluence. The results are shown in Figure 13(a). As previously observed by EPR studies of the thermally activated annealing in the dark,[2] the reaction is

$$V + O_i \xrightarrow{K} (V \cdot O)^* \to V \cdot O, \qquad (19)$$

where an excited configuration of the vacancy-oxygen pair $(V \cdot O)^*$ is first formed. Under laser excitation, no reverse break up of the $(V \cdot O)^*$ pair is observed (the conversion is complete), and *no* conversion to the ground state configuration V·O occurs. This is a remarkable result because warming in the dark to only 50K is sufficient to convert completely to the V·O

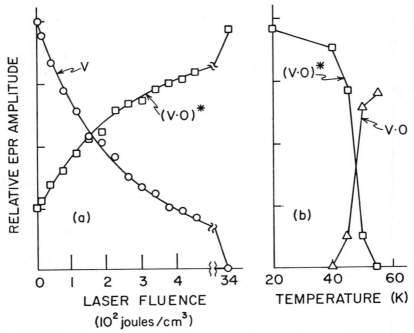

FIGURE 13. (a) Photo-induced (1.064 μ) annealing at 20.4K of vacancies in counter-doped n-type pulled silicon (P, 3×10^{16} cm^{-3}; B, 1×10^{16} cm^{-3}; O, 8×10^{17} cm^{-3}) as monitored by EPR. (b) Subsequent isochronal (15 min) anneal in the dark.

state, Figure 13(b). Subsequent photo-excitation does not regenerate (V·O)*. Apparently the electronically-enhanced processes can be very selective!

The reaction constant $K_{1\ell}$ in the n-type studies was found to be $\sim 3.5 \times 10^{-3}$ sec^{-1} watt^{-1} cm^3. With Equation (18) (now applied to oxygen trapping), an estimate of the vacancy concentration of 2×10^{15} cm^{-3}, and $N_O = 8.10^{17}$ cm^{-3}, we can estimate one jump for every ~ 20 photons incident on the sample.

The enhanced migration process for the vacancy is therefore efficient and comparable in both n- and p-type material at temperatures below carrier freeze-out, where these experiments were performed.

7.3 Mechanism for Enhanced Migration

The vacancy has five charge states in the gap. Initial speculations centered on the negative charge states as the most likely candidates for the enhanced migration. One reason was the observation that the barriers for migration were lower in n-type material (Table 3). Another good argument centered on the observation of the onset of a trigonal mode of distortion for V^- (Figure 4). Diffusion involves the motion of a neighboring atom into the vacancy site, which is a *trigonal* mode of distortion for the vacancy. V^+ and V° display only tetragonal modes, having no component of the diffusional mode. With V^- as a guide, one interesting suggestion was that $V^=$ might convert to a pure trigonal distortion with a silicon atom half way between two lattice sites.[61,62] This would be a saddle point configuration for migration and charge state change between V^- and $V^=$ would therefore provide a migration mechanism (the "Bourgoin" mechanism).

Recent results indicate that these speculations were not correct and that the enhanced migration is actually taking place when the vacancy is in a positive charged state. The first piece of evidence comes from a tentative identification of the "unknown" center X, Equation (17), as the substitutional aluminum acceptor atoms used in the p-type doping of the samples.[56,57,59] The observation in Figure 12, that $K_3/K_1 \sim I/T$ suggests that X is a Coulomb-attractive center. With a Coulomb capture radius

$$R_X \cong e^2/\epsilon kT,$$

the magnitude of K_3 implies a concentration of the center X comparable to the original doping level of the aluminum. Since the core of substitutional aluminum is negative, this implies that the vacancy is in a positive charged state when it makes its jump. A second convincing piece of evidence comes from DLTS studies in n-type silicon containing both oxygen and phosphorus.[56,58,59] In this material, thermally activated annealing at $\sim$77K produced predominantly V·P pairs (DLTS level at E_c−0.4 eV) reflecting the Coulomb attraction of the equilibrium $V^=$ charge state to the P^+ donor. Injection-enhanced annealing on the other hand produced negligible V·P pairs, the vacancies pairing instead with neutral interstitial oxygen to form the V·O pairs at E_c−0.17 eV. This provides strong evidence for Coulomb *repulsion* to P^+ in the enhanced migration state.

In order to convert electronic energy effectively into the atomic motions required for vacancy migration, a strong electron-lattice coupling is required at the defect. This suggests that the coupling derives from the Jahn-Teller instabilities of V^+ and V°. Indeed, from Table 2, the largest change in Jahn-Teller relaxation energy upon charge state change is in going from V^+ to V°, the same energy gain that is the source of the negative-U ordering.

For V° and V^+, the observed Jahn-Teller distortions are tetragonal. This raises the important theoretical question as to how the electronic energy which is dumped into the defect upon carrier capture (or direct photo-excitation) finds its way into the appropriate *trigonal* diffusion mode. Weeks et al.[63] have considered a model based on the RRK theory of molecules where energy, initially deposited into one mode, can redistribute thermodynamically amongst all of the available local modes before the energy is dissipated to the rest of the lattice. This has been applied by these authors to explain enhanced motion observed for defects in III-V semiconductors. It is not clear, however, whether this model can be justified for a vacancy where local modes are expected to be strongly coupled to the lattice and the time available for redistribution may be very short.

An alternate possibility is that upon excitation there is a strong probability of lattice relaxation directly into the trigonal mode. An indication of this is the strong trigonal *coupling coefficients* measured for the vacancy, Table 2. Classically, we can visualize two intersecting adiabatic potential energy surfaces, one associated with the trigonal mode and one with the tetragonal. The EPR observation is that the lowest energy minimum is for the tetragonal distortion. However, upon capture of a charge or upon photoexcitation, the only important question is where on the potential energy surfaces the process initiates. The resultant modes that will be excited will depend mainly on the direction of steepest descent from that point. The large V_{T2} coupling coefficient implies a steep slope in the trigonal distortion direction. The fact that this does not lead to the lowest energy minimum may not be relevant as to the *initial* "kick" given the atoms. Such simple arguments also apply statistically to the "branching ratio" where the two energy surfaces cross, as has been pointed out by Stoneham and Bartram[64] in another context. They concluded that the ratio of the transition probabilities to two vibrational modes at crossing is simply equal to the ratio of the number of vibrational quanta in each of the modes at that point, which is equivalent to the ratio of the slopes. This branching ratio argument may also be directly relevant to the vacancy, depending upon where the energy surfaces cross with respect to the point of entry, the barrier heights for motion, etc. In any case, a theoretical understanding of the dynamics of the energy release and distribution represents a formidable but worthwhile challenge to the theorist.

In the case of forward-bias injection in a junction or ionization created during high energy electron irradiation, it is evident that the energy is transferred to the vacancy via recombination at the defect of the electrons and holes produced in the bulk. In the case of 1.064μ optical injection the possibility of direct photo-excitation at the vacancy must also be considered. In Figure 12, we note that, at 20.4K, $K_{1\varrho}$ is a factor of ~100 larger for laser excitation than that, K_{1e}, for the corresponding power deposited in the sample during high energy electron irradiation, Equation (16). Each 1.16 eV photon can potentially produce an excitation, either directly at the vacancy, or indirectly via e-h generation. A high energy electron, on the other hand, is less efficient requiring on the average 3.6 eV per e-h pair generation.[65] Even with this correction, the photoexcitation is still a factor of thirty more effective per incoming photon than that resulting per e-h pair generated in the sample.

This is a strong indication that direct excitation of the vacancy dominates in the case of photoexcitation. Consistent with this interpretation, preliminary studies of the wavelength dependence of the excitation efficiency indicates the tail of an excitation band extending $\sim$0.15 eV below the band gap.[57,59] We conclude, therefore, that both e-h recombination and direct photoexcitation with near band gap light can cause the enhanced migration of the vacancy.

8. Theory of the Electronic Structure of the Vacancy

The first significant theoretical attempt at treating the electronic structure of a lattice vacancy in the diamond lattice was by Coulson and Kearsley[66] in 1957. Since that time, the lattice vacancy in the diamond lattice, and, in more recent years, specifically the lattice vacancy in *silicon,* has, in many respects, served as the principal benchmark defect for testing different theoretical approaches to the more general problem of the "deep level" in a semiconductor. As a result, a number of good reviews are available on this subject which deal specifically with the lattice vacancy in silicon.[67-72] Therefore, in this section we will present only a brief overview and summary of the present state of theory.

Broadly speaking, there have been two general approaches to the problem which can be classified as (i) finite molecular *cluster* calculations, or (ii) *perturbation* calculations on the infinite crystal states. Each approach has its advantages. The cluster methods automatically have built into them the local atomic character of the wavefunctions and can deal more naturally with local lattice relaxations and many electron effects. The perturbative methods, on the other hand, should make a better connection to the perfect crystal states, locating the defect levels with respect to the band edges, and avoiding the extraneous "surface states" of a finite cluster. Local lattice relaxations and many-electron interactions become more difficult to handle in the perturbative methods.

The original 1957 Coulson-Kearsley[66] treatment was a cluster calculation where a "defect molecule" (as they described it) was constructed simply from linear combinations of the dangling bond sp^3 atomic orbitals

(LCAO) on the four nearest neighbors. This simple approach has been used by subsequent workers (to explore many electron effects, crystal field energies, Jahn-Teller relaxations, etc.). Starting in 1970,[73] however, these calculations have tended to deal with larger clusters including many shells of atoms. Various quantum mechanical molecular orbital techniques have been tried, various methods for terminating the clusters have been explored including periodic boundary conditions, etc.[67]

The perturbative approach was first outlined in general form by Koster and Slater in 1954.[74] Although there have been many valiant attempts to treat the lattice vacancy from this approach, perhaps the first really successful treatments have appeared only relatively recently. In 1978 Bernholc, Lipari, and Pantelides,[75,76] and, independently, Baraff and Schlüter[77,78] developed a self-consistent Green's function technique, cast in an LCAO representation, which appears to give, for the first time, good single particle energies and wavefunctions for the vacancy.

Qualitatively, the cluster and perturbative approaches give essentially the same result, independent of whether the calculations are very simple or very complex. All predict for the neutral unrelaxed vacancy a partially occupied level of t_2 symmetry in the gap and a doubly occupied resonance level of a_1 symmetry in the valence band. These, of course, must correspond to the simple molecular orbitals that were used to understand the experimentally observed structure of the vacancy in section 3, and Figure 4. This is reassuring and tells us that the physics of the problem shines through somewhat independently of our feeble attempts at what is really a very difficult theoretical problem to do correctly. The different methods give results that differ in quantitative detail, however, and the best estimate for the position of the single particle t_2 level for the neutral vacancy presumably come from the recent self-consistent Green's function method and is ~ 0.7 eV above the valence band. (This agrees reasonably well with an earlier self-consistent pseudopotential band structure calculation on a "supercell" cluster by Louie et al.,[79] which gave $\sim E_v + 0.5$ eV. This technique can be visualized as the equivalent periodically extended cluster analogue to the Green's function approaches.)

8.1 Many-Electron Effects

So far, our discussion has centered on the single-particle energies. As mentioned earlier, the success of the simple one-electron models of Figure 4 in interpreting the observed structure of the vacancy in its several charge states indicates that many-electron effects are small compared to crystal field (a_1-t_2) and Jahn-Teller energies. Other more subtle details of the EPR spectra have also been cited which confirm this.[80]

On the other hand, this has not been easy for theory to come to grips with and has been the source of much controversy through the years.[69-71,80-86] The reason is simple: The neutral vacancy has four electrons localized in the molecular orbitals of Figure 4 and the configuration in its unrelaxed (T_d) state is $(a_1{}^2 t_2{}^2)$. It is therefore formally analogous to an isolated silicon atom $(3s^2 3p^2)$. It is well known that, for the silicon atom, electron-electron interactions split the configuration into terms (^3P, ^1D, ^1S) which are separated by *several electron volts*. For the vacancy one would expect the interactions to be reduced, being spread over four atoms, but it has not been at all obvious that they will be reduced enough to be ignored. In fact, the early calculations of Coulson-Kearsley and all subsequent similar calculations, have seemed to indicate that the many-electron effects are not small at all and, instead, should dominate over the crystal-field and Jahn-Teller energies.

These results appear to be in error, but why? Arguments have been presented that delocalization of the a_1 and t_2 orbitals in the solid should greatly reduce the term splittings,[80] and simple cluster calculations were performed to demonstrate that fact.[83] Subsequently, however, Surratt and Goddard[84] performed what should have been a more realistic calculation for one of the same clusters and concluded that the effects were not reduced significantly at all. Recently, Lannoo et al.[86] have developed a perturbative treatment of the many-electron effects that can be used with the single-particle Green's function results. They seem to strike a middle ground concluding that the effects are reduced significantly but not as much perhaps as was indicated in the calculations of Watkins and Messmer.[83] They concluded, however, that the reduction was great enough for the neutral vacancy that the Jahn-Teller energies would quench the many-electron effects.

In a more recent paper,[87] Lannoo, using the same approach has come to a similar conclusion for the negative vacancy (V^-), a more critical test. He was able to explain the small non-vanishing hyperfine interactions found by ENDOR[13] for sites in the nodal plane of the one-electron spin orbital (b_1 in Figure 4d) by small admixtures of excited configurations predicted as residual many-electron effects in his calculation.

It seems therefore that this controversy may be gradually being resolved. More work is required, however, before it will be understood why all of these different but presumably "realistic" approaches give such different answers. In addition, we must be particularly cautious about generalizing from what we are learning for the vacancy in silicon to other systems.

For instance, like V^- in silicon, the neutral gallium vacancy in GaP also has three electrons in a t_2 gap orbital. There, however, many-electron effects apparently dominate over the Jahn-Teller effect, and the ground state is instead the high spin $^4A_2$ state.[88] The balance between these two competing effects can therefore be a delicate one and provides a severe test for theory to deal with.

8.2 Jahn-Teller Effects

Most of these theoretical approaches have, at one time or another, been used to estimate Jahn-Teller coupling coefficients and relaxation energies. Here we will consider only the recent results of Baraff et al.[17,19] using the Green's function techniques. By expanding the LCAO basis, they were able to estimate the tetragonal Jahn-Teller coupling coefficient for a single electron in the t_2 gap orbital to be $V_E^+ = 1.12$ eV/Å.[20] These calculations are not yet capable of accurate total energy estimates so these workers instead estimated a force constant from a classical two-parameter Keating model with a large cluster of atoms surrounding the vacancy, giving $k_E = 3.7$ eV/Å^2.[20] With these values they obtained, using Equation (6), Jahn-Teller energies for V^+ of 0.17 eV, and for V°, 0.68 eV (assuming $V_E^\circ = 2V_E^+$).

Comparing these to the values in Table II, we see that their theoretical estimates for V_E are a factor of three smaller than the experimental estimates. On the other hand, the Jahn-Teller energies are only a factor of two lower. As was pointed out in section 3, significant errors are possible in extracting the experimental estimates. The question therefore is which is more correct—theory or experiment?

At the present time, the best judgment that we can make is that, at least as far as the Jahn-Teller *energies* are concerned, the theoretical calculations probably provide the best estimate. The principal reason for this conclusion is as follows: on the basis of their Jahn-Teller estimates, and with the added estimate from their calculations of a separation between single-and double-donor levels (Hubbard correlation energy U) for the *unrelaxed* vacancy of ∿0.25 eV, Baraff et al. were led to the remarkable prediction that the Jahn-Teller relaxed donor and double donor states would be inverted, in negative-U ordering, as shown in Figure 5. In addition, their predicted level positions and the magnitude of the predicted negative effective-U are remarkably close to the experimental values given in Figure 5.

Since the level positions are critically dependent upon the magnitudes of the Jahn-Teller energies, these theoretical estimates must be quite close. In fact, Baraff et al. also took this alternative view and adjusted the magnitudes of the Jahn-Teller energies and level positions for the unrelaxed vacancy for best fit to the experimental values.[18] They found that the allowed ranges of these values were indeed close to their calculated values.

It appears therefore that the Jahn-Teller energies may be being overestimated by approximately a factor of $\simeq 2$ in the stress alignment studies, Table 2. As far as the linear Jahn-Teller coupling coefficients and the corresponding force constants, it is probably premature to make a judgment. Just as there are pitfalls in interpreting the experimental stress alignment studies in terms of local-mode distortions there may be corresponding pitfalls in oversimplifying the distortion mode theoretically as simply involving nearest neighbor motions. Fortunately for both experiment and theory, the errors in estimating force constants and coupling coefficients are somewhat cancelling and Jahn-Teller energies tend to be more accurate.[2]

The apparent success of Baraff et al.[17,19] in dealing with the tetragonal Jahn-Teller distortion is encouraging. However, it is important that the other modes of distortion also be explored by these methods. For instance, earlier calculations using the Coulson-Kearsley "defect molecule" approach have estimated large Jahn-Teller trigonal coupling coefficients, which with the valence-force estimates for force constants (section 3.3) inevitably predicted the dominance of trigonal distortions[69,70,89] rather than the experimentally observed tetragonal ones. (The experimentally estimated trigonal distortion coefficients are also large, as shown in Table 2.) The approach of Baraff et al. should also therefore be applied to the trigonal distortions to test whether their approach really does avoid this same error or not. In any event, it is important to map out the trigonal energy surfaces, as mentioned in section 7.3, to help understand the recombination-enhanced processes.

Ultimately, with the recent emergence of electronic-structure calculational techniques with pseudopotentials that supply reliable total energies, it should hopefully be possible to avoid the hybrid valence-force treatment altogether and calculate the distortions directly. This is important because there is good evidence that the valence-force treatment may be grossly in error.[89] An example is the vacancy in diamond, which from analysis of phonon side bands in its optical spectra[90] reveals that the trigonal force constants are actually stiffer than the tetragonal ones, just the reverse of all valence force estimates. This could explain the trigonal vs. tetragonal error in calculations using the valence force model. If, indeed, the valence force model is in error, the question is why? Many challenges remain for the theorist before the Jahn-Teller effects for the vacancy in silicon can be considered to be understood.

9. Summary

Experimentally a great deal has been learned about the lattice vacancy in silicon. On the one hand, it can be considered a remarkably uncomplicated defect: Simple one-electron molecular-orbital models explain most of its qualitative features—the electronic structure of each of its several charges states and the origin of its Jahn-Teller instabilities. Many-electron effects, which could have made the vacancy a very complex center, apparently are not very important.

The major complexity of the vacancy arises from its Jahn-Teller instabilities. These help to crowd many charge states into the gap and even invert some of the levels into negative-U ordering. They also are undoubtedly the origin of the sensitivity of its activation barrier for migration to charge state and its remarkable *athermal* migration under electronic excitation.

The principal experimental information that is not available at present is the level position for the single- and double- acceptor states. In addition, it would be highly desirable to have optical excitation and ionization cross sections for each of the charge states as well as the carrier capture cross sections. This information is necessary in order to map out the complex energy surfaces associated with the Jahn-Teller relaxations.

The theory of the vacancy has made remarkable advances recently in the form of the single particle self-consistent Green's function calculations. Results using this method appear to indicate that wavefunctions, electrical level positions, and lattice relaxations can be reliably estimated. This first flush of success is very encouraging. The real test will be to extend these calculations to probe other critical aspects for the defect. These include (1) treating other distortion modes (trigonal and breathing as well as tetragonal), and with improved pseudopotentials that provide total energies, so that the critical energy surfaces in configuration space can be mapped out; (2) extension to the negative charged states of the vacancy; and (3) a closer more critical treatment of many-electron effects.

Finally, in the introduction to this chapter, we stated that understanding the vacancy represented a first logical step toward unraveling its role in impurity- and self-diffusion, and other complex high-temperature "processing" phenomena. Unfortunately, this has not turned out to be so easy. Experimentally we have determined that the vacancy is highly mobile with activation energies for migration of only a few tenths of an electron volt. Diffusion, on the other hand, involves activation energies of ∿5 eV. At present writing, there has been no satisfactory model that integrates these two facts. This remains for the future.

Acknowledgement. This review was made possible by support from the U.S. Navy Office of Naval Research Electronics and Solid State Program, Contract No. N00014-76-C-1097. Some of the previously unpublished results presented in this review were from research also supported under this contract.

References

1. G. D. Watkins, in *Radiation Damage in Semiconductors* (Dunod, Paris, 1964) p. 97.
2. G. D. Watkins, in *Lattice Defects in Semiconductors 1974* (Inst. Phys. Conf. Ser. 23, London, 1975) p. 1.
3. G. D. Watkins, unpublished.
4. E. G. Wickner and D. P. Snowden, Bull. Am. Phys. Soc. *9*, 706 (1964).
5. L. J. Cheng and J. C. Corelli, Phys. Rev. *140*, A2130 (1965).
6. R. E. McKeighen and J. S. Koehler, Phys. Rev. *134*, 462 (1971).
7. J. C. Bourgoin and J. W. Corbett, Phys. Lett. *A 38*, 132 (1972).
8. G. D. Watkins, R. P. Messmer, C. Weigl, D. Peak, and J. W. Corbett, Phys. Rev. Letters *27*, 1573 (1971).
9. J. C. Bourgoin and J. W. Corbett, in *Lattice Defects in Semiconductors 1974* (Inst. of Phys. Conf. Se. 23, London, 1975) p. 149.
10. G. D. Watkins, J. R. Troxell, and A. P. Chatterjee, in *Defects and Radiation Effects in Semiconductors 1978* (Inst. of Phys. Conf. Se. 46, London, 1979) p. 16.
11. G. D. Watkins, J. Phys. Soc. Japan *18*, Suppl II, 22 (1963).
12. G. D. Watkins, in *Defects and Their Structure in Non-metallic Solids* (1976), ed. by B. Henderson and A. E. Hughes (Plenum, New York 1976) p. 203.
13. M. Sprenger, S. H. Muller, and C. A. J. Ammerlaan, Physica *116B*, 224 (1983).
14. A. A. Kaplyanskii, opt. i. Spektr. *16*, 602 (1964) [English transl: Opt. Spectry. (USSR) *16*, 329 (1964)].
15. F. P. Larkins and A. M. Stoneham, J. Phys. C.: Solid State Phys. *4*, 143 (1971).
16. R. A. Swalin, J. Phys. Chem. Solids *18*, 290 (1961).
17. G. A. Baraff, E. O. Kane, and M. Schlüter, Phys. Rev. Lett. *43*, 956 (1979).
18. G. A. Baraff, E. O. Kane, and M. Schlüter, Phys. Rev. *B21*, 3563 (1980).
19. G. A. Baraff, E. O. Kane, and M. Schlüter, Phys. Rev. B22, 5662 (1980).
20. The values k = 14.8 eV/Å^2 and V$_e$ = 2.25 eV/Å given in Refs. 17–19 are not for normalized modes. The values quoted here are the equivalent ones for the normalized modes used in this paper.
21. V. Opik and M. H. L. Pryce, Proc. Roy. Soc. *A238*, 425 (1957).
22. J. Hubbard, Proc. Roy. Soc. *A276*, 238 (1963).
23. P. W. Anderson, Phys. Rev. Lett. *34*, 953 (1975).

24. R. A. Street and N. F. Mott, Phys. Rev. Lett. *35*, 1293 (1975).
25. L. C. Kimerling, in *Radiation Effects in Semiconductors 1976* (Inst. Phys. Conf. Se. 31, London, 1977) p. 221.
26. J. C. Brabant, M. Pugnet, J. Barbolla, and M. Brousseau, J. Appl. Phys. *47*, 4809 (1976).
27. J.C. Brabant, M. Pugent, J. Barbolla, and M. Brousseau, in *Radiation Effects in Semiconductors 1976* (Inst. Phys. Conf. Ser. 31, London, 1977) p. 200.
28. G. D. Watkins and J. R. Troxell, Phys. Rev. Lett. *44*, 593 (1980).
29. G. D. Watkins and A. P. Chattergee, and R. D. Harris, in *Defects and Radiation Effects in Semiconductors 1980* (Inst. Conf. Ser. 59, London, 1981) p. 199.
30. G. D. Watkins, in *Defects in Semiconductors*, ed. by J. Narayan and T. Y. Tan (North Holland, New York, 1981) p. 21.
31. G. D. Watkins, Phys. Rev. *B12*, 4383 (1975).
32. J. L. Newton, A. P. Chatterjee, R. D. Harris, and G. D. Watkins, Physica *116B*, 219 (1983).
33. J. R. Troxell and G. D. Watkins, Bull. Am. Phys. Soc. *24*, 18 (1979).
34. J. R. Troxell, PhD Thesis, Lehigh University, 1979, unpublished.
35. G. D. Watkins, IEEE Trans. on Nucl. Science *NS-16*, 13 (1969).
36. G. D. Watkins and J. W. Corbett, Phys. Rev. *121*, 1001 (1961).
37. G. Bemski, J. Appl. Phys. *30*, 1195 (1959).
38. K. L. Brower, Phys. Rev. *B4*, 1968 (1971).
39. K. L. Brower, Phys. Rev. *B5*, 4274 (1972).
40. G. D. Watkins and J. W. Corbett, Phys. Rev. *134A*, 1359 (1964).
41. E. L. Elkin and G. D. Watkins, Phys. Rev. *174*, 881 (1968).
42. G. D. Watkins, Phys. Rev. *155*, 802 (1967).
43. G. D. Watkins, Phys. Rev. *B13*, 2511 (1976).
44. G. D. Watkins, Phys. Rev. *B12*, 4383 (1975).
45. G. D. Watkins and J. W. Corbett, Phys. Rev. *138*, A543 (1965).
46. J. W. Corbett and G. D. Watkins, Phys. Rev. *138*, A555 (1965).
47. C. A. J. Ammerlaan and G. D. Watkins, Phys. Rev. *B5*, 3988 (1972).
48. J. G. deWit, C. A. J. Ammerlaan, and E. G. Sieverts, in *Lattice Defects in Semiconductors 1974* (Inst. of Phys. Conf. Ser. 23, London, 1975) p. 178.
49. J. G. deWit, E. G. Sieverts and C. A. J. Ammerlaan, Phys. Rev. *B14*, 3494 (1976).
50. E. G. Sieverts, S. H. Muller, and C. A. J. Ammerlaan, Phys. Rev. *B18*, 6834 (1978).
51. C. H. Van de Linde and C. A. J. Ammerlaan, in *Defects and Radiation Effects in Semiconductors 1978* (Inst. of Phys. Conf. Ser. 46, London, 1979) p. 242.
52. G. D. Watkins, unpublished.
53. H. C. Casey and G. L. Pearson, in *Point Defects in Solids, Vol. 2*, ed. by J. H. Crawford, Jr., and L. M. Slifkin (Plenum, New York, 1975), Chapter 2.
54. B. L. Gregory, J. Appl. Phys. *36*, 3765 (1965).
55. G. D. Warkins, J. R. Troxell, A. P. Chatterjee, and R. D. Harris, in *Radiation Physics of Semiconductors and Related Materials, 1979*, ed. by G. P. Kekelidze and V. I. Shakhovtsov (Tbilisi State University Press, Tbilisi 1980) p. 97.
56. G. D. Watkins, A. P. Chatterjee, R. D. Harris, and J. R. Troxell, Semiconductors and Insulators *5*, 321 (1983).
57. A. P. Chatterjee, PhD Thesis, Lehigh University, 1982, unpublished.
58. R. D. Harris, PhD Thesis, Lehigh University, 1982, unpublished.
59. A. P. Chatterjee, R. D. Harris, and G. D. Watkins, to be published.
60. R. D. Harris and G. D. Watkins, B. Am. Phys. Soc. *24*, 18 (1979).

61. J. W. Corbett and J. C. Bourgoin, in *Point Defects in Solids, Vol. 2,* ed. by J. H. Crawford, Jr., and L. M. Slifkin (Plenum, New York, 1975), Chapter 1.
62. J. C. Bourgoin and J. W. Corbett, Radiation Effects *36*, 157 (1978).
63. J. D. Weeks, J. C. Tully, and L. C. Kimerling, Phys. Rev. *B12*, 3286 (1975).
64. A. M. Stoneham and R. H. Bartram, Solid State Electronics *21*, 1325 (1978).
65. C. A. Klein, J. Appl. Phys. *39*, 2029 (1968).
66. C. A. Coulson and M. J. Kearsley, Proc. Roy. Soc. *A 241*, 433 (1957).
67. S. T. Pantelides, Rev. Mod. Phys. *50*, 797 (1978).
68. M. Jaros, Adv. in Physics *29*, 409 (1980).
69. A. M. Stoneham, *Theory of Defects in Solids* (Clarendon Press, Oxford, 1975), Chapter 27.
70. M. Lannoo and J. Bourgoin, *Point Defects in Semiconductors I* (Springer Verlag, Berlin 1981).
71. R. P. Messmer, in *Lattice Defects in Semiconductors, 1974,* (Inst. Phys. Conf. Ser. 23, London, 1975) p. 44.
72. M. Jaros, *Deep Levels in Semiconductors,* (A. Hilger, Ltd., Bristol, 1982).
73. R. P. Messmer and G. D. Watkins, Phys. Rev. Lett. *25*, 656 (1970).
74. G. F. Koster and J. C. Slater, Phys. Rev. *95*, 1167 (1954).
75. J. Bernholc, N. O. Lipari, and S. T. Pantelides, Phys. Rev. Lett. *41*, 895 (1978).
76. J. Bernholc, N. O. Lipari, and S. T. Pantelides, Phys. Rev. B*21*, 3545 (1980).
77. G. A. Baraff and M. Schlüter, Phys. Rev. Lett. *41*, 892 (1978).
78. G. A. Baraff and M. Schlüter, Phys. Rev. *B19*, 4965 (1979).
79. S. G. Louie, M. Schlüter, J. R. Chelikowsky, and M. L. Cohen, Phys. Rev. *B13*, 1654 (1976).
80. G. D. Watkins, in *Radiation Damage and Defects in Semiconductors* (Inst. of Phys. Conf. Ser. 16, London, 1973) p. 228.
81. A. M. Stoneham, in *Radiation Effects in Semiconductors,* ed. by J. W. Corbett and G. D. Watkins (Gordon and Breach, London, 1971) p. 7.
82. A. B. Lidiard, in *Radiation Damage and Defects in Semiconductors* (Inst. of Phys. Conf. Ser. 16, London, 1973) p. 238.
83. G. D. Watkins and R. P. Messmer, Phys. Rev. Lett. *32*, 1244 (1974).
84. G. T. Surratt, W. A. Goddard III, Sol St. Comm. *22*, 413 (1977).
85. M. Lannoo, in *Defects and Radiation Effects in Semiconductors, 1978* (Inst. Phys. Conf. Ser. 46, London, 1979) p. 1.
86. M. Lannoo, G. A. Baraff, and M. Schlüter, Phys. Rev. B*24*, 943, 945 (1981).
87. M. Lannoo, Phys. Rev. *B28*, 2403 (1983).
88. T. A. Kennedy, N. D. Wilsey, J. J. Krebs, and G. H. Stauss, Phys. Rev. Letters *50*, 1281 (1983).
89. G. G. Deleo, W. B. Fowler, and G. D. Watkins, Phys. Rev. *B29*, 3193 (1984).
90. G. Davies, J. Phys. C *15*, L 149 (1982).

CHAPTER 4

Oxygen and Oxygen Associates in Gallium Phosphide and Related Semiconductors

P. J. Dean†

Royal Signals and Radar Establishment
St. Andrews Road, Gt. Malvern, Worcestershire

1. INTRODUCTION—THE STORY UP TO DECEMBER 1966
AND SUMMARY OF LATER DEVELOPMENTS
 1.1 General Background
 1.2 Oxygen-related 0.4 eV donor
 1.3 Red luminescence—the distant DAP model
 1.4 Evidence of a related infra-red luminescence
 1.5 Observation of a deep bound excitation in GaP:Cd,O
 1.6 Capacitance spectroscopy and the second bound electron state
 1.7 Controversies and other forms of O in GaP
 1.8 Current theoretical views on the nature of the O donor ground state
 1.9 The role of O in other III–V semiconductors
2. THE NEW STORY OF O IN GaP
 2.1 A cautionary introduction
 2.2 Spectral properties of the red bound excitons—isotope effects
 2.3 Zeeman effect in the red bound exciton of GaP:Cd,O
 2.4 Isoelectronic trap description of the Cd-O and Zn-O associates
 2.5 Competitive bound exciton and distant pair recombination channels

†Deceased

185

2.6   Relationship between red BE and infra-red distant DAP
        luminescence and recombination-enhanced defect reactions
        at the Zn-O associate
2.7   Derivation of the O-donor ionization energy from IR DAP spectra
2.8   Excited states of the O donor-anticipated form
2.9   Electron capture at the isolated O donor—a radiative process
2.10 Uniaxial stress effects on electron capture luminescence
2.11 Photoluminescence excitation spectra of the electron capture
        luminescence
2.12 Derivation of the O donor ionization energy from PLE spectra
        of the electron capture luminescence
2.13 Intensity quenching of electron capture PL and PLE spectra
        by co-doping
2.14 Further consequences of the deep donor state due to O_P
3. FURTHER PROPERTIES OF THE ONE-ELECTRON STATE
3.1   Excitation spectroscopy of associate luminescence
3.2   Concentration of O_P donors and Zn_{Ga}-O_P associates
3.3   Emergence of free to bound luminescence at O
3.4   Free to bound processes and the red associate luminescence
        at 300°K
3.5   Associates of O with other impurity species
3.6   Existence and stability of the V_{Ga}-O_P associate
3.7   New identifications for Cu, O-related luminescence
4. PHOTOCAPACITANCE, DLTS AND THE SECOND ELECTRON
   STATE
4.1   Photocapacitance spectroscopy of GaP:O
4.2   Capacitance spectroscopy of the Zn-O associate
4.3   Temperature dependence of capture cross-sections
4.4   Evidence for a tightly bound state for a second electron
        at the O donor
5. CONTROVERSIAL ASPECTS OF THE PROPERTIES
   OF THE O DONOR
5.1   Can the second electron state be understood with *small* lattice
        relaxation?
5.2   Electron capture luminescence in GaP:O revisited
5.3   Zeeman and further uniaxial stress measurements on the capture
        luminescence
5.4   Further O-related luminescence and optical absorption
        and theoretical predictions of a centred, metastable state of O_P
5.5   Excited hole states in the O_P donor D^0_cXBE
5.6   Anomalous forms of D^0_cXBE absorption in PLE of distant DAP
        PLE of the O_P donor

5.7   Anomalous temperature dependence of σ^0_{PI} in distant DAP PLE spectra

5.8   Local vibronic properties of O_P in different charge states

5.9   Character of the shallow O_P^- excited states

5.10 Absence of luminescence from the O_P donor BE and Auger recombinations

5.11 Interpretation of the ODMR result of Figure 43

5.12 The new (1982) molecular-orbital model of Morgan

6. ADDITIONAL FORMS OF O IN GaP

7. FURTHER THEORETICAL CONSIDERATIONS FOR THE DEEP O DONOR

7.1   Tight-binding description of the isolated O_P donor

7.2   Tight-binding description of the associates involving O_P

8. DEEP STATES ASSOCIATED WITH O IN OTHER III-V SEMICONDUCTORS

8.1   O luminescence and EL2 trap in GaAs

8.2   Shallow and deep O-related states in InSb

1. Introduction—The Story up to December 1966 and Summary of Later Developments

1.1 General Background

Gallium phosphide is an indirect gap III-V semiconductor with energy gaps ~ 2.35 eV at $4°$K and 2.27 eV at $300°$K. These energies are just large enough to sustain reasonably efficient luminescence for red, orange, yellow and greenish light, even at $300°$K. The electrical behaviour of GaP is sufficiently amenable to manipulation by chemical dopants that efficient pn junctions can be made. This combination of properties produced a large amount of interest in GaP, growing rapidly in the early 1960s, in connection with its use in research and, eventually, in the early 1970s, in the large scale manufacture of light emitting diodes (LEDs).[1] Gallium phosphide is not the only host from which commercial LEDs have been made. However, the other main system, $GaAs_{1-x}P_x$, not only contains GaP but also depends for much of its device design upon concepts developed from work on GaP, particularly for those LEDs operating at the shorter wavelengths, in the orange $\rightarrow$ green.

The research phase of the work on GaP was rich in the discovery of entirely new concepts for the optoelectronic behaviour of semiconductors relevant for LED applications. Two of these were the first very clear demonstration of the mechanism of electron-hole recombination at distant donor acceptor pairs (DAP), with thorough evaluation of its peculiar kinetic and other attributes,[2] and the discovery of an entirely new type of impurity-related centre, the isoelectronic trap. The DAP process is strongly radiative under certain conditions. This encouraged the hope that the DAP mechanism would provide the key to the activation of efficient near-gap luminescence at $300°K$, a prerequisite for multi-color operation. This hope was not realized. The low transition rates inherent for the distant DAP recombination process gave poor luminescence performance in the face of strong competition from thermally activated detrapping processes very fast at $300°K$ for typically shallow donors and acceptors in GaP. The desired activators of efficient near-gap luminescence at $300°K$ were in fact provided by the isoelectronic traps, particularly N_P in GaP. Such centers are ineptly named in this context, since their key role is as a re-combination center rather than just a trapping center.[3]

1.2 Oxygen-related 0.4 eV Donor

Throughout this work the effects of O incorporation also proved to be of great significance. The early, controlled LEDs were often made by indiffu-sion of Zn or Cd into n-type single crystals.[4] These shallow acceptors also promote the near-gap yellow-green DAP luminescence prominent at low temperatures.[5] However, a broad red luminescence band near 1.8 eV was invariably present at 300K, suggesting the presence of a recombination center some 0.4 eV from one of the band edges.[4] A connection with shallow acceptors such as Zn was suspected at a very early stage.[6] How-ever, the persistence of this red band among non-deliberately-doped crys-tals, and an apparent enhancement in Ga-rich material, led to the initial suggestion that a V_P-related center might be involved.[4,7] Very soon after-wards, the existence of the 1.96 eV luminescence band ($77°K$) was noticed to correlate strongly with crystal growth under O-rich conditions.[8] This was observed both in GaP prepared from a vapor phase reaction of Ga_2O and P, and in systems which depend upon high-temperature quartz walls as in the early floating-zone[12] or horizontal-boat gradient-freeze tech-niques.[13] Independent evidence for the role of both Zn and O in pro-moting red luminescence in GaP grown from Ga solution was also obtained at that time.[14] These observations led to the suggestion that the 0.4 eV trap involves O. Oxygen would presumably occur as O_P according to the normal concepts of chemical substitution. It was also noticed that such crystals contained a deep donor-like trap of ionization energy between 0.2 and 0.7 eV, depending on the sample and other conditions of Hall

measurements in the range $800 \rightarrow 1100°K$.[8] In view of the variability of
the electrical activation energy, the ionization energy of this deep donor
was roughly estimated at 0.4 eV by using the simple recombination model
in Figure 1 and the ~ 0.4 eV displacement below E_g of the peak energy of
the red luminescence band, ~ 1.96 eV at 80°K. No account was taken of
the mechanism responsible for the large, nearly temperature-independent
width (~ 0.2 eV) of the red luminescence band in this estimate of $(E_D)_O$.
Strong saturation of the electroluminescence of this band at a junction
current density of only ~ 10 A cm^{-2} was used to obtain an estimate of the
concentration of these deep donors, only $\sim 10^{16}$ cm^{-3} (Ref. 8). Thermal
quenching of the luminescence intensity between 80°K and 300°K suggested an activation energy of only 20 meV. However, no understanding of
such a small value was obtained at that time.[8]

The idea that O_P might be a much deeper donor than the other group-VI
substituents such as S, which was used to promote electrical conductivity

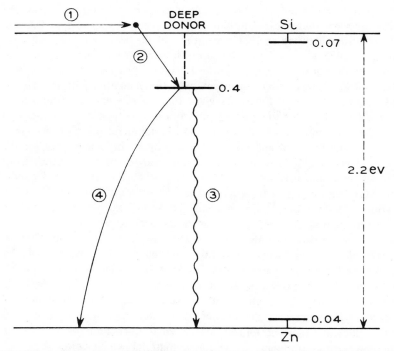

FIGURE 1 Mechanism of radiative recombination of injected minority electrons
through ~ 0.4 eV deep O donors initially believed responsible for the red electroluminescence in GaP pn junctions (1) injection, (2) trapping, (3) radiative decay,
(4) competitive non-radiative decay (Gershenzon & Mikulyak, Ref. 8).

with $E_D \sim 0.13$ eV according to early Hall analyses,[8] seemed plausible in view of its small ionic radius. Similar ideas for the behavior of O in GaAs were current at that time.[10,11] There was also some discussion of a second, interstitial role of O in the early work and of a deep state as well as shallow state associated with each acceptor studied.[8] We now know that the sharp green electroluminescence attributed to interstitial O is in fact due to the N isoelectronic trap.[3] We shall see that the "deep acceptor level" invoked to explain Zn and Cd-associated red luminescence[8] is in fact due to the Zn-O or Cd-O 'molecular' isoelectronic trap, also responsible for the '0.4 eV donor'.

1.3 Red Luminescence–The Distant DAP Model

Further analysis of the strong luminescence in terms of a 0.4 eV donor was encouraged by the observation of spectral differences between Zn and Cd-doped GaP (Figure 2). This and other properties all seemed consistent with the view that the red luminescence is produced by distant DAP recombinations between shallow acceptors and the deep O donor, now with $(E_D)_O = 0.475$ eV.[15] The energy upshift of ~ 30 meV between the red bands in Zn compared with Cd-doped crystals, observed both at $80°K$[15] and $300°K$, seemed fully consistent with the predictions of the distant DAP mechanism (inset Figure 2), given the accurate energy difference of 33 meV obtained for these acceptors from the analysis of the structured DAP luminescence that also promote at low temperatures.[18] The conclusion that the O donor and a shallow acceptor are both necessary for the red luminescence[14] was confirmed from separate doping with either Zn or Cd and simultaneously with Ga_2O_3.[15] The effects are rather less dramatic for O-doping, but this was recognized as a result of difficulties in the preparation of O-free GaP.[15] The dependence upon hole binding at shallow acceptors such as Zn or Cd, whose ionization energies were thought at that time to be, respectively 45 and 78 meV[15] explained the strong thermal quenching of luminescence intensity as well as the reduction of this quenching for p-type material. Zinc doping not only enhances the concentration of distant Zn-O DAP luminescence activators, but the increase in free hole density consequent upon the excess concentration of shallow Zn acceptors also increases the likelihood of hole recapture at each Zn-O DAP following thermal ionization. Thermal quenching of red photoluminescence in optimally Zn,O-doped material which contains $\sim 10^{18}$ cm^{-3} Zn acceptors (Figure 3) is therefore much waker than for low Zn-doped crystals.[17]

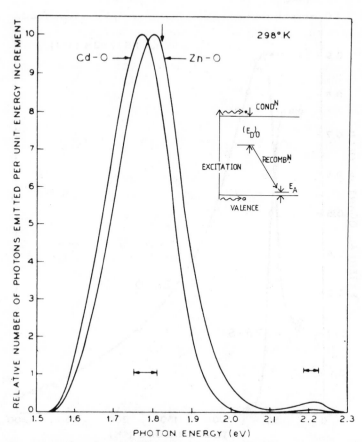

FIGURE 2 Normalized 298°K photoluminescence spectra for GaP single crystals grown from Ga solutions containing 0.01 mole % Ga_2O_3 and either 0.1 at % Zn or 15 at % Cd. The spectral resolution is indicated by horizontal arrows. The vertical arrow denotes the position of the free hole to bound electron luminescence peak (Figure 1) then expected, which should be invariant to changes in acceptor species, Zn or Cd. The inset shows the donor acceptor pair mechanism then believed relevant. (Modified from Gershenzon et al., Ref. 17).

External photoluminescence efficiencies as high as 1.5% were obtained at 300°K for optimally-doped crystals.[15] Efficiencies as high as 11% were obtained using photoexcitation photon energies adjusted to minimize the influence of surface recombination.[23] External red electroluminescence efficiencies of ~ 1.5% were reported for early GaP LEDs doped with both

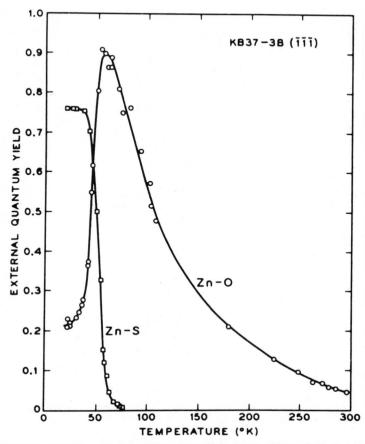

FIGURE 3 Absolute quantum efficiencies of the Zn–O (red) and Zn–S (yellow–green) photoluminescence bands as a function of temperature, showing the competition of the S and O-related electron traps near 50°K. (Gershenzon and Mikulyak Ref. 23).

Zn and O (Ref. 16). However, we shall see that the conclusion, which seemed well founded in 1965,[15,17,18] that the 300°K red luminescence is due to a distant DAP recombination process is not correct. Indeed, the mechanism does not involve electron-hole recombinations within distant impurity pairs of *any kind*. However, the deduction from the spectral shift observed between Cd- and Zn-doped materials that this luminescence cannot involve free to bound luminescence at a simple ~ 0.47 eV donor involving O alone,[15,17] at least for temperatures < 300°K,[19] remains entirely valid. The conclusion that the luminescence occurs at a center with a shallow hole trapping state is also correct.

It should be noted that the many electronic reactions involving the O_P donor are summarized for convenient reference in Table 6 in the Appendix at the end of the article.

1.4 Evidence of a Related Infra-red Luminescence

The scene was set for a series of measurements made during 1967 which transformed the earlier views of the role of O in GaP just described. We shall see that certain features, such as the appearance in GaP:Zn,O of a further infra-red electroluminescence band peaking near 1.36 eV (Figure 4), briefly mentioned in the early work[16,19,20] but essentially ignored and certainly not understood, played a vital role in the clarification of the

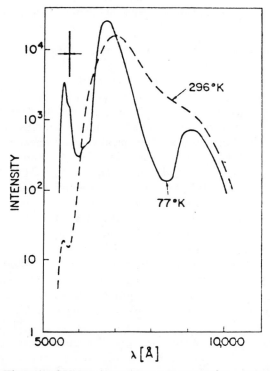

FIGURE 4 Electroluminescence spectra recorded at 77°K and 296°K from a p-n junction in a GaP crystal grown from Ga solution and containing Te donors on the n-side and Zn acceptors plus O donors on the p side. The spectra are corrected for detector response and show the green (Te-Zn distant DAP at 77°K), red (Zn-O-related) and infra-red (then unidentified, now known to be O-Zn distant DAP) characteristic luminescence bands (Lorenz and Pilkuhn, Ref. 19).

behavior of the O_P donor. Kinetic measurements, mainly at 80°K,[22] established that the infra-red and red luminescence both have the lifetime characteristics of electron-hole recombinations between distant DAP, and very deep donors were postulated for the infra-red luminescence. However, these observations could not discriminate between distant DAP and interimpurity recombination involving other types of centers. In hindsight, an additional clue to the correct interpretation can be found in the analysis of the lineshapes and thermal broadening of the broad red and infrared luminescence bands between 80°K and 500–600°K.[21] It was concluded that the centers responsible for these bands both occur on the same (P) lattice site and have similar Stokes shifts involving dominant coupling to a mode of energy $\sim$ 20 meV. Electron binding energies of 0.31 and 0.73 eV were obtained for the two centers. The former is much lower than reported in Section 1.2 just because of recognition given to the contribution of *coupling* to the luminescence lineshape and spectral positions.

1.5 Observation of a Deep Bound Exciton in GaP:Cd,O

The breakthrough in understanding of the electronic properties of O in GaP was not obtained through a proper juxtaposition of the already-existing clues, described in Section 1.4, as might have been possible. Rather, it was sparked off by essentially chance additional observations, approximately simultaneously and independently by research workers at IBM[24] and BTL.[25] Both groups observed sharp structure at the short-wavelength side of the broad red luminescence of GaP:Cd,O at low temperatures. This behavior had escaped earlier detection mainly because the circumstances of phonon coupling for the much more extensivly studied GaP:Zn,O are sufficiently different that no equivalent strong structure can be detected. The form of this structure is entirely inconsistent with a distant DAP recombination mechanism.[2] Further studies by both research teams quickly led to a new interpretation involving an associate bound exciton (BE). This new interpretation encompassed both the red and infra-red bands, their interaction with the green shallow DAP bands, and an entirely new infrared luminescence band peaking near 0.79 eV. These developments are summarized in Section 2 of this review. They lead to a much larger value for the binding energy of a single electron at the O_P donor, nearly 0.9 eV (Sections 2.7 and 2.12). Additional properties of this one-electron state of the deep O donor are described in Section 3, including a comparison of a variety of luminescent associates involving cation substituents other than Zn and Cd.

1.6 Capicitance Spectroscopy and the Second Bound Electron State

The next stage in the story involved the demonstration of the role of the deep O_P donor and associates such as Zn_{Ga}-O_P in the photocapacitancc and deep level transient spectroscopy of GaP, described in Section 4. Measurements on GaP:O strongly stimulated the pioneering development work of these deep level assessment techniques, which has subsequently led to their extensive application to the analysis of deep levels in a wide variety of semiconductors.

1.7 Controversies and Other Forms of O in GaP

Some aspects of the results on O in GaP, particularly concerning properties of a remarkable tightly bound state for a second electron, rapidly became a subject of deep controversy. This controversy, and some very recent work which has helped to clarify many aspects of it, are described in Section 5. This new work includes both experimental and theoretical evidence for a new state of O^- with moderate binding energy and slight, symmetric lattice relaxation. The experimental evidence for this new state is already as well-established as that for $O°$, including isotope shifts. The formation of this state by optical absorption is intimately linked with the neutral O donor BE, which probably has a very high Auger non-radiative recombination rate. This new state is metastable with respect to the strongly distorted, rebonded state of O^- observed through photocapacitance, which may also exhibit phonon-assisted recombination with weakly bound holes at low temperature. This process may be displaced by multiphonon emission at higher temperatures, a characteristic property of a center with very large lattice relaxation. Both processes may conspire to produce the large hole capture cross-section of the O^- state, relatively temperature independent at the higher temperatures. Section 6 describes apparently independent properties of O in GaP which are reminiscent of some well-known aspects of the behavior of O in Si.

1.8 Current Theoretical Views on the Nature of the O Donor Ground State

The early description of the ground state of the O donor was based upon a straightforward extrapolation of effective-mass concepts. The valley-orbit single state constructed from a symmetric combination of contributions from the equivalent conduction band minima near symmetry point X in the Brillouin Zone of indirect-gap GaP was believed subject to an excessively large valley-orbit interaction, caused by the strong electronegativity of the O_p donor core (Section 2.8). More recent theoretical studies based upon a quasi-molecular orbital model, more appropriate for such a deep state, suggest an entirely different description of the level near the center of the gap, with a further state deep in the valence band. This new theoretical approach, which leads to conclusions very similar to earlier pseudopotential studies with regard to the importance of valence band contributions to deep states within the energy gap,[166] is described in Section 7. A particular form of this molecular orbital description was recently introduced to explain some experimental results allegedly not well described by the classical model of O_p bonding in GaP (Section 5.12). However, a consensus from all available experimental data, including some new work expressly designed to test the new theory, together with theoretical difficulties, all described in Section 5, particularly 5.6, 5.7 and 5.12, strongly supports the original, 'classical' model used in the main body of this review.

1.9 The Role of O in Other III–V Semiconductors

Section 8 is devoted to a brief summary of our lamentably insecure understanding of the electronic properties of O in other III–V semiconductors, mainly GaAs, $A\ell_x Ga_{1-x}As$ and InSb. We shall see that for GaAs essentially no reliable consensus has emerged from a wide body of work extending over more than 20 years. This situation is considerably worse even than the difficulties which have clouded our understanding of the behavior of the transition metal Cr in GaAs![26]

2. The New Story of O in GaP

2.1 A Cautionary Introduction

We begin this section with a note of caution. The story presented here concerns the electronic behavior of just one form of O in GaP, namely the deep substitutional donor O_p. This appears to be the electro-optically dominant form of O in GaP grown from Ga solution or from the vapor at the low temperatures typical of epitaxial growth for LEDs, $\sim 1000 \rightarrow 1100°C$. We certainly cannot pretend that no other forms may occur, perhaps with quite different properties. Some of these alternate forms appear to be much more important in bulk-grown GaP, typically prepared near 1450°C, and are considered in Section 6. For the present, we are concerned only with the relatively well-understood, anion-substitutional form of O.

2.2 Spectral Properties of the Red Bound Excitons—Isotope Effects

The highly structured photoluminescence and photoluminescence excitation (PLE) spectrum of GaP:Cd, O shown in Figure 5 led to the immediate realization that the interpretation given in Section 1.3 required extensive revision. The form of this spectrum is characteristic for a bound exciton (BE) transition at a single, well-defined center. It contains at 10°K a single no-phonon line A, rather than the spectrally-distributed multitude of sharp lines expected for the DAP process, assuming that observation of the no-phonon lines is not precluded by the degree of phonon coupling. No corre-

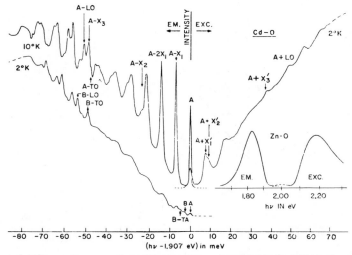

FIGURE 5 The main spectra are characteristic red photoluminescence (em) and photoluminescence excitation (exc) from a solution grown GaP single crystal doped with Cd and O. The excitation spectrum is temperature-independent below 20°K and contains a single no-phonon line A, also observed in luminescence. A second no-phonon line B appears at slightly lower energies in luminescence at the lowest temperatures. The strong phonon-assisted transitions are discussed in the text and in Table 1. The insert shows similar spectra typical of Zn plus O-doped crystals, in which the no-phonon lines cannot be detected (Morgan et al. Ref. 24).

198   P. DEAN

spondingly prominent structure can be seen in the red luminescence of
GaP:Zn, O (Figure 6). However, weak broad sub-bands do appear, characteristic of coupling to a local (in band resonance) mode of energy ~ 6
meV, slightly *lower* energy than the strong 7 meV mode in the side-bands
of the GaP:Cd, O spectrum (Table 1). This is unexpected, since $M_{Cd} \gg M_{Zn}$, and suggests that the local force constants are very sensitive to the
size of the substituent cations. On the other hand, substitution of Cd^{110}
for Cd^{114} results in an increase in energy of the 7 meV phonon in luminescence by $(1.36 \pm 0.20)\%$ (Table 1 and Figure 7) about 3/4 of the isotope
shift expected if the Cd atom carries the entire kinetic energy in this vibrational mode.

In common with virtually all such complex spectra involving BE at associates in semiconductors, no successful *detailed* analysis of the phonon
coupling has yet been made for this Cd-O center. The classical analysis of

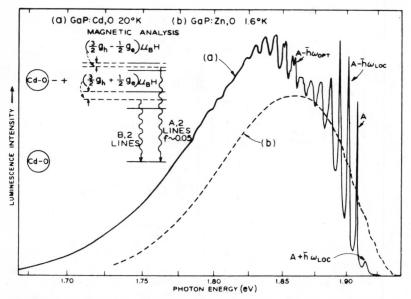

FIGURE 6 Detailed low temperature red photoluminescence spectra of GaP double
doped with Cd and O, compared with Zn and O. Weak structure can just be discerned
for the Zn plus O-doped crystal, due to multi-phonon assisted transitions. However,
the rich structure and no-phonon transitions similar to those characteristic of the
Cd-O bound exciton cannot be detected. The inset shows the Zeeman splitting and
oscillator strength f of transitions from the | 2, ± 1 > (allowed, A line) and | 2, ± 2 >
(forbidden, B line) BE states shown as $^2\Pi$ and $^2\Delta$ on the right in Figure 9, where here
we use | J, M_J > notation.

TABLE 1 Phonon energies and isotope shifts observed in O-associate spectra

Associate / Feature	$Cd_{Ga}\,O_P$ (184)b	$Zn_{Ga}\,O_P$ (25)	$Mg_{Ga}\,O_P$ (105)	$Be_{Ga}\,O_P$ (105)	$Cu_I\,Cu_{Ga}\,Cu_I$ (118)	$L_I\,L_I\,Ga\,O_P$ (107)	$L_I\,L_I\,Ga\,O_P$ Isotope Shifts(107) ^6Li–^7Li	$0^{18}\,0^{16}$
Phonon Energy (meV) and Type	7.0 ± 0.05 $\Gamma_1^\dagger$ (L.X$_1$) 7.3 ± 0.1 > (L.X$_1$) $8.6 \pm 0.1^*$ > (L.X$_2$) $16.4 \pm 0.2^*$ > $37.2 \pm 0.4^*$ > (X$_3$) 40.9 ± 0.2 $\Gamma_1^\dagger$ 45.2 ± 0.2 Γ_3 46.3 ± 0.2 Γ_3 47.3 ± 0.2 $\Gamma_1^\dagger$ (0$_1$,X$_3$) $48.6 \pm 0.2^{*,a}$ Γ_1 (0) 49.7 ± 0.2 Γ_1 (0$_2$) 49.9 ± 0.2 Γ_3 50.4 ± 0.2 Γ_3	6.0 ± 0.1 > (L)	13.2 (TA) 28.9 (L.1) 32.0 (L.2 Gap) 35.3 (L.3 Gap) 42.0 (L.4) 45.6 (TO) 50.0 (LO) 52.3 (L.6)	2.1 (L.1) 10.9 (L.2) 18.5 (L.3) 30.0 (L.4) 33.2 (L.5 Gap) 45.6 (TO) 47.4 (L.6) 49.3 (L.7) 50.2 (LO) 53.1 (L.8) 64.1 (L.9)	5.3 (L.L$_1$) 4.5 (L.A$_1$) 6.8 (5.3 + 1.5?) (L.L$_2$) 6.0 (4.5 + 1.5?) (L.A$_2$) 8.8 (5.3 + 1.5 + 2.0) (L.L$_3$) 8.7 (L.A$_3$) 12.0 (L.L$_4$) 13.5 (L.L$_5$) 14.3 (L.L$_6$) 18.1 (TA?) 27.1 (L.LA?) 39.7 (L.L$_{12}$ Gap) 45.4 (TO) 49.1 (L.Sn) 50.1 (LO)	11.1 (TA.L) 13.2 (TA.X) 13.0* (TA.X) 20.4 (L.1) 20.4* (L.1) 25.7 Γ_3 (L.A.L) 32.2 (L.2) 36.3 (L.L1– Gap) 32.2* (L.L1 Gap) 40.5 (L.L2 Gap) 36.0* (L.L2 Gap) 44.9 (L.3) 45.6 (TO.4) 45.8 (L.L3) 46.7 (L.L4) 47.5 (L.L5) 47.5* (L.L5) 47.8 – (L. unlabelled) 49.9 (L.5) 50.0 (LO) 50.3 (L.L6) 57.9 (L.L7)	–0.05 –0.75 +0.45 –0.05 –0.06 –0.10 –0.08 –0.18 –0.45 +4.23	–0.04(?) –0.33 ? –0.08 –0.07 ? 0.19 –0.10 ? –0.29
Isotope Shifts (meV)	$\Gamma_1^\perp$ Cd^{114} – Cd^{110} ↓ 0.09$_5$ NP 0^{16} – 0^{18} –0.65		NP 0^{16} – 0^{18} –0.07$_3$ 0.09		Cu^{65} – Cu^{63} (L.L$_1$) + 0.15 (L.L$_2$) + 0.27 (L.L$_3$) + 0.32 (L.L$_5$) + 0.27 (L.LA?) + 0.47	$\sqrt{P}$ Li^7 Li^7 0^{16} → Li^7 Li^7 0^{18} + 0.80 Four values of $\hbar\omega_{L7}$ appear for equal proportions of Li^6 and Li^7 in diffusion source		

L designates local mode. other symbols TA LA TO LO designate replicas from appropriate branches of host GaP lattice. Multiphonon modes not listed

* Observed in optical absorption or PLE

† May involve an electronic excited state

a Probably two unresolved components separated by ~1 meV

b Phonon replica X$_2$ energy ~ 23.4 meV (Fig 5) was not confirmed in ref (184)

c Superimposed on host phonon replica

d Phonon energies quoted for standard isotope combination. Li^7 0^{16}

e Phonon energies quoted for Cu65

f The Γ_1 replicas are receiving modes. Γ_3 are promoting modes for the forbidden transition

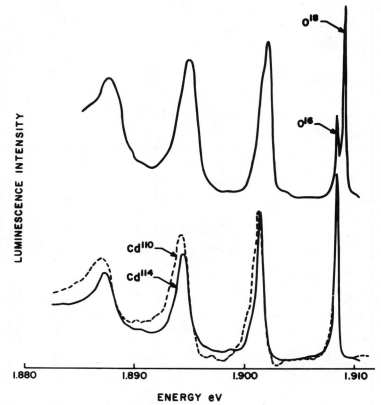

FIGURE 7 The upper part shows the shift in the no-phonon line of the Cd-O red BE photoluminescence induced by the isotopic substitution $O^{16} \rightarrow O^{18}$. The lower shows the isotope shift in the 7.0 meV phonon replica produced by the substitution $Cd^{114} \rightarrow Cd^{110}$ (Henry et al., Ref. 76).

the broad, essentially featureless red luminescence band associated with Zn-O already mentioned in Section 1.4, used the simplest type of configurational coordinate description to describe the additional broadening of the overall spectral shape with increasing temperature,[21] according to the following equation for the bandwidth ΔE:

$$(\Delta E)^2 = (2.36)^2 \, E_{FC} \, \hbar\omega \, (1 + 2\bar{n}),\qquad(1)$$

where $\bar{n}$ is the occupation number for a phonon of energy $\hbar\omega$ and E_{FC} is the Franck-Condon shift between the pure electronic transition energy and the peak energy of the luminescence spectrum. This equation assumes linear electron-phonon coupling, which we shall see is invalid for these bound excitons. The single energy $\hbar\omega \sim 20$ meV resulting from this analysis[21] represents some average value for the actual phonon coupling.[73] We see from Figure 6 and Table 1 that this is a crude approximation to the complex phonon coupling observed for the Cd-O bound exciton. We also conclude below that still further phonons, passing unresolved in the luminescence spectra, must also couple strongly in the electronic transition to account for the isotope effects observed. These latter modes have different (T_2) symmetries compared with the A_1 "receiving" modes which are involved in the linear approximation of the usual configurational coordinate analysis.

The involvement of O as well as Cd in the red luminescence has been established byond doubt through the observation of an energy increase of the no-phonon line A by 0.65 meV when $O^{16} \rightarrow O^{18}$ (Figure 7). The physical origin of this increase is a significant dependence of the energies of local modes which couple strongly to the electronic states upon the difference in local charge between the initial and final state of the electronic transition. The effect appears in the no-phonon transition through the dependence of the zero-point energies on the local vibrational modes.[27] The sense of this non-linear electron-phonon interaction is a softening effect of excess electrons and holes on the energies of the vibrational modes of the perfect lattice which form a basis for the local modes. This energy reduction causes an increase in the electronic transition energy with isotope mass according to[27]

$$\Delta E = \frac{3}{4}\hbar \sum_i \Delta\omega_i \frac{\delta M_i}{M_i}, \qquad (2)$$

where $\Delta\omega_i$ is the increase in local vibrational mode of index i on removal of the excess electron from the O donor, and δM_i is the change in vibrational mass associated with this mode when $O^{16} \rightarrow O^{18}$ assuming that the mode frequencies are proportional to $(M_i)^{-\frac{1}{2}}$. Equation (2) contains a factor of 3 since the O_p vibration involves a 3-fold degenerate mode. The softening of the phonons by trapped carriers is also related to the temperature dependence of the band gap of the pure semiconductor—that part not caused by thermal expansion, which accounts for $< 20\%$ of dE_g/dT in GaP.[27] The quantitative description of the isotope shifts listed in Table 1 shows that the hole contribution to the 'electronic' part of dE_g/dT is three to four times greater than that of the electron in GaP.[27]

Morgan et al.[24] suggested that the modes X_2 and X_3 in Figure 5 might be the local modes which control this isotope shift, since their total energy appears to decrease by ~ 27 meV between PL and PLE, that is between

TABLE 2

EXCITON AND SINGLE PARTICLE BINDING ENERGIES AT O–ASSOCIATES

Property / Associate	Exciton Localisation Energy [a] E_{BX} (meV)	Electron Binding Energy [b] E_e (meV)	E_{REP}[a] $= (E_D)_0^c - E_e$ (meV)	Pauling Radius of Cation, r (Å)	$e^2/\epsilon\,(r_C + r_0)$ [d] (meV)
$Cd_{Ga}-O_P$	420.2 [25]	400 ± 6	498	0.97	552
$Zn_{Ga}-O_P$	319 ± 5 [29]	296 ± 6	602	0.74	611
$Mg_{Ga}-O_P$	165 [105]	143 ± 10	755	0.65	638
$Be_{Ga}-O_P$	135.7 [105]	115 ± 10	783	0.31	764
$Cu_{Ga}-O_P$	~ 550 [194] [f]	~ 570 [g]	~ 860 [h]	0.72	617
$Li_{Ga}-Li_{Ga}-O_P$	238.1 [107]	215 ± 8	683	0.60 (Li⁺)	653

a Calculated for E_{GX} = 2.3285 eV

b Calculated for E_G = 2.350 eV, E_h ~42–45 meV

c Calculated for $(E_D)_0$ = 898 ± 1 meV

d Calculated for r_0 = 140 Å; $e^2/\epsilon\,(r_{Ga} + r_p)$ = 554 meV

e E_h and E_e probably comparable for $Cu_{Ga}-O_P$ associate

f Assuming 1.78 eV BE line is due to this associate

g Electron–hole energy sum $E_e + E_h$, since E_A is more comparable with E_D in this associate

h Calculated as $(E_D + E_A) - (E_e + E_h)$

emission and absorption (Table 1). However, this decrease greatly exceeds the lower limit of ~ 8 meV estimated from Equation (2) using the experimental value of the no-phonon isotope shift for O. No significant differences in the energies of X_2 and X_3 when $O^{16} \to O^{18}$ were detected directly, either in PL or PLE, so this identification seems to be misleading.[76] Probably X_2 and X_3 have A_1 (or Γ_1) symmetries rather than the $T_2(\Gamma_3)$ required for "isotope effect" modes. However, we have no direct evidence on this point (Table 1). We can only conclude that the phonons which dominate in this isotope shift do not produce well-resolved sidebands in the optical spectra, but may contribute to the large background intensities in the broad phonon wings of the PL and PLE spectra.[29]

The most important aspect of this isotope shift is the direct link it provides between O and the center responsible for the luminescence in Figure 5. This is a very valuable feature. Chemical control of O incorporation is very troublesome in semiconductors with relatively high growth temperatures such as GaP. The difficulties result from the relatively low limiting solubility and the persistence of O as a significant inadvertent contaminant. Before these isotope experiments, it was easy to conclude provisionally from results on a limited series of crystal growth runs that the addition of O *did not* correlate well with the strength of the red luminescence! This incorrect negative correlation seemed plausible in view of an attractive early model whereby the low energy mode X, (Table 2) was attributed to a Cd_{Ga}-V_P associate, with force constant softening dominated by the broken bonds associated with the vacancy.[28]

2.3 Zeeman Effect in the Red Bound Exciton of GaP:Cd,O

Further important evidence of the nature of the transitions at the Cd-O associate was obtained from a study of the Zeeman properties of no-phonon line A and an additional line B which appears ~ 2.1 meV to lower energy in spectra recorded at the lowest temperatures (Figure 8). The luminescence ratio A/B increases with T according to $\exp(-\delta E/kT)$, where $\delta E \sim 2$ meV $\sim (h\nu_A - h\nu_B)$. Thus, the splitting occurs in the initial state of the luminescence transition. The A/B intensity ratio at low temperatures is very sensitive to sample quality, decreasing in the presence of local strains or other perturbations. This ratio also decreases strongly in the magnetic field according to

$$\left(\frac{A}{B}\right)_B = \left(\frac{A}{B}\right)_{B=0} / B^2 \sin^2 \theta \qquad (3)$$

where θ is the angle between B and the symmetry axis of the Cd-O associate defined below.

The strongly anisotropic *splittings* of the B line in a magnetic field (Figure 8d) can be analyzed in terms of transitions from the $|2, \pm 2>$ state of an exciton bound at the center containing no additional electronic particles (Figure 8c). The origin of the $|J, M_J>$ description is shown in the right hand portion of Figure 9, where it is assumed that the splitting of the hole (valence band) states in the axial field of the center, δE_{AX} is large compared with the electron-hole j–j coupling, δE_{j-j} in the BE. It is also assumed that the sign of δE_{AX} is such that the Γ_4, $\Gamma_5(m_j = \pm \frac{3}{2})$ hole states become energetically lowest, equivalent to the effect of *tensional* axial strain on the Γ_8 valence band maximum of a zincblende semiconductor such as GaP.[30] This sense of the local strain is consistent with expectation for a center involving O_P, in view of the small covalent radius of O compared with the P atom it replaces. Quite different, closely isotropic magneto-optical behavior results if the BE is formed from holes of symmetry $\Gamma_6(m_j = \pm \frac{1}{2})$ (Ref. 31) (Section 2.7). The form of the Zeeman anisotropy shown in Figure 8d, and also for the A line which involves transitions from the $|2, \pm 1>$ state (Figures 8c and 9, right) and which exhibits an overall g-value about 24% of that of B,[76] establishes that the local axial field has [111] symmetry, consistent with a Cd_{Ga}-O_P associate.

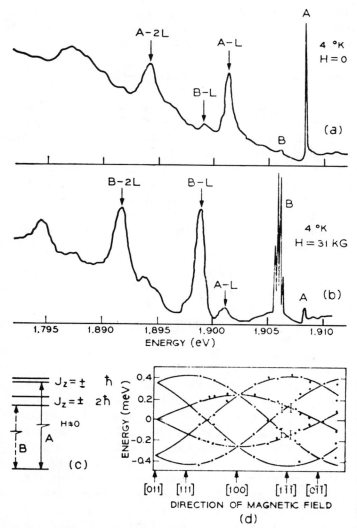

FIGURE 8   The Cd-O red BE luminescence at 4.2°K (a) with magnetic field H = O
and (b) for H = 31 kG with H lying between [011] and [111]. Six magnetic subcomponents result from the different orientations of a trigonal symmetry (C_{3v}) centre
relative to the magnetic field (part (d)), but the outer pair have very small intensity
as H approaches [111]. No-phonon transition B is forbidden for H = O but becomes
allowed by mixing with A for finite H. The level diagram for the BE states observed
is shown in (c), where J_Z is defined relative to the trigonal axis of the associate. The
magnetic field was rotated in a plane perpendicular to the [O$\overline{1}\overline{1}$] direction in (d)
(Henry et al., Ref. 25).

2.4 Isoelectronic Trap Description of the Cd-O and Zn-O Associates

These centers may be viewed as a 'molecular-type' isoelectronic trap, since the Cd-O and Zn-O associates are isoelectronic with the Ga-P molecule they replace. The Γ_6 symmetry of the electron bound to the Cd-O

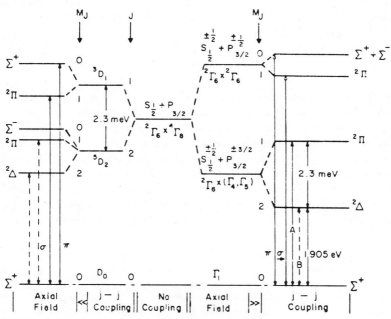

FIGURE 9 Energy levels of an exciton bound to a centre with trigonal symmetry for two cases; the j-j splitting $\delta E_{j\text{-}j} \gg \delta E_{Ax}$ where δE_{Ax} is the splitting of the hole states in the trigonal field (left side) and $\delta E_{j\text{-}j} \ll \delta E_{Ax}$ (right side). The uncoupled electron $S_{1/2}^{m_e}$ and hole $P_{3/2}^{m_h}$ states which produce the levels are shown together with the irreducible representations of the appropriate crystal field groups. Total angular momentum values J and their components M_J along the symmetry axis are given as appropriate. The states on the far left and right are labelled with the notation for a diatomic molecule and their degeneracies are indicated as superscripts. Allowed electric dipole transitions from the J-O ground state are shown by solid vertical lines while dashed lines become allowed in perturbing fields. The polarizations of emitted photons relative to the symmetry axis are indicated as π or σ. The Cd-O Be behaves according to the right hand side (Morgan et al, Ref. 24). A completely different behavior holds if the δE_{Ax} splitting has the reverse sign to that assumed here, as for associate involving Cu_{Ga} in GaP (Section 2.7), and for the hold bound to the O− state (Section 5.3).

center assumed in Figures 8 and 9 is appropriate only for an electron-at-tractive center, such as a P-site donor[32] rather than a Ga-site substituent.[33] Isoelectronic traps are classified as donor or acceptor-like according to whether binding of the exciton is dominated by interactions of the neutral center with, respectively the hole or the electron.[34] The results just de-scribed suggest that exciton binding to Cd_{Ga}-O_P is dominated by electron interaction with the O donor, consistent with the expectation that O_P will be an unusually deep donor in GaP (Sections 1.2 and 1.8). We may intro-duce the concept of an effective donor energy $(E_D)_O^{eff}$, reduced from the ionization energy of the isolated O_P donor by the repulsive effect of the neighboring ionized acceptor E_{rep}

$$(E_D)_O^{eff} = (E_D)_O - E_{rep} \qquad (4)$$

Given a sufficiently large value of $(E_D)_O$, the value of $(E_D)_O^{eff}$ will be $>$ O, that is a bound state will still exist for the electron alone, energy E_e. Once the electron is bound, the BE can be readily formed from the Coulomb interaction between the trapped electron and a relatively distant hole, energy E_h. This simple interpretation of exciton binding, in which the contributions of the electron and hole are evaluated independently, will be a good approximation if $E_e \gg E_h$. Exactly this situation exists for these remarkable centers (Figure 10).* The electronic reactions involving this neutral associate are summarized in Table 6 in the Appendix at the end of the article.

---

*The basic principle behind Equation (4), that the luminescence in a DA associate involving one deep and one shallow species is governed by an effective donor binding energy where the energy reduction factor E_{rep} is approximately the DAP Coulomb interaction energy $e^2/\epsilon R$ (Equation (6)), was first applied to visible luminescence in diamond.[197] In this case, the DA associate or BE center was held responsible for characteristic blue (band A) luminescence, while recombinations at unassociated DAP give rise to yellow-green luminescence. Both luminescence bands are spectrally very broad due to very strong phonon coupling associated with electron coupling to the $\sim 3.9eV$ donor whose binding energy was deduced from the analysis of these bands using Equations (4) and (6). The acceptor binding energy E_A is only $\sim 0.36eV$, thought due to Aℓ in 1965 but now known to be due to B.[198] Further work has con-firmed the kinetic and spectral properties of these two broad bands expected from this model. However, the remarkably deep donor is now known to be due to an N-N pair center (A center)[199,200] not isolated N which has smaller donor binding energy, $E_D \sim 1.7eV$.[201] The A center is one of the major N aggregate centers in the most common form of natural diamond, type Ia.

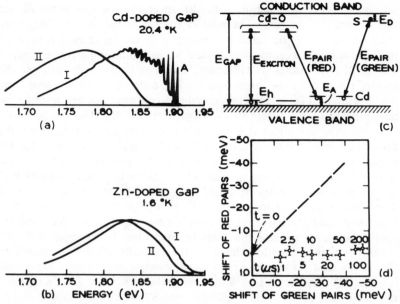

FIGURE 10   (a) Curve I Cd–O red BE luminescence, curve II (Cd–O)–Cd red distant pair luminescence. These bands may be of comparable intensity for appropriate conditions of doping and excitation at low temperatures. (b) Curve I Zn–O red BE luminescence, curve II (Zn–O)–Zn red distant pair luminescence. (c) Level diagram distinguishing the BE and distant pair recombinations for Cd, O-doped GaP, including a green DAP transition involving the shallow S donor. (d) Shift of the (Cd–O)–Cd distant pair luminescence versus the shift of the S–Cd DAP luminescence with decay time after pulsed excitation. The shifts were measured at the high energy side half-maximum of each band. The lack of shift for the red luminescence shows that the electron trap responsible for the red luminescence is uncharged before electron capture, unlike a normal donor such as S^+ or O^+ (Henry et al., Ref. 25).

2.5   Competitive Bound Exciton and Distant Pair Recombination Channels

Once the associate has trapped an electron, recombination with a hole may occur at low temperatures *either* through the BE state just described *or* from a distant neutral acceptor (Figure 10c). These two recombination channels give two luminescence bands of similar overall shape, determined by the phonon coupling. This coupling is dominated by electron binding to the associate under the strong, short range forces common to both processes. The fine structure in the BE spectra is not seen in the distant pair spectra (Figure 10a, b) because of spectral broadening resulting from small variations of the transition energy with the separation R between the associate and acceptor sites, typical for such inter-impurity processes.[2] The small but distinct overall energy increases between the pair and BE spectra result from the larger hole binding energies at the acceptor compared with E_h (Table 2). The central-cell contributions are positive for hole binding to the acceptors but *negative* in the BE. This results partly from the repulsive effect at the O_p site and, more significantly, from the delocalization of the negative charge to which the hole is bound. The BE recombination channel is favored for crystals prepared by slow cooling from $\sim 800°C$ after crystal growth, which results in a large proportion of associates, and in spectra measured at high optical excitation rates or at short time delay following pulsed excitation (Figure 11). The distant pair transitions exhibit characteristic slow time decay (Figure 10d) and are readily saturated, whereas the much larger electron-hole overlap in the BE ensures fast recombination times for the optically allowed A transition. The decay time of the BE transition is strongly temperature dependent (Figure 12). This results at low temperatures, $\leqslant 50°K$ for Cd-O, from thermalization between the forbidden and allowed transitions related to the B and A lines, (Figure 8a and at higher temperatures from the onset of thermal equilibrium between the hole states at the BE, the isolated acceptors and the valence band. The first effect may be analyzed according to the detailed balance relation

$$\tau = \tau_B \left[1 + g\exp(-\Delta E/kT) \right] \Bigg/ \left[1 + \frac{\tau_B}{\tau_A}\exp(-\Delta E/kT) \right] \quad (5)$$

where τ_A, τ_B are the lifetimes of the A and B states, g is the degeneracy ratio of the A and B states and ΔE is their energy separation. The data for the Cd-O Be in Figure 12 yields $\tau_A = 0.1$ μsec and $\tau_B = 0.6$ μsec.[29] No such behavior is observed for the Zn-O BE. It is believed that this is a consequence of a large degree of phonon-induced mixing between these states consistent with the absence of clearly resolved no-phonon structure (Figure 6). Equilibrium between the hole states above $\sim 60°K$ removes any spectral shift in time-resolved spectra (Figure 11). The evaluation of spectral form suggests that the BE recombinations predominate above $\sim 80°K$, even for crystals prepared to *minimize* the concentrations of Cd-O associates. Similar effects occur for GaP:Zn, O, although the generally smaller characteristic energies ensure that BE dominance sets in at lower

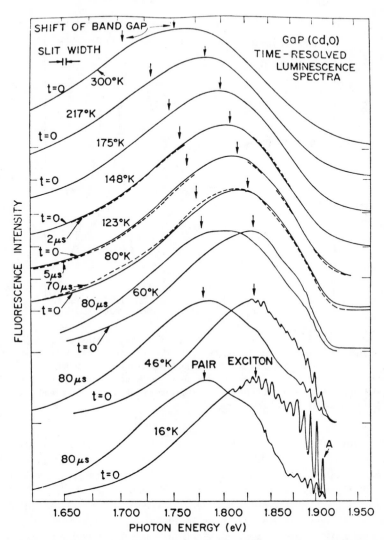

FIGURE 11 Normalized time-resolved red BE luminescence spectra of a Cd, O double-doped GaP crystal prepared so as to minimize the concentration of Cd–O associates. The BE and distant pair bands can still be clearly distinguished at low temperatures, with overall spectral peaks indicated by vertical arrows. The expected movement allowing for the temperature dependence of E_g in GaP is shown in the upper spectra. The BE process clearly predominates when the holes thermalize between the available states above ~ 80°K and the luminescence decay then becomes much faster (Figure 12), (Cuthbert et al., Ref. 29).

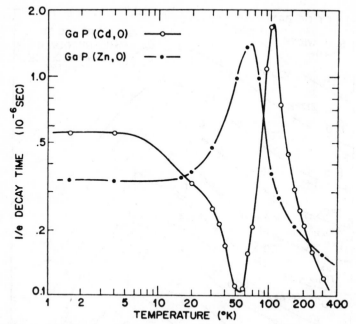

FIGURE 12 The l/e decay times ($\tau_{l/e}$) of red luminescence plotted as a function of temperature compared for Cd, O and Zn, O double-doped GaP. For Cd, O, the initial decrease in $\tau_{l/e}$ above 2°K is caused by thermal re-distribution of the BE population between the A and B states. The changes above 60°K are caused by the onset of thermal equilibrium of holes between the BE, isolated shallow acceptors and valence band. (Cuthbert et al., Ref. 29).

temperature, $\geqslant 60°$K.[29] The decrease in lifetime exhibited by both systems at $\geqslant 100°$K is readily explained in terms of a statistical analysis involving the occupation probabilities of holes at the acceptor and BE states described by standard Fermi-Dirac expressions.[29]

There is a major difference between the spectral characteristics of the distant pair recombination processes in Figure 10a, b and those at conventional DAP. Since the associate electron trap is uncharged in its ground state, there is no "Coulomb" interaction term $e^2/\epsilon R$, in contract to the equation for the electron-hole no-phonon recombination energy hυ DAP at distant DAP

$$h\upsilon_{DAP} = E_g - (E_D + E_A) + e^2/\epsilon R + f(R), \qquad (6)$$

where f(R) contains higher order terms designed to account for the polarization interaction in the excited DAP state.[2] The term f(R) is small even for very close DAP if $E_D \gg E_A$, as in the systems described here, and so E_{rep} in Equation (4) is essentially $e^2/\epsilon R$. Absence of this dominant R-dependent term in the decay at neutral associate-acceptor distant pairs removes essentially all the spectral shift during time decay of this red pair luminescence (Figure 10d). This shift is one of the most characteristic features of normal DAP luminescence and is clearly exhibited by the green DAP in GaP (Figure 10d).

This same electrostatic interaction energy $e^2/\epsilon R$ also describes the main part of the energy gain on association of the DAP. The association energy was found to be 0.54 ± 0.15 eV from measurements of the thermally-activated increase in the intensity ratio between infra-red and red luminescence bands attributed to thermal dissociation of Zn-O associated during anneals near 800°K.[35] However, later measurements resulted in an appreciably higher activation energy ~ 0.74 eV, attributed to a contribution from the diffusion activation energy of Zn in O-free GaP.[36] The enhancement effects of low temperature treatments for the intensity of the red luminescence already discovered empirically for GaP:Zn,O LEDs,[38] find a natural interpretation on this new associate-activator model.

2.6 Relationship Between Red BE and Infra-red Distant DAP Luminescence and Recombination-Enhanced Defect Reactions at the Zn-O Associate.

The BE transitions of Section 2.4 are themselves equivalent to recombinations at nearest-neighbour members of the DAP distribution, shell number $m = 1$ for a Type II spectrum.[2] Assuming E_{rep} in Equation (4) is just $e^2/\epsilon R$, we find $(E_D)_0 = (E_D)^{eff} + 0.53$ eV. The values of $(E_D)^{eff} = E_e$ given in Table 2 then indicate that $(E_D)_0 \sim 0.9$ eV, about twice the values suggested in Section 1.2. The magnitude of E_e for Cd-O is obtained directly from the energy of the A BE line, revised according to recent reappraisals of the band gap of GaP.[37] The smaller value for the Zn-O associate was obtained from the mean of the half-height energies on the high energy side of the PL spectrum and low energy side of the equally featureless PLE (absorption) spectrum, since these spectra do not exhibit significant intensity near the no-phonon line[29] unlike Cd-O (Figure 5). Table 2 also shows reasonable agreement between the experimental values of E_e and those calculated from the independently-derived values of $(E_D)_0$ (Sections 2.7 and 2.12) using values of E_{rep} with $R_{m=1}$ determined from the sum of Pauling ionic radii for the various components of the associates. Equation (4) with $E_{rep} = e^2/\epsilon R$ then predicts a significant increase in no-phonon transition energy with decrease in ionic radius of the cation component of the associate (Table 2), discussed further in Section 3.5.

Given the large energy depth just estimated for the isolated 0_p donor, distant DAP recombinations between this donor and the usual shallow acceptors became an obvious possible interpretation[24,25] for the ~ 1.4 eV luminescence mentioned in Section 1.4, at least at low temperatures. Clear proof for this identification was quickly found from the appropriate energy shifts observed in the overall infra-red spectra recorded under *weak* optical excitation between crystals doped with the different acceptors C, Zn and Cd (Figure 13).[39] These spectra show phonon coupling of quite moderate degree, considering the very large ionization energy of the 0_p donor, with a local and band mode (Table 3). However, their form is consistent with the associate spectra (Figure 10), in that the magnitude of this phonon coupling is relatively insensitive to the shallow binding of the hole between the different acceptors (Figure 13).

Yet stronger confirmation of the distant DAP model is furnished by observation of the host of sharp lines which emerge from the high energy tail of these spectra at *high* optical excitation rates (Figure 14). The involvement of 0_p in these spectra was unequivocally demonstrated by their shift to higher energy by ~ 0.71±0.02 meV when $0^{16} \rightarrow 0^{18}$, a shift in the same sense and just slightly larger than for the associates (Table 1). No consequent shift could be detected in the energies of broad phonon replicas resolved in these distant DAP spectra. Once again, it was concluded that the modes listed in Table 3 are not primarily responsible for the no-phonon isotope shift. The form of these multiline spectra depends on the lattice site of the acceptor, exactly as expected if the donor is on the P sub-lattice. In particular, gaps appear in the 0_p–C_p Type I spectrum at the predicted values of the shell number m, namely m = 14, 30, 46 et seq.[2] In addition, the detailed agreement between experiment and prediction for the fine structure in these spectra, caused by the dependence of the exact value of the DAP transition energy on the vector value of their separation R, not just on scalar R as in Equation (6), due to a sensitivity to the exact lattice environment of the DAP, provides very important evidence that both the shallow acceptors and deep donors are in undistorted tetrahedral lattice site positions. Any deviations from this condition should produce significant qualitative differences in the overall form of the fine structure between these infrared DAP spectra and the green DAP spectra involving shallow donors such as S_p and Si_{Ga}, similar to the complications observed for the axial acceptor mentioned in Section 6. No such differences have been detected.[40]

A further important point is the absence of significant red luminescence from GaP:C,0. The large decrease in transition energy of ~ 120 meV between m = 1 DAP in the Type II and Type I spectra, respectively resulting from R = 2.36 Å and R = 3.84 Å in Equation (6), removes any practical

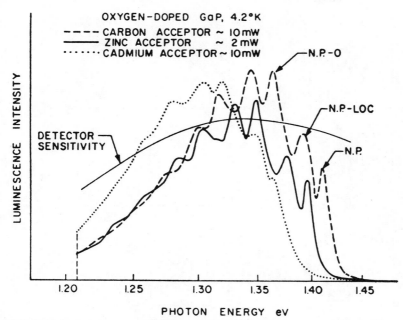

FIGURE 13 The low temperature near infrared distant DAP luminescence spectra from GaP solution-grown single crystals characteristic of doping with O donors and either C, Zn or Cd shallow acceptors. The spectra were recorded under very low excitation intensities to minimize the broadening of the no-phonon components NP caused by the dispersion of DAP energy with pair separation. Two phonons are resolved in the phonon-assisted transitions (Table 3). The spectra are mutually displaced by the indicated energies, identical to those seen in yellow-green luminescence involving the same acceptors and shallow S donors (Dean et al., Ref. 39).

interest in this system for LEDs, since the resulting luminescence overlaps very poorly with the luminosity curve for normal human vision.[1] However, this large reduction in the term $e^2/\epsilon R$ also removes any significant tendency for O_p–C_p pairs to associate at temperatures where the impurities may be adequately mobile in the GaP lattice. A further difficulty, which compounds the effect of the reduction in $e^2/\epsilon R$, is the much poorer diffusivity likely for C compared with Zn, where the interstitial-substitution diffusion mechanism ensures adequate mobility at temperatures as low as 400°C. Indeed, luminescence from the m = 1 and m = 2 DAP has not been seen for the O-C system, presumably because the probability of formation of such close DAP is small for a randomly distributed system when significant association does not occur. By contrast, luminescence from the m = 2 and 3 DAP cannot normally be seen in the O-Zn and O-Cd DAP spectra against the background of strong phono-assisted luminescence

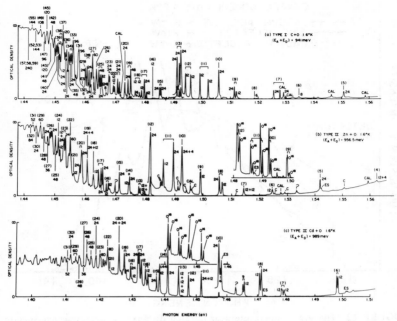

FIGURE 14   Densitometer recordings from photographic plates of the characteristic low temperature infrared luminescence spectra of GaP double-doped with (a) C+O, (b) Zn+O and (c) Cd+O. Two different exposure times and excitation intensities were used in each spectrum to adequately record the relatively weak higher energy DAP no-phonon lines. The lines marked C in spectrum (b) are due to transitions at extraneous C acceptors. CAL are calibration lines. Recombinations from excited DAP states are denoted ES. The bracketed integers are the DAP shell number m, while unbracketed integers denote the number of pair sites within a given shell. Fine structure arises from two or more inequivalent groups of DAP within many shells of given scalar separation R. The insets in spectra (b) and (c) shows the replication of DAP lines due to the 0.71 meV energy increase when some of the O^{16} is replaced by O^{18}. The linewidths are instrumental (Dean et al., Ref. 39).

from the m = 1 DAP (or BE), again because association is insignificant for m ⩾ 2 in these spectra.

Very interesting results have been reported very recently[211] for photo-degraded GaP:Zn,O, which has been subjected to intense excitation under blue-green light from an Ar^+ laser to the power levels indicated in Figure 14Aa. It is clearly seen that for excitation near the carefully chosen temperature of 470°K, the intensity of luminescence from the higher energy m = 1 Zn_{Ga} −Op associate continually decreases in favor of the

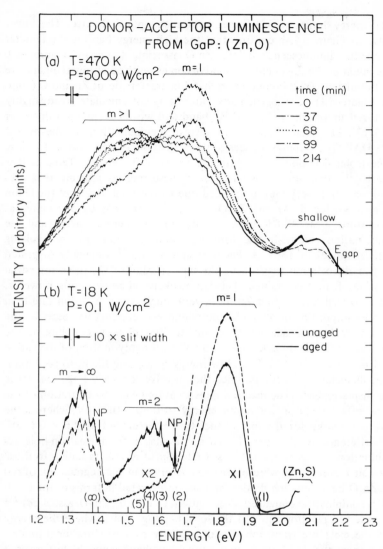

FIGURE 14A Photoluminescence recorded at various temperatures (T) and laser power densities (P) from the heavily photo-excited p regions of liquid phase epitaxial pn junctions of GaP:Zn,O; (a) Shows spectra recorded at the indicated times after the onset of intense photo excitation. (b) Shows spectra recorded at much lower excitation densities and low temperature at the beginning and end of the photo degradation process. Theoretical predictions for the O–Zn DAP no-phonon energies (Figure 15) are shown on the abscissa.

lower energy luminescence under continued optical fluence. The lower
spectra in Figure 14Ab confirm that the high energy band is the familiar
red exciton luminescence of the Zn-O associate (compare Figure 6b),
while much of the lower energy band becomes resolved at low temperature
into the distant DAP luminescence band characteristic of Zn and O (compare Figure 13). The particularly interesting intermediate band, highly
structured unlike the red Zn, O BE band and with a no-phonon line near
1.655 eV, has been shown[211] to involve recombinations at the m = 2
O-Zn DAP. The phonon replicas of this transition are very similar to those
of the isolated O donor in Figure 17, discussed in detail in Table 3. It is
suggested[211] that since the energies of these replicas indicate that substitutional O_P is very weakly bonded due to the small radius of the O ion
(see also Section 5, 4), the Zn-O m = 1 associate may be dissociated at
temperatures near 470°K by a combination of thermal energy and the
recombination energy in the form of a multi-phonon process involving
the local modes in Table 3. Feenstra and McGill[211] originally suggested
that this dissociation may occur through ejection of O_P to an intersitial
site, where it may be mobile. This is not likely on two grounds, however.
First, it is well known that Zn has a very much larger diffusion rate than
group-VI impurities in all III-V compound semiconductors. Second, the
low energy local mode which dominates the phonon coupling at the
Zn-O center (Table 1) is known to involve primarily motion of the cation
substituent (Section 2.2). Such low-energy modes are likely to predominate in dissociation reactions, since they involve relatively large real space
ionic displacements. The reaction rates measured close to saturation of the
optical effect yield an activation energy of 0.60 ± 0.07 eV, whereas the
activation energy for thermal dissociation is estimated as 2.6 ± 0.6 eV.
The difference is of order the energy available from electron-hole recombination at the m = 1 DAP associate. Spectral effects similar to those
shown in Figure 14Aa were previously observed in the degradation of red
GaP:Zn,O light emitting diodes, and also interpreted in terms of the localized recombination-enhanced dissociation (RED) of Zn-O associates.[212]
The observed reaction rates fit the local heating model, where the energy
from the elctronic transition produces highly excited vibrational modes,
better than the local excitation model involving some highly excited
electronic state with weak bonds.[252]

Despite the apparent phenomenological success of the explanation of the
effects in Figure 14A in terms of a local heating model of RED,[218] resulting in dissociation through a chosen reaction co-ordinate, some difficulties
remain unresolved. First, it is not easy to understand how a reaction involving the successful conversion of 316 local photons (1.9 eV shared
amongst ~ 6 meV local modes) into an inelastic displacement of a local

reaction co-ordinate can be efficient. Symmetry also presents problems, since the local vibrational motion involves a $<111>$ mode, whereas the total displacement in the transformation of an m = 1 Type II DAP to m = 2 is of $<110>$ type. The first problem is slightly alleviated if a much larger local phonon is considered, perhaps the one responsible for the O-induced isotope shift in the no-phonon line. We have seen that this mode is not clearly identified in the sideband spectrum of the m = 1 DAP associate (Section 2.2). However, comparison with the phonon sidebands in both the m = 2 DAP center[211] and in the optical spectra associated with the isolated O_p donor (Table 3) suggests an energy close to 25 meV. Even so, the absorption of the entire transition energy into this mode results in the production of 76 phonons in the local heating mechanism. Somehow, this energy must be efficiently coupled into the ionic displacement, with no effective energy loss other than that required to overcome the reaction barrier. There is currently no basic general understanding as to how this might occur.[219]

The reaction in Figure 14A may proceed by the impulsive relaxation of the small O ion through the triad of Ga atoms to which it remains bonded and into the tetrahedral interstitial site remote from the Zn_{Ga} to which it was bonded in the m = 1 DAP associate. If this mechanism is valid, the Zn-O associate responsible for the 1.655 eV photoluminescence line would not be the true m = 2 Type II DAP associated of $<3\overline{1}\overline{1}>$ symmetry. Rather, it would retain the $<111>$ symmetry of the m = 1 associate, but with a greatly increased value of R. The new Zn-O separattion would be comparable with that of the true m = 2 DAP state and therefore would produce a closely similar optical transition energy througn Equation 6. It may well be that the transformation between these two $<111>$ symmetry states proceeds via the localized excitation model, in which use is made of a lower barrier to the reaction in an excited electronic state of the center, rather than the vibronic mechanism via local heating. An obvious trigger for this reaction is the effect on the m = 1 DAP of localization of the additional electronic charge at the O_p member of the associate on capture of the extra electron at the reaction temperature of 470°K. We show in Section 2.2 and discuss in greater detail in Section 5.8 that the extra electronic charge produces a considerable weakening of the nearest neighbor force constants. In the present case, the weakening will be mainly concentrated on the bond to the Zn_{Ga} nearest neighbor, since this ion already holds a negative charge with respect to the GaP lattice. In this way, the RED might be included by the electrostatic effect of the additional charge. It is clearly necessary to establish the symmetry of the center responsible for the 1.655 eV˙transition, to distinguish between these two possible models. This may be achieved by the Zeeman effect,[76] or by uniaxial stress,[30] as we describe for the m = 1 associate in Section 2.3. In this case, the Zeeman experiment will be greatly hampered by the remarkable excess width of the 1.655 eV "no-phonon" line, ~ 2 meV, huge compared with the m = 1 or m $\geqslant$ 4 DAP centers, $\leqslant 0.1$ meV. This excess width strongly suggests that the 1.655 eV line is not a normal member of the DAP series. It is more consistent with the alternative model advanced above, since the exact position of the "O_I" may be much more subject to perturbations than a normal O_p center directly bonded to four nearest neighbors. Any model for the 1.655 eV line must certainly account for this excess width, ignored so far.[211] Present attempts to understand the energy of the ~ 6 meV in band resonance mode energy of the $Zn_{Ga}-O_p$ associate in terms of the Green's function theory which seems quite successful for the modes of isolated O_p (Section 5.8) has revealed the need for very small O-Ga and Zn-O force constants.[253]

It should also be noted that the technical difficulties for the local heating model of the RED process in providing an account of the data in Figure 14A is probably not unique. The difficulties are perhaps rather more apparent in the present case only because we know much more about this particular center than is the case for the majority of centers which have been found to exhibit this type of process.[218]

The transition oscillator strength of the DAP process can be described through the transition probability W(R)

$$W(R) = W_m \exp(-R/a) \tag{7}$$

where a is a characteristic length, normally one half of the Bohr radius of the least tightly bound electronic particle in the case where the binding energies are unequal.[2] This is certainly true for the O-C system. However, careful comparison with the kinetics of the S-G DAP luminescence has shown that the appropriate value of a is also half the Bohr radius of the shallow C acceptor, about 9.1 Å.[47] The parameter W_m, which represents the DAP transition rate for completely overlapping centers is found to be ~ 3 times larger for O-C compared with S-C pairs,[47] despite the reduction of nearly one order expected from the usual ν^3 factor in the transition rate for a dipole-allowed process. This indicates the extent to which binding to the very deep O_p donor augments the probability for indirect transitions even relative to a shallow P-site donor like S, which is already considerably favored in this respect by the form of the band structure in GaP.[32] We shall see below evidence that the ground state of the donor is dominated by admixture with valence band states, which can account for this large difference in recombination rates. The value of W_m estimated in this way for both O-C and O-Zn DAP corresponds to a decay time of ~ 660 nsec,[47] much larger than directly measured for the Zn-O bound exciton, ~ 100 nsec.[29] This large discrepancy simply means that Equation (7) is inappropriate for transitions in m = 1 DAP, though it can be used to compare rates for transitions through *distant* donors as we have done here.

2.7 Derivation of the O-donor Ionization Energy From The IR DAP Spectra.

The energies of these DAP spectra are well described by Equation (6) over a very large range of R, with negligible contribution from the term f(R) because of the very small polarizability of the very deep O_P donor (Figure 15). This agreement extends even to the nearest neighbor pairs, the exciton-binding associate described in Section 2.4, since $e^2/\epsilon R_{m=1} = 550$ meV if $\epsilon = 11.1$, whereas the no-phonon lines of the Cd-O and Zn-O associate lie 553 and ~ 600 meV above the R = ∞ transition energies obtained from Figure 15 (Table 2). Refinements in the analysis, considering the multipole nature of the DA interaction in the final state of the DAP luminescence transition, have led to re-assignment for a few of the shell subcomponents, particularly for the O-Cd spectrum, and an improved overall fit at low-to-moderate m values.[45,46] Vink et al[41] have studied the envelope of the Type I O-C DAP spectrum to remarkably large values of m and R, respectively ~ 330 and ~ 70 Å (Figure 16). Although individual pair lines cannot be resolved beyond m ~ 55, R ~ 30 Å (Figures 13 and 14), significant variations in the overall envelope of DAP intensities can still be observed, and accounted for in the main. These results have been used to evaluate f(R) of Equation (6) over the different ranges of R, and some origins of these additional small terms have been identified.[41] The resulting analysis provided accurate values of $\epsilon_{1.6°K} = 11.02 \pm 0.05$ and $h\nu_{R=\infty} = 1.3966$ eV. Since $E_g = 2.350 \pm 0.001$ eV,[37] the energy sum $(E_D)_O + (E_A)_C = 953.4 \pm 1$ meV. An independent value of $(E_A)_C$ can be obtained from the value of $(E_A)_{Zn} = 70.0 \pm 0.5$ meV obtained from direct infrared absorption,[42] from analysis of NN associate$-$Zn distant pair recombinations[43] and from electrical activation energies extrapolated to the infinite dilution limit.[44] The energy difference $(E_A)_{Zn}-(E_A)_C = 15.3$ meV is accurately known from energy differences in DAP spectra involving a common donor, such as in Figures 14 and 15.[18] Thus we obtain an accurate value of $(E_D)_O = 898.7 \pm 1$ meV.

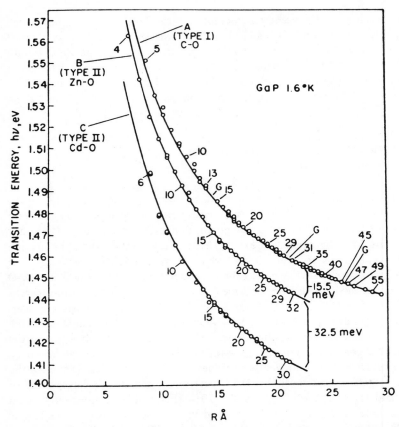

FIGURE 15 The energies of discrete infrared DAP no-phonon lines associated with deep O donors and the indicated shallow acceptors versus the DAP separations R. The experimental points lie very close to Equation (6) with the energy $(E_D)_O$ = 898 meV and $(E_A)_C$ = 54.5 meV, $(E_A)_{Zn}$ = 70 meV and $(E_A)_{Cd}$ = 102.5 meV. The shell numbers of some lines are indicated. Gaps predicted for the Type I spectrum (both impurities on the same sublattice) are indicated as G. (Dean et al., Ref. 39).

2.8 Exited States of the O Donor – Anticipated Form

The large value of $(E_D)_O$ in GaP was viewed initially simply as a large extension from the behavior of the more nearly effective-mass-like donors such as S_P. The effective triple degeneracy of the ground states of such donors is lifted by the valley-orbit interaction to give a d-like doublet $1s(E)$ and an s-like singlet $1s(A_1)$. The latter is constructed by a symmetric combination from the three equivalent conduction band minima,[48] and

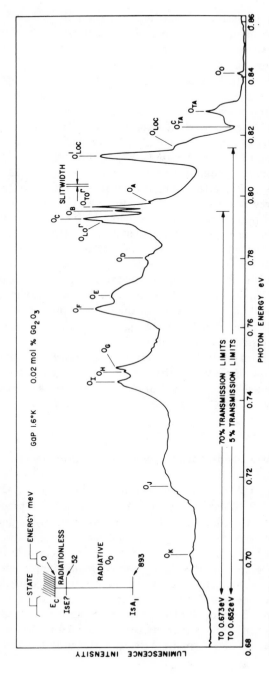

FIGURE 16 Infra-red low temperature photoluminescence characteristic of O-doped GaP recorded with a PbS detector. Despite some recent controversy, the balance of evidence associates this spectrum with electron capture by the ionized donor through the shallow excited state shown as $1s(E)$ in the inset, a transition within the single electron state of O_P. The valley-orbit splitting of this deep donor is $1s(E) \rightarrow 1s(A_1)$. The weak no-phonon line is O_O. The phonons selected in the strong phonon sideband are identified in Table 3. Transmission limits are indicated for the bandpass filter used in the luminescence excitation spectra of Figure 19 (Dean and Henry, Ref. 53).

221

alone exhibits an antinodal wave-function at the impurity core.[56] This strong core overlap leads to a large increase in the ionization energy of the $1s(A_1)$ state for donor impurities with a strongly attractive short range potential for electrons, as might be anticipated for a strongly electronegative substituent ion such as O_P. The effects of inter-valley scattering in the short range impurity potential are mediated by the repulsive effect of overlap of Bloch functions from different valleys and by the inter-valley kinetic energy term. A stable, relatively shallow bound state results from the interplay of these factors if the central cell potential is not too large, as for an isocoric impurity such as S_P in GaP.[49] In this case, the $1s(A_1) \rightarrow 1s(E)$ 'valley-orbit' splitting is known to be ~ 53.4 meV,[50] with the $1s(E)$ state relatively close to the effective mass estimate for the donor ionization energy.[51] However, the description of this valley orbit splitting is outside the scope of normal effective mass theory.[49] It was initially considered that this behavior might be extrapolated to the case of O_p in GaP, whereby the d-like character is the $1s(E)$ state would ensure little sensitivity to the central cell while the energy of the $1s(A_1)$ state became greatly enhanced by the strong attractive central cell potential of the O atom, characteristic of members of the first row of the periodic table. This view seemed well able to account for the detailed properties of the ground and excited states of the isolated O_p donor, which we now describe. However, such strong deviations from the effect mass theory should bring hybridization with the valence bands and significant distortions of local charge. We show in Section 7 that a more recent molecular-orbital description gives a much better account of such remarkably deep impurity states, where E_D is no longer negligible compared with E_G of the host semiconductor.

2.9 Electron Capture at the Isolated O Donor—A Radiative Process

On the view that $(E_D)_O^{1s(E)}$ might remain near 50 meV while $(E_D)_O^{1s(A_1)}$ increases to nearly 0.9 eV, an energy gap ΔE of ~ 17 $\hbar\omega_{LO}$ is predicted between these states, assuming that $1s(E)$ is the most tightly bound excited state. The optical spectra in Figure 13 indicate a Huang-Rhys electron-phonon coupling parameter S of only $S \sim 2$ in the electronic transition in which an electron in this $1s(A_1)$ state becomes annihilated. It therefore seemed most unlikely that the transition rate for multiphonon recombinations could be significantly competitive when an electron is captured by this ionized donor. Initial capture proceeds through the large number of shallow excited states characteristic of the long range Coulomb potential, which are effective trapping states at sufficiently low temperatures.[52] However, there is a bottleneck at the final stage of capture, the $1s(E) \rightarrow 1s(A_1)$ de-excitation. No Auger process can occur in a dilutely-doped sample at low temperatures. The only available mechanism involves a radiative transition, whose temperature-independent transition rate τ_R^{-1} may be described by the general form

$$\tau_R^{-1} = C \exp\left[-\gamma(\Delta E - S\hbar\omega)/\hbar\omega\right], \qquad (8)$$

where C may be $\sim 10^9$ sec^{-1} and the factor γ can be evaluated exactly if the dominant lattice frequency ω is independent of the electronic state, when $\gamma = \ln(\Delta E/S\hbar\omega)-(\Delta E - S\hbar\omega/\Delta E)$. The existence of the no-phonon isotope shift (Table 3) and the data in Figure 34 show that this approximation has limited validity for O_P in GaP. Thus, it was predicted that electron capture by ionized O_P donors should be strongly radiative. This strong capture luminescence was quickly found[53] very close to the predicted energy and with the expected spectral form (Figure 16). The no-phonon line O_o is very weak, consistent with an $E \rightarrow A$ forbidden transition. However, it is possible to observe a *decrease* in transition energy on isotope substitution $O^{16} \rightarrow O^{18}$, reported as -0.67 ± 0.05 meV in the early work[53] and -0.72 ± 0.02 meV more recently[141] from spectra obtained with a much more sensitive Ge photodetector (Section (5)). This shift is equal and opposite to that observed in the DPA luminescence, just as expected if the capture luminescence is its pre-cursor for the one-electron state of O and if the phonon coupling observed in these transitions is essentially independent of whether they involve an initial electron state very diffusely localized at the O donor or a final state in which the electron has been transferred to a distant acceptor. Various local phonon modes appear in the sidebands of this luminescence (Figure 17). These are identifiable on the basis of a comparison with the known form of the phonon density of states in GaP, mirrored in the phonon sidebands of the BE luminescence of the N isoelectronic trap,[3] and in several cases, through the spectral shifts they exhibit an isotope substitution (Table 3). The general expression for the change in local mode energy $\Delta \hbar\omega_{loc} > \hbar\omega_{LO}$ is for an harmonic oscillator,

$$\Delta \hbar\omega_{loc} = -\frac{1}{2} \omega_{loc} \frac{\Delta M}{M}, \qquad (9)$$

where $\frac{\Delta M}{M}$ is the change of reduced mass due to the O substitution. Use of Equation (9) assumes that the replica O_{loc} in the luminescence spectrum (Figure 17) involves a vibration in which most of the kinetic energy is contained in motion of the O atom (Table 3). This is surprising, since this

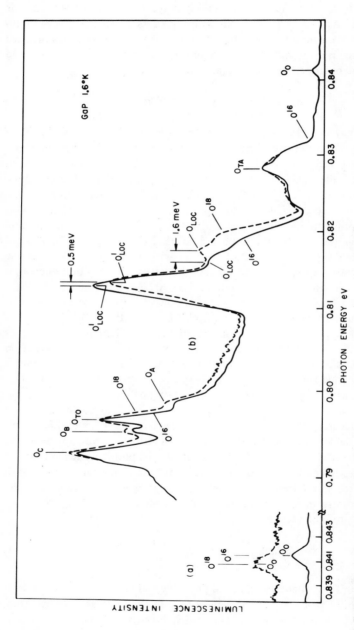

FIGURE 17 Low temperature infra-red electron capture luminescence associated with the deep O donor in GaP compared for O^{16} and for O^{18}. Part (a) shows the shift of ~ 0.7 meV to lower energy of the no-phonon line O_O due to the isotropic substitution $O^{16} \rightarrow O^{18}$. The dashed spectrum for O^{18}-doped GaP in (b) has been shifted relative to the indicated energy scale to superpose the no-phonon lines, so as to directly reveal the O isotope-related changes in the vibronic spectrum (Dean and Henry, Ref. 53).

224

TABLE 3

PHONON ENERGIES AND ISOTOPE SHIFTS OBSERVED IN O_P–RELATED SPECTRA

Technique \ Feature	Electron Capture Luminescence(53)a	PLE of Electron Capture Luminescence (53)	DAP Luminescence (39)	DAP Luminescence and PLE(72)	Configurational Co-ordinate Analysis 0^0	Configurational Co-ordinate Analysis O^-(119)d
Label and Phonon Energy (meV) (for O^{16}) and Isotope Shift $O^{16} \rightarrow O^{18}$(meV)	O_{TA} 13.1 O_{TA}^C 18.0 O_{LOC} 24.7, −1.6 O_{LOC} 28.4, −0.5 O_A 43.0 O_{TO}^{Γ} 44.8 O_B 46.1 O_C 48.7 O_{LO}^{Γ} 49.8	8.7 17.4 $\hbar\omega_B$ 46.4 $\hbar\omega_C$ 48.8	19.5 ± 1 47.0 ± 0.5	19 ± 1, 1.65 ± 0.15 b 48 ± 1, 1.1 ± 0.1 b	2 $S\hbar\omega_1$ = 170 (21) $\hbar\omega_1$ = 20 $S\hbar\omega_1$ = 90 (119) $\hbar\omega_2$ = 0.78 $\hbar\omega_1^c$	$S\hbar\omega_1$ = 1.29 eV $S\hbar\omega_2$ = 0.45 eV ω_2 = 0.65 ω_1^c
No – Phonon Isotope Shift $O^{16} \rightarrow O^{18}$ (meV)	−0.72 ± 0.02	∼ −0.7	+ 0.71 ± 0.02			

a Multiphonon involve ∼ 47.5 meV optical mode

b Huang–Rhys parameter S

c $\hbar\omega_1$ and $\hbar\omega_2$ are defined in Figs 34 and 42

d over–estimated by factor of ∼ 2 according to refs (131) and (140)

225

mode is embedded in the continuum of (longitudinal acoustic) host lattice modes. The more prominent replica O'_{loc} involves a mode with a much greater proportion of motion in the host atoms (Table 3). We shall see in Section 6 that a true local mode, defined as one where $\hbar\omega_{loc} > \hbar\omega_{LO}$ of the host lattice, is expected with energy near 60 meV. However, no trace of such a mode was found in the initial work, where all phonons of energy $\geqslant \hbar\omega_{LO} \sim 50$ meV were attributed to combination bands involving a phonon of energy ~ 47.5 meV, nor in more recent work. Some of the phonons were resolved much more clearly with the improved photo-detector, particularly in the range $TO^\Gamma \rightarrow LO^\Gamma$ near 0.795 eV in Figure 16. Electronic reactions involving the isolated O_P donor are summarized in Table 6 in the Appendix at the end of the article.

2.10 Uniaxial Stress Effects on Electron Capture Luminescence

Attempts were made to confirm the suggested identification of the excited state of the capture luminescence through measurements of the effect of uniaxial stress. Limitations of resolution and signal to noise in the early work[53] prevented observation of stress-induced splittings in the weak no-phonon line, and also ruled out any hope for Zeeman studies of this spectrum in 1968. Use of the broader but much more intense phonon replicas O_C and O'_{loc} (Figure 17) precluded the spectral resolution of stress splittings ΔE_χ at stresses sufficiently low that $\Delta E_\chi \leqslant kT$, so that only line *shifts* rather than *splittings* could be seen. These shift rates were much greater for stress $\chi \parallel \langle 100 \rangle$ than for stress $\chi \parallel \langle 111 \rangle$. The shift rate in the latter case, $\frac{dE}{d\chi} = -0.9 \times 10^{-6}$ eV cm^2 kg^{-1}, is approximately consistent with the hydrostatic stress coefficient of the indirect energy gap in GaP[54] allowing for the factor of 3 reduction to the hydrostatic component in a uniaxial stress measurement. This suggests that the ground state of the deep O donor is appreciably hybridized with the valence band. The excess shift rate for $\chi \parallel \langle 100 \rangle$ was attributed to the *splitting* expected for the 1s(E) excited state for this stress direction, which makes the $\langle 100 \rangle$ conduction band minima energetically inequivalent. The splitting of the 1s(E) state deduced from this interpretation was $(5.7 \pm 0.7) \times 10^{-6}$ eV cm^2 kg^{-1}, which should be $\frac{2}{3}$ of the splitting between the conduction band valleys for small stress in this direction.[55] The resulting prediction for the splitting coefficient of the conduction-band valleys is in good agreement with the value directly measured from piezo-absorption measurement on the indirect absorption edge,[54] $(8.6 \pm 0.8) \times 10^{-6}$ eV cm^2 kg^{-1}.

These results strongly support the suggested identification of the luminescence excited state with the 1s(E) donor state, rather than for example the $2s(A_1)$ state strongly lowered by central cell effects, the most plausible alternative for an s-like excited state of lowest transition energy (Section 7.1). The $2s(A_1)$ state should not split for any stress direction when the general stress splitting of the conduction band, ~ 10 meV for typical stress values in the experiment, is small compared with the valley-orbit splitting of the 2s state, which must be assumed no less than ~ 34 meV for this identification to be possible.[61] The binding energy of the 1s(E) state derived from this identification is ~ 58 meV, still close to the effective mass estimate for the 1s donor ground state disregarding central cell effects, but a few meV deeper than the 1s(E) states for much shallower donors such as S_p (Table 4). These results for the very deep O_p donor confirm the predicted insensitivity of all donor states other than $1s(A_1)$ to the central-cell potential of the donor impurity.[56] More detailed uniaxial stress data for the no-phonon component O_o of the luminsescence spectrum are discussed in Section 5, together with Zeeman measurements made possible with the improved infrared detectors now available. We simply note here that these results confirm the origin of the O_p donor states described above.

TABLE 4

ANALYSIS OF EXCITED STATES OF Si, O AND S DONORS

Donor / Binding Energy (meV)	Si$_{Ga}$ (59)	O$_p$ (53)	S$_p$ (61)	Theory (61)
$4p_{\pm}$	3.8	–	–	5.0
$3p_{\pm}$	6.2	7.0	6.9	7.4
$4f_o$	7.9	–	8.2	7.8
$2p_{\pm}$	10.5 [a]	10.5 [a]	10.5 [a]	10.5
$4p_o$	–	14.3	12.0	12.4
$3p_o$	18.6	20.0	17.9	18.9
$2p_o$	35.9 (37)	33.9	35.8	35.6
$3d_o (A_1)$	–	–	12.4	10.3 [b]
$3s (A_1)$	–	–	17.1	15.6 [b]
$2s (A_1)$	–	–	26.0	24.4 [b]
$1s (E)$	–	56.6	54.8	~ 62 [b]

a Fitted to theoretical value for $2p_{\pm}$ state in ref (61)
b These theoretical values are appropriate to ns (E) states

2.11 Photoluminescence Excitation Spectra of the Electron Capture Luminescence

Further strong support for the identification of the O_P donor excited state responsible for the electron capture luminescence was obtained from optical absorption. Measurements of the O donor photoexcitation through direct optical absorption[87] proved too insensitive to detect the anticipated fine structure near 0.9 eV. The overall form of this absorption is best determined through the excitation spectrum for infrared quenching of the free hole to bound electron luminescence at the O_P donor (Figure 18a).[72] These spectra can be obtained with high sensitivity only $> 60°$K, while the complementary photocapacitance technique (Section 2.1, (Figure 33) can be used only above $\sim 120°$K. Under these conditions, the pure electronic transitions to anticipated excited states are unresolvable and appear only as a weak tail to the photoionization spectrum, shown below 0.9 eV in the inset to Figure 18a. This very broad absorption spectrum, with peak energy ~ 0.4 eV above threshold and very little strength in the transitions to shallow, nearly effective mass-like excited states, is very unlike such spectra for hydrogenic donors. However, it is quite as expected for transitions from such a deep non-effective mass-like ground state as we have demonstrated for the O_P donor in GaP. One theoretical model for this process treats the effective ground state binding potential as a δ-function, and is valid for transitions to a conduction band which can be reasonably approximated by a single isotropic effective mass.[58] On this well known model, the peak absorption occurs near $2E_D$. However, the model neglects the additional structure and continuum enhancement near E_D due to the Coulomb binding universally present for donor centers. The spectral form including the effects of such shallow states at a deep center has been discussed by Messenger and Blakemore.[57] The complementary transition with threshold at $h\nu = E_G - (E_D)_O$ is shown in Figure 18b, together with absorption due to the An-O center discussed in Section 2.4.

The structured spectra in Figure 19 were recorded by the indirect PLE technique, in which the detected signal was the O capture luminescence recorded through a custom-made interference filter whose specification is included in Figure 16. A very important point is that no significant PLE component appears at O_o, the position of the luminescence no-phonon PL line, on a scale of sensitivity where strong sharp line structure appears at > 20 meV to higher energy. This observation adds considerable weight to the identification in Section 2.10 of the O_o excited state as 1s(E), rather than $2p_o$, the deepest of the p excited states. It also suggests that the 1s(E) → 1s(A) selection rule remains quite strong for this very deep donor.

This, in turn, argues against the presence of any significant de-stabilization of the O_P donor from the position corresponding to the full tetrahedral symmetry of the P lattice site, supporting the conclusions already obtained in Section 2.6 from the detailed form of the structured DAP spectra. The

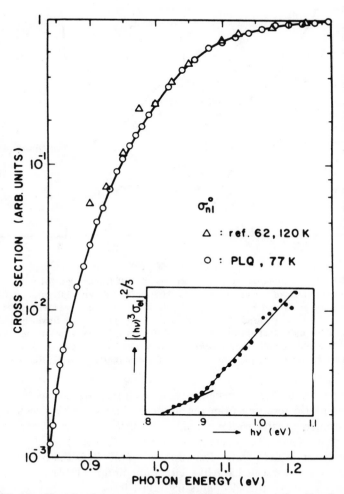

FIGURE 18a Photoluminescence quenching (PLQ) spectrum for the O-related infrared DAP luminescence, which reveals σ°_{nl} (hv), measured at 77°K in single crystals of GaP:O and compared with photocapacitance measurements of Henry et al. (Ref. 119). The insert shows the electronic component $\sigma^{\circ}_{el,nl}$ resulting from an approximate deconvolution, plotted so as to linearize the Lucovsky theoretical form which results from Equation (17) with f = 1. (Monemar and Samuelson, Ref. 72).

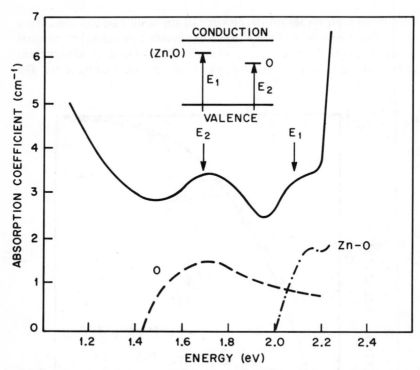

FIGURE 18b Wide-range optical absorption at 300°K recorded for a GaP (Zn,O) solution grown single crystal doped with 0.07 mole % Zn and 0.02 mole % Ga_2O_3. The maxima at E_1 and E_2 correspond to the absorption processes with Zn–O associates and isolated O donors indicated in the inset. The dashed curves indicate the shapes of the individual absorption curves determined from excitation spectra of the appropriate photoluminescence, with arbitrary relative sensitivities. Both spectral shapes are dominated by phonon assisted transitions. (Dishman et al., Ref. 87).

much greater oscillator strength of the no-phonon transitions in the PLE spectrum, also resulting from the influence of the parity selection rule, is confirmed by the weakness and very different form of the phonon coupling in Figure 20 compared with Figure 16 (Table 3).

The donor excited states responsible for the PLE sharp structure in Figure 19 are all assigned as p like, either np_o or $np_\pm$ (Table 4). The p_o–$p_\pm$ splittings are large for all donors in GaP, a feature of the large effective mass anisotropy, itself a consequence of the "camels-back" structure of the conduction band minima about the zone boundary point X in the Brillouin zone.[51] This structure produces unusually large value of the longitudinal electron effective mass parameter m_ϱ^*,[59] particularly relevant

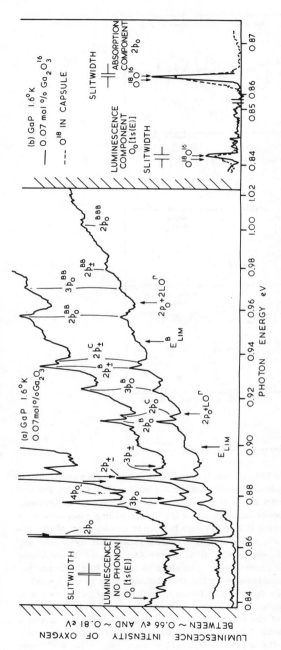

FIGURE 19 (a) Low temperature excitation spectra of the infrared photoluminescence associated with O in GaP, recorded with the detector filter indicated in Figure 16. The strongest peaks represent no-phonon excitations from the deep $1s(A_1)$ ground state to the indicated shallow p-like excited states split by the strong mass anisotropy of the conduction band in GaP. The downward pointing arrows are obtained from estimates of the excited state energies given from a rudimentary description of the effect of the camel's back conduction band structure. More precise estimates of the binding energies are included in Table 4. Peaks with superscripts B, C, BB are attributed to phonon-assisted transitions (Table 3). (b) Shows that the $1s(A_1) \rightarrow 2p_0$ absorption components exhibits a shift similar to the $1s(E) \rightarrow 1s(A_1)$ luminescence component in a crystal containing O^{18} as well as O^{16} (dashed) compared with O^{16} alone (full-line). (Dean and Henry, Ref. 53, reinterpreted by Carter et al., Ref. 59).

231

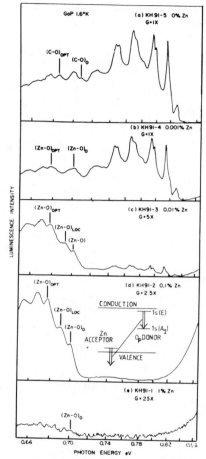

FIGURE 20 Unsaturated low temperature infrared photoluminescence spectra from a GaP single crystal grown from Ga solution containing 0.02 mole % Ga_2O_3 and the indicated at % of Zn. Spectrum (a) is mainly due to the electron capture luminescence in Figure 16, although some evidence of the DAP luminescence associated with the deep O donor and C acceptor (Figure 13) can be seen in second order of the diffraction grating below ∼ 0.7 eV. Spectrum (b) is similar, though the weak DAP are now dominantly O–Zn and the low energy tail of the red Zn–O related luminescence appears near 0.86 eV, again in second order. The electron capture luminescence becomes progressively weaker compared with the DAP luminescence in spectra (b) → (d), as the concentration of Zn acceptors increases from 0.15 through 0.5 to 1.5 × 10^{18} cm^{-3}. Finally, at the highest Zn acceptor concentration ∼ 5 × 10^{18} cm^{-3} in spectrum (e), the intensities of the infrared DAP luminescence are the red luminescence are also drastically reduced. The inset in (d) shows the inter-centre Auger process believed responsible for the capture luminescence quenching. (Modified from Dean and Henry, Ref. 53).

for excited donor states with binding energies $\leqslant 10$ meV since the height of the "camels-back" at X is ~ 3 meV.[51,60]

The analysis of the donor excited states given in Figure 19 and Table 4 has been revised from the original[53] to take account of the "camels-back" effects.[51,59,60,61] The assignments are substantially unaltered, however, except that it is now recognized that there is no need to invoke valley-orbit splittings for the p-like excited states. The original work on the electron, capture luminescence at the O donor pre-dated even the early experimental work on the excited states of shallow donors in GaP.[62] The analysis in the latter work is now known to be inadequate. The correspondence between p-like excited states of the deep O_P donor and much shallower donors such as S_P is now recognized as much closer[59] than appeared to be possible according to effective mass estimates current in 1969, when a relatively low value of m_e^* seemed appropriate.[53]

The strong central-cell effects which dominate the binding of the $1s(A_1)$ donor ground state therefore have negligible effect on the *energies* of the p-like donor excited states. By contrast, the *probabilities* of transitions to these relatively very diffuse excited states from the $1s(A_1)$ ground state are influenced very *strongly* by the tight binding of the electron in the donor ground state. The absence in the $2p_0$ wave-function of the azimuthal nodes characteristic of the $2p_\pm$ wave-function ensures much better overlap between the $2p_0$ state and the very compact $1s(A_1)$ state. Faulkner calculated that the intensity ratio $[I_{1s(A_1) \to 2p_0}]/[I_{1s(A_1) \to 2p\pm}]$ increases from ~ 0.32 for an effective mass donor in Si, or GaP, to ~ 0.85 for a deep donor such as O_P in GaP, and could reach a maximum value of ~ 1.3. The intensity ratio of the corresponding lines in Figure 19 is ~ 1.0, whereas the corresponding value for S_P, the most effective-mass-like donor for which results are available in GaP[42,59] is ~ 0.1.

2.12 Derivation of the O Donor Ionization Energy from the Excitation Spectra

Estimates of $(E_D)_O$ can be obtained both from the threshold E_O^I of a broad component observed near 0.90 eV, which appears to represent the onset of bound to continuum excitations, as well as from the energy of the absorption line attributed to $1s(A_1) \to 2p_\pm$ internal excitations (Figure 19). These two methods respectively produce $(E_D)_O = 897$ meV and 898.0 ± 1 meV, if the most recent effective mass estimate of $(E_D)^{2p\pm} = 10.5 \pm 1$ meV[61] is used in the latter case. This concordant value is in excellent agreement with the quite independent result from the DAP spectra, $(E_D)_O = 898.7 \pm 1$ meV (Section 2.7). The closeness of this agreement, to better than 1 part in 900, provides just *one* important reason to doubt the re-assignment of the capture luminescence to transitions involving the *second* electron bound at the O_p donor, which was suggested in some later work[63] (Section 5).

2.13 Intensity Quenching of Electron Capture PL and PLE Spectra by Co-doping

Three further remarkable results were obtained in the original work on the capture luminescence.[53] These have been held by some[63] to cast doubt on the assignment to transitions involving the single electron bound to the O_p donor, however, we now show that perfectly straightforward interpretations are available within the initial interpretation. Two of them were advanced in 1968, but a third depends upon knowledge not available at that date.

The first observation was the strong quenching of the intensity of the electron capture luminescence as the concentration of shallow acceptor co-dopants was increased from the 'background' contamination level, typically in the $10^{15} - 10^{16}$ cm^{-3} range, to $\geq 10^{18}$ cm^{-3} (Figure 20). The remarkable feature is the continued presence of the DAP luminescence at acceptor concentrations $\sim 2 \times 10^{18}$ cm^{-3}, even though the electron capture luminescence has been essentially completely quenched away. The existence of the DAP luminescence confirms that initially compensated O_p donors are still being photo-neutralized in this p-type material, which is doped well below the concentration of Zn acceptors at which a phase transition to the metallic conduction state occurs, near $N_{Zn} \sim 2 \times 10^{19}$ cm^{-3}.[64] The most probable explanation, given the strong evidence for the validity of the interpretation we have offered for all three luminescence bands appearing in Figure 20, is that the capture luminescence is being quenched by the inter-center Auger process also shown schematically in Figure 20. The electron-hole overlap integral which is the kernel of this Auger process[65] will be of similar type to that responsible for the DAP transition itself. However, the magnitude of this overlap, described by Equation 7, will be appreciably enhanced for the Auger process by the fact that the interaction in the initial state of the transition involves the relatively diffuse 1s(E) donor state, as well as the shallow acceptor 1s(T_2) ground state. Study of the multiline DAP transitions indicates that the acceptor ground state remains substantially discrete up to $\sim 2 \times 10^{18}$ cm^{-3}, as calculations predict. However, considerable hole delocalization due to percolation processes is expected at the highest Zn concentration in

Figure 20, $\sim 5 \times 10^{18}$ cm^{-3}. The additional quenching of the O-Zn DAP luminescence observed at this concentration is attributed to hopping of the bound holes between adjacent Zn sites, which greatly enhances the probability that the photo-created holes will migrate to luminescence quenching centers involving more tightly bound hole states before DAP luminescence can occur. This type of concentration quenching mechanism for luminescence appears to have wide validity in semiconductors and in many classes of phosphors.

The second remarkable effect is the observation[53] that the DAP luminescence saturates with increasing optical excitation intensity much faster than the electron capture luminescence at low acceptor concentrations. Saturation of the distant DAP luminescence, in itself, is no surprise because of the inherently low transition probability.[2] However, it might have been expected that this would then provide a bottle-neck for the electron capture luminescence at the O donor. The most plausible generic explanation for this discrepancy is the existence of a bound state for a second electron, whereby release of the first electron might occur on recombination with a subsequently captured hole. This interpretation was initially offered[53] in terms of the D BE state expected from an extrapolation of the behavior of shallow donors in GaP.[66] Auger recombination should provide an even more dominant channel for exciton binding at the deep O donor, because of the strong overlap of the tightly bound electrons. We shall see that the plausibility of this type of interpretation is greatly enhanced by the later observation of the O$^-$ state through capacitance spectroscopy (Section 4.4). The existence of this intra O$^-$ state Auger process has been confirmed from recent remarkable ODMR results (Section 5.2) and from recent interpretations of additional optical spectra associated with the O donor in GaP (Section 5.4).

The third phenomenon appears in two forms, related to the 'quenching' effects of additional shallow donors on both the PL and PLE spectra of the electron capture process. The capture PL is no longer observed in crystals containing concentrations of shallow donors such as Te greater than $\sim 4 \times 10^{17}$ cm^{-3}.[53] Conditions for observation of the PLE structure in Figure 20 were even more critical. A number of crystals of GaP:O were co-doped with shallow donors to a variety of concentrations, designed only to ensure that the deep O donors were neutral since no bias light was used in these PLE experiments and $(E_D)_O < E_g/2$ for GaP. The majority of these crystals failed to give any PLE response at the energies shown in Figure 19, although all gave quite strong electron capture PL under the usual above-gap photoexcitation. This effect was not understood at the time, and therefore was not reported in 1968. However, retrospective consideration indicates that the very few crystals which showed good PLE spectra were those which would have been weakly p-type in the absence of doping by O; that is the concentration ratio of O and shallow donors was > 1. Any temptation to conclude from these results that the spectrum in Figure 19 has no relation to O is removed by the observation in the lowest energy no-phonon line of an isotope shift of the same order as observed for the O_o line in the capture luminescence and in the DAP luminescence, when $O^{16} \rightarrow O^{18}$. Both of these observations relating to the effects of the equilibrium Fermi level are readily interpretable, given the current knowledge of a stable O^- state whose binding energy appreciably exceeds that characteristic of shallow donors in GaP, ie is $\gg 0.1$ eV, Section 4.4. If the concentration ratio between O plus shallow acceptors and the shallow donors is < 1, readily attainable with modest additions to the background concentration of shallow donors given the very low solubility limit of O donors, $\leqslant 5 \times 10^{16}$ cm^{-3} (Section 3.2), then all the O donors will be in the O^- state and no PLE of the O° state can be observed. The electron capture PL may still be observed under photoexcitation of moderately n-type crystals. The criterion for this may be that the concentration ratio between shallow and O donors is still sufficiently small that the recombinations of the photoexcited minority holes is not so completely dominated by electrons on shallow donors that the O centers remain in the O^- state. Evidently, this condition becomes invalidated at shallow donor concentrations $\geqslant 4 \times 10^{17}$ cm^{-3}, at least for low acceptor concentrations.

2.14 Further Consequences of the Deep Donor State Due to O_P

The observations described above provide very comprehensive and coherent support for the existence of a deep donor-like state of binding energy ~ 0.9 eV related to the one-electron state of O_P. It has been clear for a very long time that the properties of GaP are substantially influenced by very deep states which provide efficient pathways for electron-hole creation by sequential multistep excitation processes. Undoubtedly, the O_P donor is one of the most important of these for light in the energy region just above $E_g/2$, ie above $E_g - E_{D}$ or 1.45 eV, the threshold for this two-step process. One of the early observations of green ~ 2.21 eV DAP luminescence in GaP involved excitation by light of energy near 1.8 eV, probably involving the absorption processes shown in Figure 18b although additional weak, structured absorption was also reported.[67] More recently, such multistep excitation processes have been explicitly referred to the

O donor.[68] The threshold for the excitation process close to 8500 Å ($\sim$1.46 eV) agrees closely with expectation for a $\sim$0.90 eV donor. Particularly weak, volume excitation processes are possible by this means, giving narrow DAP spectra.

This PLE process has been turned around to provide a means of determining E_g from the threshold condition for the no-phonon transition $h\nu_T = Eg - (E_D)_O$ assuming $(E_D)_O$ is known. Monemar and Samuelson[69] show that the extreme threshold region contains a no-phonon part below the discontinuity due to a $\sim$19 meV phonon replica (Table 3) at $\sim$1.465 eV (Figure 21). These authors analyzed their data with an approximation to the Lucovsky formalism.[58] This accounts for the fact that, unlike the donor photoionization transition to the conduction band edge, the complementary transition from the valence band is parity-allowed assuming that momentum-conservation in the indirect transition is satisfied by scattering in the deep-state impurity potential. Such scattering provides a strong component of the bound electron wave-function at k = O, the wave-vector of the p-like valence band maximum. Assuming the valence band density of states is $\rho(E_v)$, determined from a full 6 $\times$ 6 matrix description in this analysis,[69] the no-phonon absorption cross-section σ_{P1} $(h\nu)$ can be written

$$\sigma_{P1}(h\nu) = \frac{\rho(E_v)}{[(E_c(\Gamma) - E_D) + \dfrac{m_v^*}{m_c^*}(h\nu - h\nu_T)]^2} \qquad (10)$$

This expression gives a close account of the experimental data in Figure 21. However, these authors use a value of $(E_D)_O$ which is $\sim$1.5 meV too large. The value from Table 4 gives $E_g = 2.351 \pm 0.002$ eV, in excellent agreement with the estimates $E_G = 2.350 \pm 0.002$ eV by Bindemann et al using the same technique for the O donor[70] and free to bound luminescence at shallow donors[71] and also with the estimates from the analyses of transitions involving independently determined shallow acceptor energies.[42,43] A more thorough analysis of the optical cross-section relevant for Equation 10 has been provided recently by Banks et al.[220]

The two phonon energies derived from a careful analysis of the low-level absorption for this hole photoionization edge are included in Table 4. We shall see in Section 4.1 that the PLE technique may provide the cleanest method of evaluating the low-level form of the optical cross-section between two and five orders below its peak near 1.8 eV, apparently because

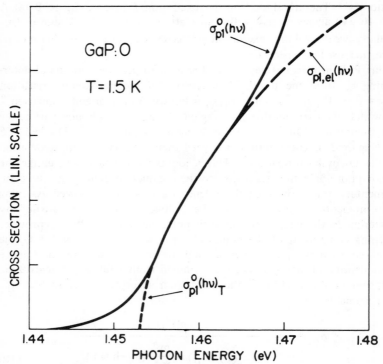

FIGURE 21 The experimental optical cross-section (—————) for the transition
to the ionized O donor in GaP denoted in the inset, determined at low temperature
from the PLE spectrum of the infrared DAP luminescence. The purely electronic
spectral component $\sigma^o_{pl,el}$ (hv) is deduced by a deconvolution procedure discussed
by Monemar and Samuelson (Ref. 72). The curve σ^o_{pl} (hv)$_T$ is theoretical (Equation
10 with $m^*_X/m^*_c \sim 8$). The onset of the first 19 meV phonon-assisted component is
clearly shown near 1.65 eV. Apart from the very weak tail below 1.453 eV, the
deconvoluted $\sigma_{pl,el}$ (hv) curve agrees both with the full experimental curve and the
theoretical form (Monemar and Samuelson, Ref. 69).

of the effect of junction electric fields inherently present in the capacitance techniques.[72] The most appropriate form of σ_{P1} (hv) might be
expected to contain the $(hv–hv)^{1/2}$ energy numerator of Equation (10)
instead of the Lucovsky form $(hv–hv_T)^{3/2}$ for a parity-forbidden transition which describes the electron-photoionization threshold to the conduction band. This is so despite evidence that the deep ground state of the
O donor appears to contain a strong admixture of valence band states
(Section 2.10). In fact, the experimental data in Figure 21 seem closer
to the exponent 1/2 than to 3/2.

3. Further Properties of the One-Electron State

3.1 Excitation Spectroscopy of Associate Luminescence

We have seen in Section 2.2 that information on the absorption of the O donor-acceptor associates can readily be obtained from excitation spectra of the associated luminescence. Only data close to threshold is shown in Figure 5, whereas Figure 22 shows a much wider range extending above the indirect energy gap. Positive contributions to the PLE are obtained from absorption components resulting in the creation of free excitons, thresholds TA^E and LA^E, and from the creation of BE at the N isoelectronic trap, responsible for the lines at A and A_x. Very similar results were reported by Welber and Morgan,[74] who also showed that the quantum efficiency for photoexcitation at the broad peak near 2.19 eV, at the A-line and at the broad band near 2.4 eV are all nearly unity for optimally doped crystals. The suggestion[74] that the strong positive PLE response at the A-line requires the existence of an excited state of the Zn-O BE essentially isoenergetic with the N BE[3] is not very plausible, however. Such assumptions are not necessary, since it is well-known that efficient excitation tunnelling occurs from the N BE to the deeper NN traps when the concentration of the latter is only $\sim 1 \times 10^{15}$ cm^{-3}.[3] This value is comparable with Zn-O associate concentrations in optimally doped crystals, while typical N concentrations inadvertently present in these crystals grown from Ga solution with no special precautions to exclude N can reach well into the 10^{17} cm^{-3} range. Significant excitation transfer is plausible at these concentrations, according to general observations of other types of PLE spectra.[75] It should be noted that the Zn-O associate concentrations quoted by Cuthbert et al.[29] are over-estimated by a factor of 9 due to a numerical error in the expression used for absorption cross-section. The corrected value of $\sim 6 \times 10^{15}$ cm^{-3} for optimally-doped crystals containing $\sim 5 \times 10^{17}$ cm^{-3} Zn acceptors is reasonably consistent with later, more accurate estimates (Section 3.2).[76] This low value represents a significant proportion of the total O donor concentration, which may be only slightly above 10^{16} cm^{-3} according to high temperature Hall data.[79]

Cuthbert et al.,[29] also noted a marked deviation between the increasing strength of the Zn-O absorption and the rapidly decreasing photoluminescence efficiency for [Zn] $> 10^{18}$ cm^{-3}. This was attributed to the influence of Auger recombination processes involving free holes or holes bound at nearly shallow acceptors. Similar effects have been reported for p-type GaP in this concentration range through room temperature electroluminescence.[77]

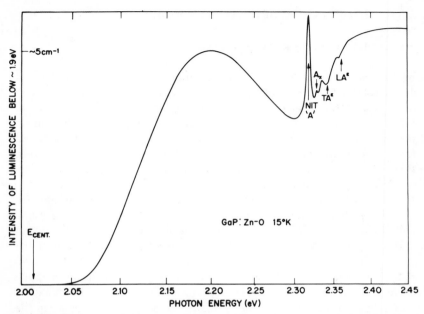

FIGURE 22 Low temperature excitation spectrum for red Zn–O photoluminescence recorded from an optimally Zn, O-doped solution grown GaP single crystal. The absorption coefficient indicated is the maximum observed in such crystals at the broad peak associated with the Zn–O BE near 2.18 eV. The structure above 2.317 eV is due to N-induced (A and A_x) and intrinsic (TA^E, LA^E) absorption processes. The no-phonon transition is estimated to fall near E_{CENT}. (Cuthbert et al., Ref. 29).

Evidence has been obtained for quantum-mechanical interference between absorption at the N A-line and the underlying Zn-O multiphonon-assisted BE absorption in the PLE spectrum of Zn-O luminescence recorded at 4.2°K for a crystal containing $\sim 10^{18}$ cm^{-3} N isoelectronic traps.[78] Even this observation does not require an independent excited state of the Zn-O BE to be resonant with the A BE. Interference can occur between N-induced absorption processes in the vicinity of a given Zn-O center and absorption due to the Zn-O center itself. Provided that strong spatial overlap occurs between these centers, creation of the N BE can be a virtual intermediate state leading to the final state of Zn-O BE+nℏω, where ℏω represents a dominant phonon coupling to the Zn-O center. Interference in this direct, non-thermally activated inter-center interaction is responsible for a dispersive form of the PLE near the N A-line. This dispersive shape is transformed towards the normal lineshape characteristic of simple additive absorption at the N and Zn-O centers with increasing temperature above $\sim 15^e$K. Then, energy transfer between N absorption and Zn-O luminescence becomes predominantly the more usual thermally-activated hopping process.

3.2 Concentrations of O_p Donors and $Zn_{Ga}-O_p$ Associates

We have already noted in Section 3.1 that the concentration of substitutional O_p donors is strictly limited in GaP compared with the shallow donors and acceptors used to promote electrical conductivity at 300°K, where concentrations well in excess of 10^{18} cm^{-3} are readily achieved.[1] There have been a large number of estimates of the O concentration and rather rather fewer, for the Zn-O concentration. Restricting attention for the moment to data for crystals grown from Ga solution near 1100°C, we observe the results for $[O_p]$ to fall into two groups. One group, utilizing indirect techniques such as analysis of the influence of O-doping on the bulk minority carrier lifetime,[80] additional compensation of Zn acceptors by O-doping, [81,82] combination of optical absorption, PLE and luminescence saturation[87] or from the high temperature Hall effect gives results near 10^{17} cm^{-3}, and as high as 3×10^{17} cm^{-3}.[80,87] The strong enhancement of [O] by Zn co-doping claimed from the lifetime and Hall analyses[79,80] has not been reproduced by some other workers.[81,83] However, results from (p, α) activation analysis[85] suggest at least a factor of ~ 2 increase in [O] for a Zn concentration up to 5×10^{17} cm^{-3}. Some Schottky barrier photocapacitance results[88] indicate a factor of ~ 3.5 enhancement between O-doped LPE GaP co-doped with Te and with Zn. The solubility of O in Ga increases continuously with temperature,[81] but the distribution coefficient decreases with temperature so that there is an optimum growth temperature near 1030° for the O-doping of GaP grown from Ga solution.[82] Problems of Ga_2O_3 co-precipitation occur when attempts are made to exceed the equilibrium O concentration in the GaP, and this has a deleterious effect on the optoelectronic properties.[84] The presence of undetected, fine precipitates may cause the total O concentration measured by activation analysis, up to $\sim 9 \times 10^{16}$ cm^{-3} in Zn-doped solution-grown materials,[85] to be slightly over-estimated, apparently a serious problem in earlier ^{3}He activation analysis where concentrations of $\sim 10^{19}$ cm^{-3} were reported.[86]

The second group of results obtained by a variety of more direct techniques, principally photocapacitance,[88-91] DLTS[92,93] and below-gap two step PLE[83] gave much lower values of [O], $\leqslant 3 \times 10^{16}$ cm^{-3}. A strong decrease in [O] near the p-n junction was reported in several cases.[89,91,93] Estimates based on below-gap PLE suggest that the value of [O] in LEC

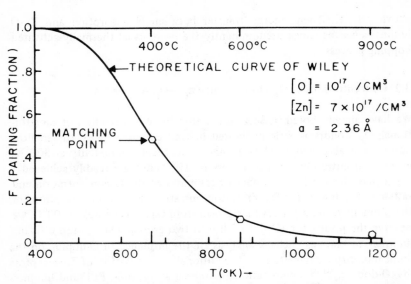

FIGURE 23 Comparison of experimental data on the fraction of deep O donors present in Zn–O associates, obtained from the integrated red (BE) to infrared (distant DAP) O-induced photoluminescence in °K (points) measured as a function of anneal temperature under saturated anneal conditions, with a theoretical curve for pair equilibrium (full line) (Hughes et al., Ref. 98).

GaP may be up to 7×10^{16} cm^{-3}.[94] However, we show in Section 6 that a significant concentration of O is present in alternative forms in the GaP lattice for this material.

Estimates of the concentration of Zn-O associates in optimally annealed material are rather more concordant. The maximum values are estimated at $\sim 10^{16}$ cm^{-3} according to photocapacitance,[89] DLTS[92,93] and junction photocurrent techniques.[95] Values nearly one order lower are often obtained, sometimes in material of high efficiency.[93,96] The theory of ion-paring for the Zn-O system has been discussed by Wiley.[97] Careful analysis of effects of annealing on the red and infra-red photoluminescence of solution grown GaP:Zn,O, sputter-coated to minimize Zn out-diffusion, revealed good agreement with the theory for [O] = 10^{17} cm^{-3} and [Zn] = 7×10^{17} cm^{-3}. These results indicate that nearly 50% of the O may be found in associates after a long anneal at 400°C (Figure 23).[98] The diode results also frequently suggest a ratio of [Zn-O]/[O] close to 0.5, though usually at significantly lower overall concentrations as we have seen.[89,92,93]

3.3 Emergence of Free to Bound Luminescence at O

The description of the infra-red O-related luminescence in terms of the distant DAP process is not expected to hold above $\sim 80°K$. The holes come into equilibrium with the valence band in this temperature range.[29,90] The relatively slow DAP recombination process then becomes non-competitive, as is well known for shallow DAP in GaP.[2,71] The dominant spectral form is expected to transfer to the radiative recombination of free holes with electrons which remain tightly bound to the deep O donor up to temperatures well above $300°K$. This is exactly analogous to the well known case of the edge luminescence of direct gap semiconductors, where the electron is the free particle and the whole transitional process occurs over much lower temperature range for cases such as GaAs.[100] The decay time of the infrared luminescence in GaP:Zn,O was observed to vary as p^{-1} above $200°K$, where the lifetime increases with decreasing temperatures as the holes become increasingly frozen out below $400°$ (Figure 24).[101] The luminescence peak energy shifts about 15 meV towards higher energy between 100 and $160°K$ as the DAP process becomes replaced by the FB process. This agrees approximately with expectation for a 70 meV acceptor, considering the likely value of the Coulomb shift $e^2/\epsilon R$ in Equation (6) at $100°K$, where transitions involving less distant pairs are emphasized compared with $4°K$ because of competition from thermal ionization.[2] Similar conclusions were obtained by Dishman,[102] who reported the emergence of structure associated with the FB band between $70°K$ and $100°K$ (Figure 25), with just the DAP band (Figure 13) at lower temperatures. The no-phonon threshold $h\upsilon_T$ energy reported at $\sim 87°K$, 1.445 ± 0.002 eV, results in an estimate $(E_D)_O = E_g\text{-}h\upsilon_T = 898 \pm 3$ meV, assuming $E_G = 2.343 \pm 0.002$ eV at $87°K$, in excellent agreement with the values reported in Sections 2.7 and 2.12. The spectral position of the infrared band at $300°K$ is independent of the binding energy of the acceptor co-dopant, as expected if the FB mechanism is then dominant. The total intensity of the infrared luminescence increases between $60°K$ and $\sim 250°K$ for light acceptor co-doping $\sim 2.5 \times 10^{-17}$ cm^{-3}, while the FB component increases even more rapidly.[102,52] This confirms the dominant influence of the increasing thermal delocalization of holes from the Zn acceptors,[101] though three-center Auger recombinations were invoked to explain the weakness of the DAP luminescence at $300°K$. The integrated infrared luminescence decreased continuously above $60°K$ in more heavily doped samples containing 10^{18} cm^{-3} Zn acceptors, allegedly due to Auger processes involving two free holes.[102]

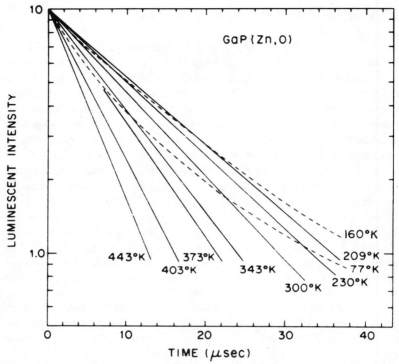

FIGURE 24 Time decay of the infrared luminescence in optimally Z,O-doped solution grown GaP single crystals after pulsed excitation with a He-Ne laser, recorded at the indicated temperatures between 77°K and 443°K. The decay remains non-exponential with very long-lived components below 77°K. (Bhargava, Ref. 103).

3.4 Free to Bound Processes and the Red Associate Luminescence at 300°K

This behavior of the infra-red related luminescence at 300°K contrasts strongly with that of the red luminescence. We have already commented in Section 1.3 that spectral shifts occur when the acceptor co-dopant is changed (Figure 2), showing that O alone cannot be responsible. However, we now recognize that this shift is caused by the sensitivity of E_e in the deep electron trap to the nature of acceptor component of the associate (Section 2.4), rather than to the operation of the distant pair mechanism at 300°K, (Figure 11). Bachrach and Jayson[103] made a careful study of the spectral evolution of this red luminescence between 80°K and 300°K, looking particularly for the presence of a component attributable to the

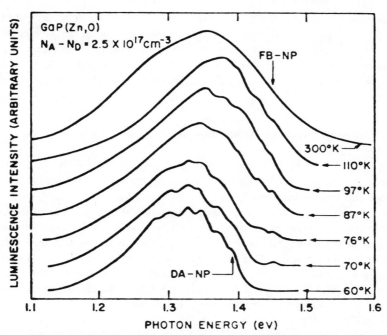

FIGURE 25 Infrared photoluminescence spectra from lightly Zn-doped GaP:Zn,O recorded as a function of temperature. The intensity-zero of each successive spectrum has been offset vertically. The no-phonon energies for free hole to bound electron and distant DAP transitions are designated by FB-NP and DA-NP. The FB process becomes significant above ~ 70°K. (Dishman, Ref. 102).

radiative recombination of free holes with electrons bound to the Zn-O associates. No such evidence was found. The peak energy remains nearly independent of temperature, with no shifts attributable to plasma screening of the thermally-generated free holes or to their kinetic energy, and no significant difference in the 300°K lineshape between crystals doped with 1×10^{18} and 3×10^{18} Zn acceptors. The relative importance of any FB process is estimated to double in this doping range.

The spectral line-shapes of the red luminescence are well represented by Gaussian functions, with a linear temperature dependence of half-width between ~ 80°K and 300°K. The holes on the Zn-O associates are thermally ionized even more readily than those at the isolated Zn acceptors in this temperature range, in view of their smaller binding energies (Table 2) ~ 35 meV compared with ~ 70 meV (Section 2.4). The strong difference between the red and infra-red luminescence as far as success of the bound to bound processes in the competition with the free to bound processes is concerned is directly due to the much greater transition probability of the red bound exciton (lifetime $\sim$ 100 nsec) compared with the average distant DAP (mean lifetime $\sim$ 1 msec). The absence of the FB component in the red luminescence is of some moment for a long-standing controversy in the early 1970s concerning the interpretation of the kinetics of time decay.[90]

3.5 Associated of O With Other Impurity Species

We have seen in Table 2 that E_e is appreciably smaller for the Zn-O associate than for the Cd-O associate. This partly encouraged the idea that the Coulomb interaction term $E_c = e^2/\epsilon R$ in Equation (6), or the term E_{rep} in Equation (4), might increase with decreasing atomic number of the cation, giving a larger value of no-phonon energy and a corresponding increase in the transition energy of the red luminescence, greatly needed for the LED applications.[1] Unfortunately, the lighter Group II atoms are very reactive, and special means are necessary to incorporate them either during crystal growth[104] or by in-diffusion. In-diffusion of Be and Mg from liquid Ga sources was found to be reasonably controllable.[105]

Doping by Mg introduces yellow BE luminescence, whose no-phonon lines near 2.16 eV (Figure 26) suggest an electron binding energy E_e of only $\sim$ 155 meV, much less than for Zn (Table 2). The involvement of O was confirmed by a small, $\sim$ 0.007 meV increase in transition energies of the no-phonon lines when $O^{16} \rightarrow O^{18}$. The local mode X_3, whose energy lies just above $\hbar\omega_{LO}$ of the GaP lattice, exhibits an energy *decrease* by $\sim$ 0.09 meV. *Three* no-phonon lines are observed (Figure 26), consistent with BE recombination at an axial center with a much smaller local crystal field splitting than the Cd-O associate.[105] The phonon coupling in Figure 26 is much weaker than for Cd-O or Zn-O (Figure 6) generally consistent with the much smaller value of E_e, which is also intimately linked with the much smaller no-phonon isotope shift.[27] There is no suggestion of the low frequency mode which we have seen is specifically related to vibrations of a heavy cation (Table 1). This yellow photoluminescence may be quite efficient at 77°K.[106] However, electroluminescence results have not been encouraging,[106] probably because of limitations in the concentrations of the Mg-O associates.

The deep level behavior of Be in GaP is even less straightforward. Two deep BE systems have been reported,[105] and neither exhibits O isotope shifts in their no-phonon lines nor in any of the local modes observed, including one of quite high energy $\sim$ 62.7 meV. However, it seems probable that the higher energy system, with weak phonon coupling and a set

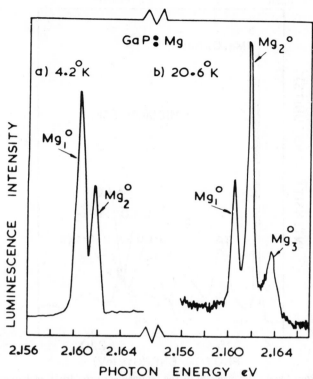

FIGURE 26 Low temperature luminescence spectra associated with the Mg–O Be in GaP, recorded from Mg-diffused solution grown single crystals of GaP:O. (a) shows the two lowest energy no-phonon lines, while (b) shows a third present at slightly higher temperature (Dean and Ilegems, Ref. 105).

of no-phonon lines near 2.15 eV (Figure 27), is the analogue of the Cd-O associate. The lower energy system, with no-phonon line near 2.12 eV, may be analogous to a number of 'additional' as yet unidentified Zn-related BE lines which appear most prominently in Zn-diffused GaP.[105] This system is also prominent in Zn growth-doped solution grown GaP LEDs, sugjected to recombination enhanced degradation after prior radiation damage by γ-rays.[192] It seems most likely that these centers involve Zn_I in some form. We therefore have an interesting progression of E_e values, which appear to scale roughly as expected from the influence of changes in the cation-O bond length on the Coulomb term in Equation (6), (Table 2). This possible Be-O associate also exhibits three or four no-phonon lines. Detailed Zeeman analysis was not possible because of the spectral

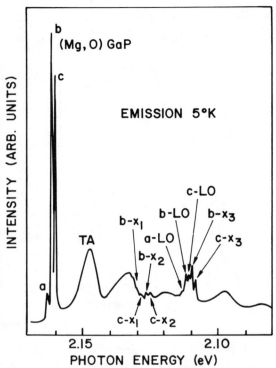

FIGURE 26c Shows the majority of the luminescence spectrum, including the major phonon replicas listed in Table 1 (Bhargava et al., Ref. 106). Note the reversal of energy scale in (c) compared with (a) and (b), also no-phonon components $M^\circ_{g1} \to M^\circ_{g3}$ are re-labelled c → a, respectively.

broadening in heavily-diffused crystals. However, the appearance of field-induced components at lower energies is consistent with expectation for a system like Cd-O (Equation (3)), although with a much lower crystal field splitting more comparable, with the Mg-O associate.[105] The crystal field and j–j splittings are expected to decrease rapidly with localization energy of these BE.

Very dramatic red luminescence has been observed for GaP double-doped with Li and O.[107] *Four* no-phonon lines have been resolved near 2.09 eV, together with very complex associated phonon-assisted transitions (Figure 28). The form of the phonon structure is more like that exhibited by isoelectronic associates with lighter cation components such as Mg-O (Figure 26) or Be-O (Figure 27) than like Zn-O or Cd-O (Figure 6). However, the strength of phonon coupling is greater and the crystal

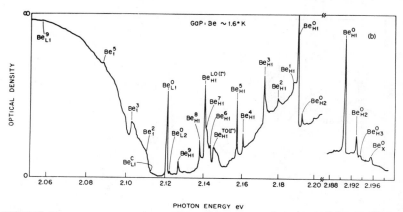

PHOTON ENERGY eV

FIGURE 27 Low temperature photoluminescence recorded photographically from Be-diffused solution grown GaP single crystals. The O point on the ordinate scale has been set arbitrarily, through scale expansion, to emphasize spectral detail. The superscripts denote whether the line is a no-phonon transition (O) or a phonon-assisted transition, as detailed in Table 1. Subscripts H and L denote luminescence components from two distinct Be-related centres. The H series is believed to involve the Be–O BE (Table 2). The L series remains unidentified, but may be associated with interstitial Be analogous to further spectra observed in Zn-diffused GaP (Dean and Ilegems, Ref. 105).

field and j-j splittings are also larger than for the Mg-O and Be-O associates, all consequences of the larger exciton localization energy (Table 2). A great deal of additonal sideband structure occurs due to local modes, (Table 1), which we shall see provides vital information on the nature of the center which binds this exciton. Thermalization and a comparison of luminescence and absorption (Figure 29) indicates that these no-phonon lines arise entirely from splittings in the excited state of the center, like the other O-related complexes described above. The lowest energy line Li_{L1}^{O} is very weak in a strain-free crystal, even at $1.6°K$ where it is strongly favored by thermalization. However, it is strongly enhanced in a magnetic field (Equation (3)), and exhibits Zeeman properties very similar to the Cd-O associate (Figure 8). The properties are consistent in detail with a $<111>$ symmetry neutral center binding an exciton, with no additional electronic particles.

Like the Mg-O and Be-O center, but unlike the still more tightly bound Cd-O associates (Table 2), the splitting of the hole states derived from the local axial field is comparable with the J-J splitting for the Li,O BE. The presence of a fifth no-phonon line representing a forbidden optical transitions can be deduced from the analysis.[107] The relationship of this center with O is again firmly established by isotope shifts of + 0.80 meV in the

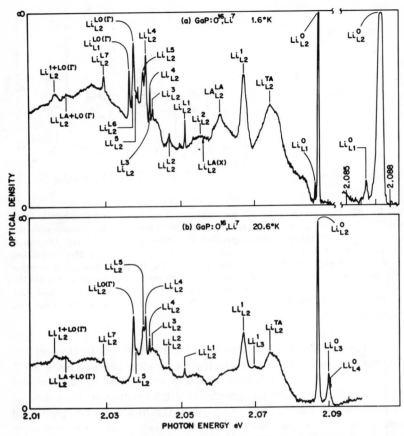

FIGURE 28 Low temperature photoluminescence recorded photographically from O-doped Li-diffused GaP grown at ~ 1100°C from Ga solution or vapor using wet H_2 transport showing detail in the Li-Li-O BE transitions. Superscripts denote no-phonon (0) or phonon-assisted transitions involving the lattice or local modes detailed in Table 1. The subscripts distinguish transitions from different electronic states of the BE formed by j-j coupling and crystal field splitting as in Figure 9. Two no-phonon lines appear at 1.6°K, shown in detail to the right in (a). Additional higher energy no-phonon lines appear at 20.6°K in (b). (Dean, Ref. 107).

no-phonon lines when $O^{16} \rightarrow O^{18}$ and *decreases* in the energies of in-band resonance modes L_3, L_4 and L_5, in gap mode L_1 and in the local mode L_7. The much larger magnitude of the no-phonon isotope shift again correlates with the stronger phonon coupling and larger E_e compared with the Mg-O and, still more with the Be-O associates. The relationship with Li hardly needs verification, in view of the strong connection of this char-

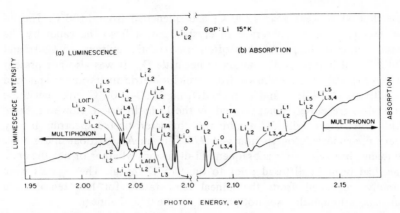

FIGURE 29   (a) **Low temperature photoluminescence of the Li-Li-O BE recorded photoelectrically. (b) Low temperature optical transmission also recorded photoelectrically. The absorption coefficient near peak Li^0_{L2} is ~ 0.5 cm^{-1}. The general decrease in transmission at the right is partly due to the spectral output of the W source. Note that the energy scale in (b) is magnified 2 times compared with (a). (Dean, Ref. 107).**

acteristic red spectrum with Li in-diffusion and its complete absence without deliberate contamination by Li. However, energy decreases when $Li^6 \rightarrow Li^7$ occur for all phonon replicas L1 → L7 (Table 1). The behavior of the highest energy local mode replica L7 in the presence of equal proportions of Li^6 and Li^7 is particularly revealing. It breaks up into not just two but into *four* sub-components. This proves beyond doubt that the associate center binding the exciton contains *two inequivalent* Li atoms, together with O. All this evidence led to the suggestion that the associate is Li_I-Li_{Ga}-O_P, where the two Li atoms both enter a V_{Ga} site adjacent to the substitutional O atom, producing a form of bi-interstitial, polarized by the adjacent O_P donor.

This three-component associate has the neutral form required for efficient luminescence.[1] It is another, slightly more complex form of 'molecular' isoelectronic substituent. The no-phonon lines again fall close to the expected position if the modification of the binding energy of the electron at the O_P donor is calculated in the elementary manner described above for the binary associates (Table 2).

A particularly interesting aspect of the Li-Li-O bound exciton stems from the fact that it can be formed under quite gentle conditions, for example, by in-diffusion at only 400° for 60 min. This strongly suggests that the V_{Ga}-O_P associates may exist prior to diffusion in the crystals used, typically those grown from Ga solution near 1100°C, from the vapor near 950°C or from non-stoichiometric melts using the liquid-encapsulated

Czochralski process near 1200°C, and quenched from 1100°C. The luminescence was *not* observed in crystals grown from the vapor by the halide process similar to that often used in the epitaxy of GaAs and InP,[108] which apparently contain rather little O_P. It was also not present in Li-diffused crystals grown from stoichiometric melts under standard conditions at 1470°C and quenched from 1200–1300°C, although these crystals were closely compensated by the Li in-diffusion. However, these LEC crystals generally did exhibit this characteristic red spectrum if they were pre-annealed for 16 hours at 975°C in an $A\ell_2O_3$ boat under a N_2 ambient, but only after subsequent Li diffusion, whether or not the crystals had been Li diffused prior to the 975°C anneal. The crystals were rapidly quenched from the anneal temperature for low temperature anneals, although this was not possible for the 975°C anneal.

3.6 Existence and Stability of the V_{Ga}-O_P Associate

These observations in Sections 2.6 suggest that the Li-Li-O associate is unstable at 975°C, as expected from the discussion of the thermal behavior of the Zn-O associate in Section 2.5. They also indicate that the V_{Ga}-O_P associate may form preferentially near 900–1000°C, though it is also unstable above 1200–1300°C. It has been suggested that V_{Ga} may be preferentially created adjacent to D_{As} in GaAs initially Li-diffused at high temperatures, ~ 800°C.[109] It is not known whether such an effect may be assumed in GaP. However, Li in-diffusion seems to form an effective method for labelling an elusive type of crystal defect in III–V semiconductors, the cation vacancy. This possibility has not been pursued for GaP in recent years, despite early proof of its usefulness in demonstrating the presence of O in LEC GaP and the existence of a strong ion exchange with the B_2O_3 encapsulant which complicates the verification of the relation with O of various infrared local mode absorption lines in LEC GaP,[107,142] discussed further in Section 6. The Li-Li-O luminescence spectrum has also been used to calibrate the O^{16}/O^{18} isotope ratio in activation analysis for O in solution-grown GaP.[85]

The evidence for the formation and high-temperature stability of V_{Ga}-O_P associates is significant in view of discussions of the role of V_{Ga} as a killer center responsible for the very poor overall luminescence efficiency of LEC GaP.[114] The strong link between decrease in minority carrier lifetime τ_m and increase in the growth temperature T_G of p-type GaP:Zn,O established by Jordan et al.[114] was believed to implicate V_{Ga} in the dominant shunt path for electron recombination according to analysis of the GaP solidus.[115] Van Vechten[116] has criticized this conclusion on the grounds that the electrical conductivity of this material was not consistent with the presence of the large concentrations of V_{Ga} predicted from this interpretation. He suggested that the overwhelming majority of the V_{Ga} should be present in neutral associates as a result of reactions with charge-conjugate defects or impurities as the crystal cools below T_G. The associate V_{Ga}-O_P is not in itself neutral, but could be an important element of such complexes. However, the relationship between τ_m and T_G^{-1} was established between $\sim$1200°C and 1470°C, a temperature range in which we have seen that the data on Li in-diffusion suggests that at least the V_{Ga}-O_P complex is not only unstable but also does not form significantly during normal schedules of cooling the crystal ingots to room temperature. Substantial annealing periods are required at 975°C in order to prepare these associates. We can only conclude that the mobilities of these constituent defects in the temperature range in which associates such as V_{Ga}-O_P are stable may be significantly smaller than assumed by Van Vechten.[116] This is consistent with recent observations on Ga self-diffusion in GaAs, which indicate very limited diffusion profiles, $\sim$1μm extrapolated depth after hr diffusion at temperatures as high as 1200°C.[184] The diffusion mechanism is believed to be of dissociative interstitial-substitutional type, so that these measurements provide an *upper limit* to the diffusivity of V_{Ga}.

3.7 New Identifications for Cu,O-Related Luminescence

Direct evidence of the electronic properties of the V_{Ga}-O_P associate itself has proved difficult in GaP. The efficient complex orange luminescence (COL) (Figure 30a) related to this complex by Bhargava et al.[110] is now known to be Cu-related and seemed not to involve O_P at all.[111] The fact that the COL is quenched in Li in-diffusion[110] is not itself sufficient to prove that COL is related to V_{Ga}-O_P, since Li is well-known to form complexes with a wide variety of centers in semiconductors. It would be particularly attracted to such sites as Cu_{Ga}, thought to form a part of the associate responsible for COL. Indeed, characteristic near gap luminescence spectra associated with both Cu *and* Li has been observed in GaP.[112] It is difficult to be sure whether any of these centers involve the COL associate, transformed by the addition of Li. Such a transformed complex could be a non-radiative center. However, an attractive postulate relates the COL to the Cu_I-Cu_{Ga}-Cu_I associate. The high transition energy, 2.177 eV, is consistent with the strong central cell potential of the Cu_{Ga} (acceptor $(E_A \sim 0.53$ eV)[193] given a reasonable value of E_D for the interstitial Cu_I-donor which 'reacts' with Cu_{Ga} to form the neutral substitutional-

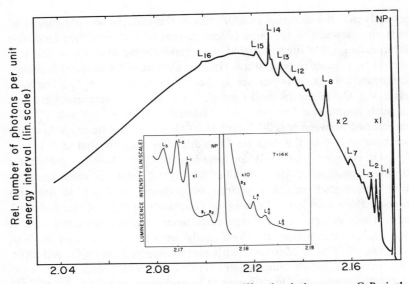

FIGURE 30a Photoluminescence from a Cu-diffused solution-grown GaP single crystal possibly related to the Cu_{Ga}-Cu_I BE, recorded at 1.8°K. Only a single no-phonon line is seen. The phonon-assisted transitions are analyzed in Table 1. The inset shows the changes in the low frequency modes $L_1 \rightarrow L_3$ at 14°K, with some evidence of a mode near $h\omega_{L1}$-$h\omega_{L2}$ in the Stokes (δ_1) and anti-Stokes (δ_3) regions.

interstitial associate, $(Cu_I$-Cu_{Ga}-$Cu_I)$, say $(E_D)_{Cu_I}$ ~0.1 eV like the interstial donors observed for Li in GaP.[185] This model can account for the unusual annealing properties of the COL luminescence[113,186] where the quenching in low temperature equilibrium anneals near 600°C is attributed[118] to completion for Cu from another center of higher dissociation while the quenching above 800°C is attributed to the dissociation of this associate, just like the Zn-O associate (Figure 23). This model also accounts for the very unusual form of the low energy ($h\omega$ ~5.3 meV, sensitive to M_{Cu}) in band resonance phonon replicas observed for COL (Figure 30a). These replicas may be more complex than for the Cd-O associate (Figure 6) because of the additional low frequency vibrational modes possible for an associate involving interstitial atoms unpinned by substitutional impurities on adjacent lattice sites.[118] Such a center is expected to form because of the long range Coulomb attraction between the ionized states of the Cu_{Ga} acceptor and the Cu_I donor, which is held responsible for the characteristically fast mass transport of Cu in most semiconductors. Once again, the neutral associate formed by the reaction between these species is expected to be strongly luminescent. *Two* Cu_I, each acting as single donors can be accommodated to neutralize the Cu_{Ga} double

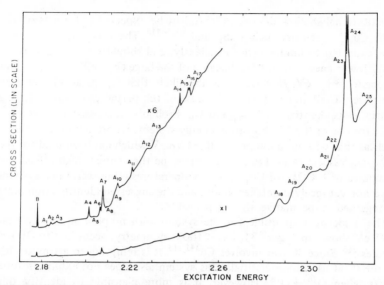

FIGURE 30b Optical absorption associated with this BE determined from the PLE spectrum of the luminescence in (a), recorded at $2°K$. B is the no-phonon line in (a), while $A_1 \rightarrow A_3$ are phonon replicas like $L_1 \rightarrow L_3$. Component A_4 is the second no-phonon line expected from j-j coupling of two $j = \frac{1}{2}$ electronic particles, while the components $A_6 \rightarrow A_{17}$ are phonon replicas of A_4. Components $A_{18} \rightarrow A_{20}$ possibly involve excited orbital electronic states of the BE, while $A_{21} \rightarrow A_{25}$ involve NN and N isoelectronic traps (Dean et al., Ref. 118).

acceptor, leaving all Cu ions in a $3d^{10}$ configuration. It is also necessary to assume *two* interstitial species to account for the large number of low energy local modes, *two* of which are allowed and two are forbidden in terms of the usual electron-phonon interactions. The concept of a neutral trap, capable of binding an exciton tightly to a center containing no additional electronic particles, and therefore with no Auger recombination possible for transitions within the complex itself, plays a central role in our understanding of the activation of efficient deep state luminescence in an indirect gap semiconductor like GaP.[1,111] The quenching of COL under Li saturation might then be attributed to the displacement of one (or more) Cu_I to give further luminescent associates according to the reaction

$$(Cu_I\text{-}Cu_{Ga}\text{-}Cu_I) + Li_I \rightarrow (Li_I\text{-}Cu_{Ga}\text{-}Cu_I) + Cu_I \qquad (11)$$

Attribution of COL to this $(Cu_I\text{-}Cu_{Ga}\text{-}Cu_I)$ associates accounts for its persistence in Cu-saturated GaP and the absence of a correlation with substitutional shallow donors which might be expected to form associates with Cu_{Ga} acceptors such as S_P and Te_P.[113] The Cu_{Ga}-O_P associate is also expected to luminescence efficiently, and should occur at a substantially lower energy than COL because of the large $(E_D + E_A)$ sum for this associate, ~ 1.5 eV. However, it is also likely that E_{rep} is also larger than the purely Coulombic value E_C because of the partial cancellation of the central cell potentials of Cu_{Ga} and O_P. Probably, this associate is responsible for one of the low transition energy systems reported by Monemar[194] such as that with no phonon line at ~ 1.78 eV which has been used for the Cu_{Ga}-O_P associates in Table 2. This may be the center which effectively competes with COL for Cu_I at low temperatures. However, these systems have not yet received detailed study, and the suggested identification must be regarded as speculative, though plausible.

The associate responsible for the system with no-phonon energy near 1.91 eV shown in Figure 31, has $<100>$ symmetry, according to ODMR studies.[213] Since it also involves $Cu^{111,113}$ it is tempting to attribute this to the $(Cu$-$Cu)_{Ga}$ bi-interstitial. This system also does not exhibit isotope shifts when $O^{16} \rightarrow O^{18}$. However, it is more plausible to identify this center as Cu_{Ga}-D_i, where D_i is an unidentified interstitial donor ion. This is so since although the 1.91 eV spectrum exhibits a well-defined local mode in the LA region of the lattice vibrational spectrum with a large Cu isotope shift, which suggests that Cu_{Ga} is involved as for the COL, the

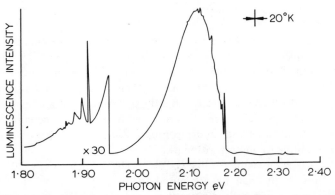

FIGURE 31 Cathodoluminescence recorded at $\sim 20°K$ from a GaP single crystal grown from Ga solution and diffused with Cu at 1050°C for 1 hr. The COL spectrum with no-phonon line near 2.18 eV and the lower energy spectrum with no-phonon line near 1.91 eV are clearly visible. (Wight et al., Ref. 113).

~10 meV low frequency mode shows *no* Cu isotope shift, suggesting that the interstitial donor is not Cu. The energy shift of ~0.27 eV between this system and COL may be due to a reduction in E_{rep} due to the removal of one of the interstitial donor ions. An increase of E_{rep} from ~0.55 to ~0.9 eV as required by this identification of the 1.78 eV system with Cu_{Ga}-O_P therefore seem plausible. The long lifetime of the COL suggests that the BE ground state is a pure spin triplet, supported by its isotropic Zeeman splitting with g close to 4.[111,117] This implies that the splitting of the hole states in the Γ_{15} valence band in the local axial field of the associate is much larger than the local spin orbit splitting Δ_{SO}. The pure spin hole state becomes highest for complete decoupling under compressive strain, in contrast to Cd_{Ga}-O_P where the local strain is tensional due to the small size of the O_P ion. Hybridization of the hole wave-function with the d-orbitals of Cu can help to reduce Δ_{SO} much below the 80 meV of the GaP lattice for reasons similar to the behavior of the Cu_{Zn} acceptor in ZnO.[187] The outstandingly large values of the J-J splitting energies of both COL and the 1.91 eV system, $\Delta = 2.32$ meV[118] and $\Delta \sim 100$ meV, respectively, compared to all other BE systems reported here imply very strong electron-hole overlap. This can be understood if, in contrast to the Hopfield, Thomas and Lynch[34] model of isoelectronic traps both the hole and the electron interact strongly with the core of the associate. This can happen if the local strain field is responsible for a significant fraction of the binding of the electron and hole. The associated local splitting of the conduction band states can also be responsible for an additional singlet-triplet no-phonon pair observed for some axial centers with appropriate strength of the local (compressive) strain field.[214]

We therefore arrise at the remarkably comprehensive set of associates involving O in GaP listed in Table 1, including the lowest energy, strongest and most persistent spectrum in Cu + Li doped crystals.[112] The Zeeman properties of this latter bound exciton, like COL and the 1.91 eV system differ markedly from all the associates involving *shallow* main-group acceptors Be → Cd. Once again, the sign of the local crystal field splitting of the hole states is reversed, bringing the $m_j = \pm\frac{1}{2}$ states lowest. The decay time for COL is ~110 μsec at 4°K and 57 μsec at 77°K, about 3 orders of magnitude longer than the allowed transition at Cd-O, consistent with a spin-forbidden recombination at the triplet state formed by J-J coupling between an $m_j = \frac{1}{2}$ hole and the $S = \frac{1}{2}$ electron. This factor enhances the decrease in oscillator strength consequent upon binding to a hole-attractive Ga-site impurity, caused by the particular form of the GaP band structure.[32]

4. Photocapacitance, DLTS and the Second Electron State

4.1 Photocapacitance Spectroscopy of GaP:O

A further dramatic development in the understanding of the electronic behavior of O in GaP came from a decision to develop spectroscopic techniques which did not depend upon the existence of radiative recombination processes at these centers. This decision was made partly with the objective of obtaining more information on non-radiative recombination processes in semiconductors.[88] The O donor in GaP was selected in view of the large amount of knowledge of this system which had been obtained from the purely optical studies described above and also because of the need to determine whether the isolated O_p donor provided an important contribution to the shunt path recombination processes which so strictly limit the efficiency of GaP LEDs.[1]

The first step, using the photocapacitance technique[88] did not abandon optical spectroscopy altogether. It can be regarded as a particularly sensitive method for the study of optical absorption at deep states. The available oscillator strength of such states is typically distributed widely in energy (Figure 18), causing direct absorption measurements to be insensitive and of limited accuracy. The basic principle in photocapacitance spectroscopy is the conversion of a small *difference* in optical transmittance to a difference in capacitance of a p-n junction or Schottky barrier. The advantage is that capacitance can be measured with much greater sensitivity than optical transmittance. The technique retains the advantage of optical absorption for the determination of *concentrations* of the deep centers responsible, since these concentrations are directly proportional to the differences in capacitance recorded under suitable conditions. The technique is superior to photoconductivity in this respect and also because the nature of the transition, electron addition or removal, can be deduced from the *sign* of the capacitance change. It is additionally superior to PLE in not being restricted to centers which promote significant luminescence. Photocapacitance is *complementary* to photoluminescence in that the energy depth of majority traps to be measured must be significantly greater than the centers which provide the bulk electrical conductivity through which the depletion layer whose capacitance is studied is defined. Similar limitations apply to the DLTS technique described below. Those centers giving the shallowest electronic states are generally best suited to analysis by photoluminescence, although we have seen that O_p in GaP is a most striking exception. The capacitance techniques also have some additional disadvantage in that they cannot be used at very low temperature to obtain the greatest freedom from thermal broadening in the spectra, since the basic conductivity in the bulk of the crystal must not be frozen out. However, they can be applied directly to any device which involves a pn junction or Schottky barrier.

The most accurate data available from photocapacitance on the two optical cross-sections σ_{n1}^o and σ_{p1}^o associated with the one electron state of the O_p donor (Figure 32)[119] is consistent with the much less precise data for σ_{p1}^o given in Figure 18b. These data for σ_{n1}^o are substantially corrected, in that a large peak just below 1.2 eV reported initially[88] is now recognized as an artifact. These data are now in good agreement with the spectral form obtained from junction photocurrent[120] and photoluminescence quenching (PLQ) (Figure 18a). Earlier, Bjorklund and Grimmeiss[258] measured absolute values of σ_{n1}^o and σ_{p1}^o at 300K, finding a broad peak for both at a level close to 10^{-16} cm^2. They also estimated $[O_p]$ at 7.5×10^{14} cm^{-3} in a solution-grown epitaxial GaP:Zn,O diode. It is crucial to measure the temperature dependence of these cross-sections, in order to determine the full configurational co-ordinate diagram which represents both the electronic energies and their variation with local lattice co-ordinate. The form of these diagrams, particularly the offsets in lattice co-ordinate between electronic states with and without an electronic particle bound at the center provide an indication of the relative importance of phonon-assisted optical transitions and of the competitive multiphonon emission (MPE) process. The MPE non-radiative recombination process has major importance and considerable generality.[119,152]

The configuration co-ordinate diagram in Figure 33 for a single electron bound at O was calculated for 526°K using luminescence data, which suggest a Huang-Rhys phonon coupling parameter S close to 2, and is in general agreement with the optical cross-section. It was necessary to use a reduction factor of ~0.78 for the mean vibrational frequency when the state is occupied in order to fit the measured value of the σ_{p1} carrier capture cross-section. However, no significant difference was seen between the energies of the two receiving mode phonons resolved in experimental spectra exhibiting these optical cross-sections (Figure 34), at least to an accuracy of $\sim \pm 5\%$.[72] This does not necessarily detract from the explanation in contrast to the statements of Monemar and Samuelson.[72] Presumably, the discrepancy gives evidence of the inadequacy of the linear form of phonon coupling in application to this particular deep center. The *thermal* cross-sections were measured using a minority carrier injection, pulse diode bias technique.[121] This ensured that carrier capture occurred in neutral material, away from significant influence of the electric field in the depletion layer of the pn junction. Multiphonon

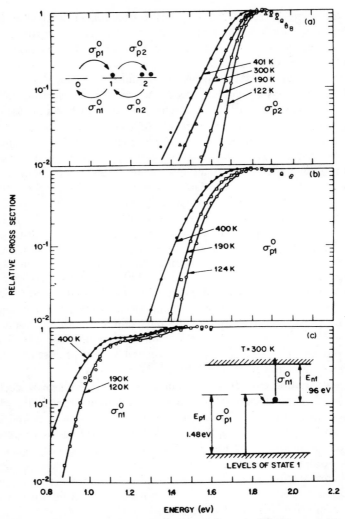

FIGURE 32 Optical absorption cross-sections σ^o_{p1} and σ^o_{n1} of state 1 (single electron, O^o) and σ^o_{p2} of state 2 (two electron, O^-) for O in GaP determined from photocapacitance measurements on p-n junctions grown by liquid phase epitaxy. Note the small broadening of σ^o_{p1} and σ^o_{n1} compared with σ^o_{p2}. The inset in (a) defines the changes in charge state of the O_p center. The insert in (c) indicates the small lattice relaxation in O^o which results from the transition of cross-section σ^o_{p1}, establishing the state from which σ^o_{n1} can be produced. The energy levels of O^o determined from luminescence are consistent with the photocapacitance data. The component E_2 in Figure 18b is the same as σ^o_{p1} here. The threshold of σ^o_{p1} is also shown in Figure 21. (Henry and Lang, Ref. 119).

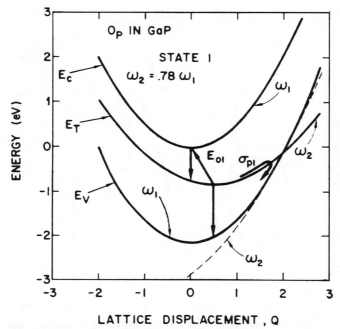

FIGURE 33 Configurational co-ordinate diagram for state 1 (O°) of O_P in GaP. The diagram was constructed from luminescence spectra involving the transitions indicated by vertical arrows. The reduction in frequency when the state is occupied of the single 'average' lattice vibration used in the analysis, $\omega_2 = 0.78\ \omega_1$, was adjusted to fit the σ_{p1} thermal cross-section. The complementary cross-section σ_{n1} is not thermally activated (Table 5).

emission carrier capture processes are important for σ_{p1} only above 450°K, and are not at all significant for σ_{n1} at least below 580°K (Figure 35). The optical cross-sections of Figure 32 are measured in the junction field, however. This is believed to account for discrepancies between the photocapacitance and PLQ or PLE spectral forms, most evident at low temperature (Figures 18a and 36a).[72]

The sensitive PLE technique has been used to obtain the accurate estimates of the threshold energies discussed in Section 2.14 (Figure 21). Derivative techniques show that phonons of similar energy couple to the transition in absorption (σ_{p1}) as well as emission (DAP) (Figure 35), with average energies ~ 19 meV and 48 meV (Table 4) and similar coupling strengths, as near as can be determined.[72] This latter point does not support the earlier contention of Dishman and Di Domenico that the oscillator strength of this transition increases $\sim$ tenfold between lumines-

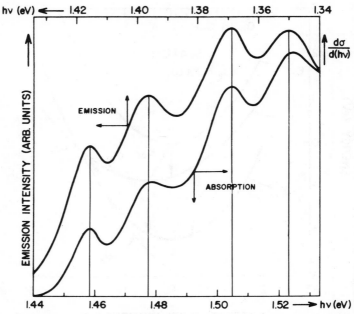

FIGURE 34 Detailed comparison of the photoluminescence (emission) spectrum and the derivative absorption of the optical spectrum (determined from PLE) measured at 4°K through infrared distant DAP luminescence involving the deep O_p donor. The edges near 1.46 eV are pure electronic (Figure 21). The agreement between these spectra suggests that the two-phonons resolved have closely similar energies in luminescence and absorption (Table 3). (Monemar and Samuelson, Ref. 72).

cence and optical absorption.[126] Careful study of the temperature dependence of the PLE and PLQ spectra (Figures 18a and 36b) confirms the earlier evidence (Section 2.10) that the O_p single electron state is pinned to the valence band at least below 175°K.[123] The electronic part of the σ_{p1}^{o} has been de-convoluted from the derivative spectra, and appears to contain two components, separated by ~80 meV. This energy is very close to the spin-orbit splitting in the valence band at the center of the Brillouin zone in GaP.[124] For this reason, and because it is difficult to account for a splitting of only 80 meV between S-like subcomponents of the ground state of the O_p donor, it is tempting to assign the higher energy component in σ_{p1}^{o} to electronic excitation from the split-off valence band. Although Samuelson and Monemar[123] reject this explanation because of difficulties with the temperature dependence observed for this structure, 'he present author believes that this is the most likely interpretation,

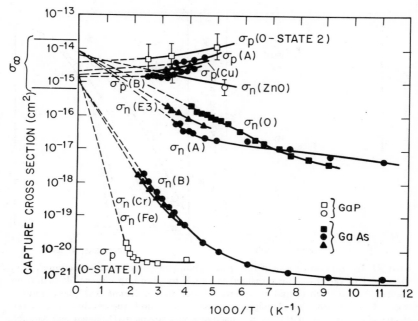

FIGURE 35 Thermal capture cross-sections for electrons, σ_n and holes, σ_p for several deep states in GaP and GaAs, including the two states of the O_p donor and the Zn–O associate in GaP. The identification of the O-related centre (DLTS state EL2) in GaAs is controversial (Section 8.1) but the states labelled Cu, Fe and Cr certainly involve these species. The E3 radiation damage defect in GaAs has been associated with V_{Ga}, while A and B remain unidentified. Capture into most of these states involves strong thermal activation, described by the multiple phonon emission (MPE) theory. The dashed lines show that these cross-sections extrapolate to a common value near the theoretical estimate in the $10^{-15} \rightarrow 10^{-14}$ cm^2 range as T $\rightarrow \infty$. (Henry and Lang, Ref. 119).

particularly since it is also difficult to understand this temperature dependence on the alternative model proposed.[123]

4.2 Capacitance Spectroscopy of the Zn-O Associate

The Zn-O electron trap can also be observed in photocapacitance spectra of appropriately-doped p-n junctions.[89] However, the electron capture cross-section and concentration of this associate, as well as of the O donor, is probably best obtained with the DLTS technique. This method of

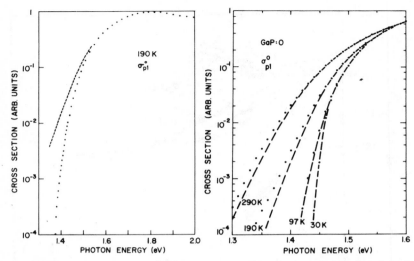

FIGURE 36 (a) Comparison of the optical photoneutralization cross-section $\sigma^o_{pl}(h\nu)$ for state 1 of the O_p donor measured at $190°K$ by photocapacitance (dashed, from Figure 32) and from excitation spectra of the infrared distant DAP photolumines-cence. The additional broadening in the dashed spectrum is apparently caused by the electric field present in the photocapacitance measurement (Monemar and Samuel-son, Ref. 72). (b) Observed (points) and theoretical (dashed) spectral variation of $\sigma^o_{pl}(h\nu)$ for the indicated temperatures. The theoretical form is obtained by fitting to the $30°K$ experimental data, since fitting with the two phonons and coupling para-meters in Table 3 systematically underestimates the experimental broadening at all temperatures. (Samuelson and Monemar, Ref. 123).

capacitance spectroscopy uses temperature as a spectroscopic parameter, through its strong influence on thermal emission rates.[92] Use of detailed balance shows that the thermal emission rate e is related to the electron trapping cross-section σ_n by

$$e = \sigma_n \langle v \rangle N_c \exp(-\Delta E/kT) \qquad (12)$$

where $\langle v \rangle$ is the mean thermal velocity of the electrons, N_c is the con-duction band density of states with the spin degeneracy factor removed for excitations from a relatively shallow donor-like state of depth ΔE below the conduction band. The DLTS technique was first developed with the study of the electronic states of O in GaP in mind. However during the past five years it has subsequently been applied to many deep states, the majority so far still unidentified (Setion 8), in a wide variety of semicon-ductors but particularly for Si and GaAs.[152] Use of this technique with p-n junctions readily allows measurement of minority carrier properties, essential for the study of electron trapping in GaP:Zn,O.

Application of the DLTS technique to the Zn-O trap in GaP immediately revealed one complication particularly significant for measurements on relatively shallow traps, namely the strong enhancement of the electron emission rate by the junction field (Figure 37). This apparently results from an influence of the field on the pre-factor in Equation (11),[92] caused by field-induced tunnelling processes.[153] However it is possible to determine this effect directly by using a clearing pulse of variable height following the forward bias (electron injection) pulse, so that the electron capture rate could be measured as a function of position (and therefore electric field) in the depletion region. The resulting extrapolated estimates of $(\sigma_n)_{ZnO}$ and of $(\sigma_{n1})_O$ are included in Table 5, and the latter agrees well with the result from photoluminescence kinetics.[125] The study of parallel recombination at O and Zn-O in the *same sample*, using previous knowledge of hole recombination times and electron thermal emission rates obtained from photocapacitance,[88,121] and from photoluminescence[125] to ensure correct identification of the DLTS signatures, enabled the relative concentrations and capture rates of these two centers to be compared directly.[92] The results showed that the recombinations through the O donor amount to only $\leqslant 4\%$ of those through Zn-O. Thus the O donor does not contribute significantly to the shunt path recombination in optimally doped p-type GaP:Zn,O. This is true even at 300°K, where σ_n for Zn-O is substantially reduced by thermal re-emission before recombination.[90,121] However, σ_n for the deep O trap is also reduced

TABLE 5

ELECTRON AND HOLE CAPTURE CROSS–SECTIONS[a] THERMAL ACTIVATION ENERGIES AND
OPTICAL CROSS–SECTIONS OF O_p DONOR

Property / Donor State	σ_n (cm^2)	σ_p (cm^2)	Electron emission Activation Energy (eV)	Hole emission Activation Energy (eV)	σ_n^0 (cm^2)	σ_p^0 (cm^2)
O_p^0	2×10^{-18}	5×10^{-21}	1.14 ± 0.06	0.46	$\sim 1 \times 10^{-16}$[c] at 1.1 eV	1.3×10^{-16}[c] at 1.8 eV
O_p^-	3×10^{-19}	6×10^{-15}	0.89 ± 0.06	~ 0	0.8×10^{-16}[c] at 2.1 eV	0.5×10^{-16}[c] at 1.8 eV
$Zn_{Ga} - O_p$	2×10^{-15}	$\sim 10^{-14}$[b]	0.285	~ 0.04		3.5×10^{-16}[d] at 2.18 eV

a Measured at 300 °K in ref (119)
b Measured at 50 °K in ref (125)
c Measured at 170 °K in ref (88)
d Measured at 80 °K in ref ( )

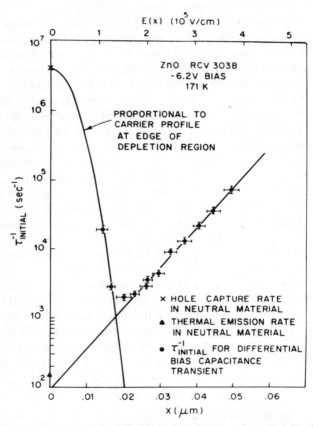

FIGURE 37 The thermal emission rate for electrons from the Zn–O associate measured in a GaP pn junction prepared of liquid phase epitaxy. The relatively fast non-exponential capacitance decay due to electron emission from the relatively shallow Zn–O traps following electron capture during a ~ 60 μsec forward bias (injection) pulse were measured with the rate-profile technique. The profile was obtained using a 'clear' reverse bias pulse V_c of variable height within the range set by the reverse bias of capacitance measurement, here −6.2 eV. The distance scale x, measured from the outer edge of the depletion layer, is obtained from V_c using the usual relation for depletion within an abrupt asymmetric pn junction, corrected for Debye tail recombination during V_c. The variation of τ^{-1} with x beyond x = 0.2 μm is dominated by the increasing influence of ionization induced by the junction field. The increase of τ^{-1} with decrease in x below x = 0.2 μm is attributed to enhanced recombination due to the hole Debye tail from the undepleted p layer. The Gaussian profile is obtained for a hole diffusion length L_D = 44 Å, while the junction is at x = 0.12 μm. (Lang, Ref. 92).

significantly between 100 and 300°K, through re-emission from the shallow 1s(E) excited state (Figure 16) which provides a large capture cross-section at low temperatures.[121] The simplified kinetic analysis of Henry et al.[90] shows that as much as 60% of the electron recombinations in optimally doped p-type GaP:Zn,O may pass through the Zn-O trap, though only about half of these are radiative. Low temperature annealing serves both to increase [Zn-O] 5–6 fold and to decrease the recombination rate through the unidentified shunt path by a factor close to 10.

The line-shape of the Zn-O luminescence band has been analyzed by Henry and Lang[119] using the expressions for the first two moments $<h\nu>$ and $<(h\nu-h\bar{\nu})^2>$

$$h\nu = h\bar{\nu} = E_o - S\hbar\omega \qquad (13)$$

and

$$<(h\nu-h\nu)^2> = <E^2> = S(\hbar\omega)^2(2\bar{n} + 1) \qquad (14)$$

where $\bar{n} = [\exp(\hbar\omega)/kT) - 1]^{-1}$ and E_o is the energy of the pure electronic transition coupled to a single (effective) phonon of energy $\hbar\omega$ in the linear approximation. The value of $S\hbar\omega \sim 0.19$ eV using the no-phonon energy in Table 2, so $S \sim 10.2$ if $\hbar\omega = 18.6$ meV as determined from Equation (13). These values agree reasonably well with the early results of Morgan et al.[21] discussed in Section 1.4. Appreciably smaller and significantly temperature-dependent values of E_e were derived by Dishman and Di Domenico[126] from a study of the temperature dependence of the half-width of the Zn-O absorption band. This study yielded $h\omega = 18.8$ meV, in good agreement with the above, but a much smaller coupling constant. They also concluded that the oscillator strength was ~ 3 times larger in absorption compared with luminescence. However, this result seem to be dependent upon removal of an explicit term for the degeneracy ratio of the initial and final electronic states from the expression for the oscillator strength.

Henry and Lang[119] also discuss the large discrepancy between the value of the electron capture cross-section σ_n *calculated* for the Zn-O trap using parameters from the low temperature luminescence and the value measured from DLTS at 190°K. The large excess in the theoretical value is shown to arise from the non-linear dependence of trap energy on local lattice co-ordinate, also needed to produce the large no-phonon isotope shift (Section 2.2). The average binding energy E_e during low temperature luminescence is ~ 0.47 eV (Figure 6), but is only ~ 0.14 eV during electron capture. The probability that the electron is within the square well used to approximate the short range potential of the Żn-O trap decreases markedly as the effective value of E_e is reduced by such a large factor. This in turn decreases $S\hbar\omega$, which increases the barrier to electron capture E_B since

$$E_B = (E_o - S\hbar\omega)^2 / 4S\hbar\omega \qquad (15)$$

Thus, the electron capture cross-section is reduced. The electron capture cross-section of the Zn-O trap σ_n actually increases with temperature (Figure 35) due to the multiphonon emission non-radiative capture process, which causes σ_n in Equation (12) itself to take an activated form in the high temperature limit[119]

$$\sigma_n = \sigma_{n_o} \exp(-E_B/kT) \qquad (16)$$

4.3 Temperature Dependence of Capture Cross-Sections

The cross-sections for the various traps in Figure 35 were obtained using the pulsed bias technique[127] in DLTS measurements for majority carrier capture in material beyond the depletion layer, made neutral during the forward bias, so as to avoid the electric field effects illustrated for the Zn-O trap in Figure 37.[92] The one exception is the very large hole cross-section σ_{p2} for the second electron state discussed in Section 4.4. This cross-section is so large that capture of photo-generated holes by O traps in the depletion layer of a semi-transparent Schottky barrier had to be employed. The cross-section was evaluated from the measured capacitance change and the junction photocurrent, which is proportional to the injected hole flux.[119] Most of the traps in Figure 35 show thermally activated cross-sections of the form of Equation (16), attributed to the MPE process. Insertion of Equation (16) into Equation (12) shows that the activation energy observed in the electron emission rate, the directly measured quantity in the DLTS method, contains two contributions. Roughly, these are the trap depth ($E_G - E_{O1}$) and the energy barrier to electron emission represented by E_B in the configurational co-ordinate diagram of Figure 33.[154]

4.4 Evidence for a Tightly Bound State for a Second Electron at the O_p Donor

The most remarkable result from the initial photocapacitance studies on GaP:Zn,O double LPE diodes [88,121] was evidence of a second electron trapping state with large binding energy at the O donor. As mentioned in

Section 2.13 such a state is expected from analogy with the behavior of shallow donors. Indeed, the hydrogenic model predicts a finite bound state for any donor, but with ionization energy only $\sim 5\%$ of E_D. The totally unexpected feature of these results on the deep O donor in GaP was the evidence that the binding energy of this second electron state E_{D2} was at least comparable with that of the first, binding energy $E_{D1} \sim 0.9$ eV (Sections 2.7 and 2.12). Earlier indirect, though through hindsight un-doubtedly correct, evidence for such a second electron state of O_p had been obtained from analysis of differential saturation in the infrared dis-tant DAP photoluminescence of GaP:Zn,O.[250] The red Zn-O associate luminescence remained linear or even superlinear. Estimates of the capture cross-section of the second electron led to a limiting value of the trap depth of > 0.4 eV, consistent with the metastable level discussed in Sec-tion 5.4 as well as the relaxed rebonded ultra deep state of O^- discussed here. Clear evidence for the presence of a second deep state is contained in Figure 38. The fact that the dominant deep states associated with the O donor occur mainly on the p side of the pn junction ensures that photo-

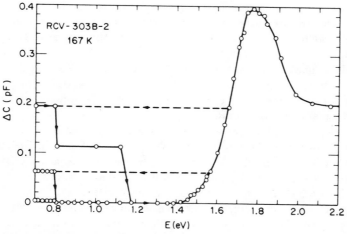

FIGURE 38 Steady-state capacitance changes versus energy of the photoexcitation in an abrupt asymmetric GaP pn junction made by liquid phase epitaxy. The diode was initially cooled from 300°K to 167°K, then reverse-biassed and irradiated with light starting from below 0.8 eV. The capacitance decrease at 0.8 eV is due to the σ_{n1}° threshold (Figure 32). Above the σ_{p1}° threshold near 1.4 eV, state 1 of the 0 donor is repopulated and the junction capacitance rises again. Above 1.58 eV, thres-hold for σ_{p2}° (Figure 32), state 2 begins to fill rapidly and the capacitance increases further. These extra electrons can also be reversibly removed, but only for photon energies above 1.15 eV, the threshold for σ_{n2}° (Figure 40). Note that the sum of the σ_{n2}° and σ_{p2}° threshold energies $\gg E_g$, unlike the sum for σ_{n1}° and σ_{p1}°. (Kukimoto et al., Ref. 88).

ionization of the O donor, electron state 1 in inset, with threshold near 0.8 eV at 167°K, corresponds to a *reduction* in junction capacitance C. The complementary photoneutralization transition, threshold near 1.4 eV, produces an *increase* in C. Provided that the photon energy $h\nu_{EXC}$ is held below 1.58 eV, these two transitions are the only significant effects in a diode which was cooled from 300°K and then reversed biassed so that the few electrons actually on the deep O donor became accessible for detection through photoionization processes in the widened depletion layer. However, large additional changes occur for $h\nu_{EXC} \geqslant 1.6$ eV, where a rapid increase in C indicates a further optical excitation process for electrons from the valence band. These additional electrons are removed only for $h\nu_{EXC} \geqslant 1.15$ eV (Figure 38). Pulse experiments readily showed that the recombination lifetime of the electron in this second state was at least four orders of magnitude shorter than the electron in state 1 of the O donor (Table 5).[121] Proof that these two deep electron traps are nevertheless both associated with the deep O donor was also obtained from these pulse experiments. Pulse reductions in reverse bias of the pn junction, designed to be much too short to remove electrons from state 1 of the O donor by hole capture in the p-type bulk material but adequate to completely remove the electrons from state 2, resulted in an equal and opposite *increase* in the population of state 1. This was revealed by the subsequent behavior under ∼1 eV infra-red light (Figure 39) which is only capable of photoionizing state 1 of the O donor (Figure 38). If these two electron traps involved different centers, removal of electrons from state 2 would leave the long-lived electron population in O donor state 1 unaffected under these conditions.

The electron capture cross-section of this second electron state is only about 5% of that of O state 1, although $\sigma_{p2} \gg \sigma_{p1}$ (Table 5). The photoionization cross section σ_{n2} is strongly temperature dependent (Figure 40), much more so than σ_{n1}^o, just as σ_{p2}^o is more temperature sensitive than σ_{p1}^o (Figure 32). All these features have been interpreted[119] in terms a very large change in energy of state on electron capture, ∼1.56 eV (Figure 39 inset), associated with the large lattice relaxation in the derived configurational co-ordinate diagram (Figure 41). The strikingly small value of the electron capture cross-section σ_{n2}, occurring despite the modest value of the activation energy, $E_A \sim 0.11$ eV in Equation (15), is semi-quantitatively explained by the possibility that the configurational curve of bound state 2 never intersects the conduction band. This is quite plausible according to Figure 41, though the behavior as these states approach one another is hard to predict with precision, using the simplified models inevitable for such configurational co-ordinate descriptions. The hole capture cross-section σ_{p2} was thought to be large and relatively temperature-insensitive

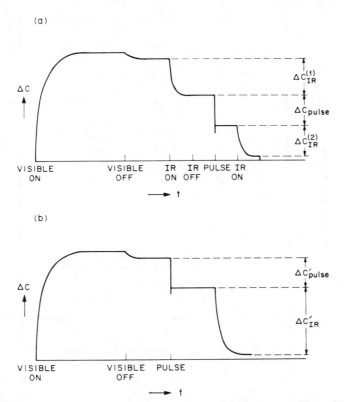

FIGURE 39 Changes in capacitance of the GaP pn junction used in Figure 38, measured at 167°K. States 1 and 2 of the O_P donor are partially filled by initial white light irradiation. Debye tail effects cause the slight capacitance decrease when the white light is removed. In part (a) electrons in state 1 were then completely removed by 1 eV infrared light, giving the capacitance decrease Δ^1_{IR}. The reverse bias was then reduced from 9V to 1V, for a time ~ 50 nsec long enough to completely empty state 2 but not state 1 (Table 5). This results in a capacitance decrease ΔC_{pulse}. However this condition again prepared electrons in state 1 removed by a further 1 eV light pulse, giving a capacitance change ΔC^2_{IR}. In part (b), the bias pulse was applied first after white light preparation, resulting in $\Delta C'_{pulse} = \Delta C_{pulse}$. Then, use of the 1 eV light gives $\Delta C'_{IR} = \Delta C^1_{IR} + \Delta C^2_{IR}$ from (a). These and decay rate experiments prove that states 1 and 2 are two different charge states of a single center (Henry et al., Ref. 121).

(Figure 35) because of the negligible energy barrier in the intersection with the valence band (Figure 41). Some finite barrier such as E_B in Figure 41, albeit small, must exist in order that the O^- state remains sufficiently stable against spontaneous dissociation at ~170°K to permit the experi-

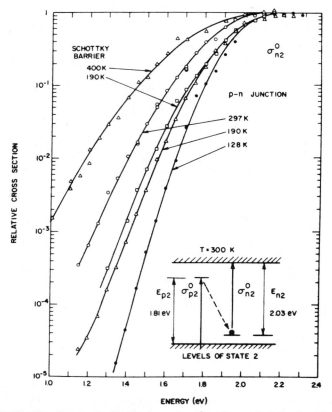

FIGURE 40 The photoionization cross-section σ°_{n2} of state 2 of the deep O_P donor in GaP, measured as in Figure 32. The strong thermal broadening was fitted to the experimental data (points) as in Figure 33a, using Gaussian profiles convoluted with an electronic form with a simple square root edge. The change of slope near 1.2 eV in the 190°K data is caused by thermal ionization of the trapped electron. Thermal effects were subtracted at higher temperatures. The energy levels of state 2 before and after lattice relaxation are shown in the insert, determined from analysis of σ°_{n2} and σ°_{p2} (Figure 32a). The exact magnitude of the electron energy decrease on lattice relaxation is controversial, but it is certainly very large. (Henry and Lang, Ref. 119).

ments described by Henry et al.[121] This contrast with the impression conveyed by the O^- configurational co-ordinate diagram as published by Henry and Lang.[119] We shall also see in Section 5.3 that an intra ($O^- + h$) bound state Auger process is also important at least at the very low temperatures of studies of magnetic resonance effects on optical spectra and for some additional effects on the low temperature luminescence. The

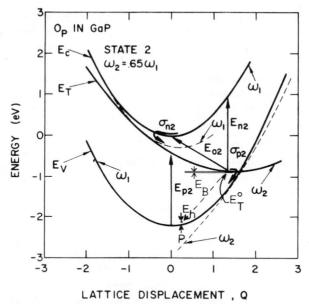

LATTICE DISPLACEMENT , Q

FIGURE 41   Configurational co-ordinate diagram for state 2(O⁻) of O_P in GaP. The diagram was determined from measurement of E_{p2}, E_{n2} and E_{O2}. The quantities ω_1 and ω_2 are the dominant 'mean' lattice vibrational frequencies before and after state 2 is occupied. The dashed curves were used to calculate thermal capture cross-sections (Table 5) with no change in vibrational frequency and strikingly fail to account for the optical cross-sections. The point at P represents the ∼ 43 meV bound hole state at O⁻ which takes part in Auger recombinations at low temperature. E_B is the residual barrier for the MPE process for electrons in the minimum of E_T near E°_T. (Modified from Henry and Lang, Ref. 119).

rapid *decrease* of σ_{n1} with increasing temperature,[121] not shown in Figure 35, is attributed to thermal ionization of the shallow excited state of the O donor (Section 2.9).

Electronic reactions involving the second electron state of the O_p donor are summarized in Table 6 in the Appendix at the end of this article.

5. Controversial Aspects of the Properties of the O Donor

A number of controversies have emerged during the past five years or so, involving closely related aspects of the two deep electron states at the O_p donor. This is probably inevitable in view of the intense interest in this system and the difficulty of obtaining a completely successful quantitative account of all aspects of these states, particularly state 2 where significant discrepancies between the theoretical and experimental widths of the photoionization and photoneutralization cross-section spectra were obtained.[119] It is unreasonable to expect a relatively simple configurational co-ordinate model to give a precise description of all aspects of the behavior of a center exhibiting such large lattice relaxation as is claimed in Figure 41. The first point of controversy concerns the very existence of this large lattice relaxation in the second state 2.

5.1 Can the Second Electron State Be Understood with Small Lattice Relaxation?

We have seen that the relatively strong temperature dependence of both optical cross-sections for state 2 of the deep O donor (Figures 32 and 40) can be given a relatively good interpretation in terms of large lattice relaxation. Theoretical calculations for these energy states also show that a large binding energy for state 2 cannot be obtained without large lattice relaxation, for example the truncated pseudopotential analyses of Jaros and Ross[128] and of Pantelides.[129] However, Jaros and Ross do not support the placement of this level in the lower half of the gap as in Figure 41, considering only a *symmetrical* distortion. Later calculations restricted to linear electron-lattice coupling and symmetrical distortions gave relaxation-induced energy shifts of ~80 meV for electron state 1, in good agreement with experiment (Figure 34) and ~0.55 eV for state 2, although the latter value was not constant for σ_{n2}^o and σ_{p2}^o suggesting the importance of non-linear interaction terms.[130] The much larger energy shift for state 2 claimed in Figure 41, with a Franck-Condon displacement E_{n2}-E_{O2} ~1.2 eV, and final energy E_{n2} over 80% of E_g cannot be accounted for exclusively by a symmetric distortion induced by the extra charge on the O atom in the O^- state. However, approximate calculations indicate that further increases in state energy, perhaps as large as 0.55 eV, can occur if the O atom also relaxes away from a nearest-neighbor Ga atom by ~35% of the Ga-P bond length.[131] Such a relaxation requires that the energy gain from the strong lattice polarization due to the second highly localized electronic particle in the O^- state be sufficient to overcome the energy barrier implied by the symmetric initial configuration of O^{-*} discussed in Section 5.4. There is no nearby p-like excited state available for hybridization to provide a distorted configuration to first order of electron-phonon coupling. The estimated energy gain is believed to be an upper limit to the total energy gain, since this first order calculation sets to zero bonding force constants outside the first Ga shell surrounding the distorted O_p site. We shall see in Section 5.2 experimental evidence that the distortion is along <110> rather than <111> crystal axes. We also show in Section 5.4 that very recent considerations suggest an intermediate, relatively shallow *symmetric* state of O^-, whose existence provides interpretations for some hitherto largely disregarded experimental properties of O 'in GaP and also yields a realistic mechanism for electron capture into the strongly relaxed ground state observed in the photocapacitance experiments.

Grimmeiss et al.[132] obtained the optical cross-section shown in Figure 42 over a very wide range in both σ and photon energy, using bulk 300°K photoconductivity in n-type material, where all the O should be in the O^- state. There is no direct evidence in this type of spectrum whether each threshold shown involves transitions to the conduction band or from the valence band. However, all the structure in Figure 42 was interpreted by photoionization transitions to the conduction band, presumably on the grounds that all donors are rapidly returned to the O^- state by thermal

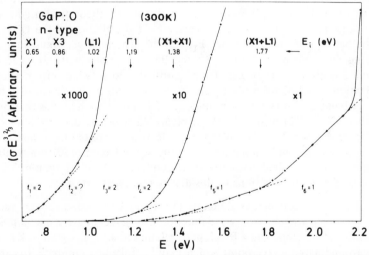

FIGURE 42 Spectral dependence of the photoionization cross-section associated with the O_p donor, measured at 300°K in an n-type GaP single crystal prepared by solution growth. All the O donors should be in the O^- charge state. The cross-section is presented as $(\sigma E^3)^{2/3}$ versus photon energy E according to predictions of the theoretical forms for different degrees of forbiddenness of the optical transition, producing different values of the index f. Seven thresholds were identified including that at E_g near 2.2 eV, in contrast to the single threshold near 1.1 eV in Figure 40. The identification of these extra weak features with absorption processes involving state 2 of the 0 donor is controversial. (Grimmeiss et al., Ref. 132).

equilibrium processes at 300°K. The symmetry of various conduction band minima should determine the form of the absorption cross-section *near to threshold*, through selection rules appropriate to a donor state of definite parity. There are two selection rules, arising from the matrix element of the impurity potential between a given band and the bound state and the matrix element of the dipole operator between dominant band functions in the impurity state and the given band. Thus, a given ionization process can be allowed, once-or twice-forbidden, f = 0, 1 or 2. Then, the form of the absorption cross-section near threshold $h\nu_o$ can be written

$$\sigma \propto (h\nu)^{-3} (h\nu - h\nu_o)^{f+\frac{1}{2}} \tag{17}$$

The analysis given in Figure 42 suggests that the O^- state has p-like symmetry and lies only 0.65 eV below the lowest conduction band minimum (X_1) with transitions to the X_3, L_1 and Γ_1 minima following up to ~1.2 eV. All these processes are weak and apparently involve doubly forbidden transitions. These low energy thresholds were all interpreted in terms of no-phonon processes, despite the strong theoretical arguments that large lattice relaxation is necessary for tight binding of the O^- state such that E_{D2} even approaches E_{D1}.[128-131] The stronger transitions involving O^-, shown in Figures 32 and 38, were not clearly assigned by Grimmeiss et al.[132] They were later identified with a 'two-electron' photo-ionization process, where both electrons are excited simultaneously, one into a band and one into a bound excited state.[133] Thus, the threshold labelled (X1 + X1) near 1.38 eV in Figure 42 which is the threshold σ_{n2}^o in Figure 40, is held to mean that one electron is excited into the lowest X_1 conduction band minimum while the second is excited to a shallow bound state associated with the same conduction band minimum. There are several serious objections to this drastic re-interpretation

1. Selection rules such as Equation (17) apply only over an energy range for which the electron wave-function can be adequately decomposed into an envelope and a Bloch part, relating to a small region of k space around a symmetry point which may be linked by compatibility relations. This would hold over an energy range of at most 0.1 eV for localized electron states in GaP.[140] However, the data in Figure 42 have been reduced by the subtraction of components of form allegedly represented by Equation (17) covering energy ranges extending between 0.4 and nearly 1.0 eV. Analysis of the data in these terms seems totally misconceived. Transitions could not be qualitatively forbidden over such wide energy ranges.[140]

2. These absorption cross-sections are strongly temperature dependent above $\sim 100°K$, as expected on the strong lattice relaxation model. Morgan[133] suggested that the 'two-electron' transitions should exhibit very much larger coupling to the lattice than either one electron transition involving state 1 or 2, but this remains entirely speculative at present.

3. The binding energy of state 2 derived from Figure 42 is 0.24 eV *smaller* than the revised value of the thermal activation energy E_{O2} (Figure 41), obtained from the thermal emission rate measured over a wide temperature range.[119] These latter measurements were made in the field of a Schottky barrier. However, the corrections for electric field effects in such a large activation energy should not be large.[92,134] In any case, electric field effects should produce a slight *underestimate* of the correct (zero-field) activation energy relevant for the low field photoconductivity measurements. It is not clear how this uncorrected thermal estimate can possibly be so much *larger* than the threshold energy for photoionization. Such a low energy for the optical threshold results partly from the view that the lowest energy photoionization processes in Figure 42 are dominantly due to no-phonon transitions even at 300°K. No attempt was made to verify this assignment through detailed measurements of the temperature dependence of the photoconductivity spectrum. Henry and Lang[119] presented strong evidence to the contrary for photoionization at $h\nu \leqslant 1$ eV (Figure 40). They have not been able to reproduce the low energy structure in Figure 42 in measurements which extend over a similar range in σ_{n2}^o, ~ 5 decades, aside from effects which they identify with extraneous impurities possessing lower photoionization thresholds. They believe that the lower energy data of Grimmeiss et al.[132] may be similarly contaminated. However Morgan[133] has made the alternative suggestion that the loss of structure may be due to the electric field inevitable in the photocapacitance measurements.

4. Morgan[133] has also suggested that the further optical absorption edge seen below 1.8 eV in Figure 42 may represent the strong process labelled σ_{p2}^o in Figure 32. However, it is not possible that these features can have the same origin, since this component has been labelled as a 'two-electron' photoionization process, whereas the sign of the capacitance changes shows clearly that it involves the complementary photoneutralization process.[88,119] The presence of such complementary optical processes is supported by the PLE and PLQ studies[72,123] mentioned in Sections 2.14 and 4.1. Only thresholds for state 1 are involved, since the measurements were made on p-type bulk material

in order to enhance the luminescence required in these techniques. These measurements are also complementary to the capacitance studies, in that they can be made at, and indeed are best suited to, the lowest temperatures $\leqslant 150°K$ where the luminescence is strongest. The photocapacitance technique is restricted to $T \geqslant 120°K$ in GaP because of freeze-out of bulk conductivity. Comparison of these techniques in the intermediate temperature range, Figures 18a and 36a suggest that the electric field causes additional *softening* in the absorption edge of state 1, rather than the *removal* of low energy features as needed to reconcile the data in Figures 32, 40 and 42. The comparison of the PLE, PLQ and photocapacitance data also suggests that these effects are very significant only for the relatively weak phonon-coupled absorption in state 1 of the O donor and at temperatures well below 300°K. The PLE and PLQ techniques and the revised form of capacitance technique described in Appendix B of Henry and Lang[119] have the great advantage relative to photoconductivity that optical absorption processes specific to the impurity of interest are discriminated relative to those of extraneous species.

5. Additional temptations to assign small lattice relaxations to the capture of the second electron at the O donor provided from unpublished 1.5°K PLE spectra of infra-red DAP luminescence involving the O donor[135] are at least partially removed by the refutation of the re-assignment of the electron capture (Section 2.9) to the second electron state of the O donor discussed in Section 5.2. The new PLE structure between ~1.75 and ~1.9 eV, (Figure 49a below), the higher energy portion of which is apparently difficult to reproduce, was interpreted in terms of 'two-electron' excitation processes involving quantum interference effects. Analysis in these terms suggests a binding energy of the second electron of 0.92 ± 0.03 eV. This value is barely consistent with the ionization energy obtained from the PLE spectrum of the capture luminescence (Section 2.12), with which it is connected on the revised interpretation advanced in Section 5.2. It is also far from the estimate of 0.65 eV obtained from Figure 42. Very small lattice relaxation was again deduced, despite the problems this brings for the existence of such a large binding energy for the second electron.[128-131] Baraff et al.[202] derive an e-e repulsion energy of ~0.5 eV for the O⁻ center, while Jaros[130] derives $0.2 \rightarrow 0.3$ eV, both far from zero! We give an alternative, very different interpretation for this interesting PLE structure in Section 5.6, where the influence of a relatively shallow, substantially unrelaxed metastable excited state of O⁻ is considered. These problems have also been recognized by Morgan.[138] He emphasized the need for an extended potential, gen-

erated by changes in the crystal parameters for the atoms surrounding the O_P donor, in order to reduce the electron-electron repulsion in the O^- state to obtain a sufficiently large binding energy for the second electron. This suggestion acknowledges the importance of lattice relaxation! The entire debate then simply reduces to whether the lattice relaxation deduced from the photocapacitance lineshapes is, or is not, "excessive".[138]

Considering points (1)-(5) above in the context of the mass of well-documented information on the O donor, frequently with rather precise and apparently successful quantitative analysis, given in the bulk of this review, the present author believes that on balance a convincing case has been made for substantial lattice relaxation in state 2 of the O donor. The current theoretical view[188] is that large lattice relaxation must be invoked to bring E_{D2} to the same order as E_{D1}. However, it is very difficult to reproduce the extreme situation shown in Figure 41, where the electron state 2 has relaxed deep into the lower half of the gap when occupied, even with a simplified theoretical model which is believed to over-estimate the electronic effect of an asymmetric relaxation. One of the main reasons for placing this level so low is to account for the large and nearly temperature-independent value of σ_{p2} within the MPE model (Figure 35). It may be that this temperature-insensitive cross-section is dominated by the Auger recombinations. A prerequisite for this process is that $E_g - E_{n2}$ in Figure 41 must be larger than E_{O1} in Figure 33, a condition which would be more easily reconciled with theory. We shall see in Section 5.3 further evidence that Auger recombinations may be important for the tightly bound O^- two-electron state, in addition to evidence discussed in Section 2.13. We now consider alternative suggestions which have been made for the interpretation of the electron capture luminescence near 0.79 eV.

5.2 Electron Capture Luminescence in GaP:O Revisited

The suggested re-assignment of the luminescence spectrum in Figure 16 to capture processes involving the second electron bound to the O donor,[136] rather than the original assignment to capture of the first electron described in Section 2.9 is intimately linked to the arguments that this state is not associated with strong lattice relaxation (Section 5.1), since the luminescence clearly indicates a modest Stokes shift. Two reasons were initially presented for this reassignment. The first noted that the stability of the O^- state would rule out the possibility of observing the PLE spectrum of this luminescence in bulk material with a Fermi level pinned within a few tenths of an eV below the conduction band by excess shallow donors. When it was pointed out that this spectrum indeed did not appear in such material, but only in crystals with a lower Fermi level consistent with the presence of significant concentrations of the O^o charge state (Section 2.13), this counter evidence was held to be inconclusive for the support of either model.[137]

Thus, the only initial evidence against the assignment of the 0.79 eV luminescence to capture of the first electron was the differences in form of the phonon coupling exhibited by the capture luminescence (Figures 16 and 17) and the infra-red pair luminescence involving the O donor (Figures 13 and 34). It was recognized that a large difference should occur because the no-phonon transition O_o in Figure 16 is parity-forbidden according to the to the assignment in Section 2.9. Coupling to non-symmetric, promoting phonons is required for a significant transition probability. The inconsistencies are therefore reduced to the question whether it is impossible that the 19 and 48 meV replicas from Figure 34, held to be a characteristic of the one electron state of the O donor (Table 3), may not play an appropriate role in the form of the complex sidebands in Figures 16 and 17. Morgan[136] asserted that they could not, though no serious quantitative attempt was made to de-convolute the complex spectra in Figures 16 and 17. No allowance was made for possible influence of the fact that the electronic transitions in Figure 34 are indirect, whereas those in Figures 16 and 17 are direct. The re-interpretation suggests alternative explanations for some of the effects described in Section 2.13, but this is not a particularly critical feature since the existing interpretations are quite feasible. We shall see that there are many strong reasons for retaining the original interpretation, in which the spectra of Figures 16 and 17 represent complementary transitions of the one electron state of the O_p donor. It seems most likely that the problems emphasized by Morgan relate to our incomplete understanding of the intricacies of phonon coupling for very different types of electronic transition at a very deep center in an indirect gap semiconductor rather than to mis-assignments of the charge states and transitions involved.

Very interesting new results emerged from optically detected magnetic resonance (ODMR) studies of the 0.79 eV and other O-related luminescence in GaP. The 0.79 eV luminescence was preferentially excited by 1.55 eV light from a Kr$^+$ laser, though as we have seen in Section (2n) all recombination processes within the band gap of GaP:O remain accessible under these conditions because of two-step excitation processes. The small changes in the intensity of the 0.79 eV luminescence induced by 16.5 GHz microwave photons as the magnetic field was increased from 0T to 1T (Figure 43) are consistent with a spin-triplet state with a D parameter of $+ 2.32 \times 10^{-2}$ meV.[139] Significantly anisotropic g values indicate a center with $<110>$ local symmetry.[139] Detection of the crossing between the

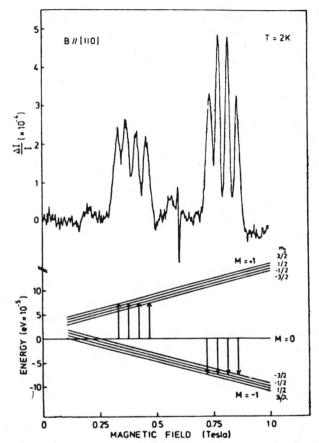

FIGURE 43 Upper, optically detected magnetic resonance measured at 16.5 GHz on the ~ 0.78 eV luminescence associated with electron relaxation within O° in Figure 16. Lower, energy level scheme for a triplet electron-state in an axial field, showing hyperfine interaction with a single Ga nucleus with nuclear spin $I = 3/2$. The angular dependence of the resonances suggest a distorted center with $< 110 >$ axial symmetry. The triplet state probably results from j-j coupling of the two electrons in the O^{-} state, and the ODMR signal may occur in luminescence from the O° state due to an Auger process such as in Figure 46 (Gal et al., Ref. 139).

$M = 0$ and $M = -1$ levels through changes in the total intensity of the luminescence confirmed the triplet nature of this state. Hyperfine interaction between the electron spin and a *single* Ga nucleus[139] was attributed to relaxation effects similar to those suggested as being necessary to obtain a very large energy for the O^{-} state.[131] Similar ODMR signals were observed on the infra-red O^{-} related DPA luminescence and on the Zn acceptor BE luminescence in samples which also showed the 0.79 eV luminescence, although the polarization properties were weaker for detection of the other types of luminescence. Mechanisms were suggested to account for these observations on the assumption that the triplet state observed on ODMR is the O^- state of the O_P donor and that this triplet state is the emitting state for the 0.79 eV luminescence in Figure 16. The second of these assumptions supports the re-assignment of Morgan.[63] However, it begs a number of serious difficulties most of which were disregarded both by Morgan[63] and by Gal et al.[139] These difficulties are

1. Analysis of the PLE of the 0.79 eV luminescence (Section 2.12) and of the IR DAP luminescence (Section 2.7), both of which have the spectral form expected for the single electron state of the deep O_P donor, yield concordant values of E_D to better than 1 part in 900. It is difficult to believe that this is merely a coincidence, or that an excitation within O^- can ape so closely the form expected for excitations to the shallow Coulomb excited states within O°.

2. We have seen that all agree that the O° state has a binding energy near 0.9 eV, with quite small lattice relaxation associated with removal of the single electron. Multiphonon emission processes are therefore unlikely to occur, consistent with the type of temperature dependence observed for the relevant capture cross-section σ_{n1}.[119,121] The only possible capture process at low temperature in dilutely co-doped material seems to be a radiative process from the shallow excited state necessary to account for the large *increase* in capture cross-section with decrease in temperature below 300°K.[53,121] However, the 0.79 eV luminescence in Figure 16 is the only one available to assign to this process, at least above ~0.35 eV where exhaustive measurements of GaP:O luminescence have been made.

3. The close agreement between the *magnitudes* of no-phonon isotope shifts in the 0.79 eV and in the infra-red O-related DAP luminescence when $O^{16} \rightarrow O^{18}$, with opposite *signs*, (Tables 1 and 3) is fully consistent with the assignment of both bands to transitions of the single electron bound to O_P, (Section 2.5 and 2.9). The agreement in the magnitudes of these shifts to well within the combined experimental errors of ~4%, must be accepted as another unexpected coincidence on the revised interpretation. The magnitude of these no-phonon isotope shifts is governed by the changes in charge distribution local to the impurity atom during the electronic transition.[27] It is physically implausible that the changes in local charge distribution could be so closely similar for reactions involving the first and second electrons bound to a deep donor, particularly if little lattice relaxation can

occur in either process because of the form of the optical spectra assigned to them both on this model.

4. The weak phonon coupling observed for the 0.79 eV luminescence and the evidence of significant lattice distortion in the triplet state responsible for the ODMR signal, makes it necessary to assume two widely separated electronic energy states of O^- with comparatively little lattice relaxation between them. This is thought to be most implausible.[140] The assumption that similar strong lattice relaxation occurs in both these energy states also removes the attractive interpretation of the rapid *increase* in σ_{n1} with decreasing temperature[53,121] in terms of thermal ionization of the shallow excited state of O^o from which luminescence occurs.

5.3 Zeeman and Further Uniaxial Stress Measurements on the Capture Luminescence

The existence of the severe difficulties for the re-assignment of the 0.79 eV luminescence to the second electron state O^- led to further attempts to resolve the two models.[141] Zeeman measurements were not entirely conclusive, since a triplet structure is expected on either model (Figure 44). The models are distinguishable only by the extent of thermalization between the magnetic subcomponents since the *full* magnetic splitting occurs in the triplet luminescence excited state for the O^- model (Figure 44b), but only *half* for the O^o model (Figure 44a). The experimental intensity ratio of the outer components, similarly circularly-polarized on both models, was consistent with the O^o model, assuming slight heating of the spin population to $\sim 3.6°K$ above the bath temperature $\sim 1.8°K$ under the typical high optical excitation conditions. Such a degree of heating was supported by changes in this ratio observed under reductions in the optical pumping rate. The decay time of the 0.79 eV luminescence is ~ 10 μsec at $5°K$,[196] amply long enough for strong thermalization to occur between the magnetic substates before luminescence occurs. This relatively long lifetime is quite expected from the forbidden nature of this transition discussed in Section 2.9. The fact that the triplet state of Gal et al[139] is unthermalized is consistent with the fast nonradiative recombination processes expected if it is assigned to the O^- state.[119,121] This situation then lessens the significance of the considerable intensity observed from the highest energy magnetic subcomponent in Figure 44 at high fields if the luminescence is assigned to the O^- state. However, the intensity ratio of the outer magnetic sub-components was observed to decrease significantly as the magnetic field increased, just as expected if

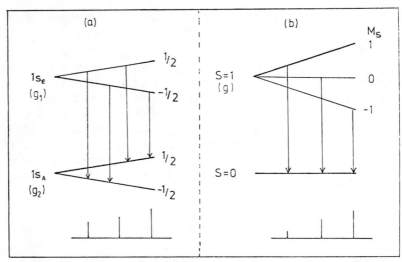

FIGURE 44 Two possible interpretations for the Zeeman splittings observed for no-phonon luminescence line O_0 in Figure 16. In (a) the model shown in the inset of Figure 16 is assumed. Equal magnetic splittings appear in the ls(E) initial and ls(A_1) final states, as expected for an s = ½ electron, with g close to 2, appropriate in GaP. In (b) the initial state is assumed to be a spin triplet, again with electron g value close to 2. The models differ only in their predictions for the relative intensities of the magnetic subcomponents, as shown at the bottom. The experimental data accord better with model (a). (Gal et al., Ref. 141).

thermalization was occurring. Thus, these Zeeman results strongly favor the O^0 model, though they do not provide an absolutely conclusive distinction.

Further strong support for the O^0 model was obtained from uniaxial stress effects, capable of demonstration for the no-phonon line with the improved detectors available in the recent work[141] (Figure 45). The O^0 model of Section 2.10 predicts only *line-shifts* under [111] stress, because the 1s(E) luminescence excited state remains unsplit[55] as the conduction band minima lie along the [100] axes of the Brillouin Zone. The uniaxial stress data in Figure 45 conform with this expectation. They are inconsistent with the triplet O^- model, where behavior similar to that observed [100] stress is also expected under < 111 > stress for a center with the < 110 > symmetry axis consistent with the ODMR results. The observed splittings for [110] and [100] uniaxial stress (Figure 45) are indicated from the O^0 model. The 1s(E) excited state splits for each of these stress directions into two subcomponents, involving different contributions from the initially degenerate conduction band minima which become energetically inequivalent under these circumstances.[55]

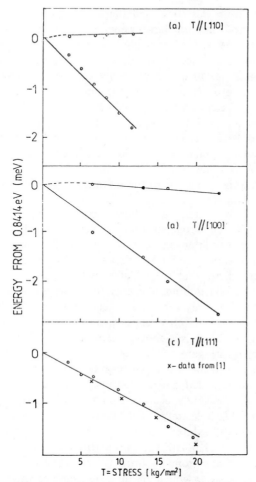

FIGURE 45 Uniaxial stress effects on the no-phonon luminescence line O_o in Figure 16, measured at temperatures $\sim 20°$ K such that thermalization of intensity between the stress-split subcomponents was not too large. The x data in part (c), stress T [111] were taken from Ref. 16. The absence of a splitting for this stress direction is in agreement with the assignment of the luminescence in Figure 16, since the conduction band valleys in GaP remain degenerate under this trigonal perturbation. (Gal et al., Ref. 141).

These stress data, combined with the magnetic data and the many other strong arguments listed above, led Gal et al.[141] to reinstate the original interpretation for the 0.79 eV luminescence given in Section 2.9. We must still account for the triplet state detected by ODMR (Figure 44). Gal et al.[141] have suggested that this is indeed related to the O^- state. The lack of luminescence from this state, and the short lifetime suggested by the weak thermalization and the large value of σ_{p2} (Table 5) are all consistent with expectation for a strongly lattice-coupled system.[119,121] The occurrence of the resonance on the 0.79 eV luminescence and the other O-related luminescense involving the single electron state suggests some fast, non-radiative process connecting the one and two-electron systems. Memory of the two-electron triplet spin state carried via the one electron emission is also required to account for the strong polarization dependences observed between the resonances obtained for detection of the oppositely circularly polarized σ_+ and σ_- components of the 0.79 eV luminescence. A possible process is shown in Figure 46. This intra-O$^-$ state process is to be preferred to the intercenter process involving a distant acceptor advocated by Gal et al.[141] since the ODMR spectra are observed in crystals containing shallow acceptor concentrations of less than $\sim 10^{17}$ cm^{-3}. This concentration is well below that required for the generically similar inter-center Auger process involving the one electron luminescence state described in Section 2.13 (Figure 20). The model in Figure 46 recognizes that the O$^-$ state will produce a shallow bound state for a hole, similar in binding energy to the acceptor-O$_P$ associates in Table 2. There will be only very small exchange interaction between this hole and the $J = 1$ two-electron state because of overlap considerations, rather as found for the Bi isoelectronic trap in InP.[195] This residual interaction is neglected in Figure 46. It is also assumed in Figure 46a that the local relaxation due to the second electron in O$^-$ causes a reversal in sense of the local crystal field, from the tensional form exhibited by the acceptor cation-O$_P$ associates (Section 2.3) to a compressive axial stress. Then, the $m_j = \pm 1/2$ hole states of the stress-split valence band lie uppermost, and the hole will exhibit magnetic properties closely similar to either electron. The form and polarization properties established in the ODMR spectra can then result from an intra O$^-$ center Auger (non-radiative) recombination process in which the hole recombines with one electron leaving the second, raised either to the 1s(E) state or to the conduction band from which recapture occurs to the 1s(E) emitting state for the 0.79 eV luminescence. The former process, occurring with no change of electron spin of the remaining electron, is presumably required to account for the strong polarization-dependence observed in the ODMR spectra for the sets of resonances occurring at the two different fields in Figure 43. Such an Auger process is highly likely for this O$^-$ + h complex in view of the very strong overlap between the two tightly-bound electrons in the O$^-$ state.

Although the vertical energy difference $E_g - E_{n2}$ is only ~ 0.25 eV in Figure 41, much less than $E_{1s(A_1)} - E_{1s(E)}$ within O^o, this Auger process

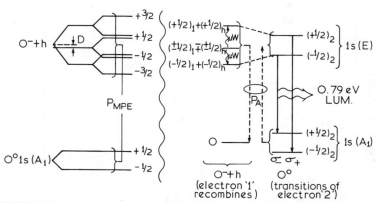

(a) MULTIPLE PHONON
       EMISSION

(b) AUGER RECOMBINATION WITHIN
(O^-+h) → LUMINESCENCE WITHIN O°

FIGURE 46 An Auger model whereby ODMR between the magnetic substates of the O^- state of the deep O donor in GaP, shown as hv to the left, can influence the recombination rate and luminescence intensity of transitions within O_0 responsible for the ~ 0.78 eV luminescence in Figure 16. One electron within O^- is assumed to recombine with a hole bound weakly to the O^- center, while the second electron to the 1s(E) state by a spin-conserving process. For simplicity only recombination from the $M_S = \pm 1$ states of O^- is shown. (Dean, Ref. 196).

can occur for the fully relaxed state of O^- since the hole effectively recombines from a value of lattice co-ordiante Q much closer to the origin in Figure 41 than the region labelled σ_{p2}. However the transition probability for this non-vertical Auger process will be strongly reduced in exactly the same way as the probability of radiative recombination. The reduction will be very large for the huge lattice relaxation shown in Figure 41. We consider in the next Section the further possibility that this Auger process involves a newly recognized metastable excited state of O^- in which very little lattice relaxation occurs as a result of capture of the second electron.

5.4 Further O-related Luminescence and Optical Absorption and Theoretical Predictions of a Centered Metastable State of O_p

Recent studies [141] have uncovered a new infra-red luminescence spectrum which appeared to be a persistent feature of O-doped GaP. These initial measurements were made on bulk GaP, and it is now clear that limitations of material quality produced excess broadening of the highest energy weak component near 0.528 eV, to a width ~ 1.6 meV approaching that of two

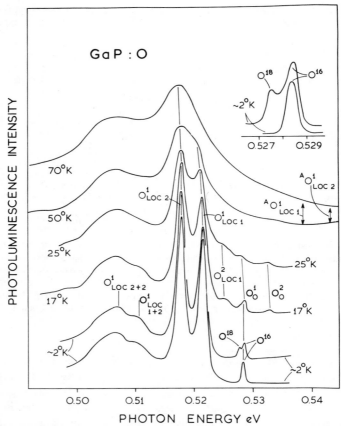

FIGURE 47 Low temperature photoluminescence in GaP:O which is only observed when the O-capture luminescence of Figure 16 is present. The no-phonon line O_0^1 near 0.528 eV, shows an energy shift when $O^{16} \rightarrow O^{18}$ as in the inset. The phonon-assisted lower energy components mainly involve two local promoting-type modes of T_2 symmetry, with energies of 6.8 (LOCI) and 10.4 meV (LOC2). A second no-phonon component O_0^2 with similar phonon replicas appears in spectra recorded at $> 20°$K, (Dean et al., Ref. 223).

stronger components near 0.520 and 0.517 eV. This encouraged the assignment of all three of these components to no-phonon transitions.[141] This interpretation would have required appropriate splittings in the *final state* of this luminescence process, to permit the observed spectral form including temperature-independent relative intensities below 10°K. No explanation was offered for these states, partly because Gal et al.[141] did not clearly establish a connection of this spectrum with O_p. More recent work,[196] with solution grown GaP single crystals, has shown that the width of the $\sim$0.528 eV component can be as low as 0.7 meV, still limited by the resolution of the spectrometer (component O_o^1 Figure 47) but much smaller than the widths of $\sim$1.7 meV for components O_{LOC1}^1 and O_{LOC2}^1. This immediately suggests that only the $\sim$0.528 eV component O_o^1 involves a no-phonon transition. The recent work also shows an O-isotope shift in this no-phonon line of -0.82 ± 0.03 meV when $O^{16} \rightarrow O^{18}$. This is distinctly larger than that for the luminescence spectrum in Figure 17, attributed to electron capture into the one-electron state of the deep O_p donor in Section 2.9, -0.72 ± 0.02 meV. Three important points should be noted. First, the observation of this isotope shift clearly establishes the relationship of this spectrum with O. Second, the *sign* of this isotope shift indicates that the transition responsible for the luminescence in Figure 47, like that in Figure 17, involves an increase in local antibonding electronic charge at the O center (Section 2.2). Finally, the *magnitude* of the shift is consistent with a transition involving the T_d-symmetry O_p center, as we discuss in Section 2.8.

Several years ago, Bachrach[205] observed a sharp optical absorption line near 1.738 eV at 4°K, with associated generally broader structure to higher energies (Figure 48). This absorption appeared to relate to the presence of O, although it was not possible to be dogmatic about this assignment in view of the difficulties mentioned in Section 2.2. Bachrach believed that this absorption might involve the creation of an exciton bound to O_p^o. We can now show strong evidence in support of this hypothesis, which was very tentative when first suggested particularly because of the obvious incompatibility of the spectrum in Figure 48 with the heavily relaxed state of O^- revealed in the photocapacitance results described in Section 4.1. However, one particular observation from the early work tended to support this assignment. The absorption in Figure 48 was observable only for those crystals with Fermi level between $\sim$0.9 eV and $\sim$0.55 eV below the conduction band.[205] This is exactly the criterion discussed in Section 2m for the observation of the structured PLE spectrum of the capture luminescence into the one-electron state of the O_p donor, consistent with the attribution of *both* these absorptions to electronic transitions involving the *neutral* donor O_p^o. The absorption line is also consistent with the negative dip observed near 1.739 eV by Samuelson and Monemar[135] in the PLE spectrum of the O-related DAP luminescence (Figure 49a) whose form suggested to them an inter-bound state transition of negligible oscillator strength, superimposed upon a relatively strong absorption continuum due to the photoneutralization of O_p donors. However, more recent PLE work[206] has shown that the complex structure above 1.77 eV reported by Samuelson and Monemar is not a straightforward feature of a pure optical absorption process involving isolated O_p,

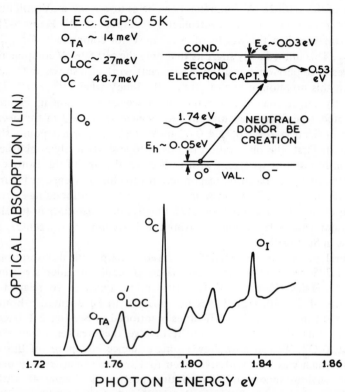

FIGURE 48 Optical absorption observed in the near infra-red at low temperature
for GaP:O crystals co-doped so as to leave the Fermi level between ~ 0.5 eV and ~
0.9 eV below the conduction band minimum (Bachrach, Ref. 205).

since these later results are consistent simply with the negative image of
the ~1.738 eV no-phonon and higher energy phonon assisted transitions
superimposed upon the featureless absorption continuum. Indeed, the
absorption measurements of Bachrach indicate an oscillator strength of
the no-phonon absorption line O_o *alone* of ~10^{-3}. Considering the magnitude of the no-phonon sidebands in Figure 48, together with the fact that
the oscillator strength of the no-phonon transition of the shallow S donor
BE, which has comparatively minor phonon replicas, is also ~10^{-3} in
GaP,[215] we conclude that the oscillator strength of the 1.738 eV transition
is not negligible but actually rather large. It is relatively easy to detect the
1.738 eV absorption line through the internal absorption it induces in the
broad band luminescence frequently present in the same spectral region[196]
(Figure 49b). Indeed, this is often the optimum way of observing this

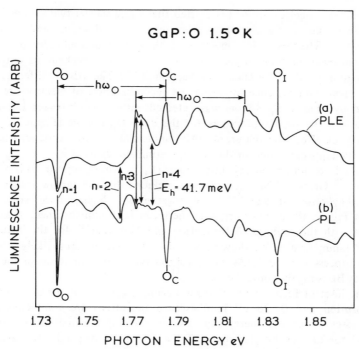

FIGURE 49 (a) Small portion of the low temperature σ°_{p1} optical absorption cross-section measured by PLE of the O–C distant DAP luminescence in semi-insulating GaP at low temperature. Details of the spectrum above the principal no-phonon line O_0 contrast strongly with the conventional absorption in Figure 48 and that shown in more detail in (b) recorded through internal absorption of underlying broad band luminescence. The arrows indicate excited states of the hole within the $D.^{\circ}X$ BE. Spectrum (a) courtesy of Ref. 135 (Dean, Ref. 196).

line, particularly for small crystals of irregular external habit. The relative *form* of the phonon sidebands in Figure 48, very different from Figure 47, also suggests an allowed transition of appreciable oscillator strength. We present in Section 5f a straightforward explanation for the striking differences in form between spectra (a) and (b) in Figure 49 above ~ 1.77 eV.

The absorption spectrum in Figure 48 was actually reported 18 years ago, in pioneering studies of long wavelength (red or near infra-red) excitation of the green DAP luminescence of solution grown GaP crystals.[216] Gross and Nedzvetskii correctly concluded that a two-step excitation process involving a deep state center must be involved, and argued against the participation of Cu. However, they misunderstood the effect on such DAP spectra of the low volume density excitation provided by this type of extrinsic mechanism, and suggested that some special two step process not involving the transfer of electrons into the conduction band must be operative. The excitation mechanism they used undoubtedly involved the one-electron state of the deep O_p donor, with mechanisms entirely analogous to those discussed in Section 2.12. The trivial difference is that green luminescence involving the recombination of electrons recaptured by shallow S_p donors was detected, rather than infra-red luminescence due to similar processes for the deep O_p donors (Figure 13). A variety of effects on the green DAP luminescence can be produced using simultaneous excitation in the red, quantum energy $h\nu > E_G - (E_D)_o$ and infra-red, quantum energy just above $(E_D)_o$ which support this viewpoint.[249] Gross and Nedzvetskii also concluded that the narrow absorption line they observed near 1.738 eV, together with higher energy structure as in Figure 48, did not promote the green DAP luminescence, in agreement with the findings of Samuelson and Monemar[135] for the infra-red DAP luminescence. We show in Section 5.10 that the most likely relaxation process for the 1.738 eV transition interpreted in terms of the O_p donor BE is highly nonradiative.

The inset of Figure 50 contains a diagram suggesting that the 1.738 eV absorption and 0.528 eV luminescence transitions are complementary processes within the energy gap of GaP, involving the first two reactions listed for the second electron in Table A1 in the Appendix (metastable centered O_p^-). This assignment is supported by six factors.

1. The energies are consistent with this assignment if we assume that both transitions involve shallow bound excited states as is required to account for the narrow, relatively temperature-independent widths of the no-phonon lines in Figures 47 and 48. Thus, if we assume that luminescence proceeds from an excited state of binding energy $E_e \sim 40$ meV, while the absorption process produces a bound hole E_h also ~ 40 meV, then the relationship from Figure 48 inset for the luminescence energy $h\nu_L$ agrees with experiment

$$h\nu_L = E_G - (1.738 + E_e^* + E_h) \qquad (18)$$

The absorption transition illustrated is effectively the neutral O_p donor BE photocreation process suggested by Bachrach[205] and for long the subject of speculation. The value of E_h assumed in Figure 48 is very reasonable, since the hole is bound to the O_p^- center only by the long range Coulomb potential of the additional electronic charge. A hole binding energy comparable with the effective mass binding energy of acceptors, $(E_A)^{EM}$ should result, slightly reduced by the

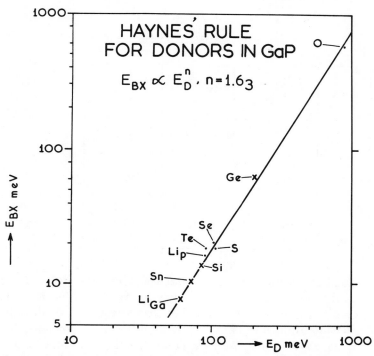

FIGURE 50 The variation of the BE localization energy with E_D for 9 donors in GaP including O_P. The x points refer to Ga-site donors, points refer to P-site donors. The inset shows the relationship between the bound exciton transition of Figure 48 and the electron capture luminescence of Figure 47 (Dean, Ref. 196).

delocalization of the charge of the excess electron compared with the core structure of a real acceptor. The value of $(E_A)^{EM}$ is $\sim$56 meV for GaP.[207] In fact, we give detailed consideration in Section 5.6 to the form of the structure near 1.77 eV in Figure 49 and thereby derive an accurate value of $E_h = 41.7 \pm 0.3$ meV, thus providing an accurate value of E_e through Equation (18), $E_e^* \sim 42 \pm 1$ meV since $E_g = 2.350 \pm 0.001$ eV.[60]

2. Sturge[206] has recently observed an *increase* in the energy of the 1.738 eV absorption line when $O^{16} \rightarrow O^{18}$. Once again, this removes any speculative element in the assignment of this absorption to O. His estimate of $\sim$ +0.65 $\pm$ 0.15 meV has been confirmed in sign but not magnitude by more recent studies of Dean et al.[223] who observed the 1.738 eV line most strongly through internal absorption superimposed upon unrelated broad band luminescence. Their isotope shift of +0.42 $\pm$ 0.05 meV (Table 6) is only about half of the value for the

TABLE 6

g Values from D^o,X BE in GaP

Donor	Ionisation Energy (meV)	Electron g_e	Isotropic Hole K	Anisotropic Hole L
Te	92.6	1.9_9	0.8_2	0.09
Se	105	1.9_9	0.7_8	0.11
S	107	1.9_8	0.7_1	0.14
Ge	204	1.9	0.58	0.15
O	898	2.0_0	0.9_2	0.02_3

0.528 eV electron capture luminescence line associated with transitions to the same O^- state according to the inset of Figure 50. However, the two transitions differ in two respects. First, there is the presence of the hole in the O_p^o donor BE believed responsible for the 1.738 eV transition. We have shown that E_h = 42 meV, while the electron binding energy in the ground state of metastable O_p is E_e = 570 meV from the D^o, X BE transition energy. The electron-hole correlation within the BE will reduce the isotope shift compared with the second electron capture PL. This reduction is expected to be only of order E_h/E_e. Therefore, it should be less than 10%, only a small proportion of the reduction observed experimentally, 49 ± 7% according to Dean et al, though it is more consistent with the apparently less accurate result of Sturge[206] which produces a reduction in isotope shift of 20 ± 18%. Further reasons for a larger isotope shift on the 0.528 eV transition compared with the 1.738 eV transition according to the interpretation of Figure 50 inset are discussed in Section 5.9. No isotope shifts are observed in the phonon sidebands of Figure 48, consistent with the expectation that symmetric receiving vibrational modes which do not involve motion of the O atom should predominate in this strongly allowed BE transition. This contrasts strongly with the behavior in the forbidden capture luminescence within O_p^- (Section 2.8).

3. The interpretations discussed in (1) and (2) are strongly supported by recent theoretical/self consistent (Greens function) calculations of Baraff et al.[203] These calculations predict a bound state of O^- with

relaxation following electron capture into the one-electron state (Figure 16) and a binding energy ~ 0.45 eV less than unrelaxed $(E_D)_O$, or ~ 0.53 eV after allowing for the ~ 0.085 eV lattice relaxation (Table 3). This value is very close to that predicted for unrelaxed O^- from the screened, rather than truly self-consistent, pseudopotential theory of Jaros,[166] $(E_t)_O - \sim 0.6$ eV. The bound state wavefunctions derived by Baraff et al and Jaros are also essentially identical. The properties of this centered state of O^- are contrasted in Figure 51 with those of the much deeper, heavily lattice-relaxed state derived from the photocapacitance experiments described in Section 4. This centered state O^- must be metastable with respect to the asymmetrically heavily-relaxed (rebonded) state. Any other form of configurational curve, for example one containing only an inflexion point or lesser perturbation rather than a true subsidiary minimum at the centered lattice co-ordinate, could not produce the type of spectra shown in Figures 47 and 48 containing sharp no-phonon lines and relatively weak phonon coupling. Thus, despite the fact that the only optical transitions seen are those in which the centered O^- is the final state, the expectation is that hot optical processes could not be confined to such a narrow range of configuration co-ordinate. Such hot processes are the only ones possible in the absence of a subsidiary minimum in the E, q curve. There are no general arguments, from symmetry or elsewhere, requiring a true potential barrier for displacements of q away from the centered value. The accuracy of the Greens function calculations is not adequate to decide whether or not such a

GaP:O

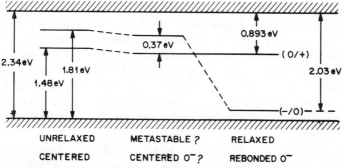

FIGURE 51 Level structure for GaP:O including the single electron state in unrelaxed (left) and relaxed (center) forms and the two electron state in unrelaxed (left), symmetrically relaxed (center) and fully relaxed (rebonded) form (right) (Baraff et al., Ref. 203).

barrier exists. The only available evidence on this point is the *positive* sign of the spring constant deduced from the extrapolation to centered O^- from the observed properties of O^0. The corrected value deduced from the analysis in Section 5.8 of the spectra described here is still smaller, yet also positive. Theory tells us little at present about the rate of reaction between the centered metastable and relaxed rebonded state.[203] Experiment merely sets an upper limit through the use of the uncertainty relationship of $\sim 10^{-12}$ sec^{-1} if the observed no-phonon linewidths of ~ 0.7 meV are wholly due to lifetime broadening.

4. The value of the BE localization energy E_{BX} derived from the no-phonon transition O_o in Figure 48 by the usual procedure $E_{BX} = E_{GX} - (h\nu)_{O_o}$ falls *remarkably* close to the trend established for *all* the 8 other donors so far identified in GaP when plotted in the Haynes' rule form,[221,185] $E_{BX} = f(E_D)$, as shown in Figure 50. A remarkable feature of Figure 50 is that this uniform trend is exhibited independent of whether the donors are Ga or P-site substituents, despite the different symmetry and strength of interaction with the impurity central cell typical for the different lattice sites.[32] The trend describes Li_{Ga} as a donor with $E_D = 61$ meV, close to the effective mass value of 59meV[51], as well as very deep donors such as O_p, E_D nearly 900 meV. The behavior is also relatively unaffected by whether the donor behaves normally with respect to the symmetry of the deepest one-electron bound state like Sn_{Ga} and Si_{Ga}, or abnormally, like Ge_{Ga}.[222] Clearly the trends in central cell effects are remarkably uniform for the first and second electrons to bind in the donor ground state, despite the large electron-electron repulsion. The effect shown in Figure 50 represents a substantial deviation from the normal form of Haynes' rule,[221,31] where the index n = 1. The conformity of the data for O_o from Figure 48 to the trend established for the other donors strongly supports the identification with the O_p donor BE. Theoretical support for the superlinear trend of E_{BX} with E_D in Figure 50 has been obtained with a modified Hartree Fock calculation in which the electrons are divided, core and conduction band, with separately adjustable effective masses and screening constants.[251] The core electrons are adjusted to account for the central cell contributions to E_D.

5. The magneto-optical properties of the 1.738 eV no-phonon line observed in Voigt configuration and B = 10T (Figure 52)[223] are very similar to those established for the S neutral donor BE[224] and fully consistent with expectation for the O neutral donor BE. Only one

component is significant in π polarization (E∥B) and two in σ polarization (E⊥B) because of strong thermalization between the magnetic subcomponents in the initial state, D^0 (Figure 52 inset nomograph). The $D^0_?X$ BE is expected to behave magnetically like a free hole, because the two electrons couple to give $J_e = 0$ as required by the exclusion principle if both electrons are to take maximum advantage of the deep short range potential on the O^+_p core in the ground state of O^-_p (metastable). The hole and $D^0_?X$ BE will have magnetic quantum number $J = 3/2$ at the T_d-symmetry O_p (metastable) site, and can be represented by the magnetic Hamiltonian[31]

$$H_{exc} = K\mu_\beta \underset{\sim}{J} \cdot \underset{\sim}{B} + L\mu_\beta (J^3_x B_x + J^3_y B_y + J^3_z B_z) \qquad (19)$$

where μ_β is the Bohr magnetron, K and L are the isotropic and anisotropic components of the hole g value. The neutral donor has a very simple isotropic Zeeman behavior in GaP,[31]

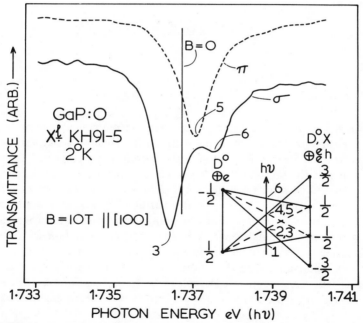

FIGURE 52 Zeeman splittings of the 1.738 eV absorption line at B = 1OT and 2°K, for B∥ < 110 >. Data are shown both for E ∥ B(π) and E ⊥ B(σ). The inset shows an interpretation in terms of transitions exclusively from the lower magnetic substate of the neutral donor, strongly favored by thermalization under these conditions (Dean et al., Ref. 223).

$$H_{gr} = g_e \, \mu_\beta \, \underset{\sim}{S} \cdot \underset{\sim}{B} \qquad\qquad (20)$$

where g_e is the electron g-value. The magnitudes of g_e, K and L derived from the fan diagram in Figure 53, taken over a range of large magnetic fields where the magnetic sub-components are very well-defined at $2°K$, are consistent with those from D^o, X BE involving the shallower P-site donors in GaP (Table 6). The larger K and smaller L for the O_p donor accords with the anticipated tighter binding of the hole according to arguments of Bimberg et al.[232] Diamagnetic terms have been neglected in Figure 19, consistent with the linear trends in Figure 53. The predicted diamagnetic shift on n 'isoelectronic acceptor' model of the $D^o X$ BE for $m_h^* = 0.5 \, m_o$ and $E_h = 42$ meV is only 0.08_3 meV even at 23T. Morgan's new molecular orbital model for the electronic configurations of O_p in GaP [233] briefly described in Section 5.12, predicts quite different symmetries for the states of O_p involved in the D^o, X BE photocreation process, namely (Figure 58 below)

$$^4T_1 + h\nu \rightarrow {}^3T_1 + h_B \qquad\qquad (21)$$

instead of

$$^2A_1 + h\nu \rightarrow {}^1A_1 + h_B \qquad\qquad (22)$$

on the classical model in which all five electrons contributed by the O ion enter the valence bonding with the surrounding Ga atoms. If we make with Morgan the further rather strong and unexpected assumption that the orbital magnetic character of the T_1 state is completely quenched by strong confinement of the antibonding electrons within different dangling bonds on the surrounding Ga atoms and also neglect the spin orbit coupling, then the two models differ magnetically only in the spin-degeneracy of the electron states. Morgan has also pointed out[234] that these two models then cannot be distinguished from observations on transitions from a single, heavily thermalized magnetic substate of D^o. He notes that transitions from a single substate with total electron spin $S = n/2$ to an $n + 1$ electron state of spin $(n - 1)/2$ plus a hole give energies and relative intensities independent of n, and similar to those of Figure 52 inset. Evidence as to the value of n, 1 as in Equation (22) or 3 as in Equation (21) requires measurement of the relative *intensities* of hot magnetic subcomponents. The *energies* of these hot absorption subcomponents are again model-independent. The requirement of very large magnetic field neces-

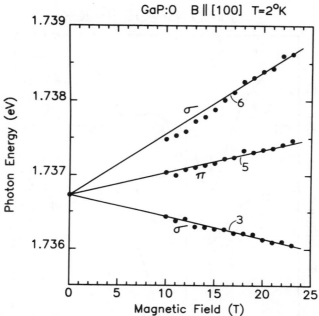

FIGURE 53   Magnetic fan diagram for the polarized Zeeman subcomponents of the O_p donor D^o, X BE no-phonon line up to B = 23T (Dean et al., Ref. 223).

sitated by the large limiting zero-field linewidth of the 1.738 eV absorption line, discussed further in Section 5.10, also necessitates a relatively high temperature for significant intensity in the hot bands. There is an optimum ratio of magnetic field and temperature, since significant increase in linewidth occurs above 10°K. Optimum conditions were found near B = 20T and T = 20°K (Figure 54).[223] Five of the six predicted magnetic subcomponents were clearly observed, and the missing σ component 1 is predicted to be weak and difficult to observe under the prevailing experimental conditions. According to a transition nomograph based upon Equation (21), the intensity of the hot to cold clearly resolved π components is $[Q/(3 + 2Q)]$ where $Q = \exp(-g_e\mu_\beta B/kT) = 0.25$ at B = 20T and T = 20°K. Thus, this predicted ratio is 0.07, very different from that given by the model in Figure 52 inset, Equation (22); ratio = Q or ~0.25. The experimental result from Figure 54 is close to 0.27, clearly in favor of the classical description of Equation (22). Further experimental attempts to distinguish between the classical and new Morgan descriptions are reserved to Section 5.12. The final important point to be noted here is that the very good fit of the D^o_pX BE Zeeman data by the nomograph of Figure 52 inset and particularly by Equation (19) provides direct

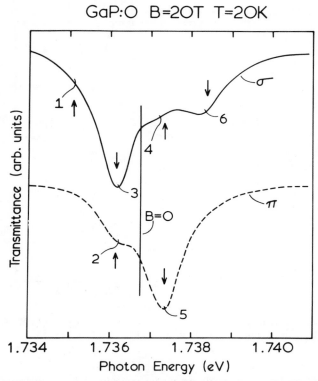

FIGURE 54 The no-phonon O_P donor D_c^oX BE absorption recorded at a combination of magnetic field and temperature optimum for the observation of hot, thermalizing subcomponents 1, 2 and 4 arising from optical absorption from the $m_s = +\frac{1}{2}$ magnetic substate of the neutral donor (Figure 52). These subcomponents are denoted by ↑ (Dean et al., Ref. 223).

experimental evidence of the centered nature of the metastable O_p^- state. Any significant axial component in the local crystal field would produce a reduction in the hole degeneracy, with either a strongly anisotropic g value or a closely isotropic $g_h \sim g_e \sim 2$, depending upon the effective sign of the splitting of the hole states under the axial field.[31] The experimental data are clearly inconsistent with such effects.

6. The luminescence in Figure 47 is observable readily only from O^- doped crystals which are not co-doped with acceptors. This is fully consistent with the interpretation of Figure 48, since the equilibrium charge state of O is O^+ in p-type material. One photo-excited minority carrier (electron) must be captured before the 0.53 eV luminescence

can result from capture of a second electron. The mean lifetime of the photoneutralized O_p donor will be considerably shortened by DAP recombination in significantly p-type GaP. The Figure 47 luminescence is found to quench much more rapidly in moderately p-type crystals, $N_A-N_D \sim$ low 10^{17} cm^{-3} than the Figure 16 luminescence according to Figure 20, involving processes described in Section 2.13. This is readily understandable if the Figure 47 luminescence results from the capture of a *second* electron at O_p. It is inconsistent with the new model of Morgan[233] in which Figure 16 and Figure 47 luminescences are both attributed to capture of a single electron at O_p^+ (Section 5.12).

5.5 Excited Hole States in the O_p Donor $D_p^o X$ BE

Weak excited states in the $D_p^o X$ BE absorption can be recorded most conveniently through the minima they produce in broad band photoluminescence which may be present near 1.8 eV, as in Figure 49b. The weak but sharp structure near 1.775 eV, about 37 meV above no-phonon line O_o is of particular interest. This structure exhibits LO($h\omega_o$) phonon replicas with coupling constants very similar to the series O_C, O_I, which are LO replicas of O_o. This is supported both by their displacement energies and the fact that their Zeeman splitting is exactly like that of O_o described in Section 5.4. This suggests that all these spectral features are an integral part of the $D_p^o X$ BE absorption, in which the phonon coupling should be dominated by the strong binding of the second electron to O_p^o.

The $D_p^o X$ BE is equivalent to D^-h where the hole is bound to D^- core by the long range Coulomb potential. The D^- core then takes on the core character of a simple acceptor. The usual (infinite) series of hole excited states must exist. The positions of the deepest of these are denoted by n = 1 → 4 in Figure 49, where n = 1 involves the $D_p^o X$ BE state described in Section 5.4. The situation is completely analogous to isoelectronic acceptors,[34] recently verified experimentally in GaP for several of the deeper NN pair traps by Cohen and Sturge.[235] The behavior is very similar to a point defect isoelectronic trap, provided the N-N separation is smaller than the radius of the bound hole. The binding energies of the hole excited states follow a simple inverse power law dependence on the principal quantum n number n, from which the series limit may be deduced as for a normal Rydberg series. Substantial deviation occurs for the ground state where the influence of the extended charge on the "acceptor" core is most significant (Table 7). Such core charge delocalization decreases the binding energy of the hole, as expected. Table 8 shows very close similarities between the deep O_p donor $D_p^o X$ BE and the NN trap data. The generally slightly larger values for this $D_p^o X$ BE are a consequence of the very tight binding at this ultra-deep donor, $E_e = 570$ meV compared with only 124 meV for NN_1^-. The fact that E_h remains so far below $(E_A)_{EM} \sim 56.3$ meV (Section 5.4) is surprising for such large values of E_e, however. It may indicate unexpectedly large electron-hole correlation, perhaps associated with an unusual form of strain contribution to the potential.[223] Such an effect would also contribute to the reduced isotope shift in the bound exciton discussed in Section 5.4 and further in Section 5.9.

The existence of this hole Rydberg series went unrecognized for some years after the original data in Figures 48 and 49 became available, largely because of the anomalously large width and consequent apparent weakness of the $n = 2$ component. This is clearly a consequence of strong coupling between the 1s–2s excitation energy of ~ 28 meV (Table 7) and the broad phonon replica, mean phonon energy near 27 meV which couples strongly to these BE electronic transitions. This coupling leads to a broadening of the electronic transition energy related to the effect noted in the photoexcitation spectrum of Ga acceptors in Si.[236] Here, it is a matter of quantum interference between transitions leading to the final states $(BE_{n=1} + \text{phonon})$ and $(BE_{n=2})$, however.

5.6 Anomalous Forms of $D_p^o X$ BE Absorption in Distant DAP PLE of the O_p Donor

The remarkable difference in spectral form between the two versions of the $D_p^o X$ BE spectrum in Figure 49 stimulated interpretations[135] in terms of the complicated 'two-electron' excitation processes already briefly mentioned in Section 5.1. It was regarded by some as significant evidence that the classical description of the properties of the O_p donor required a fundamental revision. Certainly, it also helped encourage the attempts to assign the luminescence of Figure 16 to capture of a second electron, which have already been refuted in Section 5. The salient facts are clear in Figure 49. All the structure consistently appears as negative dips in Figure 49b, equivalent to a straightforward optical absorption spectrum. However, when the same spectrum is observed superimposed upon a photoluminescence excitation (PLE) spectrum, conditions exist as in Figure 49a where many of the features appear positively while others remain negative, particularly O_o. It is very important to notice that the strength of optical phonon coupling remains constant between the spectra, irrespective of whether or not the sense of a given spectral feature becomes changed.

TABLE 7

Binding energies of holes at electron traps in GaP (meV)

Trap	$E_B{}^a$	$E_{1s}-E_{2s}$	$E_{1s}-E_{3s}$	$E_{1s}-E_{4s}$	E_h	E_e
$NN_1{}^b$	164	26.4	33.1	35.6	40.3	124
$NN_2{}^b$	159	28.1	35.1	37.4	41.7	117
$NN_3{}^b$	85	26.2	33.2	35.8	37.2	45
$NN_4{}^b$	60	24.5	30.9	33.4	34.6	22
O_p	612	~ 28.0	34.8	37.2	41.7	570

a - Binding energy relative to E_G = 2.350 eV (ref 30).

b - Hole excitation energies from ref 32.

Thus, all processes appear related to the same basic BE states. There is no evidence that the effects relate to the introduction of independent absorption processes in the PLE spectra.

Examination of the details of the PLE experiment provides straightforward interpretations for these apparently very complex properties. The absorption features are superimposed upon broad absorption from the σ_{p1}^o optical cross-section (Section 4.1, Figures 32 and 36) which dominates the PLE spectrum for O_p-Zn_{Ga} or O_p-C_p distant donor acceptor pair (DAP) luminescence, or for S_p-C_p luminescence as first discussed by Gross and Nedzvetskii.[216] Such negative features in PLE can only be seen in non p-type semi-insulating crystals for reasons already discussed in Sections 2.13 and 5; the corresponding PLE spectra in Zn-doped crystals are entirely smooth in this energy region well above the ~1.452 eV threshold discussed in Section 2.12. Studies of these PLE spectra were initiated[72,123] as a sensitive way of measuring the spectral form of σ_{p1}^o, particularly near threshold (Section 4.1). However, it is clear in the present context as well as for the effects discussed in Section 5.7 that considerable care is necessary to anticipate the complications which may result from measuring PLE rather than optical absorption directly. These complications have common in the fact the PLE spectra can strongly emphasize particular types of absorption process, namely those which most efficiently prepare the state from which the detected signal proceeds. Such effects were found to have dramatic consequences for near band gap PLE spectra of N_p isoelectronic trap-related luminescence, with a subtle evolution as a function of N concentration.[237] Here, we are concerned with optical processes which most effectively prepare neutral distant DAP.

Consider the $D_c^o X$ BE optical phonon replica O_c. This 48.7 meV phonon lies well above the 41.7 meV hole ionization limit just discussed. Optical absorption at O_c therefore may produce two distinct final states. The phonon may propagate far from the O_p donor where it is created, leaving the $D_c^o X$ BE in its ground state. However, it may interact again with the BE, liberating the hole into the valence band. The latter process evidently contributes positively to the DAP PLE spectrum, since most such holes will be efficiently captured by a nearby acceptor at 2°K and electrons are available at neutral donors. The former process can only make a negative PLE contribution, like O_o itself. The experimental results show that absorption due to the latter process predominates, while no such process is energetically available for the ground state O_o at 2°K. However, line O_o shifts ~3.9 meV to lower energy and becomes *positive* as T increases from 2°K to 77°K. This indicates an appreciable thermal ionization probability of the n = 1 hole at 77°K, as expected particularly since the effective thermal ionization energy will be significantly less than the optical value of ~42 meV due to interimpurity interactions at N_A, N_D ~ 10^{16} cm^{-3} in these semi-insulating samples. The fact that the energy shift is much less than the band gap reduction of 6.1 meV is an expected characteristic of an ultra-deep donor state. This is discussed further for the single electron state in Section 2.10 and 7. The experimental evidence from free to bound luminescence (Section 3.3)[102,241] and from distant DAP PLE spectra[123] shows that the one-electron state is even more closely pinned relative to the valence band, in agreement with uniaxial stress data.[53]

The striking reversal of sign of the n = 3, 4 . . . features between figure 49b and 49a can now be understood if the ~7 meV binding energy of the n = 3 hole state (Table 7) is inadequate to prevent hole delocalization even at 1.5°K. This can only result from significant intersite tunnelling at these impurity concentrations. This process is sufficiently strong to produce complete delocalization of electrons from 6 meV donors in GaAs at donor doping N_D ~ 10^{16} cm^{-3}.[238] The situation is less extreme here because m_h in GaP is ~7.5 × m_e in GaAs and the critical parameter, the bound state radius a^*, varies as $(m^*)^{-\frac{1}{2}}$. The quenching of the PLE contribution from the n = 2 hole state in Figure 49a suggests that doping and other conditions are such that tunnelling from this 13.7 meV hole state can just annul the negative contributions due to normal absorption competition. If this is so, the critical nature of the Mott-type insulator/metal transition ensures that tunnelling delocalization should predominate for the n = 3 state and that the contribution ratio n = 4/n = 3 should be larger in PLE than in absorption. Both of these effects appear in Figure 49. Intersite tunnelling has been shown to make an important contribution to changes in sample conductivity at very low temperatures.[239] In the present case, the contribution to positive distant DAP PLE response of hole tunnelling from excited $D_c^0 X$ BE states will be enhanced by the existence of acceptor excited states with closely similar binding energies. It is notable that the n = 2 state contribution to PLE can be seen positively in the LO phonon replica. It is evidently enhanced by the same process which makes O_c positive at $1.5°K$.

This new interpretation[196] of these old results[135] proposes a general mechanism, applicable to similar transitions involving any deep state in any semiconductor. It also accounts naturally for the appearance of similar inverted structure in photoconductivity excitation spectra from these same crystals.[240] The new interpretation seems much more plausible than the very case-specific 'two-electron' excitation processes suggested previously,[240] which have led to considerable difficulties with the general scheme of transitions and binding energies of the O_p donor in GaP.

5.7 Anomalous Temperature Dependence of σ_{p1}^0 in Distant DAP PLE Spectra

The overall form of the one electron photoneutralization optical cross-section σ_{p1}^0 has been described in Section 4.1. The photocapacitance data reported there is necessarily obtained at $T \geqslant 100°K$. The PLE technique of Monemar and Samuleson[72] can be used sensitively down to the lowest temperatures, and has been employed to derive detailed information about the form of the phonon coupling in absorption and emission processes of the one-electron state (Figure 34). The measurements are made in p-type material to avoid the effects of Section 5.6 and to ensure that the PLE response is linearly related to σ_{p1}^0, free of complications from transitions via the conduction band and effects of the two electron state.[72] Besides providing evidence on the phonon coupling strength from the extreme threshold region of σ_{p1}^0 shown in Figure 34 and providing a deconvolution of the purely electronic component of the cross-section,[72] these low temperature measurements also revealed unexpectedly large thermal broadening and quenching effects,[123] some of which extend over a wide energy range as shown in the upper part of Figure 55. It is necessary to discuss these effects in some detail, since they have been advanced as one of two major reasons which it is claimed necessitate the radical new description of the electronic configuration of O_p in GaP described in Section 5.12.

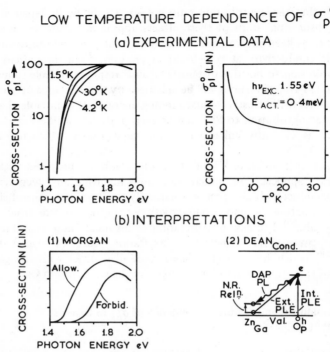

FIGURE 55 (a) Experimental form of the near threshold region of optical cross-section σ_{p1}^o determined at the indicated temperatures in p-type GaP:Zn,O through PLE spectroscopy of the O–Zn distant DAP luminescence. The rapid thermal quenching of a significant portion of the near-threshold PLE efficiency is well-described by a very low thermal activation energy, near 0.4 meV (Samuelson and Monemar, Ref. 123). (b) Contains two contrasting interpretations of (a); on the left involving absorption from two states of O_p^+ separated by 0.4 meV and on the right by thermal quenching of efficient PLE contributions via shallow concentration-perturbed excited states of the Zn acceptors (Dean, Ref. 196).

The most dramatic of these thermal effects, shown in Figure 55, involves a rapid quenching of σ_{p1}^o between threshold near 1.45 eV and the peak at ~1.8 eV.[123] The portion of σ_{p1}^o quenched between 4.2°K and 30°K has a shape reminescent of a moderately phonon-broadened spectrum with half-width ~0.22 eV. This is significantly larger than the half width of the distant DAP PLE spectrum involving the O_p donor, ~0.16 eV (Figure 13). Similar effects were observed when the green S_p donor-related DAP luminescence were detected, making it most likely that the temperature dependence results from thermalization of states near threshold of σ_{p1}^o. Analysis of the data in Figure 55, upper right, reveals a remarkably small activation energy for the thermal quenching of 0.40 ± 0.05 meV. Samuelson and Monemar[123] note that this is of the order of a typical exchange splitting. However, the classical description of the O_p^+ donor, in the initial state of the transition, admits of no exchange splitting. Morgan's new model (Section 5.12) has been tailored to produce such a splitting between $^3T_2{}^{(+)}$ and $^1A_1{}^{(+)}$ subcomponents (Figure 58 below). This model attempts to reproduce the experimental data through the assumption that the $^3T_2{}^{(+)}$ state is highest and that transitions from it to the $^4T_1{}^{(o)}$ ground state of O_p^o are forbidden (Section 5.12). The resultant threshold of the thermally activated absorption subcomponent can in principle produce the observed T-dependence of σ_{p1}^o, as shown in Figure 55, bottom left. However, there is no quantitative basis for such a very small splitting of two such states, particularly within an ultra-tight binding model. Most people regard this attribution as implausible, at least until it is *strongly* supported by experimental evidence.[242] Fortunately, an alternative model is available which, like that in Section 5.6, does not make appeal to any fortuitous peculiarities of the behavior of the O_p donor.

The general interpretation of this effect[196] recognizes that intra-DAP optical absorption processes can contribute to σ_{p1}^o in the significantly p-type crystals used in these experiments.[72] Such absorption, involving electron photoexcitation from compensated acceptors, is naturally weak compared with electron photoexcitation from the valence band. However, we recognize that these intra-DAP processes will make a disproportionately strong contribution to the PLE spectrum. They are 100% efficient in producing the detected distant DAP luminescence, unlike processes involving the photocreation of free holes in the valence band. The significance of such direct excitation processes can be judged from recent experimental work involving tunable dye lasers, when the PLE excitation efficiency permits selective detection of PL within a narrow range of $h\nu$ and therefore extending over only a small range of DAP separation R according to Equation (6). In such experiments, the selective intra DAP absorption processes result in narrow lines within the DAP spectra. The integral intensity of these lines is appreciable compared with that of the underlying broad absorption with low energy threshold at $h\nu = E_G - E_A$, which represents non-selective DAP PLE processes.[243] In the experiments of Monemar and Samuelson,[72] the PLE spectrum was recorded for DAP luminescence over a wide range of DAP separations, since the light was detected through a low energy pass filter rather than a high resolution monochromator.[243] Thus, no structure due to absorption processes at discrete DAP could be resolved. The range of DAP R which predominates in the selective processes will be defined by a balance between two conflicting aspects. Absorption at relatively close DAP will be favored by the exponential term in the overlap expression of Equation (7). However, this effect will be balanced by the well-known[2] rapid decrease of effective energy density of DAP states with decreasing R. The most effective DAP absorption will fall at some intermediate value of R comparable with a_{acc}, the radius of the relevant acceptor state. We have noted in Section 2.6 that $a_{cc} \leqslant 9$ Å for the acceptor ground state. However, we can also consider transitions to acceptor excited states in selective DAP PLE spectra.[243] Then, a_{acc} may be ~ 20 Å, or even greater if we pick sufficiently high excited states of the long range Coulomb potential. Consideration of hole creation in high excited states introduces a further trade-off, since the overlap benefits from Equation (7) becomes lost if the hole becomes delocalized before internal relaxation to the acceptor ground state can occur. Such delocalizations occur at the lowest temperature for states sufficiently extended that the Mott criterion for an insulator/metal transition is exceeded, namely for acceptor densities N_A such that

$$N_A \, a_{acc}^3 > 0.25 \qquad\qquad (23)$$

According to this interpretation of the data in Figure 55, the low activation energy $E_{ACT} \sim 0.5$ meV derived from thermal quenching of σ_{p1}^o represents an average value for the high Coulomb excited hole state which dominates the overlap term of Equation (7). This will be an effective thermal activation energy for an excited state of significantly larger optical binding energy, since the delocalization effect of the inter-acceptor interactions will be considerable at $(N_A - N_D) \sim$ low 10^{17} cm^{-3}. There will also be considerable stoichastic spatial fluctuations in this average, resulting from fluctuations in $(N_A - N_D)$ about the average value for the crystals. Judging from the experimental data on donors in GaAs,[238] the optical value of E_A^{EXC} for the dominant acceptor excited state at $N_A - N_D$ $\sim 2 \times 10^{17}$ cm^{-3} is ~ 5 meV, since $N_{LIM} \propto a_{acc}^{-3} \propto m_h^{3/2}$ from Equation (23). If optical absorption is regarded as an adiabatic process, the relevant a_{acc} is then ~ 30 Å, corresponding to shell number m ~ 60 from Figure 15. This seems a plausible region of compromise between the conflictive factors already mentioned, in view of the significant energy density of DAP sites for R ~ 30 Å.[2]

The rapid thermal quenching of a portion of σ_{p1}^o at low temperatures is thereby explained as the removal of this intra PLE excitation channel through thermal dissociation of the relevant high acceptor state before de-excitation to the ground state can occur. The indication from Figure 55 upper right that this channel can amount to $\sim 60\%$ of the near threshold PLE seems feasible for these crystals according to general experience of distant DAP PLE.[243]

Morgan[244] has commented that the mechanism proposed here cannot produce the observed T-dependence nor the lineshape of the thermally quenched contribution to σ^o_{p1}. This objection seems to be based upon fundamental misunderstandings of the dominant energies and processes in the problem. The observation that the effective E_{ACT} is relatively sample-independent follows naturally from the model, since Equation (7) forces selection of the maximum value of a_{acc} consistent with adequate thermal stability at $1.5°K$. Higher acceptor excited states can contribute to the intra DAP PLE in crystals of lower $N_A - N_D$, and this will tend to offset the reduced fractional thermal quenching which must ultimately occur as $N_A - N_D$ falls. The observation of a spectrally *broad* optical contribution from this intra-DAP PLE process also receives a natural interpretation. There are two contributions to this broadening, both significant. First, like all optical processes involving change of charge in the one-electron state of O_p, phonon broadening must be considered. We have already seen that this will contribute ~ 0.16 eV of the observed width of ~ 0.22 eV, even if the phonon coupling is constant between emission and absorption as Monemar and Samuelson[72] believe (Section 4.1). The data in Figure 55 were not recorded under conditions adequate for the observation of the weak undulations in σ^o_{p1} which appear very near threshold in derivative spectra (Figure 34) and indicate the strength of the phonon coupling. The existence of another contribution to the broadening is indicated by the excess width. The form of $\sigma^o_{p1\ el}$, the pure electronic part of σ^o_{p1} deconvoluted by Samuelson and Monemar,[123] shows that the thermally quenched contribution has a half width of rather more than 0.1 eV, perhaps as much as ~ 0.15 eV. However, despite the fact that the intra-DAP PLE process just described involves bound to bound transitions, even the pure electronic contribution is *not* expected to have the spectrally narrow form normally found for such processes at $1.5°K$. The reason is found in Equation (4). In the present problem, E_{rep} will vary over a wide range depending upon the range of R which can contribute to the absorption. We have already explained that the overlap required for the transition oscillator strength is fundamentally governed by the minimum value of E_A^{EXC} permitted by local fluctuations in $N_A - N_D$. These fluctuations will be only weakly dependent on the local N_D, since $N_D \sim 0.1\ N_A$ in these crystals. The large values of a_{acc} possible for the PLE process described encompass a wide range of R and so a wide range of E_{rep}. The *maximum* value of E_{rep} is ~ 0.6 eV for $O_p - Zn_{Ga}$ DAP (Table 2). Consideration of Figure 15 and the $R = \infty$ limit for $O_p - Zn_{Ga}$ DAP of ~ 1.38 eV shows that an energy spread of ~ 0.15 eV encompasses DAP pure electronic transitions extending from $R = \infty$ down to $R = 10$ Å, even when only ground state to ground state processes are considered. Light from DAP extending over the whole of this separation range was detectable in the PLE experiments of Monemar and Samuelson.[72]

We conclude that the suggested interpretation involves processes which ought to occur and which can reproduce all the major features of these remarkable experiments. We also note that the model suggested here predicts similar spectral forms for the temperature dependent and temperature-independent (below $\sim 77^{\circ}$K) contributions to $\sigma^{o}_{p1\ e1}$, since parity-allowed processes are available for each on the classical description of O_p. This result is consistent with the original deductions from experiment,[123] while the new interpretation of Morgan[233,244] conflicts with experiment on this point. Thus, there is no need to appeal to radical interpretations of the electronic configuration of O_p in GaP as in Section 5.12 to construct such remarkably small energy splittings in the O_p^+ state.

5.8 Local Vibronic Properties of O_p in Different Charge States

Baraff et al.[203] have used the isotope shifts derived from Figure 17 to estimate further properties of the O^- state. The 24.7 meV local mode is the only one for which motion is substantially confined to the O atom (Table 3 and discussion in Section 2.9), and may be used to estimate the force constant of Ga-O bond-stretching vibration. This constant is only about 20% of the equivalent Ga–P bond stretching constant, which would give a normal local mode energy of just over 60 meV for vibrations of the O_p atom against the surrounding Ga atoms. This reduced force constant may then be used together with the observed no-phonon isotope shift to derive the energy of this mode for O_p^+, which is 32.4 meV assuming that this three-fold degenerate (T_2) mode dominates the sum in Equation (2). This assumption is justified by the calculated charge density of the extra electron in O°, which is substantially confined to the Ga-O bonds.[203] The ratio of the calculated force constants is 0.76, agreeably close to the value of 0.78 derived in the configurational co-ordinate analysis of the capture cross-sections σ_{n1} and σ_{p1} (Figure 33). The weakening of the force constant on electron capture is consistent with the deduction that the state into which capture occurs is an antibonding state (Section 2.2) derived mainly from Ga dangling bonds (Section 7.1).

It is then argued[203] that the initial stages of electron capture into O^- should proceed via a similar symmetric (centered) state. The force constant of the Ga-P bond should decrease further by a similar magnitude, as a consequence of the repulsion of the Ga ions by the further negative charge on the O. An extremely weak force constant results, only $\sim 15\%$ of that for O^+ and $\sim 5.5\%$ of the Ga-P value! It is possible that the weakness of these Ga-O bonds in GaP is further accentuated in GaAs, as a consequence of the appreciably larger lattice constant of this host. This would produce a much reduced overall stability of the O_{As} center in GaAs. The consequences of this for optical spectra may well explain the elusive nature of evidence for the behavior of the O donor in GaAs (Section 8.1). The assumption of an equal reduction of Ga-O force constant upon capture of the second electron predicts an O-dominated T_2 symmetry vibrational mode energy 16.2 meV. However, the recent work[196] shows two modes (Figure 47) of energy 6.75 meV and 10.36 meV (Table 8). There are two T_2 modes for a tetrahedrally bonded O_p-Ga^4 molecule, one whose eigenvector primarily involves O motion and the other associated with Ga motion. The measured O-isotope shifts for these dominant promoting modes indicate a low percentage of kinetic energy for O motion in each (Table 6) as must be so if *two* such modes are present for a T_d-symmetry substituent. If the mean energy of these modes is used, Equation (2) predicts a NP isotope shift of –1.53 meV for $O^{16} \rightarrow O^{18}$, reduced to –0.61 meV when allowance is made for the 40% KE associated with O motion (Table 6). The residual NP isotope shift of –0.21 meV evidently derives from contributions of additional promoting modes to the sum in Equation (2). Some of these modes are listed in Table 8.

Feenstra and McGill[253] have recently reported briefly on a Green's function calculation of the defect vibrational modes of O_p in GaP, considering only nearest-neighbor bond stretching and bond bending effects. The model applies only when the force constants are not too weak, and therefore mainly to the reactions between O_p^+ and O_p^0. We have already noted that it is not successful in describing the low energy T_2 modes of the Zn_{Ga}-O_p associate (Section 2.6). The vibrational modes of the defect depend upon a single adjustable parameter $\Delta f/f$ (Figure 56), representing the fractional change in force constants relative to the bulk material. Defect modes of A_1 and T_2 symmetry only are found in the range $-1 < \Delta f/f < +1$, the modes considered in the above discussion. The A_1 mode always appears as a resonance in the acoustic branch of the lattice vibrational spectrum, calculated on a rigid ion model and shown at the left in Figure 56. The T_2 mode becomes a true local mode for $\Delta f/f > -0.4$ and reaches ~ 56 meV at $\Delta f/f = 0$. Using the 19 and 47 meV phonons determined for O_p^+ from Figure 34 we see that $\Delta f/f \sim -0.5$, in fair agreement with $\Delta f/f \sim -0.6_3$ deduced by Baraff et al.[203] from independent experimental data.

The electron in the O^- metastable state should posses $1s(A_1)$ symmetry, pairing with the first electron (Section 2.8) in a spin-singlet state. The

P. DEAN

TABLE 8

PHONON ENERGIES AND ISOTOPE SHIFTS IN 0^--RELATED SPECTRA[196]

Spectrum Feature	Electron Capture Luminescence	0^O_P Donor Bound Exciton
(a) Label and Phonon Energy (meV) (for 0^{16}) and Isotope Shift $0^{16} \to 0^{18}$ (meV)	0^1_{LOC1} 6.75, -0.17 0^1_{LOC2} 10.36, -0.22 0^2_{LOC1+2} 17.2 0^1_{LOC2+2} 20.9, -0.5 0^1_{LOC3} 26.9, -0.3_5 0^1_{LOC4} 30.0 0^1_{LOC5} 34.2, -0.6_6 0^1_{LOC6} 39.1 0^1_o 46.6, ~0 0^1_{LO} 50.0_1, 0	0_{TA} ~14.0 0^1_{LOC} 28.0 0_{LOC} 34.8 0_A 41.9 0_c 48.3
No-Phonon Isotope Shift $0^{16} \to 0^{18}$ (meV)	-0.82 ± 0.03	$+0.42 \pm 0.05$

(a) Refers to Figs 47 and 48

singlet state couples in first order only to S-like localized modes (breathing modes) and no Jahn Teller effect can occur from a center with such a non-degenerate state. The tunnelling transition to the strongly distorted state invoked from photocapacitance (Figure 41) must be promoted by higher-order terms in the electron-phonon interaction, which become more important when the basic force constant is low. The relaxation process may be visualized as a bound center equivalent of the familiar hole self-trapping process in crystal lattices which exhibit small polaron behavior. The fact that an exciton state involving such a hole can exhibit optical properties characteristic both of the unrelaxed as well as the fully-relaxed hole states is firmly established in certain alkali halide hosts.[204] The O_p center in GaP is the first case of a similar effect for an electronic particle bound to a defect or impurity center.

The spectral form of Figure 47 indicates that the 6.75 and 10.36 meV phonons are promoting modes, and that the no-phonon transition O_o^1 is forbidden. To this extent the spectra of Figures 16, 17 and 47 are very similar. The reason for the electric dipole-forbidden nature of the capture luminescence into the one-electron state has been discussed in Sections 2.8 and 2.9 and is consistent with the observed lifetime of ~10 μsec in lightly co-doped crystals at low temperatures.[196] We have measured the lifetime of the luminescence in Figure 47 under similar conditions, and find it to be ~25 μsec.[196] This long lifetime, and the distinctive spectral

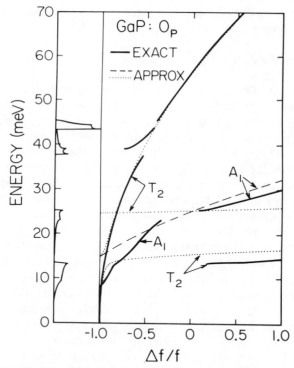

FIGURE 56 The defect vibrational modes of O_P in GaP. The mode energy is shown versus the defect local force constant $\Delta f/f$. The heavy solid lines are solutions from Green's function theory. The dashed lines are A_1 molecular modes and the dotted lines T_2 modes for an OGa_4 molecule in a rigid lattice. The bulk phonon density of states at the left is calculated from a rigid ion model (Feenstra and McGill, Ref. 253).

form in Figure 47, can be understood once the likely character of the initial state of the luminescence transition is examined (Section 5.9).

The notation in Figure 47 and Table 8 is misleading in the sense that it appears to indicate that the overtone structure involves a second inter-action with LOC2, a T_2 promoting mode. General considerations and the intensity sequence between the phonon sidebands indicate that this cannot be true. The phonon replicas caused by multiple phonon emis-sion certainly involve additional phonons of energy close to 10.5 meV, but these must be accepting modes, symmetry A_1. The existence of such a mode in Figure 47, rather than the 19 meV accepting mode held to be characteristic of transitions to the one-electron state (Figure 34 and Section 5.2) provides further evidence that the Figure 47 luminescence involves transitions of the *second* electron. The reduction in energy of the A_1 mode from 19 to 10.5 meV is qualitatively consistent with the same weakening of the Ga–O bond which reduces the T_2 mode energy from 24.7 to ~10.4 meV.

5.9 Character of the shallow O_p^- excited states

Electron binding to O° is formally equivalent to binding at an isoelec-tronic trap, with the additional complication of exchange interaction between the two electronic particles. Faulkner[156] has shown that, pro-vided the binding potential is large enough, the lowest excited state of such a center will be the E state with an antisymmetric wavefunction combination from the crystallographically inequivalent conduction band minima of GaP. Arguments equivalent to those in Section 2.8 suggest that binding energy in this E state will be very much less than for the A_1 ground state, which involves a symmetric contribution of wavefunction contributions from the individual conduction band valleys. Thus, the internal transition $E \rightarrow A_1$ will be forbidden in the same way as for the one electron capture process of Section 2.9. Electron-electron exchange cannot complicate the final state of this transition, which can only occur in an antisymmetric spin state since both electrons must have A_1-like orbital character. By contrast, in the initial state the second weakly bound has E character, while the first is tightly bound in the $1s(A_1)$ ground state of the O_p donor. Because of the orthogonality of the E and A_1 orbital states, both J = 0 and J = 1 spin-spin states may occur. The spin triplet state is ex-pected to lie at slightly lower energy, since the electron-electron inter-action is repulsive and produces a positive value of the exchange integral.

The shallow excited state of O_p^- demanded by the interpretation in inset of Figure 50 is equivalent to the states identified for shallow $D_0^\circ X$ BE in Si, a semiconductor with band structure qualitatively identical with GaP. Transition δ in Figure 57 has been seen in optical obsorption as well as in luminescence for the P D^o,X BE.[31] The excited BE state is weakly but definitely bound. There is no evidence that this binding is removed as the donor and the D_c^oX BE ground state becomes deeper with increase in the central cell potential V_{cc}, but the D_c^oX BE excited state remains close to the free exciton, relatively independent of V_{cc}. A similar D_c^oX BE excited state has been observed for donors in GaP.[245] The capture luminescence process in Figure 50 inset is equivalent to a radiative relaxation between the excited states responsible for BE transitions δ and

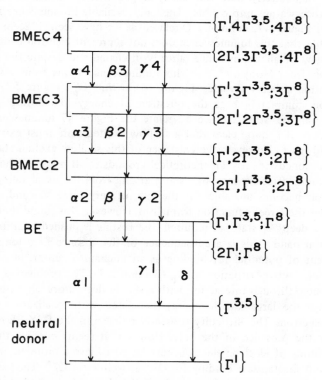

FIGURE 57 The electronic configurations from the shell model for substitutional D_c^oX BE and bound molecular exciton complexes in a semiconductor with band structure like Si or GaP, omitting the small $\Gamma_3-\Gamma_5$ electron state and electron hole J-J splittings. The electrons are placed in Γ_1 or $\Gamma_{3,5}$ states derived from valley-orbit coupling, while the holes are placed in a Γ_8 shell containing up to four particles. Possible transition series are indicated. We are particularly concerned with the absorption versions of α_1 and δ (Kirczenow, Ref. 254).

α, in Figure 57, except for the presence of the hole in the BE transitions. This will not be an efficient radiative process in Si, or for the shallow donors such as S in GaP, since the energy to be emitted falls well below the one-phonon cut-off of lattice vibrational energies. The case of the O_p $D_p^o X$ BE represents an unusual extreme of the opposite limit due to the large transition energy and reduced vibrational energies. If the relevant accepting mode energy is ~ 10.5 meV (Section 5.8), electron capture from the conduction band into the centered O_p^- state requires a 54 step multi-phonon emission process, and still a high-order 11-step process for the lattice LO(Γ) phonon. It is reasonable to expect this process to be radiative in relatively pure GaP singly doped with O, where intercenter Auger processes are improbable. The only available luminescence is that shown in Figure 47, definitely O-related as we have seen. It seems most improbable that this luminescence does not represent electron capture at $O_{p'}^o$, in view of all the evidence, already discussed and despite the recent suggestion from Morgan[233] that this luminescence stems from a second de-excitation process (besides Figure 16) *within* O_p^o (Figure 57). This suggestion completely begs the question of energy loss during second electron capture at O_p^o. If we do relate the Figure 47 luminescence to this process, it is quite clear that a shallow excited state must exist as in Figure 50 inset. However, the existence of this shallow excited state has been questioned recently on theoretical grounds.[247] It is argued that both the A_1 and E effective mass-like states arising from the usual valley-orbit interaction become unstable for the case of very large V_{cc} and can be treated by the Greens function formalism. However, it is found that while there is a deep A_1 state, as required, the E state is pushed high into the conduction band. This is a consequence of the absence of a long range component of potential for binding at an uncharged center, in contrast to the one-electron situation at O_p (Section 2.8). The problem is that it is impossible through interaction with a simple deep short range potential to balance the large kinetic energy associated with localization of the second electron. The difficulty is greatly relieved in the $D_p^o X$ BE excited state by the presence of the hole. However, it seems possible that improved forms of electron-electron correlation to give a more extended E state could substantiate binding for O^-* as well as for O_h^-*. The existence of the latter seems certain for the deep O_p donor from the behavior of the shallow $D_p^o X$ BE just discussed and a consideration of the validity of the Haynes rule trend in Figure 50. Baraff[246] has recently suggested that the O^-* state should be treated with a basis similar to the H^- ion. There, a carefully and explicitly correlated trial wavefunction is required as input to a three-parameter variational calculation in order to obtain the bound state observed in nature. Such explicit correlation in the O_p^{-*} state could help to account for the large isotope shift in Figure 47 (Section 5.5) as well as stabilizing the O_p^{-*} state itself. The existence of the E-like valley orbit state in the semiconductor, together with the large mass anisotropy of the GaP conduction band both further enhance the possibility of obtaining a bound O_p^{-*} state. The deep O_p^- ground metastable state must then be regarded as a special feature of the strong short range potential in the semiconductor, raising E_e/E_D far above the $\sim 5\%$ value permitted by the isotropic Coulomb potential for the H^- ion.

A second no-phon line O_o^2 has recently been seen[196] in luminescence spectra measured at $\sim 20°K$, and is about 4.5 meV above O_o^1 (Figure 47). These spectra suggest a ratio of oscillator strengths for no-phonon transitions from these two states of about 5:1 in favor of O_o^2. The higher component O_o^2 has phonon replicas similar to O_o^1, at least as far as LOC1 is concerned, but with much reduced coupling strength, in inverse proportion to the no-phonon oscillator strength. This is as expected if these phonons are promoting modes. The discussion in Section 5.8 suggests that at least LOC2 is a T_2-type mode, as required for a promoting mode, and also explains the marked spectral differences between the sideband structure in the luminescence spectra of Figures 17 and 47.

The properties just described are all consistent with the attribution of this ~ 4.5 meV splitting to exchange, since transitions form O_o^1 are spin-forbidden in addition to being forbidden by the orbital character of the states. There are a number of possible explanations for the evident fact that these simple selection rules are only moderately effective. Firstly, transitions to the A_1 ground state from E excited states of both the one and two-electron systems can be forced by admixtures of T_2 components in the shallow excited states. Such T_2 components are available from bound states derived from the X_3 conduction band minima split-off by the antisymmetric potential on the Ga-P bond. This mixture can be promoted by the uniaxial strain fields which are always present in real crystals. The relative intensities of the no-phonon lines exhibit a degree of specimen dependence, as expected from the dependence of the random stress field on incidental characteristics of individual samples. Second, the A_1 character of the very deep donor ground state can be weakened by strain-induced mixing with the T_2 valence band. The spin selection rule is broken down by weak spin-orbit coupling with these same T_2 states. Both effects depend upon the same order of perturbation theory. The ratio of the transition probabilities from the spin singlet and spin triplet excited states depends upon the ratio of the respective matrix elements, which require detailed calculations. The longer low temperature lifetime observed for the luminescence in Figure 47 compared with Figure 17 is also consistent with the presence of the extra selection rule.

Expectation that the triplet state should lie lowest, resulting from the positive value of the electron-electron exchange integral is inconsistent with a very recent magneto-optical study.[223] No splitting or shift significant compared with the zero field linewidth Δ of 0.65 eV was observed at B = 1OT and 2°K, whereas the lower component of the spin triplet should shift down by almost 2Δ and be strongly favored by thermalization under these conditions. We can only conclude that either the exchange splitting is negative, a possible effect of electron-electron correlation, or that the interpretation offered for this luminescence in Figure 48 inset is incorrect. However, it is very difficult to believe that this clearly O-related luminescence, falling very close to the energy predicted for electron capture at O^- from the interpretation of the 1.738 eV line in terms of the neutral O_p donor BE, must be associated with O in some different form. We have seen that the centered O^- state is predicted to involve only modest lattice relaxation. Thus, electron capture at O^o is not likely to involve predominantly multiphonon energy relaxation. No alternative luminescence spectrum is available for association with this electron capture process.

The *magnitude* of this splitting attributed to exchange is currently also a problem, since theory can accommodate the ~ 40 meV excited state of the centered O^- only if the second electron is spatially well-separated from the first, bound in the long range tail of the potential of O^o.[188] However, problems of similar origin and approximately equal magnitude occur on a possible alternative interpretation in terms of a dynamic Jahn-Teller effect, assuming a large Ham reduction factor to account for the magnitude of the splitting.[219] The only obvious further explanation in terms of a normal phonon replica also presents difficulties, since the charge of the second electron is substantially removed from the vicinity of the O_p^o donor in these states. We have already seen in Section 4.5 that the sense of the non-linear electron-phonon coupling should produce an *increase* in the mode energies for O^o plus weakly bound electron compared with the lowest state of centered, metastable O^-, whereas this assignment would require a substantial *decrease* in vibrational energy. Indeed, further broad structure observed in luminescence spectra[196] recorded respectively at $\geq 20°K$ and $> 30°K$ is consistent with anti-Stokes replicas involving modes LOC1 and LOC2 with energy *enhancements* of $\sim 23\%$ relative to the Stokes replicas listed in Table 6. This energy enhancement is remarkably similar to the value in Figure 33, giving useful support to the assumption of Baraff et al.[203] that the reductions in force constant of the Ga-O bond for $O^+ \rightarrow O^o$ and $O^o \rightarrow O^-$ may be similar (Section 5.8).

It is interesting to note that the strength of the higher order phonon coupling below ~ 0.5 eV is significantly underestimated in Figure 47 as a result of the influence of water vapor absorption. Allowing for this, the forms of Figures 16 and 47 are consistent with the expectation of similar lattice relaxation on capture of the second electron as for the first. The very different strength and form of the sidebands in Figures 47 and 48 is also consistent with the fact that the neutral O_P donor BE is created from an allowed transition between a T_2 hole and an A_1 electron state. The sidebands in Figure 48 are dominated by *receiving* rather than promoting modes, with peak energies of ~ 13, ~ 28 and ~ 48.3 meV (Table 7). These are very similar to the sidebands observed for the N_P BE in GaP.[3] supporting the case for a similarity between the O^- state and an electron trapped at an isoelectronic made at the beginning of this section.

5.10 Absence of luminescence from the O_P donor BE and Auger recombinations

No trace of luminescence corresponding to the neutral O_P donor BE has been obtained. There are two possible explanations. The first is relaxation to the rebonded state in Figure 49, with subsequent capture of the hole by multiphonon emission. At low temperatures, this would probably occur through the weakly bound state E_h in Figure 41, perhaps with contributions from a phonon-assisted Auger process (Section 5.3). The second possibility is an Auger recombination process within the centered metastable state itself. This process is energetically possible, since the available recombination energy of ~ 1.74 eV exceeds the binding energy of the neutral O_P donor, 0.9 eV. Auger recombinations in this state do not suffer the large reduction in transition rate from poor overlap due to a large lattice relaxation, in contrast to the process described in Section 5.3. The Auger transition rate may be estimated very roughly by scaling from the rates measured for shallow donor such as S_p in GaP, $P_A \sim 5 \times 10^7$ sec^{-1}.[208] The required scaling factor is $(E_{n2})_o/(E_D)_S{}^3$, since P_A is proportional to the densities of overlapping particles. We consider only the second power of the electron density. The further potential increase in P_A caused by the larger hole binding energy in the OBE will be offset by the decreased e-h overlap due to the increased electron localization within O^- compared with S^-. We therefore predict $P_A \sim 5 \times 10^9$ sec^{-1}. The hole capture rate derived from the capture cross-section σ_{p2} measured at

$200°K$ for $N_A - N_D \sim 4 \times 10^{17}$ cm^{-3}, $\sim 10^{-14}$ cm^2 (Table 5), is $\sim 4 \times 10^{10}$ sec^{-1}. Recent theoretical arguments[248] suggest that the Auger rate for this $D_p^o X$ BE may be significantly less than the crude estimate of 5×10^9 sec^{-1}. For all these reasons, it seems unlikely that the Auger process can provide the large non-radiative rate implied by the $D_p^o X$ BE no-phonon linewidth, $\sim 10^{12}$ sec^{-1}, and by the failure to observe $D_p^o X$ BE luminescence, which implies a rate $> 10^{11}$ sec^{-1} since the radiative lifetime is no longer than a few μsec[208] and the radiative efficiency is no more than $\sim 10^{-6}$ [208,223] We conclude that relaxation to the rebonded configuration establishes the lifetime of the centered metastable O_p^- state.

5.11 Interpretation of the ODMR result of Figure 43

The further question of the interpretation of the ODMR spectrum remains open. It seems most probable that the magnetic resonance is occurring in the triplet excited state of the O$^-$ center described in Section 5.9. It is possible that the suggestion of a reduced $< 110 >$ symmetry made by Gal et al[139] is an artifact of unrecognized influence of a dynamic Jahn Teller effect, since the turning points in the angular dependence of the g values would be the same for both static and dynamic Jahn Teller effect.[206] It is necessary to interpret the ODMR spectrum for crystal orientation away from the turning points to distinguish these possibilities, and this has not been done. The orbitally degenerate E state described in Section 5.6 could exhibit a dynamic Jahn Teller effect. This possibility could remove the apparent incompatibility between the ODMR result and the models for O^o and O$^-$ in Figure 50. The question of the particularly intimate link established between the magnetic resonance and the one electron state luminescence of Figure 16[139] remains. The only plausible possibility involves the postulate that the center exhibiting ODMR must decay partially by an Auger process. Any center giving Auger recombinations would do[206]. If we adhere to the excited state of O$^-$, then we must have an Auger effect with a bound hole, effectively through a high excited state of the neutral O$_p$ donor BE. Although this recombination process is rather improbable, we recognize that the radiative lifetime from O^{-*} is rather long and that the ODMR signals are weak, so a weak but observable effect might occur at the high optical pumping rates employed in the ODMR experiment. To obtain the polarization link between resonance and luminescence illustrated in Figure 46, it is necessary to assume further that the capture time of electrons at O$^+$ is spin dependent and that the electrons ejected in the Auger transition retain some spin polarization memory during energy relaxation within the conduction band. Recent measurements of hot electron luminescence in $GaAs^{217}$ suggests that this may be possible. The final condition on the model is that the wavefunction of the bound electron in O^{-*} must be sufficiently restricted to exhibit significant overlap on the nearest neighbor Ga ions to the O_p site. Comparison with the N_p BE in $GaP^{3,156}$ suggests that this may be possible, even for an excited state of binding energy only ~ 40 meV. Combination of these considerations does provide some possibility for a consistent interpretation of this ODMR result, which seems to be one of the most puzzling yet observed in any semiconductor.

5.12 The new (1982) "molecular orbital" model of Morgan

Morgan was motivated to suggest a radical new description of the electronic configuration of centered O_p in $GaP,^{233,244}$ particularly by the differences in form of the phonon coupling between the one-electron capture luminescence within O_p^o and either the PLE spectrum of this luminescence or the resultant distant DAP luminescence (Section 5.2) and by the anomalous temperature dependence of σ_{p1}^o (Section 5.7). We have already described in Section 5.7 a reasonable alternative interpretation for the second of these properties within the classical model of Section 1.8 and 2.8, as well as the anomalous structure in the PLE spectra of distant DAP luminescence involving the O_p donor when measured in semi-insulating GaP (Section 5.6), another aspect which had previously encouraged the search for novel assignments in the spectroscopy of the O_p donor. We now briefly describe the basis of the new model and the results of recent additional experiments which attempt to resolve the question as to whether the new or classical descriptions render the best account of the practical situation. The results of one of these tests have already been described in Section 5.5, strongly favoring the classical model used in the bulk of this review.

The classical description (CD), used for all conventional donor states in all semiconductors, including deep donor states in Si, puts the five $n = 2$ electrons of O_p^+ into sp^3 bonds with the four surrounding Ga atoms, forming the usual complete set of covalent bonds familiar in an $A_{III}-B_V$ compound semiconductor. The additional electrons then enter antibonding states, symmetry classification respectively $^2A_1$ and $^1A_1$ for the lowest electronic states of O_p^o and O_p^- (Sections 1.8, 2.8 and 7). Morgan uses atomic term energies to suggest that the 2p level in O^o is too deep and the radii of the O orbitals are too small to permit the O orbitals to participate with the surrounding Ga atoms in sp^3 bonding in the accepted manner.[233] He suggests that the O remains neutral, while only two of the three electrons in O_p^+ contributed from the surrounding Ga atoms are present in the dangling bonds. The motivation for this seems primarily the perceived need to obtain not one ($^1A_1$) but two electronic states of O_p^+, so as to explain the small ~ 0.5 meV splitting discussed in Section 5.7. This very small energy splitting was arbitrarily assigned to the two states $^3T_2^{(+)}$ and $^1A_1^{(+)}$ formed from the antisymmetric and symmetric combinations of the tetrahedral Ga dangling bonds in the usual way (Figure 58). The model assumes the two electrons are sufficiently highly confined on *different* dangling bonds so that the orbital moment does not change during electronic transitions, while coupling of the electron spins gives $S = 1$ and $S = 0$ for the two states. The $^1T_2^{(+)}$ state also expected from coupling two t_2 electrons is not discussed by Morgan, presumably because of the assumption of zero exchange splitting with $^3T_2^{(+)}$. The two states of O_p^+ are estimated to contain the different proportions of a_1 1-electron combinations of the four Ga dangling bonds shown in Table 9. It is there-

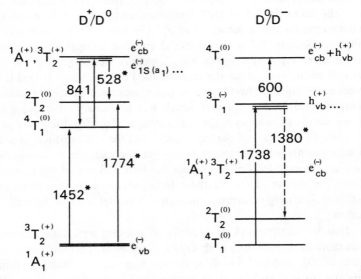

FIGURE 58 Energy levels, state configurations and oxygen transitions in GaP according to the new molecular orbital (MO) re-appraisal. The states are made up from the deep MO configurations at the left of each level plus (where shown) the effective mass electrons e^- or holes h^+ indicated at the right. Charges are in parenthesis, energies in meV. CB and VB are conduction and valence band edges. The dangling arrow at right suggests Auger transitions and asterisks mark transitions thought to excite A_1 or Γ_1 (breathing mode) phonons. Dashed transitions are unconfirmed, experiment suggests that the level marked $^5A_2^-$ is actually a spin triplet, while $^4T_1^0$ is a spin doublet (Morgan, Ref. 233).

TABLE 5. Allowed states for the positive, neutral and negative oxygen defect in GaP. n_ϵ, the coefficient of the two-center potential energy integral ϵ, depends on the number, $n(a_1)$, of a_1 electrons.

	$^m\Gamma =$	$^1A_1$	$^3T_2$	$^1T_2$	$^3T_1$	^1E
(+)	$n_\epsilon =$	4	2	0	-2	-2
	$n(a_1) =$	3/2	1	1/2	0	0

	$^m\Gamma =$	$^4T_1$	$^2T_2$	^2E	$^2T_1$	$^4A_2$
(0)	$n_\epsilon =$	1	2	0	-2	-3
	$n(a_1) =$	1	5/4	3/4	1/4	0

	$^m\Gamma =$	$^5A_2$	$^3T_1$	^1E	
(−)	$n_\epsilon =$	0	0	0	$n(a_1) = 1$

fore the more remarkable that the energies of these very different ultra-deep unscreened states should be separated by as little as 0.5 meV! The model really involves no more than a fitting procedure as far as the energies of the MO states are concerned. A large number of additional molecular-orbital states are generated by the addition of extra electrons, two of which are selected in Figure 58. These two states are considered to form the final states for the 841 meV (Figure 16) and 528 meV (Figure 47) electron capture luminescence, both assigned to transitions between the same two charge states of O_p despite the difficulties this brings, described already (Sections 5.8 and 5.9). The 528 meV transition is said to be allowed, which conflicts with the description in Sections 5.4, and 5.8, while the 841 meV transition is forbidden as required by experiment. Detailed assignments place the emitting state of the 528 meV luminescence ~9 meV above that of the 841 meV luminescence, so that it becomes very remarkable that both can be observed at 1.5°K, for weak as well as strong optical excitation rates,[196] particularly in view of the long lifetimes of emitting states in both systems (Section 5.8). However, this results from assignment of the 1.774 eV structure in Figure 49a to electron excitation from valence band to the $^2T_2^{(o)}$ final state of the 528 meV luminescence, in conflict with the alternative assignment of this spectral feature in Section 5.6.

Many of the disabilities of this radical new model have been reviewed elsewhere.[196] We give emphasis here to a few of these. An immediate theoretical difficulty is that it takes no account of the expected effects on the O_p atomic states of the strong screening expected from the valence electrons in a predominantly covalently bonded III–V compound semiconductor such as GaP. The self-consistent Green's function and other theoretical techniques, normally quite successful in the description of deep impurity as well as host states,[166,203] predict that the least tightly bound 2p electron in O_p^+ lies ~3 eV below the top of the host valence band,[246] not ~1 eV as required by the basis leading to Figure 57. Location of two electrons in the antibonding gap states rather than in the bonding states within the valence band establishes an energy deficit for O_p^+ of ~8 eV. Such a large energy cannot be established by electron–electron correlation, which is expected to be at most ~0.5 eV (Section 5.4).

All experimental tests so far designed to check the detailed assignments in Figure 58 have favored the classical model. In general, this is distinguished by low spin assignments compared with Figure 57, for example in the Zeeman studies of the O_p donor $D^0_?X$ BE already described (Section 5.4). Further Zeeman tests include a study of discrete DAP no-phonon lines. The O_p-Zn_{Ga} transitions were shown to exhibit Zeeman splittings indistinguishable from those of the S_p-Zn_{Ga} DAP, where the classical description certainly holds (S_p is an isocoric substituent), and consistent with the predictions of the classical model.[223] Finally, a careful study of the evolution of the intensities of the magnetic subcomponents of the 841 meV no-phonon line (Figure 59) showed that the integrated intensities of the σ and π transitions are both independent of magnetic field and of $Q = \exp(-g\mu_\beta B/kT)$,[223] where $g\mu_\beta B$ is the splitting of the emitting state (Figure 44a). These new results provide clear support for the model of Figure 44a over Figure 44b, also refuted in Section 5.2. They are also inconsistent with the assignment of the 841 meV transition in Figure 57.

The experimental and theoretical consensus is therefore strongly against this new model. It has already been shown that the classical description, properly interpreted, is capable of providing a very detailed description of all the manifold experimental data on O_p in GaP, with the apparent exception of the single aspect of the phonon coupling in the one electron transitions stressed by Morgan. On the other hand, many other aspects of the phonon coupling are well accounted for within the classical model. It is the author's view that the residual problem over the 19 meV phonon reflects an inadequate level of sophistication in the current theoretical treatments of phonon coupling in these spectra, rather than any need to invoke radical new descriptions for the electronic configurations of the various states and the transitions between them.

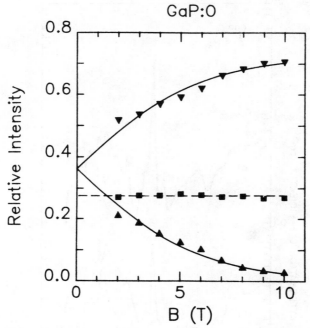

FIGURE 59 The relative intensities of the no-phonon magnetic subcomponent of the one-electron capture luminescence of O_P donors in GaP, O_0 in Figure 16. Triangles represent $\Delta m_J = \pm 1$ transitions, with the lower line corresponding to the highest energy subcomponent. Squares represent $\Delta m_J = 0$ transitions (Dean et al., Ref. 233).

6. Additional Forms of O in GaP

We have seen in Section 3.2 that a large proportion of the O in GaP may exist in Ga_2O_3 precipitates, unless care is taken not to exceed the solid solubility limit. Other forms are possible. One appears particularly in LEC material, normally contaminated with both B and O through intimate contact with the B_2O_3 encapsulant during crystal growth. The existence of this new form of O was revealed through infrared absorption measurements in the energy region where local mode effects of O may occur (Figure 60).[142] The band close to 10 μm normally appears in LEC crystals, together with several lines near 9 μm. The ~10 μm band can be related to the well-known ~9 μm band in o-doped Si, which involves a ν_3-type vibration within a "molecule" formed by an O atom which has interrupted

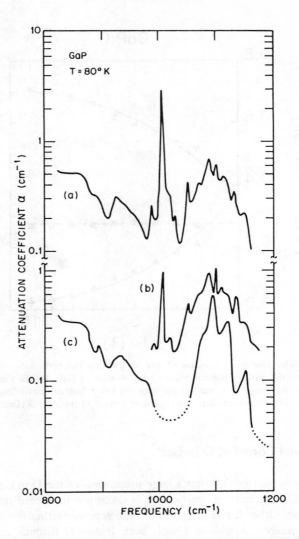

FIGURE 60 Infra-red absorption spectra of (a) an O-doped bulk (LEC)-grown GaP single crystal compared with (b) from an undoped crystal. Curve (c) is a composite obtained from several samples, chosen to exhibit low absorption. It probably represents the intrinsic absorption of the GaP lattice. Optical three-phonon processes are probably responsible for the structure above 1000 cm^{-1} in curve (c). The prominent narrow mode near 1000 cm^{-1} is believed to be caused by interstitial O which interrupts a Ga–P bond, while additional lines near 1100 cm^{-1} may arise from interstitial O associated with substitutional impurities such as B_{Ga} and C_P. (Barker et al., Ref. 142).

a Si = Si band to form an XY_2 type of structure. The ν_3 mode involves motion mainly of the O atom in a direction approximately parallel to the Si–Si axis, in which a large degree of bond stretching is involved.[143] A very low frequency vibrational type mode has also been predicted and observed.[144] Comparison of the behavior of the oxides SiO_2, GeO_2 and $GaPO_4$ suggests that a mode near 10 μm is the expected analogue of the ~9 μm mode in Si for O in a Ga-O_I-P molecular configuration in GaP. Calculations using a linear chain model[142] suggest another local mode of energy ~55 meV, only just above the cutoff energy of the lattice, ~50 meV, as well as the ~124 meV (10 μm) high frequency mode. Most of the energy is carried in the adjacent Ga and P atoms for the predicted 55 meV mode.

Attempts to verify directly the relation of this 10 μm mode with O, through O^{18}-induced shifts, failed because of excessive ion-exchange with B_2O_3 encapsulant. However, the linear chain model also gave a fair description of the O-related local modes in Si using similar ratios of force constants. Therefore, it seems very likely that the assignment is correct, besides the evidence that the strength of this mode generally increases with attempts to dope with O (Figure 60). The model gives 58.9 meV for the local mode of substitutional O_P, quite close to the 57.5 meV line attributed to this mode by Arai et al.[146] in absorption measurements on crystals grown from Ga-rich solutions in sealed silica tubes. The linear chain model also leads to a relation for the oscillator strength of the 10 μm mode, from which interstitial O_I concentrations of 0.6 to 6×10^{16} cm^{-3} were derived, with the lower values near the seed end of the LEC crystals. Berman and Barker[145] also estimated that about half of the total [O] measured by Lightowlers et al.[85] for LPE GaP in the absence of significant precipitates may be interstitial, that is about 3×10^{16} cm^{-3} for optimally Zn,O doped material grown at 1040°C.

Simple chemical considerations suggest that much of this interstitial O may be present in some associate configuration. The linear chain model suggests that the 4 cm^{-1} splitting of a weak satellite of the ~10 μm line is much too small to involve O_I on neighboring sites, though a second nearest neighbor configuration is possible. However 'molecules' of the form B_{Ga}-O_I-P or Ga-O_I-C_P are expected to generate ν_3-type vibrations near 9 μm, and may be responsible for some of the weak but sharp lines in Figure 60. These lines occur additionally to the broader intrinsic absorption, due mainly to 3 phonon processes in this region. There is little definite information concerning any electron trapping properties associated with O_I in any of these forms in GaP. The isolated Ga-O_I-P molecule is a neutral entity overall, just like the corresponding center in Si.

There is some evidence that interstitial O may be created in ZnTe by ion implantation[147] and produces BE of decreased localization energy, possibly in association with the substitutional form O_{Te} which is an isoelectronic acceptor.[148] It may be that we are dealing here with Zn-O_I-Te molecules in the environment of the O_{Te} trap. A possible luminescence system involving O_I observed in some Cd-doped GaP after Li in-diffusion has no-phonon structure near 2.217 eV,[111] with the form expected for exciton binding at a neutral center. This spectrum also exhibits a small increase in no-phonon energy by ~ 0.35 meV when $O^{16} \rightarrow O^{18}$. This system may involve the neutral complex Cd_{Ga}-O_I-Li_I-P, since it also exhibits the low energy phonon structure which seems to be a feature of heavy cation substituents (Section 3.5). It is a matter of speculation whether the Si-O defect discussed by Bachrach et al.[149] involves O_I rather than O_P, which may be preferred since the Si_{Ga} and O_P donors are both positively charged when ionized and are unlikely to form an associate. Similarly, the remarkably shallow axial acceptor described by Dean et al.[150] and thought of as possible H_{Ga}-O_P associate may instead involve O_I in some appropriate form as suggested by analogy with recent findings in H-contaminated Ge.[151] However, the obvious simple center involving interstitials O and H is a donor[151] rather than an acceptor. Unfortunately, it is very difficult to confirm any of these speculations. Isotopic substitution H → D did not give a positive result for this shallow acceptor in GaP, since the no-phonon DAP lines were far too broad (~ 0.1 meV) to detect the very small isotope shifts of $\sim 20 \rightarrow 50$ μeV observed in the ground state energies of H-related shallow donors and acceptors in Ge.[151]

7. Further Theoretical Considerations for the Deep O Donor

7.1 Tight-Binding Description of the Isolated O_P Donor

Oxygen in GaP is a well-researched example of a class of deep states in semiconductors which have a number of peculiar properties when compared with other impurities of the same class, for example the Group VI donors in GaP. Of all these impurities, only O_P has a deep state although large differences in atomic energies occur between other members of this group. An extreme tight binding description of these states[155] shows a nearly hyperbolic variation of trap (donor) depth with impurity potential V, defined as the difference between nearest–neighbor matrix elements for the impurity–host and host–host interactions. Only the central cell part of the impurity potential is retained, so donors of higher atomic number (lower s orbital energy) than Se in GaP are unbound (Figure 61). The intersection of each quasi-hyperbola with the band edges defines attractive and repulsive thresholds of potentials, one of which must be exceeded if the short range defect potential is to form a bound state. Thus far, the model is an extension of that used by Faulkner to describe isoelectronic traps such as N in GaP, long recognized to be a prototype deep state even

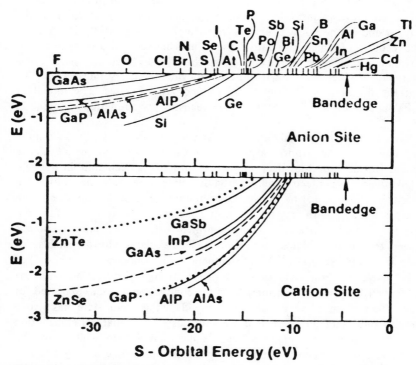

FIGURE 61 Predicted energies relative to the conduction band edge for the A_1- symmetric deep impurity levels as a function of the orbital energy of the impurity. Relevant impurities are listed at the top in the order of decreasing predicted binding energy of the trap. Each quasi-hyperbola is for a different host, as indicated. The impurities should be selected according to those likely to substitute on the anion (upper) or cation (lower) sublattice. Only major trends in the ordering are significant. Anion impurities to the right of Se are unbound in GaP according to this theory, which ignores the residual binding of a donor species due to the long range Coulomb interaction with the donor core. Where no quasi-hyperbola is given, as for anions in InP, no deep state is predicted for any substitutional impurity. (Hjalmarson, Ref. 155).

though its binding energy is small.[156] However, the important new feature is that this quasi-molecular orbital description recognizes the asymptotes of the hyperbolas as the dangling bond or ideal vacancy energies, to which the very deep traps become pinned. Thus, the energy of the $1s(A_1)$ ground state of O_P becomes pinned just above the A_1 Ga dangling bond energy. The key difference from the extrapolated effective-mass description used above (Sections 1.8 and 2.8) is, therefore, that the wavefunctions of the deep O_P donor are predominantly host-like rather than impurity like.[155] This pinning explains why differences in the impurity potential near the upper end of the range affect the trap energies rather weakly. The pinning concept is of qualitative rather than quantitative significance, however. The quasi-molecular orbital theory which underlies it ignores charge transfer effects which are known to be significant for the crystal field splitting of the dangling bond levels into T_2 and A_1 substates. The model of Hjalmarson et al.[155] does not give realistic wave-functions for O_P in GaP.[188]

The state with primarily impurity-like wavefunctions, expected from the effective mass description, also exists, but is below or within the valence band for all impurities deep enough to produce a deep state in the gap. This filled, 'hyperdeep' level is a bonding state on a molecular orbital description, orthogonal to the deep anti-bonding state in the gap. For an impurity like O_P, with very large S orbital energy relative to P, wavefunction of the hyperdeep state is mainly confined to the anion impurity, so the deep trap state is almost pure Ga-like. This explains why the analysis of stress effects (Section 2.10) and temperature dependence (Section 4.1) of the single electron state of O_P suggests strong hybridization with the valence band, rather than pinning to the conduction band, as expected for an effective mass-like donor, since the dangling bond state has strong multiband character.[155] This description also gives a natural explanation for the absence of a relatively deep 2s-like excited state, which might be expected if the deep O-donor energy states represented simply a scaled version of the effective mass states with all S-like states uniformly influenced by the central cell potential according to the relative densities of their wavefunction overlap $|\psi|^2_{r=O}$ on the donor core. This model would predict a $2s(A_1)$ state with binding energy $\sim 10\%$ of that of the $1s(A_1)$ ground state, since $|\psi|^2_{r=O} :: (1/n^3)$, where n is the orbital quantum number. The strong electron capture luminescence observed from the $1s(E)$ state of binding energy 58 meV (Section 2.10) clearly shows that no $2s(A_1)$ state can occur at ~ 90 meV below the conduction band. We have already noted in Section 5.3 that the excited state responsible for the capture luminescence cannot be the $2s(A_1)$ state, since this state would not split under any direction of uniaxial stress, in disagreement with Figure 45.

In contrast to the effective mass model, which emphasizes the infinite series of excited states characteristic of the Coulomb attractive potential of the impurity core, the tight binding description permits only a single deep s-like anti-bonding impurity state.

7.2 Tight-Binding Description of Associates Involving O_P

The model of Hjalmarson et al.[155] has recently been extended to the case of paired substitutional sp^3-bonded associates of cation–oxygen type, where the oxygen is substitutional on the anion site.[157] The matrix elements of the defect potential are again constructed from differences in the impurity-host atomic energies at each site, neglecting long-range interactions. The two states of A_1 and T_2 (p-like) symmetry predicted for a point defect substituent in the T_d point group of the zincblende lattice are replaced by an a_1 (σ-like) and e (π-like) state in the C_{3V} symmetry of the associate. The e-type molecular states should lie close to the original T_2 states, but the a_1 states involve a linear combination of the point defect A_1 and T_2 states. Thus, the A_1 state is subject to major influence by the cation-site 'spectator' member of the associate.[157] A substantial reduction in binding energy in forming the a_1 state is generally expected, since the type of cation substituent likely to best form an associate with the O_P is one which generally reduces the electron binding energy to the O_P (Table 2). This is achieved on the tight binding model by a decrease in the energy of the 'conduction band' state of the defect molecule as a result of the replacement of a neighboring Ga ion by a more electropositive ion such as Zn in the "defect molecule" discussed in Figure 2 of Hjalmarson et al.[155] Associates containing electropositive spectators produce energy levels which tend to, and are pinned by the V_{Ga}-O_P level (Figure 62). We have seen in Section 3.6 that no experimental value exists for this associate in GaP. However, this energy is expected to be close to that of the Cu_{Ga}-O_P,[157] which we have identified with the 1.868 eV system in Section 3.7. The implausible assignment of V_{Ga}-O_P to the COL luminescence by Sankey et al.[157] causes the experimental points labelled V_{Ga}-O_P to be placed 0.31 eV too high in Figure 62 if the Cu_{Ga}-O_P associate energy is to be taken as a model for V_{Ga}-O_P. However, this experimental point should not only be down-shifted in energy but also moved to the center of the abscissa scale, near the value indicated for Zn–O. As discussed in Section 3.7, the COL luminescence most probably does not involve an associate containing O_P.

Strongly electronegative spectators rapidly increase the binding energy of the a_1 molecular state towards the valence band edge (Figure 62).

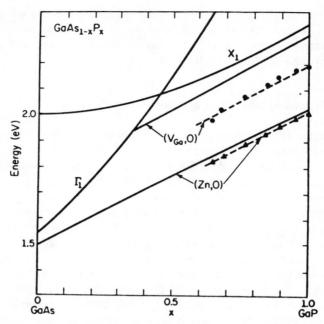

FIGURE 62 Energy levels of associates Zn–O and V_{Ga}–O as a function of alloy composition x in $GaAs_{1-x}P_x$. The solid lines are theory and the dashed lines are experimental data. The data for the V_{Ga}–0 centre are identified with the COL spectrum of Figure 30a in the limit x=1, ie for GaP. However, we have seen that this center definitely involves Cu (Table 1), so these data may be re-interpreted as involving the Cu_{Ga}–Cu_Γ–Cu_I associate. The no-phonon energies shown here are estimated from the broad experimental data by assuming that the Stokes shifts in the alloy are identical to those in GaP. (Sankey et al., Ref. 157).

Such very deep associate states would probably be very difficult to identify experimentally. However, interest in them may be largely academic. They involve spectator ions which are difficult to prepare on the cation site, such as the anti-site P_{Ga}. Even if present, such substituents would be very unlikely to associate with O_P (Section 3.5). The charge differences between the constituent members of the associates and the host atoms they replace have been disregarded in the formulation of this molecular orbital description of the host energy states.[157] However, such charge differences certainly cannot be disregarded for the statistics of the ion pairing necessary for associate formation. Charge differences between the associates have also been used quite successfully as a basis for the description of electron binding to these associates in Table 2. There, we used a very simple description, in which the energy reduction factor E_{rep} in

Equation (4) is obtained simply as $e^2/\epsilon R$ from Equation (6), considering the associate as the limiting case of nearest neighbor donor–acceptor pair (Section 3.5). The basic philosophy of this description emphasizes the neutral character, or 'molecular isoelectronic trap' nature, of all the associates for which useful experimental data exist, a feature entirely disregarded by Sankey et al.[157]

8. Deep States Associated with O in Other III–V Semiconductors

8.1 Gallium Arsenide

We have seen that there is a great deal of detailed experimental evidence on properties of O in GaP. Despite a considerable amount of controversy, generated mainly over certain aspects of these data, it has been possible to establish a remarkably coherent overall picture, described above. No comparably well-substantiated attributions exist for any other III–V semiconductor. By far the most important of these is GaAs, also the recipient of most attention concerning the properties of O. The successful growth of high resistivity "semi-insulating" bulk material, desired for device applications and apparently governed by the presence of O, was reported in the early 1960s.[158–161] Detailed analysis of Hall data taken over a wide temperature range suggested that this material contained a dominant mid-gap trapping level, about 0.75 eV below the conduction band. Deep level characterization by the thermally stimulated current[162,163] and optical transient current[164] techniques on high resistivity material and DLTS studies[165] on n-type material all confirmed such a level, the DLTS technique yielding an apparent activation energy of 0.825 eV due to the contribution of E_B (Equation (16)) to the activation energy of the thermal emission rate,[154,183] The trap depth associated with this DLTS peak EL2, normally dominant in undoped bulk-grown and VPE GaAs, was close to 0.75 eV. This energy lies near early[166] and more recent theoretical estimates[155] for O_{As} in GaAs. The latter estimate was obtained from a seemingly plausible description of the variation of this state with alloy parameter x in $GaAs_{1-x}P_x$, supported by experimental evidence that the concentration of the relevant state scaled with the dose of ion-implanted O. It has also been demonstrated that EL2 is a donor rather than a neutral electron trap.[225] The key role of EL2 in the compensation of semi-insulating GaAs has been demonstrated very clearly from a combined study of electrical (Hall) and optical absorption ([Cr]).[226]

Unfortunately, it has proved very difficult to obtain conclusive proof that this dominant electron trap is caused by the point defect deep donor O_{As}, or indeed is even O-related. Recent accurate mass-spectrometer studies show very low total O concentrations, $\leqslant 1 \times 10^{16}$ cm^{-3} even in LEC crystals deliberately doped with Ga_2O_3, and no O could be detected by local vibrational mode (LVM) spectroscopy in this material.[167] However, this negative result is contradicted by other reports,[228] where a midgap level at $\sim 10^{16}$ cm^{-3} and an interstitial center at concentration $\sim 10^{17}$ cm^{-3} was reported for bulk GaAs doped with Ga_2O_3, the interstitial center giving LVM absorption near 100 meV. Much higher [O] are frequently claimed from neutron activation and other techniques,[168] but it is clear that surface contamination and possibly effects of Ga_2O_3 inclusions could be responsible for much of these high apparent concentrations. Deep-level introduction is not the only mechanism by which O may tend to assist compensation of n-type GaAs. The control of the major shallow donor Si through a metallurgical reaction involving the formation of SiO_2 was suggested for GaAs in 1963[169] and certainly plays a major role, as it also does in GaP.[170] In strong contrast with GaP, no optical spectra containing sharp structure capable of association with O have been found in GaAs, aside from a single bound exciton line near 1.4885 eV[171] which appears to involve the recombination of an exciton bound to a neutral deep donor, such as could be caused by O_{As}. This BE line appeared to correlate with the addition of O in VPE growth, but no certain attribution to O_{As} was possible.[171] Other optical features tentatively associated with O include a broad optical absorption near 0.7 eV.[172] The correlation of these featureless optical bands with O is best described as tentative, though this may be understandable in view of the well-known difficulties in chemical control of O in III-V semiconductors (Section 2.2). Attempts to create the 0.75 eV trap in LPE GaAs by the addition of Ga_2O_3 to the Ga growth solution have been repeatedly unsuccessful[173-176] again in sharp contrast with GaP.

Apparently clinching evidence that the EL2 level cannot involve O was recently obtained from a combined secondary ion mass spectrometry–DLTS study of horizontal Bridgman bulk GaAs.[177] The EL2 trap concentration was variable and reached 2×10^{16} cm^{-3}, while the total [O] was constant near 2×10^{15} cm^{-3}. The negative conclusion is further supported by Wallis et al.[178] who failed to observe significant correlation between [EL2] and deliberate addition of H_2O or SiH_4 to VPE GaAs grown by the halide technique or by the organometallic process, where [EL2] is significantly lower. Similar conclusions were reported very recently by Martin et al.[255] who found no correlation between [EL2] and Ga_2O_3 added to, and [O] in, LEC GaAs, but a gettering of Si according to well-established ideas. The observation of apparently the same mid-gap electron trap in $A\ell_x Ga_{1-x} As$ over a wide range of $x(0 \leqslant X \leqslant 0.35)$[227] provides further evidence that it cannot be caused by a simple center involving O. However, the appearance of EL2 in O-implanted LPE layers after annealing at $870°K$ has been claimed as evidence of a relation between EL2 and O, although $[EL2] < 0.01 [O]$.[231] The reduction in N_D-N_A observed in halide VPE GaAs was entirely attributed to the suppression of Si incorporation. A similar effect observed in organometallic $A\ell_x Ga_{1-x} As$ was attributed to *compensation* of the shallow donors by an acceptor-like defect involving $A\ell$ and O.[178] The DLTS measurements suggests that this is caused by a trap ~0.41 eV below the conduction band, comparable with the associate levels discussed for $GaAs_{1-x}P_x$ by Sankey et al.[157] A considerable amount of evidence has now accumulated to suggest that the EL2 level in GaAs may involve V_{Ga}[179-182] associated with a center as yet unidentified but which may produce an electron trap 0.39 eV below the conduction band.[182] The EL2 donor concentration correlates with the square root of AsH_3 partial pressure in MOCVD growth of GaAs,[256] and with the As concentration in the melt of LEC GaAs.[257] Alternative contributions to the antisite defect As_{Ga} have been made even more recently.[230] The broad luminescence band with peak energy near 0.645 eV, originally believed to be a signature of O in GaAs[160] has recently been quantitatively related to the EL2 trap through a configurational co-ordinate type analysis.[229]

8.2 Indium Antimonide

An interesting feature of Figure 61 is the prediction that a bound state may exist for O as an anion substituent in GaAs, but not in InP. Of course, such a substitution should always give a near-effective mass-like bound state, neglected in the molecular–orbital description of Section 7.1. This description would certainly predict a deep bound state associated with the higher, large mass conduction band minima in InSb, because of the large difference in s-orbital energy for Sb and O. However, this effect will be countered for states associated with the lowest Γ_1 conduction band minimum by the very low electron mass in InSb, which makes it impossible to obtain carrier freeze-out even at $2°K$ in the purest samples available, $N_D-N_A \sim 10^{14}$ cm^{-3}. Carrier freeze-out does occur at quite low magnetic fields in such samples.[189] Recently, very remarkable central cell effects have been reported for such material.[190] These involve a shallow Γ_1-related donor state with central cell correction (chemical shift) remarkably large for InSb, and with an unusually large shift rate under increases in magnetic field or hydrostatic pressure (Figure 63). Both of these pertur-

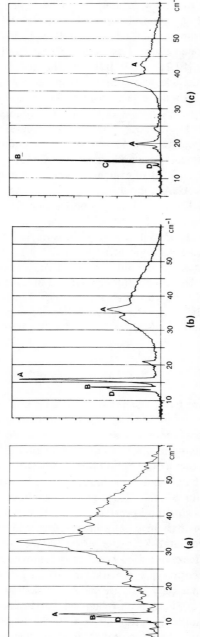

FIGURE 63 Far-infrared photoconductivity spectra of high purity InSb recorded with a Fourier-transform spectrometer at 4.2°K and at three different combinations of magnetic field and uniform hydrostatic pressure. (a) is recorded at zero pressure and B = 10T showing structure in the $1s \rightarrow 2p-$ lines due to three donors A, B and D. The separation of A and D is ~ 0.1 meV. (b) is recorded at a pressure of 4.9 kbar and B = 9T. The integrated intensity of transition involving donor A has increased to about 4 x that of B. Their separation is now 0.31 meV, resolved for $1s \rightarrow 1p_0$ as well as $1s \rightarrow 2p-$. (c) is recorded at 4.9 kbar and B = 13.5T. The contribution due to donor A has fallen dramatically to less than half of that of B, while their separation is now 0.63 meV. Donor A may be related to O_{Sb}, and it is believed that a deep state associated primarily with the large electron mass off-center conduction band minima becomes energetically lowest above ~ 5 kbar. (Davidson et al., Ref. 190).

336

bations tend to make InSb more indirect, that is they reduce the energy difference between the central Γ_1 and higher indirect X and L conduction band minima. Above a critical pressure of ~ 5 k bar, the contributions of the deviant donor to the photothermal photoconductivity process by which such shallow centers are conveniently detected and chemically discriminated[191] become removed, while signals due to other shallow donors remain strong and sharp. This is believed due to the resonance near this energy between a deep state associated with the higher conduction band minima and the shallow Γ_1-related state of the exceptional donor. There is significant evidence that this unusual donor may involve O, obtained in the usual way through correlations with attempts to add O.

This assignment can only be regarded as tentative, though plausible, at present, in view of the difficulties over the chemistry of O-doping already emphasized (Section 2.2). However, these results possibly provide the best evidence available at present for electronic activity of substitutional O in any III-V semiconductor besides GaP, and are therefore of great interest.

Acknowledgments. The author is indebted to many colleagues, a significant proportion of them co-workers, who over the years have contributed to the subject matter of this article. The emergence of O in GaP as the characterized and understood chemical substituent deep state system for any semiconductor is due to their collective efforts. There are too many names to list conveniently, and in any case I hope that all have received just citation in the reference list. However, three particular credits must be made, two personal and the third more general. The author is particularly grateful to Charles H. Henry for his collaboration on the luminescence properties of GaP:O during a most fruitful period of research in 1967–68. Chuck Henry returned to this topic 5 years later as the instigator of the greatly increased emphasis on the study of non-radiative effects through capacitance spectroscopy, which is now a very familiar feature of semiconductor research. He then conducted research which led to an important attempt to account for this admittedly 'hard problem.' This resulted in a paper, Ref. 119 in this article, which is comprehensive and yet places very clear emphasis on the essential physical principles which are involved. The author is also grateful to Tom Morgan, whose persistent inquiries into perceived limitations of understanding over many years ultimately led to a remarkable renaissance of activity on GaP:O during 1981–83, and to his other colleagues associated with the recent work on Refs. 141 and 223. The more general point concerns the thanks all the assessment scientists who have been involved in this work owe to the skilled crystal growers of a number of laboratories, who have contributed the carefully controlled semiconductor material which has been essential for the advancement of our understanding on a hard though fascinating topic.

Appendix

In this Appendix we summarize the electronic reactions involving the O_p donor in the form of a Table for easy reference. These reactions have been discussed throughout the article as indicated in the last column of the Table.

TABLE A1. Electronic reactions involving O_p donor

First or Second Electron to Bind	Centre(s) Involved	Type of Reaction	Symbolic Equation	Energy Released[X]	Section Cited
First	Isolated O_p donor	Electron capture 1S (E) → 1S(A_1)	$[\oplus e*] \rightarrow [\oplus e] + h\nu_L$	$h\nu_L = 0.841$ eV (peak)	2.9
"	"	Electron excitation 1S(A_1) – $2p_0$	$[\oplus e] \rightarrow [\oplus e*] - h\nu_A$	$h\nu_A = 0.863$ eV (peak)	2.11
"	"	Electron excitation VB → 1S(A_1)	$[\oplus] + ev_B \rightarrow [\oplus e] + h\nu_B - h\nu_A$	$h\nu_A = 1.453$ eV (threshold)	2.14
"	"	Free hole luminescence at neutral O_p donor	$[\ominus \oplus e] + h\nu_B \rightarrow [\ominus \oplus] + h\nu_L$	$h\nu_L = 1.445$ eV (peak at 87°K)	3.3
"	O_p donor and distant acceptor	Electron capture with Auger release of acceptor hole	$[\oplus e*] \rightarrow [\oplus e] + E$ $[\ominus h] + E \rightarrow [\ominus] + h*_{VB}$	–	2.13
"	"	Distant DAP luminescence	$[\oplus e] + [\ominus h] \rightarrow [\oplus] + [\ominus] + h\nu_L$	$h\nu_L \sim 1.36 - 1.41$ eV (peaks)	2.6–2.7
"	O_p and neighbouring acceptor	Donor-acceptor associate luminescence (or absorption)	$[\{ \oplus \ominus e|h] \rightarrow [\ominus \oplus] + h\nu_L$	$h\nu_L \sim 1.91 - 2.19^{\dagger}$ (peaks)	2.2–2.4 3.1, 3.5
"	"	Donor-acceptor associate dissociation reaction	$[\ominus \oplus e] + [\ominus] \rightarrow [\ominus]$ or $[\ominus \oplus e|h] \rightarrow [\oplus]$	–	2.6
"	" plus distant acceptor	Donor-acceptor associate – distant acceptor luminescence	$[\ominus \oplus e] + [\ominus h] \rightarrow [\ominus] + [\oplus] + h\nu_L$	$h\nu_L \sim 1.86 - 1.92$ eV† (thresholds)	2.5
"	Isolated O_p donor	Photoionization of O_p donor	$[\oplus e] \rightarrow [\oplus] + e_{CB} - h\nu_A$	$h\nu_A \sim 0.90$ eV (threshold)	2.2, 4.1

338

First or Second Electron to Bind	Centre(s) Involved	Type of Reaction	Symbolic Equation	Energy Released^X	Section Cited
Second	Isolated O_P donor	Electron capture $S(E) \to S(A_1)$	$[\oplus ee^*]_M \to [\oplus ee]_M + h\nu_L$	$h\nu_L = 0.528$ eV (peak)	5.4
"	"	Photocreation of neutral O_P donor bound excitation	$[\oplus e] \to [\oplus ee]_M h] - h\nu_A$	$h\nu_A = 1.738$ eV (peak)	5.4
"	"	Auger recombination of neutral O_P donor BE	$[(\oplus ee)_M h] \to [\oplus] + e_{CB}$	-	5.7
"	"	Possible explanation of triplet state ODMR	$\begin{cases}[(\oplus ee^*)_M h] \to [\oplus] + e_{CB}\\ [\oplus] + e_{CB} \to [\oplus e] + h\nu_L\end{cases}$	$h\nu_L = 0.841$ eV (peak)	5.8
"	"	Photoionization of ground state of O_P centre	$[\oplus ee]_G \to [\oplus e] + e_{CB} - h\nu_A$	$h\nu_A \sim 1.4$ eV (120°K threshold)	4.4
"	"	Photocreation of O_P^- in ground state	$[\oplus e] + e_{VB} \to [\oplus ee]_G + h\nu_{VB} - h\nu_A$	$h\nu_A \sim 1.65$ eV (120°K threshold)	4.4

Subscript M denotes metastable, centred state of O_P^-, G denotes rebonded, ground state $\oplus$ denotes the core of the O_P donor; $\ominus$ denotes the core of a substitutional acceptor such as Zn_{Ga}. e and h are electronic particles, hν is a photon, subscript L in luminescence, A in absorption.

X Negative sign means endothermic reaction.

* Denotes excited state of electronic particle.

† Denotes dependence on type of acceptor codopant.

339

References

1. A. A. Bergh and P. J. Dean 'Light Emitting Diodes', Clarendon Press, Oxford, 1976.
2. P. J. Dean, *Progress in Solid State Chemistry,* ed. by J. O. McCaldin and G. Somorjai (Pergamon, New York, 1973), 8, p. 1.
3. D. G. Thomas and J. J. Hopfield, Phys. Rev. *150*, 680 (1966).
4. M. Gershenzon and R. M. Mikulyak, J. Appl. Phys. *32*, 1338 (1961).
5. M. Gershenzon, R. M. Mikulyak, R. A. Logan, and P. W. Foy, Solid State Electron. *7*, 113 (1964).
6. G. A. Wolf, R. A. Herbert, and J. D. Broder, Phys. Rev. *100*, 1144 (1955).
7. H. G. Grimmeiss and H. Koelmans, Philips Research Repts, *15*, 290 (1960).
8. M. Gershenzon and R. M. Mikulyak, Solid State Electronics *5*, 313 (1962).
9. M. Gershenzon and R. M. Mikulyak, J. Electrochem Soc. *108*, 548 (1961).
10. L. R. Weisberg, F. D. Rosi, and P. G. Herkart, *Properties of Elemental and Compound Semiconductors,* ed. by H. C. Gatos (Interscience, New York, 1960) p. 25.
11. C. H. Gooch, C. Hilsum, and B. Holeman, J. Appl. Phys. *32*, 2069 (1961).
12. C. J. Frosch and L. Derick, j. Electrochem. Soc. *108*, 251 (1961).
13. C. J. Frosch, M. Gershenzon, and D. F. Gibbs, Symposium on the Preparation of Single Crystals of the III-V Compounds, Battelle Memorial Institute, Columbus, Ohio 1959.
14. J. Starkiewicz and J. W. Allen, J. Phys. Chem. Solids *23*, 881 (1962).
15. M. Gershenzon, F. A. Trumbore, R. M. Mikulyak, and M. Kowalchik, J. Appl. Phys. *36*, 1528 (1965).
16. H. G. Grimmeiss and H. Scholz, Phys. Lett. *8*, 233 (1964).
17. M. Gershenzon, F. A. Trumbore, R. M. Mikulyak, and M. Kowalchik, J. Appl. Phys. *37*, 483 (1966).
18. F. A. Trumbore and D. G. Thomas, Phys. Rev. *137*, A1030 (1965).
19. M. R. Lorenz and M. H. Pilkuhn, J. Appl. Phys. *37*, 4094 (1966).
20. M. Gershenzon, R. A. Logan, D. F. Nelson, and D. G. Thomas, Bull. Am. Phys. Soc. *9*, 236 (1964).
21. T. N. Morgan, M. H. Pilkuhn, and M. R. Lorenz, Proc. Int. Lum. Conf., Budapest, 1966 (Akademiai Kiado, Budapest, 1968) p. 1926.
22. J. A. W. van der Does de Bye, Phys. Rev. *147*, 589 (1966).
23. M. Gershenzon and R. M. Mikulyak, Appl. Phys. Lett. *8*, 245 (1966).
24. T. N. Morgan, B. Welber, and R. N. Bhargava, Phys. Rev. *166*, 751 (1968).
25. C. H. Henry, P. J. Dean, and J. D. Cuthbert, Phys. Rev. *166*, 754 (1968).
26. A recent conference was largely devoted to GaAs:Cr Semi-insulating III-V materials, Nottingham, 1980, ed. by G. J. Rees (Shiva, Orpington 1980); see also article by Allen, this volume.
27. V. Heine and C. H. Henry, Phys. Rev. *B11*, 3795 (1975).
28. J. J. Hopfield, private communication, 1967.
29. J. D. Cuthbert, C. H. Henry, and P. J. Dean, Phys. Rev. *170*, 739 (1968).
30. Jane van W. Morgan and T. N. Morgan, Phys. Rev. *B1,* 739 (1970).
31. P. J. Dean and D. C. Herbert, 'Bound excitons' in 'Excitons' ed. by K. Cho (Springer-Verlag, Berlin, 1979), Figure 3.10 p. 82.
32. T. N. Morgan, Phys. Rev. Lett *21*, 819 (1968).
33. P. J. Dean, R. A. Faulkner, and S. Kimura, Phys. Rev. *B2*, 4062 (1970).
34. J. J. Hopfield, D. G. Thomas, and R. T. Lynch, Phys. Rev. Lett. *17*, 312 (1966).

35. A. Onton and M. Lorenz, Appl. Phys. Lett. *12*, 115 (1968).
36. M. Toyama and A. Kasami, Japan J. Appl. Phys. *11*, 860 (1972).
37. A. A. Kopylov and A. N. Pikhtin, Solid State Comm. *26*, 735 (1978); G. F. Glinskii, A. A. Kopylov, and A. N. Pikhtin, ibid, *30*, 631 (1979).
38. R. A. Logan, H. G. White, and F. A. Trumbore, Appl. Phys. Lett. *10*, 206 (1967).
39. P. J. Dean, C. H. Henry, and C. J. Frosch, Phys. Rev. *168*, 812 (1968).
40. P. J. Dean and Lyle Patrick, Phys. Rev. *B2*, 1888 (1970); T. N. Morgan and H. Maier, Phys. Rev. Lett. *27*, 1200 (1971).
41. A. T. Vink, R. L. A. Van der Heyden, and J. A. W. Van der Does de Bye, J. Luminesc. *8*, 105 (1973).
42. A. A. Kopylov and A. N. Pikhtin, Solid State Commun. *26*, 735 (1978).
43. M. D. Sturge, A. T. Vink, and F. P. J. Kuijpers, Appl. Phys. Lett. *32*, 49 (1978).
44. H. C. Casey, Jr., F. Ermanis, and K. B. Wolfstirn, J. Appl. Phys. *40*, 2945 (1969).
45. Lyle Patrick, Phys. Rev. *180*, 794 (1969).
46. P. J. Dean and Lyle Patrick, Phys. Rev. *B2*, 1888 (1970).
47. A. T. Vink, Luminesc. *9*, 159 (1974).
48. The effect of the "camels back" form of the conduction band minima in GaP near symmetry point X in the Brillouin Zone can be neglected for states of binding energy >> 10 meV, A. C. Carter, P. J. Dean, M. S. Skolnick, and R. A. Stradling, J. Phys. C. *10*, 5111 (1977).
49. D. C. Herbert and J. Inkson, J. Phys. C: Solid State Phys. *10*, L695 (1977).
50. D. D. Manchon and P. J. Dean, Proc. Int. Conf. Phys. Semicond., Cambridge 1970 ed. by S. P. Keller, J. C. Hensel, and F. Stern (USAEC, Oak Ridge, Tenn. 1971) p. 760.
51. A. A. Kopylov and A. N. Pikhtin Fiz. Tekh. Poluprov. *11*, 867 (1977) [Engl. Transl. Sov. Phys. Semicond. *11*, 510 (1977)].
52. M. Lax, Phys. Rev. *119*, 1502 (1960).
53. P. J. Dean and C. H. Henry, Phys. Rev. *176*, 928 (1968).
54. I. Balslev, Proc. Intern. Conf. Phys. Semicond. Kyoto 1966, J. Phys. Soc. Japan *21*, Supp. 101 (1966).
55. D. K. Wilson and G. Feher, Phys. Rev. *124*, 1068 (1961).
56. W. Kohn, Solid State Physics, ed. by F. Seitz and D. Turnbull, (Academic Inc. New York 1957) *5*, 257.
57. R. A. Messenger and J. S. Blakemore, Phys. Rev. *B4*, 1873 (1971).
58. G. Lucovsky, Solid State Commun. *3*, 299 (1965).
59. A. C. Carter, P. J. Dean, M. S. Skolnick, and R. A. Stradling, J. Phys. C.: Solid State Phys. *10*, 5111 (1977).
60. R. G. Humphreys, U. Rossler, and M. Cardona, Phys. Rev. *B18*, 5590 (1978).
61. Y. C. Chang and T. C. McGill, Solid State Commun. *33*, 1035 (1980).
62. A. Onton, Phys. Rev. *186*, 786 (1969).
63. T. N. Morgan, Phys. Rev. Lett. *40*, 190 (1978).
64. H. C. Casey, Jr., F. Ermanis, and K. B. Wolfstirn, J. Appl. Phys. *40*, 2945 (1969).
65. A. R. Beattie and P. T. Landsberg, Proc. Roy. Soc. Ser. A *249*, 16 (1969).
66. D. F. Nelson, J. D. Cuthbert, P. J. Dean, and D. G. Thomas, Phys. Rev. Lett. *17*, 1262 (1962).
67. E. F. Gross and D. S. Nedzvetskii, Engl. Transl. Sov. Phys. Doklady *8*, 989 (1964).
68. A. Schindler, R. Bindemann, and K. Kreher, Phys. Stat. Solidi *b59*, 439 (1973).

69. B. Monemar and L. Samuelson, Solid State Commun. *26*, 165 (1978).
70. R. Bindemann, E. Hempel, and K. Kreher, Phys. Stat. Sol. *a52*, 201 (1979).
71. R. Bindemann, R. Schwabe, and T. Hansel, Phys. Stat. Sol. b*87*, 169 (1978).
72. B. Monemar and L. Samuelson, Phys. Rev. *B,18* 809 (1978).
73. J. A. W. van der Does de Bye, J. Phys. Chem. Solids *30*, 1293 (1968).
74. B. Welber and T. N. Morgan, Phys. Rev. *170*, 767 (1968).
75. P. J. Dean, Phys. Rev. *168*, 889 (1968).
76. C. H. Henry, P. J. Dean, D. G. Thomas, and J. J. Hopfield, 'Localized Excitations in Solids, ed. by R. F. Wallis (Plenum, New York, 1968) p 267.
77. P. D. Dapkus, W. H. Hackett, Jr., O. G. Lorimor, and R. Z. Bachrach, J. Appl. Phys. *45*, 4920 (1974).
78. P. J. Dean, unpublished data, 1968.
79. A. T. Peters and R. C. Vink, Inst. Phys. Conf. Ser. *24*, 254 (1975).
80. J. A. W. van der Does de Bye, J. Electrochem. Soc. *123*, 544 (1976).
81. L. M. Foster and J. Scardefield, J. Electrochem. Soc. *116*, 494 (1969).
82. R. H. Saul, J. Electron. Mat. *1*, 16 (1972).
83. R. Bindemann, H. Fischer, and K. Kreher, Phys. Stat. Sol. (a)*43*, 529 (1977).
84. M. Kowalchik, A. S. Jordon, and M. H. Read, J. Electrochem. Soc. *119*, 756 (1972).
85. E. C. Lightowlers, J. C. North, A. S. Jordon, L. Derick, and J. L. Merz, J. Appl. Phys. *44*, 4758 (1973).
86. C. K. Kim, Radiochem. Radioanal. Lett. *2*, 53 (1969).
87. J. M. Dishman, M. DiDomenico, Jr., and R. Caruso, Phys. Rev. *B2*, 1988 (1970).
88. H. Kukimoto, C. H. Henry, and F. R. Merritt, Phys. Rev. *B7*, 2486 (1973).
89. H. Kukimoto and M. Mizuta, Japan, J. Appl. Phys. *43*, Supp. 95 (1973).
90. C. H. Henry, R. Z. Bachrach, and N. E. Schumaker, Phys. Rev. *B8*, 4761 (1973).
91. M. Mizuta and H. Kukimoto, Japan, J. Appl. Phys. *14*, 1631 (1975).
92. D. V. Lang, J. Appl. Phys. *45*, 3014 (1974).
93. C. H. Henry and P. D. Dapkus, J. Appl. Phys. *47*, 4067 (1976).
94. R. Caruso, M. DiDomenico Jr., H. W. Verleur, and A. R. Van Neida, J. Phys. Chem. Solids, *33*, 689 (1972).
95. M. Mizuta and H. Kukimoto, Japan, J. Appl. Phys. *14*, 1617 (1975).
96. M. Mizuta, J. Yoshino, and H. Kukimoto, IEEE, Trans. *ED-26*, 1194 (1979).
97. J. D. Wiley, private communication, 1970.
98. D. L. Hughes, H. D. Pruett, and M. R. Notis, private communication, 1970.
100. D. J. Ashen, P. J. Dean, D. T. J. Hurle, J. B. Mullin, A. M. White, and P. D. Greene, J. Phys. Chem. Solids *36*, 1041 (1975).
101. R. N. Bhargava, Phys. Rev. *B2*, 387 (1970).
102. J. M. Dishman, Phys. Rev. *B3*, 2588 (1971).
103. R. Z. Bachrach and J. S. Jayson, Phys. Rev. *B7*, 2540 (1973).
104. P. J. Dean, E. G. Schonherr, and R. B. Zetterstrom, J. Appl. Phys. *41*, 3474 (1970).
105. P. J. Dean and M. Ilegems, J. Luminesc. *4*, 201 (1971).
106. R. N. Bhargava, C. Michel, W. L. Lupatkin, R. L. Bronnes, and S. K. Kurtz, Appl. Phys. Lett. *20*, 227 (1972).
107. P. J. Dean, Phys. Rev. *B4*, 2596 (1971).
108. J. R. Knight, D. Effer, and P. R. Evans, Solid State Commun. *8*, 178 (1965).
109. W. Hayes, Phys. Rev. *138*, A 1227 (1965).
110. R. N. Bhargava, S. K. Kurtz, A. T. Vink, and R. C. Peters, Phys. Rev. Letters *27*, 183 (1971).

111. P. J. Dean, J. Luminesce, 7, 51 (1973).
112. P. J. Dean, unpublished data, 1972.
113. D. Wight, J. W. A. Trussler, and W. Harding, Proc. Intern. Conf. Phys. Semicond. Warsaw (PWN, Warsaw 1972) p. 1091.
114. A. S. Jordan, A. R. Von Neida, R. Caruso, and M. DiDomenico, Jr., Appl. Phys. Lett. 19, 394 (1971).
115. A. S. Jordan, R. Caruso, A. R. Von Neida, and M. E. Weiner, J. Appl. Phys. 45, 3472 (1974).
116. J. A. Van Vechten, J. Electrochem. Soc. 122, 419, 423 (1975).
117. P. J. Dean, Solid State Commun. 9, 2211 (1971).
118. P. J. Dean, B. Monemar, H. P. Gislason, and D. C. Herbert.
119. C. H. Henry and D. V. Lang, Phys. Rev. B15, 989 (1977).
120. S. Braun and H. G. Grimmeiss, Solid State Commun. 12, 657 (1973).
121. C. H. Henry, H. Kukimoto, G. L. Miller, and F. R. Merritt, Phys. Rev. B7 2499 (1973).
123. L. Samuelson and B. Monemar, Phys. Rev. B18, 830 (1978).
124. P. J. Dean, G. Kaminsky, and R. B. Zetterstrom, J. Appl. Phys. 38, 3551 (1967).
125. J. S. Jayson, R. Z. Bachrach, P. D. Dapkus, and N. E. Schumaker, Phys. Rev. B6, 2357 (1972).
126. J. M. Dishman and M. D. DiDomenico Jr. Phys. Rev. B4, 2621 (1971).
127. D. V. Lang, J. Appl. Phys. 45, 3023 (1974).
128. M. Jaros and S. F. Ross, Proc. Int. Conf. Phys. Semicond, Stuttgart 1974, ed. by M. H. Pilkuhn (Teubner, Stuttgart 1974) p. 401.
129. S. T. Pantelides, Solid State Commun. 14, 1255 (1973).
130. M. Jaros, Phys. Rev. B16, 3694 (1977).
131. S. Brand and M. Jaros, Solid State Commun. 21, 875 (1977).
132. H. G. Grimmeiss, L. A. Ledebo, C. Ovren, and T. N. Morgan, Proc. Intern. Conf. Phys. Semicond. Stuttgart, 1974, ed. by M. H. Pilkuhn (Teubner, Stuttgart 1974) p. 386.
133. T. N. Morgan, J. Electron Mat. 4, 1029 (1975);
134. J. L. Pautrat, Solid State Electron. 23, 661 (1980).
135. L. Samuelson and B. Monemar, quoted in L. Samuelson PhD Thesis, University of Lund. (1977).
136. Ref. 63. This reassignment was later withdrawn–T. N. Morgan, Proc. 12 Int. Conf. Defects in Semiconductors, Amsterdam, 1982, Physica 116B, 131 (1983).
137. Footnote (9) of Ref. 63.
138. T. N. Morgan, Proc. Int. Conf. Phys. Semicond. Edinburgh 1978 (Inst. Phys. Conf. Ser. 43, Bristol, 1978) p. 311.
139. M. Gal, B. C. Cavenett, and P. Smith Phys. Rev. Lett. 43, 1611 (1979).
140. M. Jaros, private communication, 1980.
141. M. Gal, B. C. Cavenett, and P. J. Dean, J. Phys. C. 14, 1507 (1981).
142. A. S. Barker Jr., R. Berman, and H. W. Verleur, J. Phys. Chem. Solids. 34, 123 (1973).
143. R. C. Newman, Adv. in Phys. 18, 545 (1969).
144. D. R. Bosomworth, W. Hayes, A. R. L. Spray, and G. D. Watkins, Proc. Roy. Soc. Lond A317, 133 (1970).
145. R. Berman and A. S. Barker Jr., unpublished estimates quoted in Ref. 85.
146. T. Arai, N. Asanuma, K. Kudo and S. Umemoto, Japan, J. Appl. Phys. 11, 206 (1972).
147. J. L. Merz, Proc. Int. Conf. Phys. Semicond. Cambridge 1970, ed. by S. P.

Keller, J. C. Hensel and F. Stern (USAEC, Oak Ridge, Tenn.) p. 251.

149. R. Z. Bachrach, O. G. Lorimor, L. R. Dawson, and K. B. Wolfstirn, J. Appl. Phys. *43*, 5098 (1972).

150. P. J. Dean, R. A. Faulkner, and E. G. Schonherr, Proc. Intern, Conf. Phys. Semicond. Cambridge 1970, ed. by S. P. Keller, J. C. Hensel, and F. Stern, (USAEC, Oak Ridge, Tenn. 1971) p. 286.

151. E. E. Haller, W. L. Hansen, and F. S. Goulding, Adv. in Phys., to be published.

152. D. Lang, Proc. Int. Conf. Phys. Semicond, Kyoto (1980).

153. S. Makram–Ebeid Appl. Phys. Lett. *37*, 464 (1980).

154. G. L. Miller, D. V. Lang, and L. C. Kimerling, Ann. Rev. Mat. Sci. 377 (1977).

155. H. P. Hjalmarson, P. Vogl, D. J. Wolford, and J. D. Dow, Phys. Rev. Letters *44*, 810 (1980).

156. R. A. Faulkner, Phys. Rev. *175*, 991 (1968).

157. O. F. Sankey, H. P. Hjalmarson, J. D. Dow, D. J. Wolford, and B. G. Streetman, Phys. Rev. Letters *45*, 1656 (1980).

158. J. Blanc and L. R. Weisberg, Nature (London) *192*, 155 (1961).

159. C. H. Gooch, C. Hilsum, and B. R. Holeman, J. Appl. Phys. *32*, S2069 (1961).

160. N. G. Ainslie, S. E. Blum, and J. F. Woods, J. Appl. Phys. *33*, 2391 (1962).

161. R. W. Haisty, E. W. Mehel, and R. Stratton, J. Phys. Chem. Solids *23*, 829 (1962).

162. J. Blanc, R. H. Bube, and L. R. Weisberg, J. Phys. Chem. Solids *25*, 225 (1964).

163. S. M. Sze and J. C. Irvin, Solid State Electron *11*, 599 (1968).

164. G. M. Martin and D. Bois, J. Electrochem. Soc.

165. R. Williams, J. Appl. Phys. *37*, 3411 (1966).

166. M. Jaros, J. Phys. C. *8*, 2455 (1975).

167. M. R. Brozel, J. B. Clegg, and R. C. Newman, J. Phys. D. *11*, 1331 (1978).

168. R. A. Malinauskas, L. Ya Pervova, and V. I. Fistul' Fiz. Tekh. Poluprovodn. *13*, 2270 (1979) [Engl. Transl. Sov. Phys. Semicond. *13*, 1330 (1979)].

169. J. F. Woods and N. G. Ainslie, J. Appl. Phys. *34*, 1469 (1963).

170. C. J. Frosch, C. D. Thurmond, H. G. White, and J. A. May, Trans. Metall. Soc. AIME *239*, 365 (1967).

171. A. M. White, P. J. Dean, D. J. Ashen, J. B. Mullin, and B. Day, Proc. Intern. Conf. Phys. Semicond. Stuttgart 1974, ed. by M. H. Pilkuhn (Teubner, Stuttgart 1974) p. 381.

172. A. E. Lin, E. Owelianovski, and R. H. Bube, J. Appl. Phys. *47*, 1852 (1976).

173. M. Otsubo, K. Segawa, and H. Miki, J. Appl. Phys. Japan, *12*, 797 (1973).

174. D. V. Lang and R. A. Logan, J. Electron. Mat. *4*, 1053 (1975).

175. L. A. Ledebo, PhD Thesis, Lund (1976).

176. E. Andre, quited in Ref. 177.

177. A. Huber, N. T. Linh, M. Valladon, J. L. Debrun, G. M. Martin, A. Mittonneau and A. Mircea, J. Appl. Phys. *50*, 4022 (1979).

178. R. H. Wallis, M. A. di Forte Poisson, M. Bonnet, G. Beuchet, and J. P. Duchemin Proc. Conf. GaAs and related compounds, Vienna (1980).

179. A. Mircea, A. Mittonneau, A. Hallais, and M. Jaros, Phys. Rev. *B16*, 3665 (1977).

180. M. D. Miller, G. H. Olsen, and M. Ettenberg, Appl. Phys. Lett. *31*, 538 (1977).

181. M. Ozeki, J. Komeno, A. Shibatomi, and S. Ohkawa, J. Appl. Phys. *50*, 4808 (1979).

182. D. L. Partin, J. W. Chen, A. G. Milnes, and L. F. Vassamillet, J. Appl. Phys. *50*, 6845 (1979).
183. G. M. Martin, A. Mittonneau, and A. Mircea, Electron-Lett *13*, 191 (1977).
184. H. D. Palfrey, A. F. W. Willoughby, and M. Brown, Electrochem. Soc. Meeting Abstr. 80-1 No. 146 (May 1980) p. 388.
185. P. J. Dean, Luminescence of Crystals, Molecules and Solutions, ed. by F. E. Williams (Plenum, New York, 1973) p. 538.
186. G. M. Blom and R. N. Bhargava, J. Cryst. Growth *17*, 38 (1972).
187. D. J. Robbins, D. C. Herbert, and P. J. Dean, J. Phys. C.
188. M. Jaros, private communication, 1981.
189. R. Kaplan, R. A. Cooke, and R. A. Stradling, Solid State Commun. *26*, 741 (1978).
190. A. M. Davidson, P. Knowles, P. Makado, and R. A. Stradling, to be published.
191. Sh. M. Kogan and T. M. Lifshits, Phys. Stat. Sol. *a39*, 11 (1977).
192. P. J. Dean and W. J. Choyke, Adv. in Phys. *26*, 1 (1977).
193. B. Monemar, J. Luminesc. *5*, 472 (1972).
194. B. Monemar, J. Luminesc. *5*, 239 (1972).
195. A. M. White, P. J. Dean, K. M. Fairhurst, W. Bardsley, and B. Day J. Phys. *C7*, L35,(1974).
196. P. J. Dean, Proc. Int. Conf. Phys. Semicond. Montpellier 1982, Physica *117B*, 140 (1983).
197. P. J. Dean, Phys. Rev. *139*, A588 (1965).
198. E. C. Lightowlers and A. T. Collins, J. Phys. D: Appl. Phys. *9*, 951 (1976).
199. G. Davies, J. Phys. C.: Solid State Phys. *9*, L537 (1976).
200. A. T. Collins and S. Rafique, J. Phys. C.: Solid State Phys. *11*, 1375 (1978).
201. L. A. Vermeulen and R. G. Farrer, Diamond Res. 1975 (Suppl. to Ind. Diamond Rev.) (Ind. Diamond Inform. Bureau, 1975) p. 18.
202. G. A. Baraff, E. O. Kane, and M. Schluter, Phys. Rev. (to be published).
203. G. A. Baraff, E. O. Kane, and M. Schluter, Phys. Rev. Lett. 47, 601 (1981).
204. K. Nasu and Y. Toyozawa. J. Phys. Soc. Japan *50* 235 (1981).
205. R. Z. Bachrach, private communication, 1981.
206. M. D. Sturge, private communication, 1981.
207. A. A. Kopylov and A. N. Pikhtin, Solid State Commun. *26*, 735 (1978).
208. D. F. Nelson, J. D. Cuthbert, P. J. Dean, and D. G. Thomas, Phys. Rev. Lett., *17*, 1262 (1966).
209. D. G. Thomas, M. Gershenzon, and J. J. Hopfield, Phys. Rev. *131*, 2397 (1963).
210. P. J. Dean, J. D. Cuthbert, D. G. Thomas, and R. T. Lynch, Phys. Rev. Lett. *18*, 122 (1967).
211. R. M. Feenstra and T. C. McGill, to be published.
212. P. D. Dapkus and C. H. Henry, J. Appl. Phys. *47*, 4061 (1976).
213. S. Depinna, B. C. Cavenett, N. Killoran, and B. Monemar, to be published.
214. H. P. Gislason, B. Monemar et al. to be published.
215. P. J. Dean, Phys. Rev. *157*, 655 (1967).
216. E. F. Gross and D. S. Nedzvetskii, Doklady Akad. Nauk. SSSR *152*, 1335 (1963). [(Engl. Transl. Sov. Phys. Doklady *8*, 989 (1964)].
217. D. N. Mirlin, I. Ja Karlik, L. P. Nikitin, I. I. Reshina, and V. F. Sapega, Solid State Commun. *37*, 757 (1981).
218. L. C. Kimerling, Solid State Electronics, *21*, 1391 (1978).

346 P. DEAN

219. A. M. Stoneham, private communication, 1981.
220. P. Banks, M. Jaros, S. Brand, J. Phys. C.: ST Phys. Vol. *13*, p. 6167 (1980).
221. J. R. Haynes, Phys. Rev. Lett. *4*, 361 (1960).
222. P. J. Dean, W. Schairer, M. Lorenz, and T. N. Morgan, J. Luminesc. *9*, 343 (1974).
223. P. J. Dean, M. Skolnick, Ch. Uihlein, and D. C. Herbert, to be published.
224. P. J. Dean, D. Bimberg, and F. Mansfield, Phys. Rev. *B15*, 3906 (1977).
225. A Mircea, A Mitonneau, L. Hollan, and A Briere, Appl. Phys. *11*, 153 (1976).
226. G. M. Martin, J. P. Farges, G. Jacob, J. P. Hallais, and G. Poibaud, J. Appl. Phys. 51, 2840 (1980).
227. E. E. Wagner, D. E. Mars, G. Hom, and G. B. Stringfellow, J. Appl. Phys. *51*, 5434 (1980).
228. Z. L. Akkerman, L. A. Borisova, and A. F. Kavchenko, Fiz. Tekh. Poluprovodn *10*, 997, 1976 [Engl. Transl. Sov. Phys. Semicond. *10*, 590 (1976)].
229. A. Mircea-Roussel and S. Makram-Ebeid, Appl. Phys. Lett. *38*, 1007 (1981).
230. J. Schneider, private communication, 1981.
231. T. Ikoma, M. Takikawa, and M. Taniguchi, Proc. 1981 Int. Symp. on GaAs and Related Compounds, Oiso.
232. D. Bimberg and P. J. Dean, Phys. Rev. *B15*, 3917 (1977).
233. T. N. Morgan, Phys. Rev. Letters *49*, 173 (1982).
234. T. N. Morgan, private communication, 1982.
235. E. Cohen and M. D. Sturge, Phys. Rev. *B15*, 1039 (1977).
236. A. Onton, P. Fisher, and A. K. Ramdas, Phys. Rev. Lett. *19*, 781 (1967).
237. J. J. Hopfield, P. J. Dean, and D. G. Thomas, Phys. Rev. *158*, 748 (1967).
238. C. M. Wolfe and G. J. Stillman, Proc. IIIrd Int. Symp. on GaAs, Aachen, 1970 (Inst. Phys. Conf. Ser. *9*) p. 3.
239. A. C. Carter, G. P. Carver, R. J. Nicholas, and R. A. Stradling, Solid State Commun. *24*, 55 (1977).
240. B. Monemar, Recent developments in condensed matter physics, ed. by J. T. Devreese (Plenum, New York, 1981) p. 441.
241. R. N. Bhargava, J. Appl. Phys. *41*, 3698 (1970).
242. Consensus of opinion amongst solid state theorists at the Montpellier Semiconductor Conference, 1982, was strongly against the molecular orbital ultra tight binding model for the electronic configuration of O_p in GaP.
243. This can be inferred from data on S_p-C_p shallow DAP in GaP - R. A. Street and W. Senske, Phys. Rev. Lett. *37*, 1292 (1976), using interpretations similar to those for ZnTe where the dependence of the relative proportion of selective processes on excitation intensity has been demonstrated by S. Nakashima and A. Nakamura, J. Phys. Soc. Japan 49 supp. A. 193 (1980).
244. T. N. Morgan, Proc. Int. Conf. Phys. Semicond. Montpellier, 1982, Physica *117B*, 146 (1983).
245. K. A. Elliott and T. C. McGill, Phys. Rev. *B21*, 2426 (1980).
246. G. A. Baraff, private communication, 1982.
247. M. Jaros and S. Brand, J. Phys. C.: *15*, L743 (1982).
248. M. Jaros, private communication, 1982.
249. V. G. Gaivoron and M. I. Elinson, Fiz. Tekh. Puluprovodn. *9*, 1934 (1975) [Engl. Transl. Sov. Phys. Semicond, *9*, 1269 (1976)].
250. J. S. Jayson, R. Z. Bachrach, P. D. Dapkus, and N. E. Schumaker, Phys. Rev. *B6*, 2357 (1972).
251. A. M. Stoneham and A. H. Harker, J. Phys. C.: *8*, 1102 (1975).

252. R. M. Feenstra and T. C. McGill, Phys. Rev. *B25*, 6329 (1982).
253. R. M. Feenstra and T. C. McGill, Proc. Int. Conf. Phys. Semicond. Montpellier 1982, Physcia 117B, 149 (1983); also see R. J. Havenstein, T. C. McGill, and R. M. Feenstra, Phys. Rev. B 29, 1858 (1984); Phys. Rev. B 28, 5793 (1983).
254. G. Kirezenow, Solid State Commun. 21, 713 (1977).
253. R. M. Feenstra and T. C. McGill, Proc. Int. Conf. Phys. Semicond. Montpellier 1982, Physcia *117B*, 149 (1983); also see R. J. Havenstein, T. C. McGill, and R. M. Feenstra, Phys. Rev. B*29*, 1858 (1984); Phys. Rev. B*28*, 5793 (1983).
254. G. Kirezenow, Solid State Commun. *21*, 713 (1977).
255. G. M. Martin, G. Jacob, J. P. Hallais, F. Grainger, J. A. Roberts, B. Clegg, P. Blood, and G. Poiblaud, J. Phys. C.: *15*, 1841 (1982).
256. L. Samuelson, P. Omling, H. Titze, and H. G. Grimmeiss, J. Cryst. Growth *55*, 164 (1981).
257. D. E. Holmes, R. T. Chen, K. R. Elliott, and G. E. Kirkpatrick, Appl. Phys. Lett. *40*, 46 (1982).
258. G. Bjorklund and H. G. Grimmeiss, Solid State Electronics *14*, 589 (1971).

CHAPTER 5

The Two Dominant Recombination Centers in N-Type GaP

A. R. Peaker and B. Hamilton

The University of Manchester Institute of Science and Technology
P.O. Box 88, Manchester M601QD, UK

1. RECOMBINATION IN LIGHT-EMITTING DIODES
   1.1 Energy-Loss Processes in LED's
2. MECHANISMS FOR NON-RADIATIVE RECOMBINATION
3. CHARACTERIZATION OF POWERFUL RECOMBINATION
   CENTERS
   3.1 Deep Electron States in n-type GaP
   3.2 Minority Carrier Capture Techniques Using Schottky Diodes
   3.3 Hole Traps in n-type GaP
   3.4 Minority Carrier Trap Spectroscopy (MCTS)
4. CAPTURE CROSS SECTIONS
5. LIFETIME STUDIES
6. PHYSICAL IDENTIFICATION OF THE 0.75-eV DEFECT
7. IDENTIFICATION OF THE 0.95-eV DEFECT
8. SUMMARY

1. Recombination in Light-Emitting Diodes

Deep states in semiconductors are studied widely primarily because of their role as recombination or generation centers. In many devices, it is necessary to have close control of the minority-carrier lifetime. The best known examples are high-speed silicon diodes in which gold is added to produce deep states. These states increase the recombination rate and thus reduce the device turn-off time. In other cases, it is desirable to reduce the recombination rate via deep states and thus achieve a long minority carrier lifetime. For example, the performance of many bipolar devices is improved in this manner. Quite often, however, the deep states acting as recombination centers are present inadvertently, that is they arise from the growth or fabrication process and are the result of impurities or structural defects in the lattice and are often of unknown origin. In almost all cases in moderately-doped indirect-gap materials, the dominant recombination process is via deep states associated with centers which are present inadvertently. This has dramatic implications for the silicon device industry in terms of process repeat-bility and control, but, in the case of indirect-gap light-emitting diodes (LED), it is the crucial parameter which defines device efficiency.

Although LEDs are widely used, they are very inefficient. A typical commercial device converts less than 0.1% of the energy fed into it into visible radiation; most of the remaining 99.9% being lost as heat. In the indirect-gap green-emitting gallium phosphide and yellow- or orange-emitting gallium arsenide phosphide, the largest proportion of the nonradiative processes occur via deep states.

This article summarizes what is known about two deep levels which are the major contributors to the nonradiative recombination path in n-type gallium phosphide, particularly in crystals where the technologically controllable macroscopic recombination centers (dislocations, surfaces and interfaces) do not dominate. As will be shown later it is these deep states which control the efficiency of green emission from gallium phosphide. Closely related states are present in gallium arsenide phosphide and there is accumulating evidence that they are also the main non-radiative recombination route in the indirect gap composition range of this material. In the very widely used red emitting $GaAs_{0.6}P_{0.4}$ (a direct gap material) related states are also present but it appears that they act as only one of a number of comparable non-radiative recombination routes.

1.1 Energy-Loss Processes in LEDs

The processes leading to light emission from an LED can be conveniently regarded as falling into three stages; the first being an excitation process which raises the energy of the current carriers (holes, electrons or more usually both). In the LED, this is often called the injection process because, essentially, carriers are introduced into those regions of the semiconductor where radiative recombination can occur. Secondly, the excited carriers give up their energy in recombination processes. Some of the recombination will be radiative, producing light, and some nonradiative, serving merely to heat up the semiconductor material and its surroundings. The third process is often regarded as inevitable, namely the extraction or passage of the photons from the semiconductor material to the observer. Unfortunately, in some classes of material this is a major source of loss. The three processes can be assigned individual efficiencies: η_i the injection or excitation efficiency, η_r the recombination efficiency and η_o the extraction efficiency.

In a p-n junction in GaP, the injection efficiency can usually be made quite high, often around 60% and so in a device with an overall efficiency of 0.1% this can hardly be regarded as a major source of loss. Similarly, in this indirect gap material the self absorption is low. The absorption coefficient in typical device-grade material is less than 100 cm^{-1} and so the major source of loss in the extraction efficiency is due to the loss of the totally internally reflected light giving an overall η_o of around 20% in a normal device. This is a significant loss, but, again, is not the major deficiency.

Gallium phosphide has a band gap of 2.27 eV at room temperature which should enable emission of wavelengths longer than 560 nm to be obtained. Actually, the most efficient green emission at 565 nm is achieved by recombination via the isoelectronic impurity nitrogen. Similar radiative recombination process occur in GaAsP:N.

Basically, the situation in gallium phosphide is that good injection efficiencies can be achieved; the extraction efficiency is moderately good but the radiative recombination efficiency is low. This is essentially because the non-radiative processes compete for carriers with the radiative recombination. It is the balance between the rates at which the carrier traverse the two recombination paths which determines the efficiency

$$\eta_r = \frac{\text{radiative recombination rate}}{\text{total recombination rate}} = \tau_{nr}^{-1}/(\tau_r^{-1} + \tau_{nr}^{-1}) \qquad (1)$$

where τ_r is the radiative and τ_{nr} the nonradiative lifetime. If, as is often the case in indirect gap materials, $\tau_r \gg \tau_{nr}$ then:

$$\eta_r = \tau_{nr}/\tau_r \qquad (2)$$

Under these conditions, the measured minority carrier life-time is essentially the nonradiative lifetime. It is apparent that the efficiency can be improved either by increasing the nonradiative lifetime or decreasing the radiative lifetime. The extent to which the latter process can be undertaken depends very much on the isoelectronic trap involved. In the case of nitrogen, the fact that the exciton is very weakly bound means that it may well be released thermally from the nitrogen atom and dissociate without recombination allowing the excited carriers to follow other routes, usually nonradiative. The result is that recombination via nitrogen is extremely efficient at temperatures below 100K but is strongly quenched as the temperature is increased. This results in a typical value of η_r at room temperature in a commercially green-emitting device of around 0.4%. With a corresponding value of τ_{nr} of the order of 60 ns. Some freedom for decreasing τ_r does exist by simply increasing the concentration of the nitrogen atoms in the lattice. In liquid-phase expitaxy, this becomes difficult on the technologically desirable (100) orientation, and is limited to $[N] \cong 3 \times 10^{18}$ cm^{-3}. In vapour-phase growth systems, much larger concentrations can be used, but the yellow luminescence from the excitons bound to the nitrogen pairs then predominates. The resultant radiative lifetime is shorter than for the green emission case, but this does not dramatically change the radiative efficiency because the non-radiative lifetime tends also to be shorter in this more heavily-doped vapour phase material.

In green-emitting gallium phosphide devices, we are left with only one realistic way of substantially improving the efficiency; that is to reduce the carrier flux through the nonradiative path.

2. Mechanisms for Non-Radiative Recombination

The technological importance of elucidating the non-radiative path in GaP has led many research groups to address the problem during the last ten years. In general terms, it is clear that several mechanisms could carry the non-radiative recombination flux, acting individually or in parallel. We can identify three broad categories whch might be responsible:

1. Auger recombination
2. Recombination at macroscopic defects
3. Point defect recombination

Auger recombination, in which multi-particle collisions account for the energy (and momentum) dissipation required for recombination, can always dominate at high doping levels. For n-type GaP, the Auger recombination coefficients for band-to-band processes dictate that only at very high doping levels would the mechanism dominate; it is clear that this process is not the fundamental limiting one in GaP. Auger recombination can, of course, take place involving deep levels, but again there is no experimental evidence that this is the dominant non-radiative process. The physics of Auger recombination in GaP has been addressed by Hill and Landsberg.[1] The experimental evidence that allows us to neglect Auger recombination is simply that the minority carrier lifetime and diffusion length are independent of free electron concentration over a wide range of doping levels (N_D-N_A < 5 × 10^{18} cm^{-3}). This has been demonstrated by Young et al[2] among others. Recombination at macroscopic defects, however, is by no means an insignificant phenomenon in GaP. Indeed, mechanisms which fall into this general category can dominate the recombination physics, and it is only by carefully studying such processes that we arrive at the conclusion that microscopic, point defects, are the fundamental problem.

The importance of macroscopic centers in GaP has been graphically illustrated by using the scanning electron microscope to produce cathodoluminescence micrographs of the band edge emission. Davidson and co-workers[3] demonstrated that luminescence from a liquid phase epitaxial (L.P.E.) layer showed regions of pronounced local quenching; the now well known 'black dot' effect. Work at Oxford University combined this sort of measurement with T.E.M. studies showing conclusively that the black dots were associated with line dislocations propagating from the substrate/epitaxial interface through the epitaxial layer[4]. The local quenching of the luminescence is due to severe non-radiative recombination at or around the dislocation, though the dynamics of the non-radiative processes at the dislocation are not understood. This fact leads onto a common feature of macroscopic recombination centers, that the rate limiting process for recombination is the diffusive transport of the minority carrier to the center and not the dynamic processes occuring at the center. This concept is invaluable in modelling the physics of such systems because the problem reduces to solving the carrier transport equations with appropriate geometric and boundary conditions. For example, the non-radiative recombination rate due to dislocations can be modelled by considering the dislocations to be a uniform array of line sinks. Several workers have modelled dislocation recombination in GaP in this way; these include Suzuki and Matsumoto,[5] and Van Opdorp et al[6] who also investigated the transient minority carrier response at dislocations.

A rigorous study of diffusion-controlled recombination mechanisms in GaP has been made by Wight and co-workers at the Royal Signals and Radar Establishment (R.S.R.E.), and in one paper they review the experimental data and its theoretical interpretation.[7] These authors consider the role of plane sinks (of which surface or interface recombination are special cases); line sinks appropriate to dislocations, and point sinks. The latter case was considered in order to evaluate the behaviour of a 'giant' point defect and to make comparisons with the lifetime properties of deep-level controlled material. Although diffusion-controlled point defect lifetimes were not observed, this phenomenon is discussed in section 4 of this chapter. An important concept underpinning all diffusion-controlled mechanisms is the relationship between the size of the defect and the mean free path of the minority carrier λ. If a capture radius r_o is ascribed to the defect, within which the capture probability is infinite (i.e. the minority carrier density tends to zero), then the recombination rate will be diffusion controlled if $r_o > \lambda$. If $r_o < \lambda$ then the local depression of the minority carrier concentration is small and the particle fluxes into the recombination are ballistic. This is an underlying assumption of the Shockley-Read recombination model and is required to make the concept of a geometric capture cross section valid. The switch between 'diffusion' and 'recombination' controlled lifetimes and its dependence on capture radius is an important concept in relation to powerful recombination centers.

The two most important diffusion-controlled recombination mechanisms in GaP are surface/interface recombination, and recombination at dislocations. The surface/interface recombination problem can be studied in thin epitaxial layers, since both the free surface and the substrate epitaxial interface have recombination velocities, S, approaching infinity.[8] The diffusion controlled lifetime prediction for such a filament structure of total thickness θ, is:

$$\tau = \theta^2 / 4\pi^2 D \qquad (3)$$

where D is the diffusion coefficient of the minority carriers (holes). The agreement between theory and experiment is well borne out as can be seen from Figure 1.

These data were obtained using cathodoluminescence decay as a measurement of lifetime in a bevelled epitaxial layer. The slight roll-off of the data points for $\theta > 25$ μm is an indication that a bulk recombination process is beginning to influence the measurement as the layer thickness increases. It is characteristic of diffusion-controlled mechanisms that the temperature dependence of lifetime reflects the transport of carriers and not the dynamics of the capture process. In the present case, the variation of D with temperature accounts for the variation of lifetime. For disloca-

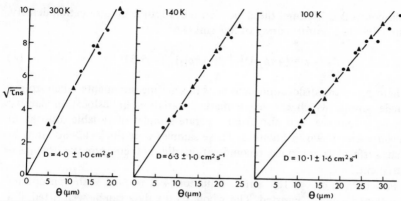

FIGURE 1 The minority carrier lifetime limited by surface and interface recombination in in-type GaP as a function of epitaxial layer thickness θ.

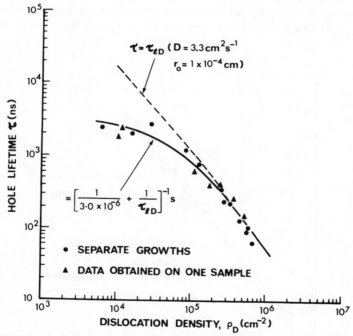

FIGURE 2 The dependence of diffusion limited minority carrier lifetime on dislocation density.

tion controlled lifetime, the dependence of lifetime on dislocation density (modelled as a uniform array of line sinks) is:

$$\tau = [4\pi\rho_D D]^{-1} \, [\ln(\pi\rho_D r_0^2) - 6/5] \tag{4}$$

where ρ_D is the dislocation density. Lifetime measurements taken from a single sample which showed a marked variation in dislocation density across its surface, and also from separate samples of variable dislocation density, are shown in Figure 2. These samples were thick ($t \geqslant 300 \, \mu$m) so that surface or interface recombination did not compromise the measurement.

For $10^5 \leqslant \rho_D \leqslant 10^6$ cm^{-2}, Equation 4 is obeyed with appropriate values of D and r_0 inserted. The whole of the data can be well fitted, if a parallel bulk recombination lifetime of $\sim 3 \, \mu$s is added to the dislocation controlled lifetime.

As for the surface/interface controlled systems, the hole diffusion coefficient dictates the temperature dependence of τ in the dislocation limited regime. Since $D \propto T^{-1.2}$, the lifetime $\tau \propto T^{1.2}$. This relationship agrees

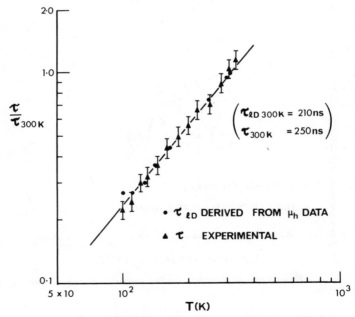

FIGURE 3 Temperature dependence of dislocation limited lifetime.

well with experiment as shown in Figure 3, in which the normalized life-times are fitted to Equation 6 at each temperature point.

This characterization of the dominant diffusion-controlled mechanisms shows that, provided an epitaxial layer is sufficiently thick and also has a low dislocation density, the macroscopic defects are not responsible for the major non-radiative path which certainly controls the lifetime. For example a layer which is $\sim 300\ \mu m$ thick with a dislocation density of $\sim 10^4\ cm^{-2}$, would have a diffusion-controlled lifetime of $\tau \cong 10\ \mu s$; a figure which has not been achieved even under the most favorable growth conditions. Clearly, 'high quality' epitaxial layers of this sort, which should form the very basis of an effort to produce efficient light emission from GaP, fall far short of the ceiling set by the macroscopic defects. We are then forced to conclude that microscopic point defects of the Shockley-Read type, control the recombination. Such defects would need to be powerful recombination centers since they operate even in the cleanest crystals. In order to detect such centers special techniques have been developed and these are described in the next section.

3. Characterization of Powerful Recombination Centers

For a point defect to carry a large fraction of the recombination traffic in a semiconductor, its electronic structure must exhibit several definite characteristics. Firstly it must produce at least one stable electron state deep in the forbidden gap; if the state is too near a band edge, then thermal re-emission of one of the bound carriers to that band will be in strong kinetic competition with the capture mechanisms which drive the recombination. Secondly the defect must have large capture cross-sections, especially for the minority carrier, which in essence means that the energy dissipation mechanisms which operate during the capture process must be efficient. This is an essential characteristic of any significant recombination center as the recombination rate is going to depend on the cross section concentration product, and the concentration of inadvertant centers will be certainly much lower than the intentionally added radiative recombination centers. An accurate characterization of a recombination center must address each of these aspects of the defect. Broadly speaking, it is necessary to measure the concentration and energy depth of a specific deep electron state, and then determine its electron and hole capture cross sections. Having isolated those defects which could be efficient recombination centers, their significance can be investigated by intentionally changing the concentration of the defect or by screening crystals with a wide range of lifetimes. This technique should establish a correlation between the lifetime and the concentration of the defect.

Our concern here is with n-type GaP. It was established in section 2 that surfaces (or interfaces) and dislocations can, in particular circumstances, control the lifetime in this material. The experiments reviewed in this section were carried out on crystals for which this was not the case and, indeed, point defects were the most likely lifetime controlling factor. The samples used in these studies were undoped or lightly doped epitaxial layers with free electron densities in the range 5×10^{14} cm^{-3} to 5×10^{15} cm^{-3}. The layers were of low dislocation density ($N_{dis} \sim 3 \times 10^4$ cm^{-2}) and were sufficiently thick to allow a measurement of true bulk lifetime without interference from surfaces or interfaces. The problems associated with lifetime measurements are discussed in detail in section 5.

3.1 Deep Electron States in n-type GaP

A typical n-type GaP crystal might contain many species of deep electron states. This statement is true for many compound semiconductors. It is often a consequence of high crystal growth temperatures and the attendant problems of contamination. Stoichiometry control is obviously another serious problem in compound semiconductors. We shall not give a survey here of the many published results for GaP, but will concentrate on those defects which are potentially efficient recombination centers even in extremely pure material. A comprehensive review of published energy levels for deep states in GaP has been given elsewhere.[9,10] The initial stages of this investigation were carried out on very pure LPE (Liquid Phase Epitaxy) crystals which were produced in an extremely clean growth furnace at RSRE. The inadvertant contamination in these crystals was low, and in many cases the concentration of deep electron traps in the upper half on the gap in this n-type material was below the detection limit ($\leqslant 10^{11}$ cm^{-3}). Nevertheless the samples appeared to exhibit point-defect controlled lifetimes. With hindsight, it is clear that the most important states cannot be detected with standard DLTS or extrinsic photocapacitance measurements using Schottky diodes. In fact, the lifetime controlling defects have electron states positioned below the middle of the energy gap Ev + 0.75 ev and Ev + 0.95 ev. For such states, $e_p^t \gg e_n^t$ (a definition of transition rate parameters is given below) and no occupancy change can be effected during the detection phase of a DLTS measurement, following a forward bias pulse applied to the Schottky diode.

The detection methods developed for these defects exploit one of the essential properties of an efficient recombination center, viz its large capture cross-section for minority carriers. The idea is to set up an excitation condition in which the electron occupied state is subjected to pulse or steady flux of minority carriers (i.e. holes in this case of n-type material). The electron states with large minority carrier capture cross-sections will then rapidly capture minority carriers. This charge exchange must result in a capacitance transient which carries information on the capture process. Subsequent thermal emission of the minority carrier can also be made to yield up the binding energy of the electron state relative to the minority band, provided the level is located in the minority half of the energy gap, i.e. $e_p^t > e_n^t$ in the present case. Such a technique, carries some very significant advantages. Being a Schottky diode method, it can explore the properties of the crystal, free of any process-induced defects which might be associated with the fabrication of a p-n junction. It is especially sensitive to just those defects which might control the lifetime. It is not restricted to a particular 'half' of the energy gap for the detection and for measurement of minority carrier capture cross-sections whereas an unambiguous measurement of thermal binding energy always relies on one of the conditions $e_n^t \gg e_p^t$ or $e_p^t \gg e_n^t$ being fulfilled. Finally the method does not depend on the detailed spectral shape of photoionization cross-sections.

The excitation of deep level capacitance transients using an optically generated particle flux to perturb the deep level electron occupancy is not a new idea. Indeed one of the first transient capacitance experiments ever published[11] used a strongly absorbed light source to drive an electron flux through a transparent Schottky barrier on n-type GaAs. It was then possible to observe electron (majority carrier) capture at deep levels. The following techniques use more weakly absorbed light to generate a flux which is predominantly minority carriers.

3.2 Minority Carrier Capture Techniques using Schottky Diodes

When a zero or reverse-bias Schottky barrier is illuminated with near-band-edge light, a short-circuit photocurrent is produced. This photocurrent is derived from two main sources. The first is from photons which are absorbed in the depletion region and generate electron-hole pairs which are separated by the field. This creates a generation current; the second is from minority carriers generated in the bulk, but within a diffusion length of the barrier which diffuse into the extracting field and then drift through the barrier to the metal interface, creating a diffusion-controlled current. In n-type material, this second current component provides a pure-hole flux through the space-charge region since the electrons generated in the bulk are repelled by the potential barrier. The diffusion current can then be regarded as a mechanism for injecting a single (minority) carrier species into the barrier. If the light is not very strongly absorbed and the diffusion length is large compared with the space-charge width, the total minority-carrier flux should be large compared with the electron component of the generation current. Deep levels in the barrier which are initially filled with electrons by applying a forward bias pulse, can respond to the hole flux by capturing holes at a rate which depends on their hole capture cross section and the density of injected holes. The hole-capture transition can then be monitored as the build up of positive space charge due to the trapped holes causes the high-frequency capacitance of the barrier to increase. This idea is the basis of the minority-carrier-capture (MCC) method.[12]

Excitation with weakly absorbed (near band edge) light could result in several distinct interactions. The non-equilibrium situation is potentially complex. If we consider a single deep-level species, then the interactions which may take place during illumination are depicted in Figure 4 and the transition rate parameters are defined for each process in Table 1.

In Table 1, standard notation has been used, with ϕ denoting the photon flux and the intrinsic absorption coefficient. The incremental values

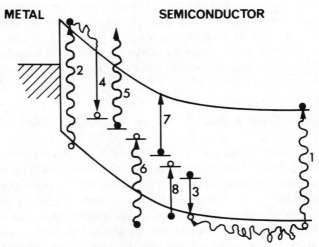

FIGURE 4 The complex excitation which could result from near band edge illumination.

TABLE 1

Transition	Process	Transition Probability unit Time
Optical absorption	1	proportional to $\phi\alpha$
Optical absorption	2	proportional to $\phi\alpha$
Hole capture	3	$C_p = \Delta p(V_{th}\sigma_p) = c_p\Delta p$
Electron capture	4	$C_n = \Delta n(V_{th}\sigma_n) = c_n\Delta n$
Electron photoionization	5	$e_n^o = \phi\,\sigma_n^o$
Hole photoionization	6	$e_p^o = \phi\,\sigma_p^o$
Electron thermal ionization	7	$e_n^t = c_n N_c \exp - [E_c-E_T]/kT$
Hole thermal ionization	8	$e_p^t = c_p N_v \exp - [E_T-E_v [/kT$

Δn and Δp have been used to denote that all carrier densities in the depletion layer are injected densities.

Two classes of information are potentially available; the response of the electron occupancy of the deep state when the light pulse is applied, and the subsequent (purely thermal) readjustment when the light pulse switches off. Applying first order kinetics to the instantaneous density of trapped electrons n_T and including all the deep level transitions in Figure 4,

$$dn_T/dt = [e_p^t + e_p^o + c_n\Delta n]\; p_T - [e_n^t + e_n^o + c_p\Delta p]\, n_T, \qquad (5)$$

where p_T is the instantaeous concentration of trapped holes at the deep level which itself has a concentration N_T, i.e. $p_T = (N_T - n_T)$. With the boundary condition $n_T = N_T$ for $t = o$, achieved with a forward bias filling pulse, the solution to (5) is

$$n_T(t)/N_T = [e_n^t + e_n^o + c_p\Delta p]\,\tau_1\, \exp\text{-}(t/\tau_1) + [e_p^t + e_p^o + c_n\Delta n]\,\tau_1, \quad (6)$$

where $\tau_1 = [e_n^t + e_p^t + e_n^o + e_p^o + c_n\Delta n + c_p\Delta p]^{-1}$. $\qquad (7)$

This simple solution depends on complete uniformity of transition probabilities and carrier densities throughout the potential barrier. The unique response time is not obtained, if any of the individual rate processes is spatially varying. When the light pulse is removed, the trapped electron density adjusts according to:

$$dn_T/dt = e_p^t\, p_t - e_n^t n_T. \qquad (8)$$

The initial condition for this equation is set by Equation 6, but, assuming that steady state (saturation) conditions were met during the optical pulse, then, using the $t = \infty$ value of n_T in 6 as the initial condition, we obtain

$$n_T(t)/N_T = [(e_p^t + e_p^o + c_n \Delta n)\tau_1 - e_p^t \tau_2] \exp-(\frac{t}{\tau_2}). \qquad (9)$$

Again, a unique response time τ_2 is obtained:

$$\tau_2 = [e_n^t + e_p^t]^{-1}, \qquad (10)$$

provided that e_n^t and e_p^t are everywhere constant, and that thermal emission is the only thermal response mechanism available to the deep center.

The charge-exchange kinetics are reflected directly in the high-frequency capacitance per unit area of the barrier:

$$C/A = q\epsilon\rho^+/2V_{TOT} \qquad (11)$$

where ρ^+ is the positive-space-charge density in the barrier and V_{TOT} is the total potential drop across the barrier. For an n-type Schottky barrier with uniformly distributed donors and donor-like deep states, $e^+ = (N_D^+ + p_T)$ where N_D^+ is the concentration of donor states which can emit their electrons in a time which is comparable with or shorter than the period of the applied rf voltage. This concentration may include donor states which are only partially ionized in the bulk of crystal. Since N_D is constant and $n_T(t)$ and $p_T(t)$ are described by Equations (5)–(8), the time dependence of the capacitance is formally determined.

If all the available optical and thermal transitions occurred, the extraction of deep level parameters from the experiment would be difficult. All transient capacitance deep level techniques involve establishing experimental conditions or exploiting deep level properties which reduce Equations like (5) – (10) to much simpler forms from which data can be extracted. By far the largest volume of published information on deep levels is derived either from purely optical extraction using sub band-gap light, which produces (essentially) zero carrier flux in the depletion layer, or by thermal transitions alone as in DLTS experiments. Reviews of these classes of technique have been given respectively by Grimmeiss[13] and Miller et al.[14] An early and remarkably comprehensive survey of possible excitation methods has been compiled by Sah.[15]

For a particular class of defects in n-type material, the hole transitions will become extremely important. Consider a defect which has an electron state in the lower half of the energy gap in n-type GaP. The thermal-ionization rate constant for electrons is likely to be small at moderate temperatures (400K) because of the large energy term $(E_c - E_T)$ in the Boltzmann factor of the detailed balance expression for e_n^t. Therefore, unless there is a massive difference in the capture cross sections, $e_p^t \gg e_n^t$ at all practicable temperatures and the thermal hole emission will dominate over the electron emission. If the hole diffusion length L_p is larger than the space-charge width, W, then the diffusion component of the photocurrent dominates and $\Delta p \gg \Delta n$. Unless $\sigma_n \gg \sigma_p$, we have $c_p \Delta p \gg c_n \Delta n$, and the hole-capture rate dominates the electron-capture rate. The electron-capture transition becomes progressively less important as the hole diffusion length increases, and is generally insignificant for defects with $\sigma_\rho > \sigma_n$. The direct excitation of the defect (photoionization) is always possible in such an optically excited barrier and this 'interference' with the capture/emission data must be deconvolved by a separate investigation of the optical properties of the defect in a rigorous study. Simple checks which tell one whether or not capture/emission transitions dominate the transient capacitance response can be made however, and rest largely on the sensitivity of the response to the magnitude of the photocurrent, rather than the photon energy. Examples are given below. The hole capture rate can in a sense be enhanced by optimizing the diffusion component of the photocurrent per incident photon. For example front face illumination through a semi-transparent diode, using a wavelength which maximized the diffusion component of the photocurrent, as determined by the spectral response, could be used. This has been dealt with for the case of GaP by Young and Wight.[2] More elegant solutions in which the excitation source does not penetrate the barrier can be envisaged, though they are generally more difficult to implement. Excitation of the back face of the diode with highly absorbed light could be used in a thin sample (e.g. $1/\alpha \leqslant$ thickness $\leqslant 3L_p$) to diffuse holes into the front face barrier. Electron beam excitation is another potential source of minority carrier diffusion currents.

In the ideal case with states deep into the 'minority half' of the energy gap, with a dominant diffusion component in the photocurrent and with weak photoionization rates, minority carrier capture and emission dominate the transient capacitance response. Equation 6 then reduces to:

$$n_T(t)/N_T = c_p \Delta p \tau_1 \, \exp -(t/\tau_1) + e_p^t \tau_1, \qquad (12)$$

where

$$\tau_1 = [e_p^t + c_p \Delta p]^{-1} \qquad (13)$$

with the optical pulse removed,

$$n_T(t)/N_T = [e_p^t \tau_1 - 1] \exp - (t/\tau_2), \qquad (14)$$

where

$$\tau_2 = [e_p^t]^{-1} \qquad (15)$$

The way in which these fractional changes in electron occupancy influence the high frequency capacitance of the diode is clear from Equation (11). For small capacitance changes $\Delta C(t)$ can be obtained from

$$C^2(t) - C^2(0)/C^2(0) \cong 2\Delta C(t)/C(0) = p_T/N_D \qquad (16)$$

and from (12):

$$p_T(t) = N_T c_p \Delta p \tau_1 \ [1 - \exp - (t/\tau_1)], \qquad (17)$$

so that

$$\Delta C(t) = I_{12} C(0) N_T/N_D c_p \Delta p \ \tau_1 \ [1 - \exp - (t/\tau_1)]. \qquad (18)$$

The increase in p_T causes a positive incremental capacitance change in n-type material, and τ_1 in Equation (13) will be referred to as the capacitance rise time τ_r. Similar reasoning shows that the capacitance decay observed on removing the light is given by

$$\Delta C(t) = \frac{1}{2} C(0) N_T/N_D c_p \Delta p \tau_1 \exp - (t/\tau_2) \qquad (19)$$

where τ_2 is given by Equation (15) and will be referred to as the capacitance decay time τ_d.

Measurement of both τ_r and τ_d give values for $c_p \Delta p$ and e_p^t. The thermal emission rate for holes e_p^t measured over a range of temperature yields the thermal binding energy (subject to certain correction factors which have been discussed by Mircea et al.[16] The risetime τ_r contains information on the hole capture cross-section. This parameter is all important in evaluating the point defect recombination physics since it exercises control of the net recombination rate at the deep level in all but conditions of high injection when the center may saturate. Combining Equations (13) and (15),

$$c_p = [\tau_r^{-1} - \tau_d^{-1}]/\Delta p \qquad (20)$$

hence:

$$\sigma_p = [\tau_r^{-1} - \tau_d^{-1}]/Vth\Delta p, \qquad (21)$$

whilst the thermal hole velocity can be reasonably estimated quantifying the hole density due to the optical excitation presents problems. Assuming that the current flow inside the barrier region is governed by drift, the injected-hole density is then only uniform if the drift velocity is a constant. However, this should not be the case for the material used in this work in which the maximum field strength was typically 10^4 V/cm, and in any event can never be true over the entire barrier region since the electric field strength E varies linearly with distance X (assuming constant $N_d - N_A$ as $E = Emax\,[1-X/W]$, where W is the total space-charge width.

It is found experimentally that transients with characteristic exponential regions can be obtained. These have been analyzed to make an estimate of Δp by assuming $\Delta p = J_{ph}/q\mu<E>$ in the depletion layer. Some nonexponential components of $\Delta c(t)$ are often observed, but these do not appear to be caused by nonuniformity in Δp, and are discussed in the next section.

So for these simple approximations, Equation (21) becomes:

$$\sigma_p = q\mu<E>[\tau_r^{-1} - \tau_d^{-1}]/Jph\,Vth., \qquad (22)$$

where $\mu<E>$ is the average drift velocity of the hole in the depletion layer. Having established the criteria for measuring e_p^t (and hence binding energy) and σ_p, the final parameter which can be extracted from the hole trapping capacitance response is the deep level concentration N_T. This can be evaluated from the low temperature limit ($e_p^t \rightarrow 0$) of equation 18 which (for a long pulse, $t \rightarrow \infty$) gives:

$$2\Delta C(\infty)/C(0) = N_T/N_D \qquad (23)$$

N_D, or strictly the ionized net donor concentration can be obtained from conventional C-V data measured under the same conditions of temperature and frequency.

The underlying physics of this optically excited Schottky diode method for evaluating the minority carrier properties of a deep level is illustrated Figure 5. Essentially, the deep level is forced by the excitation, and by virtue of its position in the energy gap, to behave as a classic hole trap exchanging charge with the valence band alone. Those defects with large

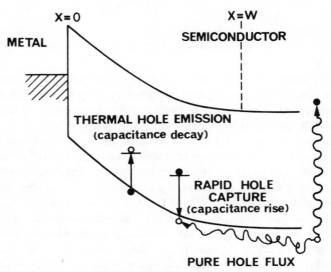

FIGURE 5 The simple excitation which should dominate in a minor hole capture experiment.

values of σ_p are selectively detected, and it is, of course, these defects which are likely to act as powerful recombination centers when situated in the bulk semiconductor. The complete classification of the recombination efficiency of the defect depends also on the measurement of the majority carrier capture cross section, σ_n in this case. Compared to the evaluation of the minority cross section, this presents few difficulties and is discussed in Section 4.

3.3 Hole Traps in n-type GaP

Two 'hole traps' have been systematically observed in undoped n-type GaP, and much of the discussion in the remainder of this chapter will center around these defects. Before detailing the experimental characterization of these defects, their properties are summarized in Table 2.

The defect at Ev + 0.75 eV can act as a lifetime controlling defect in GaP and is certainly the dominant recombination center in many of the LPE and VPE layers measured. The values of thermal hole binding energy and hole capture cross sections listed in Table 2 were calculated using the analysis of the last section. Obviously, in order to do this, the justification for neglecting the many possible, non hole-trapping interactions must be

TABLE 2

Hole Thermal Binding Energy	σ_p (300K)	σ_n (300K)	Origin
0.75 eV	5×10^{-14} cm^2	10^{-16} cm^2	Possibly V_p related
0.95 eV	2×10^{-15} cm^2	3×10^{-23} cm^2	Associated with Ni

established experimentally. This is reported in detail elsewhere.[12] Briefly, it was shown that with near-band-edge excitation the capacitance response from both defects was linked to the hole current in the barrier and not to photoionization. Interestingly, the 0.75 eV defect exhibits no measurable photoionization response. Only light of sufficient energy to generate a hole flux could change its occupancy and then the change was due to the injected holes and not the photon flux. The 0.95 eV level does exhibit photoinoization transitions but, for the excitation used to measure its hole capture properties, the influence of photoionization is weak. In order to extract usable information from (rather than merely 'detect') a hole trap it is essential to measure well-defined capacitance rise and decay times. A typical risetime measurement is given for the 0.75 eV defect in Figure 6.

At low reverse bias (or small forward bias) non-exponential transients were obtained but at moderate to high reverse bias these levelled out into a single well defined experimental response, apart from a small fast initial portion. This is typical behavior for many optically pumped capacitance experiments. The non-exponential behavior has its origin in electron trapping at the defect from the Debye tail. This, of course, combines with and speeds up, the hole-capture-induced signal. The influence of free carriers on photocapacitance measurements has been discussed by Braun and Grimmeiss.[17] Capacitance decay times for both defects show the same sort of dependence on applied bias. The influence of the barrier electric field on these measurements is evidently small since increasing the average field strength produces a more ideal response. It seems likely however that some traps will have field dependent cross sections and the necessity to undertake these measurements in a depletion field is a fundamental limitation.

The hole capture cross sections are central to the role which these defects play in the recombination physics of the material and they are discussed further in Section 4; The correlation between the concentration of these defects and lifetime is discussed in Section 5. The basic capaci-

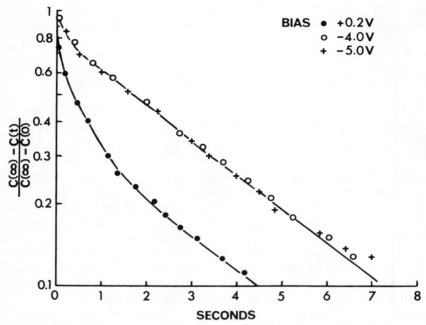

FIGURE 6 Hole trapping capacitance transients showing substantial exponential regions provided the reverse bias voltage is not too small.

tance response yields, through its temperature dependence, the thermal hole binding energy. The activation plots for both defects are shown in Figure 7.

The mean figures for binding energy quoted in Table 2 take into account correction factors due to the observed temperature variation of σ_p.

3.4 Minority Carrier Trap Spectroscopy (MCTS)

We have discussed the basis of a Schottky diode technique for detecting and characterizing recombination centers in bulk crystals, and have demonstrated its application to n-type GaP. The 'single shot' approach described, whilst it yields accurate data, suffers from the problems of low data acquisition rates. An obvious extension of the technique would be to use high optical pumping rates, giving high hole capture rates and then to apply standard DLTS signal processing [18] on the thermal release transient (capacitance decay). This would then detect the hole traps spectroscopically[19] and enable one to calculate binding energies from the peak shift

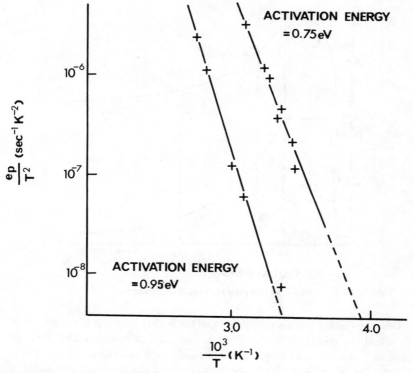

FIGURE 7 Arrhenius plots for the two hole traps.

with temperature in exactly the same way as conventional DLTS. This experiment has been performed on GaP using a He-Cd laser excitation source. The capacitance transients are shown in Figure 8.

The high intensity laser pulse provides the minority carrier equivalent of the forward bias filling pulse. Saturation of the hole occupancy is achieved for condition (b) in Figure 8 if $t_{on} \gg c_p \Delta p$. During the 'off' state the capacitance decay is sampled at two times, t_1 and t_2 which define the rate window $(t_2 - t_1)/\ln t_2/t_1$ a peak is produced on the plot when this is equal to τ_d. Since the MCTS probes hole traps rather than electron traps one would expect to see a completely different spectrum than that obtained by ordinary DLTS. This is indeed the case as is shown in Figure 9 which compares MCTS with DLTS measured on the same Schottky diode. Several hole traps appear in the MCTS spectrum which is dominated by the 0.95 eV level. Since the capacitance transients for hole traps are in the opposite sign for those of electron traps the spectra peak in opposing senses.

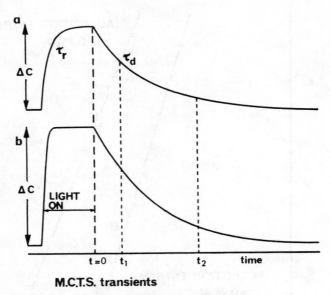

M.C.T.S. transients

FIGURE 8 Laser excited hole trapping capacitance transients.

One interesting feature of MCTS spectra is that they can be easily made to reveal the relative importance of a hole trap, as quantified by the hole capture cross section concentration product. The saturation condition $t_{on} \gg c_p \Delta p$ depends on both the magnitude of the photocurrent and σ_p.

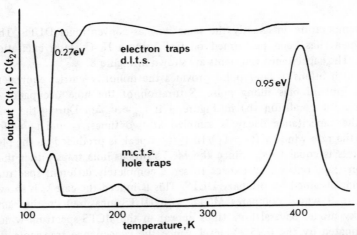

FIGURE 9 A comparison of standard DLTS and (laser excited) MCTS spectra measured on the same sample.

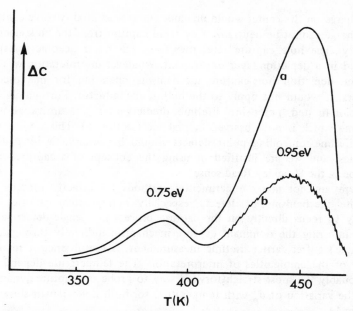

FIGURE 10 The effect of reducing the laser power in an MCTS measurement, for a) the excitation photocurrent was $30\,\mu A$ and in b) $8\,\mu A$.

For a constant t_{on}, saturation is achieved for traps with large σ_p at a correspondingly smaller photocurrent than for traps with small σ_p. Therefore, if the laser power is reduced, traps with small σ_p will tend to disappear from the spectrum. This effect is shown in Figure 10 for the 0.75 eV and 0.95 eV hole traps. Changing the photocurrent from $39\mu A$ to $8\mu A$ causes no change in the 0.75 eV peak amplitude, while that of the 0.95 eV trap, with its smaller hole cross-section is significantly attenuated. The two sprectra have been slightly offset to show this.

4. Capture Cross-Sections

The physical mechanisms which operate to dissipate the recombination energy are of interest because they may give information on the nature of the defect. For both the 0.95 eV and the 0.75 eV states, these mechanisms are efficient for hole capture, since σ_p is large in each case. Ultimately, we should expect defects with very large capture cross-sections, i.e. those with $r_o > \lambda$, to have transport-limited rates in a space-charge neutral semiconductor. In a depletion layer experiment, it is difficult to predict the capacitance response for such a center. After the initial hole capture event, the charge on the center would preclude (or at least modify) hole capture; the charge cannot be neutralized by rapid capture from the bulk electron density. The hole capture rate, therefore, can never become transport limited in a depletion-layer experiment. Whatever interpretation was put on the depletion hole-capture mechanisms operating for macroscopic centers, it would not apply to the bulk semiconductor. Fortunately, the diffusion-limited 'point-sink' lifetime, predicts a temperature variation of lifetime which is not observed in practice (section 5). This implies that the lifetime-controlling point defects should be describable by ballistic kinetics, and we are justified in using the concept of a capture cross-section in the Shockley-Read sense.

There are not many experimental methods for directly probing the capture mechanisms of defects, especially non-radiative capture. The ability to focus directly on the capture process at a single defect rather than inferring the dominant capture mechanism indirectly through (for example) a free-carrier lifetime measurement, is useful since it removes some of the ambiguities of interpretation. The temperature dependence is probably the most straightforward way to probe the capture dynamics, and the variation of σ_p with temperature for both hole traps is shown in Figure 11.

The most obvious feature of both cross sections is that they are larger at low temperatures, and obey a power law of the form $\sigma_p \propto T^{-n}$ over a substantial range of temperature. This temperature variation is in the opposite sense to that expected for a capture event in which the energy dissipation occurs by multiphonon emission.

The idea of multiphonon emission as a mechanism for non-radiative capture is well established, and has been reviewed and extended for capture into neutral centers by Henry and Lang,[20] who also illustrate that the model is appropriate for many deep level states, especially in III-V compounds. Multiphonon capture cross-sections are thermally activated and take the form.

$$\sigma = \sigma(T = \infty) \exp-(\Delta E/kT) \qquad (24)$$

The temperature activation has its origin in the need to distort the lattice locally in order to couple the free-carrier and deep-level energies. Once captured into the deep level, the latter can relax to its equilibrium value, dissipating energy by local vibrations which are ultimately converted to lattice phonons. The large-amplitude vibrations needed to carry away the energy ($\sim Eg/2$) can cause defect motion (or even damage), Kimerling[21] has reviewed the topic of recombination-enhanced defect motion in several semiconducting systems.

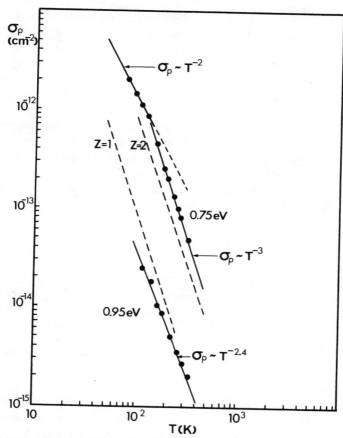

FIGURE 11 The observed temperature dependence of hole capture cross section for the 0.75 eV and 0.95 eV levels. Also shown in the prediction of equation(25) for Z = 1 and Z = 2.

The temperature dependence of σ for the two hole traps indicates that the rate limiting step in the hole capture event is not multiphonon emission. In fact a T^{-n} power law is more reminiscent of a Lax cascade process.[22] The Lax model describes capture by attractively-charged centers; the carrier descending down a quasi-continuous ladder of excited states with the emission of a lattice phonon at each step. The model works well (especially at low temperature) for capture at ionized donor or acceptor states where the coulomb field provides the necessary excited-state distribution, and the ground state is itself shallow.

In a recent re-appraisal of the Lax cascade calculation Abakumof and

Yassievich[23] argue that capture is dominated by electrons of energy, m^*S^2, where S is the velocity of sound in the crystal, into levels of similar binding energy. They derive a 'cascade' capture cross-section of the form

$$\sigma = 8/3 \; [q^2 \, Z^2 /\epsilon kT]^3 \, m^* E^2 /\rho \hbar_4 \qquad (25)$$

where eZ is the total charge of the defect (before capture), E is the deformation potential, for low energy phonons, of the band from which capture occurs and ρ is the crystal density. For hole capture we require the valence band value of E for acoustic phonons in order to apply Equation (25). Wight et al[7] argue that $E \simeq 11.8$ eV based on hole mobility measurements near 100K when the mobility can be limited by acoustic phonons ($\mu_n \sim T^{-1.5}$). In this case equation 25 predicts:

$$\sigma_p \simeq 10^{-7} Z^3 /T^3 \qquad (26)$$

and the resultant cross sections for Z = 1 and Z = 2 are plotted besides the measured cross-sections in Figure 11. The fit for the 0.95 eV state to the Z = 1 curve is reasonable. The temperature dependence of σ_p for the 0.75 eV state is T^{-3} for $130K \leqslant T \leqslant 320K$ and the absolute magnitude of σ_p agrees with equation 26 for Z = 2 to within a factor of two in the same temperature range.

The idea that hole capture into these very deep levels may be correctly described by a (modified) Lax model is an intriguing one. Certainly the Lax process cannot describe transitions into the ground states of these defects, and the implication is that the total cross section is limited by the initial capture stage into excited states. This in turn implies that the transition to the ground state is extremely efficient for these centers. The initial capture and subsequent large scale energy dissipation process occur in series, so, if the initial event is rate-limiting, then that process will be sensed by any experiment designed to probe the capture cross-section. In these circumstances, capture into the excited states of an attractively charged deep level would mask the important transition process into the ground state, which could be a multiphonon process with a very low value of ΔE. Such a process would require a large lattice shift for the transition between the excited states and the distorted ground state to occur.

Other models have been developed for the capture cross section at attractively charged centers, and all of them predict a temperature variation in the same sense as that observed for the 0.75 eV and 0.95 eV states. For example, the model of Gibb et al[24] treats two-step capture, with thermal re-emission from a single excited state back to the parent band.

The re-emission term can dominate the overall measured capture cross-section leading to a temperature dependance of the form $\sigma \sim \exp \Delta E/kT$ where ΔE is the binding energy of the excited state.

Henry and Lang[20] pointed out that for attractively charged centers, in which the capture energy is dissipated by multiphonon emission, the local enhancement of the minority carrier density (Sommerfeld factor), due to the coulomb field increases the local capture rate. The capture cross-section is calculated from a rate ($\tau^{-1} = c_p \Delta p$) which assumes that the carrier density near the recombination center is just the average bulk carrier density. This leads to a calculated cross-section which is larger than the multiphonon cross-section. If the characteristic multiphonon energy is small, then the thermalization of the local carrier density can dominate the temperature dependence of σ. A cross-section which weakens with increasing temperature may be measured, and, in the simplest case, this leads again to $\sigma \sim \exp \Delta E/kT$. The interpretation of σ in terms a specific model is difficult, but it is reasonably certain that they are attractive centers for holes in n-type material, and the observed power laws may simply reflect the domination of the early stages of capture in the total cross-sections. The major energy dissipation step(s) remain uncertain.

The majority carrier capture cross-sections for hole traps can be measured using a pulse bias Schottky diode technique.[25] The idea is to fill the deep states with holes by illumination at a temperature in which thermal hole emission is negligible. The holes are stored in the deep levels which can be re-filled by electron capture during forward bias. If forward bias pulses are applied which are short compared with the electron trapping time $\tau = (c_n n)^{-1}$ (where n is the bulk electron density), then the deep states can be made to fill in a slow and controllable manner as shown in Figure 12. The decay of capacitance takes the form

$$\Delta C \propto \exp - [\sigma_n n \, Vth].t \qquad (27)$$

where t is the total time spent in forward bias from which σ_n can be calculated. This method is precise because n is accurately known. One of the few potential problems of interpretation results from the application of too small a forward bias pulse in which case the depletion layer is not sufficiently collapsed and the hole traps see a non-uniform electron density during the capture phase.[26]

The data shown in Figure 12 are for electron capture into the 0.95 eV hole trap. The measured electron capture cross-section at 300K for this defect is remarkably small: $\sigma_n = 3 \times 10^{-23}$ cm^2. The 0.75 eV state has an electron capture cross-section (measured in the same way) $\sigma_n \simeq 2 \times 10^{-17}$

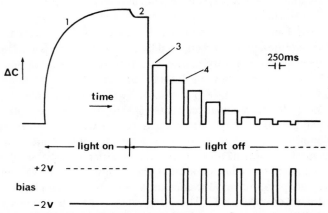

FIGURE 12 The controlled electron refilling of the 0.95 eV trap with short forward bias pulses, following the optical hole filling pulse.

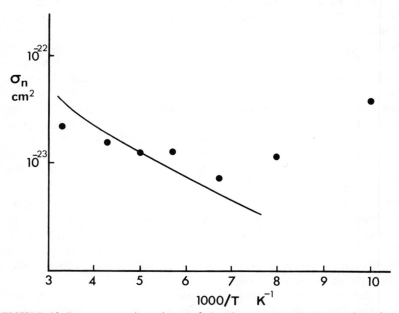

FIGURE 13 Temperature dependence of the electron capture cross section of the 0.95 eV trap. The solid line is equation (28) for capture into a repulsive center.

cm^2. Both values are small, though σ_n (0.75 eV) is reasonable for multi-phonon capture at a neutral center. A cross-section of $\sim 10^{-23}$ cm^2, however is difficult to explain, unless the center is repulsively charged i.e. negatively charged before electron capture. If the 0.95 eV center remains negative after hole capture, then it would have to be doubly negatively charged before hole capture. This would be contrary to the assignment of Z = 1 for the 0.95 eV level and Z = 2 for the 0.75 eV level, based on the data of Figure 11. On this assignment, we would expect the 0.75 eV state to remain repulsive to electrons after hole capture and therefore exhibit the smaller electron capture cross-section. Some support for the notion that σ_n (0.95 eV) is governed by repulsive capture is gained by a comparison with a model discussed by Bonch-Bruevich and Landsberg in their review of non-radiative processes.[27] According to this author, repulsive capture is strengthened by tunneling mechanisms occurring near the top of the potential barrier leading to

$$\sigma \sim \exp -[T/T_0]^{1/3} \text{ where } T_0 = 27 \pi^2 m^* Z^2 q^4 / 2\epsilon^2 h^2 k. \tag{28}$$

The measured values for σ_n (0.95 eV) are compared with this expression for 130K $\leqslant$ T $\leqslant$ 320K in Figure 13. Both the absolute magnitude and temperature dependence agree quite well in the higher temperature ranges.

5. Lifetime Studies

The capture cross-section data suggest that the 0.95 eV level should not be an efficient recombination center since its low electron cross-section would predict saturation of the recombination rate at all but the lowest injection levels. The 0.75 eV level however has all the hallmarks of a powerfull recombination center and indeed its concentration has been shown to correlate well with lifetime in undoped crystals. [12] This correlation has now been extended to S-N doped device grade material and the data are shown in Figure 14.

The slope of the data indicates that $\tau \propto 1/N_T$, and a good fit is obtained using the experimentally determined hole capture cross-section for the 0.75 eV state, $\sigma_p = 6 \times 10^{-14}$ cm^2 (300K). The tendency for the data points to fall below the fitted straight line at short values of τ, indicates that other parallel recombination paths become important in low-lifetime material.

The good agreement between theory and experiment for $\tau \geqslant 10^{-7}$ s shows that the 0.75 eV level acts as a classic Shockley-Read center in the

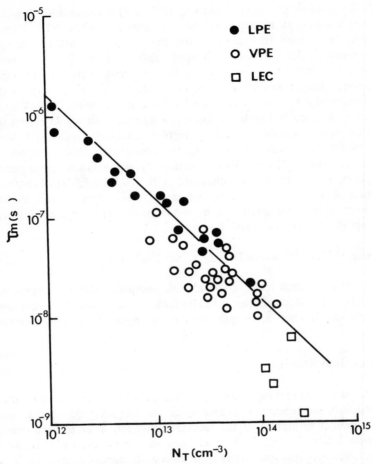

FIGURE 14 The correlation data for the minority carrier lifetime data and concentration of the 0.75 eV state. The solid line represents the predicted relationship for $\sigma p = 6 \times 10^{14}$ cm^2.

bulk crystal, and that the cathodoluminescence decay method measures the low-level lifetime of the material. Further proof that the 0.75 eV state controls the lifetime can be obtained by measuring the temperature dependence of τ. This is shown for a typical undoped L.P.E. sample in Figure 15. The lifetime as a function of the temperature, for the same sample, predicted from the measured capture cross-section of the 0.75 eV center at different temperatures and the other relevant materials proper-

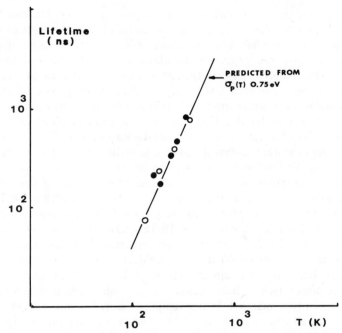

FIGURE 15 Temperature dependence of the minority carrier lifetime in 0.75 eV state controlled material; ● from c.l. decay measurements, and ○ from diffusion length measurements.

ties is also shown. The predicted behavior, $\tau \sim T^{-2.5}$ agrees well with the experimental values of lifetime.

The possibility that this very efficient hole capturing center operates by a diffusion-controlled mechanism can now be discounted since the temperature dependence of τ for diffusion-controlled point sinks can be derived from:

$$\tau = 1/4\pi D\rho_T r_o \qquad (29)$$

where D is the hole diffusion coefficient and ρ_T is the density of sinks.[7] The temperature dependence of τ, in this case is through D, predicting $\tau \propto T^{-1.2}$. Evidently, although the capture cross-section for the 0.75 eV center is large, its effective capture radius r_o is considerably smaller than the hole mean free path ($r_o/\lambda \sim 0.25$ at 300K) so the defect functions as a microscopic Shockley-Read recombination center.

It is interesting to compare these results with those of Tell and Kuijpers.[28] They undertook a comprehensive series of measurements of deep states on as-grown GaP. They described twelve states in terms of their thermal emission and capture properties and quoted typical concentrations in VPE material. Despite this, they were only able to account for 10% of the recombination traffic via the observed deep states. Some of their samples were remeasured at UMIST confirming their assessment of the deep-state populations. However, in these samples there were large concentrations of a state at 0.8 eV from the valence band (H5 in Ref. 28) and the 0.95 eV state discussed earlier (H6 in Ref. 28) and attributed to nickel. If the lifetime were controlled by the 0.75 eV state in these samples, its concentration would only need to be a fraction of that of the other two states with similar emission rates. The selectivity of DLTS as a rate window discriminator is not very good[29] and the practical manifestation of this is the width of the DLTS peaks. The inevitable result is that, although the 0.75 eV state is more important in terms of its recombination rate, it is not dominant in concentration and is therefore missed in normal DLTS in these samples. Figure 16 shows the emission plots of the three states derived from various samples using several techniques in both p and n-type material. In one sample measured (grown by VPE from PCl_3 in an all-glass system) the 0.95 eV state was not present and H5 and the 0.75 eV state could be distinguished at higher rate windows (H5 and the 0.75 eV state were present in this sample in comparable concentrations). Even so, in the p-type material of Ref. 28, this cannot account for the lifetime. In fact, the 0.75 eV state is not of crucial importance in p-type layers because of its small electron cross section. In the n-type material of Ref. 28, the 0.75 eV state could be resolved using the MCTS technique with a reduced intensity of laser excitation. This minimized the signal from H5 and H6 because of their smaller capture cross-sections (Figure 10), and so a reasonable estimate of the lifetime attributable to recombination through the 0.75 eV state could be made. On this basis, it seems that the material of Ref. 28 does conform to the general pattern, in that the 0.75 eV state dominates the recombination in n-type layers.

Because of the smaller electron cross-section, the 0.75 eV states contribution to recombination in p-type material is small. Since, however, the dominant output from commercial green LEDs is from the n-layer, in practice, it is this state which is important in light output terms. The differences in the hole and electron cross-sections, raise the interesting point that, at some excitation density, the 0.75 eV state might saturate and the devices become more efficient at high excitation densities (high

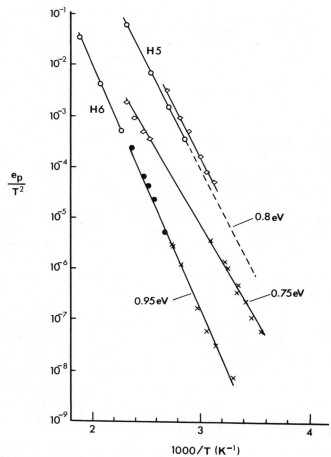

FIGURE 16 A comparison of the activation plots of hole traps observed in p and n-type gallium phosphide by Tell[28] and by Hamilton et al[12] o DLTS on p-type (Tell), ◊ DLTS on p-type (Hamilton), o MCTS on n-type (Hamilton), x MCTS on n-type (Hamilton).

currents in the LED case). In fact, the 'saturation' is progressive; when the states near the junction are handling their maximum carrier flux (defined by σ_n and n), the effective local diffusion length increases. Thus, unless the carriers find some other recombination path, they spread into a region where the recombination centers are not saturated, producing only a slowly progressing increase in efficiency.

The exact analysis is difficult because of current spreading in a real device structure and the resultant non uniform saturation. Qualitatively, however, this is exactly what is seen to happen in terms of the dependence of efficiency on current density in green LEDs. The actual power law depends on the device geometry, but the local values of current density at which the various regimes change depend on the trap parameters.

6. Physical Identification of the 0.75 eV Defect

A general problem in studies of deep states is that of ascribing a chemical and/or structural identity to an electrically-defined defect, particularly when the centers are present inadvertently. The deep states we have described so far are no exception and, very significantly, it is this information on identity which is of crucial importance both to those needing to improve LED performance and to those trying to understand deep states in a generalized sense.

Deep-state experiments measure capture or emission of carriers induced by thermal or optical processes. The way in which these parameters are affected by the microscopic structure of the defect is not known systematically. Consequently, identification tends to depend on correlation with doping experiments or in the case of structural defects with radiation damage and annealing. In some cases, it is possible to combine this work with luminescence, absorption or E.P.R. techniques which may give an unambigious identity to the defect. Such cases are rare in the literature. Apart from the sheer volume of work involved, there are many physical reasons why definitive links between the different types of experiments can only be achieved occasionally.[30]

In the case of the 0.75 eV state, optical techniques are denied us, simply because, as discussed earlier, the deep-state displays no optical activity. In addition, E.P.R. techniques would not be expected to have sufficient sensivivity in GaP to detect the rather low concentrations ($\sim 10^{13}$ cm^{-3}) of this state normally present in device grade material. This is often the case because of the broader resonance lines due to the large nuclear spins of the III-V elements. This leaves essentially only experiments where defects are introduced and a check made on whether or not the concentration of the state changes.

A comprehensive investigation of the effects of adding various impurities during L.P.E. growth has failed to generate a state with similar properties to the 0.75 eV trap.[31,32] Brunwin et al[33] diffused a selection of 3d transition metals into gallium phosphide and observed the thermal emission properties of the states generated. Despite looking specifically for a trap corresponding to that at 0.75 eV, none was found above the typical background level. These experiments included the elements considered as common contaminants such as iron, chromium and nickel.

Although such negative results can never be taken as conclusive, it appears that the dominant recombination center in gallium phosphide is not a chemical contaminant. This is compatible with the observation discussed earlier that all the GaP device-grade material grown by V.P.E. or normal cool-down L.P.E. which was examined appeared to have comparable concentrations of the 0.75 eV state. It would seem unlikely that when using reagents and reactor materials drawn from very different origins there would be such similar levels of chemical contaminants present. There exists, therefore, a distinct possibility that the 0.75 eV state is a structural defect.

Several authors have investigated what the likely behavior of lattice defects would be in gallium phosphide. Jaros and Brand[34] considered a number of structural defects in the III-V's using the pseudopotential method. Their findings supported the intuitive prediction that cation vacancies would act as acceptors and anion vacancies as donors. They calculate that the isolated gallium vacancy should be just above the valance band (0.2 eV) and that the phosphorus vacancy should be near the conduction band.

Kennedy and Wilsey[35] have interpreted E.P.R. signals from electron irradiated material (2 MeV) as being due to the gallium vacancy (V_{Ga}). This appears to have overall tetrahedral symmetry with respect to its nearest neighbors in contrast to the considerable distortion detected in silicon and the II-VI's around the vacancy. Scheffler et al[36] have questioned Kennedy's interpretation but a later study[37] goes some way to resolving these differences. Parallel DLTS measurements indicate that the state assigned to the gallium vacancy has a much smaller cross section and different emission properties[37] to the 0.75 eV defect.

There has been no positive identification of optical spectra associated the isolated phosphorus vacancy. Indeed there is some evidence that, at room temperature, the isolated vacancies are extremely mobile and consequently it would be unlikely that they would be present in significant concentrations because the charged vacancy would become trapped at other attractive defects which certainly exist in the grown crystal. Although a number of vacancy complexes have been considered theoretically, a case in which calculations have been supported by experimental work is the antisite defect P_{Ga}. Jaros[38] undertook a non-self-consistent pseudopotential calculation for this defect and placed the A_1 level at

0.2 eV from the conduction band. More recently Scheffler, et al[36] completed an entirely self-consistent calculation and determined two levels for P_{Ga} one at 0.6 eV and one at 1.1 eV from the conduction band. This is consistent with the EPR studies by Kaufmann et al.[39] which provides definitive experimental evidence of the existence of P_{Ga}.

The observation of P_{Ga} is all the more significant as the work was done on as-grown L.E.C. (Liquid Encapsulated Czochralski) material. Quite high concentrations have been observed,[40] up to 4×10^{16} cm^{-3}, which is markedly more than the highest concentrations of the 0.75 eV state seen in L.E.C. materials (5×10^{14} cm^{-3}). Although this difference could be associated with the choice of samples, a more fundamental objection is that the phosphorus antisite is a double donor while the 0.75 eV-state behavior is acceptor-like. In addition, pseudopotential calculations already quoted place the first ionization energy higer in the gap than the 0.75 eV and Kaufman et al also have some evidence that the second charged state is near mid gap.[41] There is no data in the literature ascribed to the Ga_P antisite although from thermodynamic considerations a higher concentration than that of the P_{Ga} defect would be expected in as-grown material.

Both antisites are doubly charged, with Ga_P being a double acceptor. Van-Vechten[42] estimates that, at a growth temperature of 1600K, the equilibrium concentration of Ga_P would be 5×10^{18} cm^{-3}. Clearly these are not frozen in, as there are not 5×10^{18} cm^{-3} double acceptors in undoped GaP. Indeed, the compensating acceptors in n-type material usually appear at 0.1 N_D over quite a wide range of N_D (10^{16} to 10^{18} cm^{-3}) implying that these acceptors are donor related.[43]

Van Vechten has proposed that the antisite defects react with vacancies to form complexes of the type $V_{Ga}^- P_{Ga}^{+2} V_{Ga}^-$ or $V_P^+ Ga_P^{-2} V_P^+$ which are overall neutral.[44] This could explain the absence of large concentration of compensating acceptors of the type Ga_P, but is not consistent with Kaufmann's observation of E.S.R. concentrations of isolated P_{Ga} comparable with the equilibrium values at the growth temperature as calculated by mass-action expectations.

Van-Vechten proposes that the $V_{Ga} P_{Ga} V_{Ga}$ defect should be an efficient recombination center on the basis of the ability to dispose of energy more effectively than $V_P Ga_P V_P$. There are of course many other factors to take into account in assessing the recombination rate as have been discussed earlier; not least of which is the concentration of the defect. It seems likely that $V_P Ga_P V_P$ will be present in considerably higher concentrations. An early growth experiment by Jordan[45] produced a range of gallium phosphide ingots from non-stoichiometric melts on the gallium-rich side of the phase diagram. He demonstrated a proportionality between the calculated equilibrium gallium vacancy concentration and the inverse of the minority carrier lifetime, indicating that the dominant recombination center is associated with the gallium vacancy population.

In vapor-phase material, Stringfellow and Hall[46] have changed the III/V ratio in the gas stream and observed an increase in lifetime with increasing III/V ratio. Their data are shown in Figure 17 together with that of Wessels.[47] The solid line is for a linear relation between lifetime and gallium activity, with the constant of proportionality adjusted to fit the experimental data. Four n-type layers grown by Stringfellow and two by Wessels have been measured by us and are included in the data of Figure 14. The material conforms to the pattern in which the lifetime is con-

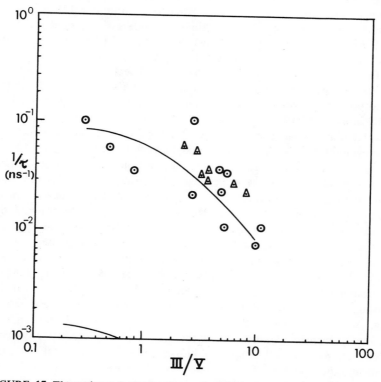

FIGURE 17 The reciprocal of minority carrier lifetime as a function of III-V ratio in the gas stream ⊙ photoluminescence decay measurements (Stringfellow[46]), △ charge storage measurements (Wessels[47]). The upper solid line was calculated assuming the lifetime is proportional to the gallium activity. The lower line used Jordan's data derived from LEC material[45].

trolled by the 0.75 eV state. These results would imply that if the dominant recombination center is an isolated point defect it must lie on the gallium site. As chemical contaminants have been ruled out previously, this leaves the V_{Ga} which, if the incorporation is thermodynamically controlled, will be proportional to the reciprocal of the gallium activity.

There are, however, problems with this interpretation. Stringfellow has pointed out that the dominant donor in his material is sulphur (which occupies the phosphorus site) and hence its incorporation should be proportional to the gallium activity. In fact a dependence of the opposite sense is observed. This suggests that the incorporation may be kinetically controlled in V.P.E. although it is possible that the dominant donor was silicon which is easy to underestimate as it is optically less intrusive than sulphur. However, if Jordan's data are taken and the equilibrium vacancy population in the V.P.E. material related to the lifetime by the same constant of proportionality, the values are two orders of magnitude too long. The predicted relationship is drawn in on Figure 17 so supporting the theory of a kinetically controlled defect system.

The formation of complexes of structural defects would be an alternative explanation but would not alone resolve the anomaly of sulphur incorporation without resort to donor complexes or defects related to the phosphorus sub-lattice.

A relationship between the 0.75 eV state and gross structural defects has been observed.[48] In this study epitaxial layers were grown by cool-down L.P.E. on L.E.C. substrates of different dislocation densitites. For identical growth conditions there is an excellent relationship with the 0.75 eV state concentration of the form $N_T \propto N_{DIS}$, the dislocation density being determined by an etch-pit count. This is shown in Figure 18. Although the etch pit density was actually measured in the substrate, it is nearly equal to the dislocation density in the epitaxial layer over the range used.

An interesting possibility is that the dislocation itself acts as a deep state. Numerically, in this sample, the measured value of N_T roughly corresponds to a defect for each occasion that the dislocation threads through a plane of atoms. However, if the growth conditions are different, e.g., V.P.E. instead of L.P.E., the ratio of defects to dislocations is substantially increased, indicative of the fact that they are not physically identical. As dislocations are well-known sinks for rapidly diffusing impurities, the increase of deep states with increasing dislocation density supports the view that the 0.75 eV defect is not of chemical origin unless we postulate that decorated dislocations act as the 0.75 eV trap. However, there are simply not enough dislocations to provide the measured concentration of deep states in V.P.E. or L.E.C. material.

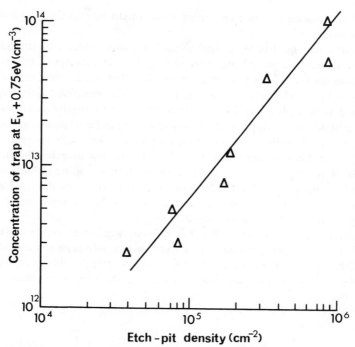

FIGURE 18 Relationship between substrate etch pit density and the concentration of the deep state at 0.75 eV as measured near the surface of the epitaxial layer.

A marked increase in the 0.75 eV state concentration is also seen at interfaces between L.E.C. substrates andV.P.E. layers.[49] At such interfaces, it is known that high concentrations of dislocation loops can occur. The overall position appears to be that the localized stress which the dislocation relieves also favors the formation of the defect structure associated with the 0.75 eV state, perhaps through the process of dislocation climb spawning point defects.

An alternative gross structural defect has been proposed by Stoneham[50] as a powerful recombination center in gallium phosphide. He has considered the possibility that colloids exist in the material consisting of clusters of $10^2 - 10^4$ gallium atoms. Much larger ($\geqslant 1\,\mu$m) gallium precipitates are often seen at interfaces but with separations which would not permit them to act as effective recombination centers. There is as yet no evidence from T.E.M. studies that gallium colloids such as those postulated by Stoneham occur in gallium phosphide. In addition, the temperature dependence of the capture cross section militates against a diffusion-limited mechanism to a neutral center which might be expected to be the case for the model.[51]

Radiation damage has been used extensively to produce structural defects in semiconductors. Lang & Kimerling[52] have catalogued electron traps resulting from 1 MeV electron irradiation. The results obtained were similar to those of Smith & Newmann[53] who also examined hole traps generated with 2 MeV and 0.8 MeV irradiation. The tentative conclusion was that states in the upper half of the gap tended to be related to gallium displacement while those in the lower half were linked with phosphorus displacement. Unfortunately the 0.75 eV state was not detected by these workers. A subsequent investigation by Brunwin et al[54] using (1 MeV and 0.5 MeV radiation to specifically investigate the effect on the 0.75 eV state was frustrated by the large concentrations of other defects generated, making sensitive capture measurements impossible.

An intriguing experiment on the 0.75 eV state was conducted by Peaker et al.[48] Two undoped epitaxial layers were cleaved into several pieces and subjected to heat treatment for various times at 800°C with effectively zero phosphorus overpressure. The effect on the 0.75 eV trap is shown in Figure 19.

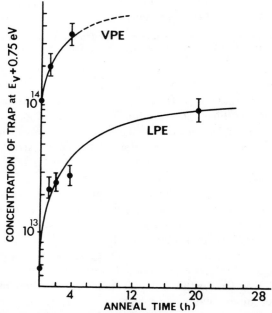

FIGURE 19 The effect of heat treatment at 800°C on the concentration of the deep state at 0.75 eV near the surface of the expitaxial layers.

It can be seen that the trap concentration (measured near the surface) increases dramatically. Subsequent work showed that, if the experiment was repeated for a 4 hour period at 800°C in closed capsules, those slices with equilibrium phosphorus overpressure or above showed no increase in the 0.75 eV state concentration (in fact in some cases a marginal decrease) while those with low phosphorus overpressure behaved like the open-tube samples. It seems probable that the dominant mechanism is phosphorus loss equating to an indiffusion of phosphorus vacancies. However, it is interesting to note that there is a substantial difference in the initial concentration of the 0.75 eV state between V.P.E. and L.P.E. layers and this difference is maintained or even enhanced during the annealing process. In addition, the electrical evidence suggesting that the 0.75 eV state is an acceptor makes the argument that the state is simply Vp unacceptable.

The cool-down L.P.E. layers described above are typical in displaying lower concentrations of the 0.75 eV state and hence longer lifetimes than layers grown by V.P.E. although the growth temperatures are very similar and hence the equilibrium vacancy population would also be expected to be similar. If the defects we observe are structure-related, then it seems to support the view that the defect incorporation is kinetically controlled. If the L.P.E. layer were grown very slowly, the defect population would tend towards the equilibrium value. Harding et al[55] developed a liquid-phase growth system which relied on a small temperature gradient to transport the phosphorus from a gallium phosphide source to the epitaxial layer. Material grown in this way was characterized by very high cathodo-luminescent efficiency, long lifetimes ($\sim 5\,\mu s$) and a 0.75 eV trap concentration below the detection limit which was in most cases considerably less than that predicted by the trap lifetime relationship shown in Figure 14. In fact, the minority carrier lifetime in this material has been shown to be limited by diffusion of carrier to surfaces, interfaces and dislocations.[51] The measurements of Harding's layers have been done on the as-grown epitaxy using Schottky diodes. Attempts to make p-n junctions in the material have resulted in the reintroduction of the 0.75 eV state presumably because of phosphorus loss during processing.

Nishizawa et al[56] have developed a L.P.E. growth system very similar to that of Harding using a temperature gradient technique. The system has been scaled up to provide commercially quantities of material for L.E.D. production by the Stanley Electronic Company. The L.E.D.'s produced by Nishizawa's technique are now commercially available and are considerably more efficient than other commercial green-emitting devices. These diodes have been tested by us, and, although they have a significant deep-state population, the 0.75 eV level was not detected. Nishizawa's technique uses a vertical melt above the slice with the bottom kept a few degrees lower than the top. The substrate is slid out from under the melt in a carrier into the furnance atmosphere. Nishizawa et al have published work relating the deep level population to the phosphorus overpressure. Figure 20 shows the concentrations of two states detected by a photo-capacitance method as a function of phosphorus overpressure in the growth chamber.

The 0.75 eV state referred to by Nishizawa is not the same as that discussed here as it exhibits a marked photoresponse but the overpressure of phosphorus appears to prevent the formation of the main recombination center as well as reducing the population of other deep states. It is difficult to understand how the phosphorus overpressure above the melt could have any real influence on the growth kinetics at the interface in a saturated solution. However, because of the annealing experiments reported earlier, it seems likely that its role is to prevent phosphorus loss from the surface of the slice while it is still in the furnace zone.

A number of factors emerge from this, it seems that there is some evidence that the incorporation of defects during growth is often kinetically controlled and is not an equilibrium process as has been assumed by many previous workers in trying to relate point defects to material perfection performance.

The 0.75 eV defect appears to be related to the loss of phosphorus, although there is a distinct possibility that this could be in an indirect way, for example by the mass action effect of phosphorus loss on gallium

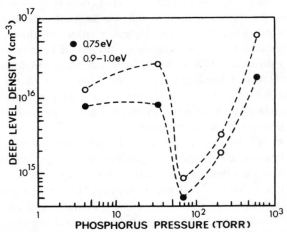

FIGURE 20 The deep level concentration in LPE layers as a function of phosphorus overpressure during growth (from Nishizawa et al[56]).

vacancies and antisites. The absence of any positive experimental observation by any technique of phosphorus-related defects supports this, although the deep-level populations we are considering are small for microscopic techniques in the III-V's ($\sim 10^{15}$ cm^{-3}). A very important factor is that the electrical measurements lead us to believe we are looking for an acceptor-like state, but we would sensibly expect simple structural defects associated with the phosphorus sub-lattice to be donors.

7. Identification of the 0.95 eV Defect

Dean et al[57] reported that an efficient luminescence band at 2 μm appeared to show an approximate relationship with the concentration of the deep state at 0.95 eV. A series of diffusion experiments suggested that both were related to nickel. More comprehensive studies of chemical impurities in GaP confirmed this[31,33] in the sense that the concentration of the 0.95 eV state was shown to rise substantially under all conditions of nickel diffusion while other diffusions (including other 3d transition metals) did not significantly change the measured concentration of the 0.95 eV state. The fact that stringent cross checks on the diffusant species producing the trap is emphasized in the light of cautionary discussions in the literature which point out that the observed deep levels can frequently be attributed to contaminants present in chemically undetected quantities in the main diffusant species or in the furnace ware. Chiang and Pearson[58] observed this in relation to annealing and more recently Ledebo et al[59] have put forward a convincing case that the deep state previously associated with nickel in gallium arsenide is in fact due to inadvertent copper contamination. It seems unlikely that this could be the case in gallium phosphide. Although this Ni related level has been referred to as the 0.95 eV state, it must be emphasized that this is its measured thermal activation energy corrected by 0.03 eV for the temperature dependance of the cross section. If, however, holes are optically excited from the state it is found that the threshold energy is anomalously low (0.6 eV). This is discussed in detail elsewhere[60] but is essentially due to the fact that the state is tied to the conduction band while the thermal processes are activated as hole emission to the valence band.

Nickel in gallium phosphide has been extensively studied using luminescence, absorption and ESR techniques. Three charged states of nickel on the gallium site have been detected in addition to luminescence signals which originate from donor acceptor associates constituted from nickel and shallow donor species.

The neutral state of the acceptor Ni^{3+} ($3d^7$) has been observed in p-type material[61] by ESR. This has been linked with the absorption and Hall measurement of Abagyan et al[62] which ascribes a thermal activation energy of 0.5 eV to the state for hole emission. This is confirmed by more recent measurements[63] on p-type gallium phosphide which indicate that a level with a thermal activation energy of 0.51 eV and an optical threshold for hole emission of 0.52 eV results from converting Ni^{3+} to Ni^{2+}. The Ni^{2+} ($3d^8$) level is not detectable by EPR because of its diamagnetic ground state. However, it can be identified by optical absorption at 0.97 due to d-d transitions.[64,65]

The Ni^{1+} ($3d^9$) state has been extensively studied by Kaufmann et al.[66] They have related the luminescence spectrum to nickel by analyzing the isotope contributions to the zero phonon line. The Zeeman splitting is consistent with EPR measurements on the same crystals and provides a very sound identity of Ni^{1+}. Strong evidence of the existence of Ni-S, Ni-Ge and Ni-Te associates on GaP has been provided by Ennen et al[67] from luminescence and absorption spectra. They have ascribed sharp line structure specifically to Ni-S, Ni-Ge and Ni-Te pairs. An estimate of the concentration derived from the absorption spectra puts them at about 10% of the total nickel content in the samples they measured. These were n-type LEC crystals diffused in closed capsules between 1000 and 1200°C, the initial carrier concentration was $10^{17} - 10^{19}$ cm^{-3}.

In the as-grown VPE material, the 0.95 eV state is present in quite large concentrations (10^{14} cm^{-3}), often more than any other defect measurable by deep-level techniques. In n-type nickel-diffused material of normal quality, the only other state ever detected in comparable concentrations is an electron trap referred to as V_1, which is not nickel specific.[33] There is however an electron trap produced by the nickel diffusion which is present at a concentration of around 10% of that of the 0.95 eV state. Photocapacitance measurements indicate a threshold of just below 0.6 eV for exciting an electron into the conduction band.[68] The form of the dependence of σ_n on wavelength is unusual although similar features have been seen in Si: Se and attributed to excited states.[69] The same form of spectrum has also been noted in GaP:Ni by Peaker and Grimmeiss.[70] Although photon energies as low as 0.6 eV could remove an electron from the state, there was no detectable thermal emission even at room temperature despite the electron capture cross section being easily measurable ($\sigma_n = 10^{-19}$ cm^2). It seems likely that the removal of electrons from the level is via a photo thermal process involving excited states. Bishop et al[71] have studied n-type GaP diffused with nickel. Their results indicate that the dominant luminescence is from Ni^{1+} in agreement with

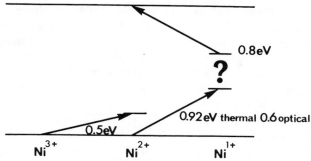

FIGURE 21 Nickel in gallium phosphide. Both the possibilities indicated for the Ni^{2+} to Ni^{1+} transitions have been observed but there is considerable evidence that the 0.92 thermal value is associated with a nickel complex rather than simple substitutional nickel.

Kaufmann's work and the unitaxial stress measurements of Hayes et al.[72] Bishop et al have examined the photoluminescence excitation spectrum so linking the previously observed absorption and luminescence but also seeing a reduction in intensity due to competing absorption from Ni^{2+}. Using this as a base, they conducted a series of dual source optical experiments which indicated that a photon energy of 0.8 eV is sufficient to convert Ni^{1+} into Ni^{2+} by removing an electron to the conduction band. The inference is that the Ni^{1+} state is less than 0.8 eV from the conduction band. Bishop et al also conclude that the electron capture cross section of Ni^{2+} is greater than the hole capture cross section of Ni^{1+}. This is a surprising result for a negatively charged center. They postulate an electron capture mechanism whereby sulphur donors play a role via a Stokes phonon assisted tunneling process. Although this could be significant at their measurement temperature of 5K, it is unlikely that it would be applicable above 80K due to the donor occupancy factor. The measurements also indicate that the optical cross sections are such that: $\sigma_e^0 > \sigma_p^0$ over the range 0.8 to 2 eV. These findings are incompatible with the state at 0.95 eV from the valence band but could reasonably relate to the nickel associated state seen by Szawelska et al and by Peaker and Grimmeiss.

If we accept the foregoing evidence, it appears that the state with a thermal activation energy of 0.95 eV is not the Ni^{1+} to Ni^{2+} transition nor Ni^{3+} to N^{2+}, partly because of the energy discrepancies but also because of the considerable differences in cross sections between the 0.95 eV state and those identified specifically as substitutional nickel. This leaves us with the distinct possibility that the 0.95 eV state is a complex, but with the anomaly that its concentration is ten times that of any other deep state in the diffused material which could sensibly be nickel related. In addition, no state has been detected by ESR, luminescence or absorption which appears to be present in sufficiently large concentrations compared to simple substitutional nickel on the gallium site already identified. Despite these reservations there is considerable other evidence which suggests that the 0.95 eV state is a complex.

Dean et al[57] diffused nickel for 24 hours at 1000°C under slight excess phosphorus pressure into LPE, VPE and LEC material. Although the samples were in the same capsule the concentrations of the 0.95 eV state were very different so that: $[N_T]_{LEC} > [N_T]_{UPE} > [N_T]_{LPE}$. Even within the same crystal, very considerable concentration changes have been noted. Hamilton and Peaker[49] measured the concentration profile of the 0.95 eV state throughout the thickness of a VPE layer and into an LEC substrate. The profile is shown in Figure 22.

The very pronounced peak in the substrate near the interface is not due to contamination immediately prior to growth. When Ni was diffused from the surface of the expitaxial layer into the substrate, the peak was replicated at an order of magnitude higher concentration. This could be due to

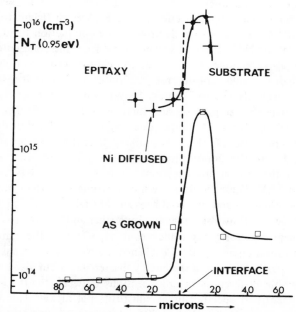

FIGURE 22 A comparison of the interface profiles for the Ni associated deep level in as grown and nickel diffused VPE material.

more sites being available for the nickel in this region (enhanced solubility due to excess gallium vacancies) or due to an increase in the concentration of the species which is paired with the nickel. This latter alternative would of course be the case only if the state were a complex. The carrier concentration was seen to be a step function at the interface, it being essentially flat in the epi-layer and also flat in the substrate. This can be taken to mirror the sulphur donor concentration and so the identity of Ni-S pairs for the 0.95 eV state seems unlikely unless the nickel concentration itself showed a peak.

Radiation damage experiments have not produced results which are definitive enough to enable us to draw any really significant conclusions. Electron irradiation at 0.5 MeV with a dose of 5×10^{15} e $-$ cm^{-2} increased the 0.95 eV state concentration by a factor of two. Irradiation at 1 or 2 MeV had no detectable effect on the 0.95 eV state.[54]

It seems that there are arguments against assigning the 0.95 eV state to any nickel related defect that has been identified to date using ESR, luminescence or absorption. This is despite the pronounced photoresponse of the state, the definite association with nickel and the high concentration in diffused material. However, despite the frustrations associated with these studies, it is apparent that the materials supplier has a chemical identification to help him eliminate one of the most ubiquitous deep states in gallium phosphide.

8. Summary

The gallium phosphide states described in this chapter illustrate problems associated with the study of deep levels which are common to many materials. Techniques have been used which enable the electrical parameters of the deep states to be measured reliably and very reproducibly. In the case of gallium phosphide and, indeed, in several indirect gap materials, it has been shown that it is possible to use the deep state data to calculate the recombination rates using Shockley-Hall-Read kinetics. Gallium phosphide has a band gap which enables us to see transitions which result in luminescence in a favorable region of the spectrum and this is a major factor accounting for a truly vast amount of work on this material. The situation with regard to electron spin resonance measurements in gallium phosphide is far from ideal, particularly from the point of view of sensitivity, but considerable success has been achieved in identifying specific structural defects and impurities. Gallium phosphide has also been favored by very considerable attention from a theoretical view point and this chapter has attempted to relate some of this work to the experimental studies.

Despite this very substantial investment of research effort the story is far from complete. We know that of the two deep states identified as important recombination centers one is associated with nickel and one with a structural defect. The wealth of studies on the topic has not been able to define a specific atomic structure to associate with the electrical parameters. We do know the activation energy of the defects, the thermal capture cross sections and their temperature dependence and their typical concentrations in n-type gallium phosphide. These results enable a minority carrier lifetime to be calculated which agrees (within a factor or two) with the actual measured lifetime.

The concentration of the 0.75 eV state has been correlated with the lifetime trend on layers grown by several techniques and orginating from many laboratories and commercial material suppliers. Although there is some scatter, the overall correlation is very good. In addition, the temperature dependencies of the cross sections have been measured and the minority carrier lifetime calculated from these data agrees well with the measured lifetime as a function of temperature. Although exact quantitative measurements of satuation of the recombination centers are not possible because of difficulties in determining the excitation density, the saturation behavior is qualitively what would be expected from the cross section measurements. We have then evidence from four different types of measurement that support the view that the 0.75 eV state controls the lifetime in most n-type gallium phosphide aided to a small degree in some material at low excitation densities by the nickel related 0.95 eV state.

Unfortunately, none of these measurements or even the optical measurements on the nickel-related state gives us a clear link with the luminescence or with the EPR data which could conceivably provide detailed structural identities. The gulf that exists between electrical and macroscopic measurement is not unique to gallium phosphide and perhaps the biggest challenge in the study of deep states today is to build a bridge between these two classes of measurement.

References

1. D. Hill and P. T. Landsberg, Proc. Roy. Soc. *347*, 547 (1976).
2. M. L. Young and D. R. Wight, J. Phys. D. *7*, 1824 (1974).
3. S. M. Davidson, M. Z. Iqbal, and D. C. Northrop. Phys. Stat. Sol. *29*, 571 (1975).
4. J. M. Titchmarsh, G. R. Booker, W. Harding, and D. R. Wight, Int. of Materials Science, *12*, 341 (1977).
5. T. Suzuki and Y. Matsumoto, Appl. Phys. Lett. *26*, 431 (1975).

6. C. Van Opdorp, A. T. Vink, and C. Werkhoven Inst. Phys. Conf. Series 33a, Edingburgh 1977, p. 317.
7. D. R. Wight, I. D. Blenkisop, W. Harding, and B. Hamilton, Phys. Rev. B. *23*, 5495 (1981).
8. I. D. Blenkisop, W. Harding and D. R. Wight, Electron. Lett. *13*, 14 (1977).
9. G. F. Neumark and K. Kosai, "Deep Levels in Wide Band Gap III-V Semi-conductors" Semiconductors and Semimetals, Vol 19, Academic Press, 1983, p. 1.
10. U. Kaufmann and J. Schneider in 'Advances in Electronics and Electron Physics' Vol. 58 Ed. C. Martin, Academic Press 1982.
11. R. Williams, J. Appld. Phys. *37*, 341 (1966).
12. B. Hamilton, A. R. Peaker, and D. R. Wight, J. Appld. Phys. *50* 6373 (1979).
13. H. G. Grimmeiss, Ann. Rev. Mater. Sci. *7*, 341 (1977).
14. G. L. Miller, D. V. Lang, and L. C. Kimerling, Ann. Rev. Mater. Sci. *7*, 377 (1977).
15. C. T. Sah, L. Forbes, L. L. Rosier, and A. F. Tash Sol. Stat. Electron. *13*, 759 (1970).
16. A. Mircea. Jnl. de Physique 38, 41 (1977).
17. S. Braun and H. G. Grimmeiss. J. Appls. Phys. *44*, 2789 (1973).
18. D. V. Lang, J. Appld. Phys. *45*, 3023 (1974).
19. R. Brunwin, B. Hamilton, P. Jordan, and A. R. Peaker, Electron, Lett. *15*, 348 (1979).
20. C. H. Henry and D. V. Lang, Phys. Rev. B. *15*, 989, (1977).
21. L. C. Kimerling, Sol. Stat. Electron. *21*, 1391, (1978).
22. M. Lax, Phys. Rev. *119*, 1502 (1960).
23. V. N. Abakumov and I. N. Yassievich, Sov. Phys. JETP *44*, 345 (1976).
24. R. M. Gibb, G. J. Rees, and B. W. Thomas, B. L. H. Wilson, B. Hamilton, D. W. Wight, and N. F. Mott, Phil. Mag. *36*, 1021 (1977).
25. A. R. Peaker, R. F. Brunwin, P. Jordan, and B. Hamilton, Electron. Lett. *15*, 663 (1979).
26. A. Zylberstein, Appld. Phys. Lett *33*, 200 (1978).
27. V. L. Bonch-Breuvich and E. G. Landsberg, Phys. Stat. Sol. *29*, 9 (1968).
28. B. Tell and F. P. J. Kuijpers, J. Appld. Phys. *49*, 5938 (1978).
29. S. Hodgarth, Electron. Lett. *15*, 724 (1979).
30. S. T. Pantelides, Reviews of Modern Physics *50*, 805 (1978).
31. 'Non-Radiative Recombination in Gallium Phosphide' University of Manchester Institute of Science & Technology, 1978 Annual Report on U. K. Ministry of Defence Contract RU 19-5 (DCVD, Fulham, London SW6 1TR).
32. Efficiency Improvements in Gallium Phosphide' Plessey Ltd. 1979 Annual Report on RP9-178 (DCVD Fulham, London SW6 1TR).
33. R. F. Brunwin, B. Hamilton, J. Hodgkinson, A. R. Peaker, and P. J. Dean, Solid State Electron, *24*, 249 (1981).
34. M. Jaros and S. Brand, Phys. Rev. *B14*, 4494 (1976).
35. T. A. Kennedy and N. D. Wilsey, Phys. Rev. Lett. *41*, 977 (1978) and also in 'Defects and Radiation Effects in Semiconductors 1978 Ins. Phys. Conf. Ser. 46, p. 375.
36. M. Scheffler, S. T. Pantelides, N. O. Lipari, and J. Benholc, Phys. Rev. Lett. *47*, 413 (1981) and M. Scheffler, J. Bernholc, N. O. Lipari, and S. T. Pantelides, Phys. Rev. B, 29, 3269 (1984).
37. P. M. Mooney and T. A. Kennedy, J. Phys. C, 17, 6277 (1984).
38. M. Janos, J. Phys. C. 11, L213 (1978).

39. U. Kaufman, J. Schneider, and A. Rauber, Appl. Phys. Lett. *29*, 312 (1976).
40. U. Kaufmann and T. A. Kennedy, J. Electronic Mat. *10*, 347 (1981).
41. U. Kaufman, J. Schneider, R. Wörner, T. A. Kennedy, and N. D. Wilsey, J. Phys. C. *14*, L951 (1981).
42. J. A. Van Vechten, J. Electrochem Soc. *122*, 1556 (1975) and J. Electronic Mat. *4*, 1159 (1975).
43. A. Mottram, A. R. Peaker, and P. D. Sudlow, J. Electrochem Soc. *118*, 318 (1971).
44. J. A. Van Vechten, J. Electrochem Soc. *122*, 423 (1975).
45. A. S. Jordan, A. R. Von Neida, R. Caruso and M. Di Domenico, Appld. Phys. Lett *19*, 394 (1971).
46. G. B. Stringfellow and H. T. Hall Jr., J. Electrochem Soc. *123*, 916 (1976).
47. B. W. Wessels, J. Electrochem Soc. *122*, 402 (1975).
48. A. R. Peaker, B. Hamilton, D. R. Wight, I. D. Blenkinsop, W. Harding and R. Gibb, Inst. of Phys. Conf. Series *33a*, 326 (1977).
49. B. Hamilton and A. R. Peaker, Solid St. Electron. *21*, 1513 (1978).
50. A. M. Stoneham, 'Report on Theoretical Study GMC/MAS/CD6500062 Killer Center Model in GaP' AERE Harwell, England 1977.
51. D. R. Wight, I. D. Blenkinsop, W. Harding and B. Hamilton, Phys. Rev. B *23*, 5495 (1981).
52. D. V. Lang and L. C. Kimerling, Appld. Phys. Lett. *28*, 248 (1976).
53. B. L. Smith and R. Newmann in 'Material and Device Technology of Gallium Phosphide' Ferranti Ltd. 1975 Annual Report on RP1-64 (DCVD Fulham, London SW6, 1TR).
54. R. Brunwin, B. Hamilton and A. R. Peaker in 'Non-Radiative Recombination in GaP', University of Manchester Institute of Science and Technology, 1978 Annual Report on RU 19-5 (DCVD Fulham, London SW6 1TR).
55. D. R. Wight, J. Phys. D. *10*, 431 (1977).
56. J. Nishizawa and Y. Okuno, IEEE Trans Electron. Devices. E022, 716 (1975) and Japan J. Appld. Phys. *19*, 25 (1979).
57. P. J. Dean, A. M. White, B. Hamilton, A. R. Peaker and R. M. Gibb, J. Phys. D. *10*, 2545 (1977).
58. S. Y. Chiang and G. L. Pearson, J. Appld. Phys. *46*, 7 (1975).
59. V. Kumar and L - Å Ledebo, J. Appld. Phys. *52*, 4866 (1981).
60. A. R. Peaker and H. G. Grimmeiss, to be published.
61. U. Kaufmann and J. Schneider, Solid State Commun. *25*, 113 (1978).
62. S. A. Abagyan, G. A. Ivanov and Yu N. Kuzentsov, Sov. Phys. Semicond. *10*, 1283 (1976).
63. A. R. Peaker, U. Kaufmann, Z. G. Wang, R. Wörner, B. Hamilton, and H. G. Grimmeiss, J. Phys. C., 17, 6161 (1984).
64. J. M. Baranowski, J. W. Allen and G. L. Pearson, Phys. Rev. *167*, 758 (1968).
65. H. Ennen and U. Kaufmann, J. Appld. Phys. *51*, 1615 (1980).
66. U. Kaufmann, W. H. Koshel, and J. Schneider, Phys. Rev. B *19*, 3343 (1979).
67. H. Ennen, U. Kaufmann, and J. Schneider, Appld. Phys. Lett. *38*, 355 (1981).
68. H. R. Szawelska, J. M. Noras, and J. W. Allen, J. Phys. C. (1981).
69. H. G. Grimmeiss and B. Scarstan, Phys. Rev. B. *23*, 1947 (1981).
70. A. R. Peaker and H. G. Grimmeiss, unpublished (1979).
71. S. G. Bishop, P. J. Dean, P. Porteus, and D. J. Robbins, J. Phys. C. *13*, 1331 (1980).
72. W. Hayes, J. E. Ryan, C. L. West, and P. J. Dean, J. Phys. C. *12*, L815 (1979).

CHAPTER 6

The Mid-Gap Donor Level EL2 in GaAs

G. M. Martin and S. Makram–Ebeid

Laboratoires d'Electronique et de Physique Appliquée
3, avenue Descartes, 94450, LIMEIL–BREVANNES (France)

1. INTRODUCTION
2. THE PROBLEM OF OXYGEN IN GaAs
    2.1 Experimental data reported in the 60's concerning the growth of bulk crystals
    2.2 Assumptions previously proposed to account for earlier data
    2.3 Recent experimental data
        2.3.1 Chemical analysis of oxygen
        2.3.2 Doping experiment in bulk materials
        2.3.3 Doping experiment during liquid phase epitaxy
        2.3.4 Doping experiment during vapor phase epitaxy
        2.3.5 Oxygen implantation
3. THE IDENTIFICATION PROBLEM
    3.1 Artificial creation of EL2
        3.1.1 Implantation and neutron irradiation experiments
        3.1.2 Electron irradiation experiments
        3.1.3 Stress induced effects
    3.2 Correlation with thermodynamical parameters
        3.2.1 Effects of stoechiometry
        3.2.2 Effects of dislocation induced stresses
        3.2.3 Effects of growth rate
        3.2.4 LPE and MBE materials

    3.2.5   Influence of free electron concentration
    3.2.6   Influence of Cr doping
  3.3 Correlation with surface equilibrium
    3.3.1   Influence of capping condition on the EL2 exodiffusion
    3.3.2   Influence of defects on the EL2 exodiffusion
  3.4 Is EL2 the arsenic antisite defect?
    3.4.1   Fast neutron irradiation experiments
    3.4.2   Uniaxial stress experiments
    3.4.3   Compared optical behaviours of EL2 and As_{Ga}
    3.4.4   Model proposals
4. ELECTRICAL AND OPTICAL CHARACTERIZATION OF EL2
  4.1 Electrical properties of EL2
    4.1.1   EL2 is a deep donor
    4.1.2   Thermal electron and hole emission and capture data
    4.1.3   Behaviour of EL2 under high electric fields and
            electron injection conditions
    4.1.4   Attempts to observe anisotropic behaviour of EL2
  4.2 Optical properties of EL2
    4.2.1   Photo-ionization and photo-absorption data
    4.2.2   Configuration-coordinate diagram
    4.2.3   The metastable state of EL2
    4.2.4   Photoluminescence bands attributed to EL2
  4.3 EL2 behaviour in GaAs related ternary alloys and in GaAs
      submitted to high pressure
    4.3.1   Constant temperature measurements (300 K)
    4.3.2   Determination of the free energy of ionization
            extrapolated at O K
    4.3.3   Discussion
    4.3.4   Comparison with theoretical predictions for various
            defects models
5. DISCUSSION ON THE STRUCTURE OF THE EL2 DEFECT
6. CONCLUSION

1. Introduction

The existence of a mid-gap deep level in GaAs was postulated as early as 1961 from assessment of high resistivity bulk materials. Further spectro-scopic experiments confirmed this assumption and gave accurate infor-mation about the electronic and optical properties of this level. It has been shown to have a marked influence on the properties of semi-insulating ma-terials and, as such, has aroused considerable interest, since these materials are used as substrates in the GaAs integrated circuits technology. But this level also presents some strange properties rarely observed for other deep levels in semiconductors, and the problem of its physico-chemical origin is a challenge which, together with the problem of understanding its observed properties, has excited the imagination of theorists and the inventiveness of experimentalists. This level is now generally referred to as EL2, a label which will be exclusively used later on, whatever label is used in all the studies which have been devoted to it.

Attention was first attracted to that level during the studies of Bridgman grown ingots which could be made to have high resistivity when oxygen was present in the growth ampoule. It was then simply supposed that this EL2 level is related to oxygen on Ga sites. Thus, during several years, the problem of EL2 was closely associated with the problem of oxygen in GaAs. In our Section 2, we will show how the clarification of that latter problem at the beginning of the 80's has allowed to reorientate research on the origin of EL2. All the recent observations concerning the control of concentration of that level during the growth of materials and further technological processes and the various hypotheses proposed for its origin will be reviewed in Section 3. Section 4 will be devoted to the detailed investigations of the unique opto-electronic properties of EL2. A comprehensive study of the properties of this level in its "normal" state will be given. The "strange" properties noticed for that level will also be presented, together with the phenomenological hypothesis of the existence of a metastable "excited" state for EL2 which can account for most of the experimental observations. Finally, a discussion of the possible origin of EL2 will be made in view of all the data gathered.

2. The Problem of Oxygen in GaAs

2.1 Experimental Data Reported in the 60's Concerning the Growth of Bulk Crystals

The effects of the presence of oxygen during the growth of bulk crystals were mentioned as early as 1956 by Willardson and Beer[1] and the observation that high resistivity materials can be obtained by adding oxygen was reported by Hilsum and Rose-innes in 1961.[2]

Actually, very high-resistivity and even semi-insulating materials were obtained later using the horizontal Bridgman (Bg) technique by introducing an oxygen partial pressure into the ampoule[3-5] or by adding Ga_2O_3 to the melt.[5-8] Different behaviors have been observed for the electron mobility μ_n as a function of the oxygen treatment. According to Woods and Ainslie[4] (see Figure 1), when the oxygen partial pressure is increased, μ_n states to increase, reaches a maximum, and then decreases, while the resistivity ρ increases continuously. Fertin et al,[5] however, only observed a decrease of mobility over a similar range of variation of oxygen partial pressure.

2.2 Assumptions Previously Proposed to Account for Earlier Data

Some conclusions were drawn, which were later considered as definitive, although rarely confirmed by chemical analysis:

— The presence of oxygen in the form of a Ga_2O gas, prevents decomposition of quartz walls and prevents Si contamination of GaAs.

— These high resistivity materials were mainly assessed by Hall effect measurements. The results show that the activation energy of the Hall constant was close to 0.75 eV, and it was thus proposed that a deep level could actually lie at this energy. Since it was a drop of the electron concentration which was recorded as a function of the oxy-

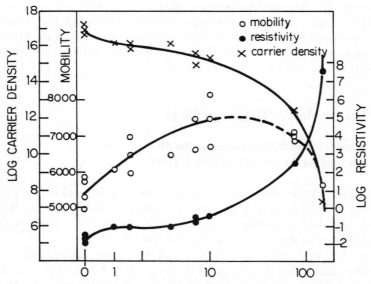

FIGURE 1 Electrical properties of GaAs grown by horizontal Bg technique, as a function of the pressure of oxygen initially added to the reaction tube. After Ref. 3.

gen treatment, it was first proposed that the deep mid-gap level is of acceptor nature,[9] compensating the shallow donors. But it was also suggested that it might well be a deep donor; the reduction of the free electron concentration could be explained by the selective chemical action of oxygen on shallow levels. This chemical action would leave us with an excess of shallow acceptors over shallow donors and the deep oxygen level would compensate the excess carrier thus leading to the observed high resistivity.[9,11] This hypothesis would have been in favor of an assignment of the level to O_{As} expected to be a donor.

— This mid-gap level was thought to be related to oxygen, but there was no experimental evidence, at that time, for the existence of the level itself and of course, no evidence for the increase of concentration of that level with the oxygen treatment either. It is only a few years later that the existence of that level was definitely established by spectro-scopic experiments such as thermally stimulated current (TSC) mea-surements in the early stage[12-14] and, later on, deep level transient spectroscopy (DLTS).[15,16] In the general classification of deep levels in GaAs, Martin et al[17] labelled this level EL2, a label which is now used widely.

Many other questions remained also unsolved, concerning, for instance, the concentration and the activity of oxygen in the solid crystal, the con-centration of the deep donor itself, the concentration and the thermal stability of Si-O complexes possibly incorporated during the growth.[6,18]

More generally, for some time, research focussed on the following four questions:

— Which is (are) the effect(s) of oxygen during Bridgman growth?
— Can GaAs be doped with oxygen?
— Are there any levels, in the band gap, related to oxygen, and, is the level EL2 related to oxygen?
— Which are the compensation phenomena taking place during the Bridgman growth in the presence of oxygen?

2.3 Recent Experimental Data

Definite progress has been achieved since new methods have been found for the quantitative determination of the concentration of oxygen, im-purities and deep levels. In the first paragraph, we will review the most confident techniques used for oxygen, while the results of attempts to introduce oxygen and to control the concentration of shallow and deep levels in bulk and epitaxial materials will be described later on.

2.3.1 Chemical Analysis of Oxygen

It is only rather recently that the problems related to the determination of oxygen concentration in GaAs have been clarified and, at least partly, solved. Early determinations of oxygen around 10^{19} cm^{-3},[4] as obtained from spark source mass spectroscopy (SSMS), have definitely been proved to be wrong, these measurements being extremely sensitive to pollution of the ion source chamber or to oxides on top of analyzed samples. When care is taken to avoid these pollution related problems,[10,11] SSMS analysis gives concentration of oxygen distinctly lower than 5×10^{16} cm^{-3} in bulk materials.[19-21] This type of measurement presents two main problems: i) the SSMS value is an average value over a large volume in which there may be oxide precipitates. Thus, the measured value may be overestimated, leading to the belief that the lowest values obtained in successive runs on the same material are the most reliable; ii) in order to provide accurate absolute values, this technique needs to be calibrated. This has been achieved using the data obtained on the same samples by another technique, the charge particle activation (CPA) technique.[22,23] The limit of detection of oxygen by SSMS has been estimated to be 2×10^{15} cm^{-3}.[21,24]

The CPA analysis is a general method which consists of generating nuclear reactions with all ions contained in a material and looking at the emission products (e$^-$, α and X rays) characteristic of a given reaction, i.e. of a given atom. In the case of the analysis of oxygen, the best way to produce the adequate nuclear reaction is to use tritons (nuclei of tritium atoms), as discussed by Debrun et al[22] and Valladon et al.[23] This technique has its own reference standard and thus provides absolute values. Pollution of surface with native oxide is eliminated, using etching after the oxygen activation: further formation of another native oxide is unimportant since the corresponding oxygen atoms are not activated and thus not taken into consideration during measurement. In the nuclear processes, the tritons penetrate over a few tenths of microns and thus the obtained concentration is an average over that thickness. The limit of detection is around 2×10^{15} cm^{-3}.[22,23]

Another nuclear reaction using ^{3}He as incident particles can also be used, but the experimental procedure of 0 analysis is by far more difficult. Emori et al[25] claimed that the detection limit can be as low as 6×10^{14} cm^{-3}.

The secondary ion mass spectroscopy (SIMS) analysis is specially dedicated to probe much thinner layers and to give profiles of concentration within the micron or sub-micron range. It has been calibrated using either the results obtained by CPA[26] or oxygen implantations.[27] When profiling oxygen in materials to be analyzed, one observes a sharp decrease immediately below the surface before reaching a plateau from which the oxygen concentration can be safely calculated. Oxide precipitates, when present, give rise to overshoots in the plateau region and can easily be detected.[26] Thus, the SIMS technique would be the most appropriate if the detection limit was low enough. In the experiments of Huber et al[26] the oxygen concentration was reported to be as low as 2×10^{15} cm^{-3}, while others give higher values around 1 to 2 10^{16} cm^{-3}, as reported by Kaminska, et al[28] for data recorded by the C. A. Evans's group.

2.3.2 Doping Experiments in Bulk Materials

The study of the influence of doping with Ga_2O_3 in the Bridgman growth has been resumed by two teams at M.I.T. (USA)[28] and LEP (France).[21,29] Both have concluded that the concentration of silicon in the melt and thus in the solid crystal is determined by chemical equilibriums which prevail at high temperature in the presence of Ga_2O gas. It has even been observed (see Figure 2) that the final value of Si in the solid depends only on that equilibrium and not on the intentional doping of GaAs with Si.[21] In that latter case, a gettering of Si purposely added into the melt has actually been observed.[21] The concentration of other donors (S, Se, Te), however, remains constant during that process.[28]

Both groups noticed that the concentration of oxygen, as determined by SIMS[28] or by CPA[21] and SSMS,[21] is low, i.e. lower than a few 10^{16} cm^{-3}

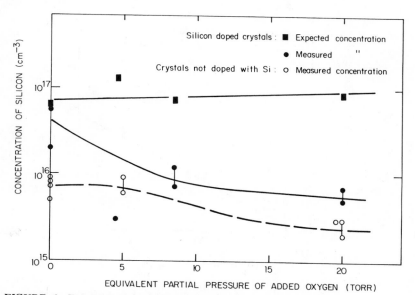

FIGURE 2 Expected and measured concentrations of silicon in various ingots grown under different Ga_2O_3 treatments. After Ref. 21.

in crystals, whatever the Ga_2O_3 addition used. A systematic study of oxygen by CPA in a quite large number of ingots grown using various amounts of Ga_2O_3 has been made by Hallais et al.[29] The results are shown in Figure 3. Most of the values are close to 10^{16} cm^{-3}. These authors think that oxide precipitates (for instance Cr oxides) could be responsible for the scattering of data and that the concentration of isolated oxygen atoms corresponds to the minimum values, distinctly below 10^{16} cm^{-3}. These minimum values are observed to increase with the Ga_2O_3 addition (see the dotted line). This is the only indication that GaAs could be "doped" with oxygen, which is not definitely proved. Emori et al[25] also concluded from a rather large number of results that intentional oxygen doping into Bg materials is quite inefficient.

Very low concentrations of oxygen (down to 2×10^{15} cm^{-3}) were also measured by SIMS in Bridgman crystals and it was noticed that the concentration of oxygen was distinctly lower than that of EL2, established to

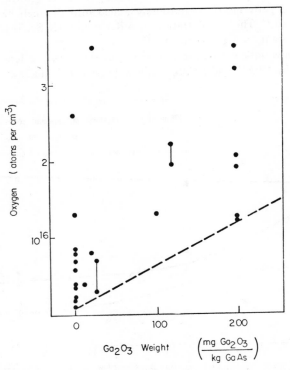

FIGURE 3 Oxygen concentration, as deduced from CPA analysis, in various ingots grown under different Ga_2O_3 treatments. After Ref. 29.

be around 1.5×10^{16} cm^{-3}.[26] It was thus concluded that this is a direct
proof that EL2 is not related to oxygen at all.[26] It was further demonstrated that the concentration of EL2, deduced from DLTS in conducting
materials and from optical absorption measurements in conducting or
semi-insulating materials, does not vary at all in crystals grown with different amounts of Ga_2O_3 added to the melt, i.e. under different partial
pressure of oxygen (see Figure 4).[21] This is quite consistent with the previous conclusion. This is also in agreement with studies made on low resistivity Bg materials which attributed the increase of EL2 with Ga_2O_3
addition to the oxygen-related decrease of Si donors, but not to the incorporation of oxygen itself.[28]

Another important conclusion has been drawn from the comparison of
the chemical concentration of Si (deduced from SSMS measurements) and
of that of Si atoms being on Ga site, in oxygen treated materials, as seen in
Table 1. Both being practically identical it has been concluded that the
Si atoms, remaining in the melt at equilibrium with Ga_2O gas, are incorporated as shallow donors and that Si-O complexes, if they exist, are in
negligible concentration, i.e. below 10^{15} cm^{-3}, in those materials.[21]

In summary, the only clearly established effect of oxygen or Ga_2O_3
partial pressure during Bridgman growth is to reduce the concentration of

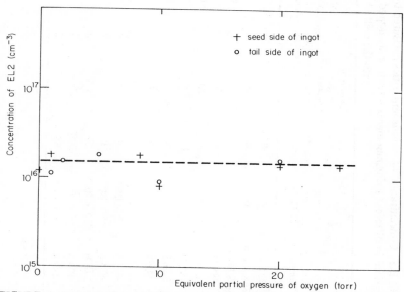

FIGURE 4 Concentration of EL2, as deduced from DLTS or optical absorption
measurements, in the seed side and tail side of ingots grown under different Ga_2O_3
treatments. After Ref. 21.

TABLE 1 Concentration of various impurities measured by S.S.M.S., C.P.A., and L.V.M. (Local Vibrational Mode Absorption) in different Bridgman ingots intentionally doped or not with Si, and in which Ga_2O_3 was or was not added into the melt. After Ref. 21.

Label of sample	Doping levels Si/GaAs	Ga_2O_3 (mg)	Equivalent oxygen partial pressure (torr)	Silicon Ex-pected	Silicon SSMS (10^{16} cm^{-3})	Oxygen LVM	Oxygen C.P.A. (cm^{-3})	Oxygen SSMS (cm^{-3})	Te SSMS (cm^{-3})	S SSMS (cm^{-3})
221 M	0	0	0		0.7				$\leqslant 5 \times 10^{15}$	$\leqslant 2 \times 10^{15}$
366 S Top	0	0	0		0.9	0.8	3.8×10^{16}		$\leqslant 5 \times 10^{15}$	2×10^{15}
366 S Bottom	0	0	0		0.5				$\leqslant 5 \times 10^{15}$	$\leqslant 1 \times 10^{15}$
399 M	0	0	0		0.5		4.4×10^{15}	$\leqslant 2 \times 10^{15}$	$\leqslant 4 \times 10^{15}$	$< 10^{15}$
369 S	0	200	20		0.3		1.2×10^{16}	$\leqslant 2 \times 10^{15}$	$\leqslant 4 \times 10^{15}$	$< 10^{15}$
437 M	0	200	20		0.2			$\leqslant 2 \times 10^{15}$	$\leqslant 5 \times 10^{15}$	$\leqslant 1 \times 10^{15}$
323 T	0	50	5		0.6	0.3	2×10^{16}	$\leqslant 2 \times 10^{15}$	$\leqslant 5 \times 10^{15}$	$\leqslant 1 \times 10^{15}$
319 S	0	50	5			0.9				
427 S	5×10^{-6}	0	0	6.5	6	5	$\leqslant 10^{15}$	$\leqslant 2 \times 10^{15}$	$\leqslant 5 \times 10^{15}$	1×10^{15}
473 S	5×10^{-6}	0	0	6.5	1.7	2	2×10^{15}	$\leqslant 2 \times 10^{15}$	$\leqslant 5 \times 10^{15}$	2×10^{15}
346 M	10×10^{-6}	20	4.5	13.0	0.3		2×10^{15}	$\leqslant 2 \times 10^{15}$		
404 S	7×10^{-6}	200	20.0	8.4	0.7	0.5	2.1×10^{16}	$\leqslant 2 \times 10^{15}$	$\leqslant 5 \times 10^{15}$	$\leqslant 1 \times 10^{15}$
487 S	6×10^{-6}	85	8.5	7.2	0.7	1.2		$\leqslant 2 \times 10^{15}$	$\leqslant 5 \times 10^{15}$	2×10^{15}

silicon incorporated in the crystal, while the concentration of EL2 remains unchanged. Since the concentration of other donors like Se or Te is extremely small, lower than 10^{15} cm^{-3} in the studied crystals,[21] it seems reasonable to trust a scheme of compensation where the deep donor EL2 is compensating a smaller concentration of shallow levels which are mainly of acceptor nature on account of the drop in concentration of shallow donors.[21,30] The same scheme is admitted for liquid encapsulated Czochralski (LEC) materials.[31] In that case, the concentration of Si in ingots have been observed to be directly related to the humidity of the boric oxide encapsulant.[32] When appropriate conditions are fulfilled, the total concentration of the main shallow donor impurities (Si, S, Se, Te) has been measured to be as low as 1.4×10^{15} cm^{-3},[33] while EL2 stands slightly above 10^{16} cm^{-3}. The background p-type concentration is, in that case, believed to be due to C.[34]

2.3.3 Doping Experiments During Liquid Phase Epitaxy

EL2 has never been mentioned to be present in thick LPE layers,[17,36] even if Ga_2O_3 was added in the growth melt. This question of the absence of EL2 in these layers will be discussed later on (see Section 3.2.4).

2.3.4 Doping Experiments During Vapor Phase Epitaxy

Introduction of oxygen during vapor phase epitaxy also leads to a decrease of the free carrier density,[35,37,39] but it seems difficult to get high-resistivity layers reproducibly. Again, Wolfe et al[35] noticed a decrease of the Si line in the far infrared photo-conductivity spectrum when Ga_2O_3 is added in the Ga source melt. These authors considered that observation as a proof that the Si concentration was decreased and that oxygen does not produce any shallow level.[35] The electron mobility in layers grown with oxygen has been reported to be distinctly lower than its usual value, and the presence of acceptor complexes related to 0 were assumed to account for that effect.[37] One may wonder whether local inhomogeneities of the resistivity of the layers are not at least partly responsible for that effect, as it has been suggested for bulk materials.[30]

As far as the level EL2 is concerned, Wallis et al[39] did not see any correlation between its concentration, or the concentration of any other deep levels, and the injection of oxygen.

2.3.5 Oxygen Implantation

The implantation of an ion is, at first glance, the surest way to introduce impurities in materials. But great care must be taken to separate the effects of defects created along the path of the implanted ion from those of the ion itself. One of the ways to do so is to compare the effects of an ion with that of others having comparable masses. In the case of GaAs, data are available for oxygen, but also for boron and neon in particular. Furthermore, a thermal treatment is necessary to cure the defects and to "activate" the implanted ions, i.e. to have them placed on the electrically active lattice sites. During the annealing, the implanted ions may also diffuse, which makes the study still more complex. It is thus necessary to study the variations of their concentration profile as a function of temperature.

In that paragraph, we will concentrate on the effects which have been clearly attributed to the oxygen itself. It was reported by Favennec et al,[40] several years ago, that GaAs layers implanted with a sufficiently large oxygen dose remain highly resistive even after annealing at temperatures as high as 900°C. Many studies[41-43] have been devoted so far to that effect often considered as especially interesting for the technology of GaAs integrated circuits. It has been recently discovered[43] that the carrier removal rate, measured after 870°C annealing, may be quite different from one material to another, as it can be noticed in the literature.[40-43] In particular, it has been measured to be about two orders of magnitude larger in silicon-doped materials than in those doped with selenium.[43] This is true if oxygen is implanted in n-type materials purposely doped with one of these two donors[44] (see Figure 5), or if it is co-implanted with one of them in a semi-insulating substrate.[43] This behavior is quite different from that of boron,[45] which is also used to insulate n-type active regions of integrated circuits.[46] Boron implantation also leads to considerable carrier removal, but almost complete free carrier recovery is observed after annealing at 550°C if the boron dose is lower or equal to 10^{12} at.cm^{-2}. After annealing at 870°C, compensation still remains noticeable only for very high doses of the order of 10^{14} at.cm^{-2} (see Figure 5), whatever the donor species (silicon or tellurium) in the implanted material.[45]

A detailed investigation[44] of deep levels observed in the oxygen-implanted layers further annealed at 870°C under Si_3N_4 capping has shown that at least one, if not two, deep levels are generated in originally selenium-doped materials. These levels, labelled Se-O and X, are not observed in silicon-doped materials, as seen in Figure 6. Using cross-correlations from results obtained for implantation of various species in both types of starting materials, it has been definitely established that, at least, one of them is associated with both oxygen and selenium. The concentration of this Se-O level linearly varies with the implantation dose while that of the level X

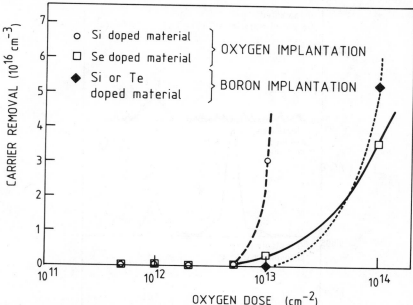

FIGURE 5   Plot of the amount of carrier removal deduced from C(V) measurements, as a function of the oxygen dose implanted in a Si– and a Se-doped material and as a function of the boron dose implanted in an Si– and a Te-doped materials. The n type materials have been implanted with oxygen at 100 keV or with boron at 380 keV and further annealed at 870°C under Si_3N_4 for 20 min. After Ref. 44.

seems to vary quadratically, as shown in Figure 7. One can notice that the amount of carrier removal noticed for larger oxygen doses, around 10^{13} at cm^{-2} (see Figure 5), is similar to the concentration of these deep levels extrapolated for the corresponding dose. It is thus very likely that the drop of the free electron concentration is due to the formation of complexes between oxygen and selenium, killing the activity of selenium as a shallow donor. The concentration of these complexes is about 100 times lower than the concentration of oxygen.[44] It is reasonable to think that same phenomena of formation of complexes take place in silicon-doped layers, between 0 and Si, even though the corresponding level(s) has (have) not yet been identified. Since the carrier removal rate observed in that case is about 100 times larger, one may assume that the formation of Si-0 complexes is more probable and that most of the oxygen atoms are then complexed with donors. The driving force for such an association is not clear. It cannot be the coulombic interaction if oxygen mainly behaves as a donor, as would be expected in GaAs, but has never been proved. Does it

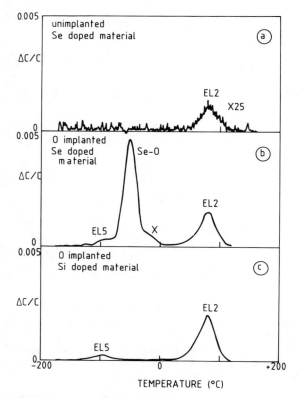

FIGURE 6 DLTS spectra recorded on (a) unimplanted and (b, c) oxygen-implanted materials (b is Se-doped; c is Si-doped) after annealing at 870°C under Si_3N_4 for 20 min. Rate window = 7.5 s^{-1}; thickness of the investigated layer = 3000 Å. After Ref. 44.

mean that oxygen is never isolated in the lattice, but always associated with a defect leading to an acceptor like complex? One may suggest a $O_{As}-V_{Ga}$ complex, a type of complex said to be acceptor by Hurle[47]

It is clear from Figure 7 that the concentration of the EL2 level remains practically constant within the implanted region whatever the oxygen dose. Again, this observation does not support the assignment of EL2 to oxygen.

For allowing the formation of these complexes between oxygen and donors, one of them must diffuse, even slightly. Donors, such as Si and Se, do not diffuse fast, as shown by various SIMS analysis.[48,49] The behavior of oxygen, again, seems rather complicated as studied by Favennec et al[50] and Deveaud et al.[51] The SIMS profile of oxygen implanted atoms is

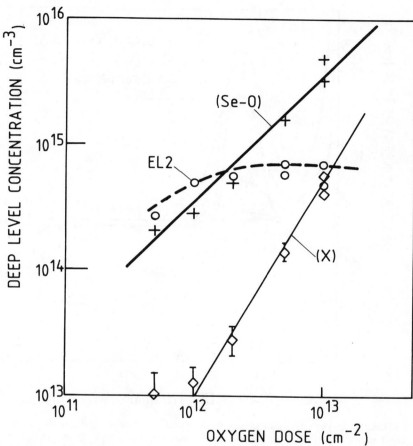

FIGURE 7 Concentration of the three deep levels (EL2), (Se-O) and (X) as a function of the oxygen dose implanted in the Se-doped material, further annealed at 870°C. The value of the concentration is averaged over the first 3000 Å below the surface. After Ref. 44.

shown in Figure 8 before and after annealing, at 900°C. When the peak concentration is larger than 10^{18} cm^{-3}, annealing does not influence too much the shape of profile which does not widen, but, on the contrary, is more restricted. When oxygen concentration is lower than 10^{18} cm^{-3}, the atoms seem to diffuse far since they are not detected any more after annealing. In the case of formation of oxygen-donor complexes studied above by DLTS measurements, the concentration of oxygen remained in the range of concentration where diffusion, and then impurity association, seem easy.

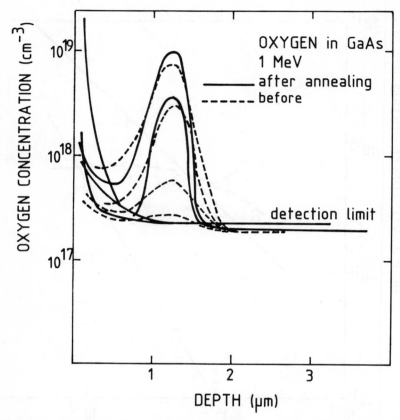

FIGURE 8   Oxygen profiles recorded by SIMS before (dotted lines) and after (solid lines) annealing at 900°C for 20 min. Implantation energy: 1 MeV. Doses: 10^{13}, 3×10^{13}, 10^{14} and 3×10^{14} at cm^{-2}. After Ref. 50.

One may also mention that impressive effects of oxygen-implantation-enhanced moving of Cr atoms have been reported.[51] According to these studies, pile-up of Cr atoms has been observed to take place in layers implanted with large doses of oxygen. This effect has also been observed in regions implanted with other ions[52] and was assumed to be due to the large concentration of defects. But it seems more efficient in the case of oxygen implantation.

In summary, the effect of oxygen implantation in n-type GaAs layers is to strongly increase their resistivity, which remains stable during annealing at temperature as high as 900°C. The carrier removal, observed after this thermal treatment, does not seem to be due to the generation of compensating acceptors, but is very likely due to the formation of chemical complexes between oxygen and donors. Depending on the probability of complexing of oxygen with the different donors, at that temperature, various carrier removal rates are obtained in materials preferentially doped with one of them. This effect is thus entirely different from the oxygen-related gettering of Si, described in previous paragraphs and which takes place during Bg, LPE or VPE growth and also allows to obtain high resistivity materials. As far as the level EL2 is concerned, its concentration has not been observed to vary appreciably with the oxygen implanted dose either.

In conclusion to this section, it turns out that the two problems of oxygen and EL2 in GaAs are clearly separated.

As far as oxygen is concerned, two *different* effects are observed to occur either during growth of materials or during annealing of oxygen-implanted layers. Both lead to rather high resistivity materials. Introducing oxygen during growth of bulk or epitaxial materials is extremely difficult since the concentration of isolated atoms has never been proved to exceed 10^{16} cm^{-3}. When it is present in the lattice, either in as grown materials or implanted ones, it has never been observed to behave as a shallow or even a deep donor and some reported results would lead to suggest that it prefers to stand complexed with defects as a V_{Ga} or acceptor like impurities, this complex behaving as an acceptor.

3. The Identification Problem

Once it became clear that EL2 is not related to oxygen, it took quite some time for evidence to accumulate that it mainly corresponds to a lattice defect. The most definite conclusions in that sense were drawn from recent experiments in which EL2 has been shown to be artificially created during a defect-generation process. These data will be presented first. Then, observations will be reviewed which relate the concentration of EL2 to various thermodynamical growth parameters. In a subsequent section, the thermodynamical equilibrium concentration of that level will be shown to be strongly influenced by surface conditions and defect concentrations nearby. All these observations provide valuable information, but the detailed nature of EL2 still remains unknown. There exist strong advocates of the notion that EL2 is associated with an As antisite defect: this assignment will be critically discussed in a last paragraph.

3.1 Artificial Creation of EL2

3.1.1 Implantation and Neutron Irradiation Experiments

Creation of a level, which was probably EL2, was reported, probably for the first time, by JERVIS et al,[53] in LPE materials implanted with a low dose of Si through a Si_3N_4 capping and annealed at 870°C. This experiment was resumed by another group[54] which checked that, actually the mid-gap level, created in very small concentration in that condition, presented at low temperature the photocapacitance characteristic of EL2 (see in Section 4).

A more significant step towards the controlled generation of EL2 has been achieved with the experiments carried out by Ikoma et al[55] and Taniguchi et al.[56,57] They observed that a deep level with electron emission rate properties similar to EL2 could be detected in noticeable concentration in LPE layers implanted with oxygen and further annealed at 600°C.[55] It would have been tempting to relate it, and EL2, to oxygen if similar effects had not been observed for other implanted ions such as As or Ga.[56,57] And, by the way, some slight differences, reported by these authors to occur in oxygen implanted materials compared to samples implanted with other ions, encouraged them to propose that there might exist two types of EL2 levels, one associated to oxygen and the other not.[56] But it still remained to prove that the created level was actually EL2.

The corresponding experiments were carried out on bulk and LPE materials.[58] Oxygen, boron and neon atoms were implanted at the same dose and the same projected range in both types of materials, and the samples annealed at 600°C. The same results were obtained whatever the implanted ion, except that the concentration of detected deep levels crudely varied as the masses of ions used. Typical DLTS spectra recorded on bulk and LPE samples is presented in Figure 9. (Prior to ion implantation, the bulk material was annealed at 870°C under Si_3N_4 to deplete the surface region of the initially present EL2 levels). Two deep levels are present in the implanted region, which look like EL2 and EL3 from their emission rate properties.[17] Profile of the concentration of the mid-gap level is presented in Figure 10, in the Ne implanted sample and in the reference one. A very large increase of concentration of that level is noticed in the implanted region. Figure 11 shows the overshoot capacitance transient under 1.06 μm illumination, which is characteristic of the transfer of EL2 from its normal state to its excited metastable state. This effect is clearly observable for the mid-gap level created in bulk material and its amplitude strictly correlated with the concentration of the level deduced from DLTS

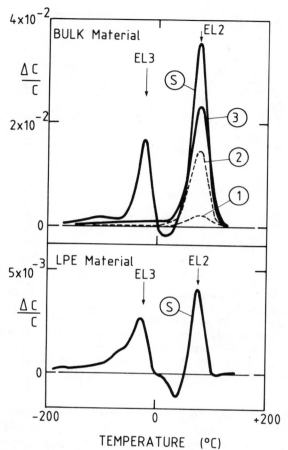

FIGURE 9 DLTS spectra recorded on bulk and LPE material implanted with neon
(125 keV, 10^{12} at.cm^{-2}) and annealed at 600°C for 15 minutes. Each spectrum corresponds to a given investigated region below the surface 0–0.27 μm for the S curves,
0.54–0.77 μm for curve 1, 1.4–1.63 μm for curve 2, 4.2–4.43 μm for curve 3. Rate
window = 7.58 s^{-1}. The arrow indicates the expected position of the level EL2 in our
experimental conditions, according to Ref. 17 (apparent activation energy = 0.82 eV,
apparent capture cross section = 1.7×10^{-13} cm^{-2}). After Ref. 58.

spectra (see Figure 10). One can definitely conclude that the defect
created under these conditions is EL2 and that it is not due to oxygen,
since the same effect occurs using neon or boron.

The overshoot capacitance transient under optical excitation was not
recorded on the LPE sample (see Figure 11). Does this correspond to a

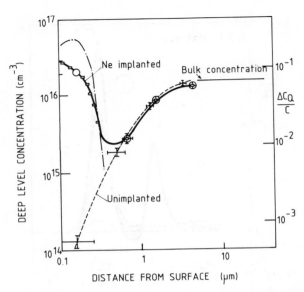

FIGURE 10 Profile of concentration of the mid-gap level in the unimplanted bulk sample (dotted curve) and in the bulk sample implanted with neon (125 keV, 10^{12} at.cm^{-2}) (full curve), both annealed at 600°C for 15 minutes. The crosses correspond to the concentrations deduced from the DLTS peak. On the other hand, the ratio $\Delta C_Q/C$, which scales the amplitude of the photoquenching effect (see text and Figure 11), is given by the open circles. The dash-dotted curve is the theoretical LSS profile of the implanted neon atoms. After Ref. 58.

fundamental effect possibly related to different concentrations of native defects in LPE, compared to bulk materials? These observations, or at least the observation of the absence of manifestation of the metastable state in Figure 11, seem better explained by the occurrence of too large a reverse current in the studied LPE diode which prevents the electron emptying of EL2, and thus prevents the formation of the metastable state, as studied by Makram-ebeid[59] (see in Section 4).

Subsequent to these experiments, we have studied the electric field and temperature dependence of the electron emission rate (e_n) related to the deep level created by the above implantation and annealing procedure. The value of e_n follows the same temperature and field dependence within the measurement accuracy for bulk and LPE materials. Quite recently, T. Ikoma (Tokyo University) has privately reported to the authors the occurrence of the EL2 overshoot behavior in LPE materials prepared as explained above, although the optical time constants seem to differ from those of EL2[60] (see Figure 12). But, according to the author's experience and to observations also reported in the literature,[61] the optical time con-

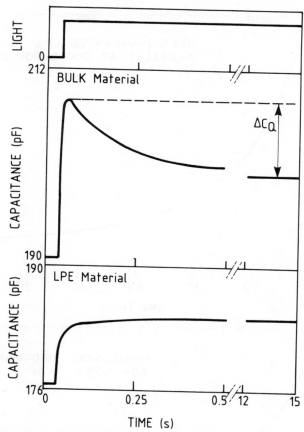

FIGURE 11 Transients of capacitance recorded at 90 K under illumination with a 1.06 μm, 100 mW laser beam, after having filled the electron traps with a long forward bias pulse. The bulk and LPE materials were both implanted with Ne (125 keV, 10^{12} at.cm^{-2}) and annealed at 600°C for 15 minutes. Investigated region below the surface: 0–0.28 μm for the bulk sample, 0–0.33 μm for the LPE sample. After Ref. 58.

stant has also been seen to vary from one sample to another in materials containing native EL2.

This is not the first time that differences in the optical properties are reported for electron levels which are electrically very close to EL2. Taniguchi and Ikoma[56,57] have reported differences in the photoioniza- tion threshold for mid-gap levels created by different methods (0-Implan- tations, Horizontal Bridgman growth or MOCVD). They also report differences in the Arrhenius activation energies of the electron emission rate. In fact, these results are interesting but difficult to interpret. We may

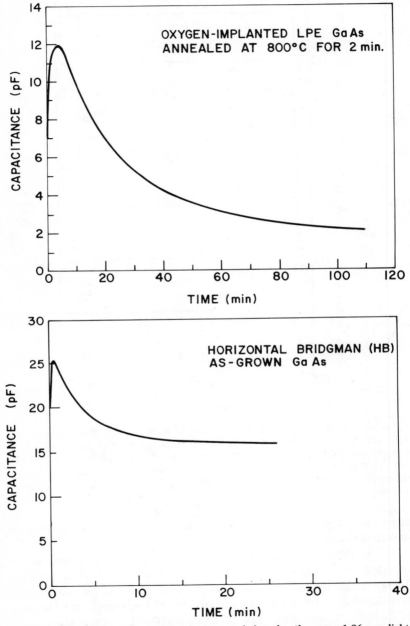

FIGURE 12 Photocapacitance transients recorded under the same 1.06 μm light power on an LPE material implanted with oxygen and further annealed at 800°C for 2 mn (figure a) and on a Bridgman bulk material (Figure b). After Ref. 60.

suggest that these types of discrepancy can be attributed to long-range interactions of the EL2 center with other defects, as for instance acceptors (on a 20–100 Å scale, say) as those which take place in GaP.[62,63] The type of interaction involved may be of electrostatic nature (Coulomb energy) or it may involve tunnelling or other effects.

At the present time, it seems established that the same mid-gap level is created in the described process, whatever the implanted ion and the implanted materials, and it has been definitely shown that it is EL2. From these data, it is clear that EL2 is a defect-associated level.

It remains to study its kinetics of formation. Some preliminary results have also been reported to Taniguchi and Ikoma[56,57] who mentioned that, accordingly to DLTS, the EL2 peak was not detected in oxygen-implanted layers further annealed at only 550°C instead of 600°C. But it was not mentioned in that case whether the material was then n-type or still highly resistive due to the presence of too many unannealed defects created by the large dose (10^{13} at cm^{-2}) used. This last hypothesis is most likely, according to implantation experiments carried out with boron,[45] which means that DLTS capacitance measurements cannot give valuable information for deep levels in this highly resistive region. Another more recent attempt has been made using neutron irradiation. When a neutron "impinges" a host atom, it ejects it from its site and everything happens as if we had implanted this atom in the surrounding matrix. Thus neutron irradiation should give similar results compared to ion implantation. And actually, DLTS spectra recorded on both type of damaged materials are quite similar,[64] the only difference being in defect creation rate due to quite different interaction cross-section of ions and neutrons respectively. Details on defects already reported to be observed in n-type materials (n = 5 X 10^{16} cm^{-3}) irradiated with a relatively low dose (10^{15} n cm^{-2}) are reported in Section 3.4.1. In a more recent experiment,[65] an n-type material has been chosen with a higher free carrier concentration (n = 1.5 X 10^{17} cm^{-3}) in order to be still conductive (n $\sim 10^{16}$ cm^{-3}) after an irradiation with a higher dose (3 X 10^{15} n cm^{-2}). The results,[65] giving the concentration of EL2 (deduced from DLTS in this still n-type material) as a function of the annealing temperature after irradiation, are reported in Figure 13. Due to the higher dose used and the larger free carrier concentration in the starting materials, the sensitivity of the experiment for the detection of EL2 has been increased and it is distinctly observed that an increase of the EL2 concentration occurs between 400 and 500°C, in a region where most of other defects (corresponding to the band U, as mentioned in 3.4.1) are annealed (Figure 13). According to these data, the creation rate of EL2 in as-irradiated material is 0.2 cm^{-1}, while it is around 1 cm^{-1} when these irradiated samples are further annealed at 600°C.[65]

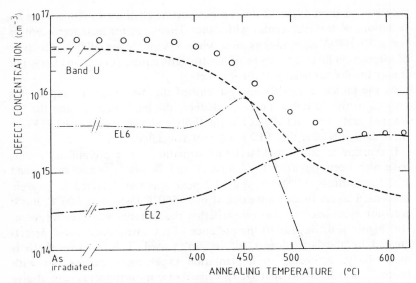

FIGURE 13 Variation of the concentration of the EL2 level and of the band U created by irradiation with a neutron dose of 3×10^{15} cm^{-2}, as a function of the annealing temperature (annealing time = 15 mn; doping level of the starting un-irradiated material = 1.5×10^{17} cm^{-3}). The material used is a bulk material previously annealed at 870°C in order to deplete the surface region from EL2 traps. After Ref. 65.

3.2.2 Electron Irradiation Experiments

Creation of EL2 has never been reported to occur during electron irradiation at room temperature and low doses,[66] in spite of the fact that a lot of different defects have been observed to be created during that process. Most of them have been shown to be simple primary defects[67] and attributed to displacement of the As sublattice (PONS et al[68]), and they were observed by many authors[69] to anneal at around 220°C. It is worth noticing that none of the detected defects could be related to a displacement of the Ga sublattice.[68] In a recent experiment, Stievenard et al[70] have repeated the same electron irradiation experiments but with the sample heated to 300°C during the irradiation in order to favor diffusion and then separation of each of the defects forming a Frenckel pair, with respect to its local recombination. Operating this way, they have observed the creation of two deep levels, one of them, present in the largest concentration, looking like EL2 from its DLTS signature. It remains to check whether this defect exhibits the metastable-state property of the EL2 level.

These observations allow to conclude that EL2 is not a simple primary defect, and it is interesting to note that, if it is EL2 which is actually created during 300°C irradiation, this is by preventing recombination of Frenkel pair defects generated in the As sublattice.

3.1.3 Stress-induced Effects

The first data recorded on plastically deformed GaAs under uniaxial stress were published by Ishida et al.[71] They show that not only EL2, but other deep levels like EL3 and HL8 have their concentration increased during that process (see Figure 14). More recently, Wosinski et al[72] repeated these experiments and also found that EL2 is increased in concentration while the concentration of other levels seem to be also enhanced and a level created in noticable concentration (very probably EL6). In the results of Ishida and Wosinski, the samples were heated at 300°C and 400°C respectively, in such a way that, when defects are generated by stress,

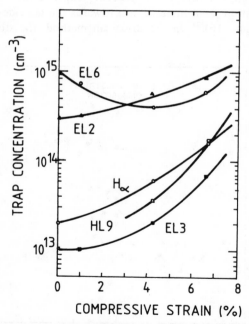

FIGURE 14 Trap concentration of three electron traps (E_α, E_β and E_δ, which are thought to correspond to EL6, EL3 and EL2 respectively) and two hole traps (H_α and H_β, H_β being expected to be HL9), as a function of compressive strain. After Ref. 72.

they have a better chance to diffuse instead of recombining as mentioned above. That is why simple primary defects, as those created by electron irradiation, are not observed in these experiments since they anneal at 220°C.

3.2 Correlation With Thermodynamical Parameters

3.2.1 Effects of Stoichiometry

The correlation between EL2 and stoichiometry has been clearly demonstrated by controlling the growth conditions of bulk and vapor phase epitaxial materials. The clearest data have been recorded by HOLMES et al[31,34,73] who noticed that the concentration of EL2, as deduced from optical absorption measurements, strongly decreases as soon as the As to Ga ratio in the starting melt is lower than 1 (see Figure 15). As a matter of fact, when the starting melt is poor in As, it will be poorer and poorer as a function of growth time which probably explains the decrease of the EL2 concentration from the seed side towards the tail side in some LEC ingots (see Figure 16).[74] In the Bridgman method, the stoichiometry is

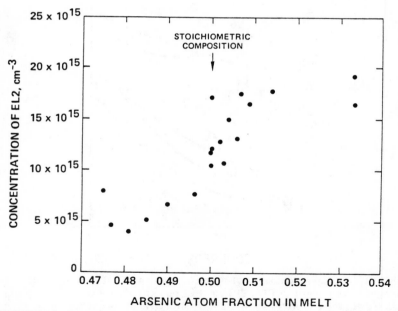

FIGURE 15 Concentration of EL2, as deduced from optical absorption measurements, versus the estimated arsenic atom fraction in the LEC melt. After Ref. 34.

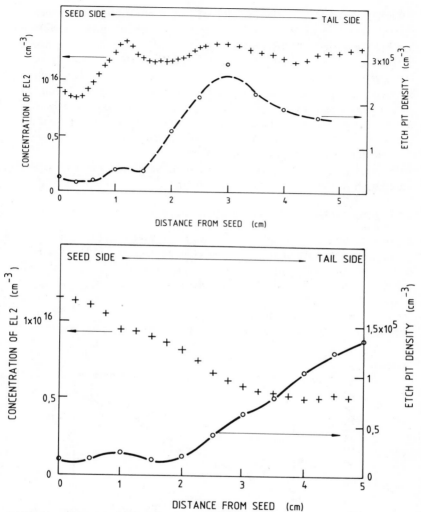

FIGURE 16 Variation of the concentration of EL2 (crosses) and of EPD (circles) from the seed side to the tail side of two different LEC ingots. After Ref. 74.

expected to be perfect due to the control of the As overpressure. This is believed to explain the high concentration ($\sim 1.5 \times 10^{16}$ cm^{-3}) of EL2 (according to Figure 15) and its homogeneity in the large majority of Bridgman ingots, as reported in Figure 4. Lagowski et al[78] also report that the EL2 concentration increases with the As pressure during Bridgman growth.

Similar observation of the decrease of EL2 concentration when decreasing the As/Ga ratio during metal-organic vapor phase epitaxy (MOVPE) has been reported by different groups,[75-77,79-81] as measured by DLTS. It has been observed that EL 2 varies as As/Ga)$^\nu$ (see Figure 17), but ν was given to be equal to 1[76], 0.5,[80] or 0.25[81] The value of 0.5 given by Samuelson et al[76] may be the most reliable since they kept constant all the growth parameters except the As mole fraction and were aware of the correction factors to be used to determine the accurate EL2 concentration from DLTS spectra. All these authors said that their results would be consistent with a V_{Ga} associated defect.

From all these data, one can be sure that the presence of EL2 is associated with As rich conditions during growth.

3.2.2 Effect of Dislocation-Induced Stresses

Figure 18 shows the variations of concentration of EL2 (as measured by optical absorption) and those of the etch pit density (EPD) in a cross-section of two different LEC ingots.[82] Both variations seem spatially correlated on a millimeter scale. They have also been observed to be correlated with the variation of resistivity of the material[83] and with the intensity of the 0.64 eV luminescence band attributed to EL2.[84] But there does not exist any direct correlation between absolute concentrations of

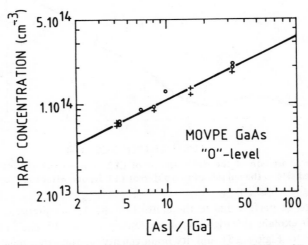

FIGURE 17 Concentration of the EL2 level as a function of the As/Ga ratio with P_{TMG} kept constant. The straight line in the figure has the slope 1/2, corresponding to $N_T \propto (P_{DASH3})^{1/2}$. The two symbols used in the figure correspond to different series of runs where different substrate materials have been used. After Ref. 80.

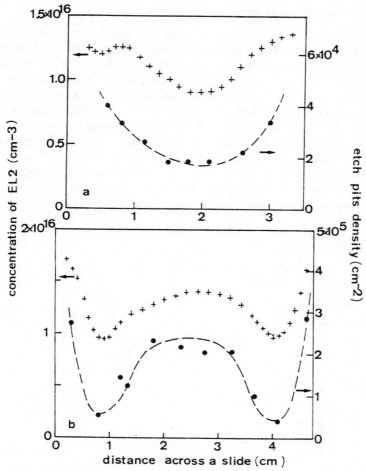

FIGURE 18 Variation of the concentration of EL2 (crosses) and of EPD (circles) on a cross section of two different LEC ingots. After Ref. 82.

EL2 and EPD, as can be already seen in the different scales of Figure 18a and b, and as reported elsewhere.[31] As a matter of fact, EL2 has been seen to be still present in a dislocation-free, high-resistivity material obtained by an LEC procedure in our laboratory.[82] But the correlation observed in Figure 18 has been assumed to result from some secondary effect as, for instance, the presence of stress in some regions of the material evidenced by the generation of dislocation therein. This would be in agreement with the observations described above in paragraph 3.1.3.

3.2.3 Effect of Growth Rate

Oseki et al[85] have published a comprehensive study of the variation of the electron concentration and of the concentration of EL2 with various growth parameters such as the $AsCl_3$ mole fraction and the growth rate. These data, obtained for growth temperatures of 710°C and 750°C in an N_2 system, are shown in Figure 19. Data are also available from Humbert et al who worked with a H_2 system[86] and at growth temperatures of 750 and 770°C. It is clear in two different systems and for different growth temperatures that EL2 is closely related to the *growth rate* and *increases with it* (see Figure 19). From these data, Oseki et al[85] already concluded that EL2 could not be a simple substitutional impurity. If so, its concentration should increase with decreasing growth rate, or with increasing growth temperature or with decreasing $AsCl_3$ mole fraction,

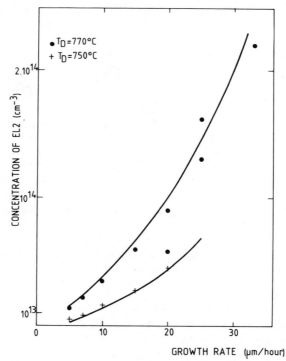

FIGURE 19 Variation of the concentration of EL2 as a function of the growth rate, for two different growth temperatures. The variation of the growth rate was obtained by varying the $AsCl_3$ mole fraction between 4×10^{-4} to almost 10^{-2}. These data were obtained in an N_2 flow growth system. After Ref. 86.

but they observed a completely opposite behavior.[85] Nevertheless, it is not yet clear whether this effects results from dynamical equilibriums controlling the formation of defects during growth or from the partial exodiffusion of EL2 at the growth temperature (see Section 3.3). In that latter case, EL2 would be large when the growth rate is high, since EL2 would not have enough time to exodiffuse. Anyway, great care must be taken to study the effects of other epitaxial growth parameters since all these results show that the study should then be made for a constant growth rate. Unfortunately, its value is rarely published in the corresponding literature.

One may remark that the EL2 concentration is largest in bulk material which corresponds to a very large growth rate (~ 1 cm per hour) while it has never been observed in MBE for which the growth rate is around 1 μm per hour. Of course, a direct comparison is impossible since the corresponding growth temperatures are extremely different.

3.2.4 LPE and MBE Materials

EL2 is usually not detected in LPE[17,36,87] and MBE[17,88,89] materials. But Day et al[90] and Xin et al[91] have observed that it can be introduced in very small concentration in MBE layers after annealing at 700°C under Si_3N_4 capping. It was not clarified whether this effect was due to the Si_3N_4-capping-induced stresses or to a diffusion of EL2 from the substrate, as noticed at the substrate-LPE layer interface by Mitonneau et al.[92] Such a diffusion from the substrate is probably responsible for the sometimes reported detection of EL2 in thin LPE layers.[93,99]

The absence of EL2 in MBE layers whatever the As or Ga rich conditions[88] could mean that the growth temperature used (600°C) is below the temperature of formation of the EL2 defect. This might also be due to the very slow MBE growth rate, as pointed out above. This cannot apply for LPE layers for which the growth rate is rather large (~ 100 μm per hour). But one may relate the absence of EL2 in these layers to the Ga rich conditions which prevail in that case.

3.2.5 Influence of Free Electron Concentration

The title of this section should be read with a question mark, and not as a definite conclusion. Observations concerning the possible dependence of EL2 on the concentration of free carriers have been reported several times in the literature. Humbert et al[86] noticed that both varied crudely in the same way under their growth conditions for growth temperatures of 750 and 770°C. In the data of Oseki et al,[85] the two concentrations varied in opposite directions, but within a different growth temperature range (650 to 750°C). Unfortunately, clear conclusions are hampered by the fact that the free carrier concentration was changed in both studies by changing the AsCl$_3$ mole fraction and not by purposely doping the material with various concentration of shallow donors. When this is done, as published by Watanabe et al[81] for MOCVD layers grown at 660°C, EL2 decreases when S doping (n = 2-3 $\times$ 10^{15} cm^{-3}) is used. All these results were measured on epitaxial layers presenting free carrier concentration lower than 10^{17} cm^{-3}. Since it was sometimes suggested that a shallow donor impurity could enter the EL2 complex composition, it may be useful to look at the results given in Table 2[33] where the concentration of EL2 from optical absorption measurements is seen to be much larger than the total concentration of the main donors deduced from SIMS measurements made by the C. A. Evans's group. These results remove any doubt left: EL2 is certainly not associated with a donor impurity.

The study published by Lagowski et al[78] is in agreement with this conclusion. They noticed in bulk materials doped with donors that the concentration of EL2, as deduced from capacitance measurements, is almost constant when the free carrier concentration n increases up to 10^{17} cm^{-3}. The same observation has been reported in VPE layers.[85] When n increases beyond 10^{17} cm^{-3}, the EL2 concentration drops sharply (see Figure 20). Before drawing any conclusion, one must evaluate the precision of capacitance-type measurement of EL2 when n is above 10^{17} cm^{-3}. This will be discussed in Chapter 4 in detail, but one can summarize here the precautions to be taken:

a) the effects of high electrical field in the investigated depleted layer leads to a decrease of the apparent DLTS peak amplitude.[95] The stronger

TABLE 2 Total concentration of donor impurities, measured by SIMS, and concentration of EL2 measured by optical absorption in different LEC high resistivity materials.

Ingot number	Total concentration of S + Se + Te + Si (cm^{-3})	Concentration of EL2 (cm^{-3})
240	2.6 10^{15}	0.9 $-$ 1.2 10^{16}
280	2.3 10^{15}	0.9 $-$ 1.1 10^{16}
330	1.5 10^{15}	8 10^{15}
400	1.4 10^{15}	7.5 10^{15}

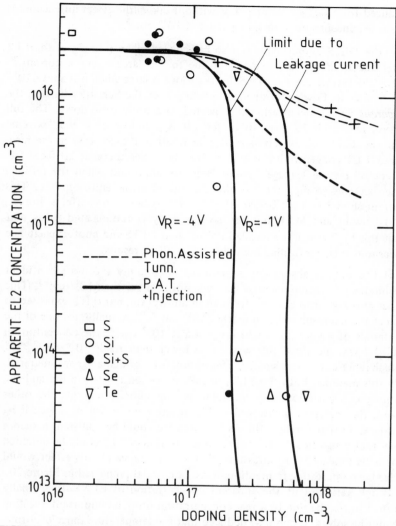

FIGURE 20 Comparison of the simulated DLTS results with the experimental dependence of the apparent EL2 concentration on free electron concentration. In the dotted curves, the simulation takes in account electrical field ionization alone while the continuous curves also includes the effect of reverse bias current injection from the metal electrode. The points are experimental after Lagowski et al.[99] and correspond to different dopant species.

the reverse bias V_R on the diode, the larger this effect. The dotted curves on the figure correspond to apparent EL2 concentrations which would be deduced for $V_R = -1$ V or -4 V without taking the effect into account, the true concentration remaining at 1.5×10^{16} cm^{-3}.

b) the DLTS amplitude may also be reduced when refilling of the EL2 level occurs in the depleted region due to too large a reverse current.[96] Since the reverse current of a Schottky diode is large when n is above 10^{17} cm^{-3}, due to Fowler-Nordheim tunneling over the Schottky barrier, the *apparent* EL2 concentration is expected to sharply drop down. The full curves in Figure 20 show this effect. It is therefore clear that extreme caution should be exercized to decide whether the decrease in the measured EL2 concentration is real or merely a measurement artifact. The effect of reverse leakage current becomes important when the refilling rate becomes comparable to the thermal electron emission rate. This current should be kept below 10^{-4} A cm^{-2} to avoid these effects. In fact, J. Lagowski and M. Kaminska have recently communicated to us that that this limit was not exceeded during their DLTS and photocapacitance measurements, confirming the reliability of those results.

c) The optical absorption experiment is certainly the most confident technique of measurement of EL2 when n is larger than 10^{17} cm^{-3}. (One must also take into account free carrier absorption, but this is expected to affect the data only when n reaches 10^{18} cm^{-3}.[97] Actually, for one of the materials of Figure 20 in which n = 4.5×10^{17} cm^{-3}, the concentration of EL2 was measured this way to be lower than 1.5×10^{15} cm^{-3},[98,99] which represents a very small concentration for a bridgman-grown material. (In this material, both the EL2 optically-measured concentration and the doping concentration were measured in our laboratories). On the other hand, the materials mentioned in Figure 20 were grown in a special Bg furnace. In that furnace, the As overpressure could be adjusted in such a way that a significant decrease of concentration of EL2 could be obtained for a particular As overpressure.[100] The importance of this effect would have been interesting to mention in the case of data reported in Figure 20.

If the variation of the apparent concentration of EL2 with n finally turned out to be due to a real concentration drop, an important question would be solved. It could be deduced that the temperature range of formation of the EL2 defect is below 780°C, since the influence of shallow donors (in concentrations lower than 10^{18} cm^{-3}) should be screened by the intrinsic free carriers present in larger concentration above that temperature.[78] On the other hand, a rather large uncertainty is expected to remain in any case concerning absolute values of concentration. In view of the difficulties of measurement, and the various assumptions concerning the structure of the EL2 associated defect, the thermodynamical models made to fit the present data are highly speculative.[78,99,101]

3.2.6 Influence of Cr Doping

The question of the influence of Cr on the concentration of EL2 in bulk
n-type or insulating materials has often been raised. Most of the time, this
question has come from electrical measurements in rather heavily compensated materials.[102] from which great care must be taken to calculate the
true deep level concentration. In particular, Schottky diode must be used
instead of Zn diffused p[+]n junction, since Zn has been shown to strongly
decrease the EL2 concentration.[17,80,103] Furthermore, it is possible but
difficult to separate the contribution of EL2 and Cr in DLTS spectra.[104,105]
Thus, the best technique is still the optical absorption. It has been shown
this way that the concentration of EL2 remains constant in Bridgman materials whatever the Cr concentration between 0 and 1.5×10^{17} cm^{-3}.[74,82]

3.3 Correlation With Surface Equilibrium

When epitaxial layers or bulk wafers are annealed, EL2 has been observed
to redistribute. Several studies have been devoted to the influence of capping conditions on the exodiffusion of the EL2 defects. It has also been
seen to be strongly modified by the presence of implantation defects.

3.3.1 Influence of Capping Condition on the EL2 Exodiffusion

Redistribution of EL2 under annealing was first reported in 1976.[106]
Mircea et al[103] then showed, in VPE layers, that this is due to an exodiffusion of that level. Okumura and Ikoma[107] also observed a similar
behavior in bulk crystals. A more quantitative assessment was further
published.[108] It was shown that the redistribution of EL2 does not follow
a simple erf curve and that it strongly depends on the capping conditions
(Figure 21). One can notice that the exodiffusion is enhanced when the
capping conditions allows more Ga outdiffusion (case of SiO$_2$). It is
slowed down when As losses are prevented using Si$_3$N$_4$ or an AsH$_3$ partial
pressure. LI and Wang[109] observed an even smaller exodiffusion in their
bulk material encapsulated with Si$_3$N$_4$. Clear conclusions are difficult to
be drawn, since these observations lead to suspect that atoms or vacancies
of both sublattices enter the physical configuration of the defect.

3.3.2 Influence of Defects on the EL2 Exodiffusion

Figure 22 plots the EL2 diffusion coefficient as a function of temperature.[108] In spite of a similar activation energy, around 4 eV, the curves differ by more than three orders of magnitude in VPE and LEC materials for
similar annealing treatment, face to face under H$_2$. It has even been found

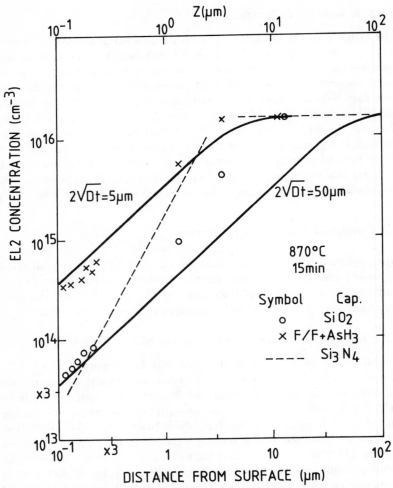

FIGURE 21 Exodiffusion of EL2 following an annealing at 870°C for 15 minutes, as a function of different capping conditions (SiO$_2$, face to face under a 3 torr AsH$_3$ overpressure, Si$_3$N$_4$). After Ref. 108.

to be different in two different LEC ingots.[108] It was checked that this is not related to the larger concentration of EL2 in bulk materials and it was suspected that stoechiometry of the host lattice or impurities therein, could influence the exodiffusion.

As a matter of fact, it has been noticed that quite different exodiffusion profiles are recorded in samples cut from the same material, annealed the same way, but previously implanted with various doses of oxygen, neon or selenium.[64,108] This seems to depend more closely on the ion species than

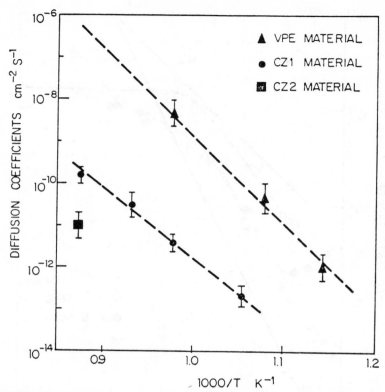

FIGURE 22 Plots of the values of the diffusion coefficient of EL2 measured in vapor phase epitaxial or bulk LEC materials. After Ref. 108.

on the implanted dose, as seen on Figure 23. Oxygen strongly reduces the exodiffusion, while the effect of neon implantation is a bit smaller.[104] On the other hand, EL2 is strongly reduced, or the exodiffusion enhanced when implanting boron.[45] It was also observed that EL2 was decreased by submitting bulk samples to H plasma at 300°C for two hours. The respective influence of implantation defects and impurities has not yet been clarified, but, following Lagowski et al,[110] one may think that either defect or impurities like H can interact with some unsaturated bonds which may form the complex EL2 defect and then help it to propagate or even to be destroyed.

3.4 Is EL2 the Arsenic Antisite Defect?

The most powerful technique to give information on the microstructural composition of defects, is certainly the electronic paramagnetic resonance

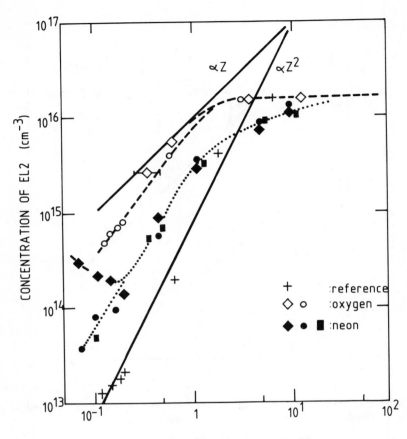

FIGURE 23 Concentration profiles of EL2 in an n type LEC material annealed at
860°C for 20 mn. + : unimplanted; open points: implanted with 0; black points:
implanted with Ne; Different doses have been used: $\diamond$: 10^{14} at.cm^{-2}; $\circ$: 10^{13}
at.cm^{-2}; $\square$: 10^{12} at.cm^{-2} . The solid lines indicate either a linear variation (α Z), or
a quadratic one (α Z^2). After Ref. 64.

(EPR) experiment. Up to recently, EPR measurements in as-grown Cr-
doped materials or in electronic-irradiated bulk GaAs materials failed to
detect any line which could correspond to a defect. The first observation
of four equivalently spaced lines in EPR spectra recorded on as-grown
undoped insulating crystals was reported by Wagner et al.[111] This signal
was attributed to the Arsenic antisite defect (i.e., Arsenic occupying normally Ga site). Then, it was demonstrated that the same lines could be
created by irradiating the materials with fast neutrons[112] or by applying
uniaxial stress at 400°C.[113] This has allowed a detailed comparison of the

electronic and optical properties of EL2 and the EPR detected defect, as well as of their annealing behavior. Hypothesis will then be proposed to account for all the reported observation.

3.4.1 Fast Neutron Irradiation Experiments

The same EPR signal, attributed to As_{Ga} has been detected in as-grown undoped LEC material and in insulating materials irradiated with fast neutrons[64,112,114] (see Figure 24). Because of the large line widths, however, it is unlikely that the effect of nearby defects cannot be resolved.[112, 115] if so, As_{Ga} is only the core of a more complex defect which gives rise to the observed EPR signal. In order to make our text as simple as possible, we will refer to that defect as "the EPR As_{Ga}", and more precisely "the EPR As_{Ga}^{4+}" since it is only detected in a given (paramagnetic) charge state.

The introduction rate of the EPR As_{Ga}^{4+} has been measured to be 10 cm^{-1} when the neutron flux is rather small, around 10^{11} cm^{-2} s^{-1}.[114] It goes down to 3 cm^{-1} when the neutron flux is larger by two or more orders of magnitudes.[112] When the neutron flux is small, an additional singlet line is detected in the EPR spectra, but its assignment is still unclear.[114] According to recent results, the temperature variation of the

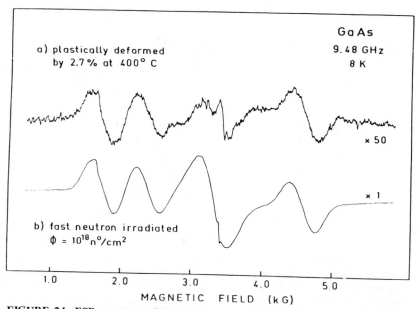

FIGURE 24 ESR spectrum of the As_{Ga} antisite defect in (a) plastically deformed and (b) fast neutron irradiated GaAs. After Ref. 115.

quadruplet hyperfine constant and the intensity, which does not follow a CURIE law, are ascribed to an exchange interaction between As_{Ga}^{4+} and an associated defect, likely to be related to the singlet line.[114]

DLTS measurements have been performed on materials irradiated with a small fluence of 10^{11} cm^{-2} s^{-1}.[64] Figure 25 shows the electron traps detected in bulk irradiated materials annealed at various temperature. The concentration of the induced deep levels, present in all the corresponding samples, is reported in Figure 26. An increase of concentration of EL2 is hardly noticeable in these conditions and, anyway, does not exceed 1 or 2 × 10^{14} cm^{-3} which corresponds to an introduction rate lower or equal to 0.2 cm^{-1}. These experiments have also been made in LPE materials by Barnes et al.[116] According to these data, the introduction rate of the EL2-like level is around 0.1 cm^{-1}.[116]

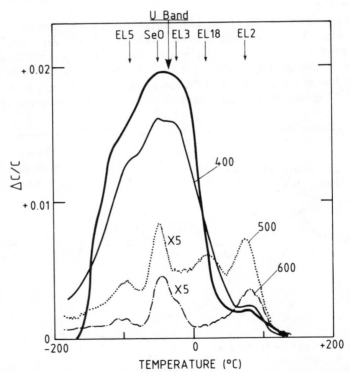

FIGURE 25 DLTS spectrum of electron traps recorded on a bulk material (previously annealed at 870°C) irradiated with fast neutrons (dose = 10^{15} .cm^{-2}). The solid line corresponds to the as-irradiated sample, the others to the material further annealed at different temperatures for 20 mn. The arrows give the expected position of well known levels. Thickness of the investigated region below the surface 3000 Å. Emission rate window: 7.5 s^{-1}. After Ref. 64.

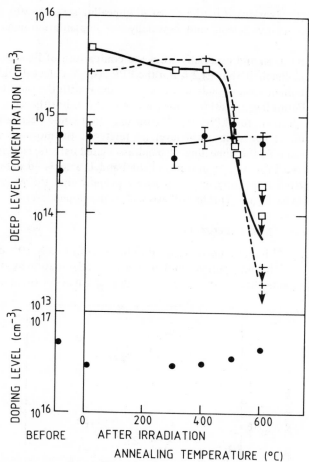

FIGURE 26 Plots of the concentration of deep levels and of the doping level, measured on the bulk LEC material (previously annealed at 870°C) before and after irradiation of fast neutrons (dose: 10^{15} .cm^2). •: concentration of EL2; □: concentration deduced from the DLTS maximum of the U band; + concentration deduced from the ODLTS maximum of the L' band; the two □ at 500°C and 600°C represent the concentration of the levels EL3 and EL18. At these temperatures, the two + represent those of HL4 and HL9. After Ref. 64.

On the other hand, two bands of defects are detected in irradiated bulk[64] or LPE[116] samples (see Figure 25). They have been labelled band U and L, and have their density of state roughly peaking at $E_c - 0.55$ eV and $E_v + 0.7$ eV respectively. The introduction rate, estimated to be 5 cm^{-1} in the case of low fluence and the annealing behavior of these bands look like that of the EPR As_{Ga}^{4+} defect created the same way, as shown in

Figure 27. It takes place in the so-called annealing stage 4 (400–500°C) observed for many defects and especially for implantation-induced defects.[45,64]

From these data, and especially from the comparison of Figures 26 and 27, we could conclude that EL2 is not the EPR As_{Ga}^{4+} defect if DLTS and EPR measurements were made in the same sample with the same conductivity and Fermi level position. Unfortunately, this is not the case. In particular the drop of the EPR As_{Ga}^{4+} concentration could be related to a change of the charge state of the level, i.e. related to the movement of the Fermi level due to the annealing of band associated defects. Nevertheless, since both the EPR As_{Ga}^{4+} defect and the band U associated defect have similar creation rates, one may reasonably propose that the As_{Ga} defect is involved in each case. This hypothesis will be developed later on.

3.4.2 Uniaxial Stress Experiments

The EPR As_{Ga}^{4+} has also been observed to increase from 3×10^{15} cm^{-3} to 2×10^{16} cm^{-3} in bulk samples undergoing a 4% deformation at 400°C.[72] In that case, annealing of the defects is not as sharp as for the neutron ir-

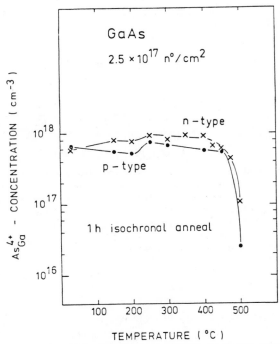

FIGURE 27 Isochronal annealing decay, in 1 h steps of the As_{Ga} antisite defect in originally n-type and p-type GaAs, fast neutron irradiated at T = 50°C. After Ref. 112.

radiated samples since 40% of the EPR As_{Ga}^{4+} concentration still remains after annealing at 550°C.[117]

As mentioned above in Section 3.1.3, the concentration of some deep levels has been observed to increase during the stress-induced deformation. Precise determination of concentration of all these levels was not always given, but, for instance, in a recent experiment, EL2 was estimated to increase from 7×10^{15} cm^{-3} to 3.4×10^{16} cm^{-3} in consequence of a plastic deformation of 5.5% under stress at 400°C. The other electron traps, the concentration of which was also increased, are probably the EL18, EL3 and EL6 levels, in rather good agreement with previously reported data[71] (see Figure 14).

These experiments are not as conclusive as those of neutron irradiation. There are two main reasons for this: i) EL2 was too large in the starting materials, i.e. of the same order of magnitude as the concentration of the introduced EPR As_{Ga}^{4+} ; ii) it is difficult to compare DLTS measurement of defects in a thin (typically 1 μm) layer at the surface and EPR measurements which makes an average over a large (typically 1 mm) thickness. This was possible in the case of neutron irradiation because fast neutrons have a mean free path of a few centimeters in GaAs. But this is expected to be very speculative in the case of stress effects which combine the generation of defects at the surface and their movement towards the bulk of samples.

3.4.3 Compared Optical Behaviors of EL2 and As_{Ga}

There are two main optical properties of EL2 which can be considered as the optical signature of EL2: its photoionization properties and its optical transfer from its normal state to a metastable state giving rise to photo-quenching effects. Both properties are extensively discussed below.

Figure 28 shows the variation of the EPR As_{Ga} under monochromatic illumination.[113,115] The overall shape of that variation, with the two steps starting at 0.75 and 1 eV, closely looks like the shape of the EL2 electron photoionization cross-section (see Figure 37).

The drop of the EPR As_{Ga}^{4+} signal under illumination, noticed in Figure 28, has been reported to be persistent in the dark, at low temperature. Figure 29[117] presents a complete scheme of that variation, just after illumination, and after further annealing at various temperature. The behavior is not the same as that of EL2, for which a complete quenching of signal is noticed under illumination and a complete recovery of the initial signal is recorded after annealing at 150 K (see below).

Thus, the EPR As_{Ga}^{4+} signal does not have exactly the same optical behavior as EL2. Furthermore, we should notice that the similarities could

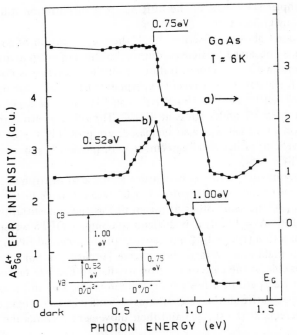

FIGURE 28 Intensity of the As$_{Ga}$ ESR signal as a function of monochromatic in-situ illumination. Curve a: semi-insulating GaAs sample; curve b: weakly p-type sample. The transitions shown in the insert correspond to the thresholds of photo-quenching and photo-enhancement of the As$_{Ga}$ ESR. After Ref. 113, 115.

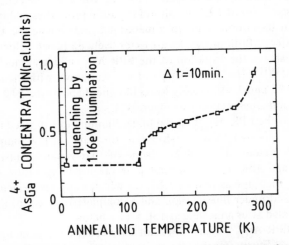

FIGURE 29 Isochronal annealing of the photo-quenching effect of As$_{Ga}$, as measured by EPR at 7 K. After Ref. 117.

of the Fermi level in these semi-insulating materials induced, for instance, by the presence of EL2.

3.4.4 Model Proposals

Two identification problems seem to be related. The first problem is the identification of the crystalline defect detected by EPR. The second problem is that of the identification of the defect responsible for the electrically measured DLTS EL2 peak. We call the first defect the EPR "antisite" defect and the second defect will be referred to as the EL2 defect.

As far as the EPR antisite defect is concerned, it is generally identified as with the As_{Ga}^{4+}. However, it has also been noted that the EPR experiments by themselves cannot exclude more complex structures where As_{Ga}^{4+} is associated with some second-neighbor defects. These defects might correspond to the singlet line detected in neutron irradiated samples. Therefore we shall leave this problem to more specialized investigations.

One may wonder whether the EPR defect is related to a DLTS peak. According to the observations reported so far, the defect in question is certainly related to near mid-gap levels. One must still decide whether EL2 or the U-band is associated to the EPR defect. In fact, the two assumptions can boil down to the same if one adopts the assumption that the U-band is derived from the EL2 defect interacting electrically with acceptors.

Those acceptors may be situated at distances of tens to hundreds of Å and may electrically interact with the EL2 defect by means of tunnelling or Coulombic effects. The result is of course a severe distortion of the EL2 DLTS peak and a shift to lower temperatures i.e. to the U-band. This appears particularly plausible since one expects the fast-neutron irradiation defects to be introduced in clusters with a very high defect density inside each cluster.

In order to summarize, we propose the following:

- the EL2 defect corresponds to the As_{Ga} EPR detected defect, i.e. As_{Ga} or a more complex defect with As_{Ga} as a core.
- the electrical and optical properties of that donor defect are modified by the presence of surrounding acceptors. Depending on their distance and concentration, the effect might be large, as noticed in irradiated materials (where many defects are created), or relatively small as observed in irradiated and annealed materials or as-grown materials (see below).

These hypotheses would allow one to explain the data mentioned in this paragraph and gathered in Figure 30. The band-U associated defects, which are present in a concentration similar to the EPR As_{Ga}^{4+} defect, could cor-

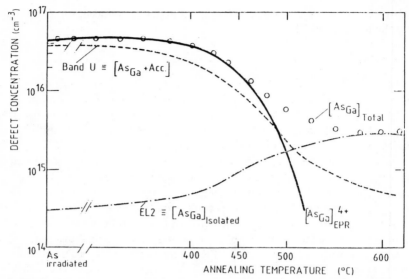

FIGURE 30 Variation of concentration of different defects (detected either in DLTS or in EPR) as a function of annealing temperature, together with an assignment proposed for each of them (part of the data come from Figures 13 and 27.

respond to (the As_{Ga} EPR defect + defect acceptor), the distance of these acceptors from the EPR As_{Ga} defect itself varying in a wide range. The perturbation of the electrical property of the EPR As_{Ga} defect (supposed to be the EL2 defect) is expected to be so large that even the activation energy of the emission rate of the level is modified and, furthermore, it is modified in a continuous way, leading to a shift of the EL2 DLTS peak and to its widening. On the other hand, we must suppose that the presence of acceptors does not change too much the EPR spectrum of the EPR As_{Ga} defect itself in that case, as suggested by EPR specialists. Thus, one may expect that, after neutron irradiation, most of the As_{Ga} are created in the As_{Ga}^{4+} charge state (in the case of insulating materials) and that most of them are under the influence of the large concentration of acceptors located at various distances giving rise to the band U in DLTS spectra recorded on n-type materials while very few are still isolated to give the usual EL2 peak.

On the other hand, we suggest that the total concentration of As_{Ga} defects decreases in the annealing stage 4 (between 400 and 500°C) from the initial concentration of created defects to a value close to the equilibrium value at the corresponding temperature. Tentatively, this expected variation is shown by open circles in Figure 30. During the annealing stage 4, most of the acceptor defects, thought to be associated with As_{Ga} to give the band U, have diffused or are annealed. Thus, above 500°C, most of the remaining As_{Ga} defects are expected to be isolated, and detected as the EL2 peak. The difference of annealing behavior of the EPR As_{Ga}^{4+} signal between 400 and 600°C, noticed in neutron irradiated materials and stressed materials, would be explained by an abrupt variation of the Fermi level in the neutron irradiated samples due to the annealing of acceptor-like defects corresponding to neutron damages. In stressed materials, the EPR As_{Ga}^{4+} signal seems to follow the same variation as the total concentration of As_{Ga} reported in Figure 30.

Conclusive experiments to test these three assumptions have yet to be done. Nevertheless, it is reasonable to think that As_{Ga} is involved in the EL2 defect, as the EL2 and the As_{Ga} EPR defects occur in similar concentration in as-grown semi-insulating GaAs.

4. Electrical and Optical Characterization of EL2

4.1 Electrical Properties of EL2

4.1.1 EL2 is A Deep Donor

The donor or acceptor nature of EL2 has been a subject of speculation even before it was directly observed and since the first successful attempts to grow semi-insulating Ga_{As} ingots without chromium doping (see Section 2). The donor nature of this level was established in a direct and unambiguous manner by Mircea et at.[103] Their experiment consisted in studying the EL2 outdiffusion under a 600–750°C annealing in VPE materials where the electron concentration (around 10^{15} cm^{-3}) was comparable with the EL2 concentration (1 to 3×10^{14} cm^{-3}). Profiling of EL2 was obtained by differential DLTS (DDLTS) while carrier profiling was obtained by a classical C–V technique. In-depth profiling was achieved by combining the above method with chemical etching. The electron concentration profile observed at low temperature, with all the EL2 centers occupied by electrons, proved flat and unaffected by the annealing process. The high-temperature carrier profile (with all the EL2 centers ionized) closely followed the outdiffusion profile of EL2. This experiment thus clearly showed that the EL2 centers are electrically neutral when occupied by electrons and are positively charged when releasing these electrons. However, these experiments of Mircea et al.[103] do not give any information on the single or double donor nature of this level. To the best knowledge of the authors, no electron trapping level has yet been definitely assigned as another charge state of EL2 in the upper half of the Ga_{As} energy gap.

The determination of the donor nature of EL2 had direct implications on the understanding of the mechanism responsible for the high resistivity of semi-insulating ingots either doped with chromium or of high purity. In fact, the donor nature of EL2 and the knowledge of its free energy for thermal ionization yield a coherent picture explaining Hall-effect data measured on a large number of semi-insulating wafers and at different temperatures.[30]

4.1.2 Thermal Electron and Hole Emission and Capture Data

The temperature dependence of the thermal electron-emission rate of EL2 has been studied by many authors.[17] All the measured emission rates scatter about the Arrhenius type relationship:

$$e_n = \alpha_n T^2{}_{na} \exp(-E_n/kT) \qquad (1)$$

where $\alpha_n = 2.28 \times 10^{20}$ cm^{-2}s^{-1}K^{-2} for GaAs

$\delta_{na} = 1.5 \pm 0.5 \times 10^{-13}$ cm^2 and

$E_n = 0.825 \pm 0.01$ eV.

In fact, even with VPE samples fabricated in the same laboratory in similar experimental conditions up to a factor of 2 dispersion in this emission rate has been reported by Mircea et al.[118] As suggested by these authors, this dispersion in emission rate may well be due to different stress conditions from sample to sample. However, there may be other reasons like the possible occurrence of ionized acceptor in the neighborhood of EL2 (less than 100 Å distance say).

As pointed out by Mircea et al,[119] the electron emission rate data are not sufficient to deduce the free energy of ionization as a function of energy. The complementary data required are those of the electron capture cross-section as a function of temperature. These data have been determined in a temperature range of 50 K to 273 K by Mitonneau et al.[120] and can be written as:

$$\sigma_n = \sigma_{n\infty} \exp(-E_{\sigma_n}/kT) + \sigma_{no}$$

where $\sigma_{n\infty} = 0.6 \times 10^{-14}$ cm^2 and $\sigma_{no} = 5 \times 10^{-19}$ cm^2 and $E_{\sigma_n} = 0.066$ eV.

The term σ_{no} in the above expression for σ_n is only important at temperatures lower than 100 K and is otherwise negligible. Assuming that the free energy for ionization E_T varies with temperature in a linear manner in the range of temperatures where both electron emission and capture data are available (T = 300 to 400 K), i.e.:

$$E_T = E_{T_o} - \alpha T$$

the principle of detailed balance yields:

$$E_{T_o} = E_n - E_{\sigma n} = 0.760 \pm 0.010 \, eV$$

and $\alpha + k \, \ln(g_1/g_0) = k \ln(\sigma_{no}/\sigma_{n\infty}) = 2.4 \times 10^{-4} \pm 0.4 \times 10^{-4} \, eV \, K^{-1}$

In fact, α represents the lattice contribution to the deep-level entropy while $k \ln(g_1/g_0)$ is the electronic contribution due to the difference in degeneracy between the filled (g_1) and empty level (g_0). Since these two contributions cannot be separated, one has to make additional assumptions to determine α. Since EL2 is a donor, Mircea et al.[97] have assumed that $g_1/g_0 = \frac{1}{2}$ following the theoretical arguments of Van Vechten and Thurmond.[121] Hence, one obtains: $\alpha = (3 \pm 0.4) \, 10^{-4} \, eV \, K^{-1}$. Rather than assuming a totally linear dependence of E_T with T, it would be more suitable to describe the very low temperature behavior of EL2 by using a temperature dependence similar to that of the band gap[122] and thus:

$$E_T = E_T(0) - \alpha \, T^2 / (T + 204) \qquad (3)$$

where $\alpha = (3.5 \pm 0.6) \, 10^{-4} \, eV \, K^{-1}$ and $E_T(0) = 0.722 \pm 0.020 \, eV$.

The electron capture behavior of EL2 has been studied by Zylberstejn[123] who has shown that when proper care is taken to avoid "zone edge" related problems, the capture kinetics of EL2 is very close to a single exponential behavior.

The temperature dependence of the electron capture cross-section of EL2[120] (Equation 2) is a very interesting feature since it may yield information about lattice relaxation properties of this level.[124,125] If one describes the lattice relaxation of EL2 by a characteristic phonon frequency $\hbar\omega$ and a linear Huang-Rhys coupling constant S, a closest fit of the experimental σ_n versus T results (Equation 2) using the multiphonon capture theory of Ridley,[125] one gets:[126]

$$\hbar\omega = 20 \pm 3 \, meV \quad and \quad S\hbar\omega = 115 \pm 50 \, meV$$

In other words, in order to accurately account for the low-temperature asymptotic behavior of σn, one obtains relatively precise information on the characteristic phonon $\hbar\omega$ but not for $S\hbar\omega$. It seems vain to try to identify the above phonon energy with peaks in the density of state of the non-localized GaAs phonon modes. The large lattice relaxation related to EL2 can only be due to local phonon modes probably associated to a Jahn-Teller effect. Moreover, the phonon energy deduced above should be considered as an average value if many phonon modes are involved. This "average phonon energy" deduced above is, however, too small to account for the relatively large absolute value of σ_n below 80 K ($\sigma_{no} \cong 5 \times 10^{-19}$ cm^2). To account for this, one may assume that the electron capture by EL2 may occur via an intermediate shallow hydrogen-like electron state. In this case, the apparent capture cross-section after Rhees et al.[127] should be written in the form:

$$\sigma_n = \frac{R}{N_c v_n} \cdot \exp(E_1/kT)$$

where R is the multiphonon transition rate from the excited shallow hydrogen-like bound state to the fundamental state of EL2, E_1 is the binding energy of that shallow electron state, while N_c and v_n are the effective

If E_1 small ($E_1 = 15 - 30$ meV), the best fit procedure described above to obtain $\hbar\omega$ is only slightly altered, but one expects a large apparent capture cross-section because the electronic part of the transition matrix describing the multiphonon transition to the ground state may be strongly enhanced by the localization of the electronic wave function in the shallow level E_1. In fact, the assumption of an excited shallow state (E_1) is also suggested by other experimental observations as will later be discussed and appears natural since EL2 is a donor.

An attempt was made by Mitonneau et al.[120] to measure the hole capture cross-section of EL2. These authors quote a hole capture cross-section of $\sigma_p = 2 \times 10^{-18}$ cm^2. Their experiment consisted in separating the DLTS signal due to EL2 from that due to a chromium impurity in a bulk p-type sample doped in the $p = 10^{17}$ cm^{-3} range. This result however should be considered with some caution in view of the investigation on the electric field response of the chromium-related deep level[105] which has been shown to transform from a hole trap at low electric fields to an electron trap at fields higher than 2×10^5 V cm^{-1}. In fact, the DLTS peak contribution attributed by Mitonneau et al.[120] to EL2 may be confused with the chromium-related signal which appeared as an electron trap in the high electric field region of their n$^+$p junction ($p = 10^{17}$ cm^{-3}, $V_R = -4$V). Recently, however, the ambiguity about this point was eliminated by Prinz and Rechkunov[128] who measured the hole capture cross-section of EL2 in

VPE layer which contained no chromium. These authors have confirmed
the value $\sigma_p = 2 \times 10^{-18}$ cm^2 near room temperature obtained by MIT-
ONNEAU et al.

4.1.3 Behavior of EL2 Under High Electric Fields and Electron Injection Conditions

Mircea and Mitonneau[129] have reported a strong anisotropy in the EL2
capacitive transient observed in Schottky barrier diode made on n-GaAs
(n = 1.4 to 1.7 $\times$ 10^{17} cm^{-3}, reverse bias - 4 V) (see Figure 28). Very different amplitudes and average time constants were observed when the
Schottky diode was fabricated on the (111) As surface, the (100) surface
or the (111) Ga surface (Figure 31). These authors attributed this very
surprising behavior to anisotropy in the potential barrier related to EL2
or to the electron wave function of the trapped electron.

Makram-Ebeid[96] has subsequently made a detailed investigation of the
electric field EL2-related transient by a differential transient technique.
The typical apparent emission rates $< e_n + e_p^* >$ related to different region of a reverse biased Schottky barrier are shown in Figure 32. It is clear
that the emission rate goes through a minimum when the electric field increases from zero and then increases very rapidly. The plateau value of

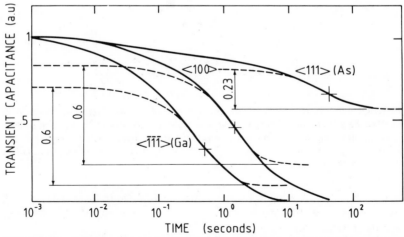

FIGURE 31 Capacitance transients recorded at 253 K after a 100 μs bias pulse of
duration from –4 V to –2 V and back to –4 V. Prior to the bias pulse the deep levels
were emptied by heating the samples at 300 K. Dashed lines are pure exponential
variations fitted to the experimental transients and corresponding to time constants
of 0.5 s, 1.4 s, and 41 s respectively. After Ref. 129.

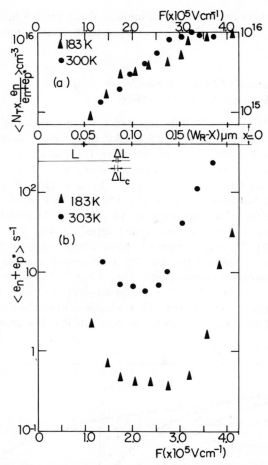

FIGURE 32 Result deduced from the differential transient technique applied to analyze EL2 under high electric field and injection conditions.[96] N_T is the levels concentration, e_n the electron emission rate (thermal + field assisted + impact ionization) and e_p^* is the net electron capture rate (from injected electrons or from Debye tail of bulk electron gas).

$< e_n + e_p^* >$ appearing in Figure 32 was also shown to be correlated with the reverse bias Fowler-Nordheim tunnelling current of the different Schottky barrier studied (Figure 33).

These data then suggest that the EL2 related transients for electric field in the range $F = 1.5$ to $2.5 \; 10^5 \; V \; cm^{-1}$ reported above may be due to the reverse bias current. Capture of electrons by empty EL2 level and impact ionization by the electron injected from the reversed Schottky barrier of

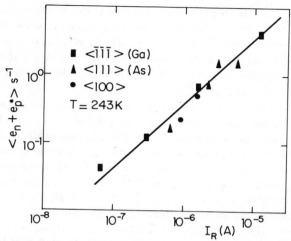

FIGURE 33 Experimental correlation between the total apparent emission rate at fields between 2 to 3 10^5 V cm^{-1} and the reverse bias current. Area of the Schottky diode = 6 10^{-3} cm^2. After Ref. 96.

the filled level may be involved in the observed anomalies. In fact, the apparent capture and impact ionization cross-sections do not seem to differ much from the low field capture cross-section of EL2.[96] At higher electric fields (F $\sim$ 2.5 10^5 V cm^{-1}), the EL2-related kinetics seem to be dominated by a phonon-assisted tunnel emission of electrons. The anisotropy (if any) for this electron emission does not seem to induce more than a factor of 4 change in the emission rates between diode fabricated on different crystallographic surfaces. These differences are small and may be exclusively due to error in evaluating the electric fields.

The quantum-mechanical theory needed to account for the phonon-assisted tunnelling emission from different levels in GaAs and GaP has been developed by Makram-Ebeid and Lannoo[95] and a best fit procedure yields for the lattice relation parameters S and $\hbar\omega$:

$$S\hbar\omega = 140 \pm 10 \text{ meV} \quad \text{and} \quad \hbar\omega = 20 \pm 5 \text{ meV}$$

The above values are consistent with electron capture cross-section data (preceding section) and will be seen to be consistent with optical data (next section). Comparison of the experimental results with theoretical predictions of the phonon assisted tunnelling model is shown in Figure 34.

The above combination of injected carrier capture, impact ionization and phonon-assisted tunnelling leads to a close and coherent description of the EL2-related transients. Very severe distortion in the shape and amplitudes of the DLTS peaks may result.[59] These results may explain at least in part the apparent decrease in the EL2 concentration with the free electron density (Figure 20 in Section 3.2). At very large dopings (n ~ 3 10^{17} cm^{-3}), the reverse-bias-injected current becomes a very serious limit to the measurement of the EL2 concentration by DLTS, ODLTS or photocapacitance techniques.[59]

The effect of the reverse bias electron injection from the metal counter electrode is visible even in the free electron profiles measured at high temperature. In fact, the comparison of the free electron profile at high temperatures (400 K say with EL2 ionized) with that measured at low temperature (below 300 K with EL2 neutral) allows the determination of the EL2 profile when the concentration of that level is comparable or larger than that of shallow donors.[26] Figure 35 shows the high-temperature and low-temperature electron profiles measured in this purpose.

The apparent EL2 concentration profile is the difference between the 400 K and 300 K carrier concentration profile. The fall in the apparent EL2 concentration profile near the surface is related to zone-edge effect,[123]

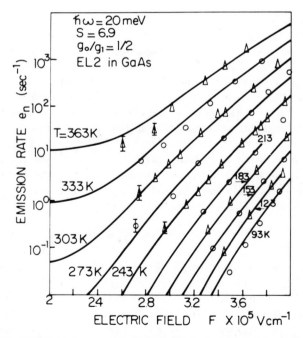

FIGURE 34 High-field electron emission rate for EL2 after substraction of injection component. The points are experimental and the curves theoretical. Ref. 90.

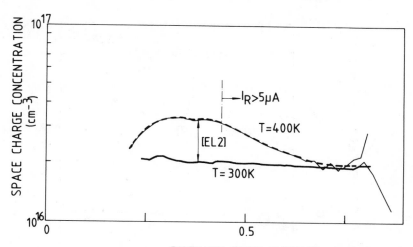

FIGURE 35 The apparent EL2 concentration deduced from the difference between the free electron profile at 20°C and 130°C. This apparent concentration falls near the surface due to zone edge effects[123] and becomes vanishingly small in the bulk because of the recapture of reverse bias injected electrons by EL2.

but the fall at large depths is due to the reverse current and results in vanishingly small values of [EL2] for $\chi \geqslant 0.7$ μm. The effect of the reverse bias current becomes important when its density exceeds 10^{-4} A cm^{-2}. This threshold current is very small and considerable measurement error may result at higher current densities.

Prinz and Rechkunov[128] have recently reported detailed measured electron and hole capture cross-section of EL2 under different electric field conditions (Figure 36). It seems that the electron and hole capture rate can be increased at high electric field (F $\geqslant 10^4$ V cm^{-1}) by up to 5 orders of magnitude over their zero field values. This result can be of considerable importance in understanding EL2-related transients, more particularly the effect of reverse bias electron injection mentioned above.

The increase of the electron capture cross-section with electric field has also other implications: it explains for example the current instabilities observed[130] while measuring the thermally stimulated current in bars of high-purity semi-insulating GaAs material and thought to be related to EL2 by Kaminska et al.[133] The observed oscillation frequency[133] was found to be proportional to the electron emission rate from EL2 for temperatures ranging from 320 to 400 K.

The large increase of electron capture cross-section with electric field seems to be in apparent contradiction with the lattice relaxation model of

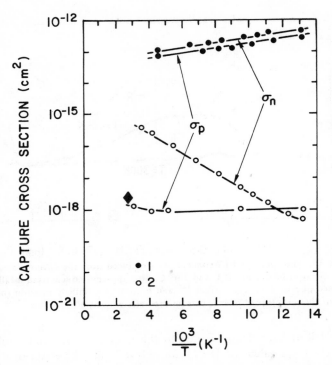

FIGURE 36 Experimental data of the capture cross-section measurements (Level EL2) in the strong electric field $F \geqslant 10^4$ V cm^{-1} (1) and in the zero electric field (2). The rhomb in the figure denotes the value of hole capture cross-section after Mitonneau et al.[120] and after Prinz and Rechkunov.[128]

EL2. In fact, the capture of hot electrons requires the release of more phonons to reach the EL2 ground state. Therefore, the multiphonon capture probability is expected to be smaller for hot electrons than for electrons which are in thermal equilibrium with the lattice (at low electric fields). One way out of this difficulty has recently been suggested by R. Pässler.[134] This author suggests that the hot electron capture may occur via an excited state of EL2 which is strongly coupled to the lattice. In this case, the capture rate may well be dramatically enhanced for hot electrons.

4.1.4 Attempts to Observe Anisotropic Behavior of EL2

As mentioned above, Mircea and Mitonneau[129] have observed a very strong anisotropic behavior in the high-field ($F = 1.5$ to 2.5×10^5 V cm^{-1}) EL2-related capacitive transients.

This observation suggested that one could reveal anisotropy in the EL2-related defect structure. The subsequent work on that subject[96] showed that the anisotropic transients could be correlated to reverse bias current injection and this would in turn suggest the involvement of hot-carrier capture and impact ionization in the process. A detailed investigation of the anisotropy in these hot-carrier related phenomena remains to be done but great care should be exercised to deconvolute problems related to the non-uniformity of the reverse current injection over the Schottky diode surface.

Another attempt to reveal possible EL2 related anisotropy has been undertaken using uniaxial stress experiments. The results however are not conclusive as will be shown in Section 4.3.

4.2 Optical Properties of EL2

4.2.1 Photo-ionization and Photo-absorption Data

The photo-ionization cross-sections σ_p° and σ_n° of EL2 with the centers initially empty (σ_p°) or filled (σ_n°) with electrons have been studied in great detail by Chantre et al.[135] Their technique consists in essence in comparing the photocapacitance transients with a given level initially empty with those transients where the level is initially filled with electrons. They use electrical or optical pulses to alter the initial electronic populations and they make their measurements at different temperatures. Comparing their data at high and low temperatures for a given level, they can determine the σ_p° and σ_n° spectra. Chantre et al.[135] have termed their technique the Deep Level Optical Spectroscopy (DLOS). The spectra obtained for σ_p° at 305 K and for σ_n° at 85 K corresponding to EL2 are shown in Figure 37. The spectra reported for σ_n° and σ_p° are drawn with an arbitrary-units vertical scale in view of the difficulty in determining the absolute values of light fluxes used. The ratio $\sigma_n^\circ/\sigma_p^\circ$ at the YAG laser photon energy (1.17 eV) has been independently found to be near 1.6 by Vincent,[136] near 3.3 by Martin[137] and about 10 by Mitonneau and Mircea.[144]

Chantre et al.[135] have made a detailed analysis of the σ_p° and σ_n° spectra using a theoretical model they have developed. Their model takes into account lattice relaxation effects, using an adjustable Franck-Condon shift Δ_{FC}, and also calculates the contribution to the photo-ionization cross-section of the optical transition rates from the bound state to the Γ, L and X conduction-band valleys. The transition matrix elements are calculated assuming allowed transition to or from the conduction-band minima or

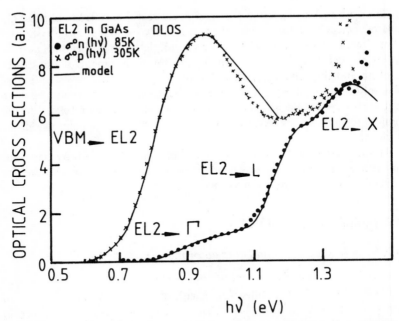

FIGURE 37 Spectral dependence of $\sigma_n 0$ (lower curve) and $\sigma_p 0$ (upper curve) after Ref. 135.

valence-band maxima respectively. This assumption does not seem to be critical in their model and furthermore allows to reduce the number of parameters involved. The only adjustable parameters in their model are the Franck-Condon shift Δ_{FC} together with the relative weights of the square of the optical transition matrix elements to each of the conduction-band valleys. An important feature in the σ_n^0 spectrum is the relatively small contribution of the transition to the Γ band minimum which is due in fact to the very low density of states (small effective mass) in this band.

An interesting feature of the above work of Chantre et al. is the unambiguous attribution of different portions of the photo-ionization spectrum of σ_n^0 of EL2 transition from the bound state to each of the Γ, L and X conduction bands. This is particularly valuable when comparing optical data related to EL2 with thermal ionization data. An important optical method used to study EL2 is the Double Source Differential Photocapacitance technique (DSDP) of White et al.[138,139] In fact, Chantre et al. have simulated the DSDP spectrum of EL2 using their σ_n^0 and σ_p^0 data. The result is shown in Figure 38. The negative peak at $h\nu = 0.8$ eV can be attributed to a valence-band-maximum-to-bound-state transition while the

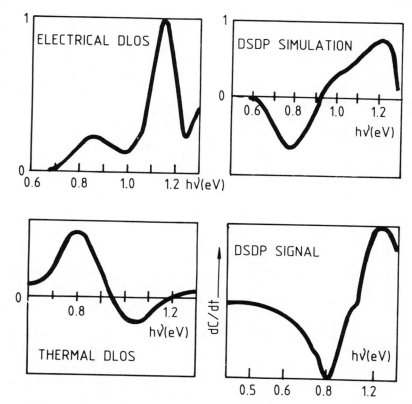

FIGURE 38 Spectral distribution of $d/dh\nu$ (σ_n°) obtained from electrical DLOS of $d/dh\nu$ (σ_p°) obtained from thermal DLOS. Also shown in the simulated DSDP signal as obtained from a linear combination of the mentioned spectra. The results are compared with the experimental DSDP spectrum of EL2 by White et al.[138] After Ref. 135.

positive peak at $h\nu = 1.2$ eV can be unambiguously attributed to a bound-state-to-L-conduction-band-minimum transition.

Figure 39 shows the extrinsic optical absorption spectrum observed in undoped semi-insulating bulk GaAs and in n-type bulk GaAs. The similarity between the optical absorption spectrum and σ_n^0 for EL2 is remarkable. The identification of this extrinsic absorption band has been used by Martin et al.[82,137] for optical non-destructive assessment of EL2 in bulk GaAs materials. The identification of the above mentioned absorption band and σ_n^0 spectra to EL2 has recently been made even more unequivocal by Martin[98] who observed a drastic change in the low temperature (10 K) extrinsic optical absorption spectrum of an undoped semi-

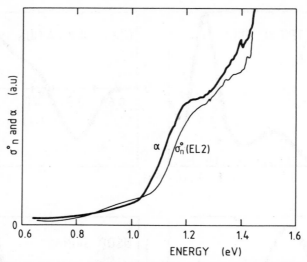

FIGURE 39 Comparison of the extrinsic absorption spectrum in high purity S.I. GaAs with the σ_n° spectrum. After Ref. 137. The difference between both curves is given by the dotted line (see also Figure 45 and text).

insulating GaAs sample after a few minutes exposure to illumination from a tungsten filament lamp. Complete disappearance of the extrinsic optical absorption features was observed (see Figure 40). This behavior can be directly related to the photocapacitance quenching behavior which is a unique finger print of EL2 as will be discussed in Section 4.2.3.

The above determination of σ_n° and σ_p° also yields a good understanding of the photoconductivity spectra observed by different authors for oxygen-doped semi-insulating GaAs. In all cases the reported photoconductivity spectra have shapes closely resembling that of the σ_n° spectrum.

Litty[140] has reported a peculiar behavior observed at different temperatures in the σ_p° spectrum which exhibits a sharp increase for photon energies above 1.2 eV. This behavior may be related to the existence of a shallow excited state for EL2 near the Γ conduction band minimum.

4.2.2 Configuration-coordinate Diagram

The optical threshold energies (for σ_n° and σ_p°) together with the thermal ionization and capture data previously reviewed yield a picture of the lattice relaxation configuration diagram of EL2. This diagram is shown in Figure 41. The electronic energy of the bound state is assumed to change with the distortion of the surrounding lattice represented by the generalized coordinate Q. The total energy (electronic plus elastic) is represented

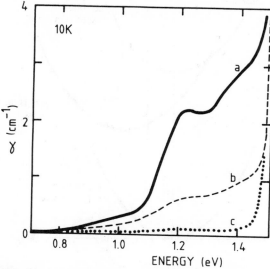

FIGURE 40 Bleaching of the EL2 induced optical absorption. Curve (a): after cooling in the dark; curves (b) and (c): after white light illumination for 1 and 10 min, respectively. After Ref. 98.

by parabolas (with the bound state empty or filled). In the configuration diagrams of Figure 41, all the optical transition energies are referred to zero K temperature taking into account the temperature dependence of the free ionization energy E_T and of the band-gap and assuming a temperature independent Franck-Condon shift Δ_{FC}. This last assumption is not justified since, in fact, Δ_{FC} is pressure dependent[141] as will be seen in a coming section and, by analogy, one expects it to be also temperature dependent. This may introduce an extra uncertainty in the O K optical transition energies of the order of 20 meV. The total uncertainty in the optical transition energies including measurement errors, can be expected to be around 30 meV.

One interesting feature of the configuration diagram is the non-linear dependence of the electronic energy of the bound state with the generalized coordinate Q. One, in fact, expects the coupling of the level with phonons to be weaker when the level is empty. However, if one takes the dependence of electronic energy on the lattice coordinate Q as a second degree polynomial, the photon energies with the empty level is expected to be about 15% larger than the photon energy corresponding to a filled level. If one believes the best-fit evaluation of the phonon energy for an empty EL2 level as obtained from the temperature dependence of the electron capture cross-section (Section 4.1.2), the phonon energy for the

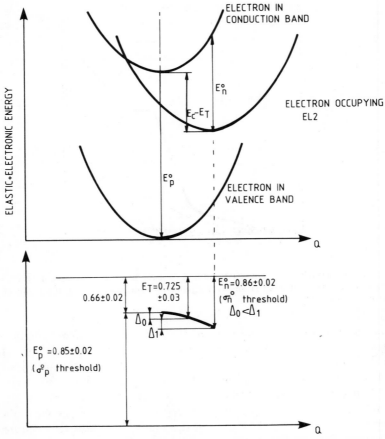

FIGURE 41 Lattice relaxation configuration diagram of EL2 as deduced from thermal and photo-ionization data. A non-linear coupling with the lattice is suggested by the fact that the thermal free energy of ionization is not midway between the ionization threshold energies for $\sigma_n{}^*$ and $\delta_p{}^\circ$. All energies are referred to 0 K.

empty and filled level should be (20 ± 3) meV and (18 ± 3) meV respectively. The above figures for the phonon energies together with the Franck-Condon shift evaluated from optical data $(100–130 \text{ meV})$ are in agreement with the values obtained (for an initially filled EL2 level) from the phonon-assisted tunnelling emission (Section 4.1.3).

The lattice relaxation picture of EL2 would be incomplete if one did not take into account a unique feature of that level which can show a meta-stable behavior related to strong lattice relaxation effects (see below).

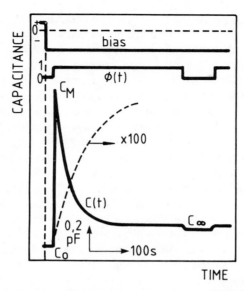

TIME

FIGURE 42 Transient photocapacitance observed in an n-GaAs Schottky barrier diode at 77 K after a direct electrical impulse and under an illumination at 1.17 eV photon energy. After Ref. 143.

4.2.3 The Metastable State of EL2

A) Capacitance Data, Thermal Transition Rates. Bois and Vincent[142] have reported a very peculiar transient photocapacitance behavior associated with EL2. The typical behavior is depicted in Figure 42. These photocapacitance transients can be observed in Schottky-barrier diodes made on n-GaAs ($n = 10^{15}$ to 1.5×10^{17} cm^{-3}) which are cooled while forward biased from ambient temperature to below 140 K, then reverse biased and then subjected to infrared illumination ($h\nu = 0.9$ to 1.5 eV). The photocapacitance transients observed consist of a fast increase of capacitance from the initial capacitance C_O to a maximum value C_M followed by a decrease to a final value C_∞, the fall time is about 100 times larger than the rise time. For VPE samples where EL2 has a concentration larger than that of any other trap, the value C_∞ and C_O are nearly equal. This indicates that the phenomenon involved is EL2 related. Another proof of this assertion, which excludes in particular possible contributions from hole traps, is the experiment performed by Vincent et al.[143] who first reverse biased their diode at room temperature for a long enough time to empty EL2, then cooled it to 77 K; the photocapacitance transient observed was decreasing. Some though shows that if a hole trap was involved in the process, an increasing photocapacitance transient could have been observed regardless of the bias sequence used.

After completion of the photocapacitance transient of Figure 42, switching the light off at 77 K and illuminating again does not result in any sizable change in capacitance (except for small changes due to zone-edge effects). The regeneration of the full overshoot behavior visible in Figure 42 depends on the time spent in the dark and on the temperature. The return of the overshoot in the photocapacitance transient is exponentially related to the time spent in the dark and occurs at a rate r which has been determined independently by Mitonneau and Mircea[144] and by Vincent et al.[143] as a function of temperature. The phenomenological explanation of this behavior which is unique among all known GaAs defects has been given by Bois and Vincent.[142] These authors are led to attribute to EL2 a state EL2M having a much stronger relaxation then the fundamental normal state EL2F. The corresponding configuration diagram would be then be as shown in Figure 43a. The charge states for EL2F and EL2M are identical (neutral EL2). The total energy of state EL2M is larger than that of EL2F, so that EL2F is the only stable state but at low temperatures, the system can stay for a very long time in the cusp of EL2M of the total energy diagram of EL2. State EL2M is therefore referred to as the metastable state. One may also refer to EL2M as an excited state, but this should not be confused with the hydrogen-like excited state mentioned in Section 4.1.2 and 4.2.1 above. *The very large lattice relaxation associated with EL2M makes it insensitive to extrinsic light excitation.*

The different optical and thermal transition rates from stable EL2F to metastable state EL2M and inversely are sketched schematically in Figure 43b.

The deexcitation rate r (return from metastable to normal state) can occur through two different mechanisms either by thermal deexcitation,[143,144] or by an Auger like process[144] in which case the deexcitation rate is very much enhanced by the presence of a free electron density n. The total deexcitation rate r can, in general, be written as:

$$r = 2 \times 10^{-11} \exp(-0.3 \text{ eV}/kT) + 1.9 \times 10^{-14} \times n \times v_n \exp(-0.107 \text{ eV}/kT) \sec^{-1}$$

where n is the free electron density in cm^{-3} and v_n the thermal electron velocity in cm/sec at the temperature T. In practice, above 140 K, the transfer to the normal state is very fast even in the absence of free carriers and one only observes the fundamental normal state above this tempera-

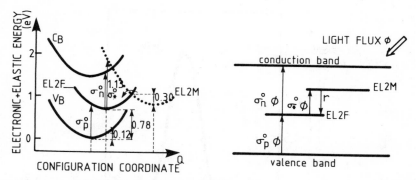

FIGURE 43 (a) Configurational diagram of EL2 including the fundamental EL2 F and the metastable state EL2 M. (b) Schematics of the different optical and thermal transition rates between the different states involved after Ref. 143.

ture. The enhancement of the deexcitation of EL2 by free electrons is an Auger type of process, but the details of the mechanism are not understood.

Figure 43b suggests that the transition of EL2 to the metastable state can occur from the normal state without transiting in the semi-conductor band. In fact, Mircea and Mitonneau have checked that EL2 can be transferred from the normal filled state (in the neutral region of the diode) to the metastable state provided the temperature is low enough (50 K say) to bring the Auger deexcitation rate to extremely small values.

B) Optical Transition Rates. As for the different optical and thermal transition rates depicted in Figure 43b, σ_n^0 and σ_p^0 are the usual photoionization cross-sections of Section 4.2.1. The cross-section σ^* corresponds to the photo-induced transition rate from normal to excited state and was shown by Vincent et al[143] to have the spectral dependence shown in Figure 44. The peak of σ^* occurs at 1.13 eV which is very near 1.17 eV (the YAG laser photon energy). The relative values of the different cross-sections at the YAG laser photon energy are:[143]

$$\sigma_n^0 = 1\sigma_n^0, \, \sigma_p^0 = 0.6\sigma_n^0, \, \sigma^* = 0.08$$

The absolute value of σ_n^0 for the same photon energy is, after Mitonneau and Mircea:[144]

$$\sigma_n^0 = 1.5 \times 10^{-16} \text{ cm}^2$$

More recently, Kaminska et al.[145] have made an extremely interesting remark. They have noted a systematic difference (see Figure 45) between

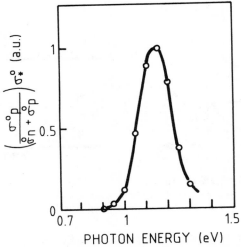

FIGURE 44 Spectral dependence of the cross-section $\sigma*$ corresponding to the optically induced transition rate from 1 EL2F to EL2M. After Ref. 143.

the shape of the electron *photo-ionization spectrum* of EL2 and that of the EL2-associated optical *absorption* in n-type materials (see Figure 39 for example). When properly normalizing the two curves so that they agree in the high and low energy parts of the spectrum, there appears to be in the optical absorption spectrum an excess absorption band centered about 1.15 eV. The shape and position of this band coincides within measurement accuracy with the band depicted in Figure 44 and which represents the optical transition rates from the fundamental to the metastable state of EL2.[143] Therefore, the optical-induced transition to the metastable state is readily visible in the optical absorption data of GaAs. Although the photo-ionization and absorption data were available in the literature (see Figure 39), this is the first time that such a connection is made. Kaminska et al.[145] have examined in great detail and with a high resolution the optical absorption of n-GaAs at 8 K around the 1.15 eV band. They revealed a fine structure in the absorption band (Figure 46). These authors interpret the fine structure as different optical transitions involving 0, 1, 2, 3, 4, or 5 phonons. The zero phonon line lies at 1.0395 eV and the associated phonon energy is 11 ± 1 meV. Although the position of the zero-phonon line at the low-energy side of the intra-center absorption band is a good point in favor for the attribution of these phonon replicas to EL2, it seems that more work has to be done to fully validate this attribution. Unfortunately, the phonon energy involved is probably related to an internal Jahn Teller effect and it is difficult to compare this phonon energy

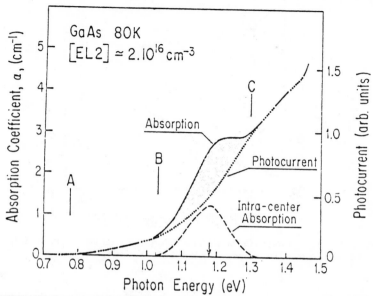

FIGURE 45 Optical absorption and photocurrent spectra measures on the same sample of melt grown GaAs with EL2 concentration of about 2×10^{16} cm^{-3}. The difference between two spectra defines the intra-center absorption band and is quite similar to the difference which may also be noticed between optical absorption and optical capture cross-section in Figure 39. It is also similar to the shape of the optical transition from the fundamental to the metastable state (Figure 44). After Ref. 145.

with other phonon energies involved in the multiphonon capture of electrons by EL2, for example. In fact, quite different local phonon modes may be involved in each case.

The symmetry of properties of the above zero phonon line was subsequently analyzed by uniaxial stress experiment by Skowronski et al.[146] In spite of some problems with the interpretation (non-linear dependence with stress in certain directions), these authors prove that there is a strong influence of the trigonal field and almost no effect from the tetragonal field. They assume that the transition to the metastable state occurs via a strongly relaxed excited state having an electronic energy resonant in the conduction band, separated from this band by a purely vibronic barrier. The crystal field symmetry of this excited state, as revealed by their work is T_2 whereas that of the fundamental initial state is A_1.

C) Metastable State Associated Effects. Vincent et al.[143] have used the above phenomenological model to account for different photo-memory effects observed by different authors in non-chromium doped semi-insulat-

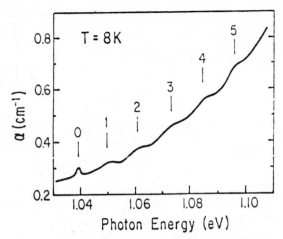

FIGURE 46 Fine structure in the 1.15 eV absorption band attributed to an EL2F to EL2M "intra-center" transition. After Ref. 145.

ing GaAs. The case of photoconductivity is particularly instructive. Lin et al.,[147] for example, reported that the spectral photoconductivity they observed has a peak at 1.04 eV photon energy which only appeared with increasing photon energy while another peak at 1.4 eV was independent of the sense of the photon-energy scan. However, a pre-illumination with photon energy 1.18 eV and/or 1.08 eV at 82 K yielded a decrease of as much as three orders of magnitude in photoconductivity and even a change in the sign of the majority photocarrier (from electron to holes). This change was found to persist when the temperature was scanned from 82 K to 120–140 K. It is clear that the transition of the electron from its normal state to the metastable state (which is optically inactive and electrically neutral) neutralizes the EL2 centers present and may yield to a type change in the photoconductivity.

As already mentioned in paragraph 4.2.1, the quenching of near-infrared optical absorption[98] in semi-insulating material can also be directly related to the transfer of the EL2 from its ground state, having rather large optical cross-sections, to its metastable state which does not present any significant optical activity under sub-band-gap light excitation.

The properties of the stable-metastable system related to EL2 are unique features of that center in GaAs. It becomes very tempting to use these features to unambiguously decide whether a phenomenon like luminescence[148,149] or EPR[112] response is related or not to EL2. However, when one applies this method to different phenomena in semi-insulating GaAs, one should keep in mind that the stable-to-metastable transition of EL2 may induce large changes in the Fermi level of this type of material that may even show as a change in sign of the majority carriers. One should therefore be aware that any phenomenon, which is sensitive to the exact position of the Fermi level in the gap, may undergo changes in intensity which are very similar to those initially observed in the very peculiar transient-photocapacitance behavior related to EL2.

D) Microscopic Model for the Metastable State. A microscopic model for the metastable behavior of EL2 is difficult to develop in view of the unknown nature of that center. However, Vincent et al.[143] propose that the two possible states EL2F and EL2M of EL2 are related to an interstitial impurity atom which is capable of sitting in two tetrahedral interstitial positions with As atoms as nearest neighbors for the first position and Ga atoms for the second.

4.2.4 Photoluminescence Bands Attributed to EL2

The configuration diagram of Figure 43a suggests that one may expect free-carrier-to-bound EL2 luminescent transitions at $h\nu = 0.65 \pm 0.04$ eV. In fact, unstructured luminescent bands peaking at 0.62 to 0.68 eV photon energies have been observed by many authors by photoluminescence[150,153] and by cathodoluminescence.[154] According to Shanabrook et al.[149,155] and to Yu et al.,[151,156,157] the luminescence band around 0.65 eV, seen by many authors, may be due to a combination in different proportions of two wide bands with peaks at 0.635 eV and 0.68 eV photon energy. Both bands are not seen in n-type GaAs and they are both observed with different proportions in both high purity and in chromium doped semi-insulating GaAs.

A) The 0.63/0.65 Photoluminescence Band. Mircea-Roussel and Makram-Ebeid[152] have observed a Photoluminescence (PL) band around 0.645 eV having a halfwidth of 130 meV at 4 K and rapidly decreasing in intensity for temperatures higher than 20 K. The position and width of this band were observed by these authors to agree with the attribution of this band to a photoluminescent transition from a free hole at the valence band maximum radiatively recombining with an EL2 center filled with an electron. The peak position and second moment for such a transition were predicted from the experimentally measured phonon-assisted tunnel emission (Sections 4.1.3 and 4.2.1 above). Nevertheless, it must be admitted that the 0.645 eV band could also possibly be associated with a conduction-band-to-bound-state transition. Considerable uncertainty is associated with the peak position that one expects for this transition, particularly in view of the non-linearity observed in the configuration diagram reported in Section 4.2.2. However, another independent observation confirms the attribution of the 0.645 eV luminescence band to EL2. The same authors[152] have reported that after annealing their semi-insulating GaAs samples at 870°C for 15 minutes under Si_3N_4 cap, the band almost completely disappears as would be expected for EL2, which was shown previously to outdiffuse during this type of heat treatment.[103,108]

Leyral et al.[148,153] approached the same problem by studying the 0.645 eV photoluminescence band using a YAG laser excitation (1.17 eV) at 4 K. The intensity of the P.L. band intensity was observed to start from a high value and then decrease with a slow time constant to a steady state value of about 10% of the original intensity. A close analysis of this photoluminescence "fatigue" behavior of the photoluminescence showed that the kinetic phenomena involved are identical to those observed for the metastable state of EL2 (previous section). This observation of photoluminescence fatigue agrees well with the attribution of the 0.645 eV PL band with EL2. However, as noted in the previous section, we should keep in mind that indirect phenomena may be involved in the above observations like the change in the photo-excited carrier density when EL2 goes from its stable to its metastable configuration. In fact, any deep center photoluminescence due to a transition from a conduction band state to a similar bound state may yield a "fatigue" behavior in that case. Leyral and Guillot[148] have also studied the photoluminescence excitation (PLE) spectrum of the 0.645 eV band taking great care to avoid luminescence fatigue phenomena. The observed PLE spectrum was observed to resemble the σ_n^0 photo-ionization spectrum of EL2. This suggests that the observed PL band is probably due to a conduction-band-to-EL2 transition. However, other possibilities remain since the photo-excited carrier density for extrinsic photon excitation is known to have a spectrum dependence similar to that of σ_n^0, as was mentioned in Section 4.2.1 above. Therefore, one cannot categorically discard the involvement of mid-gap levels other than EL2. However, since EL2 is the only mid-gap level known to occur in undoped GaAs in sizable concentration, all the combined information mentioned above give us good confidence for the attribution of the 0.645 eV PL band to EL2.

B) PLE Spectra of the 0.63/0.65 and of the 0.68 eV Bands. Two research groups[148,149] have simultaneously observed a rich structure in the near band gap Photoluminescence Excitation Spectrum of the (0.63/0.65) eV band (referred to also as the 0.635 or 0.64 eV band). Shanabrook et al.[149] made a detailed PLE study of the 0.635 eV and 0.68 eV bands. To observe differences in behavior of the two bands, they detected different luminescence wavebands where the relative importance of the two PL bands was different. The 0.635 eV band exhibited a very special type of near band photo-excitation spectrum with a sharp resonance for photon excitations about 10 meV below band gap. Above band gap, an oscillatory behavior in the PLE spectrum was observed revealing a resonant cooling of photo-excited electron by longitudinal optical phonons. This type of resonant cooling behavior was never observed before for deep level lumin- escence and indicates that a shallow level is involved as an intermediate step in the photoluminescence process.

Shanabrook et al.[149,155] attributed the 0.635 eV PL band more specific- ally to a transition from an electron trapped in the EL2 center to a neutral shallow acceptor level. This type of explanation also accounts for the sharp onset of the photoluminescence spectrum for extrinsic near band gap excitation (Figure 47). The temperature dependence of the extrinsic PLE peak suggests an acceptor activation energy of 30 ± 10 meV. To fur- ther elucidate details of this donor-acceptor transition, the above authors have made a more in-depth analysis of the 0.635 eV band by observing it in samples where it is exclusively present. They made time-resolved photo-

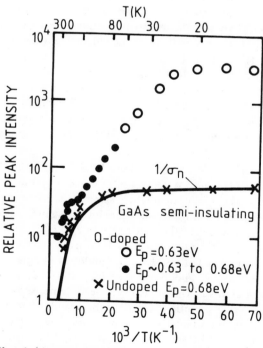

FIGURE 47 Photoluminescence excitation spectra for the 0.635 eV luminescence band at different temperatures after Ref. 149.

luminescence measurements to detect possible shifts in the peak position that could reveal different spatial separations of the donor-acceptor pair. The PL peak seemed time independent within the limit of their experimental accuracy and they concluded consequently for a fixed donor acceptor separation. Thus, the donor-acceptor pair assumption leads one to believe that EL2 is always combined with an acceptor probably co-existing as closest neighbors. This attribution, however, presents serious difficulties since one of its logical consequences is that, in n-type materials, the EL2 defect should have a negative charge state (deep level occupied and shallow acceptor ionized). This negative charge state has to be postulated to account for the presence of a shallow (hydrogen like) acceptor level. Therefore, this attribution would contradict the experimental fact[103] that the EL2 level, when filled with an electron, is neutral, i.e. is of donor nature.

In fact, the very remarkable experimental piece of work of Shanabrook et al.[149,155] can be given an alternative explanation which avoids the above-mentioned difficulties. One can assume that the 0.635 eV luminescence transition is due to the internal transition from an electron trapped in the shallow excited donor state of EL2 (see Sections 4.1.3 and 4.2.1) to the fundamental state of that defect center. Since the transition is now supposed to occur from the excited to the fundamental state of the same defect, it is only natural to observe a time-independent photoluminescence spectral shape (fixed spatial separation between initial and final states). In fact, an excited state of the EL2 level (10 to 30 meV below the conduction-band minimum) would be expected to have a large orbit of about 50 Å radius and to be relatively insensitive to phonon coupling. This would then explain the resonance cooling structure observed in the PLE spectrum since the excited intermediate level is not phonon broadened.

C) The 0.68 eV Band. Yu and Walters[24] have studied in detail the 0.63 eV and 0.68 eV PL bands. Attempts were made by these authors to correlate the presence of oxygen in their samples with these bands. It was found that, when comparing an "oxygen doped" sample with an undoped one, the oxygen content in the first one was 2×10^{15} cm^{-3} while, in the second, it was lower than the detection limit estimated to be 10^{15} cm^{-3}. The 0.63 eV band was not visible in the undoped sample. As it has been discussed in Section 2, and more especially in 2.3.2, incorporation of isolated oxygen atoms in GaAs has never been demonstrated, while it can be present via oxide pollution. On the other hand, effect of oxygen has been shown to be important from the point of view of the incorporation of silicon (in Bg and LEC ingots) or of boron (in LEC ingots); thus "oxygen treated materials" are distinctly different from the point of view of the concentration of these two inpurities. These PL results, recorded on so-called oxygen doped or undoped materials, cannot be considered as a proof that the 0.635 eV band is related to oxygen itself.

The temperature dependence of each of these two PL bands was studied by Yu and Walters[24] who observed the intensity dependence shown in Figure 48. The 0.63 eV band is seen to decrease in intensity when the temperature exceeds 25 K (activation energy about 15 meV) while the 0.68 eV band is much less sensitive to temperature. The width, shape and position of the 0.68 eV band was also studied by Yu.[157] He accounts for the increase in the half width of this band with temperature by a coupling with phonons having a characteristic energy of 20 ± 2 meV and a linear Huang-Rhys coupling constant S of 5.5 ± 0.5. These values are in good agreement with the values previously reported[126] for EL2 (see also Section

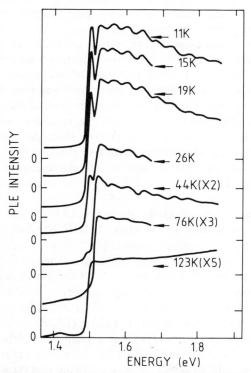

FIGURE 48 Temperature dependence of the 0.63 eV and 0.68 photoluminescence band intensities after Ref. 151. The continuous curves represent the temperature dependence of the inverse of EL2 electron capture cross-section as measured by Mitonneau et al.[120]

4.2.3 above). Yu also reports that the position of the 0.68 eV PL peaks is temperature independent for T = 5 K to 245 K.

The above experimental results of Yu[157] and Yu et al.[24,156] seem to indicate that the 0.68 eV band could well be due to an occupied-EL2-to-valence-band transition, but no evidence contradicts a possible association of this PL band with a conduction-band-to-empty-EL2 transition. This last suggestion is furthermore reinforced by a close look at the temperature dependence of the 0.68 eV peak intensity. Figure 48 follows closely the relative temperature dependence of the reciprocal of the electron capture cross-section of EL2 determined by Mitonneau et al.[120] Such a behavior would indeed be expected if the 0.68 eV PL band was due to a Γ-conduction band electron being trapped by an empty EL2. The non-radiative multi-phonon capture rate responsible for the observed electron capture cross-section of EL2 would enter in competition with the radiative (photoluminescent) transition and one would expect the 0.68 eV intensity to change with temperature in proportion to $1/\sigma_n$.

The stronger temperature dependence of the 0.635 eV band intensity agrees quite well with its attribution to a transition from a shallow excited donor level of EL2 to its fundamental state. In this case, two non-radiative mechanisms may compete with the photoluminescent transition for temperatures above 20 K. The first of these mechanisms is the thermal ionization from the shallow donor level while the second is the non-radiative multiphonon recombination to the ground state. At low temperature, the first mechanism strongly prevails and the observed 10-to-15 meV activation energy could well be near the thermal ionization energy of the shallow excited EL2 level.

D) Discussion of a Proposed Model. Instead of the models proposed previously by Mircea-Roussel and Makram-Ebeid[152] [0.64 eV band related to (EL2F) − VB trans.], by Shanabrook et al.[149,155] [0.64 eV band related to (EL2F)-acceptor-pair trans, 0.68 eV band related to (EL2F) − VB trans.], and by Yu et al[156,157] [0.64 eV band related to oxygen, 0.68 eV band to EL2], we suggest that the 0.68 eV band corresponds to CB − (EL2F) transitions and the 0.64 eV band to an internal transition from the shallow excited state to the fundamental state (EL2E) → (EL2F).

One problem remains: how does one explain the widely changing proportions of the 0.635 eV and 0.68 eV band from sample to sample? One possible way out of this difficulty is to postulate that the excited state of EL2, EL2E, cannot occur if an ionized acceptor is present within 50 Å say of the EL2 defect. The Coulomb repulsion from the ionized acceptor would push the shallow excited state into the conduction band. The association of EL2 even with relatively remote acceptor can thus account for the occasional disappearance of the excited donor level of that center and thus to the prevalence of the 0.68 eV band which is in fact much more rarely observed in undoped semi-insulating GaAs than the 0.635 eV band.

Another question arises about the above attribution of the 0.68 eV PL band to a conduction-band − EL2 transition, which should be consistent with the known thermal ionization data for that level (Section 4.1.2) and with the observation of Yu that the 0.68 eV peak is temperature independent from 5 K to 215 K. Since one assumes a free-carrier-to-bound-state transition, one should take into account a Boltzmann distribution of the photo-excited free carriers in the bands. This is particularly important on account of the relatively large temperatures involved (up to 245 K). Assuming parabolic bands, the non-zero kinetic energy of the photo-excited carrier would shift the first moment (nearly peak positions) to higher energies by an amount nkT where n = 1.5 for allowed transitions. The first moment $< h\nu >$ of the photoluminescence peak is therefore expected to be:

$$< h\nu > = E_C - E_T - \Delta_{FC} + n\,k\,T = E_C - E_T(0) - \Delta_{FC} - \alpha\,T^2/(T + 204) + 1.5\,kT$$

In Section 4.1.2, the temperature coefficient α for the free ionization energy of EL2 was shown to be $\alpha \sim (3 \pm 0.5) \times 10^{-4}$ eV K^{-1}. Consequently, one expects a maximum change of about 9 meV of $< h\nu >$ in the observed temperature range (5 K to 215 K), which is probably too small to be detectable. Of course, nearly identical conclusions would have been drawn if one assumed the 0.68 eV band to be due to a bound-state-to-valence-band transition.

Tajima and Okada[158] have made detailed micro-mappings of the 0.65 eV luminescence band together with a 0.8 eV band they observed in undoped semi-insulating GaAs wafers. They report for the 0.65 eV band intensity mapping a W-profile across a wafer diameter closely resembling the etch pit density distribution. This is very similar to the EL2 distribution previously reported by Martin et al[82] (see 3.2.2) and is probably a further argument for the association of the 0.65 eV band (identical to the 0.63-to-0.645 eV band reported by other authors) with EL2. The 0.8 eV band observed by Tajima and Okada seems to have an intensity profile decreasing with increasing EPD and vice-versa, a behavior contrasting completely with that of the 0.65 eV band.

4.3 EL2 Behavior in GaAs-related Ternary Alloys and in GaAs Submitted to High Pressure

In order to shed light on the nature of deep level defects of unknown origins, experimentalists are led to design experiments where the lattice environment of the deep level to be studied can be altered in a controllable manner. For EL2, this type of experiments have been conducted using two different approaches. In the first approach, EL2 has been studied in ternary alloy systems such as $Ga_{1-x}Al_xAs$ (Refs. 159–161) and $Ga_{1-x}In_xAs$ (Ref. 162) with x starting from very small values. Similar experiments were also performed by Woolford[162] for the $GaAs_{1-x}P_x$ system but results for small x values are not available. For the $Ga_{1-x}Al_xAs$ system, the EL2 level coexists with other deep levels so that the existing data are incomplete and the results of different laboratories do not agree. In the second approach, isotropic or uniaxial pressure is applied to a GaAs sample and the optical[138] or electrical[163] properties of EL2 were studied as functions of pressure.

Both alloying or applying large pressures introduce changes in the average lattice spacing and alter consequently the position of the different conduction-band minima relative to the valence band maximum. In the following, we shall attempt to compare the different sets of results obtained by referring the observed changes in ionization energy in each set of experiments to the corresponding change of the direct band gap energy. As reference to this comparison, we shall use the experimental results of Mircea et al.[118] for EL2 in the $Ga_{1-x}In_xAs$ system which provides the most complete set of data.

4.3.1 Constant Temperature Measurements (300 K)

If the experiments are performed at a constant temperature, the changes in the free energy ionization E_T caused by a change in composition or pressure can be studied by simultaneously determining changes in the thermal electron emission and capture rates. The equation to be used is from the principles of detailed balance:

$$e_n = \sigma_n \, v_n \, N_C g_1 \, \exp \left((E_T - E_C)/kT \right),$$

where e_n is the electron emission rate, σ_n the electron capture cross-section, N_c the effective density of state at the Γ conduction band minimum, v_n the thermal electron velocity and g_1 the degeneracy ratio for a filled electronic EL2 level. One can safely neglect variations in the factor (v_n $N_C g_1$) and, consequently, we may write:

$$\Delta (E_C - E_T) = kT \ [\Delta (\ln \sigma_n - \ln e_n)] \ .$$

If the room temperature results of Mircea et al.[118] are presented in this fashion, the dependences of $\Delta (\ln \sigma_n)$, $\Delta (\ln e_n)$ and $\Delta (E_C - E_T)$ on the indium content x are shown in Figure 49a. It is quite clear that one cannot neglect isothermal changes in the capture cross-section σ_n for different compositions x. Even though the changes kT. $\Delta (\ln e_n)$ and kT. $\Delta (\ln e_n)$ vary in a non-linear manner with x, their difference $\Delta (E_C - E_T)$ seems to be linearly related to x, with:

$$\Delta (E_C - E_T)/ \Delta x = - 12 \pm 1 \text{ meV for a 1\% change in x at T = 300 K,}$$

while, near x = 0,

$$kT \ \Delta (\ln e_n)/ \Delta x = 4.8 \text{ meV} \qquad \text{for a 1\% change in x at 300 K,}$$
and
$$kT \ \Delta (\ln \ _n)/ \Delta x = -7.27 \text{ meV} \quad \text{for a 1\% change in x at 300 K.}$$

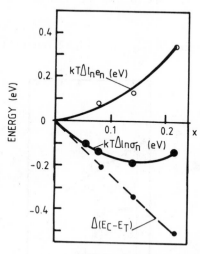

INDIUM FRACTION X

FIGURE 49b   Results of Mircea et al.[118] for the indium fraction dependence of the free energy of ionization extrapolated at 0 K.

These figures should be compared with the rate of change of the direct band gap:[118]

$$\Delta (E_C - E_V)/\Delta x = -12.3 \text{ meV for a 1\% change in x at 300 K.}$$

4.3.2 Determination of the Free Energy of Ionization
Extrapolated at O K

Rather than observing changes at a constant temperature, the alternative method of Mircea et al.[118] is to deduce the free energy of ionization projected at the absolute zero temperature, by subtracting the activation energy of the electron capture cross-section from the activation energy of the thermal electron emission rate. The result is shown in Figure 49b and one has:

$$\Delta(E_C - E_T)/\Delta x = -8.6 \text{ meV for a 1\% increase in x referred to O K.}$$

One thus reaches different conclusions when working isothermally or using differences in activation energies. At room temperature, the free energy of ionization of EL2 follows the valence top while the projected ionization energy E_T at absolute zero follows the valence band top within about 70%, more precisely:

$$\Delta(E_T - E_V)/\Delta (E_C - E_V) = 0.32, \text{ referred to 0 K.}$$

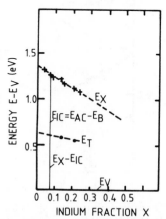

FIGURE 49a Dependence of the room temperature emission and capture rates of EL2 on the indium fraction[118] together with the deduced change in thermal ionization energy relative to the conduction band.

4.3.3 Discussion

In other words, the energy E_T referred at 0 K with respect to the valence-band top changes at about one third of the rate of the energy gap. The free ionization energy projected at absolute zero temperature is probably the most meaningful quantity to compare with theoretical predictions. However, the isothermal method described above is useful when comparing with the isothermal data reported by Zylberstejn et al.[141] and by White et al.[138] for the hydrostatic pressure dependence of EL2.

Zylberstejn has measured the room temperature thermal emission rate e_n of EL2 as a function of hydrostatic pressure P from zero to 5 K bar and finds an exponential dependence in the pressure range investigated which yields:

$$kT \cdot \Delta(\ln e_n)/\Delta P = 3.8 \pm 0.3 \text{ meV/K bar around 300 K.}$$

This figure should be compared with the rate of change of the direct band gap energy with hydrostatic pressure which, according to different authors,[164] is about 11.5 ± 1 meV/K bar. *Consequently a change at 1 K bar is roughly equivalent to a compositional change of − 0.9% in the In content of the $Ga_{1-x}In_xAs$ system.* In other words, both 1 K bar changes and a −0.9% In compositional change induce about 11.5 meV change in the band gap energy. Therefore, the above figure for kT. $\Delta(\ln e_n)/\Delta P$ is equivalent to (4.2 ± 0.3) meV change for a − 1% change in the In compotion and should be compared with the value of 4.8 meV for a 1% change in In content as deduced from the results of Mircea et al.[118]

White et al.[138] were the first to study EL2 at different hydrostatic pressures. They used a Double Source Differential Photocapacitance (DSDP) method. As was seen in section 4.2.1, the peaks exhibited in the DSDP spectrum of EL2 at 0.85 eV and 1.2 eV can be unambiguously attributed, after Chantre et al.,[135] to a valence-band-maximum-to-EL2 and to an EL2-to-L-band-minimum transitions respectively. White et al. report that at 100 K temperature, the DSPP negative peak position initially at $h\nu = 0.85$ eV changes at a rate of 1.7 meV per Kbar whereas the positive peak initially at $h\nu = 1.2$ eV changes at a rate of 1.2 meV per K bar. We thus deduce that:

$$\Delta(E_T + \Delta_{FC} - E_V)/\Delta P = 1.7 \text{ meV/K bar around 100 K,}$$

$$\Delta(E_L - E_T + \Delta_{FC})/\Delta P = 1.2 \text{ meV/K bar around 100 K,}$$

where E_V is the energy at the valence band maximum, E_L the energy of the L-conduction band minimum and Δ_{FC} the Franck-Condon shift.

Since we know the rate of change of $E_L - E_V$ with pressure to be 5.5 meV per Kbar after Aspnes,[164] we deduce,

$$\Delta(\Delta_{FC})/\Delta P = -1.3 \text{ meV/K bar}$$

and

$$\Delta(E_T - E_V)/\Delta P = 3 \text{ meV/K bar at } T = 100 \text{ K}.$$

Hence, the Franck-Condon shift is seen to decrease when the pressure increases, whereas E_T changes with respect to the valence band maximum at about ¼ of the direct band gap rate. This result must be situated midway between the results of Mircea et al.[118] when using their room temperature isothermal data and those results corresponding to the zero absolute temperature ionization energy. This is understandable since the observation of White et al. was made at an intermediate temperature (100 K).

Bastide et al.[163] have studied the thermal emission rate of EL2 as a function of an applied uniaxial stress. They observe practically identical stress dependence for the (100) and (111) directions. After proper scaling, their results are seen to be in total agreement with the isotropic pressure dependence of Zylberstejn et al.[141]

4.3.4 Comparison With Theoretical Predictions
Various Defect Models

Ren et al.[165] have attempted to use the pressure dependence of the EL2 ionization energy to get a hint about the physical nature of that level. However, there seems to be too many possible candidates which would yield similar pressure dependence as EL2 within experimental error. One of the difficulties in exploiting the experimental results is the pressure dependence of the Franck-Condon shift. Nevertheless, Ren et al.[165] argue that the existing pressure dependence data are not in contradiction with the attribution of EL2 to either substitutional oxygen on the As site or to an As atom on the Ga site. Buisson et al.[165] have made theoretical predictions for the compositional dependence of different defect-related levels in the alloy system $In_{1-y}Ga_yAs_{1-x}P_x$. Applying their numerical result for the special case $x = 0$ and y near unity, one finds that their A_1 symmetry level for an anion on cation site (such as As on Ga site) is a near mid-gap level (0.64 eV above valence band maximum compared to 0.77 eV for EL2). For the $Ga_{1-x}In_xAs$ system, their results predict that the energy level position changes relative to the valence band maximum at a rate of

— 1.7 meV for each 1% increase in the In content. The corresponding rate deduced above from the experimental results of Mircea et al.[118] is in fact nearer to − 4 meV for a 1% increase in Indium content.

In conclusion to this section, it would seem that the available experimental data are incomplete, except for the results on GaInAs of Mircea et al. who measured both capture and emission rates as a function of temperature. It is well established that the EL2 level practically follows the valence band top. The Franck-Condon shift has also been shown to be pressure dependent. One can infer that it probably also depends on the composition of the GaInAs material. This behavior has not yet been explored neither from a theoretical nor from an experimental point of view. In view of these unsolved problems, it is perhaps too early to draw definitive conclusions concerning the defect structure.

5. Discussion on the Structure of the EL2 Defect

It is useful to summarize here the main experimental evidences concerning the origin and the properties of EL2.

It has been shown in Section 2 that EL2 is not related to oxygen. All the data reported in Section 3 prove that EL2 is actually a lattice defect, the concentration of which increases in the case of As rich growth conditions. This defect does not correspond to a dislocation related or associated defect. It is certainly a point defect or a point defect complex. The thermal stability of this defect has been the object of several studies which established that the annealing threshold temperature varies from 700 to 850°C and is material dependent. The separate role of stoichiometry and of defects or impurities on the annealing process is nevertheless to be clarified. Furthermore, it seems also established that the temperature of formation of that defect lies below 780°C. One can also assert that none of the usual donor impurities (S, Si, Se, Te) is included in that possible complex defect. Among all the electrical properties reported in Chapter III, the most remarkable one is certainly the existence for that deep level of a strongly relaxed metastable state. Nevertheless, this information is not yet sufficiently precise and detailed. This explains why several models have been proposed for the structure of that defect, which all have pros and cons.

In 3.4.4 we have proposed two hypotheses which could allow to account for the large amount of reported observations:

— the EL2 defect itself corresponds to the As_{Ga} EPR detected defect, i.e. As_{Ga} or a more complex defect with As_{Ga} as a core

— when surrounded by acceptors or shallower defects at various distances, the properties of the mid-gap EL2 level are modified.

These two hypotheses allow to separate the problem of the identification of the, say, isolated EL2 defect from the problem of the "fluctuation" of its observed properties in various samples. Coming back to the former one, the simplest model proposed so far for the structure of the isolated EL2 defect has been the arsenic antisite defect As_{Ga}.[78,113] This is the simplest defect we may expect to be created in As-rich conditions, if we keep in mind that still simpler ones, like interstitials or vacancies, are excluded in view of electron irradiation experiments (see Section 3).

Furthermore, the existence of a metastable state has sometimes been supposed to correspond to an alternative position of an interstitial atom near a central point defect. More generally, the metastable state could correspond to a different arrangement of a non-symmetric complex defect. Several possibilities have been proposed: either $As_{Ga} V_{As}$,[167] which would look like the structure of DX centers in GaAlAs ternary alloys,[168,169] or $V_{As}As_{Ga}V_{Ga}$.[101] This last defect was first proposed a long time ago as the native defect which should have the best chance to remain trapped in bulk ingots in noticeable concentration.[170] As a matter of fact, the recent observation of a zero-phonon line in the optical absorption transition by Kaminska et al[145] and their study under uniaxial stress by Skowronski et al[146] could bring further insight on the symmetry of EL2 and thus on its structure.

Will the theory help to clarify the problem? So far, most of the calculations have concentrated on the energy position of defect associated levels in the gap. A recent paper shows that the assignment of EL2 as $V_{As}V_{Ga}$ could also well be possible.[171] On the other hand, it seems that the energy position of the As_{Ga} level should stand rather high in the upper half of the band gap, which would still suggest that EL2 is not As_{Ga} alone. Very rare theoretical papers have taken advantage of data available on the variation of EL2 with pressure or composition and, to our knowledge, none has tried to explain the strongly relaxed metastable state, nor the pressure dependent Franck-Condon shift. There is no doubt that many defect configurations still have to be modeled before all the theoretical results could be usefully exploited.

6. Conclusions

In this paper, we have tried to review all the properties of the main deep donor EL2 on Ga_{As}. Each time, this has been possible, we have separated those data, which allow to definitely attribute given properties to EL2, from other data which are not yet sufficiently precise or confirmed and which may obscure the already very complex EL2 problem.

As far as electrical properties are concerned, one may say that the electronic transitions between the level and the conduction band have been extensively studied, leading to a precise and detailed scheme of the configuration coordinate diagram of that level. Unfortunately, almost nothing is definitely established concerning the electrical transition between the valence band and the level. On the other hand, optical investigations of EL2 have led to a good understanding of all the optical transitions involved in the presence of EL2. But one must admit that the present state of knowledge does not allow to definitely explain the photoluminescence bands thought to be related to EL2.

It is now established that EL2 has two possible configurational states: these are the normal fundamental state and a strongly relaxed metastable state insensitive to extrinsic light stimuli. The stable-metastable system distinguishes this level among different levels in GaAs. It still remains to definitely establish that this level has also an excited state lying close to the conduction band, as is suggested from electronic capture, photoionization and photoluminescence data.

A big step towards the identification of that level has also been made recently. It has been definitely established that the problem of oxygen, and that of EL2 are clearly separated in GaAs. More generally, it is very likely that this level is not related to any given impurity, and it has been suggested to be probably a lattice defect. As discussed in the last chapter, we are still far from the definite assignment of that defect.

Acknowledgments. The authors would like to thank Andrei Mircea for his encouragement and for his interest in this work which he has pioneered. They are grateful to both him and S. Gourrier for critical reading of the manuscript. Thanks are also due to A. Mitonneau and our other colleagues from LEP, in particular J. P. Farges, J. P. Hallais, G. Jacob, A. Mircea-Roussel and C. Schemali, for their helpful collaboration. Very special thanks are also due to our colleagues G. Guillot, P. Leyral and A. Nouaillat from Insa Lyon (France), and A. Goltzene and C. Schwab from ULP Strasbourg (France), for our long-term scientific association. We also acknowledge C. E. Barnes (Sandia Labs), S. Bishop and B. V. Shanabrook (NRL), H. C. Gatos, M. Kaminska, J. Lagovski (MIT), D. G. Holmes (Rockwell), D. C. Look and P. W. Yu (Univ. Dayton), M. Lannoo and D. Stievenard (Isen Lille), T. Ikoma and M. Taniguchi (Univ. Tokyo), L. Samuelson (Univ. Lund), E. R. Weber (Pi Koln), and Zou Yuanxi (Shanghai) for interesting and useful discussions. Sims data, reported in Table 2, have been obtained by C. A. Evans and Associates.

References

1. R. K. Willardson and A. C. Beer, Elec. Mfg (January 1956).
2. C. Hilsum and A. C. Rose-Innes, Semiconducting III-V Compounds, Pergamon Press, 142 (1961).
3. N. G. Ainslie, S. E. Blum, J. F. Woods, J. Appl. Phys. 33, 2391 (1962).
4. J. F. Woods and N. G. Ainslie, J. Appl. Phys. 34, 1469 (1963).
5. J. L. Fertin, J. Lebailly, E. Deyris, Proceedings of the 1966 Symposium on GaAs, Inst. Phys. Conf. Ser. 3, 46 (1967).
6. J. M. Woodall and J. F. Woods, Solid State Comm. 4, 33 (1966).
7. J. M. Woodall, Trans. Metall. Soc. AIME, 239, 378 (1967).
8. T. Shimoda and S. Akai, Japan J. Appl. Phys. 8, 1352 (1969).
9. J. W. Allen, Nature, 187, 403 (1960).
10. J. Blanc, L. R. Weisberg, Nature, 192, 155 (1961).
11. R. W. Haisty, E. W. Mehal, R. Stratton, J. Phys. Chem. Solids, 23, 829 (1962).
12. J. Blanc, R. H. Bube, L. R. Weisberg, J. Phys. Chem. Solids, 25, 225 (1964).
13. S. M. Sze, J. C. Irvin, Solid State Electron. 11, 599 (1968).
14. G. M. Martin, D. Bois, Proceedings of the Topical Conference on Characterization Techniques (Electrochemical Society, 78-3, 32 (1978).
15. R. Williams, J. Appl. Phys. 37, 3411 (1966).
16. A. Mircea, A. Mitonneau, Appl. Phys. 8, 15 (1975).
17. G. M. Martin, A. Mitonneau, A. Mircea, Electron. Lett. 13, 191(1977).
18. M. E. Weiner and A. S. Jordan, J. Appl. Phys. 43, 1767 (1972).
19. M. R. Brozel, J. B. Clegg, R. C. Newman, J. Phys. D: Appl. Phys. 11, 1331 (1978).
20. G. W. Blackmore, J. B. Clegg, J. S. Hislop, J. B. Mullin, J. Electron. Mat., 5, 401 (1976).
21. G. M. Martin, G. Jacob, J. P. Hallais, F. Grainger, J. A. Roberts, J. B. Clegg, P. Blood, G. Poiblaud, J. Phys. C: Solid State Phys. 15, 1841 (1982).
22. J. L. Debrun, M. Valladon, A. M. Huber, N. T. Linh, Proceedings of the 7th Int. Vac. Congress and 3rd Int. Conf. of Solid Surfaces, Vienna, (R. Dobrozewsk et al., Vienna), 2645 (1977).
23. M. Valladon, G. Blondiaux, C. Koemmerer, J. P. Hallais, G. Poiblaud, A. M. Huber, J. L. Debrun, J. of Radioanalytical Chemistry, 58, 165 (1980).
24. P. W. Yu, D. C. Walters, Appl. Phys. lett. 41, 863 (1982).
25. H. Emori, M. Umehara, M. Takeya, K. Nomura, Y. Terai, T. Nozaki, Inst. Phys. Conf. Ser. 63, 47 (1982).
26. A. M. Huber, N. T. Linh, J. C. Debrun, M. Valladon, G. M. Martin, A. Mitonneau, A. Mircea, J. Appl. Phys. 50, 4022 (1979).
27. P. N. Favennec, M. Gauneau, H. L'Haridon, B. Deveaud, C. A. Evans, B. J. Blattner, Appl. Phys. Lett. 38, 271 (1981).
28. M. Kaminska, J. Lagowski, J. Parsey, K. Wada, H. C. Gatos, Int. Conf. on GaAs and Related Compounds, Oiso (Japan), (1981), Inst. Phys. Conf. Ser. 63, 197 (1982).
29. J. P. Hallais, G. M. Martin, G. Jacob, C. Koemmerer, M. Valladon, J. L. Debrun, M. L. Verheijke, J. B. Clegg, F. Grainger, J. A. Roberts, G. Poiblaud, Final Report (1981) (D.G.R.S.T. Contract 79.7.0193).
30. G. M. Martin, J. P. Farges, G. Jacob, J. P. Hallais, J. Appl. Phys. 51, 2840 (1980).

31. D. E. Holmes, R. T. Chen, K. R. Elliott, C. G. Kirkpatrick, P. W. Yu, IEEE Trans. *E. D. 29,* 1045 (1982).
32. J. R. Oliver, R. D. Fairman, R. T. Chen, P. W. Yu, Electron. Lett. *17,* 840 (1981).
33. G. Jacob, G. M. Martin, Unpublished results.
34. D. E. Holmes, R. T. Chen, K. R. Elliott, C. G. Kirkpatrick, Appl. Phys. Lett., *40,* 46 (1982).
35. C. M. Wolfe, G. E. Stillman, D. M. Kom, Proceedings of the 1976 Conf. on GaAs and Related Compounds, Inst. Phys. Conf. Ser. *33b,* 120 (1977).
36. D. V. Lang, R. A. Logan, J. Electron. Mat. *4,* 1053 (1975).
37. L. Palm, H. Bruch, K. H. Bachem, P. Balk, J. Electron. Mat. *8,* 555 (1979).
38. H. Bruch, L. Palm, F. Ponse, P. Balk, Electron. Lett. *15,* 246 (1979).
39. R. H. Wallis, M. A. Di Forte-Poisson, M. Bonnet, G. Beuchet, J. P. Duchemin, Proceedings of the 1980 Symp. on GaAs and Related Compounds, Inst. Phys. Conf. Ser. *56,* 73 (1981).
40. P. N. Favennec, J. Appl. Phys. *47,* 2532 (1976).
41. T. Itoh, T. Tsuchiya, M. Takeuchi, Jap. J. Appl. Phys. *15,* 2277 (1976).
42. S. Gecim, B. J. Sealy, K. G. Stephens, Electron. Lett. *14,* 306, (1978).
43. M. Berth, C. Venger, G. M. Martin, Electron. Lett. *17,* 873 (1981).
44. G. M. Martin, S. Makram-Ebeid, Ngo Tich Phuoc, M. Berth, C. Venger, Proceedings of the 2nd Int. Conf. on Semi-insulating III-V Materials (Evian, France), (Edited by S. Makram-Ebeid and B. Tuck, Shiva Ltd, U.K.), 275 (1982).
45. G. M. Martin, P. Secordel, C. Venger, J. Appl. Phys. *53,* 8706, (1982).
46. M. Berth, M. Cathelin, G. Durand, Tech. Digest IEDM, Washington DC, 201 (1977).
47. D. T. Hurle, Inst. Phys. Conf. Ser. *33a,* 113 (1977).
48. J. Kasahara, N. Watanabe, Proceedings of the 2nd Conf. on Semi-insulating III-V Materials (Edited by S. Makram-Ebeid and B. Tuck, Shiva Ltd, U.K.), 238 (1982).
49. A. Lidow, J. F. Gibbons, V. R. Deline, C. A. Evans Jr., Appl. Phys. Lett. *32,* 149 (1978).
50. P. N. Favennec, B. Deveaud, M. Salvi, A. Martinez, C. Armand, Electron. Lett. *18,* 203 (1982).
51. B. Deveaud, P. N. Favennec, H. L'Haridon, Proceedings of the 2nd Conf. on Semi-insulating III-V Materials (Edited by S. Makram-Ebeid and B. Tuck, Shiva Ltd., U.K.), 269 (1982).
52. P. K. Vasudev, R. G. Wilson, C. A. Evans Jr., Appl. Phys. Lett. *36,* 837 (1980).
53. T. R. Jervis, D. W. Woodward, L. F. Eastman, Electron. Lett. *20,* 620 (1979).
54. S. Makram-Ebeid, G. M. Martin, J. C. Bouley, A. Mircea, Unpublished results, (1980).
55. T. Ikoma, M. Takikawa, M. Taniguchi, Int. Conf. on GaAs and Related Compounds, Oiso (Japan) (1981), Inst. Phys. Conf. Ser. *63,* 191 (1982).
56. M. Taniguchi, T. Ikoma, Proceedings of the 2nd Conf. on Semi-insulating III-V Materials (Edited by S. Makram-Ebeid and B. Tuck, Shiva Ltd, U.K.), 283 (1982).
57. M. Taniguchi, T. Ikoma, Proceedings of the Inst. Conf. on GaAs and Related Compounds, Albuquerque (U.S.A.)(Sept. 1982), Inst. Phys. Conf. Ser. No 65, 65(1983)
58. G. M. Martin, P. Terriac, S. Makram-Ebeid, G. Guillot, M. Gavant, Appl. Phys. Lett., *42,* 61 (1983).

484   G. MARTIN AND S. MAKRAM-EBEID

59. S. Makram-Ebeid, Unpublished results.
60. M. Taniguchi and T. Ikoma, Private communication.
61. G. Vincent, D. Bois and A. Chantre, J. Appl. Phys., 53, 3643 (1982).
62. J. D. Wiley, J. Phys. Chem. Solids, 32, 2053 (1971).
63. M. Toyama, A. Kasami, Jap. J. Appl. Phys., 11, 860 (1972).
64. G. M. Martin, S. Makram-Ebeid, Proceedings of the 12th Int. Conf. on Defects in Semiconductors, Amsterdam (Holland) (Sept. 1982) Physica 116B, 371 (1983)
65. G. M. Martin, E. Esteve, S. Makram-Ebeid, P. Langlade, J. Appl. Phys. 56, 2655 (1984).
66. A. Mircea, D. Bois, Inst. Phys. Conf. Ser. 46, 82 (1979).
67. D. Pons, P. M. Mooney, J. Bourgoin, J. Appl. Phys. 51, 2038 (1980).
68. D. Pons, Proceedings of the 12th Int. Conf. on Defects in Semi-conductors, Amsterdam, (Holland) (Sept. 1982) Physica AA6 B, 388 (1983).
69. See, for instance, D. Pons, A. Mircea, J. Bourgoin, J. Appl. Phys. 51, 4150 (1980).
70. D. Stievenard, J. C. Bourgoin, D. Pons, Proceedings of the 12th Int. Conf. on Defects in Semi-conductors, Amsterdam (Holland) (Sept. 1982), Physica 116B, 394 (1983).
71. T. Ishida, K. Maeda, S. Takeuchi, Appl. Phys. 21, 257 (1980).
72. T. Wosinski, A. Morawski, T. Figielski, to be published in Appl. Phys. A, 30, 233 (1983).
73. D. E. Holmes, K. R. Elliott, R. T. Chen, C. G. Kirkpatrick, Proceedings of the 2nd Conf. on Semi-insulating III-V Materials (Edited by S. Makram-Ebeid and B. Tuck, Shiva Ltd, U.K.), 19 (1982).
74. G. M. Martin, Physica Scripta, T1, 38 (1982).
75. M. D. Miller, G. H. Olsen, M. Ettenberg, Appl. Phys. Lett. 31, 538 (1977).
76. P. K. Bhattacharya, J. W. Ku, S. J. T. Owen, V. Aebi, C. B. Cooper, R. L. Moon, Appl. Phys. Lett. 36, 304 (1980).
77. E. E. Wagner, D. E. Mars, G. Hom, G. B. Stringfellow, J. Appl. Phys. 51, 5434 (1980).
78. J. Lagowski, H. C. Gatos, J. M. Parsey, K. Wada, M. Kaminska, W. Walukiewicz, Appl. Phys. Lett. 40, 342 (1982).
79. M. Ozeki, J. Komeno, S. Ohkawa, Fujitsu Sci. Tech. J. 15, 83 (1979).
80. L. Samuelson, P. Omling, H. Titze, H. G. Grimmeis, J. Cryst. Growth, 55, 164 (1981).
81. M. O. Watanabe, A. Tanaka, T. Nakanisi, Y. Zohta, Jap. J. Appl. Phys. 20, L429 (1981).
82. G. M. Martin, G. Jacob, A. Goltzene, C. Schwab, G. Poiblaud, Proceedings of the 11th Conf. on Defects and Radiations Effects in Semiconductors (Oiso, Japan, 1980), Inst. Phys. Conf. Ser. 59, 281 (1981).
83. R. T. Blunt, Proceedings of the 2nd Conf. on Semi-insulating III-V Materials (Edited by S. Makram-Ebeid and B. Tuck, Shiva Ltd, U.K.), 107 (1982).
84. M. Tajima, Y. Okada, Proceedings of the 12th Int. Conf. on Defects in Semi-conductors, Amsterdam (Holland)(Sept. 1982) Physica AA6B, 404 (1983).
85. M. Oseki, J. Komeno, A. Shibatomi, S. Ohkawa, J. Appl. Phys. 50, 4808 (1979).
86. A. Humbert, L. Hollan, D. Bois, J. Appl. Phys. 47, 4137 (1976).
87. S. Subramanian, B. M. Arora, S. Guha, Solid State Electron, 24, 287 (1981).
88. D. V. Lang, A. Y. Cho, A. C. Gossard, M. Ilegems, W. Wiegmann, J. Appl. Phys. 47, 2558 (1976).
89. J. H. Neave, P. Blood, B. A. Joyce, Appl. Phys. Lett. 36, 311 (1980).

90. D. S. Day, J. D. Oberstar, T. J. Drummond, H. Morkoc, A. Y. Cho, B. G. Streetman, J. Electron. Mat. *10*, 445 (1980).
91. S. H. Xin, W. J. Schaff, C.E.C. Wood, L. F. Eastman, Appl. Phys. Lett. *41*, 742 (1982).
92. A. Mitonneau, J. P. Chané, J. P. André, J. Electron. Mat., *9*, 213 (1980).
93. H. J. Stocker, J. Appl. Phys. *48*, 4583 (1977).
94. Zhou Binglin, Wang Le, Shao Yongfu, Chen Qiyu, Acta Physica Sinica, *28*, 350 (1979).
95. S. Makram-Ebeid, M. Lannoo, Phys. Rev. *B 25*, 6406 (1982).
96. S. Makram-Ebeid, Proceedings of the MRS Meeting (Nov. 1980), Defects in Semiconductors (Editors J. Narayan, T. Y. Tan), (North Holland), 495 (1981).
97. J. S. Blakemore, J. Appl. Phys. 53, R123 (1982).
98. G. M. Martin, Appl. Phys. Lett, 39, 747 (1981).
99. J. Lagowski, J. M. Parsey, M. Kaminska, K. Wada, H. C. Gatos, Proceedings of the 2nd Conf. on Semi-insulating III-V Materials (Edited by S. Makram-Ebeid and B. Tuck, Shiva Ltd, U.K.), 154 (1982).
100. Y. Nanishi, J. M. Parsey, J. Lagowski, H. C. Gatos, to be published in J. Electrochem. Soc.
101. Zou Yuanxi, Zhou Jicheng, Mo Peigen, Lu Fengzmen, Li Liansheng, Shao Jiuan, Huang Lei, Proceedings of the Int. Conf. on GaAs and Related Compounds, Albuquerque (U.S.A.) (Sept. 1982), to be published.
102. See, for instance, F. Hasegawa, N. Iwata, Y. Nannichi, Jap. J. of Appl. Phys. *21*, 1479 (1982).
103. A. Mircea, A. Mitonneau, L. Hollan, A. Brière, Appl. Phys. *11*, 153 (1976).
104. G. M. Martin, A. Mitonneau, D. Pons, A. Mircea, D. W. Woodward, J. Phys. C: Solid State Physic, *13*, 3855 (1980).
105. S. Makram-Ebeid, G. M. Martin, D. W. Woodard, Proceedings of the 15th Conf. on Phys. of S. C., J. Phys. Soc. Japan, *49*, Suppl. A, 287 (1980).
106. F. Hazegawa, A. Majerfeld, Electron. Lett. *12*, 52 (1976).
107. T. Okumura, T. Ikoma, Annual Research Report of the Solid State Electronics Laboratories (Institute of Industrial Science, Tokyo, Japan), Part C (1977).
108. S. Makram-Ebeid, D. Gautard, P. Devillard, G. M. Martin, Appl. Phys. Lett. 40, 161 (1982).
109. G. P. Li, K. L. Wang, Proceedings of the Elect. Mat. Conference (June 1982) U.S.A., to be published, and J. Appl. Phys., 53, 8653 (1982).
110. J. Lagowski, M. Kaminska, J. M. Parsey Jr., H. C. Gatos, M. Lichtensteiger, Appl. Phys. Lett., 41, 1078 (1982).
111. J. R. Wagner, J. J. Krebs, G. H. Strauss, A. M. White, Solid State Commun. 36, 15 (1980).
112. R.Worner, U. Kaufmann, J. Schneider, Appl. Phys. Lett. 40, 141 (1982).
113. E. R. Weber, H. Ennen, U. Kaufmann, J. Windscheif, J. Schneider, T. Wosinski, J. Appl. Phys. 53, 6140 (1982).
114. A. Goltzene, B. Meyer, C. Schwab, J. Appl. Phys., 54, 3117, (1983); see also A. Goltzene, B. Meyer, C. Schwab, Revue Phys. Appl. *18*, 703 (1983).
115. J. Schneider, Proceedings of the 2nd Conference on Semi-insulating III-V Materials (Edited by S. Makram-Ebeid and B. Tuck, Shiva Ltd, U.K.), 144 (1982).
116. C. E. Barnes, T. E. Zipperian, L. R. Dawson, Electronic Material Conference, June 1982, to be published in the J. of Electron. Mat.
117. E. R. Weber, J. Schneider, Proceedings of the 12th Int. Conference on Defects in Semiconductors, Amsterdam (Holland)(Sept. 1982) Physcia 116B, 398 (1983).

118. A. Mircea, A. Mitonneau, J. P. Hallais and M. Jaros, Phys. Rev. B 15, 3665 (1977).
119. A. Mircea, A. Mitonneau and J. Vannimenus, J. de Phys. Lett., 38, L41 (1977).
120. A. Mitonneau, A. Mircea, G. M. Martin and D. Pons, Rev. Phys. Appl., 14, 853 (1979).
121. J. A. van Vechten and C. D. Thurmond, Phys. Rev. B14, 3539 (1976).
122. C. D. Thurmond, J. Electrochem. Soc., 122, 1133 (1975).
123. A. Zylberstejn, Appl. Phys. Lett., 32, 200 (1978).
124. C. Henry and D. V. Lang, Phys. Rev. B15, 989 (1977).
125. B. K. Ridley, J. Phys. C, 11, 2323 (1980).
126. S. Makram-Ebeid, Appl. Phys. Lett., 37, 464 (1980).
127. C. J. Rhees, H. G. Grimmeiss, E. Jantzen and B. Skarstam, J. Phys. C, 13, 6157 (1980).
128. V. YA. Prinz and S. N. Rechkunov, presented at the 4th Int. Conf. on Deep Impurity in Semiconductors, Eger, Hungary (May-June 1983).
129. A. Mircea and A. Mitonneau, J. de Phys. Lett., 40, L31 (1979).
130. A. Barraud, C. R. Acad. Sci. Paris, 256, 3632 (1963).
131. B. K. Ridley, Proc. Phys. Soc. London, 82, 954 (1963).
132. B. K. Ridley, J. Phys. C7, 1169 (1974).
133. M. Kaminska, J. M. Parsey, J. Lagowski, H. C. Gatos, Appl. Phys. Lett. 41, 989 (1982).
134. R. Pässler, work presented at the 4th Int. Conf. on Deep Impurity Centers in Semiconductors, Eger, Hungary (May-June 1983).
135. A. Chantre, G. Vincent and D. Bois, Phys. Rev. B23, 5335 (1981).
136. G. Vincent, Appl. Phys., 23, 215 (1980).
137. G. M. Martin, Proceedings of the 1st Conf. on Semi-insulating III-V Materials, Edited by G. J. Rhees, Shiva Publishing Ltd, 13-28 (1980).
138. A. M. White, P. Porteous, W. F. Sherman and A. A. Stadtmuller, Solid State Phys., 10, L 4731 (1977).
139. A. M. White, P. Porteous and P. J. Dean, Electron Materials, 5, 91 (1976).
140. F. Litty, Doctorat Thesis, INSA Lyon (France), (1982).
141. A. Zylberstejn, R. H. Wallis and J. M. Besson, Appl. Phys. Lett., 32, 764 (1978).
142. D. Bois and G. Vincent, J. Phys. Lett., 38, 351 (1977).
143. G. Vincent, D. Bois and A. Chantre, J. Appl. Phys., 53, 3643 (1982).
144. A. Mitonneau and A. Mircea, Solid State Comm., 30, 157 (1979).
145. M. Kaminska, M. Skowronski, J. Lagowski, J. M. Parsey and H. C. Gatos, Appl. Phys. Lett., 43, 302 (1983).
146. M. Skowronski, M. Kaminska, W. Kuzko and M. Godlewski. Presented at the 4th Conf. on Deep Impurity Centers in Semiconductors, Eger, Hungary, (May-June 1983).
    and also M. Kaminska, M. Skowronski, W. Kuzko, J. Lagowsky, J. Parsey and H. C. Gatos, to appear in Czech Journal of Physics.
147. A. L. Lin, E. Omelianovski and R. H. Bube, J. Appl. Phys., 47, 1852 (1976).
148. P. Leyral, G. Guillot, Proceedings of the 2nd Conf. on semi-insulating III-V Materials, Edited by S. Makram-Ebeid and B. Tuck, Shiva Publishing Ltd, 166-171 (1982).
149. B. V. Shanabrook, P. B. Klein, E. M. Swiggard and S. G. Bishop, J. Appl. Phys., 54, 336 (1983).
150. H. Nakashima, M. Matsunaya and Y. Shiraki, "GaAs and Related Compounds," Inst. Of Phys. Conf. Series, 63, 203 (1982).

152. A. Mircea-Roussel and S. Makram-Ebeid, Appl. Phys. Lett. *38*, 1007 (1981).
153. P. Leyral, G. Vincent, A. Nouaillat and G. Guillot, Solid State Comm., *42*, 67 (1982).
154. D. R. Wight, I. D. Blenkinsop and S. J. Bass, Proceedings of the 1st Conf. on Semi-insulating III-V Materials, Edited by G. J. Rhees, Shiva Publishing Ltd, 174–182 (1980).
155. B. V. Shanabrook, P. B. Klein and S. G. Bishop, presented at the 16th Conference on the Physics of Semi-conductors, Montpellier (1982), to be published.
156. P. W. Yu, D. E. Holmes and R. T. Chen, GaAs and Related Compounds, Inst. of Phys. Conf. Series 63, 200 (1982).
157. P. W. Yu, Solid State Commun., *43*, 953 (1982).
158. M. Tajima and Y. Okada, 12th Int. Conf. on Defects in Semi-conductors, C.A.J. Amerlan Editor, Amsterdam, (1982). Physica 116B, 404 (1983).
159. E. E. Wagner, D. E. Mars, G. Hon and G. B. Stringfellow, J. Appl. Phys., *51*, 5434 (1980).
160. N. M. Johnson, R. D. Burnham, D. Fekett and R. D. Yunfling in Defects in Semiconductors, J. Narayan and Y. Tan Editors, North Holland, 481 (1981).
161. T. Matsumoto, P. K. Bhattacharya and M. J. Ludowise, Appl. Phys. Lett., *41*, 662 (1982).
162. D. J. Woolford, Ph. D. Thessis, Univ. of Illinois (1979).
163. G. Bastide, G. Sagnes and C. Merlet, Rev. Phys. Appl., *15*, 1517 (1980).
164. D. E. Aspnes, Phys. Rev. B, 5331 (1976).
165. S. Y. Ren, J. D. Dowand, D. J. Wolford, Phys. Rev. B25, 7661 (1982).
166. J. P. Buisson, R. E. Alenand, J. D. Dow, Solid State Commun., *43*, 883 (1982).
167. J. Lagowski, M. Kaminska, J. M. Parsey, H. C. Gatos, W. Walukiewicz, Proceedings of the Int. Conf. on GaAs and related Compounds (1982, Albuquerque, U.S.A.) Inst. Phys. Conf. Ser No 65, 41 (1983).
168. D. V. Lang and R. A. Logan, Inst. Phys. Conf. Ser., *43*, 433 (1979).
169. D. V. Lang, J. Phys. Soc. Japan, *49*, Suppl. A, 215 (1980).
170. J. R. Van Vechten, J. Electrochem. Soc., *122*, 423 (1975).
171. P. J. Lin-Chung, Proceedings of the MRS meeting (Symposium on Defects in semi-conductors), November 1982 (Boston, U.S.A.), to be published.

CHAPTER 7

DX Centers in III-V Alloys

D. V. Lang

A.T. & T. Bell Laboratories
Murray Hill, New Jersey 07974

1. INTRODUCTION
2. LARGE LATTICE RELAXATION
   2.1 MPE Capture in the Strong-Coupling Regime
   2.2 Optical Lineshape in the Strong-Coupling Regime
   2.3 Extrinsic Self-Trapping
3. EXPERIMENTAL METHODS
   3.1 Thermally Stimulated Capacitance (TSCAP)
   3.2 Isothermal Capacitance Transients
   3.3 Deep Level Transient Spectroscopy (DLTS)
   3.4 Majority Carrier Capture Measurements
   3.5 Photocapacitance
4. SUMMARY OF EXPERIMENTAL RESULTS FOR DX CENTERS
   IN AℓGaAs AND GaAsP
   4.1 Ionization and Activation Energies
   4.2 Band Structure Effects
   4.3 Structural Information
5. CONCLUDING REMARKS

1. Introduction

The DX centers in III-V alloys such as AlGaAs and GaAsP are prototypes
of an interesting new class of semiconductor defects which is characterized
by the dominant role of lattice relaxation effects in determining carrier
capture and emission transitions. Defects exhibiting large lattice relaxation
effects have been reported in a wide variety of semiconductor systems and
have been reviewed by Langer.[1] It is not unusual to find such effects in a
semiconductor such as CdF_2,[1,2] since lattice relaxation effects are well
known to play an important role in strongly ionic materials. It is quite surprising, however, that one of the earliest reports of such defects in semiconductors was by Porowski and coworkers[3,4] in the covalent material
InSb. It is now generally accepted that lattice relaxation effects can be
quite substantial at defects in covalent semiconductors. Indeed, following
the theoretical work of Baraff, Kane, and Schluter,[5] even the perfectly
covalent material silicon has been shown by Watkins and coworkers to
have large lattice relaxation effects leading to the negative-U behavior of
the vacancy[6] and interstitial boron.[7]

The importance of large lattice relaxation effects in covalent materials is
less surprising when one notes that in all cases where the defect structure is
known, the relaxation is thought to be associated with weak or broken
bonds of the defect and would not occur in the undisturbed host lattice.
This is to be contrasted with the situation in the alkali halides where even
intrinsic free holes are self trapped due to the electron-lattice interaction.
This situation was explained by Toyozawa in terms of the concepts of
intrinsic as well as extrinsic self trapping,[8,9] and will be discussed later in
this chapter.

The DX centers are a good example of Toyozana's extrinsic self trapping
case. They have been extensively studied in AlGaAs[10-16] and in GaAsP.[17-20]
The name "DX center" was originally proposed[12] to account for the
dual observations that the defect concentration was nearly always proportional to the shallow donor concentration (hence the relation to the donor
impurity D) and the defect behaved in a decidedly non-effective-mass way
(hence the supposed association with an unknown defect X.) Since the detailed microscopic structure of this type of defect is still unclear, it is useful to continue the DX center notation for this class of behavior in AlGaAs
and GaAsP. The close similarities among the donor-related deep centers
exhibiting large lattice relaxation effects in these III-V alloys strongly
suggests that the underlying defect structure is of the same generic type
in all cases. Whether this type of defect is also closely related to those
showing similar behavior in other semiconductors is an open question at
this point.

One of the more striking phenomena associated with DX centers is persistent photoconductivity. Indeed, in the area of III-V alloys Craford et al.[17] had reported this effect in S-doped GaAsP long before DX centers were shown to be responsible.[10,12,18] This persistent photoconductivity effect, which has been seen in a number of other semiconductors as well,[1,10] was originally explained by several different models based on purely electronic effects not involving lattice relaxation. Namely, (a) forbidden capture transitions due to band structure effects[17,21,22] (b) repulsive Coulomb barriers to capture at multicharged defects,[17,23,24] and (c) macroscopic internal Coulomb barriers due to electric fields between high and low resistivity regions of inhomogenous material.[25,26] For most cases the first two models can be readily dismissed by a number of observations which have been discussed in the literature[1,10,25] although relatively weak persistence effects (< 200 sec) at low temperature (< 30 K) have been adequately explained by a double acceptor model (b).[27]

The macroscopic barrier model (c), on the other hand, cannot be dismissed as lightly. Indeed, there is solid evidence that such an effect is responsible for the persistent photoconductivity observed in thin ($<$ 5 μm) epitaxial films of GaAs on semi-insulating substrates.[26,28] In this case, the barrier is at the interface between the n-type film and the semi-insulating substrate. The more random bulk inhomogeneities envisioned by Shik and Vul,[25] such as defect clusters, doping fluctuations, precipitates, or grain boundaries, may also give rise to persistence effects in some cases. These effects are particularly likely in impure, closely compensated bulk crystals, such as those on which much of the GaSb and CdS work was done.[25] The DX center work in III-V alloys, however, was all done on very recent state-of-the-art epitaxial crystals grown by a number of different methods, such as liquid phase epitaxy (LPE), chemical vapor deposition (CVD), and molecular beam epitaxy (MBE). Since CVD and MBE produce very homogeneous films, it is not likely that random macroscopic inhomogeneities could explain the trapping behavior attributed to DX centers in these AlGaAs and GaAsP films. The semi-insulating substrate effect discussed by Queisser and Theodorou[26] clearly will not affect junction capacitance measurements on films grown on n^+ substrates. As we shall discuss, all of the data except photoconductivity were taken by junction techniques.

It is generally believed by all workers who have studied DX trapping centers in AlGaAs and GaAsP that the large lattice relaxation model gives the best overall agreement with the large body of data that is now available. This model will be discussed in detail in Section 2 of this chapter. In Section 3 we will briefly review the various junction capacitance techniques and measurement methods used to study DX centers. The data will

be summarized in Section 4. Finally, in Section 5 the results will be critically discussed in an effort to shed some light on possible microscopic
models for DX centers. While it is still premature to settle on a definitive
model, there are nevertheless a number of general microscopic features
which any successful model must incorporate. It is hoped that the critical
review presented here will help in the ultimate understanding of the detailed microstructure of this interesting class of defects.

2. Large Lattice Relaxation

Nearly all defects in solids are coupled to the lattice to some extent. The
case of DX centers provides one of the most dramatic examples of this
coupling. These defects are in the limit where the energies associated with
lattice relaxation are on the order of or larger than the binding energies
associated solely with the electronic part of the defect Hamiltonian. Thus,
the appropriate zeroth order view of such defects is that of a defect molecule embedded in a solid rather than a screened hydrogen atom, as in the
case of the effective mass theory.[29]

The effect of lattice relaxation is manifested in two major ways, namely,
(1) carrier capture via multiphonon emission (MPE), and (2) a Stokes shifts
of the transition energy for optical ionization relative to the energy necessary for thermal ionization. For the particular case of large lattice relaxation, which is appropriate to DX centers, these effects dominate the properties of the defect. In order to better understand the basic properties of
DX centers it is therefore important to briefly discuss lattice relaxation
effects in general. In this section we will consider three major areas: (a)
MPE capture in the strong-coupling regime; (b) Optical lineshape in the
strong-coupling regime; (c) Extrinsic self trapping.

2.1 MPE Capture in the Strong-Coupling Regime

In general, we may express[30,31] the MPE capture cross section σ as

$$\sigma = Af(0) \tag{1}$$

where A is a term involving only the electronic matrix elements of the
transition and $f(h\nu)$ is the well-known optical lineshape for phonon-assisted
absorption or emission transitions. The lineshape function $f(h\nu)$ has been
evaluated in closed form for arbitrary coupling strength and temperature

by Huang and Rhys,[32] Kubo and Toyozawa,[33] and Lax.[34] Since nonradiative transitions involve the overlap of vibronic states at the same energy, it is the function f(hν) evaluated at hν = 0 which dominates the temperature dependence of the MPE capture rate. The contribution of Henry and Lang[30,35] to this problem was an approximate calculation of the electronic term A for free-to-bound transitions between the continuum and bound states, respectively, of a spherical square well potential whose radius was modulated by the lattice. Considering the simplicity of the model, their value of A was in surprisingly good agreement with experimental data for deep levels in GaAs and GaP, most of which were within an order of magnitude of the Henry and Lang result, A_{HL}= 1.5 × 10^{-14} cm^2 eV for a neutral center. The previous work by Lax[34] did not consider the breakdown of the adiabatic approximation, which was a key ingredient of the HL theory, and therefore underestimated A by many orders of magnitude. As a result of Lax's work, MPE capture was considered to be unimportant is semiconductors for over 20 years.

For our purposes here of discussing the overall trends in the temperature dependence of MPE capture it is most instructive to focus attention on f(0) and to treat A, which is only weakly temperature dependent, as an adjustable parameter which may vary within approximately an order of magnitude of A_{HL}. In the limit of strong electron-phonon coupling, f(hν) approaches a Gaussian lineshape[31,34] and f(0) may be written[31] as

$$f(0) = (4\pi E_R kT^*)^{1/2} \exp(-E_B/kT^*) \qquad (2)$$

where $E_R = S\hbar\omega$ is the lowering of the energy of the bound state due to lattice relaxation, $\hbar\omega$ is the average energy of the phonons to which the defect is coupled, and E_B is the classical barrier height at the crossing of the configuration coordinate (cc) curves as shown in Figure 1. The effective temperature T^* is defined as

$$kT^* = \frac{\hbar\omega}{2} \coth \frac{\hbar\omega}{2kT} . \qquad (3)$$

Thus, at high temperatures $T^* = T$ and Eq. (2) shows the classical thermally activated capture first predicted by Mott.[36] At low temperatures, $kT^* = \hbar\omega/2$ and the zero-point vibrations of the lattice play the role of temperature in promoting MPE transitions. Thus, at low temperatures, the log of the capture cross section is proportional to the number of phonons necessary to reach the classical barrier E_B. This behavior is reminiscent of the

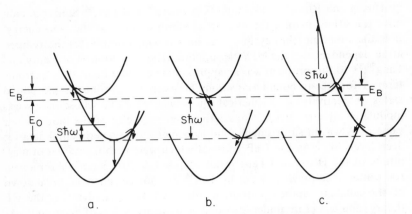

FIGURE 1 Configuration coordinate (cc) diagrams for progressively larger amounts of lattice relaxation energy, E_R . In all three cases the lowest curve corresponds to the total energy of the system in its ground state; the upper curve is the threshold energy of an e.h. pair plus an empty defect; the shifted central curve is for a free h plus an occupied defect. (a) "normal" defect; (b) extrinsic self trapping with no barrier; (c) extrinsic self trapping with same barrier as in (a). [Ref. 14]

so-called "energy-gap law" characteristic of the low-temperature weak-coupling regime where the log of the capture rate is proportional to the number of phonons needed to equal to the *depth* of the level E_0.[31,37] It has been recently proposed[37] that this weak-coupling "energy-gap law," which is valid for large molecules[31] and for narrow-line emissions in solids,[38,39] ought to also roughly describe the low temperature behavior of even more strongly coupled defects. However, as we shall discuss below, such an expectation is not supported by the deep-level capture data in GaAs and GaP.[30] For example, in Figure 2 σ_n (EL2) is larger than σ_n (E3) even though EL2 is over twice as deep as E3. Also, σ_{p2} (GaP:0) is one of the largest capture cross sections known, in spite of the fact that the oxygen level is deeper than all other levels shown in Figure 2.

The coupling strength parameter G is approximately given by[31]

$$G = S \coth \frac{\hbar\omega}{2kT} \qquad (4)$$

where $S = E_R/\hbar\omega$ is often called the Huang-Rhys factor.[32] The strong-coupling regime is then defined[31] as $G \gg 1$; the weak-coupling case corresponds to $G \leqslant 1$. However, in spite of the fact that these two regimes are well-defined mathematically, it is often difficult to determine which is more appropriate to a particular set of experiment data. A case in point is the so-called level B in GaAs.[40] The electron capture data[30,40,41] for this

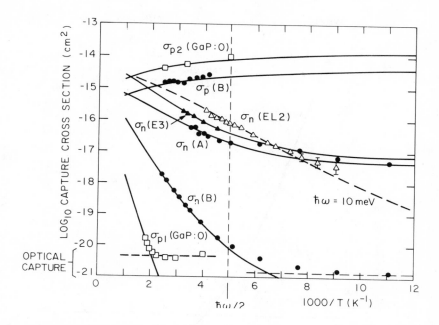

FIGURE 2 Fit of the strong-coupling MPE expression to a range of typical capture cross sections for deep levels in GaAs and GaP: The specific values of the fitting parameters are less important than the fact that the overall trends of the data are easily accounted for by this expression. [Ref. 14]

level have been fit by various MPE expressions with three widely different values of S, namely, S = 0.5–1,[42] S = 3.0,[43] and S = 7.7.[44] These values range from the weak (S ≤ 1) to strong (S > > 1) coupling regimes. In Figure 3 we show our own fit to the level-B data[45] using Equations (1–3) with A = 0.07A_{HL}, S = 5.9, and $\hbar\omega$ = 34 meV. Since E_0 = 0.72 eV in this case,[40] the classical barrier height in Figure 2 is E_B = 0.33 eV. This is to be compared with the apparent asymptotic slope of the data, namely 0.25 eV. About half of the 80meV difference between E_B = 0.33eV and the apparent value of 0.25 eV can be accounted for by the T^{-1} behavior of the exponential prefactor, with the balance due to the fact that the system is not totally in the high-temperature limit even for the highest temperature data.

The sharp transition at about 1000/T = 6 between nearly temperature independent and nearly activated behavior in Figure 3 is far too abrupt to be fit by MPE capture. Therefore, as has been suggested by Burt,[43] another capture mechanism must be operative. Since Henry and Lang[30] clearly ruled out Auger capture for this particular case, and since the center is most likely neutral before electron capture, the only possible

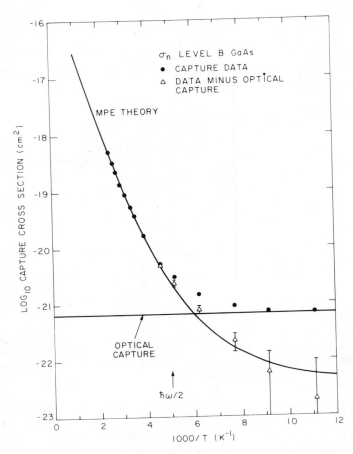

FIGURE 3 Fit of the strong-coupling MPE expression to the electron capture cross section for level B in GaAs: Adjustable parameters are A, E_B, and σ_{op}. ($\hbar\omega$ is set at 34 MeV). [Ref. 14]

mechanism is optical capture by emission of a photon. It is possible to estimate the effective cross section σ_{op} for optical capture from experimental values for the photoionization cross section σ° (E) as measured by junction photocurrent or photocapacitance. From the well-known Einstein relations between absorption and spontaneous emission[46] we have for the case of electron capture

$$\sigma_{op} = 3.5 \times 10^{19} \, \frac{\bar{n}^2 E^2 T}{vN_c} \, \sigma_n^{\circ}(E) \qquad (5)$$

where $\bar{n}$ is the refractive index, E the photon energy (eV) at the threshold of σ^0(E), v is the thermal velocity (cm/s) of free carriers and $N_{c(v)}$ is the effective-mass density of states in the conduction (valence) band (cm^{-3}). For typical values[47] of σ^0 within kT of threshold ($10^{-17} \leqslant \sigma^0 \leqslant 10^{-16}$ cm^2) we have $10^{-21} \leqslant \sigma_{op} \leqslant 10^{-20}$ cm^2. The optical capture cross section needed to fit the data in Figure 3 is in this approximate order of magnitude range.

Figure 2 shows the capture data from Ref. 30 where the solid lines are calculated using Equations (1–3) with $\hbar\omega$ = 34 meV and with $0.03 \leqslant$ A/$A_{HL} \leqslant 1$ and $0 < E_B < 0.6$ eV treated as adjustable parameters. It is clearly possible to fit the general trends of the data with the strong coupling result of Equation 2. Note in particular that the low temperature data are totally unrelated to the depths of the traps and hence cannot possibly be described by the weak-coupling "energy-gap law." For example, σ_{p1} and σ_{p2} for GaP:0 differ by more than six orders of magnitude, yet both states have $E_0 \sim 1.3$ eV. The differences in cross sections are due to the difference in the barrier to capture E_B, *not* a difference in depth. Similarly σ_n(A) $>> \sigma_n$(B), while E_0 = 1.0 eV for level A and E_0 = 0.72 eV for level B. Indeed, it is one of the major triumphs of the strong-coupling MPE theory that $\sigma_{p2} = 10^{-14}$ cm^2 for GaP:0 can be fit by the theory in spite of the fact that this is one of the deepest states ($\sim$1.3 eV) studied.

From Equation (3) it is evident that the average energy of the phonons that couple most strongly to the defect is the determining factor in separating the high and low temperature regimes. The midpoint of the transitional temperature range is at kT = $\hbar\omega$/2. In Figure 2, this corresponds to T = 200K for $\hbar\omega$ = 34 meV. Note, however, that the capture cross section of the EL2 trap[48] does not have a "knee" at 200K but is thermally activated to much lower temperatures. This fact has been cited as evidence that the activation barrier for this trap is electronic in origin rather than due to MPE.[42] Note, however, that the EL2 data in Figure 2 can be fit quite well by Equations (1)–(3) with a phonon energy of 10 meV. The very strong dependence on phonon energy is best illustrated in Figure 4 where we show theoretical curves for three different phonon energies all with E_B = 0.25 eV. The phonon energies of 10 and 34 meV were specifically chosen to match the regions of maximum density of states in the phonon spectrum of GaAs as shown in Figure 5.[49] Defects which couple to the LO or TO phonons would tend to have $\hbar\omega$ = 34 meV while those which couple to TA phonons would have $\hbar\omega \simeq 10$ meV. The phonon spectrum of silicon[49] has the same shape as that in Figure 5 except that the energy scale is a factor of two larger for Si. Thus, defects in silicon which couple to optical phonons would tend to have rather large, very weakly temperature dependent cross sections below 300K in agreement with the experimental results.[50]

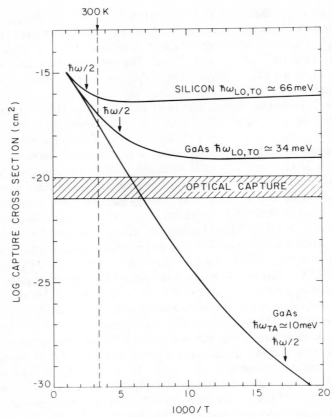

FIGURE 4   Equations (1)–(3) plotted for various phonon energies. [Ref. 14]

The behavior of $\sigma(T)$ for $\hbar\omega \cong 10$ meV in Figure 4 is qualitatively very similar to that observed for DX centers and for the decay of persistent photoconductivity.[10,11] It is clear from the discussion of MPE capture that such behavior can occur only if two conditions are met, namely, (1) the energy of the local mode vibration of the defect must be relatively small ($\hbar\omega \leqslant 10$ meV), and (2) the lattice relaxation occuring after carrier capture must be large enough so that optical capture is not possible and E_B is large ($\geqslant 0.2$ eV) (e.g. Figure 1c). The first condition makes MPE capture by zero-point vibrations very weak ($< 10^{-30}$ cm^2) and moves this regime to low temperatures (< 50K). The second condition is necessary for this weak capture regime to be observed. If optical capture were possible (e.g. Figure 1a), the cross section would drop no lower than approximately 10^{-21} cm^2 even for a very large E_B in Equation (2), as shown for

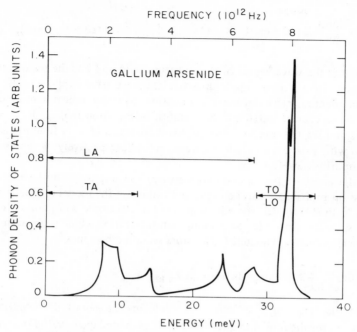

FIGURE 5 Phonon density of states in GaAs. [Ref. 14]

"normal" deep level defects in Figures 2 and 3. As we shall see, the large relaxation limit illustrated schematically in Figure 1c is also necessary to explain the very large Stokes shifts observed in the optical properties of DX centers. Thus large Stokes shifts and weak, thermally activated capture at low temperatures, which are the essential features of this class of defects, each lead independently to the large relaxation model.

2.2 Optical Lineshape in the Strong-Coupling Regime

Just as the strong-coupling regime leads to relatively simple relationships to describe MPE, so too does the expression for the lineshape of optical ionization transitions reduce to a rather simple form in this limit. We will follow here the lineshape theory of Jaros.[12,51] The number of free parameters in the part of this theory relating to the purely electronic part of the matrix element is somewhat larger than one would like for definitive curve fitting. However, in the strong-coupling limit, the lineshape is relatively insensitive to the purely electronic part.

In the absence of an electron-phonon interaction, it is a standard approximation to write the cross section σ as

$$\sigma(h\nu) = \frac{\text{const}}{h\nu} \sum_{n,\vec{k}} |\langle\psi \mid \exp(-i\vec{k}_\lambda \cdot \vec{r})\,\vec{\epsilon}_\lambda \cdot \vec{p} \mid \phi_{n,\vec{k}}\rangle|^2 \times \delta(\epsilon\theta + E_{n,\vec{k}} - h\nu), \quad (6)$$

where $\vec{k}$ is the wave vector of the radiation field and λ is the polarization direction. In the usual dipole approximation we have $\exp(-i\vec{k}_\lambda \cdot \vec{r}) \sim 1$. The momentum matrix element in Equation (6) really indicates an average over all degenerate initial and final states. In Equation (6), ϵ_θ, $E_{n,\vec{k}}$, and $\phi_{n,\vec{k}}$ stand for the impurity energy, band energy, and wave function assc ciated with a band n and reduced wave vector k, respectively; Ψ represents the impurity wave function.

In the event of strong coupling between the impurity and lattice, the transition probability can be expressed following the model of Huang and Rhys.[32] In this model, the equations for the electronic and phonon functions separate. Only the electron-phonon interaction which is linear in the lattice coordinates is included. The cross section σ becomes

$$\sigma(h\nu) \sim \frac{1}{h\nu} \sum_{n,\vec{k}} |\langle\Psi|\exp(-i\vec{k}\cdot\vec{r})\vec{\epsilon}_\lambda \cdot \vec{p}|\Psi_{n,\vec{k}}\rangle|^2 J_{n,\vec{k}} \tag{7}$$

where the function $J_{n,\vec{k}}$ carries the information about the vibrational states and for the model in question can be evaluated exactly.[52] For strong electron-phonon coupling, the expression for $J_{n,\vec{k}}$ simplifies to

$$J_{n,\vec{k}} = (\pi U)^{-\frac{1}{2}} \exp - h\nu - (|E_n| + E_{n,\vec{k}})^2/U ,$$

where

$$U = 2S(\hbar\omega)^2 /\tanh(\hbar\omega/2kT). \tag{8}$$

Here $\hbar\omega$ refers to the phonon energy, k is the Boltzmann constant, S is the Huang-Rhys factor, and the terms $S\hbar\omega$ and $E_n = E_0 + S\hbar\omega$ are defined in the configuration coordinate (CC) diagram shown in Figure 6. The preexponential term in Equation (8) obviously does not affect the shape of the optical cross section, and for our purposes can be omitted.

In Ref. 51 a series of simplifying approximations were introduced which, for a sufficiently localized Ψ, allow one to express Equation 7 in the form

$$\sigma(h\nu) \sim \frac{1}{h\nu} \int_0^\infty dE\rho(E) \times \left| \frac{(1\pm\eta)E^{\frac{1}{2}}}{|E_n|+E} + \frac{(1\mp\eta)\sqrt{E_F}}{|E_n|-E-(E_g+E_A)/2} \right|^2$$
$$\times U^{-\frac{1}{2}}\exp - \left[\frac{[h\nu-(|E_n|+E)]^2}{U} \right] , \tag{9}$$

where $\rho(E)$ represents the density of electron states, E_F is the free-electron

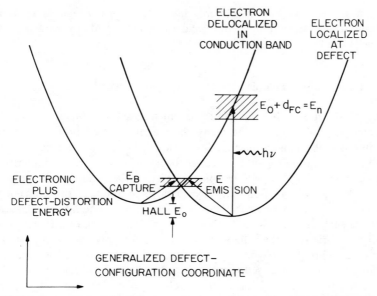

ELECTRON
DELOCALIZED
IN
CONDUCTION BAND

ELECTRON
LOCALIZED
AT
DEFECT

$E_0 + d_{FC} = E_n$

$h\nu$

ELECTRONIC
PLUS
DEFECT-DISTORTION
ENERGY

E_B
CAPTURE

E
EMISSION

HALL E_0

GENERALIZED DEFECT–
CONFIGURATION COORDINATE

FIGURE 6   Typical configuration coordinate (cc) diagram for DX centers in AlGaAs and GaAsP. This figure is drawn to scale for the case of Te-doped AlGaAs where the range of uncertainty in the data are shown as the shaded regions. [Ref. 12]

Fermi energy, E_g is the forbidden band gap, and E_A is the average or Penn gap.[51] The function $\eta(E)$ interpolates between its apparent values at $E = 0$ and $E \to \infty$, and was chosen simply as $\eta(E) = \exp(-2E/E_A)$. The required choice of + or − depends on the nodal character of Ψ and corresponds to valence-band-like or conduction-band-like deep states for the upper lower signs, respectively. In addition, $\rho(E)$ and E_n may be functions of temperature as well.

In the limit of a large S, the formula for $\sigma(h\nu)$ in Equation 9 is dominated by the lattice-relaxation Gaussian and the details of the electronic part in Equation (6) become unimportant. The approximations in Ref. 51 concerning Equation (6) are therefore perfectly acceptable in this case. A difficulty may arise, however, at $h\nu \geq 1.5$ eV, since above that energy the drop in the crystal density of states[53] at around 2 eV above the conduction-band edge begins to affect the cross section. In order to represent the cross section well at $h\nu \geq 1.5$ eV one would have to abandon this simple model and perform a more rigorous calculation of the electronic part in which the true variation of the electronic matrix element and the density of states farther from the band edge is better accounted for. However, the major use of this lineshape theory for the DX center case is to determine S, $\hbar\omega$, and E_0 from the thermal broadening and composition dependence of $\sigma(h\nu)$. Therefore, one is primarily interested in σ between its apparent threshold and maximum. In this regime it is not a bad approximation to let $\rho(E) \propto E^{1/2}$.

2.3 Extrinsic Self Trapping

The configuration coordinate models which have been constructed to phenomenologically explain the many thermal and optical properties of DX centers are all of the same generic class as the large relaxation cases shown in Figure 1c and in Figure 6. These cc diagrams bear a striking similarity to those used to explain self trapping of free carriers and excitons in ionic solids.[9] This observation led Toyozawa[8,9] to propose the concept of extrinsic self trapping to explain in general terms large lattice relaxation defects such as CdF_2:In and DX centers in III-V alloys. In this section we will very briefly outline the essential ideas of extrinsic self trapping. The reader is referred to Toyozawa's excellent review[9] for further details.

To describe extrinsic self trapping it will be helpful to first briefly review the concepts of normal (or intrinsic) self trapping. Basically, self trapping of an electron, for example, can be viewed as the result of the competition between the localization tendencies of the electron-phonon interaction at a particular lattice site and the delocalization tendencies due to overlap of atomic orbitals on nearest neighbor sites. In this very simple picture the energy for localization is the lattice relaxation energy $E_{LR} = S\hbar\omega$, while the delocalization energy B is equal, in the tight-binding approximation to one half of the width of the conduction band. Toyozawa defines an electron-phonon coupling parameter $g \equiv E_{LR}/B$ which separates the regime of free carriers ($g < 1$) from that of self-trapped carriers ($g > 1$).

The basic physics of self trapping is perhaps more clear in the cc diagrams (a, b) and band diagrams (c) shown in Figure 7 for the example of electron trapping. Part (a) of this figure shows the effect of lattice relaxation alone without the delocalizing effects of the conduction band, i.e. the excitation energy E_A corresponds to an atomic transition which experiences only the lattice relaxation effects of its environment. Part (c), on the other hand, shows the delocalizing effects of wave function overlap in a rigid lattice. Part (b) shows the combined effects of lattice relaxation and delocalization. The self-trapped state S is clearly the lowest energy state when $E_{LR} = S\hbar\omega > B$, i.e. $g > 1$.

The complete treatment of self trapping must consider the different nature of long range (optic phonon) and short range (acoustic phonon) interactions as well as the degeneracy of the self trapped state and zero point lattice vibrations.[9] The essential features of Figure 7 remain, however, with only a slight renormalization of the energy scales.

The case of *extrinsic* self trapping has to do with defects rather than the intrinsic host crystal. The particular case of most interest for DX centers is

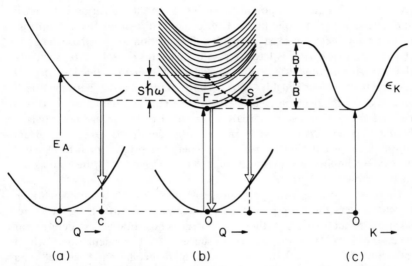

FIGURE 7 Configuration coordinate models for localized excitation (a) and exciton (b), and excitonic band in rigid lattice (c). The upper half of the Figure applied also to an electron or a hole. [Ref. 9]

a defect for which the short range electronic part of the potential is too weak by itself to bind an electron. If this defect is in a lattice where $g < 1$ so that intrinsic self trapping cannot occur, one might think at first that such a defect could have no bound states (except, perhaps, for shallow coulombic states if the defect is charged). The novel feature of Toyozawa's extrinsic self trapping idea is that a bound state can, in fact, exist even in this case due to the combined effect of the short range defect potential U and short range (acoustic) electron-phonon coupling $g_s = E_{LR}^{(ac)}/B$.

Figure 8b is a phase diagram showing the regions where the free state (F) and extrinsic self-trapped state (S) are the lowest energy state for a neutral defect with only a short range potential. For $U/B \geqslant 0.7$ the defect potential is strong enough to bind an electron without lattice relaxation, while for $g_s \geqslant 0.9$ intrinsic self trapping will occur even without a defect present. Part (a) of Figure 8 shows the sequence of cc diagrams corresponding to U/B increasing along the dashed line ($g = 0.6$) in the phase diagram below the intrinsic self trapping threshold. Note that the solid line separating the F and S regions of the phase diagram corresponds to the minima of the cc curves centered at $Q = 0$ and $Q = C$ being at exactly the same energy. For $U/B \geqslant 0.3$ in this example ($g_s = 0.6$), the combined effect of the impurity potential and the electron-phonon interaction give rise to a stable bound state S. Note, however, that the extrinsic self trapped state is separated from the free state by a potential barrier. Thus the transition between F and S behavior at the solid line in Figure 8(b) is discontinuous. In addition, metastable effects can exist near this F-S transition in the phase diagram. If the potential barrier is large (e.g. $\geqslant 0.2$ eV) this metastability is exactly the sort of long time effect typically seen (1) in many semiconductors at temperatures below roughly 100K. The potential barrier between the F and S states is an intrinsic property of short range interactions.[9] This is because for short range potentials, as opposed to long range potentials, a bound state appears only when the potential depth $\propto Q$ exceeds a certain threshold. Since the lattice distortion energy is proportional to Q^2, this necessarily implies the existence of a potential barrier in the cc diagram such as shown in Figure 8.

These extrinsic self trapping ideas give us a qualitative framework in which to view the basic physics of DX centers and other large lattice relaxation phenomena, but they do not give quantitative predictions for specific defects. However, the recent success of defect calculations[5] which take into account both the self-consistent defect potential as well as a realistic model of the lattice hold the promise that this problem will soon be solved for the large lattice relaxation case as well.

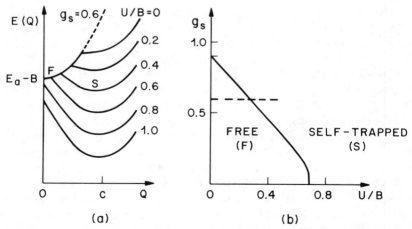

FIGURE 8 Configuration coordinate model for extrinsic self-trapping (a), and the stable state as a function of short-ranged impurity potential: U and electron-phonon coupling: g_s. [Ref. 9]

3. Experimental Methods

The primary experimental methods used to study DX centers are junction capacitance techniques. These measure three types of properties for these centers: (1) the rate of thermal emission of trapped electrons, (2) the rate of optical emission of trapped electrons, and (3) the thermal capture rates of electrons and holes. Other measurements, such as photoluminescence, the optical emission of holes, or electron spin resonance have been unable to detect DX centers. The negative results for photoluminescence and optical hole emission are consistent with the large relaxation cc diagram determined from the capacitance results. Electron spin resonance studies have not been attempted due to an expected lack of sensitivity for wide lines in thin epitaxial films ($< 10\mu m$ thick). Structural information about the symmetry of DX centers has been obtained, however, by the phonon scattering experiments of Narayanamurti et al.[54]

Since the junction capacitance measurements are rather similar for nearly all of the work published on DX centers, we will discuss in this section the techniques and give representative data to illustrate the analysis of each type of measurement. In Section 4 we will summarize the results in a way which focuses on the ionization energies and chemical trends rather than on the raw data.

There are five basic types of junction measurements used for DX centers:

1. Thermally stimulated capacitance (TSCAP).

2. Isothermal capacitance transients.

3. Deep level transient spectroscopy (DLTS).

4. Majority carrier capture measurements.

5. Photocapacitance.

These methods have all be reviewed in the literature.[47,55,56] Since some aspects are peculiar to DX centers, however, it is worthwhile to briefly discuss the techniques here as well.

3.1 Thermally Stimulated Capacitance (TSCAP)

In most cases the DX center concentration N_{DX} is a large fraction of the net donor concentration N_D and is proportional to N_D even to rather high doping levels, (i.e. $N_D \cong 10^{18} cm^{-3}$). Thus the relative capacitance change $\Delta C/C$ due to carrier trapping at (or emission from) DX centers in the depletion region of a pn junction or Schottky barrier is of the order of unity. In this large-signal limit the usual simple analysis methods used to obtain defect concentration from DLTS spectra are inadequate. TSCAP is a somewhat better diagnostic tool to study concentrations and slow meta-stable effects in such a system.

We show typical DLTS and TSCAP curves in Figures 9a and 9b, respective-ly. These techniques correspond to the same physical phenomena observed on different time scales. Thus, the DLTS peaks in Figure 9a correspond to thermal electron emission following a zero-bias voltage pulse (negative peak) and electron capture following a forward-bias injection pulse (posi-tive peak), both with a 2.7-msec time constant. The TSCAP data in Figure

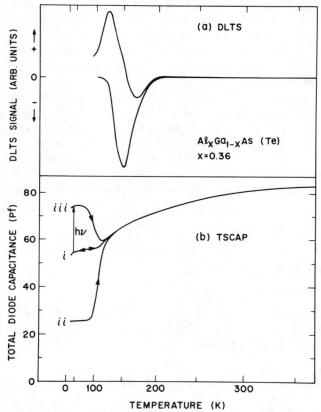

FIGURE 9 DLTS and TSCAP data for a typical sample of Te-doped $Al_xGa_{1-x}As$ with x = 0.36. The DLTS spectra correspond to a rate window of 366 sec^{-1}; the TSCAP heating rate was ~1 K/sec, Increases in C correspond to fewer trapped elec-trons, while decreases in C imply more trapped electrons. The photocapacitance tran-sition is indicated by the arrow labeled hv. [Ref. 12]

9b, on the other hand, correspond to time constants of the order of seconds. This explains the shift of the DLTS data to higher temperatures.

The procedures used in establishing the initial conditions for the three C(T) curves in Figure 9b are as follows: Curve *(i)* is the steady-state zero-bias capacitance recorded as a function of temperature. This curve is reversible for increasing or decreasing temperature scans. Curves *(ii)* and *(iii)*, on the other hand, are irreversible thermal scans corresponding to initial conditions at the lowest temperature of completely filled or completely empty DX centers, respectively. Initial condition *(ii)* is obtained by cooling the sample from about 200 K to about 50 K with +1 V bias. This bias corresponds to a narrowing of the junction space-charge layer so that nearly all DX centers (which are donors in AlGaAs[11]) are below the Fermi level and hence filled with electrons (neutral charge state). At the lowest temperature the bias is returned to 0V where the filled DX centers in the space-charge region constitute a nonequilibrium state which is metastable because the electron thermal-emission rate is vanishingly small at 50 K. Thus, as shown in Figure 10a, the space-charge layer of width W_{ii} is made up only of ionized "normal" donors of net concentration N_D. When the temperature is increased to the vicinity of 100 K, the DX centers begin to thermally emit their trapped electrons and hence become positively charged. Since $N_{DX} \gg N_D$ in most cases, this corresponds to a drastic rearrangement of the space charge at constant bias which finally results in the equilibrium width W_i, shown in Figure 10a as the step-wise charge distribution with shaded boundaries. The step at $W_i-\lambda$ corresponds to the point where the DX energy level passes through the Fermi level in the edge region of the space-charge layer. Thus between 0 and $W_i-\lambda$, the DX centers are above E_F and are empty in equilibrium so that the positive space charge is $q(N_{DX}+N_D)$. In the edge region from $W_i-\lambda$ to W_i, the energy level of the DX centers is below the Fermi level so that the equilibrium space charge is only qN_D. The steady-state capacitance change of curve *(i)* as a function of temperature corresponds primarily to the temperature dependence of the Fermi level, and consequently of λ. As λ changes with temperature at constant-bias voltage, the space-charge distribution, and hence W_i in Figure 10a, must change accordingly.

Initial condition *(iii)* in Figure 9b corresponds to all DX centers empty. This is illustrated in Figure 10a with the space-charge distribution $q(N_{DX}+N_D)$ from 0 to W_{iii} with no edge region. Condition *(iii)* is obtained at low temperature from the steady-state condition *(i)* by emptying the DX centers either *optically,* by exciting the electrons to the conduction band, or *electrically,* by recombination of the electrons with injected holes under

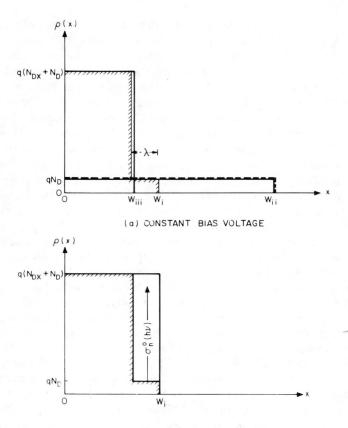

FIGURE 10 Charge density vs. distance in the abrupt-depletion approximations (a) for the junction space-charge layer corresponding to the three TSCAP initial conditions in Figure 9, and (b) for the constant-capacitance condition of the photoionization measurements. [Ref. 12]

forward bias. The optical emptying path is illustrated in Figure 9b and is utilized in the photocapacitance method to measure the electron photoionization cross section $\sigma_n^\circ(h\nu)$.

The positive-going TSCAP step, corresponding to initial condition *(ii)* is the same physical phenomenon (thermal electron emission) as the negative DLTS peak. From the magnitude of the TSCAP step one can calculate the concentration of deep levels which are emitting electrons with rates of the order of seconds in the temperature range of the step. Similarly, the negative going TSCAP step corresponding to initial condition *(iii)* arises from the same effect as the positive DLTS peak. This has been shown to be due to electron capture in the case of DX centers.

Usually a positive DLTS peak or a negative TSCAP step is due to minority-carrier emission (holes in this case). However, the fact that initial condition *(iii)* can be established in n-type Schottky barriers by illumination with photons of energy as low as 0.6 eV for Te DX centers in AlGaAs totally rules out the possibility of hole emission in this temperature range, since $\geqslant$ 1.5 eV light would be needed to empty hole traps close enough to the valence band to emit holes at the same temperature as the positive DLTS peak. The fact that electron capture can give a signal that looks so much like hole emission is due to the peculiar nature of the DX center, i.e., its electron-capture cross section is very small and thermally activated at low temperatures.

The capacitance values corresponding to conditions *(i)*, *(ii)*, and *(iii)* in Figure 9b can be put on a more quantitative basis in order to determine the DX center and net normal-donor concentrations. In Ref. 12 it was shown that the concentration of DX centers could be given by

$$\frac{N_{DX}}{N_D} = \left[\frac{C_{iii}}{C_{ii}}\right]^2 - 1 \tag{10}$$

using the abrupt-depletion approximation illustrated in Figure 10. This is reasonably accurate for all conditions except forward bias. For the data in Figure 9b, we find $N_{DX}/N_D = 8$ with $N_{DX} + N_D \sim 10^{18}\,\mathrm{cm}^{-3}$. The charge-density diagrams in Figure 10 correspond to these values.

3.2 Isothermal Capacitance Transients

Instead of recording a capacitance transient during a temperature scan as in TSCAP, it is often very useful to record such a transient at a fixed temperature. For many defects such an isothermal capacitance transient is a simple exponential function of time. In some cases, however, the capacitance transient is not exponential. Such nonexponentiality can be due to a variety of effects: such as the nonlinearity of the junction response for large signals ($\Delta C/C > 0.1$), the dependence of the thermal emission rate on the spatially varying electric field in the space charge layer, a distribution of emission rates from different defects, or a variable Coulomb barrier to capture and/or emission which depends on the occupancy of an extended

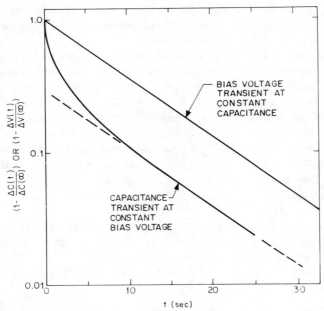

FIGURE 11 Transients due to the photoionization of DX centers as observed for the same experimental conditions by two methods of measurement: (a) capacitance transient at constant bias and (b) bias-voltage transient at constant capacitance. [Ref. 12]

defect or defect cluster. All of these effects have been seen in various defect systems.

For the DX-center case one expects nonexponential behavior because of the large value of $\Delta C/C$. This is shown for a photocapacitance transient in Figure 11a. As pointed out by Goto et al[57] this large-signal nonlinearity problem can be solved by recording the bias-voltage transient necessary to maintain a constant junction capacitance in a feedback measurement circuit. This is shown in Figure 11b. Note that even in the constant-bias case, however, the transient is exponential at long times, when the signal has become small enough ($\Delta C/C \ll 1$). Most data to be discussed in the next section were taken at constant bias with the capacitance transient characterized either by this asymptomatic slope or by the time at the ½ signal point.

Because nearly all capacitance transient data on DX centers has been taken with constant bias rather than constant capacitance, it had been assumed by nearly all workers that the observed nonexponentiality was due to the large-signal nonlinearities[56,57] such as shown in Figure 11. However, Ferenczy[58] has recently shown that the DX centers in GaAsP have *intrinsically* nonexponential thermal-emission transients and exponential optical-emission transients. This is shown in Figure 12 where the nonexponential nature of the thermal-emission transient becomes more pronounced at lower temperatures. At 119K, the transient was shown to be proportional to the logarithm of time over 4 decades (1 sec to 10^4 sec.)[58] In this case all other spurious sources of nonexponential behavior were carefully ruled out. This intrinsic nonexponentiality of the thermal emission process for DX centers at low temperatures can also be seen in the anomalous width of the DLTS lineshape, as will be discussed in the next subjection.

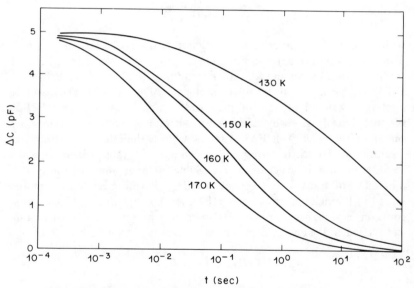

FIGURE 12 Single shot capacitance transient measurements on a n-Au-GaAs$_{.62}$P$_{.38}$:TE Schottky diode at four different temperatures. The filling width was chosen to be long enough at each temperature to reach the quiescent capacitance value. [Ref. 58]

3.3 Deep Level Transient Spectroscopy (DLTS)

The DLTS technique[55,56,60] is a convenient way to rapidly obtain a spectral fingerprint of the trap distribution in a particular sample. In fact, it can be shown[61,62] that, in the small-signal limit, the DLTS signal divided by kT gives a spectrum versus temperature which is proportional to the density of gap states versus energy, g(E). The energy-temperature relationship is given by

$$E = kT \ln(\nu \tau_m) \tag{11}$$

where ν is the exponential prefactor of the thermal emission rate

$$\tau^{-1} = e = \nu \exp(-E/kT) \tag{12}$$

and τ_m^{-1} is the DLTS rate window defined by

$$\tau_m = \frac{t_2 - t_1}{\ln(t_2/t_1)} \tag{13}$$

with t_1 and t_2 the times of the midpoint of the DLTS sampling gates.[63] The true proportionality in Equation 11 holds only for cases where ν is the same for all traps and for rather broad trap distributions ($\Delta E/E \geqslant 0.1$). For DLTS peaks due to traps with well-defined energies, i.e. g(E) given by a series of δ-functions, the full peak width at half maximum (FWHM) ΔT is approximately given by $\Delta T/T \simeq 0.1$ where T is the temperature of the center of the peak. This FWHM relationship is the inherent instrumental resolution of DLTS. It corresponds to an emission rate variation of about an order of magnitude across the DLTS line for an exponential transient.[55]

Families of traps with a spread in energy E will have a corresponding spread in thermal emission times $\tau(E)$ given by Equation 12. The thermal emission capacitance transient C(t) will thus be nonexponential in time and related to the energy distribution of traps g(E) by a Laplace transform

$$C(t) = \int g(E) \exp\left[-t/\tau(E)\right] dE. \tag{14}$$

For $\Delta t = t_2 - t_1 \leqslant t_1$, the DLTS spectrum versus temperature S(T) is given by[61]

$$S(T) \simeq \frac{\Delta t}{<t>} kT\, g(E(T)) \tag{15}$$

Where $<t> = (t_1 + t_2)/2$ and E(T) is given by Equation (11). It is clear from Equations (14 and 15) that nonexponential transients are not a serious problem for DLTS measurements. Indeed, if properly analyzed,[61,62] the DLTS spectrum gives direct information about the energy distribution of gap states which is easier to analyze than the time dependence of the transient response. Both measurement regimes are important, however, in order to obtain a detailed understanding of the system.

To illustrate the relationship between transient response and DLTS, let us compare the DLTS linewidth reported by Ferenczi[58,59] with that expected from his nonexponential transients in Figure 12. Since a factor of

10 change in τ is related to a DLTS linewidth of $\frac{\Delta T}{T} \simeq 0.1$, each decade in

time roughly corresponds to an additional 0.1T increase in ΔT (as long as $\Delta T \ll T$). Thus the transients in Figure 12 which are logarithmic in time over three to four decades should be expected to give DLTS linewidth

$\frac{\Delta T}{T}$ of between 0.3 and 0.4. According to Figure 1 of Ref. 58, Ferenczi

observes a DLTS linewidth of $\Delta T/T \simeq 0.4$, in agreement with this rough argument. It is difficult to be more precise than this without being able to account for the nonlinearities inherent in the large $\Delta C/C$ regime[56,62] which are also present in the DX center spectra.

In spite of the wide DLTS linewidth ($0.2 < \frac{\Delta T}{T} < 0.4$) which is typical of

all DX centers and indicates nonexponential thermal transients, the Arrenhius plot obtained from varying the DLTS rate window shows simple thermally activated behavior. An example is the Te DX center in AlGaAs shown in Figure 13. The thermal energies quoted in Section IV are all obtained from similar data. In all cases the activation energies will be quoted in that section with the standard T^2 correction factor. This is obtained from the slope of a plot of $\log(e_n/T^2)$ vs $1000/T$ or from the slope of $\log e_n$ vs $1000/T$ minus $2kT$. The values in Figure 13 have not been corrected.

3.4 Majority Carrier Capture Measurements

The majority carrier (electron) capture rate for DX centers can be measured in a number of different ways. Perhaps the easiest way, and historically the first, is the decay rate of the persistent photoinduced conductivity.[11,17] Electron capture may also be measured by junction capacitance methods in three different ways: (1) majority-carrier pulse-width variation, (2) positive DLTS peak position, or (3) decay of the photoinduced TSCAP state.

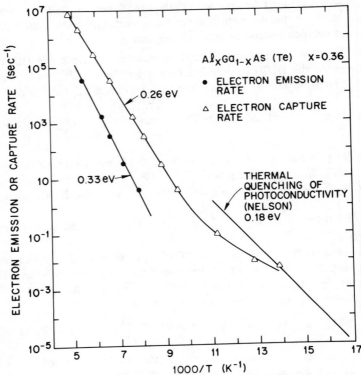

FIGURE 13 Electron-emission and -capture rates vs. inverse temperature for a typical sample of Al_xGa_{1-x}-As (Te). [Ref. 10, 11]

The majority-carrier pulse method is the standard junction technique used for measuring capture rates.[55,56,60] The analysis is straight-forward for large bias voltages ($V \gg V_{bi}$) and small trap concentrations ($N_T \ll (N_D - N_A)$). In this case the time constant for electron capture τ_e is related to the carrier concentration n and the capture cross section of the defect σ_n by

$$\tau_e^{-1} = \sigma_n <\nu_n> n \tag{16}$$

where $<\nu_n>$ is the thermal velocity of electrons.

For most cases, the DX center concentration is large relative to the carrier concentration and only relatively small bias voltages can be applied to the sample because the doping is typically quite high ($10^{17}-10^{18}\,cm^{-3}$).

Thus Equation 16 does not hold exactly and, furthermore, the capture is found to be nonexponential in time. According to Ferenczi,[58] part of this nonexponential capture behavior is an intrinsic property of σ_n which is unrelated to the large signal nonlinearities. Therefore, it is very difficult in most cases to obtain values for σ_n which are more accurate in an absolute sense then an order of magnitude estimate. It is possible, however, to obtain an activation energy from the temperature dependence of the capture rate by using some arbitrary but consistent measure of the time dependence, e.g. the time to reach the half-signal point. Data such as this were shown for Te doped AlGaAs in Figure 13.

The other two junction measurements of capture are indirect methods which are unique to DX centers and can only be properly interpreted with the aid of the more direct majority-carrier capture results. As discussed above in connection with TSCAP and DLTS, the apparent "hole emission" features in Figure 9 (i.e., the positive DLTS peak in (a) and the decay from state *(iii)* in (b)) are actually due to electron capture in the case of DX centers. This type of behavior is unique to defects which cause persistent photoconductivity and is due to the larger thermal activation barrier to electron capture, coupled with a large Stokes shift which quenches optical capture, as discussed in Section 2.

The data in Figure 13 is a compilation of all four types of electron capture measurements for Te DX centers in AlGaAs. In view of the significant uncertainties in quantitatively relating these different methods, the agreement in Figure 13 is quite remarkable. Indeed, the three junction methods overlap almost exactly when the positive DLTS peak position is used along with the half-signal point for the other methods. This makes intuitive sense, since for a symmetric DLTS line the peak corresponds exactly to the half signal condition.

Although it is difficult to obtain precise values for σ_n, we may obtain an approximate order of magnitude estimate for the cross section scale in Figure 13 from Equation 16. If we take $n = 10^{18} \text{cm}^{-3}$ and $\langle v \rangle = 10^7 \text{cm/sec}$, then at $\tau_e^{-1} = 10^{-5} \text{sec}^{-1}$ we have $\sigma_n \sim 10^{-30} \text{cm}^2$. This is an extremely small cross section and has only been reported for defects such as DX centers which exhibit persistent photoconductivity. As discussed in Section 2.1 (cf. Figure 4), however, such values are a natural consequence of MPE capture in the large lattice relaxation regime for small vibrational frequencies ($\hbar\omega \leqslant 10 \text{meV}$).

Hole capture at DX centers is not easy to measure accurately since it is a minority-carrier process. However, the fact that a positive DLTS peak due to electron capture be generated by a forward bias injection pulse in a pn junction necessarily implies that $\sigma_p \gg \sigma_n$. Craven and Finn[18] report

$\sigma_p > 10^{-12} \, \text{cm}^2$ and $\sigma_n \leqslant 10^{-19} \, \text{cm}^2$ at $T \cong 200K$ for Sulfur-related DX centers in GaAsP. Such large values of σ_p are qualitatively consistent with the cc diagrams typically used to describe DX centers.

3.5 Photocapacitance

A key feature of DX centers is the large difference between the energy for *thermal* emission as opposed to *optical* emission to the conduction band. According to the cc diagrams for DX centers (e.g. Figure 6), this is the same lattice-relaxation phenomenon which is responsible for the well-known Stokes shift between optical absorption and emission in radiative defects. However, since DX centers are nonradiative defects, it is not possible to observe the Stokes shift in the usual way. Therefore, one uses the shift between the thermal activation energy and the optical absorption threshold as measured by photocapacitance to determine the Stokes shift. The analysis of the photocapacitance lineshape to obtain the parameters of a cc diagram was discussed in Section 2.2. In this section we will discuss the aspects of photocapacitance measurements which are unique to DX centers. As we shall see, photocapacitance measurements are closely related to TSCAP measurements in the DX case.

In the low-temperature limit, where thermal emission and capture rates are negligible, the concentration of occupied DX centers $n_{DX}(t)$ is given by

$$n_{DX}(t) = N_{DX} \exp(\sigma_n^{\circ} \Phi t), \qquad (17)$$

where Φ is the optical intensity in photons/cm^2 sec and N_{DX} is the total DX center concentration. In the limit where the DX concentration is much larger than the net donor concentration ($N_{DX} \gg N_D$), the capacitance is a complicated function of $n_{DX}(t)$; thus the time dependence of C from *(i)* to *(iii)* in Figure 9b is far from the simple exponential relationship of Equation 17. This was dramatically illustrated by the typical photocapacitance transient at constant bias voltage shown in Figure 11. Thus as discussed above, it is necessary to measure the bias voltage change needed to maintain a constant capacitance. The resulting voltage transient has the simple exponential form of Equation 17.

The experimental sequence is as follows. The sample is first cooled at zero bias to reach condition *(i)*. When the monochromator is turned on, the DX centers in the edge region are emptied according to Equation 17, as shown in Figure 10b. The photoionization transient is measured by recording the feedback bias voltage as shown in Figure 11. The optical cross section σ_n° can then be obtained from the time constant of the voltage transient if Φ is known.

A peculiar property of DX centers, which is the cause of persistent photoconductivity, is that when emptied at low temperatures it is impossible to refill the centers without warming the sample. The thermal barrier due to lattice relaxation essentially stops all electron capture below about 77K, and the fact that the empty DX state is not in the gap makes it impossible to optically refill the level from the valence band. Thus in order to measure σ_n° at a different photon energy after the system is in state *(iii)*, it is first necessary to warm the sample to some temperature above the negative-going electron-capture TSCAP step in Figure 9b. This electron-capture step is the same physical phenomenon as the thermal quenching of persistent photoconductivity. State *(iii)* in Figure 9b, therefore, corresponds to the persistent-photoconductivity state seen by photo-Hall or photoconductivity measurements. As a consequence, in the measurements of $\sigma_n^\circ(h\nu)$, the temperature cycle from the measurement temperature T up to 150–200 K along curve *(iii)* and back to T along curve *(i)* is required for *each* value of $h\nu$. This is a very time consuming experiment.

A typical spectrum for Te DX centers in AlGaAs measured in this way is shown in Figure 14. Note the broadening of the threshold observed when the measurement temperature T is charged from 44K to 78K. Data of this sort give a qualitative indication of the magnitude of the Stokes shift, but must be analyzed via Equation (9) to obtain accurate quantitative results. For the example in Figure 14, the fitting parameters used in Equation (9) to obtain the solid lines shown in the figure are as follows[12]: $S\hbar\omega = 0.75 \pm 0.1$ eV, $E_0 = 0.10 \pm 0.05$, and $\hbar\omega = 10$ meV. The other parameters in Equation (9) were fixed at the values for GaAs ($E_A = 5.2$eV, $E_F = 11.5$eV) and $Al_{0.37}Ga_{0.63}As$ ($E_g = -2.05$ eV). Note that the proper optical ionization energy to use in a cc diagram such as Figure 6 is $E_0 + S\hbar\omega = 0.85$ eV for this example, not the lowest possible energy for which transitions are observed (~0.6eV). The threshold results of the photocapacitance data presented in Section 4 will all be those obtained from Equation (9).

4. Summary of Experimental Results for DX Centers in AlGaAs and GaAsP

One of the striking features of DX centers in AlGaAs and GaAsP is the qualitative similarity of their large lattice relaxation behavior, as discussed in general in the preceding sections. One should not assume from this, however, that all DX-like behavior is due to the same defect. The charac-

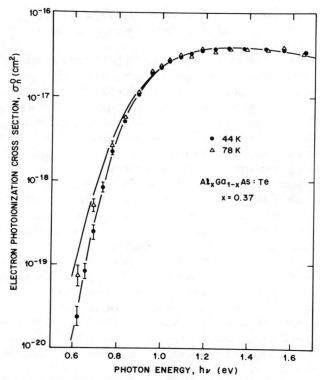

FIGURE 14 Normalized electron-photoionization cross-section $\sigma^{\circ}n$ (hν) for two temperatures. The solid lines are theoretical fits to the data. [Ref. 12]

teristic energies of the six different centers which have been reported in AlGaAs (Se,Te,Si,Sn) and GaAsP (S,Te) are sufficiently different to be easily distinguishable. In this section we will review the known facts about these different DX centers. The data can be divided into three general areas: Ionization and Activation Energies, Band-Structure Effects, and Structural Information.

4.1 Ionization and Activation Energies

As discussed in the previous sections, the most convenient way to visualize the large differences between thermal and optical energies of DX centers is with the aid of a configuration coordinate (CC) diagram. The prototypical CC diagram is shown in Figure 15. There are three types of energy "depths"

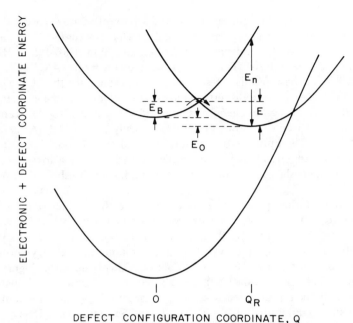

FIGURE 15 Typical cc diagram for DX centers showing the notation used in this paper.

for such a center. Namely, (1) the optical depth E_n corresponding to transitions at constant lattice coordinate (e.g., a photocapacitance measurement), (2) the thermal equilibrium depth E_0 corresponding to the activation energy of the equilibrium carrier concentration when the DX center is the dominant donor (e.g., a Hall measurement), and (3) the thermal activation depth E corresponding to the energy needed to surmount the energetic barrier between the bound and free state (e.g., a DLTS measurement). In addition, we have the activation barrier E_B to the thermal capture of a free electron. It is obvious from Figure 15 that

$$E = E_B + E_0 . \tag{18}$$

The energy of lattice relaxation $S\hbar\omega$ is related to these other energies by

$$E_n = E_0 + S\hbar\omega. \tag{19}$$

The parabolic energy curves $E(Q)$ are described analytically by the relation

$$E(Q) = E(Q_0) + \tfrac{1}{2} kQ^2 \tag{20}$$

where Q_0 is the equilibrium value of Q and k is the effective force constant.

The most complete optical and thermal data on DX centers is primarily from three research groups, namely, Craven and Finn (GaAsP:S),[18] Henning and Thomas (GaAsP:Te),[20] and Lang and coworkers (AlGaAs:Se,Te, Si,Sn).[10,12,13] These results are summarized in Table 1 for the CC diagram in Figure 15. Note the substantial shifts in optical ionization energy E_n as a result of changing the chemical species of the donor in the DX center. This chemical trend is shown rather dramatically for the AlGaAs defects in Figure 16. The optical cross sections of the defects in GaAsP fall at the extremes of the trend in Figure 16, i.e., GaAsP:Te has a lower energy threshold than AlGaAs:Te and GaAsP:S has a higher threshold than AlGa As:Si.

The trend in the optical energy is correlated with a similar trend in the thermal ionization energy, E. However, in order to consistently describe some of the defects in a CC diagram such as Figure 16, it is necessary to allow for a change in the curvature of the parabolic total energy surfaces for the different charge states of the defect. This is shown in Figure 17 for the AlGaAs system. In this figure the energy curves, given by Equation 20, have all been normalized to a force constant of $k = 1.0$ in the ionized charge state (equilibrium at $Q = 0$). The Te and Se centers (i.e., DX centers with As-site donors) have the same force constant in both charge states while the deeply bound state of the Si center has $k = 1.35$ and the Sn center has $k = 0.32$. Using the same convention, the GaAsP centers have force constants in the bound state of $k = 1.0$ for Te and $k = 0.56$ for S.

It has been argued[1,64,65] that the nearly ideal thermally activated behavior of the emission rate over the temperature range between about 77K and 120K (see e.g., Figure 13) is evidence that a simple, single coordinate CC model such as Figure 15 is inadequate to describe the data in Table 1. The basic argument is that the MPE process[30] in this temperature range is dominated by tunneling between vibronic states of the defect and therefore is not in the so-called Mott limit of purely activated behavior.[36] As we discussed in Section 2, this is true for defects which have local vibrational modes with frequencies in the range of optic phonons, e.g., $\hbar\omega_{LO,TO} \cong$ 34 meV in GaAs. However, as we showed in Figure 4, defects with vibrational frequencies in the range of the transverse acoustic phonon modes, i.e., $\hbar\omega \leqslant 10$ meV, are in the thermal activation Mott limit for temperatures as low as 100K. For materials such as InSb where $\hbar\omega_{TA} \simeq 5$ meV the activated behavior could extend to as low as 50K.[64]

In the case of the DX centers the low vibrational energy has been independently determined from the analysis of the optical lineshape, such as in Figure 14. The thermal broadening of this lineshape seen in AlGaAs: Te[12] and GaAsP:Te[20] at temperatures below 77K can only be explained

TABLE 1 Summary of DX Centers in AlGaAs and GaAsP

DONOR IMPURITY	E (eV)	E_B (eV)	E_o (eV)	E_n (eV)	k	REFERENCE
AlGaAs:						
Se	0.28 ± 0.03	0.18 ± 0.02	0.10 ± 0.05	0.85 ± 0.1	1.0	13
Te	0.28 ± 0.03	0.18 ± 0.02	0.10 ± 0.05	0.85 ± 0.1	1.0	12, 13
Si	0.43 ± 0.05	0.33 ± 0.05	0.10 ± 0.05	1.25 ± 0.1	1.35	13
Sn	$0.19 + 0.02$	< 0.1	$0.10 + 0.05$	1.1 ± 0.1	0.32	13
GaAsP:						
S	0.35	0.15 ± 0.03	0.20 ± 0.03	1.53	0.56	18
Te	0.19 ± 0.02	0.12 ± 0.03	0.07	0.65 ± 0.05	1.0	20

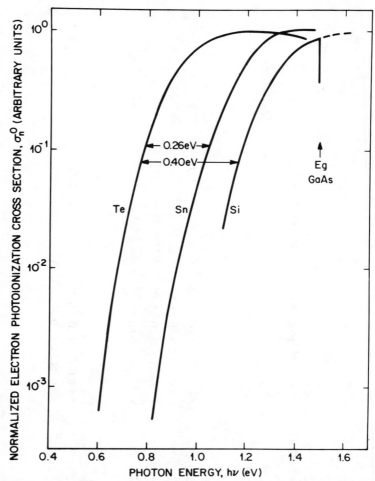

FIGURE 16 Photoionization lineshapes of Te-, Sn-, and Si-related DX centers in the n-type Al$_{0.4}$GA$_{0.6}$As. [Ref. 13]

by $\hbar\omega \leq 10$ meV. Local vibrational frequencies in the optic phonon range would not give rise to a measurable effect in this temperature range.[30] Thus, contrary to the above mentioned arguments,[1,64,65] there is no reason to exclude *a priori* a relatively simple CC diagram in the case of DX centers. This does not mean, however, that a more complex local lattice relaxation process could not be taking place. Clearly, the need to drastically change the force constants for the different charge states in some cases is

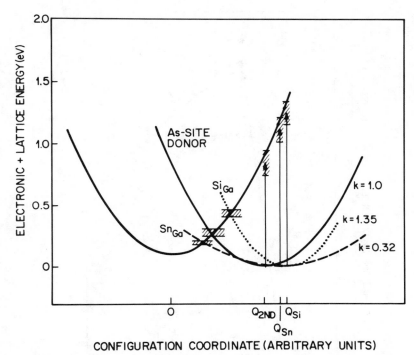

FIGURE 17 Configuration coordinate diagrams for the three types of DX centers in AlGaAs. The shaded areas show the range of uncertainty of the data. [Ref. 13]

already a departure from the simple CC model. We will discuss further the implications of these results in section 4.3 when we summarize the data relevant to the microscopic structure of DX centers.

4.2 Band-Structure Effects

A striking feature of DX centers which has been well documented for the case of AlGaAs:Te[12] and GaAsP:S[18] is the dramatic drop in apparent DX concentration as the alloy composition of the host is varied from the indirect gap region into the direct gap region of composition. This is shown for the AlGaAs:Te system in Figure 18. Hydrostatic pressure studies of the DX signal for various alloy compositions show that the precipitous drop below $x_{Al} \simeq 0.35$ in $Al_x Ga_{1-x} As$:Te and below $x_P \simeq 0.35$ in $GaAs_{1-x} P_x$:S is primarily due to band structure effects and not to sudden drop in the actual concentration of DX centers. The pressure effect is shown in

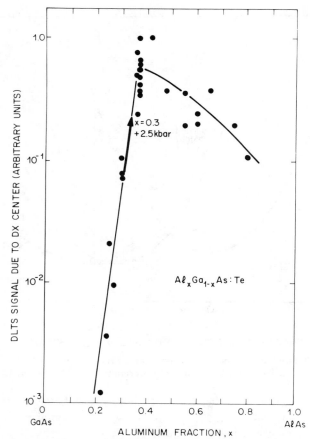

FIGURE 18 DLTS signal magnitude due to DX centers in various $Al_xGa_{1-x}As$ samples vs. the aluminum fraction x. The heavy arrow indicates the signal increase in an x = 0.3 sample induced by the application of a 2.5 kbar stress. The discontinuous drop in DLTS signal below x = 0.36 corresponds to E_0 crossing the Fermi level, as explained in the text. [Ref. 12]

Figure 18 for the case of AlGaAs:Te. It was found[12] that a pressure of 2.5 kbar increased the DX signal by a factor of 3 for x = 0.30. The slope of the arrow in Figure 18 indicates this factor of 3 along with the composition change Δx which would give rise to the same shift of the Γ conduction band minimum as does 2.5 kbar of hydrostatic pressure. This indicates that the composition dependence of the signal is primarily due to a shift of the conduction band at Γ, not a change in the DX center concentration.

A similar pressure effect was seen at x = 0.25, but not at x = 0.37; i.e., the thermal emission signal of electrons trapped at DX centers in AlGaAs:Te is pressure dependent for x < 0.35. Similar results hold for GaAsP:S.[18]

The energy level scheme which results from the pressure and composition experiments in AlGaAs:Te is shown in Figure 19. The composition (x ≃ 0.35) in Figure 18 below which the electron trapping signal begins to drop exponentially with x corresponds to the composition in Figure 19 where the DX energy level crosses the conduction band edge at Γ. The physical meaning of this picture is that the DX energy appears to be constant relative to the X minimum of the conduction band and is relatively unaffected by the Γ minimum as the band structure is changed by either pressure or composition. The Fermi level, however, follows the lowest band edge so that at and below the crossing between the DX energy level and the Γ minimum the DX center will be *above* the Fermi level in equilibrium. Thus, at equilibrium, the DX center will be only partially occupied by the Boltzmann tail of the Fermi function. Since it is this occupation that is measured in an electron trapping experiment such as TSCAP or DLTS, the apparent concentration appears to be significantly reduced.

On the other hand, the less dramatic reduction in signal as x approaches 1 in $Al_xGa_{1-x}As$:Te and in $GaAs_{1-x}P_x$:S must be due to a true change in the concentration of DX centers, since the conduction band minimum is unambiguously X-like in this composition range. In fact, centers of the sort summarized in Table 1 have not been seen in related pure binary III-V semiconductors such GaP. Apparently, some aspect of the alloy must be essential to the occurrence of these centers.

Wolford et al.[66,67] have shown that the electron-phonon coupling for nitrogen isoelectronic traps in GaAsP increases dramatically in the alloys as opposed to pure GaP. According to the extrinsic self trapping ideas reviewed in Section 2.2, we might expect that a threshold value for the local electron-phonon coupling strength would have to be reached before the shallow-deep lattice relaxation instability characteristic of DX centers could occur. According to Figure 8 this is a sharp transition, whereas the data in Figure 18 and in Ref. 18 indicate a rather gradual increase in the concentration of DX centers as one approaches the center of the alloy composition range from either GaP or AlAs. However, Toyozawa's extrinsic self trapping model of the shallow-deep instability considers a very localized electron-phonon coupling strength at the defect. Local alloy disorder would be expected to give rise to a large number of inequivalent local arrangements in the vicinity of a defect. Therefore, one might expect a wide range of coupling strengths between the trapped electron and the lattice. This could give rise to a rather gradual dependence on overall alloy

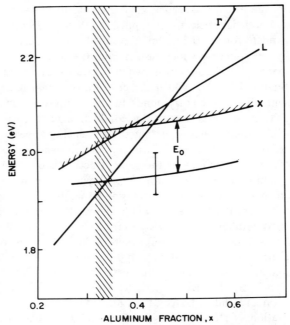

FIGURE 19 Equilibrium energy E_O of DX centers in $Al_xGa_{1-x}As$ as deduced from the fit to the optical data. The shaded region $0.32 < x < 0.35$ indicates the range of composition where E_O crosses the Fermi level. The lowest high-density-of-states minimum relevant for optical transitions is shown shaded. [Ref. 12]

composition of the number of defects which have the appropriate local configuration to meet the discontinuous self trapping criterion illustrated in Figure 8. Thus it might be possible for the concentration of defects to be more or less constant with x but for the fraction of these meeting the extrinsic self trapping criterion to vary with x. Such an idea is consistent with the data, but rather speculative at this point.

Another interesting question to consider is the behavior of DX centers in the alloy composition range ($x < 0.35$ in $Al_xGa_{1-x}As$:Te and $GaAs_{1-x}P_x$:S) where the bound state is resonant with the conduction band at Γ. One might expect such a resonant state to autoionize into the conduction band continuum. However, as with similar resonant states in CdTe:Cl,[1] the auto-ionization process is quenched by the large lattice relaxation between the states of the occupied and empty defect. The physical basis for this quenching is easy to understand with the aid of Figure 20. In this figure we have superimposed upon the CC diagram in Figure 15 an additional energy curve (shown dashed) corresponding to the Γ conduction band minimum

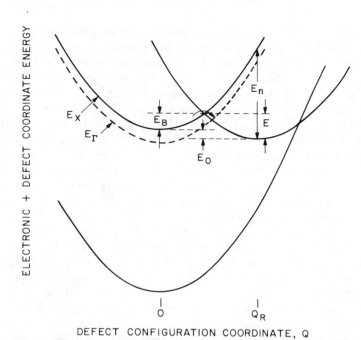

ELECTRONIC + DEFECT COORDINATE ENERGY

DEFECT CONFIGURATION COORDINATE, Q

FIGURE 20   Configuration coordinate diagram for DX centers in the direct gap regime where the total energy of the bound state is greater than that of an ionized DX center with the free electron at the Γ conduction band minimum.

for the resonant condition at x = 0.30 in AlGaAs. Note that the bound state energy at $Q = Q_R$ is below X and above Γ at $Q = 0$. The bound state is resonant with the Γ conduction band in terms of *total* energy, but it is actually very deep in the gap when viewed with the lattice frozen at $Q = Q_R$. From this point of view one would not expect autoionization to occur unless there is sufficient overlap between the vibronic wavefunction centered at $Q = 0$ and $Q = Q_R$. However, from the electron emission and capture results for these centers we know that such tunneling is much weaker than thermal emission over the classical barrier E_B.

The experimental results for AlGaAs:Te and GaAsP:S in the resonant-state limit indicate surprisingly little influence of the Γ conduction band minimum on the thermal and optical transitions of the DX center. In both systems[12,18] the DLTS peak, which is indicative of the energy E, changes only slightly in position as a function of host composition. In GaAsP:S the DLTS peak seems to move to slightly higher temperatures as x decreases below x = 0.35. Transitions to the conduction band at Γ, on the other hand, would reduce the activation energy. Thus the DX centers appear to make thermal transitions primarily to the X minimum with only very weak coupling to Γ. This sort of symmetry selection rule for thermal transitions was originally proposed[21,22] as the dominant mechanism for persistent photoconductivity. We see now that this earlier idea, which has been discounted as a mechanism for the persistence effect,[1,10] nevertheless has validity in our overall understanding of this type of defect. The idea that thermal transitions from certain types of deep traps to the Γ point of the conduction band may be forbidden by symmetry selection rules has also been proposed to explain the temperature dependence of the capture cross sections of some traps in GaAs.[68] However, such a model may be carried to extremes by attempting to explain data at compositions far from the direct-indirect crossing in the alloy where the selection rules need to be far more strict than is physically plausible.

The effect of the Γ point of the conduction band in the resonant state limit may be observed as a weak perturbation of the threshold of photoionization transitions. This is shown for AlGaAs:Te in Figure 21. This shows data from five samples of different aluminum content ($0.27 \leqslant x \leqslant 0.60$) all at the same temperature (44K). The heavy line drawn through the data is the 44K theoretical fit shown in Figure 14. The lighter lines for $x = 0.27$ and $x - 0.30$ are not fit but are intended to show how these samples, well into the direct gap composition range, have distinguishable shifts in the threshold region for $\sigma_n^{\circ} < 10^{-2} \sigma_{max}$. From Figure 20 it is clear that subthreshold optical transitions to Γ at $Q = Q_R$ are at least energetically possible when $E_\Gamma < E_X$. In fact, the slight broadening of the spectrum in Figure 21 for $E_\Gamma < E_X$ is consistant with the band structure shifts for these compositions and the much lower density of states at Γ compared to X. This indicates that there is very little symmetry selection in the optical transitions to the conduction band.

4.3 Structural Information

The basic microscopic structure of DX centers is not yet known in detail. However, various experiments give us important clues. The most definitive structural information comes from the absorption of ballistic phonons by DX centers in AlGaAs as measured by Narayanamurti et al.[54] In this experiment, phonon time of flight spectroscopy was used to separate the longitudinal acoustic phonons from the transverse phonons. By measuring the attenuation due to DX centers of ballistic phonon pulses propagating in different crystalline directions, it was possible by using the selection rules for phonon absorption to determine the symmetry of certain DX centers. Typical results are shown in Figure 22 for the case of propagation

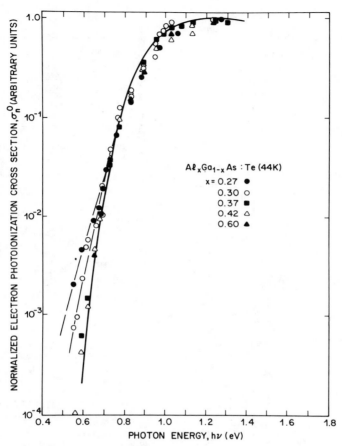

FIGURE 21 Normalized electron-photoionization cross-section $\sigma_{\bar{n}}^{\circ}$ at 44 K for five samples with different Al fractions. The heavy curve is the same theoretical fit as in Figure 14. [Ref. 12]

along the [110] direction for three samples of $Al_{0.5}Ga_{0.5}As$: (a) nominally undoped, (b) $\sim 10^{18} cm^{-3}$ Sn doped, and (c) $\sim 10^{18} cm^{-3}$ Te doped. The measured quantity is the light-induced change in the intensity of the fast transverse (FT) phonons as compared to the slow transverse (ST) phonons. The measurements were made at $\sim 1.5K$ where the light ionizes the DX centers and puts the sample in the persistent photoconducting state. The results are therefore characteristic of the ionized state of the Te and Sn DX centers.

530   D. LANG

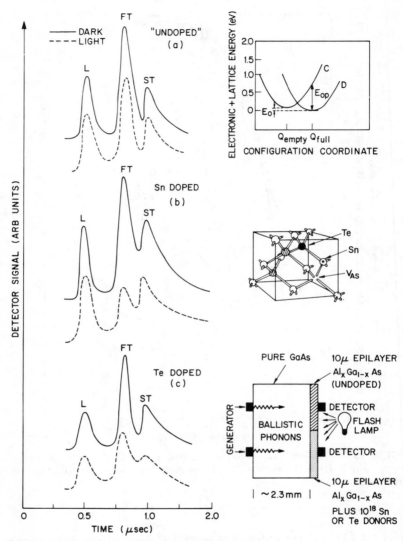

FIGURE 22 Ballistic phonon intensities for 3 different samples of $Al_{0.5}Ga_{0.5}$ As:
(a) nominally undoped material, (b) with ~ 10^{18} Te donors per cubic centimeter.
Solid lines are data taken in the dark while the dashed lines are taken after photoexcitation. Propagation direction [110]. The intensities of the ballistic phonons reveals
selective attenuation of certain groups of transverse modes depending upon the symmetry of the donor-related photoconductivity center. The right-hand side of the figure shows a typical configuration-coordinate diagram for such centers. A crystal
model of GaAs and a schematic of the experimental arrangement are also shown.
[Ref. 15]

The phonon absorption results can be interpreted as follows. The fact that the Sn-related center dramatically reduces the FT intensity without changing the ST intensity indicates an ionized defect of < 111 > trigonal symmetry. In the case of the Te-doped sample, on the other hand, the strong attenuation of both the FT and ST modes indicates an ionized defect of < 110 > orthorhombic symmetry. These assignments are corroborated by comparisons of different selection rules with absorption data in the [100] and [111] directions.[54]

As shown in the crystal model inset in Figure 22, these defect symmetries are consistent with a donor-plus-arsenic-vacancy model for DX centers which has been proposed earlier on much weaker evidence.[12,13] In spite of the consistency of the defect symmetry with this V_{As}-based model, there are still puzzling problems which lead us to believe that the issue is not yet settled. In particular, as has been pointed out by Henning and Thomas,[20] the fact that DX centers exist in GaAsP grown by chemical vapor deposition (CVD) under anion-rich (As or P) conditions makes it seem unlikely that a substantial number of anion vacancies would be available to make such a donor-vacancy complex.

If we are to seek an alternate structural model, however, the phonon results force us to consider only those defects which have the same general symmetry as a donor-V_{As} complex, namely a Ga (or Al)-site donor such as Sn is a first-nearest neighbor to X along a < 111 > direction, and an As (or P)-site donor such as Te is a second-nearest neighbor to X along a < 110 > direction. While it is difficult to imagine substitutional defects of these symmetries without invoking vacancies, impurities, or antisite defects, it is on the other hand quite plausible to imagine < 111 > or < 110 > configurations of interstitial atoms. We note in particular that pairs of atoms in adjacent tetrahedral interstitial sites would tend to have either < 111 > or < 110 > symmetry axes.

The notion of DX centers being interstitial in nature was not seriously considered originally[12] because of the lack of a reliable theoretical understanding of the bonding of interstitial atoms in covalent solids. However, the recent results of Baraff et al.[69] on Al interstitials in silicon give us important new insight into this problem. They find that in the tetrahedral (T_d) interstitial site the Al impurity makes weak tetrahedral sp^3 bonds with its nearest neighbors by weakening the bonds in the host lattice. This configuration gives rise to an $E_v + 0.17$ eV double donor state (Al^{++}/Al^{+}) as determined by ESR and DLTS measurements.[70] The calculations of Baraff et al., shown in the lower half of Figure 23 for the Al interstitial, agree with these data and in addition predict a strong p-like non-bonding resonance of T_2 symmetry slightly above the bottom of the conduction band. When the Al interstitial is moved to the lower symmetry (D_{3d}) hexagonal site between adjacent tetrahedral sites, this T_2 resonance splits and a one-fold degenerate boundstate (p_z) drops into the gap. This state can be occupied by zero, one, or two electrons and gives rise to two levels in the gap corresponding to the transitions (Al^+/Al^0) and (Al^0/Al^-). The silicon self interstitial in the upper part of Figure 23 has even a more dramatic level shift. This, however, has not been confirmed by experiment.

The fascinating aspect of these interstitial calculations from the point of view of DX centers is the strong interaction between the local bonding symmetry of the defect and the resulting states in the gap. The prototypical DX CC diagram in Figure 15 implies a shallow-deep transition similar to that calculated for the Al and Si interstitials in silicon. Namely, at $Q = 0$ the electron-trap state of the DX center is unoccupied and located in the conduction band (analogous to the T_d resonance of the tetrahedral interstitials in silicon), while at $Q = Q_R$ the DX state has dropped into the gap as a result of the change in the local environment parameterized by Q (analogous to the boundstates of the hexagonal interstitials derived from the

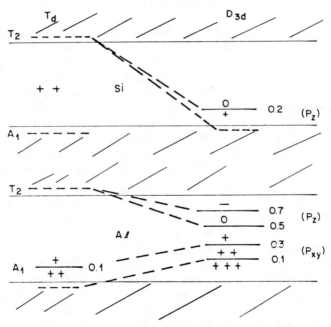

FIGURE 23 Results of self-consistent Green's function calculations of the energies of Si and Al interstitials in silicon for the tetrahedral (T_d) and hexagonal (D_{3d}) sites. [Ref. 69]

T_d resonance). While the results in Figure 23 are specifically relevant to the case of Al and Si interstitials in silicon, one might nevertheless expect rather similar behavior of interstitials in other tetrahedrally coordinated covalent solids such as III-V and certain II-VI semiconductors.

Several other aspects of DX centers also suggest the likelihood of interstitial character. We pointed out that low energy ($\hbar\omega \leqslant 10$ meV) local vibrational modes are implied by the low temperature behavior of the optical and thermal emission from DX centers. Such vibrations are characteristic of shear-like bond-bending modes such as TA phonons. From the work of Baraff et al.[69] we might expect that the weaker bonds associated with interstitial atoms would very likely give rise to the appropriate local mode vibrational energies to be consistent with the data on DX centers. An interstitial-like DX center would also help explain some very puzzling results on the recombination-enhanced interactions between DX centers and dislocation climb in AlGaAs.[71] This experiment studied electron trapping *in situ* in a scanning electron microscope and found that the growth of dislocation networks which were clearly of interstitial character was correlated with a substantial decrease in the concentration of DX centers within a few microns of the dislocation network. The most straightforward interpretation is that the DX centers were the source of the interstitial atoms needed to generate the dislocation network. However, with the donor-vacancy model dominating the thinking at that time, it was difficult to explain these results without a very speculative defect reaction model. If, on the other hand, the DX centers were actually interstitial in character, the DX-dislocation interaction would be much easier to explain.

A final point about the structure of DX centers is the fact that they do not appear to be well defined defects with exactly the same configuration in all samples, but rather a closely related family of defects for each of the six types in Table 1 with energies varying at random on the order of 20% from sample to sample. This is shown most graphically in Figure 24 where we show selected DLTS electron-emission spectra which illustrate the range of possible Te-related DX centers in $Al_x Ga_{1-x} As$.[12] Note that some samples actually show two resolved DLTS peaks, while most show a single peak of varying width located somewhere between the extremes of the double-peak examples. The range of peak positions in Figure 24 corresponds to a shift of about 60 meV in the thermal-emission activation energy of ~0.3 eV. There is no systematic correlation of these peak positions, and hence of thermal-emission energies, with mixed-crystal composition. Apparently, the dominant type of DX center in any given sample is determined more or less at random. The only possible correlation with x is that the very few double-peak examples seem to occur more readily at the extremes of the composition range, i.e., in the x = 0.20 − 0.30 or x ⩾ 0.6 range. Perhaps the relative probabilities of particular Ga or Al arrangements around the defect play a role in the slight shifts of its properties.

As we discussed in connection with Figure 12, thermal emission and capture at DX centers is nonexponential in time. A simple explanation for this result is that there is a closely related distribution of defects such as in Figure 24 which gives rise to a broadened DLTS line and, as we discussed, a nonexponential transient. Ferenczi and coworkers[58,59] have suggested that this nonexponential behavior is of a more fundamental nature. They point to the well-known fact that for systems with a large number of correlated states the transition probabilities tend to zero at the threshold energy of the transition. This is often referred to in many-body theory as an infrared divergence.[72] In the case of a thermal emission transition where many correlated lattice phonons are necessary to generate the appropriate local distortion of the defect, Ferenczi[58] shows that for long times the occupation of the defect N(t) is given by

$$N(t) \propto \exp(-Kt^\alpha/\tau_0) \tag{21}$$

where K is a constant, τ_0 is the transition time constant without many-body effects, and $0 < \alpha < 1$ is a parameter related to the particular many-body system. This infrared divergence only occurs near the threshold for the transition. Therefore at higher temperatures where sufficient thermal energy is available or for optical transition the normal exponential behavior of N(t) will be seen. As Ferenczi points out, this is generally consistent with the data in Figure 12.

We are thus left with two pictures for the nonexponential behavior of thermal emission and capture at low temperatures: (a) a distribution of defect energies, i.e., an *inhomogeneous* broadening of the DLTS line, or (b) an inherent many-body effect related to the need to assemble many correlated phonons in exactly the correct combination to drive the thermal transition, i.e., a temperature dependent *homogeneous* broadening of the DLTS line. Either of these mechanisms for the experimentally observed DLTS line broadening give us important insight into the structure of these defects. The data in Figure 24 indicate that inhomogeneous broadening is certainly present (case (a)), however, we cannot rule out the possibility that the homogeneous broadening in case (b) is present as well.

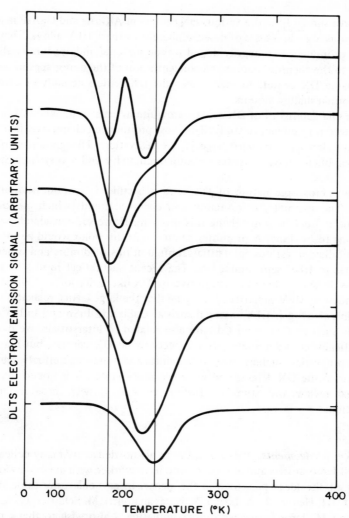

FIGURE 24   DLTS spectra of the DX centers in six Te-doped $Al_xGa_{1-x}As$ samples selected to show the typical range of peak positions and line shapes commonly observed. The rate window was 3.7×10^5 sec^{-1}, with a 2-V reverse bias and a 3-V majority-carrier pulse of 10-μsec duration. [Ref. 12]

5. Concluding Remarks

We have shown how the various DX centers in AlGaAs and GaAsP may be viewed as a general class of defects inherent to these III-V alloys. They are most pronounced in heavily-doped n-type material and are intimately related to the chemical nature and concentration of the donor species, hence the name DX center. In many cases the DX centers actually outnumber the normal shallow donors.

The characteristics of DX centers are quite unusual and make this family an interesting subject to study. The most pronounced characteristic of this class of defects is its very large lattice relaxation. This gives rise to two striking phenomena: persistent photoconductivity and a very large Stokes shift.

The microscopic nature of DX centers is still an open question at this time. The symmetry information and chemical trends which are known will be a great help in solving this problem. However, considerable work remains to be done. A promising approach at this point would be theoretical studies of various interstitial configurations of donors and other impurities in III-V semiconductors. The recent theoretical results on interstitials in silicon are very suggestive of DX-like behavior and should be extended to III-V materials. It may be that the large lattice relaxation centers in narrow gap III-V material such as GaSb and InSb and in II-VI materials such as CdTe and CdS are also related to interstitials. We did not specifically consider these letter defects to be DX centers, but many of their properties, such as large Stokes shifts and metastable effects, are very similar to the DX defects considered in this chapter. It is hoped that this critical review will stimulate further work to answer these and other questions.

Acknowledgements. It is a pleasure to acknowledge my many colleagues at Bell Laboratories and elsewhere who have worked with me in developing many of the ideas presented in this paper. In particular, my associations with C. H. Henry, R. A. Logan, V. Narayanamurti, M. Schlüter, G. A. Baraff, and M. Jaros have been most stimulating. I also wish to thank those who sent me preprints of their work in this area well before the publication date, especially R. A. Craven, H. Thomas, G. Ferenczi, and J. M. Langer.

References

1. J. M. Langer, in *New Developments in Semiconductor Physics,* edited by F. Beleznay, G. Ferenczi, and J. Giber (Springer-Verlag, Berlin, 1980), p. 123.
2. U. Peikara, J. M. Langer and B. Krukowska-Fulda, Solid State Commun. *23,* 583 (1977).
3. S. Porowski, M. Konczykowski and J. Chroboczek, Phys. Stat. Sol. (b) *63,* 291 (1974).
4. L. Dmowski, M. Konczykowski, R. Piotrzkowski and S. Porowski, Phys. Stat. Sol. (b) K131 (1976).
5. G. A. Baraff, E. O. Kane and M. Schlüter, Phys. Rev. Lett. *43,* 956 (1979).
6. G. D. Watkins and J. R. Troxell, Phys. Rev. Lett. *4,* 593 (1980).
7. J. R. Troxell and G. D. Watkins, Phys. Rev. B *22,* 921 (1980).
8. Y. Toyozawa, Solid State Electronic *21,* 1313 (1978).
9. Y. Toyozawa, in *Relaxation of Elementary Excitations,* edited by R. Kubo and E. Hanamura (Springer-Verlag, Berlin, 1980), p. 3.
10. D. V. Lang and R. A. Logan, Phys. Rev. Lett. *39,* 635 (1977).
11. R. J. Nelson, Appl. Phys. Lett. *31,* 351 (1977).
12. D. V. Lang, R. A. Logan, and M. Jaros, Phys. Rev. B*19,* 1015 (1979).
13. D. V. Lang and R. A. Logan, in *Physics of Semiconductors 1978* (Inst. Phys. Conf. Ser. No. 43, 1979), p. 433.
14. D. V. Lang, *Proc. 15th Int. Conf. Physics of Semiconductors,* J. Phys. Soc. Japan *49,* Suppl. A, 215 (1980).
15. V. Narayanamurti, R. A. Logan and M. A. Chin, Phys. Rev. Lett. *43,* 1536 (1979).
16. B. Ballard, G. Vincent, D. Bois and P. Hirtz, Appl. Phys. Lett. *34,* 108 (1979).
17. M. G. Craford, G. E. Stillman, J. A. Rossi and N. Holonyak, Jr., Phys. Rev. *168,* 867 (1968).
18. R. A. Craven and D. Finn, J. Appl. Phys. *50,* 6334 (1979).
19. G. Ferenczi, in *New Developments in Semiconductor Physics,* edited by F. Beleznay, G. Ferenczi, and J. Giber (Springer-Verlag, Berlin, 1980), p. 116.
20. I.D. Henning and H. Thomas, Solid-State Electron. *25,* 325 (1982).
21. A. Ya Vul', L. V. Golubev, L. V. Sharonova, and Yu. V. Shmartsev, Fiz. Tekh. Poluprovodn. *4,* 2347 (1970) [Sov. Phys. Semicond. *4,* 2017 (1971)].
22. G. W. Iseler, J. A. Kafalas, A. J. Strauss, H. F. MacMillan and R. H. Bube, Solid State Commun. *10,* 619 (1972).
23. M. R. Lorenz, B. Segall, and W. H. Woodbury, Phys. Rev. *134,* A751 (1964).
24. B. C. Burkey, R. P. Khosla, J. R. Fischer, and D. L. Losee, J. Appl. Phys. *47,* 1095 (1976).
25. A. Ya. Shik and A. Ya. Vul, Fiz. Tekh. Poluprovodn. *8,* 1675 (1974) [Sov. Phys. Semicond. *8,* 1085 (1975)].
26. H. J. Queisser and D. E. Theodorou, Phys. Rev. Lett. *43,* 401 (1979).
27. R. J. Keyes, J. Appl. Phys. *38,* 2619 (1967).
28. J. W. Farmer and D. R. Locker, J. Appl. Phys. *52,* 5718 (1981).
29. A. M. Stoneham, *Theory of Defects in Solids,* (Clarendon Press, Oxford, 1975).

30. C. H. Henry and D. V. Lang, Phys. Rev. B15, 989 (1977).
31. R. Englman and J. Jortner, Mol. Phys. 18, 145 (1970).
32. K. Huang and A. Rhys., Proc. Roy. Soc. 204, 406 (1950).
33. R. Kubo and Y. Toyozawa, Prog. Theor. Phys. 13, 160 (1955).
34. M. Lax, J. Chem. Phys. 20, 1752 (1952).
35. C. H. Henry, in Relaxation of Elementary Excitations, eds. R. Kubo and E. Hanamura (Springer-Verlag, Berlin, 1980), p. 19.
36. N. F. Mott, Proc. Roy. Soc. London A167, 384 (1938).
37. N. F. Mott, Solid-State Electronics 21, 1275 (1978).
38. C. W. Struck and W. H. Fonger, J. Lum. 10, 1 (1975).
39. D. J. Robbins and A. J. Thomson, Phil. Mag. 36, 999 (1977).
40. D. V. Lang and R. A. Logan, J. Electronic Mat. 4, 1053 (1975).
41. D. V. Lang and C. H. Henry, Phys. Rev. Lett. 35, 1525 (1975).
42. B. K. Ridley, J. Phys. C11, 2323 (1978).
43. M. G. Burt, J. Phys. C13, 1825 (1980).
44. R. Pässler, Phys. Stat. Sol. (b), 86 K39 (1978).
45. As has been pointed out in Ref. 44, there is a slight discrepancy between the level B data in Refs. 30 and 41 as compared with that in Ref. 40 due to the values of v_{th} used. Reference 40 and the current Figure 2 are the most accurate.
46. See e.g., H. C. Casey, Jr. and M. B. Panish, Heterostructure Lasers (Academic Press, New York, 1978) p. 120.
47. H. G. Grimmeiss, Ann. Rev. Mater. Sci. 7, 341 (1977).
48. G. M. Martin, A. Mitonneau, and A. Mircea, Electronics Letters 13, 191 (1977).
49. G. Dolling and R. A. Cowley, Proc. Phys. Soc. 88, 463 (1966).
50. E. P. Sinyavskii and V. A. Kovarskii, Fiz. Tverd. Tela 9, 1464 (1966) [Sov. Phys. Solid State 9, 1142 (1967)].
51. M. Jaros, Phys. Rev. B16, 3694 (1978).
52. T. H. Keil, Phys. Rev. 140, A601 (1965).
53. J. R. Chelikowsky and M. L. Cohen, Phys. Rev. B14, 556 (1976).
54. V. Narayanamurti, R. A. Logan, and M. A. Chin, Phys. Rev. Lett. 43, 1536 (1979).
55. G. L. Miller, D. V. Lang and L. C. Kimerling, Ann. Rev. Matl. Sci. 7, 377 (1977).
56. D. V. Lang, in Thermally Stimulated Relaxation in Solids, Vol. 37 of Topics in Applied Physics, edited by P. Braunlich (Springer, Berlin, 1979), p. 93.
57. G. Gato, S. Yanagisawa, O. Wada, and H. Takanashi, J. Appl. Phys. 13, 1127 (1974).
58. G. Ferenczi, Lund III Conference on Deep Level Impurities, 1981 (unpublished).
59. G. Ferenczi, in New Developments in Semiconductor Physics, edited by F. Beleznay, G. Ferenczi, and J. Giba (Springer, Berlin, 1980), p. 116.
60. D. V. Lang, J. Appl. Phys. 45, 3023 (1974).
61. J. D. Cohen and D. V. Lang, Phys. Rev. B25, 5321 (1982).
62. D. V. Lang, J. D. Cohen, and J. P. Harbison, Phys. Rev. B25, 5285 (1982).
63. D. S. Day, M. Y. Tsai, B. G. Streetman, and D. V. Lang, J. Appl. Phys. 50, 5093 (1979).
64. L. Dmowski, M. Baj, P. Ioannides, and R. Piotrzkowski, Phys. Rev. B26, 4495 (1982).
65. J. M. Langer, Proc. 15th Intl. Conf. Physics of Semiconductors, J. Phys. Soc. Japan 49, Suppl. A, 207 (1980).

66. D. J. Wolford, B. G. Streetman, W. Y. Hsu, and J. D. Dow, J. Lumin. *18/19*, 863 (1979).
67. D. J. Wolford, B. G. Streetman, and J. Thompson, *Proc. 15th Intl. Conf. Physics of Semiconductors*, J. Phys. Soc. Japan *49*, Suppl. A., 233 (1980).
68. A. Majerfeld and P. K. Battacharya, Appl. Phys. Lett. *33*, 259 (1978).
69. G. A. Baraff, M. Schlüter, G. Allan, Phys. Rev. Lett. *50*, 739 (1983).
70. J. R. Troxell, A. P. Chatterjee, G. D. Watkins, and L. C. Kimerling, Phys. Rev. B *19*, 5336 (1979).
71. D. V. Lang, P. M. Petroff, R. A. Logan, and W. D. Johnston, Jr., Phys. Rev. Lett. *42*, 1353 (1979).
72. K. L. Ngai, Comments on Solid State Physics *9*, 141 (1980).

CHAPTER 8

Iron Impurity Centers in III-V Semiconductors

S. G. Bishop

Naval Research Laboratory
Washington, DC 20375

1. INTRODUCTION
2. HISTORICAL PERSPECTIVE
  2.1 Electron Spin Resonance Studies
  2.2 Optical Studies: Crystal Field Splitting
  2.3 Acoustic Paramagnetic Resonance
  2.4 Transport Measurements
3. EXPERIMENTAL STUDIES OF InP:Fe
  3.1 Optical Absorption and Photoluminescence Studies of InP:Fe
  3.2 ESR Studies of InP:Fe
  3.3 Photoluminescence and Photoluminescence Excitation
    Spectroscopy in InP:Fe
  3.4 Photocapacitance and DLTS Studies of InP:Fe
  3.5 Fe-Related Complexes in InP:Fe
4. EXPERIMENTAL STUDIES OF GaAs:Fe
  4.1 Optical Absorption and Photoluminescence Studies of GaAs:Fe
  4.2 Photoconductivity, Photocapacitance, and Charge Transfer
    Absorption in GaAs:Fe
  4.3 ESR Studies of GaAs:Fe
  4.4 Mössbauer Spectroscopy in GaAs:Fe
5. EXPERIMENTAL STUDIES OF GaP:Fe
  5.1 Optical Absorption and Photoluminescence Studies of GaP:Fe
  5.2 ESR Studies of GaP:Fe

  5.3 Transport Measurements, Optical Absorption, and Photoionization
        Spectroscopy in GaP:Fe
6. NOTES ON THEORETICAL METHODS
7. SUMMARY
8. WORK IN PROGRESS AND NEW DIRECTIONS

1. Introduction

The study of the deep impurity levels introduced in III-V semiconductors
by the presence of Fe impurities is of great interest for several reasons. For
example, Fe-doped InP is a technologically important high-resistivity substrate for devices which are based on the epitaxial growth of InP and
quaternary III-V alloys (e.g. InGaAsP). A more general concern is the fact
that 3d transition metal impurities can act as efficient recombination centers which determine the minority carrier lifetime and emit radiation
strongly in the near infrared[1-4] (wavelengths ~2-3 μm). The competing
pathways (shunt paths) introduced by these centers for the recombination
of minority carriers limit the performance of light emitting diodes which is
determined by the efficiency of near-band edge emission.[2] A dramatic
illustration of the efficiency of such shunt paths is shown in Figure 1
where the ratio of the efficiency of the ~3.5 μm Fe^{2+} emission band to
the efficiency of the band-edge emission for InP:Fe is plotted as a function of Fe concentration.[5] As the Fe concentration is increased by a factor of ~10^3, the strength of the Fe^{2+} emission relative to the band edge
emission increases by nearly *five* orders of magnitude. Such efficient deep
traps can also have deleterious effects upon the majority carrier properties
which govern the performance of unipolar devices such as field-effect
transistors.

  Transition metals in general and Fe in particular are pervasive inadvertent impurities in III-V semiconductor compounds, a circumstance which
further emphasizes the practical importance of the study of the fundamental properties of the associated deep impurity levels. For example, the
Fe^{2+} luminescence band which will be discussed later in detail is observed
in essentially all bulk InP crystals studied, including semi-insulating Fe-
doped and undoped material.[6,7] Recent mass spectroscopic analysis[8] of
high purity indium raw material used in the growth of InP crystals revealed
the presence of Fe at the 3 ppma level (in addition to Mn, Cu, and Cr at
somewhat lower levels). Hence, the indium raw material is viewed as the
likely source of the ubiquitous Fe contaminants in InP. In GaAs, the Fe^{2+}
luminescence is not so universally evident in undoped bulk crystals, possibly because of the higher purity of the gallium raw material relative to
that of the available indium.

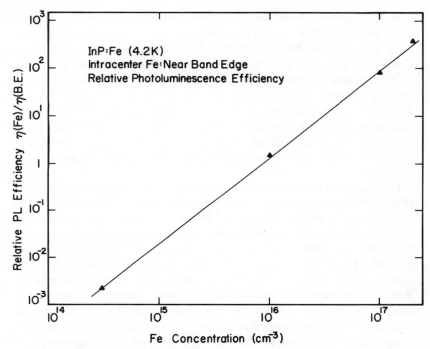

FIGURE 1 The ratio of the efficiency of the 0.35 eV Fe^{2+} emission band to the efficiency of the band edge emission for InP:Fe plotted as a function of Fe concentration. Bishop et al., Ref. 5.

The example of Fe in InP is particularly important since, as mentioned previously, Fe-doped InP is used as a semi-insulating substrate material for microwave devices.[9] Consequently, redistribution of Fe dopants from the substrate during epitaxial growth or annealing after ion implantation can introduce Fe as a contaminant in the active device. Furthermore, recent studies of the effects of capped and uncapped annealing have demonstrated that in *undoped* GaAs[10] and InP[11] in which experimental evidence indicates that the Fe concentration in the as-grown material is $\leq 10^{15} \, cm^{-3}$, Fe accumulates in readily detectable concentrations ($\sim 10^{17} \, cm^{-3}$) at the sample surface during the anneal, as evidenced by photoluminescence (PL) and secondary ion mass spectroscopy (SIMS). Thus, it would appear that at the present level of materials purity, the effects of Fe impurities in III-V semiconductors are essentially inescapable.

In a fundamental context, the study of Fe-related deep levels in III-V's is important because of its potential contributions to the physical understanding of the incorporation of transition metal ions in these semiconductors. Transition metals as substitutional impurities in III-V semiconductors form deep acceptor levels for which the effective mass description applied to main-group shallow acceptors is inappropriate.[12] These impurities have open shell core (d-band) configurations which are characterized by intra-center d-d optical transitions between atomic-like core states perturbed by the tetrahedral crystal field.[2-7,12] Consequently, the optical spectra of transition metal impurities which show such sharply structured intracon-figurational d-d transitions at energies below the host-lattice band gap provide no measure of the positions of the impurity energy levels within the forbidden energy gap. However, they can provide readily identifiable optical signatures which reveal the presence of particular transition metal impurities.

The transition metals in general and Fe in particular often exhibit variable valency when acting as substitutional impurities. Their most stable oxidation state may be strongly influenced by the chemical environment of the host lattice and it is not determined solely by the ionization of potential of the transition metal itself.[12] For example, in a semiconductor, the position of the Fermi energy in the band gap can determine whether a transition metal impurity ion is incorporated in the neutral state where it matches the valence of the host atom it has replaced, as a negatively-charged electron trap, or as a positively-charged hole trap. These variable valence states can be denoted as follows: the electrically-neutral, trivalent charge state of a transition metal acceptor on a group III (cation) site is $A^{3+} (3d^n)$ while the negatively-charged, singly-ionized acceptor state is $A^{2+} (3d^{n+1})$.

Much of our discussion of the impurity centers in III-V compounds will be concerned with substitutional Fe on the group-III (cation) site in the tetrahedral lattice. In this substitutional model, the Fe has been shown to exist in the neutral $Fe^{3+}(3d^5)$ state (sometimes referred to as "isoelec-tronic") in which it gives up three of its valence electrons to the bonds; the singly-ionized acceptor or one-electron trap state, $Fe^{2+}(3d^6)$; or, occasion-ally, the doubly-ionized acceptor or two-electron trap state, $Fe^+(3d^7)$.

In general, experimental studies of deep levels in semiconductors cannot provide a full chemical and structural description of the localized electronic states, especially if only electrical measuresments are possible. However, for the 3d transition metal impurities such as Fe, the powerful combination of optical and electron spin resonance spectroscopy can often identify the charge state and structural environment of the substitutional impurity centers. In addition, interconfigurational charge-transfer transitions which involve the excitation of electrons from the valence band to the transition-metal impurity, or from the impurity to the conduction band, give rise to broad charge-transfer absorption bands (photoionization spectra) which can reveal the position of the impurity energy levels relative to the band edges.

The present article reviews and summarizes recent progress in the experimental identification of deep levels associated with Fe impurities in the III-V semiconductors InP, GaAs, and GaP. An introductory survey of early experimental studies of Fe-doped III-V semiconductors is followed by more detailed discussions of recent work which is categorized by material and experimental technique.

2. Historical Perspective

A brief historical perspective will be presented which outlines the early work concerning Fe impurity centers in III-V compounds. Detailed discussion will be reserved for the more recent work which comprises the current state of our understanding.

The principal, experimentally observable signatures of substitutional Fe in the tetrahedral crystal lattice are the characteristic electron spin resonance (ESR) spectrum of the spin 5/2 S-state Fe^{3+} ion and the optical transitions (absorption or emission) between the $^5T_2$ excited state and ^5E ground state of the ^5D term of Fe^{2+} ions in the tetrahedral crystal field. The early studies of Fe in III-V crystals can be outlined conveniently in terms of the experimental observation and theoretical treatment of these signatures of Fe^{2+} and Fe^{3+}.

2.1 Electron Spin Resonance Studies

Electron spin resonance is one of the few experimental techniques which yields information about the immediate environment of impurity centers in solids and it has been widely applied to the study of transition metals in III-V crystals, beginning in the early 1960's. The first ESR study of transition metal ions in a III-V compound was carried out by Woodbury and Ludwig,[13] in which they observed Mn and Fe impurities in GaP in the $3d^5$ configuration. This work was closely followed by the first ESR measurements on Fe in GaAs by deWit and Estle[14,15] and Fe in InAs by Estle.[16] (For a complete bibliography of early ESR studies of transition metal ions in III-V compounds see the 1972 review article by Bashenov.[17]) In all of these studies, the observed ESR spectra were interpreted as arising from a spin 5/2 system in cubic symmetry and were attributed to $Fe^{3+}(3d^5)$

ions, presumably substituted on cation (Group III) sites. While these studies concentrated primarily on the $\Delta M = 1$ transitions in the vicinity of $g = 2$, deWit and Estle[15] did observe nine of the ten possible "quasi-forbidden" transitions for which $\Delta M > 1$.

The conventional spin Hamiltonian for the $S = 5/2$, ^6S ground state of Fe in an environment having cubic symmetry is given by

$$H = g\beta\vec{H}.\vec{S} + \frac{a}{6}(S_x^4 + S_y^4 + S_z^4 - \frac{707}{16}). \tag{1}$$

The ESR spectra are then describable in terms of two parameters, the g-value (the spectroscopic splitting factor) and the zero-field splitting, or fine-structure parameter, a. Early studies of Fe in III-V compounds[13-16] and a study of Fe in ZnTe by Hensel[18] revealed some interesting trends in the parameters g, a, and the ESR linewidths in cubic covalent crystals such as the III-V's as compared to their values in more ionic cubic crystals. The g shift is large and positive, the zero-field splitting is larger than in ionic crystals, and a large hyperfine interaction with neighboring nuclei is often observed which manifests itself either as resolved structure or as inhomogeneous broadening of the lines. All of these effects were attributed to the greater covalency of the III-V compounds. The large positive g-shift was explained on the basis of a mechanism first proposed by Fidone and Stevens.[19] They suggested that in addition to interactions with excited states of the impurity ion, one must also consider interactions with charge-transfer states in which electrons are transferred from the impurity to the ligands. For the case of Fe, Watanabe[20] showed that the positive g-shift is attributable to electron transfer from the surrounding ligands onto the Fe^{3+}. Azarbayejani et al.[21] showed that this effect can contribute to the zero-field splitting parameter a, although Estle[16] pointed out that there is apparently no simple relationship between the g-shift and the zero-field splitting parameter for the III-V compounds.

Estle[16] also discussed the large ESR linewidths for Fe in III-V compounds which arise from unresolved hyperfine structure due to interactions between the d-electrons of the transition metal impurity and the magnetic moments of the neighboring nuclei of the host lattice. He pointed out that the ratio of the linewidth observed in gallium arsenide and indium arsenide is approximately equal to the ratio of the weighted means of the magnetic moments of the gallium and indium isotopes. This means that if the hyperfine interaction is due primarily to the *next*-nearest neighbors, the required delocalization of the d-electrons is roughly constant for the two compounds. This is consistent with results obtained for Mn in II-VI compounds where only the next-nearest neighbor hyperfine structure is observed and roughly constant delocalization is indicated.[22,23]

It will become evident in the succeeding Sections that the more recent ESR studies of Fe in III-V's have confirmed, complemented, and extended rather than contradicted these early results.

2.2 Optical Studies: Crystal Field Splitting

While ESR techniques probe the Zeeman splitting of the ground state multiplet in an applied magnetic field at energies usually ~ 1 cm^{-1}, optical techniques (emission and absorption) are used to study the much larger crystal-field splittings of the ground state and excited state multiplets which occur for many 3d-electron configurations. The $Fe^{3+}(3d^5)$ S-state ion constitutes a special case from the point of view of crystal field theory in that, to first order, there can be no interaction with the crystal field since there is no orbital angular momentum.[24] However, as is evident from Equation (1), the sixfold degeneracy of the ^6S multiplet is partially lifted by a cubic crystal field,[24] but the splitting is far below the optical region (the zero-field splitting parameter a is in the range 0.02 to 0.04 cm^{-1} for the III-V compounds[25-27]). Hence, while the Fe^{3+} $(3d^5)$ ion evidences a readily observable and characteristic 5-line ESR spectrum, it exhibits no sharply structured intracenter d-to-d transitions in the optical spectral range. The converse is true for $Fe^{2+}(3d^6)$ (the one electron trap in III-V's); while an ESR spectrum characteristic of Fe^{2+} in cubic symmetry has persistently evaded detection in the III-V's, sharply structured intracenter optical transitions attributable to Fe^{2+} have been observed in a variety of II-VI and III-V semiconductors (see ref. 28 and refs. therein, also refs. 25 and 27). In addition, the $Fe^{2+}(3d^6)$ state of Fe impurities in GaAs has been studied by acoustic paramagnetic resonance (APR) techniques as will be discussed later.

In 1960 Low and Weger[29] pointed out that the optical and paramagnetic resonance spectra of the $3d^6$ configuration in the crystal field of cubic symmetry had not been observed experimentally; they calculated the energy level splittings of the ground state of the d^6 configuration in cubic and axial crystal fields as well as the Zeeman splittings of the various levels. The tetrahedral crystal field lifts the orbital degeneracy of the ^5D term of the free ion producing an orbital triplet $^5T_2$ ($^5\Gamma_5$) and an orbital doublet ^5E ($^5\Gamma_3$), with the ^5E level lying lower (Figure 2).[29,30] The ground state ^5E multiplet is further split into five equidistant levels by

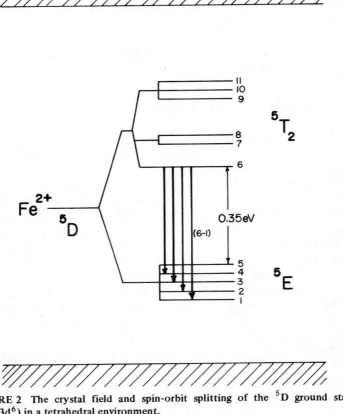

FIGURE 2 The crystal field and spin-orbit splitting of the ^5D ground state of Fe^{2+} ($3d^6$) in a tetrahedral environment.

second order spin-orbit and first order spin-spin interactions as shown in Figure 2 (third order spin-orbit effects lead to slightly unequal spacings for these levels[31]). Electric dipole transitions are allowed between the lowest state, T_2, of the $^5T_2$ excited state multiplet and the lowest four states of the ground state ^5E pentet. These transitions are now routinely observed and comprise the four-line optical signature of Fe^{2+} ions in tetrahedral symmetry.

Crystal-field theory[32] describes the energies of these transitions in terms of the crystal-field splitting $\Delta = 10D_q$ and the spin-orbit parameter λ, which may be different in first and second order effects (λ_1, λ_2). The $^5T_2$ and ^5E levels of Fe^{2+} in the tetrahedral crystal field lie at energies of $-4D_q$

and $+6D_q$, respectively, relative to the free-ion value[29,30] (D_q is negative). The highest energy of the four allowed transitions between the T_2 excited state and the ^5E ground state is given by the expression[33]

$$E_{T_2 \longleftrightarrow A_1} = \Delta + 3\lambda_1 + \frac{18}{5}\frac{\lambda^2}{\Delta^2} + 4K', \qquad (2)$$

where K' is the spacing of the five levels of the ^5E ground state given by [33]

$$K' = Kq = 6q[(\lambda_2^2/\Delta) + \rho]; \qquad (3)$$

ρ is an effective spin-spin interaction parameter and q is a reduction factor which represents the effect of a possible dynamical Jahn-Teller distortion of the ^5E ground state. The equidistant spacing of these levels is often written in the simplified form $K = 6\lambda^2/\Delta$.

Subsequent to the 1960 theoretical work of Low and Weger,[29] Slack and coworkers[34,35] studied the characteristic Fe^{2+} near infrared absorption spectrum in cubic ZnS and some related II-VI compounds and subjected the results to a detailed theoretical analysis. Slack and O'Meara[36] observed Fe^{2+} luminescence in ZnS, and a revised interpretation of the absorption and luminescence results was published by Ham and Slack[33] in 1971. During the same time period the only study of $Fe^{2+}(3d^6)$ and $Co^{2+}(3d^7)$ were studied in several II-VI compounds and, for the first time, in GaAs and GaP. Characteristic Fe^{2+} absorption spectra were observed with the strongest zero-phonon lines (ZPL) occurring at 3344 cm^{-1} and 3002 cm^{-1} in GaP and GaAs, respectively. This classic study of the crystal-field spectra of $3d^n$ impurities in a wide variety of tetrahedral crystals provided a graphic demonstration of how useful the crystal-field description is even in covalent materials.

2.3 Acoustic Paramagnetic Resonance

Apparently, no further detailed studies of the Fe^{2+} optical transitions in III-V compounds were carried out until the mid 1970's. However, these later optical studies of Fe^{2+} in the III-V's were preceded by APR investigations which can detect the $3d^6$ configuration of Fe. Ganapolskii[37(a)] discovered a strong resonance absorption of hypersound in GaAs:Fe at 4.2K in an applied magnetic field, which was found to be attributable to Fe atoms in the $3d^6$ configuration. As described earlier, the combined effects of crystal-field splitting and second-order spin-orbit splitting for $Fe^{2+}(3d^6)$

in tetrahedral symmetry produce a ground-state ^5E multiplet of five al-most equally spaced levels separated by $\cong 15$ cm^{-1} in GaAs:Fe. The lowest of these levels is a singlet and the next higher level is a triplet which is split by a slight axial distortion of the crystal field into a singlet and doub-let. The degeneracy of the doublet is lifted by an external magnetic field and these levels, characterized by a strong electron-phonon interaction, are involved in a strong resonance absorption of hypersound. The strength of this APR absorption line is proportional to the concentration and the Fe^{2+} centers and APR measurements of GaAs:Fe crystals with different concentrations of shallow compensating donors can determine the relative concentration of $Fe^{2+}(3d^6)$ ions even if this concentration is quite low in comparison to the total Fe concentration.

Ganapolskii and coworkers[37(b)] used such APR measurements on GaAs:Fe co-doped with different concentrations of Te $(1.6 \times 10^{16} -1.7 \times 10^{17}$ cm$^{-3})$ to demonstrate that the concentration of $Fe^{2+}(3d^6)$ was proportional to the TE concentration. These results were indicative of a simple compensation mechanism in which the concentration of the $Fe^{2+}(3d^6)$ centers is governed entirely by the concentration of compensat-ing donors and is not related to a proposed self-compensation mechanism. Additional APR absorptions occur within the ^5E ground state manifold. Ganapolskii[37(a)] studied the temperature dependence of the APR absorp-tions of Fe^{2+} in GaAs and obtained values for the energy separations be-tween the ground state of the pentet and the first three excited states of 1.6, 33, and 50 cm^{-1}, respectively.

Recently, Rampton[38] and Wiscombe reported the first APR study of Fe^{2+} in InP:Fe. They, too, studied absorptions within the ^5E ground state manifold and found the temperature dependence in the 4-to-14K range to be consistent with a Zeeman-split triplet lying 15 cm^{-1} above the singlet ground state, consistent with the optical data.[25]

2.4 Transport Measurements

Indium phosphide has been conspicuously absent from our discussions of early ESR and optical studies of Fe in the III-V's. This material was simply not widely available until the high-pressure crystal-growth technology reached its present state of development.[39,40] This circumstance is also reflected in the chronology of transport studies of Fe-doped III-V's. For example, Haisty and Cronin[41] established that semi-insulating GaAs could be prepared by doping with Fe in 1965, ten years before Mizuno and Watanabe[9] produced semi-insulating Fe-doped InP.

In their original study of the conductivity of GaAs:Fe, Haisty and Cronin found an acceptor level associated with Fe which lies 0.52 eV above the valence band at $0°K$. However, while calculations of the room-temperature hole concentration for an acceptor level at 0.52 eV showed that a room-temperature resistivity of $>10^6$ ohm-cm should be obtainable, the highest resistivity observed in the Fe-doped GaAs was about 2×10^5 ohm-cm. This discrepancy was addressed in a subsequent paper by Haisty[42] in which he concluded that the 0.52 acceptor level associated with Fe in GaAs stays fixed with respect to the *conduction* band as a function of temperature so that at room temperature it would be only 0.38 eV above the valence band. He also suggested the alternative possibility that there are two Fe levels fairly close together which merge at high temperatures and shift the effective thermal activation energy to lower values.

This early work was closely followed by a study of electroluminescence in GaAs:Fe by Strack[43] which was interpreted in terms of an acceptor level 0.38 eV above the valence band, and additional electrical measurements and impurity photoconductivity by Allen[44] which confirmed an acceptor level at 0.52 eV above the valence band. Measurements of electrical conductivity, Hall coefficient, and thermally-stimulated current by Kadyrov and coworkers[45] yielded an acceptor level 0.49 eV above the valence band, also in good agreement with the earlier work. Likewise, Sze and Irwin[46] found Fe-related acceptor levels 0.52 and 0.37 eV above the valence band in GaAs:Fe. The temperature dependences of the Hall coefficient, electrical conductivity, optical absorption, and steady state photoconductivity were studied by Omel'yanovskii et al.[47] Hall-coefficient data again confirmed the acceptor level with the 0.5 eV thermal activation energy at $0°K$. The apparent low energy onsets of both absorption and photoconductivity spectra shifted to lower energy with increasing temperature consistent with Haisty's original observation[42] of a reduction in the separation of the 0.5 eV Fe level from the valence band with increasing temperature. Both the optical absorption and photoconductivity spectra in the spectral range >0.8 eV indicated the presence of a second, deeper level. The samples employed in these studies were found to be so strongly compensated that the authors suggested[47] that an "automatic compensation" had taken place during growth; that is, the concentration of the compensating centers increased approximately proportionally to the Fe concentration. In a subsequent study Omel'yanovskii and coworkers[48] established the existence of an acceptor level with an activation energy of 0.2 eV through an investigation of thermally-stimulated current and the temperature dependence of the Fe^{3+} concentration as monitored by ESR in GaAs:Fe crystals.

Measurements of carrier lifetimes in GaAs:Fe by Kolchanova et al.[49] and Lukicheva et al.[50] demonstrated that non-equilibrium photo-injected carriers recombine at the same iron-related deep level which governs the equilibrium carrier density. This identification of Fe impurities as dominant recombination centers is obviously consistent with the implications of the radiative recombination data for InP:Fe shown in Figure 1.

Plesiewicz[51] studied the photo-Hall effect and photoconductivity in GaAs:Fe and found, in addition to the familiar deep acceptor level 0.52 eV above the valence band, a high concentration of shallow acceptor levels at 0.03 eV above the valence band. He regarded these shallow acceptor levels as "sensitizing centers" which are responsible for the high photosensitivity of GaAs:Fe at low temperatures. Finally, with the deep-level-transient-spectroscopy (DLTS) studies of Lang and Logan[52] which detected a deep acceptor level with an activation energy of 0.57 eV in GaAs:Fe, the chronology reaches the first study of transport in InP:Fe by Mizuno and Watanabe.[9] It is at this point that we begin the discussion of the more recent ESR, optical, luminescence, and transport studies of Fe in the III-V semiconductors.

3. Experimental Studies of InP:Fe

With the increasing availability of single crystals of InP grown by the liquid encapsulation Czochralski (LEC) method[39,40] during the 1970s, this material became the subject of intensive study. Iron-doped InP assumed particular importance when Mizuno and Watanabe[9] reported that this material could be produced with a resistivity in excess of 10^7 ohm-cm. This semi-insulating character of InP:Fe makes it superior to InP:Cr or GaAs:Fe (both have resistivity in the 10^3–10^5 ohm-cm range) as a substrate for III-V microwave devices such as FETs and planar TEDs. Consequently, in recent years, studies of the properties of InP:Fe have, with a few exceptions, totally supplanted the studies of GaAs:Fe which dominated our discussion of the early work.

While the initial studies[9,53] of the transport properties of InP:Fe demonstrated its high resistivity and detected an Fe-associated deep level 0.66 eV below the conduction band edge, they did not provide values for the effective segregation coefficient or the solubility of Fe in InP. Henry and Swiggard[40] have reported a study of the growth and properties of InP:Fe in which they used ESR measurements[26] of the Fe^{3+} concentrations in their samples to calculate an effective segregation coefficient for Fe in InP of 1.6×10^{-3}. Spectrochemical analysis was used to provide an approximate verification of the segregation coefficient.

Recently Debney and Jay[54] have reported the observation that in some cases only a small percentage of the Fe contained in InP:Fe crystals is electrically active as an electron acceptor. These workers have raised the possibility that the Fe^{3+} concentration as measured by ESR may not be a reliable measure of the total Fe concentration and, consequently, the accuracy of the effective segregation coefficient based on ESR measurements might be questionable. They suggest that the solubility of Fe as a deep acceptor may be a function of growth conditions, and that the effective segregation coefficient of Henry and Swiggard (1.6×10^{-3}) is not universally applicable. It should be pointed out that Debney and Jay's measurements were carried out for Fe concentrations at least an order of magnitude lower than the 2.5×10^{17} cm^{-3} value in the InP:Fe sample for which the effective segregation coefficient was determined by ESR. For those experiments, Henry and Swiggard[40] had produced semi-insulating InP:Fe with resistivity greater than 10^{7} ohm-cm by adding 0.15–0.3 wt % Fe to the melt, with resultant Fe concentrations in the 2×10^{16} to 2×10^{17} cm^{-3} range. More recently, these workers and others have achieved comparable resistivities with much lower Fe concentrations ($\sim 5 \times 10^{15}$ cm^{-3}) and routine procedures for growth of semi-insulating InP now employ $\sim 10^{16}$ cm^{-3} or less Fe.[55]

3.1 Optical Absorption and Photoluminescence Studies of InP:Fe

A sharply structured near infrared (0.35 eV) PL spectrum attributable to crystal field transitions at Fe^{2+} ($3d^{6}$) in a tetrahedral lattice site was first reported in Fe-doped InP by Koschel et al.[6] (1976). This PL spectrum and the corresponding near-infrared absorption spectrum as reported in greater detail by Koschel et al.[25] in 1977 are shown in Figure 3. The PL spectrum consists of four well-resolved and nearly equally spaced zero-phonon lines and a weak phonon sideband at lower energy. Similar PL[6,56] and absorption[57] spectra have been observed in GaAs:Fe and will be discussed later. All of the zero phonon spectra are explicable in terms of transitions between the excited $^{5}T_{2}$ manifold and the ^{5}E ground state manifold shown in the energy level diagram of Figure 2 for Fe^{2+}($3d^{6}$) in a tetrahedral crystal field as discussed previously. Photo-excited electrons are assumed to thermalize non-radiatively to the lowest lying state of the $^{5}T_{2}$ manifold, from which radiative transitions occur to the four lowest levels of the ^{5}E ground state pentet which has been split by second order spin-orbit effects. The transition to the highest lying level of the ^{5}E manifold ($T_{2} \rightarrow A_{2}$) is electric dipole forbidden and is not observed in the emission spectrum.

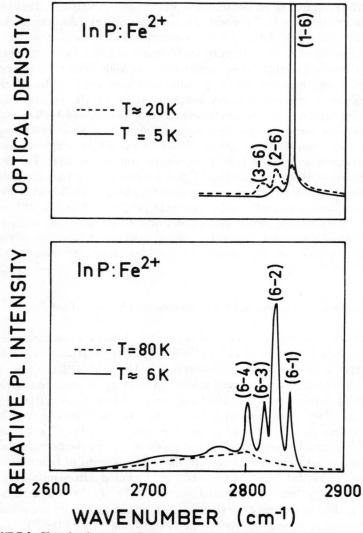

FIGURE 3   Photoluminescence (lower part) and optical absorption (upper part) of
InP:Fe^{2+}. The optical transitions of the 3d^6 state of Fe are given in parentheses
according to Figure 2. After Koschel et al., Ref. 25.

Koschel et al.[25] used the experimental values for the highest energy PL transition ($T_2 \rightarrow A_1$ = 2845 cm^{-1}) and the splitting of the PL lines (K' = 14 cm^{-1}), and the theoretical expression given in Equation (2) to arrive at a value for Δ = 10 Dq of 3040 cm^{-1}. Their calculation assumed $\lambda_1 = \lambda_2 = k\lambda_0$, where λ_0 is the free ion value of the spin-orbit coupling constant (λ_0 = −103 cm^{-1}) and k is a covalency reduction factor (k $\approx$ 0.8–0.9).

The near infrared absorption spectrum of Figure 3 taken at 5K also shows a strong zero phonon line at 2845 cm^{-1} which corresponds to the transition $A_1 \rightarrow T_2$, the inverse of the PL transition. Only the lowest level (A_1) of ^{5}E is populated at 5K and therefore only the $A_1 \rightarrow T_2$ transition is observed in absorption. At 20K the next highest levels of ^{5}E, (T_1, and E) are thermally populated and transitions from these levels to the T_2 excited state are seen in the absorption spectrum for this temperature. Ippolitova et al.[27] reported similar absorption spectra for several temperatures in InP:Fe. From their highest resolution spectra taken at 20K they derived values of the crystal field splitting parameter Δ = 10 Dq and the spin-orbit coupling constant λ of 3025 and −87 cm^{-1}, respectively. Comparison of the published absorption spectra indicates that the value of the highest energy transition resolved lies ~15 cm^{-1} lower in the data of Ippolitova et al.[27] than in the spectrum of Koschel et al.[25] It seems probable that Ippolitova et al. did not resolve the highest energy $A_1 \rightarrow T_2$ transition.

While the Fe^{2+} PL spectrum is broad and weak at 80K and unobservable at 300K, both Ippolitova et al.[27] and Iseler[58] observed a broadened but identifiable Fe^{2+} absorption spectrum near ~0.4 eV in InP:Fe at 300K. This absorption spectrum can serve as a highly useful signature of the presence of iron in the Fe^{2+} state. ESR studies of semi-insulating crystals of InP:Fe with low donor concentrations have shown that most of the iron is in the Fe^{3+} or neutral charge state. It is to be expected that back-doping with shallow donors will convert some of the Fe^{3+} to Fe^{2+}, the one electron trap state, and that n-type crystals will contain mostly Fe^{2+}. Iseler's 300K absorption data[58] and the corresponding low-temperature absorption data of Bishop et al.[5,7] and recently Tapster et al.[59] demonstrated that this is indeed the case as the characteristic Fe^{2+} absorption spectrum becomes much stronger as shallow donors are added to semi-insulating InP:Fe. Eaves et al.[60] pointed out that since all of the absorption results in semi-insulating and n-type InP:Fe can be explained in terms of the Fe^{2+} and Fe^{3+} charge states, the existence of an Fe^{1+} level within the band gap is not required. However, the most definitive statement has been provided by Tapster and coworkers.[59] Their Fe^{2+} absorption experiments on InP:Fe with increasing donor concentrations showed that after all of the Fe ions have been converted to Fe^{2+}, the addition of more shallow donors simply

increases the free carrier concentration. They conclude[59] that this demonstrates that the $Fe^+(3d^7)$ double electron trap does not exist in InP as it
does for Fe in GaP.[3,65]

In contrast to the effects of shallow donors on the strength of the Fe^{2+}
absorption, the 0.35 eV Fe^{2+} luminescence is found[7,59] to be strongly
quenched in n-type InP:Fe which has been back-doped with donors. This
was first observed by Bishop et al.[5,7] and is evident in the comparison of
the Fe^{2+} PL spectra for semi-insulating and n-type InP:Fe and for

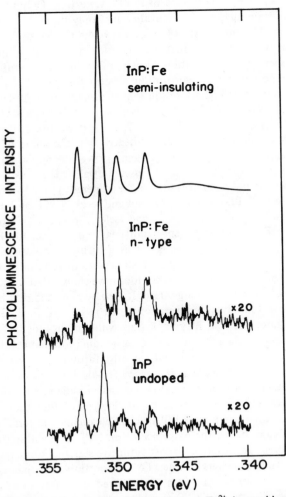

FIGURE 4   Photoluminescence spectra attributed to Fe^{2+} in semi-insulating and
n-type InP:Fe and in undoped InP. T = 4.2 K. After Bishop et al., Ref. 7.

undoped InP in Figure 4. These observations have been confirmed recently by Tapster et al.[59] On the basis of these results, it would appear that the strength of the Fe^{2+} PL is correlated with the equilibrium concentration of Fe^{3+}. Bishop et al.[5,7] suggested an excitation mechanism in which inter-band photo-excitation of free electrons and holes is followed by capture of a conduction band electron by an Fe^{3+} ion which results in the forma-tion of Fe^{2+} in the $^5T_2$ excited state. The center then relaxes to the ^5E ground state by the emission of a 0.35 eV photon, and subsequent hole capture restores the Fe to its equilibrium Fe^{3+} charge state.[7,61] This mech-anism will be considered again later in our discussion of photoluminescence excitation spectroscopy and photoconductivity in InP:Fe.

In two very recent papers Tapster et al.[59] and Leyral and coworkers[62] point out that the quenching of the Fe^{2+}-PL band in n-type InP:Fe could also be caused by non-radiative Auger processes between Fe^{2+} ($^5T_2$) and the free electrons which become probable at high electron concentrations. These two papers[59,62] also mark the first appearance in the literature of the ~0.5 eV PL band which has been observed[5] in many samples of InP:Fe and is apparently Fe-related (see Figure 5). Three possible explan-

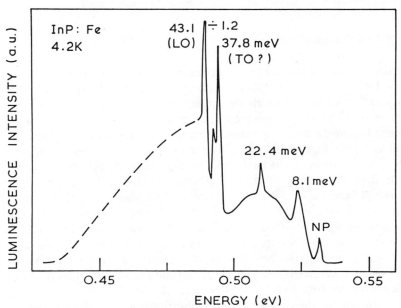

FIGURE 5 Detailed photoluminescence spectrum of InP:Fe near 0.5 eV. The energy separations of the phonon replicas from the 0.5332 eV no-phonon line are indicated. The dashed region of the spectrum is partially obscured by sharp atmospheric ab-sorption lines. The peak labelled TO(?) could also be a local phonon mode replica. After Tapster et al., Ref. 59.

ations of this band have been proposed: (1) that it is attributable to an Fe complex[59,62] rather than isolated Fe; (2) that it arises from a transition-metal impurity[62] other than Fe; and (3) that it could be due to a charge-transfer transition[59,62] of the type Fe^{2+} $(^5T_2)$ + h_{VB}^+ → Fe^{3+} + $h\nu$ (0.5 eV). To be more exact, for the charge-transfer or hole-capture process (3), Tapster et al.[59] actually suggest $[Fe^{2+}]h_b^+$ → Fe^{3+} + $h\nu$ (0.5 eV), where h_b represents a hole bound to the negatively charged Fe^{2+} in the initial state of the luminescence transition. Such transitions have been proposed previously to explain the excitation of PL bands in ZnO:Cu[63] and ZnSe:Ni.[64] Tapster et al. cite the strong phonon coupling of the 0.5 eV band (see Figure 5) as evidence for the involvement of a particle with an extended band-like wavefunction in the initial state of the transition.[64]

Leyral et al.[62] achieved particularly efficient below-gap excitation of the Fe^{2+} PL spectrum through the use of YAG laser (1.06 μm) excitation. In addition to the 0.5 eV and 0.35 eV PL bands, they observed previously reported TA zone-boundary phonon side bands 8.3 and 6.7 meV below the ZPL's of the Fe^{2+} band and a new band at 0.315 eV with the same structure as the 0.35 eV band. They suggested that this band could be due to a local-phonon-mode replica with energy separation of 36 meV.

3.2 ESR Studies of InP:Fe

A major contribution to our understanding of the incorporation of Fe in the InP lattice was provided by three parallel but independent ESR studies[25-27] of InP:Fe which were published in 1977. The Fe^{3+} ESR spectra obtained by these three groups, and the spin Hamiltonian parameters derived from the spectra are in excellent agreement and, for the most part, the interpretations of the results by the three groups are remarkably consistent. Furthermore, the results and interpretations are entirely consistent with those reported much earlier for GaAs,[14,15] GaP,[13,65] and InAs.[16]

The X-band Fe^{3+} ESR spectrum obtained by Stauss et al.[26] in a semi-insulating crystal of InP:Fe at 4.5K with H ∥ [100] is shown in Figure 6. Because of the relatively large linewidth of the spectrum, $\Delta H \cong 120$ G, the usual five-line fine-structure pattern expected for an S = 5/2 center is not fully resolved. However, the angular dependence of the positions of the spectral features (shown in ref. 27) unambiguously identifies the spectrum as attributable to Fe^{3+} $(3d^5)$ in cubic symmetry. These results do not indicate whether the Fe is located on a substitutional or an interstitial lattice site since, in the absence of distortion, both sites have tetrahedral symmetry. In fact, electron-nuclear-double-resonance (ENDOR) studies[66,67] of Fe in III-V compounds led Teuerle and Hausmann to the unexpected con-

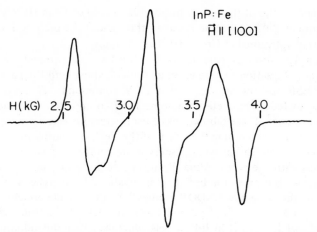

FIGURE 6   X-band EPR spectrum of Fe^{3+} in InP at 4.5 K with $\vec{H}$ // [100]. (Stauss et al., Ref. 26.)

clusion that the Fe atoms were not located on Ga sites but in interstitial sites surrounded by group V atoms in tetrahedral symmetry. Stauss et al.[26] state in a footnote concerning the ENDOR results that they consider the question unresolved, but assume in their analysis that the Fe is substitutional in Group III sites. Ippolitova and coworkers[27] assert that detailed analysis of the ENDOR capabilities shows that the available data do not provide a sufficient basis to determine the position of the Fe^{3+} ions in a lattice with T_d symmetry. However, these workers discussed two factors which they believe do provide sufficient justification for the conclusion that the Fe is substitutional on the Group-III sites: 1) In the substitutional model, an Fe atom on the Group-III site gives up two 4s and one 3d electron to form bonds. These valence electrons do not contribute to electrical conduction and the Fe impurity acts as an acceptor. Thus the material will be semi-insulating, as observed, if the Fe concentration exceeds that of the residual donors. For an Fe atom at an interstitial site, it is not clear what happens to the Fe valence electrons; the Fe impurity should behave as a donor rather than an acceptor. 2) Semi-insulating GaAs:Fe crystals are highly stable under heat treatment. Such thermal stability is more characteristic of materials in which the dopant forms a substitutional solid solution. Dopants in interstitial states are usually more strongly affected by annealing.

Although the first justification of Ippolitova et al.,[27] i.e., the acceptor behavior of Fe in GaAs and InP, is quite valid and supports the case for substitutional Fe, the second argument has been seriously undermined by more recent studies of the redistribution (diffusion) of Fe in III-V's during heat treatment.[10,68] It will be assumed throughout the present discussion that Fe is substitutional at Group-III sites.

The spin Hamiltonian parameters for $InP:Fe^{3+}$ determined by Stauss et al.[26] using Equation (1) are given in Table 1, along with the parameters for Fe^{3+} ESR spectra in some other III-V compounds. A similar table which also includes parameters for some II-VI compounds is provided by Ippolitova et al.[27] These tabulations of parameters for various compounds give some insights concerning the origin of the Fe^{3+} linewidths and g values. On the basis of the earlier work discussed previously,[15-22] superhyperfine interactions with the surrounding (ligand) nuclei are expected to make the principal contribution to the Fe^{3+} ESR linewidth ΔH. Stauss et al.[26] point out that in this case the ENDOR studies[66,67] do provide conclusive evidence that this is true for GaP and GaAs. In each of the three reported studies[25-27] of Fe^{3+} ESR in InP, it was concluded that the indium hyperfine interaction makes the dominant contribution to ΔH. That is, the width at the Fe^{3+} spectrum is due primarily to the superhyperfine interaction not with the Group V (nearest neighbor) atoms but with the Group-III (second-nearest neighbor), i.e. In, atoms. The superhyperfine interaction with the nearest-neighbor coordination sphere is apparently weak and has little effect[27] upon ΔH. Ippolitova et al.[27] point out that these results are especially significant because it is usually assumed that the wave functions of deep impurities are strongly localized, yet the ESR data indicate that the wave function of the Fe^{3+} ion is sufficiently extensive to allow a large superhyperfine interaction with the second nearest neighbor coordination sphere.

The Fe^{3+} g-value observed in InP:Fe exhibits a substantial positive shift relative to the free electron value $g_e = 2.0023$. This is consistent with the positive shifts found for Fe in the other III-V compounds shown in Table 1. Following the work of Watanabe[20] discussed previously Stauss et al.[26] and Ippolitova et al.[27] attribute the positive g-value shift in this covalent compound to electron transfer onto the central ion from the

TABLE 1 EPR parameters of Fe^{3+} $(3d^5)$ in III-V compounds (after Stauss et al., Ref. 26).

Compound	g	a $(10^{-4}\ cm^{-1})$	ΔH (G)	Refs.
GaAs	2.046	+340	50	15
GaP	2.025	+390	38	13, 66, 67
InAs	2.035	+421	130	16
InP	2.0235(10)	+221(2)	118(2)	26

neighboring ligands. Some significant trends in the g-values of Table 1 were also pointed out. Stauss et al. noted that the increase in g for the arsenides relative to the phosphides is consistent with the decrease in the band gap and the increase in the group-V ligand spin-orbit parameter. Ippolitova et al. observed that, with the exception of GaAs, the reduction in the shift of the g-value is accompanied by a reduction in the zero-field splitting parameter, a. In Figure 5 of Ref. 27, Ippolitova et al. plot a as a function of the shift Δg of the g-value relative to g_e for Fe^{3+} in a variety of III-V and II-VI compounds and demonstrate that they exhibit a nearly linear interdependence. They point out that, while a quantitative analysis of the variation of a and Δg for Fe^{3+} is not possible, the qualitative explanation of their behavior can be found in the variation of the ionicity (covalence) of the bonds in this series of compounds.

Stauss et al.[26] also studied the effect of near-infrared light of varying wavelengths on the intensity of the Fe^{3+} ESR spectrum in their InP:Fe samples. The difference in the ESR intensity for light-on and light-off conditions is plotted in Figure 7 as a function of photon energy $h\nu$. In effect, the spectrum is a quenching spectrum for the Fe^{3+} ESR. Principal spectral features include an onset of quenching near 0.75 eV, a shoulder near 1.13 eV, and a cutoff at the band edge, E_g. The observed quenching

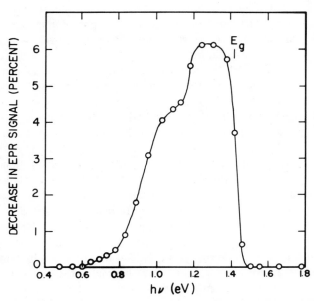

FIGURE 7 Change in Fe^{3+} EPR signal in InP:Fe during application of light with photon energy $h\nu$. E_g is the band gap energy of InP. (Stauss et al., Ref. 26.)

of the Fe^{3+} ESR results from the light-induced transfer of electrons either on or off the Fe^{3+} ions (charge transfer transitions). Stauss et al. point out that the 0.35-eV optical absorption and PL bands associated with the $^5T_2 \rightarrow {}^5E$ transitions of Fe^{2+} ions in InP:Fe do not fix the location of these levels relative to the band edges. However, they suggested a "speculative model" which is consistent with the PL spectrum and their quenching spectrum. The 0.75 eV onset in Figure 7 is attributed to a direct transfer of electrons from the valence band to Fe^{3+} which converts it to the ground state (^5E) of Fe^{2+}. Correspondingly, since the $\sim$1.1 eV shoulder lies just 0.35 eV higher in energy, they proposed that it represents a similar charge transfer from the valence band to the *excited* state ($^5T_2$) of Fe^{2+}. Clearly, this second charge-transfer transition could be followed by radiative relaxation to the ground state of Fe^{2+}, and we will return to this point in subsequent discussion.

3.3 Photoconductivity and Photoluminescence Excitation Spectroscopy in InP:Fe

The photo-induced quenching spectrum for the Fe^{3+} ESR reported by Stauss et al.[26] provided the first experimental basis for suggesting the location of Fe-related localized levels in the gap of InP:Fe. We turn now to the discussion of recent photoconductivity and photoluminescence excitation (PLE) studies in InP:Fe which have begun to provide a consistent picture of Fe-related levels in InP.

In 1979–80, Look[69,70] reported for the first time a below-gap or extrinsic peak in the room temperature photoconductivity of semi-insulating InP:Fe which is shown in Figure 8 along with Iseler's[58] absorption spectrum. The energy position ($\sim$0.44 eV peak) and shape of the photoconductivity peak are similar to those of the $\sim$0.4 eV near-infrared absorption peaks observed by Ippolitova et al.[27] and Iseler[58] in InP:Fe at room temperature. On the basis of the sharply structured optical absorption[7,25,27] and PL spectra[7,25] obtained in the same spectral range, the $\sim$0.4 eV absorption band was attributed to transitions between the ^5E ground state and $^5T_2$ excited state of Fe^{2+} ions in the tetrahedral crystal field as discussed previously in Section 3.1. Accordingly, Look[69,70] attributed the 0.44 eV peak in the photoconductivity spectrum to such Fe^{2+} intracenter optical transitions followed by thermal excitation to the conduction band. The sharply structured zero-phonon lines observed in the low temperature absorption and PL spectra apparently involve only the lowest of the spin-orbit split $^5T_2$ excited states of Fe^{2+}. Look explained the broad room-temperature absorption and photoconductivity spectra in InP:Fe on the basis of a moderately strong electron-lattice interaction in the Fe^{2+} $^5T_2$ excited state. He proposed that, in analogy with the case of Fe^{2+} in II-VI compounds,[34,71] the lattice mode could be non-symmetric, resulting in a dynamic Jahn-Teller effect. A dynamic Jahn-Teller effect has been invoked in ref. 57 to explain the Fe^{2+} absorption spectrum in GaAs:Fe which will be discussed later.

Look found a thermal activation energy for the carrier concentration in his InP:Fe samples of 0.64 eV, which agrees well with previously reported values.[9,27,53,58] In discussing the interpretation of his photoconductivity spectra, he defined three spectral regions: region 1, the "resonance" re-

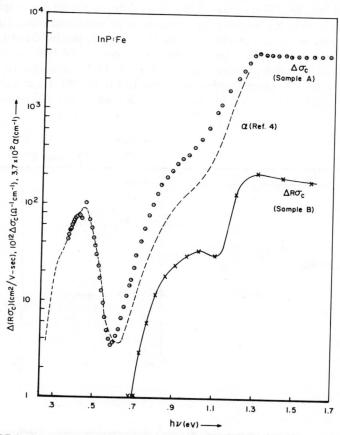

FIGURE 8 Spectral dependences of the photoconductivity $\Delta\sigma_c$ for sample A, photo-Hall mobility change $\Delta(R\sigma_c)$ for sample B, and absorption coefficient α for an InP:Fe, Sn sample from Ref. 58. The curves for α and $\Delta\sigma_c$ are normalized to each other at 0.50 eV. After Look, Ref. 69.

gion, $0.27 < h\nu < 0.59$ eV; in this region, as described above, the absorption is due to the transition $^5E \to {}^5T_2$, and photoconductivity involves the subsequent thermal excitation $^5T_2 \to$ conduction band (CB); region 2, $0.59 < h\nu < 0.70$ eV; in this region the absorption and photoconductivity are due to the optical transitions $^5E \to$ CB (the photoconductivity in both regions 1 and 2 is due to photo-excited electrons, a key point); region 3, $0.70 < h\nu < 1.34$ eV (the 297K band gap); in this region the optical transition valence band (VB) $\to {}^5E$ becomes important as evidenced by the 0.7 eV onset in hole conducitvty seen in the photo-Hall spectrum ($\Delta R\sigma_c$) of Figure 8. The dark conductivity exhibits a negative (n-type) Hall coefficient, R, and the 0.7 eV onset of increasing $\Delta R\sigma_c$ in Figure 8 corresponds to R_σ becoming less negative, an indication of photoexcited holes. The resulting placement of the Fe^{2+} ^5E and $^5T_2$ states in the InP band gap is shown in Figure 9. There are two pieces of experimental evidence which directly support this placement of the ^5E level: 1) the thermal activation energy, 0.64 eV, which should correspond to $E(CB) - E(^5E)$, and 2) the 0.7 eV onset of photo-excited hole conduction which should correspond

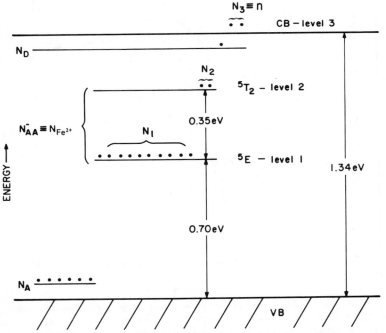

FIGURE 9   Proposed room-temperature energy diagram for Fe^{2+} in InP. After Look, Ref. 69.

to $E(^5E) - E(VB)$. (Note that $0.64 + 0.70 = 1.34$ eV, the room temperature band gap.) Clearly the level scheme of Figure 9 is consistent with the model suggested by Stauss et al.[26] to explain their photo-induced quenching spectrum for the Fe^{3+} ESR. In particular, the ~0.7 eV onset of photo-excited hole conduction corresponds to the Fe^{3+} ESR quenching transition

$$(h\nu \gtrsim 0.7 \text{ eV}) + Fe^{3+} \rightarrow Fe^{2+}(^5E) + h^+(VB), \tag{4}$$

proposed by Stauss et al.[26]

Fung et al.[72] studied the photoconductivity of InP:Fe at both room temperature and low temperature (77K and 6K). Examples of their AC photoconductivity spectra for a semi-insulating sample of InP:Fe are shown in Figure 10. These workers proposed the existence of three charge states for Fe in InP, Fe^{1+} ($3d^7$), Fe^{2+} ($3d^6$), and Fe^{3+} ($3d^5$), and they explained the features of their photoconductivity spectra in terms of optical

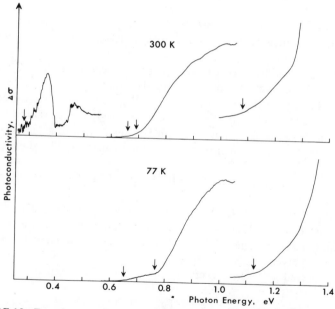

FIGURE 10   Experimental recordings of the AC photoresponse of the semi-insulating RSRE sample (L677) at 300 and 77K. The recordings taken in different regions of the spectrum were taken using different gratings and optical filters, so that there are changes in gain at 0.6 and 1.0 eV. The recordings are normalized to the incident-light intensity within each spectral region. After Fung et al., Ref. 72.

transitions between these charge states and the conduction and valence bands. They proposed that the two electron trap Fe^{1+} $(3d^7)$ is the stable charge state in n-type material. Because the low-energy onset at 0.34 eV (6K) or 0.28 eV (300K) was the only feature observed in the photoconductivity spectrum for a heavily n-type sample, it was attributed to photo-ionization of the Fe^{1+} state into the Γ conduction band minimum represented here as

$$h\nu \,(\gtrsim 0.34 \text{ eV}) + (Fe^{1+} \to Fe^{2+} + e^- \,(\text{CB})). \qquad (5)$$

The threshold which they observed at 0.64 eV, which is dominant for samples with more donors (high Fe^{2+} concentration) and only weakly observed for samples containing mostly Fe^{3+}, they attributed to transitions from the one electron trap, Fe^{2+} (^5E ground state), to the conduction band. This assignment of this threshold is, of course, consistent with that of Look,[69,70] $E(\text{CB}) - E(^5E)$, restated here as

$$h\nu \gtrsim 0.64 \text{ eV} + Fe^{2+} \to Fe^{3+} + e^- \,(\text{CB}). \qquad (6)$$

The 0.64 eV onset was not observed by Fung et al. in n-type samples, a circumstance which they attributed to the absence of Fe^{2+} in n-type material.

Fung et al. regarded[9] a weak self-quenching feature in the AC photoresponse and a stronger self-quench in the DC photoresponse at 0.78 eV as further evidence for the location of the Fe^{2+} ground state near the center of the gap. These quenching features in the strongly n-type samples and a photoconductivity onset at 0.78 eV in the semi-insulating sample were attributed to the generation of free holes in the valence band by the process

$$h\nu \gtrsim 0.78 \text{ eV} + Fe^{3+} \to Fe^{2+} \,(^5E) + h^+ \,(\text{VB}). \qquad (7)$$

This is the equivalent of Look's hole excitation transition $E(^5E) - E(\text{VB})$ observed in the photo-Hall spectrum. Obviously, the process represented in Equation (7) also coincides with the Fe^{3+} ESR optical-quenching onset at ~ 0.7 eV observed by Stauss et al.[26] The second Fe^{3+} ESR quenching feature observed by these workers at ~ 1.13 eV can be stated as

$$(h\nu \gtrsim 1.1 \text{ eV}) + Fe^{3+} \to Fe^{2+} \,(^5T_2) + h^+ \,(\text{VB}), \qquad (8)$$

in accord with their suggested model. This same process is one of the alternatives invoked by Fung et al. to explain an additional photoconductivity onset which they observed near 1.1-1.2 eV in InP:Fe. The schematic energy level diagrams proposed by Fung et al. to represent all these transitions at 6K and 300K are shown in Figure 11.

Very similar photoconductivity spectra have been obtained for InP:Fe by Yu.[73] He invoked Look's[69,70] explanation for the 0.44 eV peak in the photoconductivity which involves the Fe^{2+} intracenter optical excitation followed by thermal excitation to the conduction band. The features at 0.6 eV and 0.78 eV were also observed and ascribed to the transitions involving Fe^{2+} (^5E) as proposed by Look[69,70] and Fung et al.[72] However, the 1.1-1.2 eV photoconductivity threshold and a ~1.1 eV PL band observed in these samples were explained by Yu[68] in terms of transitions involving an Fe-related defect center (e.g. $(Fe)_{In}$-$(Vacancy)_P$). Such Fe-related complexes will be discussed in more detail in Section 3.5.

Eaves and coworkers[60] have carried out independent photoconductivity measurements on InP:Fe with somewhat different results and interpretation. The photoconductivity spectra obtained by these workers in InP:Fe samples with a wide range of electrical characteristics are presented in

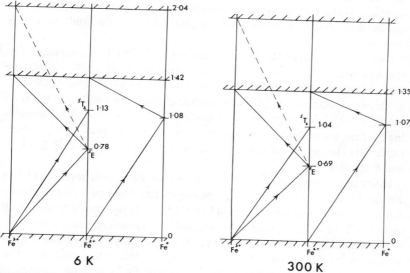

6 K   **300 K**

FIGURE 11 A schematic picture of the energy level scheme for Fe impurities in InP at (a) 6K and (b) 300K. The arrows indicate possible transitions and show the initial and final charge states of the Fe ions. Possible transitions to higher lying satellite valleys are also shown. The peak energies of the different levels with respect to the valence band edge are given in eV. After Fung et al., Ref. 72.

Figure 12. Working at 4.2 K, they found the photoconductivity threshold at ~0.65 eV (transition (6)) in *all* Fe-doped samples including two heavily n-type samples, thereby establishing the presence of the Fe^{2+} charge state in both strongly n-type and semi-insulating material. Hence, one cannot justify the assumption that Fe^{1+} is the stable charge state in heavily n-type material. Furthermore, Eaves et al. did not observe the low-energy (0.34 eV (6K) or 0.28 eV (300K)) photoconductive threshold reported by Fung et al. in n-type InP:Fe and attributed to transitions from Fe^{1+} to the CB (see Equation (5)). Eaves et al. explain the data of Fung et al. on the basis of bolometric effects involving the 0.35 eV internal Fe^{2+} transitions; the bolometric effects are not present when the photoconductivity samples are immersed in liquid He as in the experiments of Eaves et al.[60] Apparently, Look's explanation[69,70] of his photoconductivity threshold at 0.35 eV in terms of the internal transition of Fe^{2+} followed by thermal excitation to the CB is to be preferred to an explanation which requires the presence of Fe^{1+} in the material.

In their n-type samples, Eaves et al.[60] found no other sharp onsets in the photoconductivity between the 0.65 eV edge and the near-band-edge region. The second threshold at 1.15 eV was observed only in high-resistivity samples. There was no sign of quenching at this feature, indicating that the carriers produced at the 0.65 eV and 1.15 eV thresholds are of the same type.

Eaves et al.[60] suggested the level scheme in Figure 13 to explain all of the absorption, luminescence, photoconductivity and ESR results. The Fe^{2+} ^5E ground state (level Fe(I) in Figure 13) has its apparently well-established position ~0.65 eV below the conduction band, and the 0.35 eV PL band, 0.65 eV photoconductivity edge, and 0.75 eV ESR quenching feature are indicated as transitions A, B, and C, respectively. These aspects of the model are consistent with those of Stauss et al.,[26] Look,[69] and Fung et al.[72]

Eaves and coworkers[60] proposed a second Fe-related level, Fe(II) in Figure 13, located 1.15 eV below the conduction band edge to explain the strong photoconductivity threshold at 1.15 eV. It has also been suggested that the 1.15 eV threshold be explained by transitions out of the Fe^{2+} (^5E) ground state (level Fe(I) in Figure 13) to a higher conduction-band minimum.[72,74] However, Eaves et al. argued that the strength of this threshold in semi-insulating material (mostly Fe^{3+}) and its absence in n-type samples (mostly Fe^{2+}) suggest that it is related to the Fe^{3+} charge state. They speculated that the level Fe(II) in Figure 13 might be identified with Fe^{3+}, and the 1.15 eV threshold (transition D in Figure 13)

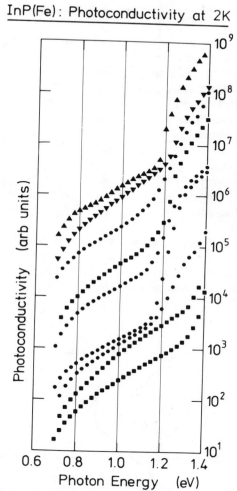

FIGURE 12 The chopped photoconductivity response of several Fe-doped InP samples at liquid helium temperature between 0.6 eV and the intrinsic edge. The vertical axis is plotted logarithmically and the vertical displacement of the curves is arbitrary. The curves are arranged as follows with the RSRE samples in order of decreasing resistivity. (a) MR 302 R 4, (b) NRL 1-74 H, (c) L 808-4, (d) L 781-3, (e) L 808-3, (f) L 808-2, (g) L 808-1, (h) 1 781-2, (i) L 781-1. The optical resolution is about 20 meV. After Eaves et al., Ref. 60.

could be represented as

$$(h\nu \gtrsim 1.15 \text{ eV}) + Fe^{3+} \rightarrow Fe^{4+} + e^- \text{ (CB)}. \qquad (9)$$

The free electrons produced by transition D can recombine at the neutral Fe^{3+} sites to form Fe^{2+} thereby quenching the Fe^{3+} ESR. It would also be possible[60] for transition D (Equation (9)) to provide extrinsic excitation of the PL transition A, if the capture of photo-excited electrons by Fe^{3+} leads to Fe^{2+} in the excited $^5T_2$ state as suggested by Bishop et al.,[5,7] and discussed in Section 3.1. However, recent studies[61] of the dependence of the optical DLTS spectrum upon thermal annealing have demonstrated that the presence of the level 0.25 eV above the valence band is highly dependent upon the thermal history of the sample. This makes the assignment of this center to an isolated Fe^{3+} ion very unlikely and appears to eliminate the possibility of associating the $\sim$1.13 eV onset in PLE and photoconductivity with $Fe^{3+} \rightarrow Fe^{4+}$ transitions represented by Equation (9).

Bishop et al.[7] proposed that transitions represented by Equation (8) could account for the extrinsic excitation of the 0.35 eV PL band which they observed in PLE spectra for InP:Fe. The PLE spectra for the 0.35 eV PL Fe^{2+} band in two semi-insulating crystals of InP:Fe are shown in

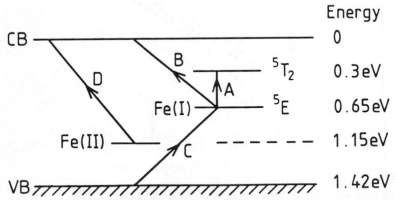

FIGURE 13 An energy level diagram for Fe in InP at liquid helium temperature. Transitions going upwards and right correspond to hole excitation in the valence band while those going upward and left are electronic transitions into the conduction band. Internal transitions are vertical. Level Fe (I) corresponds to Fe^{2+}, Level Fe (II) is unidentified. The transitions shown are described in the text. After Eaves et al., Ref. 60.

Figure 14a. In PLE spectroscopy, the intensity of the PL band is measured as a function of the wavelength of the exciting light. The PLE spectra represent a measure of the relative efficiency of various absorption processes in the excitation of the selected PL band.

Both of the PLE spectra in Figure 14a exhibit a band of extrinsic excitation extending from a low energy onset of $\sim$1.13–1.15 eV up to the onset of strong band-edge absorption which occurs near 1.4 eV. The strength of this extrinsic PLE band for the Fe^{2+} PL is correlated with the concentration of Fe^{3+} as determined by ESR measurements. Bishop et al.[7] suggested that the extrinsic excitation of the Fe^{2+} PL might be attributed to a ligand-to-metal charge transfer transition in which a valence electron from an adjacent phosphorus atom is transferred to an Fe^{3+} ion, thereby creating an Fe^{2+} ion in the excited $^5T_2$ state. The excited Fe^{2+} ion then relaxes to the ^5E ground state by the emission of a 0.35 eV photon. This excitation process is represented by Equation (8), and the $\sim$1.13 eV threshold is consistent with the energy position of the Fe^{2+} levels stipulated in the models discussed above as shown in Figures 9, 11 and 13. Of course, the transition from the Fe(II) level (located 1.15 eV below the conduction band) to the conduction band proposed by Eaves et al.[60] has a threshold similar to that for the $Fe^{3+} \rightarrow Fe^{2+}$ ($^5T_2$) transition. However, Tapster et al.[59] point out that the latter process seems more likely for the PLE since it leads directly to the Fe^{2+} PL in a one-step process. Furthermore, Leyral et al.[62] have also obtained PLE spectra for the Fe^{2+} PL band and find that they are well correlated with the 1.14 eV onset and the shape of the photoionization spectrum $\sigma_p^o(h\nu)$ (hole creation) of the Fe^{3+} to Fe^{2+} ($^5T_2$) transition. This indicates that the extrinsic absorption process which excites the 0.35 eV Fe^{2+} PL band involves the creation of free holes (as does the process suggested by Bishop et al.[7]), and argues against an extrinsic excitation mechanism which involves creation of free electrons which are captured by Fe^{3+} ions. PLE spectra have also been obtained[5,61,62] for the apparently Fe-related PL band at $\sim$0.5 eV observed in InP:Fe. As shown in Figure 14b this PLE spectrum has an onset at energy roughly corresponding to the $\sim$1.15 eV onset for the PLE spectrum of the 0.35 eV PL band. In addition it has a weak threshold extending to lower energy ($\lesssim$0.9 eV). Both Leyral et al.[62] and Skolnick speculate that the 0.5 eV PL band may be attributable to a charge-transfer transition of the type Fe^{2+} ($^5T_2$) + $h_{VB} \rightarrow Fe^{3+} + h\nu$ (0.5 eV). Leyral et al.[62] point out that this possibility is supported by the fact that the PLE spectrum of the 0.5 eV PL band has about the same onset as the hole photoionization spectrum for Fe^{2+} in InP:Fe.

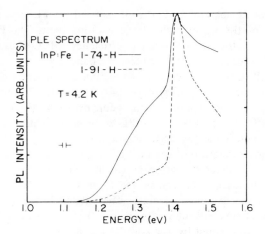

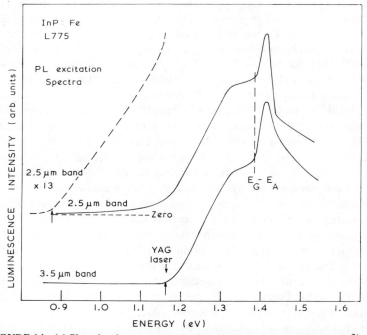

FIGURE 14 (a) Photoluminescence *excitation* spectra of the ~0.35 eV Fe^{2+} luminescence band in semi-insulating InP:Fe. The Fe concentrations are 1×10^{16} cm^{-3} for 1-91-H and 1×10^{17} cm^{-3} for 1-74-H. After Bishop et al., ref. 7. (b) Photoluminescence excitation spectrum of the ~0.5 eV luminescence band in semi-insulating InP:Fe (unpublished data provided by M. S. Skolnick).

3.4 Photocapacitance and DLTS Studies of InP:Fe

Deep impurity levels in InP have been studied by several workers using photocapacitance[59,75,76] and deep-level transient spectroscopy (DLTS).[59,77-81] A variety of deep levels have been reported with energies scattered throughout the forbidden gap of InP, but, until quite recently, no deep level had been related to a specific impurity or defect. The present discussion will be limited to recent studies of Fe-doped InP and studies of undoped material in which deep levels attributed to inadvertent Fe impurities have been observed. In particular, we will focus on three independent studies which have been carried out on the same bulk LEC n-type InP:Fe grown at R.S.R.E., Malvern, labeled L781. This sample contained sufficient residual compensating shallow donor impurities for n-type conduction to dominate at all temperatures.

White and coworkers[77] carried out DLTS measurements on bulk and epitaxial InP and detected several deep electron traps in the upper half of the band gap. Their work included measurements on the n-type bulk Fe-doped sample of InP, L781. In this sample (as well as others) they observed a trap with activation energy of 0.52 eV and capture cross-section $\sigma_\infty = 3 \times 10^{-14}$ cm^2 which they labeled R. (σ_∞ is the extrapolated high-temperature capture cross-section.) While these workers gave no identification for this trap, subsequent DLTS studies of the sample L781 by Tapster et al.[59] have identified the trap R as Fe-related.

Bremond et al.[80(a)] performed comparative DLTS measurements on several undoped samples of n-type InP and the Fe-doped L781 sample. DLTS spectra for this sample exhibited two electron traps. The temperature dependence of the emission time constant for the dominant trap yielded an activation energy of 0.63 eV and a capture cross-section of $\sigma_\infty = 3.5 \times 10^{-14}$ cm^2. It was proposed that this level is related to the one-electron trap state Fe^{2+} (3d^6) of Fe on an In site. This assignment was made primarily on the basis of the abundant evidence from transport[9,53,58] and photoconductivity[60,69,72] measurements that the Fe^{2+} ground state is located ~0.6–0.7 eV below the conduction band. In addition, Bremond et al.[80(a,b)] measured the optical photo-ionization cross-sections of this electron trap to the conduction and valence bands by deep-level optical spectroscopy. The onsets for photo-ionization agree quite well with the energy-level scheme deduced from photoconductivity in InP:Fe (Figures 9, 11 and 13). Finally, the high concentration of this 0.63 eV electron trap in the Fe-doped material compared with the concentration of deep traps observed in bulk undoped InP strongly suggests that the trap is related to the Fe-doping.

A dominant electron trap with similar DLTS and photo-ionization signatures was observed by Bremond et al.[80(a)] in undoped n-type crystals of InP, and they concluded that it was the same defect observed in Fe-doped InP. This is consistent with the fact that Fe is known to be a ubiquitous residual contaminant in InP.

Tapster and coworkers[59] have studied DLTS, photocapacitance, optical DLTS, optical absorption, and PL in the Fe-doped InP sample, L781. (Their PL results have been discussed in Section 3.1.) Their published DLTS spectrum for L781 is nearly identical to that of Bremond et al.,[80(a)] and they deduced an activation energy of 0.59 eV and $\sigma_\infty = 4 \times 10^{-14}$ cm^2 for the dominant electron trap, in reasonable agreement with Bremond et al. The photocapacitance spectrum obtained by Tapster et al. in L781 is presented in Figure 15. Initially, the device was forward biased to fill all the levels with electrons; it was then illuminated with increasing photon energy. Positive and negative thresholds near 0.7 and 0.85 eV were attributed to electron and hole emission, respectively, from the Fe level near mid gap. These transitions correspond to B and C in Figure 13. Clearly, these results are consistent with the photoconductivity data which indicate that the Fe^{3+} to Fe^{2+} transition is 0.78 eV above the valence band at 6K.

Tapster et al. suggested that the strong positive threshold in the photocapacitance spectrum near 1.2 eV might be related to the photoconductivity threshold observed in the 1.1–1.2 eV region by Fung et al.[72] and

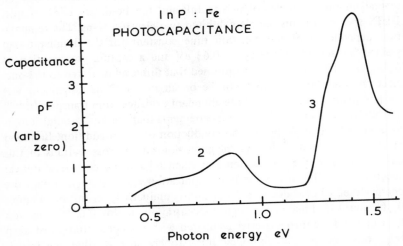

FIGURE 15 Photocapacitance spectrum of InP:Fe sample L781. After Tapster et al., Ref. 59.

Eaves et al.[60] Since the photocapacitance threshold clearly corresponds to electron emission, one can eliminate transitions from Fe^{3+} to Fe^{2+} ($^5T_2$) (Equation (8)) or transitions from the Fe^{2+} ground state to an Fe^+ charge state as possible assignments for the 1.1 eV feature in these spectra. Tapster et al.[59] suggested instead either a transition from Fe^{2+} to Fe^{3+} which produces an electron in the L-point conduction band valley (as Bremond et al.[80(b)] concluded, or a transition from a different level (0.2-0.3 eV from the valence band) to the Γ minimum of the conduction band.

Tapster and coworkers[59] also carried out an informative series of optical DLTS experiments in which the usual electrical filling pulse is replaced by a pulse of monochromatic light. The ensuing capacitance transient is analyzed by the usual DLTS technique while the temperature is scanned. The amplitude of the resulting DLTS peaks is determined by the optical ionization cross sections of the corresponding levels for the photon energy being applied. Optical DLTS spectra for 1.3 eV illumination were dominated by a thermal hole emission peak at 107K, characteristic of a shallow level 0.2-0.3 eV above the valence band. For 1.2 eV illumination the 107K peak was reduced by a factor of 10, and it became essentially negligible for lower photon energies. These results indicate that the 1.2 eV threshold in the photocapacitance spectrum is associated with the 107K optical DLTS peak. This means that this photocapacitance threshold and the photoconductivity edge at ~1.1-1.2 eV must be attributed to a transition such as D in Figure 13, and the possibility of a transition from a mid-gap level to the L conduction band minimum is eliminated. (This contradicts the interpretation of Bremond et al.[80(b)] who attributed a second strong 1.1 eV onset in the photoionization spectrum for electrons to transitions from the Fe^{2+} (^5E) ground state level to the L conduction band minimum.)

Tapster et al. suggested that the 107K optical DLTS peak might be due to a transition from Fe^{3+} to Fe^{4+}, involving a level 0.25 eV above the valence band similar to the Fe(II) level of Figure 13. They pointed out that Eaves et al.[60] have independently interpreted their 1.1 eV photoconductivity threshold in terms of the $Fe^{3+} \rightarrow Fe^{4+} + e$ process of Equation (9) (transition D in Figure 13). However, as mentioned previously, more recent optical DLTS studies[61] have shown that the presence of the level 0.25 eV above the valence band in InP:Fe is highly dependent upon the thermal history of the sample. This makes the assignment of this center to an isolated Fe^{3+} very unlikely and a more probable explanation for the level 0.25 eV above the valence band is an Fe-related complex.[59] Such complexes are the subject of the next Section, 3.5.

3.5 Fe-Related Complexes in InP:Fe

Recent experimental studies of some transition-metal doped III-V semi-conductors have demonstrated the existence of defect complexes involving the transition-metal dopants, such as near-neighbor transition metal-donor complexes. Perhaps the most celebrated example of this was the discovery through Zeeman spectroscopy[82] and optically detected magnetic resonance (ODMR) studies[83] that the well known 0.839 eV Cr-related PL band in GaAs:Cr is not attributable to isolated Cr^{2+} ions, but to recombination at a near-neighbor chromium-donor complex[84,85] or a chromium interstitial coupled to an acceptor on an As site.[85] In addition, PL and optical absorption studies[86] of Ni-diffused samples of n-type chalcogenor group IV-doped GaP and GaAs have been interpreted in terms of donor-acceptor pair radiative recombination at near-neighbor Ni-acceptor-shallow donor complexes.

In the case of InP:Fe, there is indirect evidence from PL spectra[74,81,87,88] for the existence of an Fe-related complex and a recent ESR study[89] of electron-irradiated InP:Fe has detected a nearest-neighbor complex formed by the pairing of Fe with radiation-induced defects.

The existence of vacancy-impurity complexes such as $(V)_{Ga}$-donor and $(V)_{As}$-acceptor in GaAs was proposed on the basis of PL studies[90,91] of appropriately doped crystals of GaAs. A broad PL band near 1.1–1.2 eV reminiscent of these supposedly complex-related PL bands in GaAs has been reported in a wide variety of InP crystals, both doped and undoped. Complete bibliographies of the studies are found in refs. 73 and 88; the discussion here will center on several recent studies of this PL band in which it has been attributed to an Fe-related complex.

Demberel and coworkers[87] observed two PL bands at 1.12 eV and 1.07 eV at 77K in InP:Fe crystals. The latter band was also found in undoped crystals while the 1.12 eV band appeared only in Fe-doped crystals. Without detailed justification, these authors attributed the 1.12 eV band to a neutral Fe-acceptor-Si-donor near-neighbor complex which they referred to as an isoelectronic trap. They inferred from the PL energy that the acceptor level of this complex lies about 0.28 eV above the valence band, and associated a 0.3 eV threshold in the optical absorption spectrum with the photo-ionization of this level. However, as discussed previously, most currently available evidence ascribes this threshold in the absorption and photoconductivity of InP:Fe to the Fe^{2+} "resonance" model of Look.[69,70]

Subsequent to the work of Demberel et al., Yu[73,88] studied the PL bands near ~1.1 eV in a variety of epitaxially grown and bulk crystals of

InP at 4.2K. These low-temperature experiments revealed evidence of phonon structure in the 1.1 eV PL band as shown in Figure 16a, which contrasted with the broad featureless spectra reported in previous studies. In addition, Yu found that the PL peak position varies from sample to sample over the range 1.06 to 1.15 eV, and that the PL band takes on a more symmetric shape as its peak position shifts to a lower energy. He explained the variations in the peak position and shape of these PL bands on the basis of phonon sidebands and a variable coupling strength for the localized center-phonon interaction.[88] By fitting the PL lineshape with a configuration coordinate model,[88,92] Yu determined the zero phonon energy $E_o = 1.18$ eV and the energy of the coupled phonon $\hbar\omega = 38 \pm 2$ meV, which is equivalent to the TO_Γ phonon energy for the InP lattice. The Huang-Rhys factor or coupling strength S ranges from 1.7 to 3.5 for the observed spectra. On this basis Yu concluded that the 1.06 eV PL band and all the PL bands observed at $\sim$1.10 eV originate from the same radiative recombination center. On the basis of correlations of the intensity of the $\sim$1.1 eV emission with Fe-doping of the InP crystals he proposed that the localized center responsible for this PL band is a nearest-neighbor, molecular-like, $(Fe)_{In}$-phosphorous vacancy complex.

Yu[88] also measured the temperature dependence of the intensity of the $\sim$1.10 eV PL band and determined a single activation of 0.24 eV from these data. This energy corresponds well to the difference between the ZPL energy 1.18 eV and the InP band-gap energy, 1.423 eV, and was therefore regarded as the activation energy, for thermal release of a carrier bound to the center. Yu[73] ascribed the strong 1.1-1.2 eV onset in the photoconductivity of InP:Fe to transitions involving the $(Fe)_{In}$-$(V)_p$ complex.

Yamazoe et al.[81] have also studied the near-band-edge and 1.1 eV PL bands in InP at 4.2K, and performed parallel DLTS experiments on the same samples. They found an association between the 1.1 eV PL band and a DLTS peak (labeled E_5) which has a 0.42 eV emission activation energy. This activation energy was regarded as being in good agreement with the difference between the 1.42 eV band gap energy and the 1.1 eV PL band energy. In addition, the trap E_5 and the 1.1 eV PL band exhibit similar responses to heat treatment. Both the E_5 DLTS signal and the 1.1 eV PL band were enhanced by heat treatment at $T_H = 350°C$ for 1 hr under vacuum. Correspondingly both phenomena were suppressed by heat treatment of the InP crystals under excess phosphorous pressure. These DLTS and PL features are maximized for $T_H = 500°C$, and they both decrease for $T_H > 500°C$. Yamazoe and coworkers inferred from these results that E_5 and the 1.1 eV PL band are closely related and might arise from a complex involving Fe and P vacancies. They reasoned that if an isolated P vacancy were responsible one would not have expected E_5 and the PL band to be suppressed for $T_H > 500°C$. A complex would be more likely to dissociate at the higher temperatures and, in fact, Yu[88] also observed a complete quenching of the 1.1 eV PL after a 1 hr heat treatment at 700°C under vacuum which he attributed to the dissociation of the proposed $(Fe)_{In}$-$(V)_P$ complex.

More recently, Eaves and coworkers[93] have reported a study of the 1.1–1.2 eV PL spectra in a series of Fe-doped, Mn-doped, and undoped InP samples (see Figure 16b). They observed in InP:Mn a strong broad band peaking at 1.15 eV with phonon structure at 1.184, 1.145, and 1.107 eV. This same band is observed in undoped InP and is greatly enhanced by thermal annealing. This suggests that Mn is an inadvertent impurity in InP which can diffuse to the surface during heat treatment. Equivalent observations have been made for Mn in GaAs.[10(a)] In InP:Fe, Eaves et al. observed a broad PL band peaking at 1.10 eV with weak phonon structure at 1.135, 1.098 and 1.062 eV. This work demonstrates that the variation in the peak energy of the 1.1–1.2 eV PL spectra observed[73,81,87,88] in a variety of Fe-doped and undoped InP samples is due to the fact that the observed spectra are composites of two distinct PL bands whose relative intensities vary from sample to sample. This seems to be a far more plausible explanation than the variable phonon coupling strength proposed by Yu.[88]

The most direct evidence for the formation of Fe-related complexes in InP:Fe is provided by the ESR studies of electron irradiated semi-insulating InP:Fe carried out by Kennedy and Wilsey.[89] These workers found that following irradiation with 2 MeV electrons to a fluence of 1×10^{17} cm^{-2}, an LEC grown crystal of InP containing 2×10^{16} cm^{-3} Fe showed no evidence of the usual spin 5/2 Fe^{3+} ($3d^5$) cubic ESR spectrum shown in Figure 6. A new anisotropic ESR spectrum was observed whose rotational pattern exhibited the trigonal [111] symmetry of a spin $\frac{1}{2}$ center (Fig. 17). However, the ESR lines are observed to be sharper and stronger when the magnetic field is oriented along certain high-symmetry directions and there are large departures from $g = 2$. These observations suggest that the defect has spin greater than one half with a large zero-field splitting. Since Fe is the dominant transition metal impurity, Kennedy and Wilsey concluded that Fe is involved in the observed complex.

The observed trigonal symmetry implies that the Fe is paired with a defect along a [111] axis, such as the vacancies and interstitials known to be produced by the irradiation. Possible candidates are an iron-phosphorous vacancy pair or an iron-interstitial pair with the interstitial opposite a phosphorous atom. Following the arguments first presented by Ludwig and Woodbury,[94] Kennedy and Wilsey suggested that spin-spin coupling of the S = 5/2 of Fe and the S = 1 of In$^+$ could result in S = 7/2 and the observed g$'_\perp$ value of 8. Furthermore, since the In interstitial is expected to be mobile at room temperature, the formation of iron-indium interstitial pairs is quite likely.

The new trigonal spectrum induced by the electron irradiation is completely quenched by *in situ* irradiation with 900 nm light. The fact that

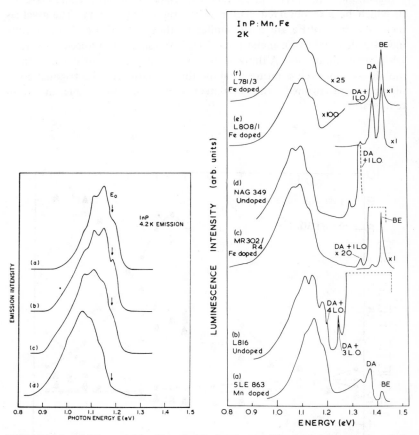

FIGURE 16 (a) The ~1.10 eV emission band at 4.2K obtained from four InP crystals. The peak position does not change with the excitation intensity. The peak position depends on the sample. The details of samples used are given in Ref. 88. E$_O$, the zero-phonon energy at 1.18 eV is indicated by arrows. After Yu, Ref. 88. (b) Low temperature (2K) PL spectra of a series of InP samples taken in the range from 0.8 to 1.42 eV. The figures "X25," etc. indicate the increased amplifier gain used for monitoring the deep level PL relative to that used for the stronger near-band-edge PL. Curves (a) and (b) show the Mn band and curves (c) to (f) the Fe band. After Eaves et al., Ref. 93(a).

the cubic Fe spectrum does not reappear during this illumination indicates that its absence after electron irradiation is due to a change in the Fe defect itself rather than a charge-state change caused by a radiation-induced change in the Fermi level. A new ESR spectrum appears during the 900 nm irradiation. Although the spectrum has not been analyzed completely, it appears to be attributable to a second Fe associate.

Ingelmund[95] has performed similar experiments on GaP:Fe and GaAs:Cr in which he also observed centers with trigonal symmetry. The g-values measured in GaP:Fe are quite similar to those of InP:Fe. These results indicate that transition metal impurities trap radiation induced defects in III-V semiconductors. Although the defects are apparently sufficiently mobile at room temperature to allow the formation of the trigonal complexes, once formed, the Fe-defect complexes are stable at room temperature.

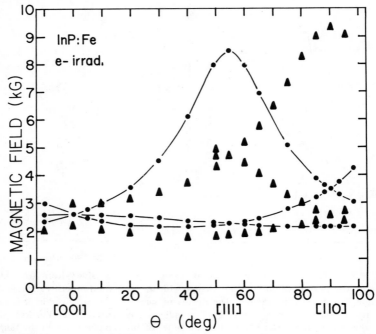

FIGURE 17   EPR line positions versus angle in the (1̄10) plane. The spectrum observed in the dark is shown as circles, the spectrum after 900 nm illumination as triangles. After Kennedy and Wilsey, Ref. 89.

4. Experimental Studies of GaAs:Fe

The early transport studies of GaAs:Fe discussed previously established the existence of a dominant Fe-related acceptor level located about 0.5 eV above the top of the valence band. In recent years a few additional DLTS studies have been carried out which have confirmed these earlier results. Mitonneau et al.[96] surveyed many samples of v.p.e., l.p.e., m.b.e., and bulk grown GaAs by DLTS techniques and found the expected hole trap with 0.52 eV activation energy in both Fe-doped l.p.e. material and as-grown m.b.e. GaAs. These two traps and a trap with 0.59 eV activation energy observed in Fe-diffused v.p.e. material were all interpreted as being Fe-related. Takikawa and Ikoma[97] used a photoexcited DLTS technique to study the Fe-related level in GaAs as a minority carrier trap. Again the Fe hole trap was found to have an activation energy of 0.51 eV, in excellent agreement with some of the previous results.

Some of the early studies also detected a level with an activation energy of 0.38 eV which was interpreted as either a second acceptor level[46] or a temperature shift of the ~0.5 eV level.[42] In addition, optical-absorption and photoconductivity spectra[47] with onsets ~0.8 eV indicated the presence of another, deeper level. We turn now to a more detailed discussion of the Fe-related levels in GaAs:Fe which will draw upon the early studies discussed previously as well as more recent optical-absorption, PL, PLE, photoconductivity and photocapacitance studies.

4.1 Optical Absorption and Photoluminescence Studies of GaAs:Fe

The sharply structured near infrared (0.37 eV) absorption spectrum characteristic of the internal d-d transitions of Fe^{2+} $(3d^6)$ in a tetrahedral crystal field was first reported for GaAs:Fe by Baranowski et al.,[28] as discussed previously. More recently, Ippolitova and Omel'yanovskii[57] have carried out a more detailed and higher resolution study of the GaAs:Fe^{2+} absorption spectrum at temperatures ranging from 4.2 to 40K (see Figure 18). As in the case of InP:Fe, the four sharp lines near 3000 cm^{-1} in Figure 18 are identified as the four allowed transitions between the ground state ^5E pentet and the lowest state of the $^5T_2$ excited state multiplet of Fe^{2+} in the tetrahedral crystal field (see Section 2.2 and the energy level diagram of Figure 2). Spectral features attributed to the participation of TA, LA, TO, and LO phonons and their combinations are also shown in Figure 18. In addition, Ippolitova and Omel'yanovskii identified purported transitions to higher lying excited states in the $^5T_2$ multiplet (labeled 2-7, 2-8, and 1-9 in Figure 18).

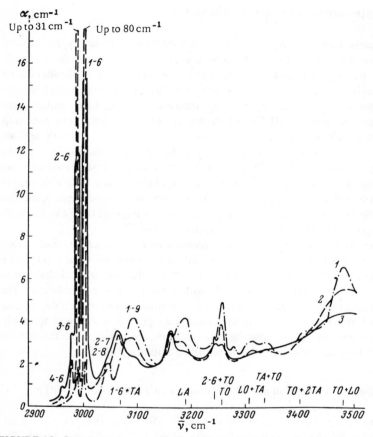

FIGURE 18 Low-temperature optical absorption spectra of GaAs:Fe. The abscissa gives the energies of the transitions 1-6 involving the participation of TA, LA, TO, LO phonons and their combinations, as well as the phonon replicas of the transition 2-6; curves 1, 2 and 3 represent the results obtained at 4.2, 10, and 20K. After Ippolitova and Omel'yanoskii, Ref. 57.

In the limit $T \rightarrow 0$, only the lowest singlet level of the five $\sim$ equally spaced levels of the ^5E ground state is populated. At higher temperatures the higher levels of ^5E are thermally populated and this is reflected in the appearance of all four allowed zero-phonon transitions in the 20K spectrum of Figure 18. Using the theoretical expressions[33-35,71] for the energies of the observed transitions (similar to Equation 2) these workers found for GaAs:Fe a crystal field splitting (Δ = 10 Dq of 2995 ± 10 cm^{-1}

and first and second order spin-orbit coupling constants $\lambda_1 = 18 \pm 2$ cm^{-1} and $\lambda_2 = 77 \pm 6$ cm^{-1}. They point out that their data indicate that the levels of the ^5E term in GaAs:Fe are not exactly equidistant and that the dynamic Jahn-Teller effect may have a slight influence on this splitting.[71]

In the high-resistivity GaAs:Fe crystals studied by Ippolitova and Omel'yanovskii, the concentration of Fe atoms in the Fe^{2+} ($3d^6$) state is governed by the concentration of shallow Te donors present. (Fe concentrations in their samples ranged from 5×10^{17} to 2×10^{18} cm^{-3} and the concentration of back-doped Te shallow donors ranged from 6×10^{14} to 1×10^{18} cm^{-3}.) In overcompensated samples, $N_{Fe^{2+}} = N_{Fe}$. A linear relationship was observed between the optical-absorption coefficient and the Fe^{2+} concentration. These results are in agreement with the APR studies[38] which also found that the Fe^{2+} concentration is governed entirely by the compensation of shallow donors rather than a self-compensation mechanism.

Luminescence spectra which correspond to the Fe^{2+} absorption spectra of Figure 18 have been reported in GaAs:Fe by Bykovskii et al.,[98] Koschel et al.,[6] and, more recently, by Nordquist et al.[10(a)] The characteristic four-line structure observed in the GaAs:Fe PL spectrum is shown in Figure 19. This ~0.37 eV spectrum is nearly identical to the ~0.35 eV PL spectra for InP:Fe shown in Figures 3 and 4 and, accordingly, is attributed to transitions from the lowest level of the $^5T_2$ excited multiplet to the four lowest levels of the ^5E ground state multiplet of Fe^{2+} ($3d^6$) in a tetrahedral crystal field. As in the case of InP:Fe, the spacing of the four sharp zero phonon lines is about 14 cm^{-1}. Bykovskii et al. studied these PL lines as a function of temperature and found their positions to be independent of temperature in the 6-20K range. Above about 20K the lines broaden and can no longer be resolved. These workers also observed a broad lower energy PL band (~0.34 eV) in GaAs:Fe which they attributed to a defect complex, not necessarily directly related to Fe.

The characteristic 0.35 eV four line PL "signature" of Fe^{2+} is often observed in nominally undoped GaAs, indicating that Fe is a pervasive impurity in this III-V compound. In addition, Nordquist and coworkers[10(a)] first reported and Yu[10(b)] later confirmed that Fe impurities which are present in not intentionally-doped GaAs at very low concentrations ($\sim10^{15}$ cm^{-3}) undergo redistribution during heat treatment which produces an accumulation near the surface of the crystal. A thin surface layer 1-3 μm thick is formed which contains Fe at concentrations $\sim10^{17}$ cm^{-3}. The initial evidence for the redistribution of inadvertent Fe impurities was provided by PLE spectra of heat-treated undoped GaAs.[10(a)] The Fe^{2+} PL spectrum for bulk Fe-doped GaAs and its corresponding PLE spectrum

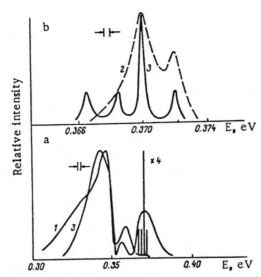

FIGURE 19 Photoluminescence spectra of Fe-doped GaAs. $T(°K)$: 1) 80; 2) 23; 3) 6. After Bykovskii et al., Ref. 56.

are shown in the top of Figure 20. In addition to excitation of the Fe^{2+} PL by highly absorbed above-gap light ($h\nu > 1.52$ eV), the PLE spectrum exhibits an efficient extrinsic (below-gap) excitation band which extends from an 0.86 eV onset to the band gap, E_g. Leyral et al.[99] have subsequently confirmed this 0.86 eV onset in extrinsic excitation. Nordquist et al.[10(a)] suggested that the excitation mechanisms proposed by Bishop et al.[7] to explain the above-gap and extrinsic PLE processes for Fe^{2+} PL in InP:Fe could explain the corresponding PLE processes for Fe^{2+} PL in GaAs:Fe. The level diagram shown in Figure 20 indicates the two types of excitation mechanisms (as discussed in Sections 3.1 and 3.3 for InP:Fe); note that the 0.86 eV onset of extrinsic excitation (compared to ~ 1.15 eV in InP:Fe) places the Fe^{3+} to Fe^{2+} ($^5T_2$) charge-transfer transition lower in the forbidden gap for GaAs:Fe than in InP:Fe (compare Figures 11 and 20).

The corresponding PL and PLE spectra for a heated, undoped GaAs sample, which exhibited no Fe^{2+} PL before heating, are shown in the bottom of Figure 20. (Heat treatment at 800°C for 30 min was carried out under SiO_2 or Si_3N_4 encapsulants, or with the close contact annealing technique.) The PLE spectrum for the heated, undoped sample shows much less efficient excitation for below-gap light than its Fe-doped counterpart, indicating that the Fe concentration is significant only where the light is highly absorbed, i.e., near the surface. SIMS measurements of the Fe concentration profile in the heated undoped crystal verified the ex-

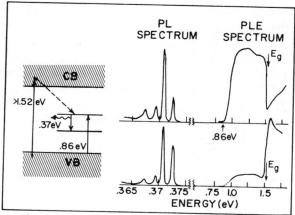

FIGURE 20 Fe^{2+} PL and PLE spectra of bulk GaAs:Fe (top) and an annealed, undoped GaAs sample (bottom). The level diagram indicates possible optical excitation mechanisms for above- and below-gap excitation of the Fe PL. After Nordquist et al., Ref. 10(a).

istence of a thin surface layer with Fe concentration $\sim 10^{17}$ cm^{-3}, while the overall concentration in the bulk is $\sim 2 \times 10^{15}$ cm^{-3}. The best estimate of the bulk concentration of Fe was provided by Fe^{3+} ESR measurements which established the Fe^{3+} concentration at $\sim 2 \times 10^{15}$ cm^{-3} in the undoped crystals.

4.2   Photoconductivity, Photocapacitance, and Charge-Transfer Absorption in GaAs:Fe

The strong extrinsic excitation band with 0.86 eV onset exhibited by the GaAs:Fe PLE spectra of Figure 20 indicates that a corresponding extrinsic optical-absorption band occurs in this spectral range. If the interpretation of the PLE spectrum in terms of $Fe^{3+} \rightarrow Fe^{2+} + h^{+}$ (VB) charge-transfer transitions is correct, this extrinsic absorption should be correlated with the Fe^{3+} concentration and should lead to photoconductivity due to the free holes generated.

Optical absorption and photoconductivity for this spectral range in GaAs:Fe have been studied fairly extensively by Soviet workers. Omel'yanovskii et al.[47] reported absorption and photoconductivity measurements on semi-insulating GaAs:Fe in the 0.3–1.5 eV range at various temperatures (Fe concentrations of 4×10^{16}–10^{18} cm^{-3}). Both the absorption and photoconductivity spectra at low temperatures exhibited onsets at ~ 0.5 eV and at ~ 0.8 eV. Measurements of the Hall coefficient of the illuminated samples indicated that the photoexcited carriers were free holes. This precludes explanations based on electron transitions from Fe levels to the conduction band; the ~0.8-eV onset for hole photoconductivity is consistent with the proposal of Bishop et al.[7] and Nordquist et al.[10(a)] to explain the 0.86 eV onset in the PLE spectrum in terms of the $Fe^{3+} \rightarrow Fe^{2+}$ ($^5T_2$) + h^+ (VB) charge-transfer transition (counterpart of Equation (8)). Although it may be fortuitous, the energy separation between the two absorption onsets at ~0.5 eV and 0.86 eV is roughly equivalent to the photon energy of the Fe^{2+} PL band, the same circumstance which led Stauss et al.[26] to suggest their level scheme for Fe^{2+} in InP (see Figure 11).

Fistul et al.[100] also studied absorption (Figure 21(a)) and photoconduc-

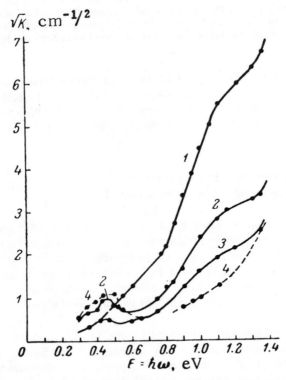

FIGURE 21   (a) Absorption spectra (T = 300°K): 1, 3) samples with a random background of compensating donors; 2, 4) samples doped with iron and tellurium (curve 2 represents a p-type sample and curve 4 an n-type sample). After Fistul et al., Ref. 100.

tivity (Figure 21b) for GaAs:Fe samples with varying degrees of compensation in the 0.2 to 1.5-eV spectral range. (Fe concentrations ranged from 3.5×10^{16} to 1×10^{18} cm^{-3}, and Te concentrations from 1×10^{16} to 3×10^{17} cm^{-3}.) They observed a broad room-temperature absorption band near 0.4 eV whose strength was proportional to the concentration of Fe^{2+} $(3d^6)$ ions in the samples. This band corresponds to the broadened Fe^{2+} intracenter transitions observed at higher resolution at low temperature by Ippolitova and Omel'yanovskii (see Figure 18). Similar room-temperature, broadened versions of the Fe^{2+} absorption spectrum were observed by Iseler[58] and Ippolitova et al.[27] in InP:Fe. In addition, Fistul et al. found that the strength of the absorption at 0.65 eV and 0.95 eV (Figure 21a) was correlated with the concentration of Fe^{3+} $(3d^5)$ ions. This, too, is consistent with the proposed excitation model in which the absorption and photoconductivity onsets at 0.5 eV are attributed to

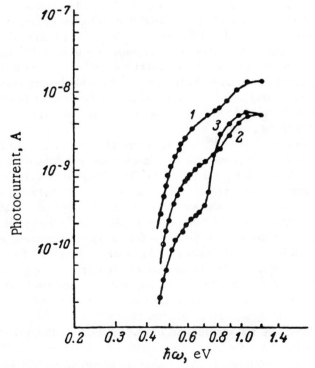

FIGURE 21 (b) Photoconductivity spectrum of iron-doped semi-insulating gallium arsenide samples 1-3 with different degrees of compensation. After Fistul et al., Ref. 100.

$Fe^{3+} \rightarrow Fe^{2+}$ (^5E) + h^+ (VB) (Equation (7)) and the onsets at ~0.8 eV are attributed to $Fe^{3+} \rightarrow Fe^{2+}$ ($^5T_2$) + h^+ (VB) (Equation (8)). These workers also obtained photoconductivity spectra (Figure 21b) which were essentially equivalent to those reported by Omel'yanovskii et al.,[47] with onsets at ~0.45–0.5 eV and ~0.8 eV. The 0.5 eV onset is clearer in photoconductivity spectra than in absorption since it is not obscured by the ~0.4 eV Fe^{2+} intracenter absorption which does not contribute to photoconductivity. This is consistent with the fact that these Fe-related levels lie lower in the gap in GaAs than in InP and the thermal excitation to the conduction band which follows optical excitation to the excited $^5T_2$ state of Fe^{2+} in InP:Fe as proposed by Look[69,70] apparently does not take place in GaAs:Fe.

The interpretation of the ~0.8 eV absorption, photoconductive, and PLE onsets in GaAs:Fe is confused somewhat by the observation in lightly Fe-doped[101] and undoped[102] GaAs crystals of an extrinsic absorption with onset near 0.78 eV. Andrianov et al.[101] ascribed this absorption to the presence of oxygen, but pointed out that, at higher Fe concentrations, this "oxygen-related" absorption was supplemented by an obviously Fe-related absorption. As mentioned previously, Fistul et al.[100] found the strength of the absorptions with onsets at ~0.5 eV and ~0.8 eV in moderately to heavily doped GaAs:Fe to be correlated with the concentration of Fe^{3+}. It seems reasonable to assume that, for Fe concentrations in the range 5×10^{16} to 10^{18} cm^{-3}, the observed extrinsic absorption onsets are attributable to the presence of the Fe.

Lebedev and coworkers[102] investigated the photocapacitance of GaAs:Fe epitaxial diodes and found two Fe-related levels with ionization energies of E_V + 0.38 eV and E_V + 0.45–0.5 eV. The change in the concentration of the charged centers produced by illumination with 0.45–0.5 eV light ($\Delta N = 3.3 \times 10^{17}$ cm^{-3}) was almost an order of magnitude greater than that produced by 0.38 eV light ($\Delta N = 4 \times 10^{16}$ cm^{-3}). For photon energies >0.5 eV, there was no further change in the capacitance, i.e., there was apparently no manifestation of an onset at ~0.8 eV. Their photocapacitance spectra were obtained in the temperature range 77–170K and led them to conclude that there are two Fe-related levels rather than one level whose activation energy shifts from 0.38 eV at 300K to 0.52 eV at 0°K, a possibility suggested earlier by Haisty.[42]

The relative spectral distribution of the photo-ionization cross section σ^0 (hν) in Fe-doped GaAs has been studied by Kitahara et al.[103] They employed the photoconductivity technique of Grimmeis and Ledebo[104] in which the photo-current is kept constant by adjusting the light source intensity, the photo-current is always much greater than the dark current, and the intensity of the light is low enough to ensure that the concentration of photoexcited holes is much less than the concentration of the deep levels. The relative spectral distribution of σ° for a VPE layer of GaAs containing $\sim 10^{16}$ cm^{-3} Fe is shown in Figure 22 for photon energies greater than 0.5 eV. The solid line represents the photo-ionization cross section calculated from the model of Lucovsky[105] with a threshold energy of 0.5 eV. Kitahara et al. pointed out that the good agreement of the σ° spectrum for the 0.5 eV threshold with the Lucovsky model suggests that the ground-state wave function for the impurity center is highly localized since the model assumes that the wave function is determined solely by a short range ion core potential.

The deviation of the data from the Lucovsky curve at ~ 0.8 eV in Figure 22 is consistent with the features in the extrinsic absorption and photo-conductivity spectra at this energy discussed previously. However, Kitahara et al. also regard the interpretation of this threshold with some uncertainty since they often observed a similar level in high resistivity undoped samples. These workers also observed a lower energy level with

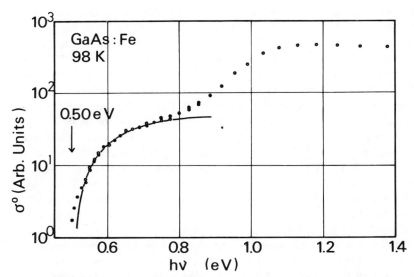

FIGURE 22   Relative spectral distribution of σ° in the $h\nu > 0.5$ eV region

(•) 1×10^{-8} A constant photocurrent
(○) 3×10^{-8} A constant photocurrent

The solid line is the calculated values using the threshold energy of 0.57 eV. After Kitahara et al., Ref. 103.

threshold of 0.37 eV, but the conditions of the Grimmeis and Ledebo technique were not fulfilled in this spectral range because the photocurrent was less than the dark current. This is consistent with the observation of Lebedev et al.[102] that ΔN for their 0.38 eV threshold was nearly an order of magnitude weaker than ΔN for their ~0.5 eV threshold.

More recently, Leyral et al.[99] have studied the spectral distribution of the photoionization cross section for *holes* in GaAs:Fe. (Their deep level optical spectroscopy technique[80(b)] allows the photoionization cross sections for holes and electrons to be measured separately.) Their spectra exhibited thresholds at ~0.5 eV and ~0.85 eV, which are consistent with transitions from the valence band to the ground state and excited-state levels of Fe^{2+} as shown in Figure 20. These photoionization-cross-section spectra lend further support to the PLE mechanism proposed for Fe^{2+} PL in GaAs:Fe by Nordquist et al.[10(a)]

4.3 ESR Studies of GaAs:Fe

In the case of Fe-doped GaAs, very little ESR work has been carried out in recent years, and the most complete study published to date is the early work of de Wit and Estle[14,15] which was discussed briefly in Section 2.1. The 9.2 GHz ESR spectrum[15] arising from the $\Delta M = 1$ transitions of Fe^{3+} in GaAs at T = 1.3K with the magnetic field aligned along the [001] direction is shown in Figure 23. Unlike the Fe^{3+} spectrum for InP:Fe (Figure 6) the five-line fine structure pattern expected for an $S = 5/2$ center is fully resolved in theGaAs:Fe spectrum of Figure 23. Again the conventional spin Hamiltonian for the $S = 5/2$, ^6S ground state of Fe in cubic symmetry given in Equation (1) is applicable. The spin Hamiltonian parameters obtained from the 1.6K data are g = 2.0462 ± 0.0006 and a = $(+339.7 \pm 0.3) \times 10^{-4}$ cm^{-1}. These correspond to the values given for GaAs:Fe in Table 1. The temperature independent line width of 54 G, which is attributable to hyperfine interactions with neighboring nuclei, is less than half that observed[25-27] for Fe^{3+} in InP (112 G) and results in the fully resolved five-line spectrum of Figure 23.

DeWit and Estle also measured the complete angular dependence of the magnetic field positions of the $\Delta M = 1$ transitions and their data are compared with the calculated positions in Figure 24. (Detailed theoretical expressions for the angular dependence of the spin Hamiltonian with the magnetic field in the $(1\bar{1}0)$ plane are given in Ref. 15.) Within the experimental uncertainty, there is complete agreement between the observed field positions and the theory. There are, however, some unexplained discrepancies between the calculated and observed angular dependences of the intensities of the transitions.

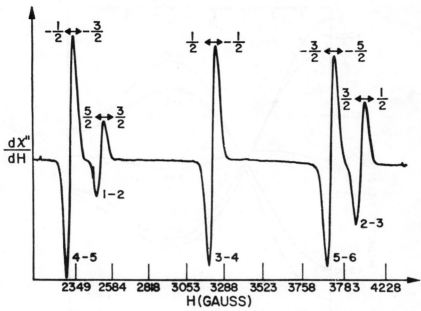

FIGURE 23 The derivative of the absorption versus magnetic field for the $\Delta M = 1$ transitions of Fe in GaAs, when $T = 1.3°K$, $\nu = 9.2$ GHz/sec, and H is along the [001] direction. The pattern of the amplitudes corresponds to a positive field splitting parameter a. The levels between which the transitions occur are designated by the strong magnetic field quantum number M in the upper part of the figure or by the rank in order of decreasing energy of the level in the lower part. After deWit and Estle, Ref. 15.

The g shift $\Delta_g = g - 2.0023 = +0.044$, is large and positive as is the case for Fe^{3+} in all the III-V compounds listed in Table 1. As discussed in Section 3.2 in connection with the Fe^{3+} ESR in InP:Fe, the large positive g-value shifts for Fe^{3+} in the III-V's is attributed[20,26,27] to electron transfer onto the central ion from the neighboring ligands in these covalent compounds.

DeWit and Estle attributed the rather large observed line width (54 G) to unresolved hyperfine structure resulting from interactions with neighboring Ga and As nuclei. On the basis of the assumption that the linewidth is due entirely to the super-hyperfine interaction with the As ligands, which occur with a 100% abundant isotope of spin $3/2$ (^{75}As), they calculated a hyperfine interaction constant of about 10 G. However, as discussed in Section 3.2 and refs. 25-27 and evidenced in the linewidths listed in Table 1, the size of the Fe^{3+} ESR linewidths in the III-V compounds is correlated with the Group III (cation) atoms rather than the Group V (anion) atoms.

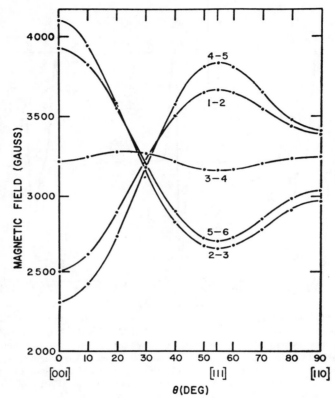

FIGURE 24 The magnetic field positions versus θ of the $\Delta M = 1$ transitions, when $T = 1.3°K$ and $\nu = 9.2$ GHz/sec. The levels between which the transitions occur are designated by their rank in order of decreasing energy. The dots are the measured values, the lines are the calculated ones. The experimental uncertainty is less than the size of the dots. After deWit and Estle, Ref. 15.

This led to the conclusion[25-27] that the linewidth of the Fe^{3+} ESR is determined primarily by the super-hyperfine interaction with the cations (second-nearest neighbors) rather than with anions (nearest neighbors).

Omel'yanovskii et al.[48] studied the effects of illumination on the intensity of the Fe^{3+} ESR in GaAs:Fe. They found an optically induced increase in the ESR signal intensity during irradiation with below-band gap light only in highly compensated samples. The spectral dependence of this optically induced increase exhibits an onset with increasing photon energy at about 0.7 eV and reaches a maximum above 0.8 eV. This effect indicates an optically-induced *increase* in the Fe^{3+} concentration due to charge transfer. Stauss et al.[26] observed an optically induced *quenching* of the Fe^{3+} ESR signal in InP:Fe (see Figure 7) which was explained in terms of the transition $Fe^{3+} \rightarrow Fe^{2+} + h^+$ (VB). The seemingly contradictory result obtained by Omel'yanovskii and coworkers in GaAs:Fe is probably explicable in terms of the photoionization of a level which is not associated with the Fe dopants. In Section 4.2 the existence of an extrinsic optical absorption[101] with onset near ~0.78 eV in lightly Fe-doped and undoped GaAs and its possible relationship to the presence of oxygen[101] were discussed. In samples containing higher concentrations of Fe, the Fe-related absorptions apparently overwhelm the effects of this 0.78 eV absorption in the absorption and photoconductivity spectra.[101] It is possible that the same level which gives rise to the "confusing" 0.78 eV absoprtion is responsible for the photo-induced increase in the Fe^{3+} concentration in the highly compensated samples studied by Omel'yanovskii et al.[48] In support of this hypothesis Andrianov et al.[101] observed that in samples whose absorption spectra are dominated by the Fe-related features (with low energy onset at 0.5 eV) there is *no* optically induced increase in the Fe^{3+} ESR intensity. In addition, no optically-induced increases were observed[106] in the Fe^{3+} ESR signal from the Fe-doped GaAs sample in which Nordquist et al.[10] studied the PLE spectrum of the Fe^{2+} PL.

Very recently, Wagner et al.[107] have obtained submillimeter ESR spectra in GaAs:Fe and InP:Fe (unpublished at this writing) in which the Fe^{2+} charge state has been observed for the first time. Their experiments are carried out at high magnetic fields (~10T) and employ a molecular gas laser as the source of submillimeter (11.2 to 23.0 cm^{-1}) radiation. The Fe^{3+} spectrum is clearly observed, and additional ESR lines are observed which are assigned to transitions within the ^5E ground state multiplet of Fe^{2+} in GaAs. The levels of the ^5E multiplet and their splitting in a magnetic field were discussed in Section 2.3 in connection with the APR studies. In the submillimeter region transitions can be induced between the lowest lying state of ^5E which is a singlet and the higher-lying states of ^5E as their positions are adjusted by the magnetic field. Initial data have involved transitions between the ground state singlet and the first excited state (a triplet) with $\vec{H} \parallel$ [111], but a more detailed study with $\vec{H}$ [100] is in progress.[107]

4.4 Mössbauer Spectroscopy in GaAs:Fe

Where applicable, Mössbauer spectroscopy provides a means to determine the charge state of impurity atoms in solids. In the case of semiconductors, a particularly interesting opportunity is presented to observe the change in the charge state of impurities as a function of the position of the Fermi level in the forbidden energy gap. For Fe in the III-V compounds the relevant parameter is the isomer shift of the ^{57}Co (^{57m}Fe) Mössbauer emission spectrum which should depend upon the position of the Fermi level. In addition, the Mössbauer spectra can reveal the electronic structure of the impurities, their position in the lattice, the symmetry of the local environment, and the formation of impurity-defect associates.

Earlier studies[108] detected an abrupt change in the quadrupole splitting of Mössbauer emission spectra of ^{57}Co (^{57m}Fe) in III-V compounds resulting from a change from n-type to p-type material. In this case the change in the splitting was attributed to a change in the charge state of Fe impurity atoms which formed associates with vacancies in the lattice. The more recent work to be discussed here involves the study of substitutional Fe impurity atoms on Ga sites in GaAs.

Seregin and coworkers[109] have investigated Mössbauer emission spectroscopy in GaAs and GaP in which the isotope ^{57}Co is diffused into the samples and the radioactive transformation of ^{57}Co (electron capture) results in the Mössbauer level ^{57m}Fe. They studied Te-doped n-type samples in which all of the Fe centers are ionized (the Fe^{2+} one electron trap) and Zn-doped p-type samples in which all of the Fe centers are in the neutral or isoelectronic Fe^{3+} state. The emission Mössbauer spectra for two such samples of GaAs are compared in Figure 25, and it is evident that the isomer shifts for the n-type and p-type samples differ significantly. Seregin et al. concluded that the isomer shift for the p-type sample corresponds to the neutral or isoelectronic Fe^{3+} centers while that of the n-type sample corresponds to the ionized Fe^{2+} centers. The absence of any quadrupole splitting in the Mössbauer spectra indicates that the symmetry of the local environment of the Fe atoms is cubic.

Interestingly, the isomer shift of the Mössbauer spectra of ^{57m}Fe in p-type GaAs differs from that expected for pure $3d^5$ electronic configuration. The observed shift is indicative of an increase of the electron density at the ^{57}Fe nuclei relative to that of $3d^5$. Seregin et al. used the isomer shift to determine the electron configuration of the Fe^{3+} center in GaAs to be $3d^5 4s^{0.52} p^{1.56}$. For the case of the ionized Fe impurity center Fe^{2+}, they found that if one assumes that the population of the valence shell (4 sp) of the Fe^{2+} is the same as that of the Fe^{3+} center and the 3d shell population is increased from $3d^5$ to $3d^6$, the expected isomer shift for Fe^{2+} would be of the order of 105 mm/s rather than the observed 0.63 mm/s for GaAs:Fe. Seregin et al. suggested that this discrepancy might be explained on the basis of a larger atomic orbital radius for the sixth 3d electron in the Fe^{2+} center in GaAs than is observed for a free ion.

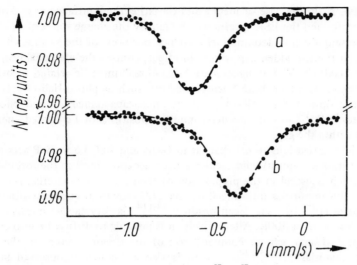

FIGURE 25 Emission Mössbauer spectra of $^{57}Co(^{57m}Fe)$ in GaAs at 295 K.
(a) N(Te) = 9.0 × 10^{17} cm^{-3},   N(Co) = 10^{17} cm^{-3};   (b)   N(Zn) = 2.2 × 10^{18} cm^{-3},
N(Co) = 10^{17} cm^{-3}. After Seregin et al., Ref. 109.

In summary, the observed change in the isomer shift of the Mössbauer
spectrum of ^{57m}Fe in GaAs which accompanies the change from p-type
to n-type material is consistent with the expected conversion from the isoelectronic Fe^{3+} charge state to the one electron trap Fe^{2+} state.

5. Experimental Studies of GaP:Fe

We begin our discussion of the properties of Fe-doped GaP with the classic
optical-absorption studies of Baranowski et al.[28] A sharply structured
spectrum centered near 0.41 eV was observed at 4.2K in GaP:Fe. This
was interpreted as the by now familiar transitions from the ^5E ground
state to $^5T_2$ excited state of Fe^{2+} ($3d^6$) in tetrahedral symmetry. In subsequent discussion we will treat this experimental signature of Fe^{2+} in
greater detail as observed in high-resolution absorption[110,111] and luminescence spectra.[31,110,112] The presence of the neutral or isoelectronic
Fe^{3+} ($3d^5$) charge state in GaP:Fe was demonstrated by the early ESR
measurements of Woodbury and Ludwig,[13] which were mentioned in the
Introduction. More recent ESR studies[65,13] to be discussed in Section 5.2
have also detected the Fe^+ ($3d^7$) charge state or "two-electron trap" in

GaP:Fe. As in the case of InP:Fe and GaAs:Fe, these intracenter d-d optical spectra and the ESR spectra which identify the charge states of Fe that are present do not determine the energy positions of the various charge states in the forbidden gap which, in effect, control the electrical properties of GaP:Fe. These charge state "signatures" must be related to other measurements which have been carried out such as photo-Hall effect,[113] extrinsic optical absorption,[111,114-117] photocapacitance,[118] and photo-excited ESR[119] in order to determine the positions of Fe-related energy levels within the gap.

Such investigations reveal that, as in GaAs and InP, Fe in GaP acts as a substitutional impurity and creates a deep acceptor level in the forbidden gap which is useful in the preparation of semi-insulating material. For example, Fe impurities introduced during LEC growth at concentrations in the range 10^{17}-10^{18} cm^{-3} can produce[111-114] GaP crystals with resistivity $\sim 10^9$ ohm-cm at 300K. Alternatively, it is possible to diffuse Fe into crystals of GaP and achieve compensation of the donors present in the as-grown material.[112,115,120,121] Iron is also a frequently observed inadvertent impurity[121] in as-grown GaP which can form non-radiative recombination centers that control minority carrier lifetimes and limit luminescence efficiency.

While the review of the experimental studies of GaP:Fe in the following Sections closely parallels those for InP:Fe and GaAs:Fe in many respects, it will also feature several studies which have been carried out only in GaP:Fe. These include piezospectroscopic and Zeeman studies[31] of the characteristic four-line Fe^{2+} PL spectrum, and the observation of ESR spectra attributable to the Fe$^+$ charge state,[65] which has been observed only in GaP:Fe.

5.1 Optical Absorption and Photoluminescence Studies of GaP:Fe

Subsequent to the original observation by Baranowski et al.[28] of the d-d absorption spectrum characteristic of Fe^{2+} (3d^6) in GaP:Fe, Soviet workers have reported higher-resolution studies of the same spectrum. Vasil'ev et al.[110] reported absorption spectra obtained at 7K in Czochralski grown crystals of GaP containing an Fe concentration of 8×10^{17} cm^{-3}, with background shallow donor concentrations $\sim 2 \times 10^{17}$ cm^{-3}. The crystals were p-type with resistivity of $\sim 10^9$ ohm-cm at 300K. Four sharp lines were observed which were identified as the familiar four allowed transitions between the ground state ^5E pentet and the lowest state of the $^5T_2$ excited state multiplet of Fe^{2+} in the tetrahedral crystal field (see Figure 2). The reported energies of these transitions, 3304 cm^{-1}, 3320 cm^{-1},

3330 cm^{-1}, and 3343 cm^{-1}, agree within one cm^{-1} with the energies of the same transitions measured in PL spectra of GaP:Fe reported by Vasil'ev et al.[110] and other workers.[31,112]

Andrianov et al.[111] were the first workers to carry out a comprehensive, although briefly described, study of electrical conductivity, optical absorption, and ESR of semi-insulating Fe-doped crystals of GaP with Fe concentrations in the range 5×10^{17}–2×10^{18} cm^{-3}. They obtained a four-line absorption spectrum at 20K which was also interpreted as intra-center zero-phonon transitions between the ground ^5E and excited $^5T_2$ states of the ^5D term of Fe^{2+} split by the tetrahedral crystal field. While no exact numbers were reported for the four transitions, they were described as being in the range 3300–3350 cm^{-1}, and the crystal-field splitting parameter $\Delta = 10\,Dq$ and the spin-orbit parameter for Fe^{2+} in GaP calculated from the spectrum were 3345 cm^{-1} and 82 cm^{-1}, respectively.

Photoluminescence studies of Fe-doped GaP were carried out in 1968 by Gershenzon et al.[122] and, more recently, by Vasil'ev et al.[110] and by Clark and Dean.[112] Vasil'ev et al.[110] reported a rather-low-resolution PL spectrum for GaP:Fe in which the energies of the four $^5T_2$–^5E transitions in emission were in complete agreement with the energies measured in absorption and given above. Clark and Dean[112] published the spectrum shown in Figure 26 which exhibits the familiar four-line spectrum near 0.41 eV with a prominent phonon replica about 41 meV lower in energy. In their analysis, they established that Δ lies between 0.405 eV and 0.438 eV (3270 cm^{-1} and 3530 cm^{-1}), which is consistent with the 3345 cm^{-1} value obtained by Andrianov et al.[111] from their absorption spectra. Clark and Dean also found that calculated values of the relative intensities of the four zero phonon lines near 0.41 eV are in good agreement with experiment. They also remind the reader that the Fe^{2+} optical spectra are relatively easy to understand or interpret because significant Jahn-Teller coupling occurs only in the upper ($^5T_2$) term[28,34,35] and, as a first approximation, this can be accounted for as a partial quenching of the first order spin-orbit coupling in the $^5T_2$ term. Ippolitova et al.[57] also discussed this point in conjunction with the Fe^{2+} optical absorption spectra in GaAs:Fe.

The most thorough and detailed study by far of the four-line optical spectrum of Fe^{2+} in the III-V's is that of West et al.,[31] in which the piezospectroscopic and Zeeman splittings of the four-line PL spectrum were investigated in GaP:Fe. For these experiments, Fe-doping was carried out by diffusing Fe at high temperature into n-type sulfur-doped crystals of GaP. High resolution GaP:Fe^{2+} PL spectra observed by these workers at 10K in the absence of externally applied fields are shown in

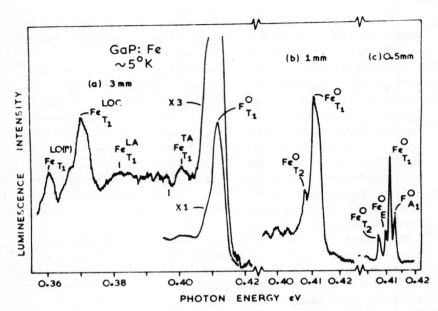

FIGURE 26 Photoluminescence spectrum of Fe^{2+} in GaP:Fe. After Clark and Dean,

Figure 27. These spectra are consistent with those reported by Vasil'ev et al.[110] and Clark and Dean.[112] The crystal-field-theoretic description of the splitting of the ^5D term of the Fe^{2+} ($3d^6$) in the tetrahedral environment is discussed in Section 2.2, and the splitting into the $^5T_2$ ($^5\Gamma_5$) excited multiplet and ^5E ($^5\Gamma_3$) ground multiplet is shown in Figure 28, which is analogous to Figure 2 except that the double group notation has been used to be consistent with the treatment of West et al.[31] In this notation, electric-dipole PL transitions are allowed from the lowest state, Γ_5, of the $^5\Gamma_5$ multiplet to the Γ_5, Γ_3, Γ_4, and Γ_1 states of $^5\Gamma_3$, but not to the Γ_2 state. West et al. ascribe the observed four-line spectrum to these electric dipole transitions, and have solved the full Hamiltonian, including crystal-field and spin-orbit effects (the latter to third order) in order to fit the data and obtain values of $\Delta = 10$ Dq and λ, the spin-orbit parameter The observed frequencies of the four lines and their calculated values are given in Table 2. They found the values $\Delta = -3559.4$ cm^{-1} and $\lambda = -93.5$ cm^{-1} to be consistent with their experimental observations.

The schematic representation of the calculated splittings for uniaxial

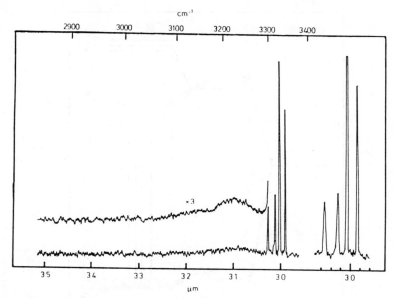

FIGURE 27 The near-infrared photoluminescence spectrum of GaP:Fe^{2+} at 10K. After West et al., Ref. 31.

stress $\vec{P}$ // [110] presented in Figure 29 demonstrated that stress along this direction removes all the degeneracy remaining in the system. Allowed transitions are shown as vertical lines in Figure 29 which indicate that transitions from the three stress-split components of the $^5\Gamma_5$ excited state to the Γ_1 ground state level are electric dipole allowed. These splittings allow one to determine the values of the parameters B and C for the excited state given in Table 3, which are consistent with the earlier measurements by Vasil'ev et al.[110] Only the differences between the A-values (Table 3) for the excited and ground states can be determined from the stress-splittings of the spectra. (The parameters $A(A_i)$, $B(B_i)$, and $C(C_i)$ determine the response of the level Γ_i to hydrostatic, tetragonal, and trigonal stresses, respectively.) The observed splitting of the four zero-phonon lines of GaP:Fe^{2+} PL for $\vec{P}$ [110] are shown in Figure 30 and are compared with solid curves calculated with the values of A_i, B_i, and C_i given in Table 3.

The effects of the Zeeman interaction upon the crystal-field split levels of Fe^{2+} ($3d^6$) were calculated by West et al.[31] by solving numerically the

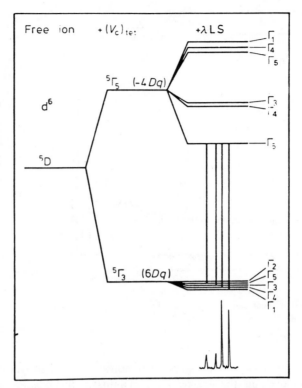

FIGURE 28 Effect of the crystal-field and spin-orbit coupling on the ^5D ground state of Fe^{2+} (d^6) in a tetrahedral environment. After West et al., Ref. 31.

TABLE 2 Photoluminescence peaks observed in the spectrum of GaP:Fe^{2+} at 10K. The theoretical positions of these lines have been calculated with and without Jahn-Teller coupling (after West et al., Ref. 31).

	Observed position cm^{-1}	Calculated position (cm^{-1}) without the Jahn-Teller effect	including the Jahn-Teller effect
Transition			
$\Gamma_5(^5\Gamma_5)-\Gamma_1(^5\Gamma_3)$	3343.5	3343.5	3343.5
$\Gamma_5(^5\Gamma_5)-B_4(^5\Gamma_3)$	3330.7	3331.0	3330.9
$\Gamma_5(^5\Gamma_5)-\Gamma_3(^5\Gamma_3)$	3319.6	3318.4	3319.6
$\Gamma_5(^5\Gamma_5)-\Gamma_5(^5\Gamma_3)$	3303.6	3303.5	3303.5

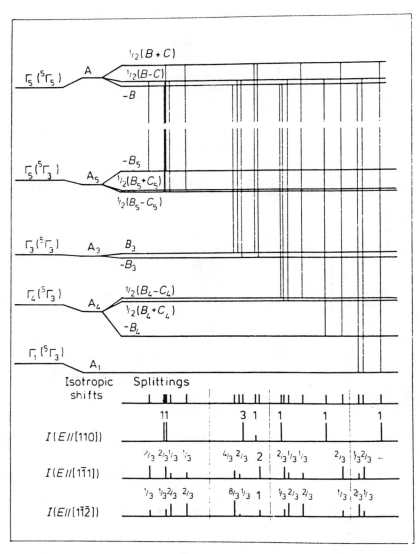

FIGURE 29 Schematic representation of calculated splittings for uniaxial stress
P ∥ [110]. Allowed transitions are shown as vertical lines. The relative polarizations
within each ZP line are shown below the transitions for the electric vector E of the
luminescence parallel to [110], [1$\bar{1}$1] and [1$\bar{1}$2]. Comparison between different ZP
lines is not meaningful in this diagram. After West et al., Ref. 31.

TABLE 3 Stress parameters for the levels involved in the near infrared luminescence of GaP:Fe^{2+}. All entries are given in units of cm^{-1}/(kg mm^{-2}) (after West et al., Ref. 31).

Level	–	B	C	A
$\Gamma_5(^5\Gamma_5)$	–	0.09	0.17	–
	i	B_1	C_1	$A-A_1$
$\Gamma_1(^5\Gamma_3)$	1	–	–	0.19
$\Gamma_3(^5\Gamma_3)$	3	0.02	–	0.11
$\Gamma_4(^5\Gamma_3)$	4	0.26	−0.03	0.18
$\Gamma_5(^5\Gamma_3)$	5	−0.14	0.01	0.14

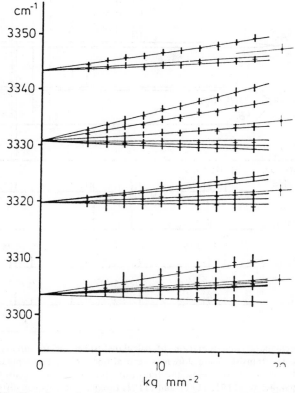

FIGURE 30 The observed splitting of the four ZP lines of GaP:Fe^{2+} for P ∥ [110]. The solid lines are calculated using the values of A_p B_i and C_1 of Table 3. After West et al., Ref. 31.

original Hamiltonian (which contained crystal-field and spin-orbit terms) with the addition of the Zeeman term

$$H_z = \mu_B \, \vec{B} \, (k\vec{L} + 2\vec{S})$$ (10)

where k is an orbital reduction factor, μ_B the Bohr magneton, and the g-value associated with spin has been set equal to 2. Figure 31 shows the splitting of the four Fe^{2+} zero phonon lines under the influence of a magnetic field applied along the [001] direction. The measured splitting of the 3343.5 cm^{-1} $\Gamma_5 (^5\Gamma_5) - \Gamma_1 (^5\Gamma_3)$ zero phonon line provides a value of the orbital reduction factor k of about 0.2. West et al. then find that the parameters λ, Δ, and k are sufficient to describe the magnetic field dependence of the PL lines for any field direction. The solid curves of Figure 31 represent the calculated magnetic field dependences including the effects of a weak Jahn-Teller coupling upon the zero field positions of the $\Gamma_5 - \Gamma_3$ transitions.

West et al. also discussed the effects of John-Teller coupling upon the $^5\Gamma_5$ and $^5\Gamma_3$ levels of Fe^{2+} in GaP. They point out that the low value of the orbital reduction factor, k = 0.2, suggests the presence of a Jahn-Teller coupling. While a simple crystal field model gives the positions of the zero-phonon lines to within 1.2 cm^{-1} (Table 2), West et al. show that the inclusion of a small Jahn-Teller coupling within the $^5\Gamma_5$ level (E_{JT} = 0.0024 cm^{-1}, Jahn-Teller reduction factor Q $(^5\Gamma_3)$ = 0.9998) improves the fit of the line positions (Table 2).

The reader is referred to West et al.[31] for a complete description of their detailed experimental studies and their calculations of crystal-field splitting, stress effects, and Zeeman interaction, as this work apparently represents the most complete available discussion of the optical signature of Fe^{2+} in a III-V compound.

Klein and Weiser[123] have reported a study of the time dependence of the 0.41 eV $^5T_2 \rightarrow {}^5E$ luminescence transition of Fe^{2+} in GaP:Fe, for the case of below-gap excitation. They measured the decay of the 0.41 eV luminescence intensity after pulsed dye laser excitation (10 nsec pulse width at 2.14 eV) and the decay curve obtained at 4.2 K is shown in Figure 32; the data exhibit an exponential decay rate of 6.6 $\mu sec)^{-1}$. These results were compared to dynamical calculations based on alternative models of the excitation-recombination process for below-gap (extrinsic) excitation. Excitation proceeds by the charge transfer process $Fe^{3+} + h\nu \rightarrow Fe^{2+} (^5T_2) + h^+$. The excited Fe^{2+} ion can then relax radiatively to its ^5E ground state with a radiative lifetime τ_o, and the Fe^{2+} ground state then recombines nonradiatively with a hole, reverting to its

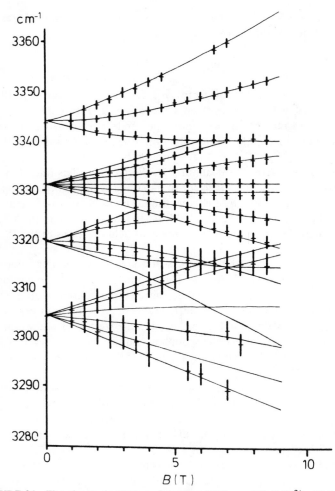

FIGURE 31 The observed splitting of the four ZP lines of GaP:Fe^{2+} with B ∥ [001]. The solid lines are a theoretical fit using $k = 0.2$, $\lambda = 93.5$ cm^{-1}, Dq -355.95 cm^{-1} and including Jahn-Teller coupling. Some of the predicted Zeeman components are too weak for detection. After West et al., Ref. 31.

Fe^{3+} charge state. Alternatively, the excited Fe^{2+} could recombine non-radiatively with a hole before a radiative transition can take place. This hole capture process may be viewed as the Coulombic capture of a free hole by the negatively-charged Fe^{2+} center, or the hole may be regarded as trapped (bound) by the negatively charged Fe^{2+} center with a character-istic lifetime τ_h.

Klein and Weiser[123] separately analyzed the rate equations for the charge-transfer process with each of the alternative hole capture mechanisms. They found that the localized-hole-capture process is expected to result in an exponential decay of the Fe^{2+} PL, whereas the free-hole capture will produce a decidedly non-exponential decay. The obviously exponential decay of Figure 32 ($1/\tau_o + 1/\tau_h = (6.6\ \mu sec)^{-1}$) favors the idea of hole localization at the Fe^{2+} site, with a characteristic lifetime for this photoinduced hole. The alternative process predicts a lifetime dependent on the time-varying free-hole and Fe^{2+} concentrations and implies a non-exponential decay.

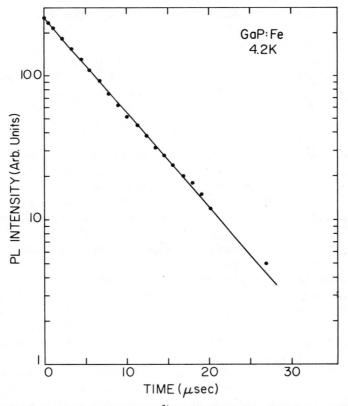

FIGURE 32   Decay of the 0.41 eV Fe^{2+} PL in GaP:Fe. The solid line is a fit of the data to an exponential decay. The measured exponential decay rate is $(6.6\ \mu sec)^{-1} = 1/\tau_o + 1/\tau_h$. After Klein and Weiser, Ref. 123.

5.2 ESR Studies of GaP:Fe

The Fe^{3+} ESR spectrum for Fe-doped GaP was first reported in the early work of Woodbury and Ludwig[13] which was mentioned in Section 2.1. This work and the studies of Suto and Nishizawa[113,119] are apparently the only reported ESR studies of GaP:Fe until the more recent work of Kaufmann and Schneider[65] and Andrianov et al.[111] Woodbury and Ludwig,[13] Suto and Nishizawa,[119] and more recently Andrianov et al.[111] observed in semi-insulating crystals of GaP:Fe a characteristic five-line ESR spectrum which was interpreted as arising from a spin 5/2 system in cubic symmetry and was attributed to Fe^{3+} $(3d^5)$ ions, the neutral or isoelectronic charge state of Fe ions substituted on cation (Ga) sites. The spin Hamiltonian parameters determined by Woodbury and Ludwig, g = 2.025, a = 390 × 10^{-4} cm^{-1}, and Δg = 38 G, are compared with the same parameters for Fe^{3+} in some other III-V compounds in Table 1 of Section 3.2. Bashenov[17] points out that the g-value decreases from GaAs to InAs to GaP, i.e., with decreasing ionicity. The large positive g-value shift is attributed to electron transfer onto the central ion from the surrounding ligands;[26,27] the linewidth of the Fe^{3+} ESR spectrum is due primarily to the superhyperfine interaction[25-27] with the *second* nearest-neighbor shell, i.e. the Ga atoms. The reader is referred to Section 3.2 for a more complete discussion of these interpretations.

Andrianov et al.[111] found a somewhat larger value for the Fe^{3+} ESR g-value, g = 2.01 ± 0.01, although with a rather large uncertainty in the measurement. More importantly, they found a linear correlation between the concentration of Fe ions in the $3d^5$ state as measured by the Fe^{3+} ESR intensity in several different GaP:Fe crystals, and the optical absorption coefficient in a below band gap charge transfer absorption band with 0.7 eV onset. They attributed this absorption to the optical transfer holes from the Fe^{3+} centers to the valence band (Equation (7)). In a closely related study, Masterov and Sobolevskii[124] observed a 1.5 eV onset in the spectrum of a photoinduced increase in the intensity of the Fe^{3+} $(3d^5)$ ESR. They attributed this photoenhancement to the charge transfer transition Fe^{2+} $(3d^6) \rightarrow Fe^{3+}$ $(3d^5) + e^-(CB)$ (Equation (6)) involving the photoionization of an electron from Fe^{2+} $(3d^6)$ to the conduction band followed by the capture of the electron by a deep trapping level. The energy of the threshold of this photo-ionization should correspond to the depth of the Fe^{2+} $(3d^6)$ level below the conduction band. The resulting position of this level relative to the conduction band, E_c–1.5 eV, is quite consistent with the 0.7 eV absorption threshold measured by Andrianov et al.[111] for photo-ionization of holes from Fe^{3+} ($3d^5$) to the valence band.

Masterov and Sobolevskii[124] also studied the thermal quenching of the photoinduced Fe^{3+} ESR signal. They found that beginning at 140K there was a reduction in the photoinduced ESR signal with increasing temperature. This was interpreted as thermal ionization of the electron trap levels, leading to the reverse charge exchange between the Fe centers, Fe^{3+} ($3d^6$) → Fe^{2+} ($3d^6$). The depth of the trapping level as determined from the thermal quenching of the ESR was $\Delta E = 0.27$ eV below the conduction band.

It should be pointed out that the results of these photoinduced ESR studies[111,124] do not appear to be consistent with the photoexcitation and photoquenching spectra for the Fe^{3+} ESR reported earlier by Suto and Nishizawa.[119] One might conclude from the fact that Suto and Nishizawa observed an apparent Fe^+ ($3d^7$) ESR spectrum under optical excitation and Andrianov et al.[111] did not, that the positions of the Fermi levels, nature and concentrations of background impurities, etc. were quite different for the crystals employed by the two groups.

As mentioned previously, the one-electron trap state of Fe in the III-V semiconductors, Fe^{2+} ($3d^6$), has an orbitally degenerate ^5E ground state and has therefore not been detected by conventional microwave ESR techniques,[65] although it has recently been observed by Wagner et al.[107] in far-infrared ESR studies of GaAs:Fe and InP:Fe (Section 4.3). In addition, the Fe^{2+} charge state has been detected in GaAs:Fe by acoustic paramagnetic resonance.[37,38] However, the *two*-electron trap state of Fe in GaP, Fe^+ ($3d^7$), has been detected by ESR. Haraldson and Ribbing[125] first reported, without identification of its origin, an ESR signal at $g = 2.131$ in n-type GaP which later proved to be the Fe^+ ($3d^7$) ESR spectrum. Subsequent ESR studies of GaP by Suto and Nishizawa[113,119] demonstrated that this resonance was associated with Fe and suggested the $3d^7$ configuration, but still provided no evidence that it was attributable to the $S = 3/2$, Γ_8 ground state of Fe^+ ($3d^7$). It remained for Kaufmann and Schneider[65] to show through the characteristic angular dependence of the Zeeman splitting of its Γ_8-ground state that this spectrum was attributable to the two-electron trap, Fe^+ ($3d^7$) (Fe^{--} in the notation of Ref. 65).

The ESR spectrum of Fe-doped n-type GaP obtained by Kaufmann and Schneider at 35 GHz and 20K is shown in Figure 33 for H // [100]. The slightly anisotropic low-field signal is the Fe^+ ($3d^7$) ESR spectrum, and the high-field ESR signal ($g = 1.999$) which is isotropic is probably attributable to shallow donor centers.[65] The low-field ESR spectrum, with

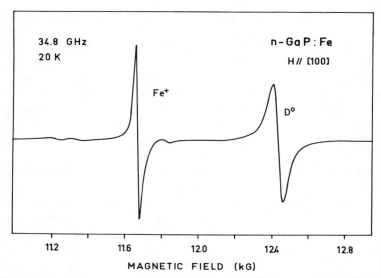

FIGURE 33 ESR-spectrum of iron-doped n-type GaP, recorded at 35 GHz and 20K. The slightly anisotropic low-field signal arises from the $3d^7$-state of Fe; the high-field isotropic signal is due to unspecified donor centers. After Kaufmann and Schneider, Ref. 65.

linewidth ≈ 28 G, was shown by Kaufmann and Schneider to be described by the Hamiltonian[126]

$$H = g\beta\vec{H}.\vec{S} + \mu\beta \, \{H_x S_x^3 + H_y S_y^3 + H_z S_z^3 - \frac{1}{5} (\vec{H}.\vec{S}) \times [3S(S+1) - 1] \}.$$ (11)

The $3d^7$ ESR spectrum exhibits a characteristic angular dependence for rotation of the external magnetic field in a (110)-plane of the crystal. In this case the effective g-value of the $M_s = -\frac{1}{2} \leftrightarrow +\frac{1}{2}$ ESR transition of the $3d^7$ state of Fe as derived from Equation (11) becomes

$$g_{eff} = g - \frac{9}{5} \mu p$$ (12)

where

$$p = 1 - \frac{5}{4} \sin^2\theta (1 + 3\cos^2\theta).$$ (13)

The angle θ is measured between $\vec{H}$ and the [001] direction. In Figure 34 the angular dependence of the low-field ESR spectrum is shown along with the fit to Equation (13) for rotation of the magnetic field in the $(1\bar{1}0)$ plane of the crystal. On the basis of the excellent agreement of experiment and theory in Figure 34, Kaufmann and Schneider concluded that the ground state of this ESR center has Γ_8 symmetry and spin $S = 3/2$, and that an Fe center with these characteristics must have the $3d^7$ configuration.

Suto and Nishizawa[113] had earlier studied the Fe^+ $(3d^7)$ ESR and Hall effect in n-type GaP crystals. They showed that the ESR spectrum later identified conclusively[65] as attributable to Fe^+ $(3d^7)$ appeared only in n-type Fe-doped crystals of GaP back-doped with Te, Sn, or Zn. On the basis of their Hall-effect data and the temperature dependence of the Fe^+ ESR intensity, these workers concluded that the Fe^+ level is close to the Te donor level at 90 meV below the conduction band. This conclusion was later confirmed qualitatively by Ennen and Kaufmann[120] who measured the annealing temperature dependence of Fe^+ and Te-neutral donor ESR signals in Te-doped GaP (n = 1.7×10^{17} cm^{-3}) into which Fe was diffused.

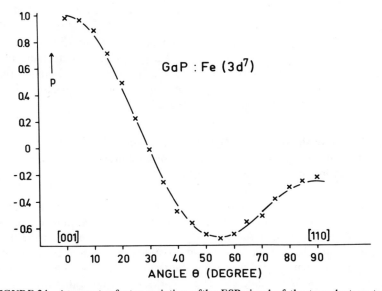

FIGURE 34 Apparent g factor variation of the ESR-signal of the two-electron trap state, $3d^7$, of iron in GaP. The external magnetic field is rotated in a $(1\bar{1}0)$-plane of the crystal. The solid line is the theoretically expected curve according to equation (12). After Kaufmann and Schneider, Ref. 65.

They found that for annealing (diffusing) temperatures above 1000°C, there was a strong decrease in the Te-neutral donor ESR signal, which was interpreted as a dropping of the low-temperature Fermi level below the Te-donor level. If, as concluded by Suto and Nishizawa,[113] the Fe^+ level lies close to the Te-donor level, a drop of the Fermi level below this donor level must also be accompanied by a decrease in the Fe^+ concentration and ESR signal. Such a decrease was observed by Ennen and Kaufmann[120] for annealing temperatures above 1000°C, confirming the approximate position of the Fe^+ level, near the bottom of the conduction band. Ennen and Kaufmann[120] noted that Tell and Kuijpers[127] observed an electron trap at E_c-0.23 eV in GaP which seemed to be related to Fe. The seemingly unusual properties of this trap reported by Tell and Kuijpers could be explained in the opinion of Ennen and Kaufmann if the filled trap corresponds to Fe^+ ($3d^7$).

One final point was made by Kaufmann and Schneider[65] on the basis of the light sensitivity of the Fe ESR signals. In not intentionally Fe-doped, semi-insulating GaP samples, and in some semi-insulating Fe-doped GaP samples both the Fe^{3+} (neutral Fe) and Fe^+ (two-electron trap) ESR spectra were observed. Under the influence of infrared or visible illumination the intensities of both signals changed as reported by Suto and Nishizawa.[119] However, the n-type Fe-doped GaP samples exhibited only the $3d^7$ Fe-related signal. Furthermore, illumination did not change the intensity of the Fe^+ ($3d^7$) ESR signal and it produced no detectable Fe^{3+} ($3d^5$) ESR spectrum. Kaufmann and Schneider therefore concluded that the one-electron Fe^{2+} ($3d^6$) and two-electron Fe^+ ($3d^7$) trap states are the most stable charge states of Fe in n-type GaP.

5.3 Transport Measurements, Optical Absorption, Photoionization Spectroscopy and Photoluminescence Excitation Spectroscopy in GaP:Fe

As part of the comprehensive study of GaP:Fe carried out by Andrianov et al.[111] and discussed in the preceding two Sections, measurements of the electrical resistivity as a function of temperature were reported for their p-type crystals of GaP:Fe, which had resistivity $\sim 10^9$ ohm-cm at 300K. The ionization energy deduced from the thermally-activated hole concentration was $E_i = 0.70 \pm 0.02$ eV. These workers concluded that this energy corresponded to the ionization energy of deep acceptor states associated with the Fe impurities.

The room-temperature optical-absorption spectra for three of the

GaP:Fe samples of Andrianov et al. are shown in Figure 35. The "resonance" band located at 0.4–0.6 eV is the Fe^{2+} ($3d^6$) intracenter d-d absorption spectrum discussed in Section 5.1 whose sharp structure, visible at low temperatures, has been greatly broadened at 300K. The amplitude of this band was clearly a function of the background donor concentration which determined the Fe^{2+} concentration. The very broad band with threshold near 0.70–0.75 eV is the absorption described in Section 5.2 as attributable to charge-transfer transitions from the valence band to the Fe^{2+} ($3d^6$) acceptor states. This threshold energy is in good agreement with the value of E_i = 0.70 eV determined from the electrical measurements, and can be interpreted as the photoionization transition $h\nu \gtrsim 0.7$ eV + Fe^{3+} ($^3d^5$) → Fe^{2+} ($3d^6$) (^5E) + h^+ (VB) (Equation (7)). In Section 5.2, it was pointed out that this position for the Fe^{2+} level, E_{VB} + 0.7 eV is consistent with the observation[124] of a 1.5 eV onset in optically induced enhancement of the Fe^{3+} ESR due to transition $h\nu \gtrsim 1.5$ eV + Fe^{2+} → Fe^{3+} + e^-(CB) (Equation (6)).

Abagyan and coworkers[114] also reported room temperature extrinsic optical absorption spectra in GaP:Fe crystals containing 1×10^{17}–3×10^{18} cm^{-3} Fe impurities. These workers also observed the Fe^{2+} intracenter transitions near 0.5 eV, but the dominant feature in their higher-energy

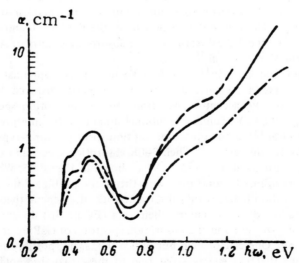

FIGURE 35 Absorption spectra of GaP:Fe at 300°K. After Andrianov et al., Ref. 111.

absorption is a broad band with apparent onset near 1.2 eV. Their absorption data do exhibit a weaker tail below 1.2 eV with onset near 0.7 eV, which they chose to ignore. This closely corresponds to the 0.7 eV onset shown in Figure 35. Correspondingly, the data of Andrianov et al. in Figure 35 exhibit a pronounced kink near 1.2 eV, a feature which these authors chose to ignore. Abagyan et al. identified their 1.2 eV onset as the threshold for excitation of electrons from the valence band to the Fe acceptor level. However, the 1.5 eV photoenhancement threshold for Fe^{3+} ESR[124] and the linear relationship[111] between the Fe^{3+} ESR intensity and the optical absorption coefficient at 0.95 eV (i.e. in the lower energy band with 0.7 eV threshold) appear to corroborate the interpretation of Andrianov et al. Furthermore, Abagyan et al. found a thermal activation energy for electrical conductivity in their GaP:Fe crystals of 0.3 eV, which must be compared with the 1.2 eV threshold for photoionization which they inferred from their absorption data. They suggested that this large discrepancy might be due to the electron-phonon interaction. In contrast, the photoionization energy deduced from the absorption data of Andrianov et al., ~0.75 eV, is in good agreement with their 0.70 eV thermal activation energy for the hole concentration. Parenthetically, Abagyan et al.[114] point out that the 1.7–1.8 eV ionization energy for the Fe level in GaP given by Okuno et al.[118] corresponds to the *maximum* of the hole photoactivation spectrum, and that for deep levels an energy close to the *onset* or threshold of the spectrum is a more accurate representation of the ionization energy. The *thresholds* exhibited by the data of Okuno et al. are consistent with the 1.2 eV features in the absorption spectra of Andrianov et al.[111] and Abagyan et al.[114]

Shishiyanu and Georgiu[115] diffused Fe into n-type epitaxial layers of GaP and observed type conversion. These workers also deduced the energy position of the Fe acceptor level from optical absorption spectra and found $E_{VB} + 0.84$ eV. Lending additional support to the interpretation of Andrianov et al.[111] is the optical-absorption spectrum for p-type GaP:Fe reported by Kirillov et al.[116] Their 300K absorption spectrum exhibited a long wavelength band at 0.7–2.2 eV, as shown in Figure 36. While these authors were primarily concerned with fitting the profile of the impurity optical absorption bands within the framework of a general theory of the photoionization of deep centers, their data (Figure 36) do confirm the existence of *two* onsets in the absorption spectrum of GaP:Fe at ~0.7 eV and ~1.2 eV. One cannot help but notice the, again, possibly fortuitous circumstance that the energy separation of these two thresholds corresponds closely to the 0.41 eV energy of the Fe^{2+} ($^5T_2 - {}^5E$) optical transition. A similar circumstance was pointed out for GaAs:Fe in Section 4.2

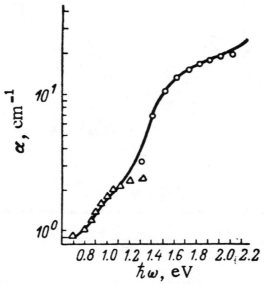

FIGURE 36 Optical absorption spectrum of p-type GaP:Fe recorded at T = 300°K. The absorption background, amounting to 0.9 cm⁻¹, is subtracted from the long-wavelength band. After Kirilov et al., Ref. 116.

(0.5 eV and 0.86 eV thresholds, 0.37 eV PL) and in Section 3.2 for InP:Fe (0.78 eV and 1.13 eV thresholds, 0.35 eV PL).

Brunwin et al.[121] carried out DLTS studies of Fe-diffused GaP. They observed a 0.27 eV electron trap in GaP samples before Fe diffusion whose concentration was greatly increased by the in-diffusion of Fe, but a similar increase was observed as a result of a control diffusion in which no Fe was present. In one Fe diffusion, a trap with activation energy of 0.64 eV was introduced in both LPE and VPE crystals. Diffusion with other transition metals (Cr, Mn, Co, Ni) did not produce this trap, but the authors describe its association with Fe as tentative. Brunwin et al.[121] also briefly reviewed some of the principal conclusions of the Soviet work which we have discussed in this section, and they suggested the level scheme shown in Figure 37 for GaP:Fe. This scheme shows the 0.7 eV onset for the transition $Fe^{3+} \rightarrow Fe^{2+} + h_{VB}^{+}$ which is inferred from the data of Andrianov et al.,[111] Abagyan et al.,[114] and Kirillov et al.[116] The 0.41 eV Fe^{2+} PL transition is also indicated and clearly implies that a second onset should be expected for transitions from the valence band to the Fe^{2+} excited state, $Fe^{3+} \rightarrow Fe^{2+*} + h_{VB}^{+}$, at an energy of 0.7 eV + 0.41 eV $\cong$ 1.1 eV. As pointed out above, such a threshold is observed at ~1.2 eV in

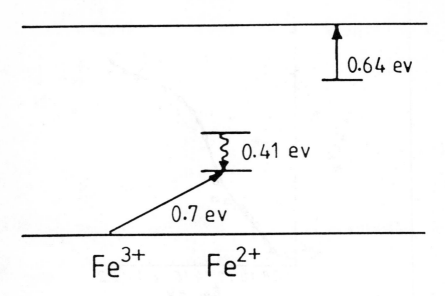

the absorption data of Refs. 111, 114, and 116. It might also be expected that the ~1.2 eV absorption band in GaP:Fe should manifest itself in the PLE spectrum of the 0.41 eV Fe^{2+} PL band, in analogy with the PLE spectra of the Fe^{2+} PL bands in InP:Fe (Figure 14) and GaAs:Fe (Figure 20). This expectation has been verified by Shanabrook et al.[128] who have carried out PLE spectroscopy on the 0.41 eV Fe^{2+} PL band in GaP:Fe. The PLE spectrum (Figure 38) at 4.2 K exhibits a strong, broad, below-gap band with the expected low energy onset at ~1.27 eV. This extrinsic PLE band corresponds to the $Fe^{3+} \rightarrow Fe^{2+} + h^+$ charge transfer absorption band invoked in the discussions of PLE for InP:Fe (Section 3.3) and GaAs:Fe (Section 4.1) and proposed by Klein and Weiser[123] in their time resolved study of Fe^{2+} PL in GaP:Fe.

We close this discussion of GaP:Fe with the observation that three groups[121,124,127] have independently observed an Fe-related electron trap in GaP with activation in the range 0.23 to 0.27 eV and a fourth group[120] has argued that this trap, when filled, corresponds to Fe^+ ($3d^7$). While the evidence for this assignment is not so convincing as that for the Fe^{3+} to Fe^{2+} transition, it might be argued that a 0.27 eV electron trap, tentatively ascribed to Fe^+, should be added to the GaP:Fe level scheme of Figure 37.

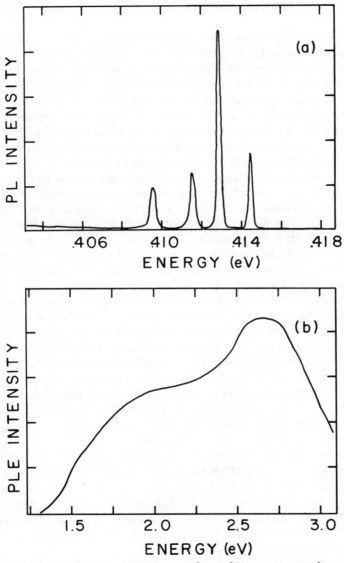

FIGURE 38   (a) PL spectrum of the 0.41 eV $^5T_2 \rightarrow \, ^5E$ transition of Fe^{2+} in GaP:Fe.
(b) PLE spectrum of the 0.41 eV PL transition (T = 4.2K). After Shanabrook et al.,
Ref. 128.

6. Notes on Theoretical Methods

The scope of this review has been confined primarily to experimental studies of Fe impurities in the III-V semiconductors. While various models or energy-level schemes have been discussed, these are empirical models whose validity, if any, rests solely on consistency with the results of a range of optical, electrical, and magnetic-resonance experiments. Crystal-field-theoretic calculations[29-36,71] have been discussed briefly in connection with the sharply structured intracenter d-d transitions of the Fe^{2+} optical spectra, but these calculations do not address the question of the position of the Fe energy levels relative to the band edges.

For a comprehensive discussion and bibliography relating to theoretical methods for the study of the states introduced by impurities and other defects in semiconductors the reader is referred to the extensive review of Pantelides.[129] For theoretical treatments specifically of transition-metal impurities in III-V semiconductors, the review of Masterov and Samorukov[130] and the recent work of Hemstreet[131] and Fazzio and Leite[132] provide a fairly complete coverage of the relatively limited number of theoretical studies which have been reported.

Hemstreet[131(a)] and Fazzio and Leite[132] have carried out closely related cluster calculations of the electronic structure of neutral 3d transition metal impurities replacing Ga atoms in the GaAs lattice. The calculations indicate that the neutral transition metal impurities produce acceptor levels deep in the fundamental gap in agreement with experimental observations. Hemstreet[131(a)] found that, on the basis of his results, the transition-metal impurities can be divided into two categories on the basis of whether or not their 3d states play an active role in determining the active electronic states of the defect. The active impurity states in the gap for Cr, Mn, Fe, and Co have substantial d character and there is evidence of bonding between the impurity and the neighboring ligands. This could be described as the "standard" behavior for structured 3d transition-metal impurities.[12] However, Ni and Cu appear to behave as simple (main-group) acceptors and their active defect levels have little d character. Hemstreet found that the d states of the Ni and Cu impurities appear as "resonances" in the valence band of the host lattice and have little effect upon the electronic properties of the material.

More recently, Hemstreet[131(b)] has applied his calculational procedure to the problem of Fe at an In site in InP. His results for the positions of the Fe^{2+} energy levels in the forbidden gap are in remarkably good agreement with experiment. He calculates 0.37 eV for the $^5E \sim {}^5T_2$ separation and onsets for photoexcitation of electrons from the valence band to the

^5E ground state and from the trap level to the conduction band of 0.80 eV and 0.62 eV, respectively. Thus his energy-level diagram would be in excellent agreement with those of Figures 9, 11, and 13.

Partin et al.[133] have estimated the binding energy of a hole at a transition metal relative to the valence band edge on the basis of a simple model which relates the binding energy to elastic stress and therefore to the radius of the impurity atom. Binding energies have been obtained from the model for the 3d transition elements from Sc to Cu in the GaAs lattice and are compared to collected experimental data. For Fe in GaAs the model yields a value of 0.44 eV. Somewhat less complete results have been obtained for the transition metal impurities in GaP. For GaP the predicted Fe binding energy is 0.55 eV.

7. Summary

For the case of Fe in the III-V semiconductors the work reviewed here has shown how the powerful combination of ESR and intracenter d-d optical spectroscopy has unambiguously identified the neutral Fe^{3+} ($3d^5$) center through the ESR spectrum[13-17,22,25-27] of this spin 5/2 S-state ion, the Fe^{2+} ($3d^6$) or one-electron trap on the basis of the $^5T_2 - {}^5E$ crystal-field-induced optical spectra,[6,7,25-28,31,57,110-112] and, in GaP:Fe, the two-electron trap, Fe^+ ($3d^7$), through the ESR spectrum[65] of its spin 3/2, Γ_8 ground state. In InP:Fe and GaAs:Fe combined ESR and optical studies[7,15,25-27,57,110,111] have shown that, in semi-insulating crystals, Fe exists in both the Fe^{3+} and Fe^{2+} states, with the Fe^{2+} concentration determined by the concentration of compensating shallow donors; in n-type InP and GaAs the Fe^{2+} state predominates. On the other hand, while semi-insulating GaP:Fe behaves similarly, in n-type GaP the Fe^{2+} and Fe^+ states are the most stable charge states.[65] The ESR study by Kaufmann and Schneider[65] which demonstrated this fact provides the first direct evidence that neutral substitutional Fe impurities in GaP can trap two electrons. In contrast, it has been concluded on the basis of optical studies of InP:Fe that the Fe^+ charge state or two-electron trap does not exist in this material.[59,60]

While the ESR and intracenter optical transitions have identified the charge states in which substitutional Fe occurs in the III-V compounds, photoconductivity, PLE spectroscopy, photo-quenching of ESR, extrinsic optical-absorption spectra, photocapacitance spectra and transport measurements have been used to establish an empirical basis for the Fe-related energy-level schemes shown in Figures 11, 20, and 37 for InP, GaAs, and GaP, respectively (see Section 3.3, 4.2, and 5.3). Optical-absorption thresholds for the $Fe^{3+} \rightarrow Fe^{2+}$ (^5E) charge transfer transition (Equations (4) and (7)) have been determined for GaAs:Fe and GaP:Fe from onsets for hole photoconductivity and for InP:Fe from the onset for hole conductivity observed in the photo-Hall effect spectrum. These thresholds fix the position of the Fe^{2+} (^5E) ground state relative to the top of the valence band for each of these III-V semiconductors. In each case, additional higher-energy onsets in photoconductivity, absorption, PLE, or photocapacitance are observed which correspond to the $Fe^{3+} \rightarrow Fe^{2+}$ ($^5T_2$) (excited state) (Equation (8)) charge-transfer transition. The most detailed studies of the charge-transfer transitions have been carried out in InP:Fe (see Section 3.3) where the onset of electron photoconductivity due to the Fe^{2+} (^5E) $\rightarrow Fe^{3+}$ (Equation (6)) charge-transfer transition has also fixed the position of the Fe^{2+} (^5E) ground state relative to the conduction band.

Eaves et al.[60] and Tapster et al.[59] have suggested the possibility of an $Fe^{3+} \rightarrow Fe^{4+}$ (Equation (9)) charge-transfer transition to explain a photo-conductivity onset near 1.1 eV in InP:Fe. (The possible existence of the Fe^{4+} charge state in an Fe-doped III-V semiconductor was proposed earlier by Okuno et al.[118] and Suto and Nishizawa[113] to explain the absence of an Fe^{3+} ESR signal in p-type samples of GaAs:Fe.) However, as pointed out at the end of Section 3.4, the dependence of this transition upon the thermal history[61] of the InP:Fe sample indicates that an Fe-related complex is a far more reasonable explanation for the level lying 0.25 eV above the valence band than an isolated Fe^{3+}. The evidence for the existence of the Fe^{4+} charge state and the charge transfer transition of Equation (9) is far from conclusive. A systematic study of Fe-related optical absorption, photoconductivity, PL, and ESR spectra in a series of InP:Fe samples ranging from semi-insulating to highly-p-type might shed some light on this question.

The observation that the efficiency of the characteristic Fe^{2+} PL band in InP:Fe when excited by above-gap light is correlated with the Fe^{3+} concentration led Bishop et al.[5,7] to propose an excitation mechanism whereby photoexcited electrons in the conduction band are captured by Fe^{3+} ions resulting in the formation of Fe^{2+} in the $^5T_2$ excited state (see Section 3.1). This proposal represents a solution to a specific case of the general problem of explaining how the recombination energy of photoexcited electron-hole pairs is transferred to the localized d-band states of transition metal impurities in semiconductors to produce the observed intracenter d-d PL transitions. Robbins and Dean[12] have discussed this problem in terms of the formation of bound "exciton" states at isoelectronic transition-metal impurity centers followed by a defect Auger recombination (DAR) process which transfers recombination energy to the core electrons of the impurity. For the interband excitation of Fe^{2+} PL in the III-V's, Klein and Weiser[123,134] have suggested a modification of the DAR model in which localization of an exciton at the isoelectronic Fe^{3+} center is followed by a DAR process which results in the "core" hole on the Fe^{3+} being driven deep into the valence band, leaving the center internally excited in the Fe^{2+} state. The excited Fe^{2+} ($^5T_2$) relaxes radiatively to Fe^{2+} (^5E) and is restored to the equilibrium Fe^{3+} state by hole capture. Exciton localization or bound-state formation at main-group (as opposed to transition-metal) isoelectronic impurities in semiconductors can be viewed in terms of sequential carrier capture.[135] The first carrier (e.g. an electron) is captured by the short-range central-cell potential of the isoelectronic impurity; the second carrier (e.g. a hole) is then captured by the Coulomb potential of the now charged center. Similarly, the isoelectronic charge state of a transition metal impurity in a semiconductor, such as Fe^{3+} ions on the group III site of the III-V's, can bind an exciton by this sequential capture process.

Whatever the actual process of carrier localization is which precedes the excitation of the Fe^{2+} PL by above-gap light, the data of Figure 1 indicate that the cross section for the process is quite significant. It is perhaps relevant that the ESR data for the isoelectronic Fe^{3+} center discussed in Section 3.2 indicate that the wavefunction of the Fe^{3+} ion is sufficiently extensive to allow a significant superhyperfine interaction with the second nearest neighbor coordination sphere.[14,15,25,27]

The efficient below-gap excitation observed in the PLE spectra of the Fe^{2+} PL in InP:Fe,[7] GaAs:Fe,[10] and GaP:Fe[128] are consistent with a simple ligand-to-metal charge-transfer process as discussed in Section 3.3. Of course, the assignment of the extrinsic PLE bands in these Fe-doped materials to this mechanism cannot be made with certainty on the basis of the PLE spectra alone. However, the author believes that the observation in all three of these III-V compounds of efficient extrinsic PLE bands for the Fe^{2+} PL with threshold energies that are in each case consistent with the valence band to Fe^{2+} ($^5T_2$) energy separation deduced from the independent experimental measurements cited previously justifies the conclusion that the extrinsic PLE is attributable to the valence band to Fe^{3+} charge transfer transitions which produce the excited state ($^5T_2$ of Fe^{2+}, which relaxes radiatively to the ground state (^5E) of Fe^{2+}. This conclusion allows one to construct the schematic level diagrams[128] of Figure 39 which compares the positions of the Fe^{2+} levels relative to the band edges in (a) GaP:Fe, (b) GaAs:Fe, and (c) InP:Fe.

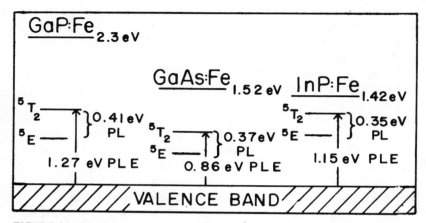

FIGURE 39 Scheme for the position of the Fe^{2+} levels relative to the band edges in (a) GaP:Fe, (b) GaAs:Fe, and (c) InP:Fe for T = 4.2K. After Shanabrook et al., Ref. 128.

8. Work in Progress and New Directions

As discussed in Section 4.3, Wagner et al.[107] have recently initiated high-magnetic-field (~10T) ESR studies of III-V semiconductors with a sub-millimeter molecular gas laser (11.223 cm^{-1}). With this technique, these workers have observed ESR lines attributable to transitions within the ^5E ground state multiplet of Fe^{2+} in GaAs and InP. This represents the first observation of the Fe^{2+} ($3d^6$) one-electron trap state in a III-V semiconductor by ESR spectroscopy. The technique has also achieved the first observation of a native defect, the As antisite, in GaAs[136] and is now established as a valuable adjunct to conventional microwave ESR. The studies of GaAs:Fe and InP:Fe are continuing.

The Fe^{2+} charge state in the III-V's is also being studied by time resolved PL techniques. Klein and Weiser have studied the time decay of the Fe^{2+} intracenter PL bands in GaP:Fe[123] (see Section 5.1), GaAs:Fe[134] and InP:Fe.[134] The radiative lifetimes of the excited Fe^{2+} ($^5T_2$) states in these III-V hosts are found to be $\approx 15\,\mu sec$, in agreement with lifetimes inferred from Fe^{2+} optical absorption measurements. Experimental results are consistent with the simple charge-transfer mechanism for below-gap excitation and with the defect Auger recombination (DAR) mechanism described previously for above-gap excitation. Nonradiative hole capture times are found to be $\approx 10\,\mu sec$. A highly interesting outgrowth of these time-resolved studies has been the recent observation by Klein and coworkers[137] of laser oscillations at 3.53 μm due to the internal Fe^{2+} ($^5T_2 \rightarrow {}^5E$) transitions in n-type InP:Fe at $\sim 2K$.

The broad range of experimental studies reviewed here has demonstrated that the deep levels found experimentally in Fe-doped semiconductors cannot be assigned to Fe impurities simply on the basis of the correlation of some property with the Fe impurity concentration. An identifiable ESR or optical signature is required for an unambiguous assignment. Arguments based on correlation with impurity concentration are not convincing because dopants can generate intrinsic lattice defects or form complexes as discussed in Section 3.5. The possible role of Fe-related complexes in controlling the properties of III-V semiconductors establishes such complexes as an important topic for continued investigation. For example, it is hoped that the recent work of Eaves and coworkers[93] which demonstrated that the 1.1–1.2 eV PL band in InP comprises two separate components attributable to Fe and Mn will stimulate additional efforts to identify the detailed nature of the Fe-related complex. Future studies of the formation or dissociation of Fe-related complexes as a function of thermal annealing conditions must take into account the fact that the Fe impurities in both undoped InP[11] and GaAs[10] and Fe-doped InP[68] have been shown to redistribute during annealing. The accumulation of Fe-rich layers at surfaces of III-V crystals during annealing[10,11,68] must be accounted for in the interpretation of, for example, PL data which are representative only of the surface of the sample.

In general, the best opportunities for obtaining improvements in our understanding of Fe impurity centers in III-V semiconductors will be provided by improved control of crystal-growth conditions and the application of new experimental techniques.

The availability of higher-purity crystals with carrier type and concentration (Fermi level) tailored for particular experimental requirements would remove many of the ambiguities of interpretation which presently afflict experimental investigations, e.g. PL studies, especially those concerning complexes. Examples of experimental techniques which can be applied profitably to the study of Fe impurities in III-V semiconductors include the high field, submillimeter ESR measurements now in progress[107] deep-level transient spectroscopy (DLTS)[59] and deep-level optical spectroscopy (DLOS),[80(b),99] optically detected magnetic resonance (ODMR), and electron nuclear double resonance (ENDOR). The photocapacitance and DLOTS experiments of Tapster and coworkers[59] and the DLOS work of Bremond et al.[80(b)] have clearly demonstrated (Section 3.4) the advantages of experiments which measure both the thermal activation energy for a trap (and determine whether hole or electron emission ensues), and the spectral threshold for photoionization of the same trap.

Studies similar to those of Refs. 59 and 80(b), for InP:Fe, and Ref. 99, for GaAs:Fe, are to be recommended for GaP:Fe. The 1.1–1.2 eV (Section 3.5) and 0.5 eV (Section 3.1) PL bands observed in InP:Fe are obvious examples of emission processes of unidentified origin to which ODMR could be applied. Finally, the work of Teuerle and Hausmann[66,67] remains as the only ENDOR study carried out on these materials and one hopes for further application of this powerful technique.

While the application of the specialized techniques listed above will certainly advance our understanding of Fe centers in III-V's, it may still be true that the most effective experimental tools are comprehensive studies in which a specifically tailored sample or series of samples is subjected to a coordinated investigation by the standard optical, magnetic resonance, photoconductivity, photocapacitance, transport, and DLTS techniques.

Acknowledgements. I would like to thank L. Eaves, G. Guillot, R. G. Humphreys, P. B. Klein and P. R. Tapster for providing preprints of their work prior to publication and for helpful discussions. In addition, I gratefully acknowledge valuable discussions with P. J. Dean, L. A. Hemstreet, R. L. Henry, P. R. Jay, T. A. Kennedy, J. J. Krebs, D. C. Look, B. D. McCombe, R. J. Nicholas, D. J. Robbins, J. Schneider, M. S. Skolnick, G. H. Stauss, R. J. Wagner, N. D. Wilsey, and P. Won Yu.

References

1. A. G. Milnes, *Deep Impurities in Semiconductors*, New York, Wiley (1977) and references therein.
2. P. J. Dean, A. M. White, B. Hamilton, A. R. Peaker, and R. M. Gibb, J. Phys. D: Appl. Phys. *10*, 2545 (1977).
3. U. Kaufmann and J. Schneider, Festkörperprobleme *Vol. XX*, J. Treusch (ed.), Vieweg, Braunschweig (1980), p. 87.
4. S. G. Bishop, P. J. Dean, P. Porteous, and D. J. Robbins, J. Phys. C: Solid St. Phys. *13*, 1331 (1980).
5. S. G. Bishop, R. L. Henry, P. B. Klein, and B. D. McCombe, Extended Abstracts of the Electrochem. Soc. Meeting, Seattle, Washington, 21–26 May 1978, *Vol. 78*-1,320; J. Electrochem. Soc. *125*, 95C (1978).
6. W. H. Koschel, S. G. Bishop, and B. D. McCombe, in *Proc. 13th Int. Conf. on Physics of Semiconductors*, E. G. Fumi (ed.), Tipografia Marves, Rome (1977), p. 1065.
7. S. G. Bishop, P. B. Klein, R. L. Henry, and B. D. McCombe, in *Semi-Insulating III-V Materials*, G. J. Rees (ed.), Shiva, Orpington (1980), p. 161.
8. E. Kubota and K. Sugii, J. Appl. Phys. *52*, 2983 (1981).
9. O. Mizuno and H. Watanabe, Electron. Lett. *11*, 118 (1975).
10. (a) P. E. R. Nordquist, P. B. Klein, S. G. Bishop, and P. G. Siebenmann, in *Gal-*

lium Arsenide and Related Compounds (1980), Inst. Phys. Conf. Ser. *No. 56*, 569 (1981); (b) P. W. Yu, J. Appl. Phys. *52*, 5786 (1981).
11. B. V. Shanabrook, P. B. Klein, P. G. Siebenmann, H. B. Dietrich, and S. G. Bishop, *Proc. 2nd Conf. on Semi-Insulating III-V Materials*, S. Makram-Ebeid and B. Tuck (eds.) Shiva, Nantwich (1982), p. 310.
12. D. J. Robbins and P. J. Dean, Adv. Phys. *27*, 499 (1978).
13. H. H. Woodbury and G. W. Ludwig, Bull. Am. Phys. Soc. *6*, 118 (1961).
14. M. deWit and T. L. Estle, Bull. Am. Phys. Soc. *7*, 449 (1962).
15. M. deWit and T. L. Estle, Phys. Rev. *132*, 195 (1963).
16. T. L. Estle, Phys. Rev. *136*, A1702 (1964).
17. V. K. Bashenov, Phys. Status Solidi A*10*, 9 (1972).
18. J. C. Hensel, Bull. Am. Phys. Soc. *9*, 244 (1964).
19. I. Fidone and K. W. H. Stevens, Proc. Phys. Soc. (London) *73*, 116 (1959).
20. H. Watanabe, J. Phys. Chem. Solids *25*, 1471 (1964).
21. G. H. Azarbayejani, H. Watanabe, and C. Kiskuchi, Bull. Am. Phys. Soc. *9*, 38 (1964).
22. See, for example, J. Schneider, S. R. Sircar, and A. Räuber, Z. Naturforsch. *18a*, 980 (1963); H. Kimmel, ibid. *18a*, 650 (1963); R. S. Title, Phys. Rev. *133*, A1613 (1964).
23. G. W. Ludwig and M. R. Lorentz, Phys. Rev. *131*, 601 (1963).
24. J. W. Orton, *Electron Paramagnetic Resonance*, Gordon and Breach, New York (1969), pp. 20, 61.
25. W. H. Koschel, U. Kaufmann, and S. G. Bishop, Solid State Commun. *21*, 1069 (1977).
26. G. H. Stauss, J. J. Krebs, and R. L. Henry, Phys. Rev. B*16*, 974 (1977).
27. G. K. Ippolitova, E. M. Omel'yanovskii, N. M. Pavlov, A. Ya. Nashel'skii, and S. V. Yakobson, Sov. Phys. Semicond. *11*, 773 (1977).
28. J. M. Baranowski, J. W. Allen, and G. L. Pearson, Phys. Rev. *160*, 627 (1967).
29. W. Low and M. Weger, Phys. Rev. *118*, 1119 (1960).
30. G. A. Slack, S. Roberts, and J. T. Vallin, Phys. Rev. *187*, 511 (1969).
31. C. L. West, W. Hayes, J. F. Ryan, and P. J. Dean, J. Phys. C: Solid St. Phys. *13*, 5631 (1980).
32. C. J. Ballhausen, *Introduction to Ligand Field Theory*, McGraw-Hill, New York (1962).
33. F. S. Ham and G. A. Slack, Phys. Rev. B*4*, 777 (1971).
34. G. A. Slack, F. S. Ham, and R. M. Chrenko, Phys. Rev. *152*, 376 (1966).
35. G. A. Slack, S. Roberts, and F. S. Ham, Phys. Rev. *155*, 170 (1967).
36. G. A. Slack and B. M. O'Meara, Phys. Rev. *163*, 335 (1967).
37. (a) E. M. Ganapol'skii, Sov. Phys.-Solid State *15*, 269 (1973); (b) E. M. Ganapol'skii, E. M. Omel'yanovskii, L. Ya. Pervova, and V. I. Fistul, Sov. Phys. Semicond. *7*, 1099 (1974).
38. V. W. Rampton and P. C. Wiscombe, Acta Phys. Slov. *32*, 35 (1982).
39. J. B. Mullin, R. J. Heritage, C. H. Holliday, and B. W. Straughan, J. Cryst. Growth *3/4*, 281 (1968).
40. R. L. Henry and E. M. Swiggard, J. Electron. Mat. *7*, 647 (1978).
41. R. W. Haisty and G. R. Cronin, Proc. 7th Int. Conf. on Physics of Semicond., M. Hulin (ed.), Dunod, Paris (1964), p. 1161.
42. R. W. Haisty, Appl. Phys. Lett. *7*, 208 (1965).
43. H. Strack, Transact. Metallurg. Soc. AIME *239*, 381 (1967).

44. G. A. Allen, J. Phys. D: Appl. Phys. *1*, 593 (1968).
45. M. A. Kadyrov, E. M. Omel'yanovskii, L. Ya. Pervova, N. N. Solov'ev, and V. I. Fistul, Sov. Phys. Semicond. *2*, 713 (1968).
46. S. M. Sze and J. C. Irvin, Solid State Electron. *11*, 599 (1968).
47. E. M. Omel'yanovskii, L. Ya. Pervova, E. P. Rashevskaya, N. N. Solov'ev, and V. I. Fistul, Sov. Phys. Semicond. *4*, 316 (1970).
48. E. M. Omel'yanovskii, N. M. Pavlov, N. N. Solov'ev, and V. I. Fistul, Sov. Phys. Semicond. *4*, 439 (1970).
49. N. M. Kolchanova, D. N. Nasledov, and G. N. Talalakin, Sov. Phys. Semicond. *4*, 106 (1970).
50. N. I. Kukicheva, O. V. Pelevin, and L. Ya. Pervova, Sov. Phys. Semicond. *5*, 169 (1971).
51. W. Plesiewicz, Phys. Status Solidi A*16*, 485 (1973).
52. D. V. Lang and R. A. Logan, J. Electron. Mat. *4*, 1053 (1975).
53. K. P. Pande and G. G. Roberts, J. Phys. C.: Solid St. Phys. *9*, 2899 (1976).
54. B. T. Debney and P. R. Jay, in *Semi-Insulating III-V Materials*, G. J. Rees (ed.) Shiva, Orpington (1980), p. 305.
55. R. L. Henry, private communication.
56. V. A. Bykovskii, V. A. Vil'kotskii, D. S. Domanevskii, and V. D. Tkachev, Sov. Phys. Semicond. *9*, 1204 (1976).
57. G. K. Ippolitova and E. M. Omel'yanovskii, Sov. Phys. Semicond. *9*, 156 (1975).
58. G. W. Iseler, in *Gallium Arsenide and Related Compounds*, C. M. Wolfe (ed.) Inst. Phys. (London) Conf. Ser. No. 45 (1979), p. 144.
59. P. R. Tapster, M. S. Skolnick, R. G. Humphreys, P. J. Dean, B. Cockayne, and W. T. MacEwan, J. Phys. C: Solid State Phys. *14*, C5069 (1981).
60. L. Eaves, A. W. Smith, P. J. Williams, B. Cockayne, and W. R. MacEwan, J. Phys. C.: Solid State Phys. *14*, C5063 (1981).
61. M. S. Skolnick, private communication.
62. P. Leyral, G. Bremond, A. Nouailhat, and G. Guillot, J. Luminescence, *24/25*, 245 (1981).
63. R. Dingle, Phys. Rev. Lett. *23*, 579 (1969).
64. S. G. Bishop, D. J. Robbins, and P. J. Dean, Solid State Commun. *33*, 119 (1980).
65. U. Kaufmann and J. Schneider, Solid State Commun. *21*, 1073 (1977).
66. W. Teuerle, E. Blaschke, and A. Hausmann, Z. Phys. *270*, 37 (1974).
67. W. Teuerle and A. Hausmann, Z. Phys. B*23*, 11 (1976).
68. D. E. Holmes, R. G. Wilson, and P. W. Yu, J. Appl. Phys. *52*, 3396 (1981); J. D. Oberstar, B. G. Streetman, J. E. Baker and P. Williams, J. Elec. Chem. Soc. *138*, 1814 (1981).
69. D. C. Look, Phys. Rev. B*20*, 4160 (1979).
70. D. C. Look, Solid State Commun. *33*, 237 (1980).
71. F. S. Ham, Phys. Rev. *138*, A1727 (1965).
72. S. Fung, R. J. Nicholas, and R. A. Stradling, J. Phys. C: Solid St. *12*, 5145 (1979).
73. P. Won Yu in *Semi-Insulating III-V Materials*, G. J. Rees (ed.) Shiva, Orpington (1980), p. 167.
74. G. Bremond, A. Nouailhat, G. Guillot, and B. Cockayne, Third "Lund" Int. Conf. on Deep Level Impurities in Semiconductors, Southbury, Conn., 26–29 May (1981) published abstracts, p. 72.
75. N. S. Grushko and A. A. Gutkin, Sov. Phys. Semicond. *8*, 1179 (1975).

76. S. H. Chiao and G. A. Antypas, J. Appl. Phys. *49*, 466 (1978).
77. A. M. White, A. J. Grant, and B. Day, Electron. Lett. *14*, 409 (1978).
78. A Majerfeld, O. Wada, and A. N. M. M. Choudbury, Appl. Phys. Lett. *33*, 957 (1978).
79. A. N. M. M. Choudbury and P. N. Robson, Electron. Lett. *15*, 277 (1979).
80. (a) G. Bremond, A. Nouailhat, and G. Guillot, Electron. Lett. *17*, 55 (1981); (b) G. Bremond, A. Nouailhat, G. Guillot, and B. Cockayne, Solid State Commun. *41*, 477 (1982).
81. Y. Yamazoe, Y. Sasai, T. Nishino, and Y. Hamakawa, Japan J. Appl. Phys. *20*, 347 (1981).
82. L. Eaves, T. Englert, T. Instone, C. Uihlein, P. J. Williams, and H. Wright, in *Semi-Insulating III-V Materials*, G. J. Rees (ed.) Shiva, Orpington (1980), p. 145.
83. N. Killoran, B. C. Cavenett, and W. E. Hagston, ibid., p. 190.
84. A. M. White, Solid State Commun. *32*, 205 (1979).
85. G. Picoli, B. Deveaud, and D. Galland, in *Semi-Insulating III-V Materials*, G. J. Rees (ed.) Shiva, Orpington (1980), p. 254; J. Physique *42*, 133 (1981).
86. H. Ennen, U. Kaufmann, and J. Schneider, Appl. Phys. Lett. *38*, 355 (1981).
87. L. A. Demberel, A. S. Popov, D. B. Kushev, and N. N. Zheleva, Phys. Status Solidi A*52*, 341 (1979).
88. P. Won Yu, Solid State Commun. *34*, 183 (1980).
89. T. A. Kennedy and N. D. Wilsey, in *Defects and Radiation Effects in Semiconductors*, 1980, R. R. Hasiguti (ed.) Inst. Phys. (London) Conf. Ser. No. *59*, 257 (1981).
90. E. W. Williams, Phys. Rev. *168*, 922 (1968).
91. C. J. Hwang, Phys. Rev. *180*, 827 (1969).
92. T. H. Keil, Phys. Rev. A*140*, 601 (1965).
93. (a) L. Eaves, A. W. Smith, N. S. Skolnick, and B. Cockayne, J. Appl. Phys. *53*, 4955 (1982); (b) L. Eaves, A. W. Smith, M. S. Skolnick, C. R. Whitehouse, and B. Cockayne, *Proc. 2nd Conf. on Semi-Insulating III-V Materials*, S. Makram-Ebeid and B. Tuck (eds.) Shiva, Nantwich (1982), p. 208.
94. G. W. Ludwig and H. H. Woodbury, in *Solid State Physics*, F. Seitz and D. Turnbull (eds.), Academic, New York and London (1962), Vol. 13, p. 223.
95. A. Ingelmund, unpublished thesis, Univ. Aachen (1979).
96. A. Mittoneau, G. M. Martin, and A. Mircia, Electron. Lett. *13*, 666 (1977).
97. M. Takikawa and T. Ikoma, Japan J. Appl. Phys. *19*, L436 (1980).
98. V. A. Bykovskii, V. A. Vil'kotskii, D. S. Domanevskii, and V. D. Tkachev, Sov. Phys. Semicond. *9*, 1204 (1976).
99. P. Leyral, F. Litty, G. Bremond, A. Nouailhat, and G. Guillot, *Proc. 2nd Conf. on Semi-Insulating III-V Materials*, S. Makram-Ebeid and B. Tuck (eds.) Shiva, Nantwich (1982), p. 192.
100. V. I. Fistul', L. Ya. Pervova, E. M. Omel'yanovskii, E. P. Rashevskaya, N. N. Solov'ev, and O. V. Pelevin, Sov. Phys. Semicond. *8*, 311 (1974).
101. D. G. Andrianov, E. M. Omel'yanovskii, E. P. Rashevskaya, and N. I. Suchkova, Sov. Phys. Semicond. *10*, 637 (1976).
102. A. A. Lebedev, F. A. Akhmedov, and M. M. Akhmedova, Sov. Phys. Semicond. *10*, 1028 (1976).
103. K. Kitahara, K. Nakai, M. Ozeki, A. Shibatomi, and K. Dazai, Japan. J. Appl. Phys. *15*, 2275 (1976).
104. H. G. Grimmeiss and L-A. Ledebo, J. Appl. Phys. *46*, 2155 (1975).
105. G. Lucovsky, Solid State Commun. *3*, 299 (1965).

106. J. J. Krebs and G. H. Stauss, unpublished results.
107. R. J. Wagner, J. J. Krebs, and G. H. Stauss, Bull. Amer. Phys. Soc. 27, 277 (1982).
108. B. I. Boltaks, A. A. Efimov, P. P. Seregin, and V. T. Shipatov, Sov. Phys.-Solid State 12, 1592 (1970).
109. P. P. Seregin, F. S. Nasredinov, and A. Sh. Bakhtiyarov, Phys. Status Solidi B91, 35 (1979).
110. A. V. Vasil'ev, G. K. Ippolitova, E. M. Omel'yanovskii, and A. I. Ryskin, Sov. Phys. Semicond. 10, 713 (1976).
111. D. C. Andrianov, P. M. Grinshtein, G. K. Ippolitova, E. M. Omel'yanovskii, N. I. Suchkova, and V. I. Fistul, Sov. Phys. Semicond. 10, 696 (1976).
112. M. G. Clark and P. J. Dean, in Proc. 14th Int. Conf. on Physics of Semicond., Edinburgh (1976), B. L. H. Wilson (ed.) Inst. Phys. (London) Conf. Ser. No. 43, 291 (1979).
113. K. Suto and J. Nishizawa, J. Appl. Phys. 43, 2247 (1972).
114. S. A. Abagyan, G. A. Ivanov, and Yu. N. Kuznetsov, Sov. Phys. Semicond. 10, 1283 (1976).
115. F. S. Shishiyanu and V. G. Georgiu, Sov. Phys. Semicond. 10, 1301 (1976).
116. V. I. Kirillov, N. N. Pribylov, and S. I. Rembeza, Sov. Phys. Semicond. 11, 1190 (1977).
117. V. F. Masterov, Yu. V. Mal'tsev, and I. B. Rusanov, Sov. Phys. Semicond. 14, 330 (1980).
118. Y. Okuno, K. Suto, and J. Nishizawa, J. Appl. Phys. 44, 832 (1973).
119. K. Suto and J. Nishizawa, J. Phys. Soc. Japan 27, 924 (1969).
120. H. Ennen and U. Kaufmann, J. Appl. Phys. 51, 1615 (1980).
121. R. F. Brunwin, B. Hamilton, J. Hodgkinson, A. R. Peaker, and P. J. Dean, Solid State Electron. 24, 249 (1981).
122. M. Gershenzon, R. A. Logan, D. F. Nelson, and F. A. Trumbore, Proc. Int. Conf. on Luminescence, Budapest, 1966, Akademiai Kiado, Budapest (1968), p. 1737.
123. P. B. Klein and K. Weiser, Solid State Commun. 41, 365 (1982).
124. V. F. Masterov and V. K. Sobolevskii, Sov. Phys. Semicond. 13, 965 (1979).
125. S. Haraldson and C. G. Ribbing, J. Phys. Chem. Solids 30, 2419 (1969).
126. F. S. Ham, G. W. Ludwig, G. D. Watkins, and H. H. Woodbury, Phys. Rev. Lett. 5, 468 (1960).
127. B. Tell and F. P. J. Kuijpers, J. Appl. Phys. 49, 5938 (1978).
128. B. V. Shanabrook, P. B. Klein, and S. G. Bishop, Proc. 12th Int. Conf. on Defects in Semiconductors, Amsterdam, Physica 116B, 444 (1983).
129. S. T. Pantelides, Rev. Mod. Phys. 50, 797 (1978).
130. V. F. Masterov and B. E. Samorukov, Sov. Phys. Semicond. 12, 363 (1978).
131. (a) L. A. Hemstreet, Phys. Rev. B22, 4590 (1980); (b) L. A. Hemstreet, Proc. 12th Int. Conf. on Defects in Semiconductors, Amsterdam, Physica 116B, 116 (1983).
132. A. Fazzio and J. R. Leite, Phys. Rev. B21, 4710 (1980).
133. D. L. Partin, J. W. Chen, A. G. Milnes, and L. F. Vassamillet, Solid State Electron. 22, 455 (1979).
134. P. B. Klein and K. Weiser, to be published.
135. J. J. Hopfield, D. G. Thomas, and R. T. Lynch, Phys. Rev. Lett. 17, 312 (1966).
136. R. J. Wagner, J. J. Krebs, and G. H. Stauss, Solid State Commn. 36, 15 (1980).
137. P. B. Klein, J. E. Furneaux, and R. L. Henry, Appl. Phys. Lett. 42, 638 (1983); Phys. Rev. B29, 1947 (1984).

CHAPTER 9

Chromium in GaAs

J. W. Allen

Wolfson Institute of Luminescence
Department of Physics
University of St. Andrews
Fife, Scotland

1. INTRODUCTION
2. CONFIGURATIONS
   2.1 The d^2 configuration
   2.2 The d^3 configuration
   2.3 The d^4 configuration
   2.4 The d^5 configuration
3. IONIZATION LEVELS
   3.1 The d^3-d^2 level
   3.2 The d^4-d^3 level
      3.2.1 Introduction
      3.2.2 Hall effect and resistivity
      3.2.3 Photoionization
      3.2.4 Capture and thermal emission
      3.2.5 Summary
   3.3 The d^5-d^4 level
4. LUMINESCENCE
   4.1 Introduction
   4.2 The 0.84 eV luminescence
   4.3 The 0.56 eV luminescence
5. OTHER PROPERTIES
6. CONCLUSIONS

1. Introduction

Since Haisty and Cronin[1,2] first observed that semi-insulating gallium arsenide could be made by doping with chromium, many investigations have
been made of the properties of GaAs:Cr. One motive is that material with
resistivity close to the theoretical maximum for GaAs can be made and
this is used as a substrate for GaAs devices, especially field effect transistors. It must be admitted, however, that the detailed scientific studies of
GaAs:Cr have so far had little effect on the manufacture of semi-insulating
material, which is still largely dependent on empirical processes. A second
motive is that chromium can exist in GaAs in a number of different charge
states, each with a set of excitation levels, and as a result there are levels in
the gap corresponding with ionization from one charge state to another.
This raises interesting problems concerning a system in which the single-
particle approximation, the basis of most semiconductor physics today,*
breaks down. A third motive, perhaps trivial yet still strong, is that GaAs:Cr
presents intellectual puzzles in the unravelling of near-coincidences. One
ionization level is near the center of the gap, so transitions to the valence
and conduction bands occur over the same energy range; the electron-spin
resonance signal expected for the d^2 and d^5 configurations are almost
identical; an excitation level and an ionization level both lie close to the
conduction band edge. These properties have led to confusion which has
required patient and ingenious work to sort out.

Incorrect application of ideas from one-electron models leads to error
and apparent contradictions. Therefore the terminology used here and its
basis will be described in some detail. For illustrative examples we use a
model in which there is a group of electrons on the impurity with properties sufficiently similar to those in the d-shell of a free atom for them to
be called d-electrons. All other electrons are in states which form the valence and conduction bands of the host crystal. A free chromium atom has
an outer electron configuration $3d^5 4s^1$. When the atom replaces a gallium
atom in GaAs, three electrons go into bonding orbitals, i.e. into the valence
band, leaving a $3d^3$ configuration. The experimental data can be interpreted in terms of changes of occupancy of the d-shell and of excitations
within the d-shell. In practice, there is mixing between the d-electrons and
the valence electrons, so the d-electrons no longer have an exact orbital
angular momentum $\ell = 2$, and some states in the conduction and valence
bands are modified by d-admixture. (Formal theoretical models illustrating these points have been constructed by Haldane and Anderson[3]
and by Fleurov and Kikoin.[4-6]) Despite this it is still useful to retain the
nomenclature pertaining to d-electrons. The model described above was
proposed by Ludwig and Woodbury[7] to explain some properties of tran-

sition metal impurities in silicon. George Watkins has suggested that in other impurity-host combinations a more appropriate model might be one in which removal of the host atom leaves a vacancy with a certain set of energy levels and wavefunctions. The impurity sits in the vacancy and, although it modifies the vacancy levels and wavefunctions, it does so to a sufficiently small extent that they are still identifiable. The impurity introduces its own levels and wavefunctions: possibly none of the levels lies above the top of the valence band. So far, it has been easier to interpret data on GaAs:Cr in terms of the d-electron/bonding-electron model than of the modified vacancy model.

For brevity, the configuration will be denoted by the occupancy of the d-shell, e.g. d^3. There are several possible states of a configuration, with a ground state and excited states. Sometimes here the word "state" with no further qualification is used to denote the ground state of a configuration. Different levels correspond to different states within a configuration. Table 1 shows alternative nomenclatures for the configurations. In semiconductor notation the charge on the impurity has an operational significance: sufficiently far from the impurity there is an additional Coulomb field (usually screened) corresponding to the charge on the impurity. This charge is shown in Table 1. Unfortunately, it has become common to use an ionic notation in describing spin resonance experiments, founded on the fiction that a GaAs crystal can be divided naturally into volumes containing either a Ga^{3+} or As^{3-} ion. Because of covalency, such a division is artificial. The ionic notation is also shown in Table 1 but will not be used here, as it leads to illogicalities such as "the neutral Cr^{3+} ion." Clark[8] suggests the use of the oxidation state, as in inorganic chemistry. This is not unambiguous, as for example Cr(III) means different configurations in atomic spectroscopy and inorganic chemistry, and the number has no immediate operational significance in semiconductor physics. Hence, in this review, the d-shell occupancy is used, if necessary with the charge as in $(d^2)^+$, even though this implies a model of bonding. A notation just show-

TABLE 1 Various ways of denoting different configurations of Cr in GaAs.

Configuration	Charge state	Ionic notation	Oxidation state	Kaufmann/ Schneider	Donor/ acceptor
d^2	Cr^+	Cr^{4+}	IV	A^+	D^+
d^3	Cr^o	Cr^{3+}	III	A^o	N^o
d^4	Cr^-	Cr^{2+}	II	A^-	A^-
d^5	Cr^{2-}	Cr^+	I	A^{--}	A^{--}

ing whether the state has donated or accepted an electron or electrons, also
shown in Table 1, has the advantage of not implying a model of bonding
and is especially useful in discussions where the validity of a model has not
yet been established.

In atomic physics, it is customary to take the state with lowest total
energy of a particular configuration as ground state, and to relate all
possible eigenenergies to this ground state. Different continua thresholds
exist, corresponding with different ionization energies. For example, if the
$'S$ ground state of neutral helium is taken to have zero energy, there is a
first ionization threshold, leaving He^+ls, and a second at higher energy
leaving He^{2+}. In semiconductor physics, it is more convenient to keep say
the bottom of the conduction band continuum as a single reference level,
irrespective of the state of the impurity, so that transport properties may
be more easily discussed. A proper definition of ionization energies, i.e.
of the energy levels of an impurity with respect to the conduction or
valence edge, is now required. The experimentally accessible energies are
differences in total energies of the system. (When the Hartree-Fock approximation is a good one, Koopmans' theorem states that the difference
between total energies is the same as the difference in single-particle energies, but for transition metal impurities the Hartree-Fock approximation is
inadequate.) When an electron is removed from an impurity with configuration i^n (e.g. d^4), resulting in a configuration i^{n-1} (e.g. d^3) together with
an electron at the bottom of the conduction band far from the impurity,
then we define the energy level as having energy

$$E_c - E_i \equiv E(i^{n-1}, e_{cb}) - E(i^n) \qquad (1)$$

below the conduction band edge.[9-11] An analogous expression defines the
position of the energy level above the top of the valence band. If interaction with lattice vibrations is not negligible, it is necessary to specify a
vibrational state as well as an electronic state in (1). There is no longer a
one-to-one correspondence between states and levels as there is in Hartree-
Fock systems: instead, the level is associated with two states. For instance,
one cannot speak of the d^4 energy level without specifying an additional
convention of nomenclature, but one can speak of the d^4-d^3 level. There
is confusion in the semiconductor literature as some authors denote a level
i^n-i^{n-1} by the single configuration i^n, while others denote the same level
by i^{n-1}, often but not always depending on whether the author is
considering a transition to the conduction band or from the valence band.
In this review the notation i^n-i^{n-1} will be used to avoid ambiguity.
Figure 1 shows schematically the relation between states and levels. It is

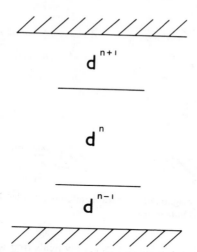

FIGURE 1 Schematic diagram of the relation of states to ionization levels.

clear that the number of levels is one fewer than the number of corresponding states.

There are two complications in describing ionization levels and excitation levels simultaneously, introduced by the use of the band edges as reference states. The first is that in the independent-particle band diagram electron energies increase upwards and hole energies increase downwards, so excitations within a configuration can be depicted as energy increasing upward or downward, depending on whether electrons or holes are being considered. The second is that the ionization level defined by (1) is characteristic of two configurations and therefore not necessarily equatable with the ground state energy of one configuration. The problems are particularly important when excitation and ionization are considered together, as in autoionization or photothermal ionization. An example is the discussion by Gloriozova and Kolesnik[12] of hole excitations in GaP:Cr and GaAs:Cr. A consistent scheme is achieved if excitations of the d^n configuration are reckoned upwards (i.e. as electron energies) from the d^n-d^{n-1} level, or if the excitations of the d^{n-1} configuration are reckoned downwards (i.e. as hole energies) from the same level. Figure 2 illustrates this for an example which, though hypothetical, could be GaAs:Cr. If the impurity is in the d^4 configuration, then, in the example, an excitation within d^4 requires more energy than is required to remove an electron from the configuration, so autoionization into the conduction band occurs, but it is insufficient to raise an electron from the valence band onto the impurity, so d^4 excitation does not lead to a hole autoionization.

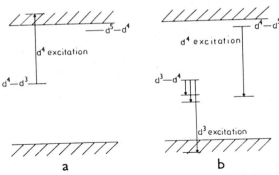

FIGURE 2   Schematic diagram of excitation and ionization levels in terms of
(a) electron energies, (b) hole energies.

In accordance with these ideas, this review deals first with configurations
and their excitations, then with the levels associated with changes in the
charge state. Although diffusion experiments show that interstitial
chromium cannot be neglected in GaAs:Cr, no firm information is as yet
available about its electronic properties. Therefore in the earlier sections
only isolated substitutional chromium is discussed. A later section describes
some optical properties attributable to a complex including substitutional
chromium, and another describes a luminescence whose origin, other than
that it apparently involves chromium, has not been definitely established.
To keep the review within reasonable bounds, only the energy levels of
GaAs:Cr and transitions between them are discussed in any detail.
Thermodynamic properties such as solubility, or diffusion properties, are
omitted although a fair amount is now known of them. Only brief reference will be made to properties like electrical conductivity which depends
on the energy levels. Theories of transition metal impurities in semiconductors require a full review and their application to GaAs:Cr cannot be
treated properly here.

2. Configurations

2.1 The d^2 Configuration

The ground state of the d^2 configuration in $T_d(\overline{4}3m)$ symmetry is an orbital singlet $^3A_2$ (Figure 3). An isotropic esr signal with $g \simeq 2$ is seen in
the dark in p-type GaAs:Cr when the Fermi level is near the valence

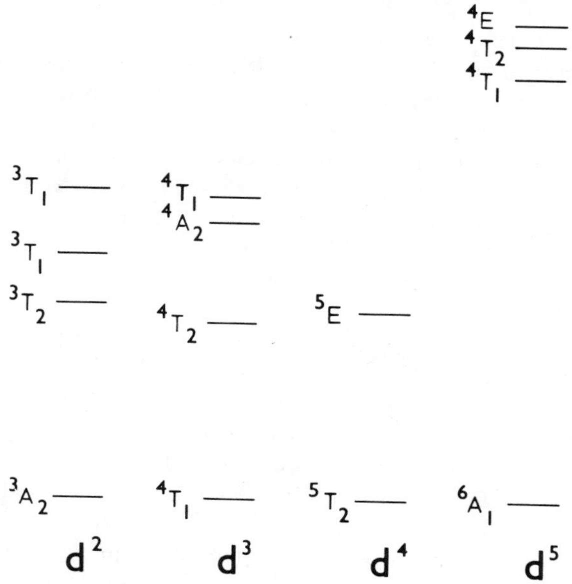

FIGURE 3 Crystal-field levels of the configurations d^2, d^3, d^4, d^5 in the absence of spin-orbit coupling. Except for d^5, only levels of states of maximum spin are shown. (See for example J. S. Griffith, "The Theory of Transition-Metal Ions," Cambridge University Press 1964.)

band.[13,14] It is also seen in semi-insulating GaAs:Cr optically excited with $h\nu > 0.75$ eV. The concentration of d^2 centers determined from the strength of the esr signal is equal to the chromium concentration in p-type GaAs:Cr. It is therefore fairly well established that the spin resonance is that of Cr d^2.

At 4.2 K the g-value is 1.994 ± 0.001 and the width of the resonance is 110 gauss, too wide for hyperfine structure to be resolved. So far there have been no reports of optical transitions between the spin-triplet levels shown in Figure 3 so the crystal field and other parameters are unknown.

Earlier, the esr signal of Cr d^2 had been observed after photoexcitation in semi-insulating GaAs and had been attributed to Cr d^5.[15,16] One possible mechanism of photoexcitation is photoionization to the conduction band of an electron from a Cr d^4 impurity followed by capture at another Cr d^4 impurity, i.e.

$$d^4 + d^4 + h\nu \rightarrow d^5 + d^3. \qquad (2)$$

Doubts about the assignment were raised by Frick and Siebert,[17] who found the esr signal even when the Fermi level was low enough for the Fe d^5 esr signal to be seen. The observation of the signal in low-resistivity p-type material by Kaufmann and Schneider and by Stauss et al. showed definitely that the assignment to Cr d^5 is incorrect. Among a number of possible photoionization mechanisms in semi-insulating material, a simple one is the raising of an electron from the valence band to a d^3 center, followed by capture of the hole at another d^3 center, i.e.

$$d^3 + d^3 + h\nu \rightarrow d^4 + d^2. \tag{3}$$

The threshold photon energy is the energy of the d^4-d^3 level above the valence band.

Although Cr d^2 in GaAs has lost an electron and is positively charged, no reports of its donor action have yet appeared.

It is unsatisfactory that our knowledge of Cr d^2 in GaAs comes almost entirely from a single unresolved esr line. More work, for example on optical excitation within the configuration and on the donor action, is desirable. Until more data are available, there is still room for alternative explanations of existing data, for example that a low chemical potential might drive substitutional chromium into an interstitial site with d^5 configuration acting as a donor.

2.2 The d^3 Configuration

The ground state of substitutional Cr d^3 in T_d(43m) symmetry is an orbital triplet $^4T_1$ and is therefore susceptible to Jahn-Teller distortion. Krebs and Stauss[18,19] observed an esr signal with orthorhombic C_{2v}(mm) symmetry in semi-insulating GaAs:Cr when the Fermi level was not too high. The signal is weak in unstrained crystals but becomes strong when uniaxial stress is applied.

Krebs and Stauss interpreted their esr spectra in terms of an orthorhombic center with principal axes [001], [110] and [1$\bar{1}$0]. A simple model is that the chromium atom is displaced along a cube axis direction from the center of the tetrahedron of nearest neighbors (Figure 4). There are six equivalent distortions, corresponding with the six cube axis directions. The six energy minima are separated from each other by low saddle points so that reorientation of the center can occur in a time shorter than experimental times even at 1.8 K. Uniaxial stress removes the equivalence of the six minima. Reorientation becomes a thermally activated process and the system goes from a dynamic to a quasi-static Jahn-Teller state.

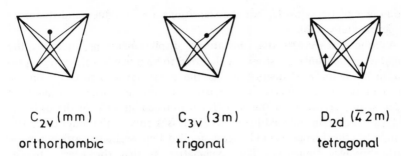

C_{2v} (mm)

orthorhombic

C_{3v} (3m)

trigonal

D_{2d} ($\bar{4}$2m)

tetragonal

FIGURE 4   Simple models of symmetry-lowering distortions of a regular tetrahedral environment.

An orthorhombic Jahn-Teller distortion of a T_i state is not energetically favored in a linear coupling theory. It is necessary to go to a non-linear theory,[20] and then coupling to E and T_2 vibrational modes simultaneously is possible. Clearly a theory with linear and non-linear Jahn-Teller coupling, spin-orbit coupling, magnetic field and uniaxial stress is not easy. Krebs and Stauss used a model with a static distortion as a first approximation. A T_1 orbital triplet is split into three orbital singlets by a C_{2v} (mm) crystal field. If this splitting is much greater than the spin-orbit splitting then the spin Hamiltonian is

$$\mathcal{H} = \mu_B \sum_i g_i H_i S_i + D[S_z^2 - (1/3) S(S+1)] + E(S_x^2 - S_y^2). \qquad (4)$$

The zero-field terms in D and E split the lowest level into two doublets. If the spacing of the doublets is much greater than the maximum photon energy causing the spin-resonance transitions, then the spin resonance is within the lowest doublet, which behaves as an effective spin $S' = \frac{1}{2}$ level with spin Hamiltonian

$$\mathcal{H} = \mu_B \, H.g'.S'. \qquad (5)$$

The effective g-value components found are 5.15, 2.37 and 1.64 along [001], [110] and [1$\bar{1}$0] respectively. From these values it is estimated that the ground state doublet splitting is roughly 10 cm^{-1} (1.2 meV). A rough estimate of 8500 cm^{-1} (1.0 eV) for the Jahn-Teller splitting of the $^4T_1$ level was obtained from the stress experiments, but this could be substantially wrong because of the assumptions which had to be made in the estimate. This value is greater than the expected value of the $T_d(\bar{4}3m)$ crystal field splitting Δ. Optical absorption measurements to clarify this point would be valuable, but the transitions are forbidden and therefore experimentally weak.

A curious feature is that the atomic displacements produced by the highest applied uniaxial stress are deduced to be much less than those produced by Jahn-Teller distortion, yet the stress has a large effect on the reorientation rate. If macroscopic elastic constants are used to estimate the microscopic strain, then the Jahn-Teller induced strain is of the order of 0.1 and the strain induced by external stress is only of the order of 0.001. Similarly, the highest [111] stress applied had negligible effect on the orthorhombic symmetry. The explanation is that the energy barrier between equivalent minima is much less than the total Jahn-Teller displacement energy of an individual minimum.

An alternative explanation of the orthorhombic symmetry, that it is produced by the effect of another impurity in a $<100>$ direction is ruled out by two observations. The first is the rapid reorientation rate, the second is the conversion by light of Cr d^3 into Cr d^4 with different symmetry. Both occur at helium temperatures at a rate far too high for the movement of an impurity between the requisite atomic positions.

2.3 The d^4 Configuration

The properties of Cr d^4 in II-VI compounds have been extensively studied both by electron spin resonance and by optical spectroscopy.[21-31] (See the review by J. Baranowski in this volume.) This has greatly helped the study of Cr d^4 in GaAs. Although Cr d^4 is a neutral center in a II-VI compound and is singly negatively charged in a III-V compound, the pattern of levels of internal transitions should not be affected by the charge, but the quantitative energy splittings will be.

In T_d ($\overline{4}3m$) symmetry the d^4 configuration is split by the crystal field into a $^5T_2$ ground state and ^5E excited state (Figure 3), together with lower spin states at higher energy. Both $^5T_2$ and ^5E are subject to Jahn-Teller distortion. Because of the Ham effect, the first-order spin-orbit interaction is quenched. An E-mode coupling with the $^5T_2$ ground state leads to a static Jahn-Teller effect. The distortion is tetragonal with three equivalent minima along the cube axes. Within a minimum the local symmetry is D_{2d}($\overline{4}2m$) and $^5T_2$ (T_d) splits into an orbital single $^5\hat{B}_2$ and an orbital doublet $^5\hat{E}$. Figure 5 shows a cross-section of the energy surfaces in normal coordinate space in the absence of spin-orbit coupling. The ^5E (T_d) excited state has a much weaker Jahn-Teller coupling and the energy surfaces have the "Mexican hat" form. Figure 5 also shows the cross-section of this surface. The orbital doublet ^5E splits into two singlets $^5\hat{A}_1$ and $^5\hat{B}_1$: which is the lower in the cross-section shown depends on the sign of

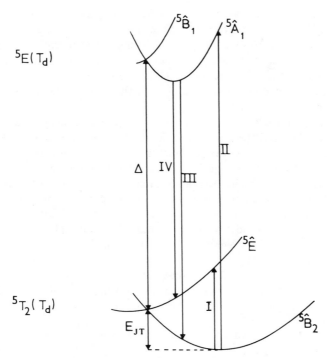

FIGURE 5 The effect of Jahn-Teller distortion on the quintet levels of d^4. Optical transitions I and II can occur in absorption and III and IV in luminescence.

the Jahn-Teller coupling coefficient. Electron spin resonance demonstrates the tetragonal symmetry of the ground state and the structure of the spectrum is determined by second-order spin-orbit and by spin-spin splittings.

Transitions I-IV of Figure 5 have been observed in the case of Cr in II-VI compounds (see Baranowski, this volume). There is an optical absorption (transition I) at ~ 1000 cm^{-1} (0.12 eV) between the Jahn-Teller split levels of $^5T_2$. As the transition is at $3E_{JT}$, where E_{JT} is the Jahn-Teller energy, this gives a simple method of determining E_{JT}. Another optical absorption (transition II) between $^5\hat{B}_2$ and $^5\hat{A}_1$ occurs at about 5500 cm^{-1} (0.68 eV). (The $^5\hat{B}_2$-$^5\hat{B}_1$ electric dipole transition is forbidden.) Its energy depends on both Δ and E_{JT}. Luminescence transitions III and IV terminate on the two branches of the split ground state and, in compounds with not too large a spin-orbit coupling, produce a distinctive double-peak spectrum. A detailed analysis of the zero-phonon lines and vibronic structure of the luminescence has not yet been made, but Grebe et al.[28] have shown that in ZnS:Cr and ZnSe:Cr the zero-phonon absorption and emis-

sion coincide. For this reason, in Figure 5, $^5\hat{A}_1$ is drawn at a lower energy
than $^5\hat{B}_1$. Grebe et al. also found that the emission has two closely-spaced
zero-phonon lines corresponding to the splitting of the ground state found
by esr. Kaminska et al.[31] have given a semi-classical treatment of the transitions in which the energy surfaces are calculated and then the emission
and absorption spectra are calculated with the classical Franck-Condon
approximation.

Krebs and Stauss[32,33] investigated the electron spin resonance of
GaAs:Cr d^4. At 4.2 K the line-width is large, $\simeq 100$ G, and above 12 K the
width increases rapidly so that a signal is not observed above 20 K. The
angular dependence of the signal is consistent with there being tetragonal
centers with three different orientations, namely the cube axes. A uniaxial
stress > 400 kg cm^{-2} along [100] aligns all the centers alone this axis in
less than 1s at 4.2 K. At lower stresses there is a thermal distribution between centers with [100] orientation and those with [101] or [001]
orientation. Thermal reorientation leads to a broadening of the esr lines.
Krebs and Stauss fitted the broadening to an Arrhenius plot with activation energy 67 ± 5 cm^{-1} (8.3 ± 0.6 meV) at zero stress, but the ultrasonic
absorption measurements of Tokumoto and Ishiguro[34] were fitted to
Raman broadening with

$$\tau^{-1} = 4.7 \ T^7 + 1.0 \times 10^7 \ T \ (s^{-1}). \tag{6}$$

Over the accessible range of temperature the experiments are not accurate
enough to distinguish between the two laws. If the Jahn-Teller distortion
at low temperature is taken to be static, then the spin Hamiltonian for a
particular orientation for the orbital singlet $^5\hat{B}_2$ with S = 2 is, when small
terms are neglected

$$\mathcal{H} = g_{/\!/} \ \mu_B S_z H_z + g_\perp \mu_B (S_x H_x + S_y H_y) + D S_z^2 + \frac{1}{6} a \ (S_x^4 + S_y^4 + S_z^4). \tag{7}$$

The values found for the coefficients are given in Table 2 for Cr d^4 in
GaAs and ZnSe. The similarity between the compounds can be seen.
Wagner and White[35] have confirmed the esr data by using far infrared laser
sources for the radiation field instead of a microwave cavity. If, in (7), a
is neglected with respect to D, then the zero-field energy levels of the split
ground state are as shown in Figure 6.

Krebs and Stauss also attempted to estimate the Jahn-Teller energy from
the esr parameters by using a cluster model with force constants taken from
the bulk elastic moduli. The value for E_{JT} obtained is 1720 ± 350 cm^{-1}
(0.21 ± 0.04 eV). This is probably a considerable overestimate and is about
three times the value deduced for ZnSe. It is also greater than the lattice

TABLE 2 Electron spin resonance quantities for Cr d^4. The quantities $g_{//}$, $g_{\perp}$, a, and D are defined in Equation (7). V_E is a Jahn-Teller coupling coefficient.

	ZnSe	GaAs
$g_{//}$	1.961 ± 0.002	1.974
$g_{\perp}$	$1.98 \ \pm 0.02$	1.997
$a(cm^{-1})$	± 0.024	$+0.031$
$D(cm^{-1})$	-2.48 ± 0.01	-1.860
$V_E(eVÅ^{-1})$	> 0.40	-0.85 ± 0.09 [a]
		$-1.2 \ \ \pm 0.2$ [b]
Reference	(25)	(32)

[a] Ref. 33
[b] Ref. 34

relaxation energy in the d^4-d^3 transition. However, the route from the experimental parameters to the value of E_{JT} is tortuous and full of approximations. Hennel et al.[36] re-interpreted the data with other approximate methods and deduced that 360 cm^{-1} $< E_{JT} <$ 660 cm^{-1} (0.045 eV $<$ $E_{JT} <$ 0.082 eV).

Further information about the fine structure of the lowest state comes from the attenuation of mechanical vibrations. Narayanamurti et al.[37] studied the absorption of ballistic phonons in semi-insulating GaAs:Cr at 1.5 K. The symmetry again was consistent with a tetragonal distortion. The phonons were not monochromatic but had a black-body spectrum from a heat pulse. By varying the temperature in order to sweep the peak of the spectrum through any energy splittings of the centers, it was found that the splitting is $\simeq 14$ K, i.e. 10 cm^{-1}, which is not far from the value of

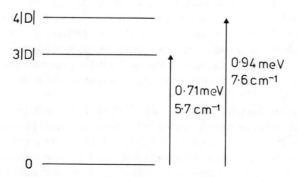

FIGURE 6 Structure of the ground state $^5\hat{B}_2$ of GaAs:Cr d^4.

7.6 cm^{-1} from esr. Tokumoto and Ishiguro[34] looked at the absorption of monochromatic ultrasonic waves and also found tetragonal distortion. A problem arises, however, with the acoustic paramagnetic resonance spectrum, reported for example by Ganapol'skii[38], Tokumoto and Ishiguro[34] and Bury et al.[39] The spectra found in apr are more complex than in esr and indicate that some of the chromium centers may form a variety of complexes. Because the selection rules for apr and esr are different, a particular complex may give a prominent spectrum in apr but not in esr. The situation has led Abhvani et al.[40] to challenge the model of Cr d^4 as having a static Jahn-Teller effect. They point out that there are always terms in the full Hamiltonian which mix crystallographically equivalent distorted orientations, so a pure static Jahn-Teller effect never occurs for an isolated impurity. In order to account for the observed esr spectra they postulate that there are strain fields at the chromium centers, produced for example by nearby (but not nearest-neighbor) impurities. Some fraction of the centers then have tetragonal symmetry because of these strains, and these centers give rise to the esr signals. Other centers will not be detectable by esr but will be by apr. There are a number of points to be resolved here. Although it is true that a strictly static Jahn-Teller distortion may be impossible, the matrix elements between different orientations may be sufficiently small that they are irrelevant in a particular experimental situation. A resolution of the conflict between the two concepts could come from experiment. If Abhvani et al. are correct, the esr signal comes from only a portion of the Cr d^4 present. So if many different samples are investigated there should be no exact relation between the concentration of Cr d^4 and the strength of the esr signal. Goltzene et al.[41] have in fact reported a correlation between the strength of the esr signal and the concentration of Cr d^4 deduced from electrical measurements. However, there is considerable scatter in the data and the samples are not made under a wide range of preparation conditions, so these experiments are not sufficiently conclusive. On the other hand, there is as yet no means of relating the strengths of the various apr signals to the concentrations of the centers producing them. It could turn out, therefore, that the apr spectra are something of a red herring, as was the 0.84 eV luminescence (section 4), in that they may arise from a very small fraction of the total chromium present.

No absorption between the $^5\hat{B}_2$ and $^5\hat{E}$ Jahn-Teller split levels (transition I, Figure 5) has yet been reported in GaAs:Cr, although Hennel et al.[36] searched for it over the range 900–5000 cm^{-1} (0.11–0.62 eV). This is a pity because the abosrption position would give a more direct measurement of the Jahn-Teller stabilization energy E_{JT} than other methods.

A broad peak in the optical absorption of GaAs:Cr d^4 at 0.90 eV (Figure 7) superposed on a rising background, has long been interpreted as the $^5T_2$-^5E (T_d, $\bar{4}3m$) internal transition. Taking into account the Jahn-Teller effect, this is the $^5\hat{B}_2$-$^5\hat{A}_1$ (D_{2d}, $\bar{4}2m$) transition (II in Figure 5). Abagyan et al.[42] and Ippolitova et al.[43] recognized that the excited level lies above the conduction band edge. The peak position at 7200 cm^{-1} (0.90 eV) was taken to be a measure of the crystal field splitting Δ but this ignores the Jahn-Teller effect. If the Jahn-Teller distortion of the ^5E state is small, the peak position is at approximately $\Delta + 2E_{JT}$, where E_{JT} is the stabilization energy in the ground state. In the absence of an accurate value of E_{JT} for GaAs:Cr d^4, an accurate value of Δ cannot be found, but it is about 1.5 times the value for the II-VI compounds. At the peak, the absorption cross-section is 1×10^{-17} cm^{-2}[43,36] at low temperature. From the width and the peak cross-section, the product $(\epsilon_{eff}/\epsilon)^2$ f is 2×10^{-2} where $(\epsilon_{eff}/\epsilon)$ is the effective field ratio and f is the oscillator strength. Abagyan et al.[42]

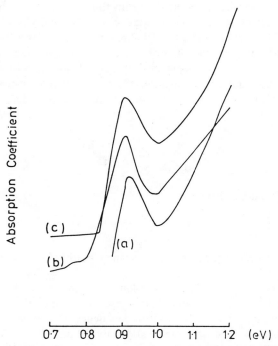

FIGURE 7 Absorption to the conduction band, in arbitrary units, of Cr d^4 in GaAs at liquid helium temperature. The curves are displaced vertically for clarity. (a) Abagyan et al.[42], (b) Ippolitova et al.[43] (c) Bois and Pinard[44].

and Ippolitova et al.[43] give oscillator strengths 2×10^{-4} and 3×10^{-3} respectively, but, in the absence of a proper theory for the effective field ratio, these values must be treated with caution. The width of the band increases with temperature. Bois and Pinard[44] fitted the variation to the theoretical form for linear coupling with equal force constants in the ground and excited states

$$W(T) = W(0) \left[\coth(\hbar\omega/2kT)\right]^{\frac{1}{2}}. \qquad (8)$$

From the fitting they found $W(0) = 90$ meV and $\hbar\omega = 13$ meV. Because a background absorption has to be subtracted to find the form of the band, there is considerable experimental uncertainty in these values. If W is the full width at half height, then in the linear theory

$$W(0) = \hbar\omega\, S^{\frac{1}{2}}\, (8\ \ell n\ 2)^{\frac{1}{2}} \qquad (9)$$

where S is the Huang-Rhys factor. From this one has $S \simeq 8$, which is an exceptionally large value for an excitation within the $3d^n$ shell for a transition metal impurity in a II-VI or III-V compound. However, the value obtained is misleading because it is obtained from a simplified model of the vibronic interaction which replaces the Jahn-Teller system of Figure 5 by two displaced parabolae of equal curvature. Hennel et al.[36] give the width Γ at liquid helium temperature as 340 ± 5 cm^{-1} but they define the width by the Gaussian shape

$$I(E) = \frac{1}{(\pi\Gamma^2)^{\frac{1}{2}}} \exp - \frac{(E - \overline{E})^2}{\Gamma^2}. \qquad (10)$$

The full width at half height is then

$$W = (4\ \ell n\ 2)^{\frac{1}{2}}\ \Gamma \qquad (11)$$

and $W(0)$ is then 70 meV in fair agreement with the value of Bois and Pinard.

Lightowlers[46] made high-resolution absorption measurements of GaAs:Cr and found, in material expected to contain Cr d^4, a line at 821 meV (6620 cm^{-1}) at 2 K. It showed evidence of containing three components (Figure 8) and had total width only 0.1 meV. At 6.5 K, another line appeared, about 1 meV lower in energy, accompanied by more fine structure. Clerjaud et al.[45] saw the same spectrum, but at lower resolution. They pointed out that the absorption could be explained by thermalization between levels of a ground state split by the same amount as that found by

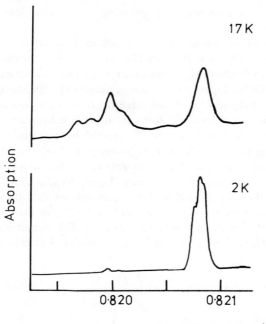

FIGURE 8 Absorption in the zero-phonon line region of the excitation of GaAs:Cr
d^4, after Ref. 46.

esr for Cr d^4. Consequently they attributed the lines to zero-phonon transitions between $^5\hat{B}_2$ and $^5\hat{A}_1$. Williams et al.[46] used high-resolution spectra to make the assignment quantitative. They deduced an energy level
scheme in which the ground state is split into three levels at 0, |3D| and
|4D|, as in Figure 6, while the excited state is split into five equally spaced
levels with spacing K. The experimental values are

$$D = -0.240 \pm 0.010 \text{ meV}, \quad K = 0.060 \pm 0.005 \text{ meV}$$

and the value of D agrees within experimental error with that found from
esr. Five approximately equally spaced levels in the excited state result
from second-order spin-orbit and spin-spin interactions in a $^5E(T_d)$
state. Hennel et al.[36] estimated the strength of the zero-phonon line with
respect to the 0.9 eV band to be 3.3×10^{-4}. Since in a linear coupling
model this is e^{-S}, a value of $S \simeq 8$ is obtained, in agreement with that from
the temperature broadening of the band. Again the result must be treated
with caution because of the experimental difficulties in finding the relative

strength and also because the linear coupling model is not strictly applicable here.

If the zero-phonon transition for excitation is at 0.82 eV and the threshold for ionization is at 0.75 eV (Section 3.2), then the excited state lies roughly 70 meV above the conduction band edge. Luminescent transitions (III and IV on Figure 5) then have to compete with autoionization transitions. Thus they will be of low intensity (unless phonon coupling is sufficiently strong to reduce the autoionization rate by many orders of magnitude).[47] Deveaud and Martinez[48] applied hydrostatic pressure to increase the band-gap of GaAs, thereby pushing the conduction band edge above the excited state, and thus were able to see the luminescence. At 12.6 k bar the spectrum has two peaks (Figure 9) like that found for ZnSe:Cr. Using the semiclassical model of Kaminska et al,[31] they deduced that the Jahn-Teller energy E_{JT} is 500 cm^{-1} (62 meV) in the $^5T_2$ ground state and 60 cm^{-1} (7 meV) in the ^5E excited state. The former is much less than the value deduced by Krebs and Stauss from esr, but over-simplified

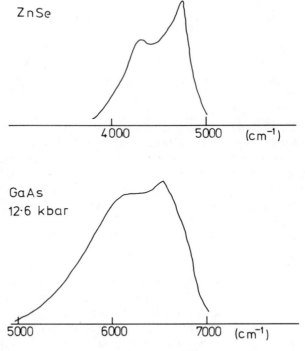

FIGURE 9 Luminescence of Cr d^4 in ZnSe[31] and in GaAs under hydrostatic pressure.[48]

models are used in interpreting both the esr and the luminescence data. At low pressures the spectrum is distorted by the presence of another luminescence band, so a sharp threshold pressure for the onset of luminescence was not seen. The resolution was not adequate for the zero-phonon structure to be seen, but the high-energy edge of the band is about 7700 cm^{-1} instead of the 6620 cm^{-1} expected at ambient pressure from the zero-phonon absorption. This implies a pressure shift of the internal transition of about 30 cm^{-1}/kbar (4 meV/kbar) which is high[49] but not unreasonably so. Deveaud et al.[50] also increased the energy gap by making a mixed crystal of $Al_xGa_{1-x}As$:Cr in the range $0.11 < x < 0.38$. Again a luminescence similar to that in ZnSe:Cr is found. It is sufficiently dissimilar to the 0.84 eV luminescence in GaAs:Cr (Section 4) for it not to be confused, although it is not clear why both luminescence transitions are not seen simultaneously. The luminescence is seen in the sample with the lowest aluminium concentration used, 11%. There is some contradiction here with the interpretation of Kocot et al.[51] of photoconductivity data. These authors observed a peak in the photoconductivity of $Al_xGa_{1-x}As$:Cr crystals for $x \leqslant 0.23$ but not for $x \geqslant 0.38$. They interpreted the peak as being due to a transition to an excited state within the conduction band, and, therefore, put the composition at which the excited level drops below the conduction band edge at a much higher aluminium concentration than do Deveaud et al. The apparent contradiction may not be real, as the idea of a purely electronic excited level varying in relation to a band edge is a simplification. Vibronic interaction must be taken into account. Figure 10 shows a schematic configurational coordinate diagram. When the excited level is not too far above the conduction band edge, the autoionization rate is reduced by a vibrational overlap integral which varies with the relative energies. There can be a range of relative energies where the autoionization rate is sufficiently small compared with radiative rates for luminescence to be seen, yet sufficiently large compared with the rate of recombination of electrons with the ionized centers for photoconductivity to be seen. Certainly the widths of the zero-phonon lines in the absorption spectrum of GaAs:Cr observed by Williams et al.[46] show the autoionization rate to be small, $\lesssim 10^{11}$ s^{-1}, for the level spacing at ambient pressure. It will decrease further as the excited level approaches the bottom of the conduction band. The smallness of the rate compared with that for normal s-d interaction is evidence that quenching of autoionization by vibrational overlap effects is nearly complete.

In summary, the optical and spin resonance properties of GaAs:Cr d^4 can be explained by a model in which the $^5T_2$ ground state is strongly split by a static Jahn-Teller distortion and the ^5E excited state by a much weaker distortion. The excited state lies above the bottom of the conduc-

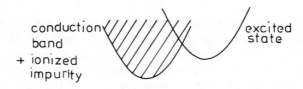

FIGURE 10 Configurational coordinate diagram showing how autoionization from an excited level above the conduction band edge can be reduced by vibrational effects.

tion band. Second-order spin-orbit and spin-spin effects produce small splittings of the lowest level, which can be seen in spin resonance and in the zero-phonon spectra. Figure 5 shows a cross-section of the energy surfaces in the absence of spin-orbit coupling. Approximating the surfaces by a configuration coordinate diagram with a single coordinate and linear coupling may be possible, but requires care in interpretation. The energy levels of the $^5T_2$ level split by spin-orbit and Jahn-Teller coupling simultaneously have not been established either experimentally or theoretically except for the fine splitting of the lowest level: in this respect our knowledge of Cr d^4 in GaAs is less than that of Cr d^4 in II-VI compounds. The sign of the Jahn-Teller coupling in the excited ^5E state is not firmly established. Although most of the optical and esr data are consistent with a model of static Jahn-Teller distortion of $^5T_2$, the experiments using phonon absorption suggest that dynamic Jahn-Teller effects cannot be ignored, and that a model in which most of the chromium is on isolated substitutional sites unaffected by other centers may be incorrect.

2.4 The d^5 Configuration

The ground state of Cr d^5 is $^6A_1$. There should be a spin-only electron spin resonance which is isotropic with $g \simeq 2$. The first excited state should be $^4T_1$. The $^6A_1$-$^4T_1$ spacing is probably greater than the energy gap, so the $^4T_1$ level would not be seen in normal absorption or luminescence experiments.

No observations of the properties of the Cr d^5 state have been reported. The problem is that the level which is occupied when d^5 exists lies near the conduction band edge, either just above or just below it (Section 3.3).

When the d^5 state exists, there is therefore a high concentration of free electrons and esr measurements become difficult or impossible. The level can be brought lower into the gap by hydrostatic pressure and, despite the experimental difficulties, an esr experiment may then give results. It is possible that luminescence from the $^4T_1$-$^6A_1$ transition, analogous to the yellow emission in ZnS:Mn but shifted to lower energy by covalency effects, could be seen in $Al_xGa_{1-x}As$:Cr crystals with high enough aluminium concentration ($x \gtrsim 0.3$) and therefore large enough band-gap.

3. Ionization Levels

3.1 The d^3-d^2 Level

When the Fermi level is near the valence band, a spin resonance attributable to Cr d^2 is seen, but when it is near mid-gap the signal is that of Cr d^3. The level corresponding with d^3-d^2 must therefore lie in the lower half of the band gap. No precise information about it is yet available. It is surprising that there have been no detailed studies of GaAs:Cr in which shallow acceptors are compensated by the chromium donor level to produce moderately high-resistivity p-type material. Nor has the level been noticed in photocapacitance experiments. If the Cr d^2 state is produced in semi-insulating material by illumination, then the d^2 spin resonance signal is decreased by further illumination in a spectral band with threshold at about 0.45 eV.[14,15] A possible mechanism is that electrons are raised from the valence band onto Cr d^2 to produce Cr d^3:

$$Cr\ d^2 + h\nu \rightarrow Cr\ d^3 + p. \tag{12}$$

The best evidence so far is therefore that the d^3-d^2 level is at 0.45 eV above the valence band at liquid helium temperature, but more work on this point is needed.

3.2 The d^4-d^3 Level

3.2.1 Introduction

The level corresponding with the transition d^4-d^3 lies near the center of the energy gap in GaAs:Cr. When the level is partly filled, the Fermi level is close to it and thus one has semi-insulating material with room temperature resistivity in the range 10^6-10^8 ohm-cm. Because this material is commercially important as a device substrate, much effort has gone into investigating the properties of the level. The fact that the level is near the gap center raises problems, however. For example, optical transitions to the valence band and the conduction band are in strongly overlapping spectral ranges, and electrical measurements are affected by the simultaneous presence of holes and electrons.

3.2.2 Hall Effect and Resistivity

Early measurements[1,2,42,52] of the variation of resistivity or Hall effect with temperature gave activation energies in the range 0.7–0.8 eV. The measurements give the free electron or hole concentration. There is no simple relation between these and the exact position of a chromium level. In equilibrium, there is a statistical distribution of occupancies of the different charge states (i.e. configurations), of the levels within a configuration, and of the conduction and valence bands. Lock[53,54] has attempted to take into account the effect of simultaneous transport in the conduction and valence bands. Clark[9] has applied statistical methods to take into account different charge states and excited levels. A complete treatment is not possible at present because one needs to know the various levels in order to perform the calculations. Even a purely electronic model of the levels, as used by Clark, is inadequate because of the lattice displacements present, and because the vibronic levels have to be known. It is clear that, although in principle one can calculate the electron distribution and hence the Hall effect or resistivity from the vibronic energy levels, the reverse process of determining the levels from these measurements is not possible with high accuracy.

Mullin et al.[55] have dissented from the usual interpretation of the thermal measurements. Their analysis of the levels uses a comparison of the temperature dependence of ohmic and space-charge-limited currents. It is not sufficient, as these authors appear to have done, to take a current varying with the square of the voltage as evidence of space-charge-limited current: one must also investigate the thickness dependence. Jimenez-Lopez et al.[56] have given an example of this. The conclusions of Mullin et al. can therefore be ignored.

3.2.3 Photoionization

The photoionization spectrum for transitions from a localized level to the conduction or valence bands can be measured by optical absorption, photoconductivity, photocapacitance or photo-excitation of esr. Each method has its advantages and disadvantages. Absorption measurements can be made in the absence of an electric field, but absorptions due to different mechanisms, as for example impurity to conduction band, impurity to valence band, excitation without ionization and free carrier absorption, are superposed and are difficult to identify and to separate. An illustration of the problems is the work of Jones and Hilton.[57] These authors made careful measurements of the absorption spectrum of semi-insulating GaAs:Cr at different temperatures, and deduced an ionization threshold energy. Without further information, however, it is impossible to state whether the transition seen was to the conduction band, from the valence band, or an admixture of both. Photoconductive responses depend not only on the photoionization rates but on recombination and transport processes. With semi-insulating material, large photoconductivity signals are possible but ambiguity of interpretation is correspondingly great. Photocapacitance is a versatile technique because different levels can be filled and emptied selectively, but the photoionization transition occurs in a region where the field is typically 10^4-10^5 V cm^{-1} and non-uniform. The spectrum in such high fields may be different, both qualitatively and quantitatively, from that without a field. Photo-excitation of esr gives an identification of the centers whose charge states are changed as a result of a photoionization transition. It is often performed with a scanned excitation source. Only thresholds can then be obtained, not the shape of the spectrum, for the occupancy of centers changes as the spectrum is scanned. (A similar problem arises with absorption in high-resistivity samples if the sample is placed at the exit slit of a spectrometer instead of the input.) Schwab and his colleagues[58] have developed a modification of photo-excited esr measurements using concepts developed for photocapacitance, which may give more precise information.

The absorption spectrum for transitions from Cr d^4 to the conduction band consists of a band rising from a threshold, on which is superposed the approximately Gaussian band from the d^4 internal transition. Figure 7 shows some reported spectra. There is some variation between spectra reported in different papers. Gutkin et al.[39] noted that part of this variation comes from a superposition of transitions from the valence band and to the conduction band, the relative magnitudes of the two components depending on the compensation ratio. It follows that transitions to the conduction band are best studied in fairly low-resistivity n-type material, where the Cr d^4 state is fully occupied, as was done by Ippolitova et al.[43] A striking feature of the spectrum is that the internal transition is strong with respect to the rising band, the latter presumably being due to transitions to free electron states in the conduction band. In the approximation that the electrons in the d-shell of the impurity are atomic-like d-electrons, both the internal d-d excitation transition and the d-s ionization transition to the bottom of the conduction band are forbidden. Insufficient knowledge about the symmetry-breaking mechanisms makes it difficult to predict the relative strengths of the transitions in a tetrahedral semiconductor. Hennel et al.[36] have gone so far as to say that the ionization transition to the free electron states is sufficiently forbidden as not to be observed. They interpret the rising absorption as being due to transitions from the valence band to residual Cr d^3 giving Cr d^4, with the peak at 0.9 eV being the Cr d^4 internal transition. The material available to these authors was very non-uniform and even in material with the Fermi level nominally in the upper half of the energy gap the Cr d^2 and d^3 esr signals were seen. It is therefore quite possible that in these particular samples transitions from the valence band were seen, but the evidence presented by these authors is not adequate to show that transitions to free electron states are not observed in better material.

No accurate value of an optical ionization energy for the $d^4 \rightarrow (d^3 + n)$ transition has been determined from absorption measurements. The problem is that the range of energy between the transition being obscured by background absorption or reflectivity and being obscured by the d^4 internal transition is much less than 0.1 eV, too small a range from which to extrapolate with accuracy to a threshold energy.

There have been fewer investigations of the absorption transition from the valence band, $d^3 \rightarrow (d^4 + p)$. Again there are problems of obtaining uniform samples with the Fermi level at a position such that a substantial fraction of the chromium is in the d^3 charge state, and of separating the required absorption from other loss mechanisms. Martinez et al.[60,61] made measurements on two samples, one made by doping in the melt and the other by diffusion. The diffused sample was high-resistivity p-type, with a resistivity such that the Fermi level should be below the d^4-d^3 level. Nevertheless, esr signals from Cr d^2, d^3 and d^4 were seen at helium temperatures, manifesting the non-uniformity of the sample. In such material the transitions

$$d^2 \rightarrow d^3 + p, \quad d^3 \rightarrow d^4 + p, \quad d^4 \rightarrow d^3 + n, \quad d^3 \rightarrow d^2 + n$$

can all occur, but Martinez et al. believed that the second predominated. Measurements were made at 4.5, 77 and 300 K, and at pressure up to 10 kbar at 77 and 300 K. Spectra at 77 and 300 K are shown in Figure 11: the 4.5 K spectrum is little different from that at 77K. A slight structure was seen and is indicated on the figure. Martinez et al. attributed this to transitions to the ground state $^5\hat{B}_2$ and lower excited state $^5\hat{E}$ of Cr d^4 (see Figure 5). The steepening of the absorption spectrum at lower temperature shows that there is phonon broadening of the edge. It is not possible in general to unravel the electronic and vibrational contributions to the shape of the absorption spectrum without additional knowledge, unless the coupling is weak. However, Martinez et al.[60,61] attempted to do so by assuming particular forms for the shape of the electronic spectrum and the form of the phonon broadening. They attempted to take into account the

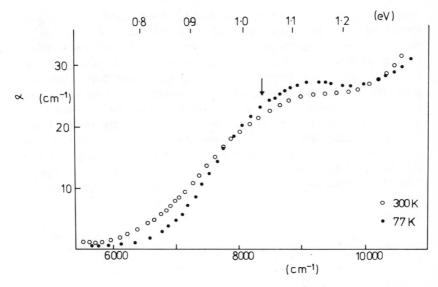

FIGURE 11 The absorption coefficient of the transition d^3–d^4, i.e. of an electron from the valence band onto d^3 to give d^4, for GaAs:Cr after Martinez et al.[61].

valence band structure of GaAs, in a simplified model, and the Jahn-Teller distortion of the chromium. It is doubtful whether the experimental data are sufficient to bear the weight of a theoretical analysis which contains a certain number of assumptions, approximations and disposable parameters. A thermal energy (i.e. the energy of a zero-phonon transition from the valence band to the level) of 0.74 eV and an optical energy of 0.92 eV, independent of temperature within the approximations used, were deduced. The difference was attributed to a combination of a symmetrical lattice relaxation and the Jahn-Teller relaxation. The thermal energy is largely determined by the threshold behavior of the absorption and should therefore be reasonably accurate irrespective of the merits of the analysis, while the optical energy deduced depends on the details of the model and is therefore less reliable. Martinez et al. also measured the pressure dependence of the absorption and found a shift of $+3 \pm 0.7$ eV/kbar. In addition, from the measured chromium concentration they found the optical cross-section: at 9000 cm^{-1} (1.12 eV) it is 9×10^{17} cm^{-2}. This value is an upper limit as the esr. measurements show that not all the chromium is initially in the d^3 state.

Semi-insulating GaAs:Cr is strongly photoconductive. G. A. Allen[52] and Broom[62] made measurements of the spectral response as a function of temperature. Spectra are shown in Figure 12. Above the threshold the response rises to a peak, beyond which it decreases again. Part of this behavior must be due to the shape of the absorption (Figure 7) with the d^4 internal peak superposed on the transitions to the conduction band continuum. However, the steepness of the drop beyond the peak at low temperatures, and the strong temperature dependence of this dip, show that this cannot be the sole cause as the absorption spectrum does not show such drastic variation. An explanation lies in the fact that the Cr d^4-d^3 level lies slightly above mid-gap. Just above threshold, electrons are ionized to the conduction band and give photoconductivity. (Allen used the change in Hall coefficient with illumination at 300 K to show that at 0.68 eV electrons are ionized to the conduction band.) At higher energies.

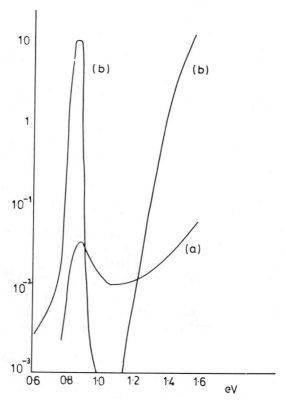

FIGURE 12   Photoconductivity spectra of GaAs:Cr at liquid nitrogen temperature. (a) Allen[52] in arbitrary units, (b) Broom[62] with photoconductive gain as the vertical axis.

electrons are raised from the valence bands into Cr d^3 centers. The holes produced recombine rapidly with photogenerated electrons and quench the photoconductivity. The relative strengths of photoconductive and quenching transitions depend on the initial compensation of the chromium, i.e. the ratio of d^4 to d^3, and therefore vary from material to material. For three samples, the threshold energies found by Allen are

| 77 K | 0.73, | 0.75, | 0.76 eV |
| 300 K | 0.56, | 0.56, | 0.60 eV |

although no analysis of the spectral shape was attempted.

Capacitance methods of investigating deep levels rely on the fact that the width and hence capacitance of a junction depletion region depends on the space charge of impurities. If the charge state of the impurities is changed by optical or thermal ionization, the resultant electrons or holes are swept out of the depletion region by the field there. The consequent change of space charge is recorded as a change of capacitance. (One can also measure the current produced by the sweeping out of the carriers.) From the sign of the capacitance change one can tell if transitions are to the conduction band or from the valence band. For qualitative work, or for rapid comparison of materials, it is possible to use continuous scanning, e.g. by measuring the capacitance as light on the junction is spectrally scanned. For accurate quantitative work, data must be taken point by point. (Although because of the high speed of each measurement the DLTS method appears to be one with a continuous scan, it is really a point by point method.) The occupancy of the deep impurities is first brought to an initial reproducible condition. For example, in n-type material all centers can be filled by forward-biasing a Schottky junction for a sufficient time in the flat-band condition. Alternatively light of suitable photon energy can be used to empty levels in the upper half of the gap, to fill levels in the lower half, or more generally to produce a partial occupancy. In that part of the depletion region where the free carrier concentration is sufficiently small, if the temperature is sufficiently low for thermal emission rates to be negligible compared with optical ones, the ratio of occupied centers n_I to the total number N_I is

$$\frac{n_I}{N_I} = \frac{e_p^o}{e_n^o + e_p^o} \, . \tag{13}$$

Here e_n^o, e_p^o are optical emission coefficients (Figure 13a) with $e^o = \sigma^o \phi$ where σ^o is the optical cross-section and ϕ the photon flux. When a repro-

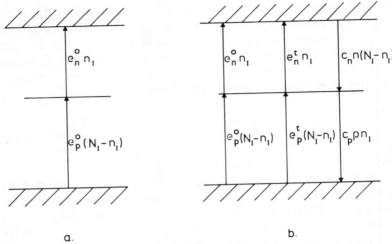

FIGURE 13 (a) Optical transition rates leading to Equation 13, (b) optical, thermal and capture transition rates.

ducible steady-state condition is reached, the bias voltage or light producing it is switched off. If the temperature is low enough, then the occupancy of the levels will stay practically constant. Another monochromatic source of light is next used to remove electrons into or out of the levels. For example, if the level is in the upper half of the gap, and the photon energy is such that

$$E_C - E_I < h\nu < E_I - E_V$$

then the level is emptied with a time constant

$$\tau^{-1} = e_n^o = \phi\sigma_n^o \tag{14}$$

From the time constant of the resulting capacitance change and the measured photon flux one can find σ_n^o. If the photon energy is large enough to give transitions to both the conduction and the valence bands, then

$$\tau^{-1} = \phi(\sigma_n^o + \sigma_p^o) \tag{15}$$

Next, the system is returned to the reproducible initial condition, and a measurement is then made at another photon energy, and so on. A problem that arises in junction capacitance measurements of deep levels is illustrated in Figure 14 for a Schottky barrier on n-type material in equi-librium. It is usually an adequate approximation to take shallow donors as fully depleted, i.e. up to a width W the density of free carriers is negligible compared with that of shallow donors, while beyond W the material is electrically neutral. As a first approximation, the deep levels are empty to a width W-x_0 and filled beyond this. In the region x_0 thermal filling is un-important for shallow centers but not for deep ones. More accurately, the occupancy at x is given by

$$\frac{n_I}{N_I} = \frac{c_n n + e_p}{c_n n + c_p p + e_n + e_p} \tag{16}$$

where c_n and c_p are capture coefficients and the emission coefficients e_n and e_p are sums of optical and thermal coefficients (Figure 13b). The free carrier densities n and p are functions of x, varying with the potential by a

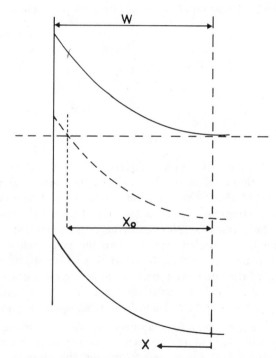

FIGURE 14 The depletion region of a Schottky junction at zero bias. The free-carrier depletion width is W and x_0 is the distance from the edge of the depletion region to the point where the Fermi level crosses the deep impurity level.

Boltzmann factor. The time constant for, say, emptying a level to the conduction band is only given by Equation (14) for that region of the depletion region where e_n^o is larger than any other term in Equation (16). In a region of width $\sim x_o$ the thermal capture is non-negligible and the time constant therefore varies with x. The capacitance change is no longer purely exponential, although it approaches an exponential with time constant given by Equation(14) or (15) at long times. It is possible to minimize the effect of thermal filling in a number of ways, for example by applying a large reverse bias so that the thermal filling region is a small fraction of the total depletion width.

Gutkin et al.[59] made photocapacitance measurements on GaAs:Cr at 120 and 295 K. The photocapacitance change was non-exponential, but from the later part of the decay curve they found $e_n^o + e_p^o$ and thus $\sigma_n^o + \sigma_p^o$. They also measured the total change of capacitance $\Delta C(\infty)$, as in Figure 15. In the absence of the thermal filling region, the initial concentration of filled centers in the depletion region is N_I and the final concentration would be given by Equation (13), so $\Delta C(\infty)$ is proportional to the difference, i.e.

$$\Delta C(\infty) \propto \frac{\sigma_n}{\sigma_n + \sigma_p} \tag{17}$$

and hence

$$\sigma_n \propto \tau^{-1} \Delta C(\infty). \tag{18}$$

Although the thermal filling makes Equations (17) and (18) inaccurate, the results of Gutkin et al. show that σ_n has a threshold at about 0.7 eV and σ_p at about 0.9 eV, although precise values are not determinable from their data. A spectrum of $\sigma_n^o + \sigma_p^o$ is shown in Figure 16. Szawelska and Allen[63] also made transient photocapacitance measurements on n-type GaAs:Cr diodes but restricted themselves to the range below 0.8 eV at liquid nitrogen temperature. Their objective was to investigate transitions to the bottom of the conduction band without the complications of the internal d^4 transition and of transitions from the valence band, both of which occur just above 0.8 eV. Their data are shown in Figure 16. They plotted $(\sigma \, h\nu)^{\frac{2}{3}}$ against $h\nu$, a form appropriate when the major contribution to the electron dipole matrix element comes from the evanescent wave region of the impurity wavefunction and the phonon coupling is small. The intercept on this plot, 0.74 ± 0.01 eV, is the optical energy, i.e. the energy of a vertical transition on a configuration coordinate dia-

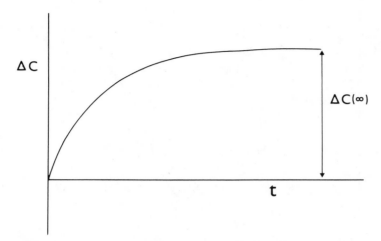

FIGURE 15   The form of the signal in a transient photocapacitance experiment. The sign of ΔC tells if centers are being filled or emptied, the time constant gives the optical cross-section and the final value ΔC(∞) gives the deep impurity concentration.

gram. A detailed analysis of the tail below the intercept in order to obtain the thermal energy was not attempted and would not have been justified by the data available. Szawelska and Allen also found the level to be 0.81 ± 0.01 eV above the valence band from the threshold energy for a photocapacitance signal in an n^+p diode. Chantre et al.[64] made transient photocapacitance measurements on n-type GaAs:Cr with a different technique for minimizing the effect of thermal filling. The rate of change of occupancy of the deep levels is

$$\frac{dn_I}{dt} = -(e_n^o + e_n^t)\, n_I + c_n n(N_I - n_I) + (e_p^o + e_p^t)\,(N_I - n_I) - c_p p N_I \quad (19)$$

where the transitions are shown in Figure 13b, the emission coefficients e being sums of optical and thermal coefficients e^o and e^t. In Equation (19), the quantities n_I, n and p are functions of position as well as time. If initially the levels are all occupied, so $n_I = N_I$, and if the hole capture rate and the thermal emission rates are small compared with the optical emission rate e_n^o, then the initial slope is

$$\left(\frac{dn_I}{dt}\right)_{t=0} = -e_n^o\, N_I. \qquad (20)$$

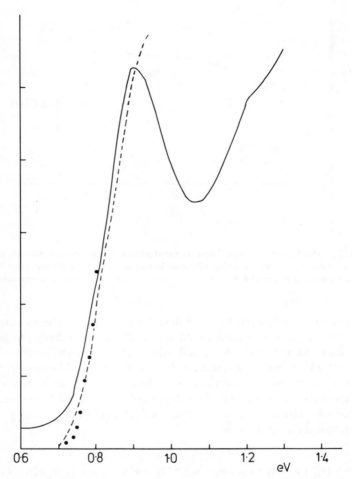

FIGURE 16 Normalized spectra of photoionization from GaAs:Cr d^4 to the conduction band as determined by photocapacitance, $-\sigma_n$ at 86K after Chantre et al.[64], ... σ_n at 84K after Szawelska and Allen,[63] - - - $\sigma_n + \sigma_p$ at 120K after Gutkin et al.[59]

Thus e_n^o and hence the optical cross-section σ_n^o can be found from the initial slope of the capacitance change instead of from the time constant. Even in the photon energy range where optical transitions from the valence band can occur one obtains σ_n^o instead of the sum $\sigma_n^o + \sigma_p^o$. In practice one cannot find the initial slope, but one has to measure a change of capacitance over a time interval sufficiently large that an adequate signal-to-noise ratio is obtained. The method therefore has advantages but is still not completely free of the effects of the thermal filling region. Chantre et al. made measurements on GaAs:Cr at 86, 121, 181 and 302 K: the 86 K spectrum is shown in Figure 16. The spectrum is similar to the absorption spectrum (Figure 7) with a peak at 0.9 eV which broadens with increasing temperature. Chantre et al. did not analyze the spectrum in detail.

The occupancy of charge states can be detected by esr, and the change of occupancy with illumination can in principle give information about optical transitions between charge states.[13-15,32] In the experimental spectra there is a threshold feature at about 0.75 eV at which the concentration of Cr d^2 increases, of Cr d^3 decreases and of Cr d^4 increases. A possible model is that, if much of the chromium is initially in the d^3 configuration, light can raise electrons from the valence band to the chromium causing an increase of Cr d^4 and a decrease of Cr d^3. The liberated holes can recombine with other Cr d^3 centers, giving Cr d^2 and a further decrease of Cr d^3. Clearly there is a problem that the occupancies are governed not only by optical emission rates but also by recombination. Von Bardeleben et al.[58] have described an experimental procedure to overcome some of the difficulties. They adopt a point-by-point method. As in a version of photocapcitance, the system is brought to a reproducible state by one light source of appropriate photon energy, and the occupancies are perturbed by another monochromatic light source. In principle, this is a powerful method of connecting optical transitions with the configurations between which they occur, but it has not yet been reported to have been applied to GaAs:Cr.

3.2.4 Capture and Thermal Emission

Junction capacitance methods can be used to measure the capture cross-sections of electrons and holes, and thermal emission rates as well as optical ones. For example consider a Schottky diode in n-type material with free electron concentration n_B in the bulk region outside the depletion layer. Suppose that initially the deep levels in the region W-x_o (Figure 14) are to an adequate approximation empty, so $n_I \simeq 0$, and suppose that the temperature is sufficiently low that in this region the thermal emission rates are very small. If the diode is biased in the forward direction with a voltage large enough to remove practically all the depletion region, then Equation (19) reduces to

$$\frac{dn_I}{dt} = c_n \, n_B \, N_I. \tag{21}$$

If the forward bias is applied in a short pulse there will be a partial filling of the levels which can be measured as a change of capacitance, and hence the value of c_n can be found. Conversely, if the levels are all filled initially by a forward bias or by optical transitions, subsequently they can be emptied thermally if the temperature is not too low (in analogy with optical emptying in photocapacitance). The time constant for emptying gives the thermal emission rate. The DLTS method in its simplest form is a variant of this second method.

Lang and Logan[65] in an early investigation of the properties of deep level impurities in GaAs by DLTS found that the thermal emission coefficient e_n^t in GaAs:Cr varied with temperature with an apparent activation energy 0.84 eV, and also found another level with activation energy 0.92 eV which may not be due to chromium.

Mitonneau et al.[66] measured the capture cross-section of electrons and holes at the d^4-d^3 level. In p-type material, one might expect the d^3-d^2 level to interfere with the measurements at low temperatures because chromium impurities initially in the d^4 configuration can capture a hole to become d^3, then another to become d^2. However, if the d^3-d^2 level is not too deep and the temperature not too low, any holes captured to give d^2 will rapidly be re-emitted to the valence band to give d^3, so the effect of the second level may be negligible. Mitonneau et al. expressed their results in terms of capture cross-sections: these can be written as

$$\sigma_n = 1.3 \times 10^{-17} \exp\left(-\frac{0.117 \, eV}{kT}\right) cm^2, \quad 250 \leqslant T \leqslant 400 \, K,$$

$$\sigma_p = 1.0 \times 10^{-16} \exp\left(-\frac{0.020 \, eV}{kT}\right) cm^2, \quad 150 \leqslant T \leqslant 400 \, K. \tag{22}$$

The measured quantities are capture rates, the cross-sections being derived quantities defined by

$$c = \sigma \bar{v} \tag{23}$$

where $\bar{v}$ is the mean carrier velocity. From the information given by Mitonneau et al. one has

$$c_n = 3.3 \times 10^{-11} \, T^{\frac{1}{2}} \exp\left(-\frac{0.117 \, eV}{kT}\right) cm^3 \, s^{-1},$$

$$c_p = 1.0 \times 10^{-10} \, T^{\frac{1}{2}} \exp\left(-\frac{0.020 \, eV}{kT}\right) cm^3 \, s^{-1}. \tag{24}$$

Although (24) is in an Arrhenius form, this does not necessarily imply that the capture process is in fact thermally activated, nor do the apparent activation energies necessarily correspond with any real energies. The temperature dependence of the capture coefficients is not very strong; for example, one has

T	200 K	400 K
c_n	0.05×10^{-11}	2.2×10^{-11} cm^3 s^{-1}
c_p	44×10^{-11}	110×10^{-11} cm^3 s^{-1} .

Martin et al.[67] measured the thermal emission rates by a method which in principle is the same as that used by Gutkin et al. to find the optical rates, although the detailed procedure is different. They filled the centers and then observed the emptying. The time constant τ for emptying and the total capacitance change $\Delta C(\infty)$ were obtained. (Instead of recording the whole transient they used the DLTS sampling procedure. The quantities τ and $\Delta C(\infty)$ were deduced from measurements of the capacitance at two times, with the assumption that the transient is purely exponential.) In analogy with Equations (15) and (17), if the centers are initially filled with electrons and if capture processes during the transient are ignored, the time constant of the capacitance change is

$$\tau^{-1} = e_n^t + e_p^t \tag{25}$$

and the total capacitance change is

$$\frac{\Delta C(\infty)}{C} = \frac{N_I}{2n_B} \frac{e_n^t}{e_n^t + e_p^t} , \tag{26}$$

where it is assumed that $N_I \ll n_B$. If N_I and n_B are measured, then from Equations (25) and (26) the coefficients e_n^t and e_p^t can be found.

Carrier capture, ignored in the derivation of Equations (25) and (26), complicates matters. Because the effects occur in other experiments some detail is given here. The occupancy of deep levels at steady state is no longer

$$f \equiv \frac{n_I}{N_I} = \frac{e_p^t}{e_n^t + e_p^t} \tag{27}$$

but is given by Equation (16). The time constant therefore varies with x, as in photocapacitance, so Equation (25) is not exact, and also Equation (26) must be modified. As an approximation, one can divide the depletion region into two regions as was done in connection with Figure 14. In one

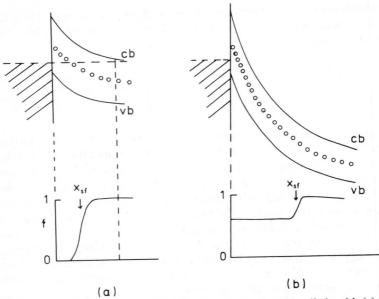

FIGURE 17 The upper part of the diagram shows a Schottky diode with (a) zero
bias and (b) large reverse bias. The lower part shows the fractional occupancy f of the
deep levels and the position x_{sf} of a step-function approximation to f.

region, capture processes are taken to be negligible, and, in the other, they
are taken to be fast enough to maintain thermal equilibrium. The occupancy is, therefore, approximated by a step function with a discontinuity
at x_{sf}. Figure 17 illustrates two situations: thermal equilibrium and large
reverse bias, both for a Schottky barrier on n-type material. In thermal
equilibrium, the occupancy of the deep centers is given by a Fermi distribution and of the free carriers by a Boltzmann distribution

$$n = n_B \exp\left(-\frac{qV(x)}{kT}\right) = n_B \exp\left(-\frac{x^2}{2L_D^2}\right), \qquad (28)$$

$$p = p_B \exp\left(+\frac{x^2}{2L_D^2}\right), \qquad (29)$$

where L_D is the Debye length

$$L_D \equiv \left(\frac{\epsilon kT}{4\pi q^2 n_B}\right)^{\frac{1}{2}} \qquad (30)$$

in Gaussian units. It is convenient to approximate the function $f \equiv n_I/N_I$ by a step function at the point at which $f = \frac{1}{2}$. Here, one has, by detailed balance,

$$c_n n = e_p \,, \quad c_p p = e_p \tag{31}$$

and from Equations (28) and (29) the step function position x_{sf} is

$$
\begin{aligned}
x_{sf} &= L_D \left[2 \left| \ln\left(\frac{c_n n_B}{e_n} \right) \right| \right]^{\frac{1}{2}} \\
&= L_D \left[2 \left| \ln\left(\frac{c_p p_B}{e_p} \right) \right| \right]^{\frac{1}{2}} .
\end{aligned}
\tag{32}
$$

(The position x_{sf} is not exactly the same as x_o in Figure 14 where the Fermi level crosses the impurity level if the occupied and unoccupied states have different degeneracies, but the difference is usually less than L_D.) Instead of Equation (26) one has

$$\frac{\Delta C(\infty)}{C} = \frac{N_I}{2n_B} \left(1 - \frac{x_{sf}}{W} \right)^2 . \tag{33}$$

At large reverse bias in steady state, both the electron and hole concentrations are reduced below the equilibrium values.[68] If the reduction is large enough, capture is negligible and the occupancy of deep levels is given by Equation (27) instead of the Fermi function. (Martin et al. incorrectly use the terms "equilibrium" and "Fermi function" when referring to a non-equilibrium steady-state situation.) In the free carrier tail region the field is sufficiently small that $n(x)$ and $p(x)$ have approximately their equilibrium values. Martin et al. make a step-function approximation to the occupancy with x_{sf} taken at the point where, in n-type material, the occupancy is midway between unity and that given by Equation (27). Equation (26) is now modified to

$$\frac{\Delta C(\infty)}{C} = \frac{N_I}{2n_B} \frac{e_n^t}{e_p^t + e_p^t} \left(1 - \frac{x_{sf}}{W} \right)^2 \tag{34}$$

They find x_{sf} (which they call λ) by an inaccurate approximation in which minority carrier capture in the region x_{sf} is neglected while the corresponding emission is not, but if the reverse bias is sufficiently large that (x_{sf}/W) is small the error introduced is not great.

For their experiments, Martin et al. used liquid epitaxy material, one layer with $n = 6 \times 10^{14}$ cm^{-3} and another with $p = 9 \times 10^{15}$ cm^{-3}. The low carrier concentration means that the depletion layer field is not very high, $\sim 5 \times 10^4$ V cm^{-1} at 10 V reverse bias. In interpreting their experimental data on the level emptying they ignored the effect of capture on τ but corrected for its effect on the magnitude of the capacitance change. They found $e_p^t \gg e_n^t$, and expressed the temperature variation as

$$e_n^t = 8.0 \times 10^4 \; T^2 \; \exp\left(-\frac{0.883 \text{ eV}}{kT}\right) \text{cm}^3 \text{ s}^{-1}, \; 40\text{-}200°C \qquad (35)$$

$$e_p^t = 4.3 \times 10^6 \; T^2 \; \exp\left(-\frac{0.858 \text{ eV}}{kT}\right) \text{cm}^3 \text{ s}^{-1}. \qquad (36)$$

The latter value was found in n-type material: in p-type they found

$$e_p^t = 1.7 \times 10^7 \; T^2 \; \exp\left(-\frac{0.886 \text{ eV}}{kT}\right) \text{cm}^3 \text{ s}^{-1}. \qquad (37)$$

Over the temperature range used, the two values of e_p^t differ by a factor $\simeq 2$, i.e. the accuracy is of the order of $\pm 50\%$. The form in which (35)-(37) are written arises from a particular model of the emission process: a different power law together with a different energy value would equally well represent the data. It is possible to find the Gibbs free energy for the electron and hole transitions from these data. By detailed balance one has

$$\frac{e_n}{c_n} = \frac{g_e}{g_f} N_c \; \exp\left(-\frac{E_n}{kT}\right) \qquad (38)$$

and

$$\frac{e_p}{c_p} = \frac{g_e}{g_f} N_v \; \exp\left(-\frac{E_p}{kT},\right) \qquad (39)$$

where g_f, g_e are the degeneracies of the filled and empty level. Martin et al. argue from the published esr data (Sections 2.2, 2.3) that g_f/g_e, i.e. $g(d^4)/g(d^3)$, is 5/4. Values of N_c and N_v are given by Blakemore.[69] The results are given in Table 3: the 300 K values are extrapolated out of the

TABLE 3 Properties of the d^4-d^3 level from the experimental data of Mitonneau et al.[66] and Martin et al.[67]. The energy gap is from Thurmond[70].

	300K	400K	
N_c	4.23×10^{17}	6.34×10^{17}	cm^{-3}
N_v	8.63×10^{18}	1.33×10^{19}	cm^{-3}
e_n	1.1×10^{-5}	9.6×10^{-2}	s^{-1}
e_p	1.5×10^{-3}	11	s^{-1}
c_n	6.2×10^{-12}	2.2×10^{-11}	$cm^3 \ s^{-1}$
c_p	8.0×10^{-10}	1.1×10^{-9}	$cm^3 \ s^{-1}$
E_n	0.672	0.641	eV
E_p	0.760	0.734	eV
$E_n + E_p$	1.432	1.375	eV
E_G	1.423	1.376	eV

temperature range of the emission experiments. The Gibbs free energies add closely to the energy gap as given by Thurmond.[70] This shows the capture and emission data to be consistent. Blakemore[69] has used the same data to find the intrinsic carrier concentration n_i from the combination of Equations (38) and (39), i.e.,

$$\frac{e_n e_p}{c_n c_p} = n_i^2 \qquad (40)$$

and also concludes that the data are consistent with data on n_i from other types of measurement. Martin et al. used an approximate method of obtaining the energies from their data instead of the method given here, but their results are not greatly different from those in Table 3.

By using Schottky diodes with a higher shallow donor concentration (to 2×10^{17} cm^{-3}), Makram-Ebeid et al.[71] were able to investigate electron emission at higher electric fields, up to 4.5×10^5 V cm^{-1}. They used a differential measuring technique which is equivalent to moving the position of x_{sf} in Figure 17b a small amount by means of a voltage pulse, and then observing the relaxation of the electron concentration on the impurity when the pulse ends. This has the advantage that the emission rate is measured in a small region where the field is nearly constant, rather than in a substantial length of the depletion region where the field varies by a factor nearly two. It has the disadvantage that the interpretation of the results depends critically on a proper evaluation of x_{sf}. Makram-Ebeid et al. made the same approximations as Martin et al.[67] with the same error, but, whereas this was not serious for the latters' measurements, it is for the formers'. (An analogous, though not precisely similar, situation is the demonstration by Braun and Grimmeiss[72] that an apparent field-dependence of the thermal emission rate in gold-doped silicon resulted from an incorrect treatment of the thermal-filling region). Makram-Ebeid et al. found an apparently strong field and temperature dependence of e_n^t. For example, they deduced that at 300K the value of e_n^t increases by three orders of magnitude between 2×10^5 V cm^{-1} and 4×10^5 V cm^{-1}. Pons and Makram-Ebeid[73] have given a theory of phonon-assisted tunnel emission of electrons from deep levels in semiconductors. It is not a theory of the second-order process in which tunneling and phonon emission or absorption occur simultaneously. Instead, the impurity is considered to be in a particular vibrational state from which an electron tunnels elastically to the conduction band: a temperature dependence arises from averaging over the thermal occupation of the vibrational states. By fitting theory and experiment, it is deduced that $S\hbar\omega$ is 195 ± 15 meV and $\hbar\omega$ is 35 ± 10 meV, so S is 5.6 ± 1.6. These values are close to those found by Hennel et al.[60,61] from absorption measurements, i.e. $S\hbar\omega = 170$ meV, $\hbar\omega = 28$ meV. It is not clear what weight can be put on the results of Makram-Ebeid et al. A proper analysis of the data would require a better treatment of filling and emptying rates at the edge of the thermal filling region, and experimental verification that transients assumed to be exponential really were so (in the experiments, the time for the transient to decay to its half-value was used as a measure of the decay rate[74]). Grimmeiss et al.[75] have shown that substantial errors can arise from neglect of certain terms in the rate equations in a related kind of experiment.

3.2.5 Summary

In summarizing the evidence concerning the position of the d^4-d^3 level in the gap, one notes that there is a discrepancy between these authors who deduce that the difference $S\hbar\omega$ between optical and thermal energies is large, of the order of 0.15 eV, and those who deduce it to be an order of magnitude smaller. (Here the "optical energy" is the vertical energy on a configuration coordinate diagram.) The small shift results from an overly simple interpretation of data, while the large shift results from interpretation in terms of more complex theories containing unsubstantiated assumptions. The data of Martin et al. (Table 3), expressed as Gibbs free energies, give an energy level just above mid-gap with a temperature variation between 300K and 400K which, not unexpectedly, shows that the level follows neither the conduction nor the valence band. Extrapolation of these values to 77K gives a value close to the optical energy determined by Szawelska and Allen by photocapacitance, which is consistent with these authors' claim that lattice relaxation is small. The emission and capture coefficients given in Table 3 are self-consistent within the limits of experimental error. It remains to be seen whether more detailed experiments confirm the claim by Makram-Ebeid and his colleagues that these quantities are strongly field-dependent at higher electric fields.

3.3 The d^5–d^4 Level

Ippolitova et al.[43] grew crystals of GaAs containing chromium and tin. The concentrations of the two impurities were found by the radioactive tracer method. When the chromium concentration exceeded the tin concentration the material was semi-insulating: when the tin concentration was the greater the material was low-resistivity n-type. Data for the latter situation are shown in Table 4. In these experiments the chromium must be acting as a single acceptor, for if there were two deep acceptor levels then since $2[Cr] > [N_D]$ the material would not be low resistivity. It follows that the d^5–d^4 level must lie above the Fermi level, which from the data of Table 4 is about 0.05 eV below the conduction band. Supporting evidence for this conclusion comes from the observation of the internal d^4 absorption band in the low-resistivity material. If the d^5–d^4 level lay substantially below the Fermi level, the chromium would be in the d^5 configuration and the d^4 absorption would not be seen.

Contrary evidence came from two sources. First, a photo-excited electron spin resonance signal was interpreted as that of Cr d^5, which implies that the d^5–d^4 level must be sufficiently deep in the gap for thermalization to be small at the measurement temperature. However, the identification was later found to be incorrect, as discussed in Section 2.1. Second, Brozel et al.[76] made electrical measurements similar to those of Ippolitova et al. and found a different result. They used crystals containing chromium and silicon. The chromium concentration was determined by mass spectrometry, the silicon concentration on the gallium and arsenic sites by

TABLE 4 Data from Ippolitova et al.[43] for GaAs containing chromium and tin shallow donors. All concentrations are in units of 10^{17} cm^{-3}.

Chromium concentration	Donor concentration	Resistivity (300 K)
1.36	1.55	0.04 ohm-cm
1.07	1.2	0.13
1.8	2.2	0.06

local mode absorption, and the electron concentration at 300K from the Hall effect. Silicon acts as a donor on the gallium site and an acceptor on the arsenic site, so the effective donor concentration $[N_D^{eff}]$ is the difference of the two concentrations. If chromium is a single acceptor, then

$$n = [N_D^{eff}] - [Cr] \tag{41}$$

while if it is a double deep acceptor then

$$n = [N_D^{eff}] - 2\,[Cr]. \tag{42}$$

Table 5 shows the data of Brozel et al.[76] (excluding two high-resistivity specimens). It is clear that Equation (42) is obeyed more closely than Equation (41). This implies that chromium has a second acceptor level below the Fermi level in these materials, i.e. below about 0.08 eV beneath the conduction band. The contradiction between the data of Tables 4 and 5 suggests an error in calibration of the concentrations by one set of workers.

Goswami et al.[77] measured the electron spin resonance of n-type GaAs:Cr. Initially the material was low-resistivity n-type and no resonance was seen because of the high loss in the microwave cavity. The carrier concentration was reduced in stages by electron irradiation, and as the resistivity increased the spin resonance signal of Cr d^4 was seen, with no signal attributable to Cr d^5. Unfortunately the experiment was not a quantitative one, i.e. the appearance of the d^4 signal was not correlated with the position of the Fermi level at successive radiation stages. However, these experiments show that the d^5–d^4 level cannot lie deep in the energy gap.

Challis and Ramdane[78] observed that the thermal conductivity at low

TABLE 5 Data from Brozel et al.[76] for GaAs containing chromium and silicon. All concentrations are in units of 10^{17} cm^{-3}.

Chromium concentration	Effective donor concentration	Electron concentration	$[N_d^{eff}] - 2\,[Cr]$	$[N_d^{eff}] - [Cr]$
0.73	1.6	0.08	0.14	0.9
1.0	3.5	1.5	1.5	2.5
1.7	4.7	2.25	1.3	3.0
3.0	6.9	3.9	0.9	3.9
1.7	14	9.8	10.6	12.3
2.5	17	11.5	12.0	14.5
5.3	34	24	23.4	28.7

temperatures of low-resistivity n-type GaAs:Cr was high and showed negligible scattering of phonons by chromium, whereas semi-insulating GaAs:Cr showed strong scattering. They deduced that in the low-resistivity material the chromium is in the d^5 configuration with a $^6A_1$ ground state with no close-spaced levels which could scatter phonons, for they believed that the Jahn-Teller ground state of Cr d^4 has such levels. Consequently, the d^5-d^4 level lies below the conduction band edge at low temperatures. Subsequently Challis et al.[79] gave reasons for believing that only a fraction of Cr d^4 would contribute to phonon scattering and withdrew their previous interpretation, even though some quantitative aspects of the concentration of scattering centers remained unresolved.

Hennel et al.[80,81] measured the absorption, resistivity and Hall effect of GaAs:Cr under hydrostatic pressure. They found a most interesting result at 77K in one sample where the free carrier concentration was 1.5×10^{17} cm^{-3} and the chromium concentration was slightly higher, 1.8×10^{17} cm^{-3}. At atmospheric pressure the absorption shows the peak at 0.9 eV from the Cr d^4 internal transition. As the pressure is increased, this peak decreases in strength and another absorption, slowly rising with increasing photon energy, appears. At 10.5 kbar, the maximum pressure reached, the Cr d^4 peak has almost disappeared, as shown in Figure 18. The resistivity increases between atmospheric pressure and 10.5 kbar by two orders of magnitude. The data were interpreted as showing the d^5-d^4 level to lie above the conduction band edge at atmospheric pressure and 77K. All the chromium atoms are then in the d^4 configuration and show the corresponding internal transition. With increasing pressure, the conduction band edge rises above the d^5-d^4 level and most of the chromium is in the d^5 configuration. Consequently the d^4 absorption decreases, and an absorption presumably associated with photoionization of Cr d^5 appears. The free electron concentration drops as electrons go into the d^5-d^4 level and hence the resistivity increases. Quantitative interpretation of the data is not possible because of the gross inhomogeneity of the sample for which spectra were presented. Electron spin resonance shows that, in this sample, roughly 10% of the chromium is in the d^3 configuration and some in the d^2 configuration.[61] Hennel and Martinez[81] attempted a quantitative fit of Hall measurements on a sample where the free electron concentration at 1 bar was apparently substantially greater than the chromium concentration so that at 12 kbar only 28% of the electrons have frozen onto d^5-d^4 levels. They deduced the level to be 55 meV above the band edge at atmospheric pressure at 77K, and to move down at a rate of 6.5×10^{-6} eV/bar with respect to the band edge. No information was given to show that the material was more uniform than their other samples, and as the Hall effect is affected by inhomogeneities it is not pos-

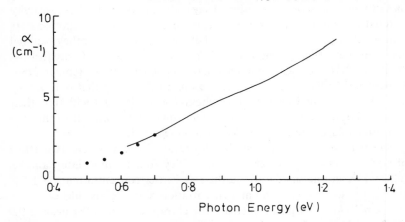

FIGURE 18 The absorption spectrum of a transition observed by Hennel et al.[80] in GaAs:Cr under hydrostatic pressure, here tentatively ascribed to ionization from Cr d^5 to the conduction band, compared with photocapacitance data of Szawelska and Allen[63]. The two sets of data are scaled at one point, although the difference in temperatures makes this an approximate procedure.

sible to give an estimate of the accuracy of the deduced values. The mobility did not behave as predicted and Hennel and Martinez postulated impurity pairing to explain this, but sample inhomogeneity is at least as likely an explanation.

The experiments described in this section so far use measures of the equilibrium occupancy of states (i.e. measures of the relative positions of the Fermi level and the impurity level) to get information about the position of the d^5-d^4 level. Szawelska and Allen[63] instead studied photoionization transitions, by using transient photocapacitance in a Schottky diode. Levels are filled electrically by forward biasing, then emptied optically. At liquid nitrogen temperatures and above, they saw the d^4-d^3 level, while at liquid helium temperature, they saw a shallower level with the same concentration as the d^4-d^3 level. The photoionization spectrum had the slow rise typical of strong phonon coupling. This is illustrated in Figure 19, which compares the spectrum of this second level with the ZnSe:Cr photoionization spectrum[82] for d^5-d^4 which also has strong phonon coupling, and with the GaAs:Cr d^4-d^3 spectrum which has weak coupling and rises much more sharply. Szawelska and Allen deduced that the energy of a vertical transition on a configuration coordinate diagram for the second level

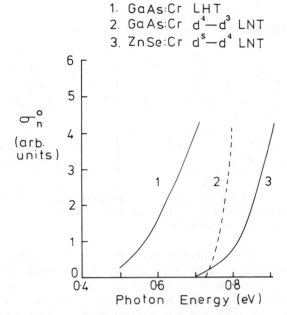

1. GaAs:Cr LHT
2. GaAs:Cr d^4—d^3 LNT
3. ZnSe:Cr d^5—d^4 LNT

FIGURE 19 A comparison of the shapes of the photoionization spectra for (1) a transition possibly ionization from d^5 to the conduction band in GaAs:Cr as in Figure 18, (2) ionization from d^4 to the conduction band in GaAs:Cr as in Figure 16 and (3) ionization from d^5 to the conduction band in ZnSe:Cr.

is about 0.5 eV, and unwisely represented this energy on an ordinary energy level diagram, i.e. as lying deep in the band gap. Such a scheme is inconsistent with their observation that electrons in the level emptied rapidly to the conduction band at liquid nitrogen temperature. They identified the level with the d^5-d^4 level of chromium because its concentration equalled that of the d^4-d^3.

The shape of the spectrum is certainly like that for d^5-d^4 in ZnSe:Cr, and, as Figure 18 shows, it is consistent with that observed for what is probably the d^5-d^4 transition in GaAs:Cr seen by Hennel et al. at 77K under hydrostatic pressure.

In the II-VI compounds, the difference between the thermal and optical ionization energies of d^4-d^5 are typically 0.3-0.4 eV,[83] apparently from interaction with a totally-symmetric (A_1) mode. One therefore expects the same order of magnitude for the GaAs:Cr d^5-d^4 transition. The levels must then, at the least, be plotted on a configuration coordinate diagram

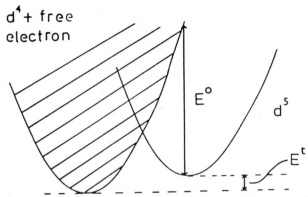

FIGURE 20 A schematic configurational coordinate diagram for ionization from d^5 to the conduction band in GaAs:Cr. The optical energy is E^o and the thermal energy is E^t, here taken to be negative.

as in Figure 20 rather than an ordinary energy-level diagram. If, in accordance with the data presented above, E^o is $\simeq 0.5$ eV and E^t is small, then the barrier for ionization is $\simeq 0.1$ eV. With this magnitude, thermal passage over the barrier would be fast at 77K on the time scale involved in, say, hydrostatic pressure experiments, but the level could be stable or metastable at helium temperatures depending on the sign of E^t.

In summary, the weight of the evidence is that the d^5-d^4 level of GaAs:Cr lies within ± 50 meV of the conduction band edge at liquid helium and liquid nitrogen temperatures, and that there is strong phonon coupling in the ionization transition. A repetition of the hydrostatic pressure measurements[36,80,81] with more homogeneous material would be valuable in giving a more precise value of the thermal energy, and time-dependent filling and emptying experiments at low temperatures would determine if there is an energy barrier to be surmounted.

4. Luminescence

4.1 Introduction

A number of luminescence bands have been associated with the presence of chromium in GaAs. Luminescence from an internal transition of Cr d^4, when the conduction band edge is raised above the excited level by hydrostatic pressure, has already been described in Section 2.3. Here only two other luminescent bands will be discussed. A few centers with large oscillator strength may give strong luminescence while a large number of centers with large non-radiative rates may give negligible luminescence. It is therefore not always easy to determine whether a luminescence band comes from a deliberately added impurity or from an unwanted impurity, nor whether the luminescence comes from a small fraction of the impurities with some special position. This is why some luminescence bands claimed to be connected with chromium will be ignored when the evidence is too slight for rational discussion.

4.2 The 0.84 eV Luminescence

Considerable confusion has been caused in the past by a luminescent band in GaAs:Cr with a peak at 0.8 eV. Attempts to fit the excited level into an energy level scheme for isolated substitutional chromium led to contradictions. It is now clear that the luminescence comes from only a small fraction of the chromium atoms, which are associated with a neighboring defect.

Luminescence at low temperatures at 0.8 eV was reported briefly by Turner and Pettit[84] in 1964 and by Allen[52] in 1968. At the international Conference on the Physics of Semiconductors at Rome in 1976 two groups[85,86] reported better-resolved spectra at liquid helium temperatures. Optical excitation with $h\nu > E_G$ was used. Both groups found a narrow line at 0.839 meV with a phonon band peaking at about 30 meV lower energy (Figure 21). A slight additional peak appears at 0.75 eV on the tail of the phonon band. The suggestion that this is produced by another impurity is disproved by the fact that several authors have since reported the spectrum and find no significant variation of the 0.75 eV feature between samples. Structure on the phonon band was attributed to TA(X) and LO(Γ) phonons, and it was noted that interaction with the TA phonons was stronger than that with LO phonons. Koschel et al.[87] used Hopfield's model of phonon coupling with two modes and found the mean number of phonons (i.e. the Huang-Rhys factor S) to be 2.1 for the TA mode and 0.9 for LO. Although the fitting is not a correct procedure, these numbers are a fair indication of the relative strengths of the coupling. Stocker and Schmidt[85] also measured the splitting of the narrow line under uniaxial stress.

Stocker and Schmidt noted that the narrow line was near the threshold of the absorption and photoconductivity bands which had been attributed to the ionization of an electron from Cr d^4 to the conduction band. They therefore proposed that the luminescence comes from the inverse transition, i.e. recombination of a conduction band electron with Cr d^3 to give

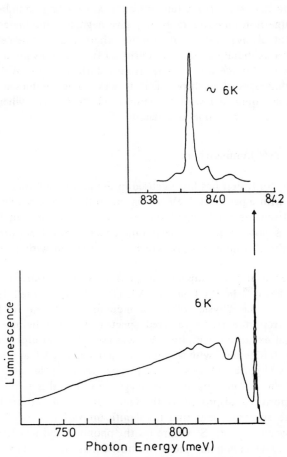

FIGURE 21   The 0.84 eV luminescence of GaAs:Cr and some of the fine structure
associated with it, after Lightowlers et al. (Refs. 90, 91).

Cr d^4. There are two serious objections to this attribution. The first is
that although 0.893 eV is near the threshold energy for photoionization
of Cr d^4 at helium temperatures it is in fact greater than it (see Section
3.2.3.). The second is that the spectrum of the recombination of free carriers with an impurity has an asymmetric shape and a width approximately 1.8 kT, because of the thermal distribution of the free carriers.
Ushakov et al.[88] showed that, at 20.4K, the width of the 0.84 eV line is
less than 1 meV whereas kT at this temperature is 1.7 meV. Koschel
et al.[86,87] attributed the luminescence to the ^5E–$^5T_2$ transition within the

Cr d^4 configuration and ignored the problem that, because the excited state would be in the conduction band, autoionization might be sufficiently fast to make radiative transitions have low probability. (A sufficiently strong phonon coupling could, however, decrease the autoionization rate and make a radiative transition non-negligible). The $^5E-^5T_2$ transition of isolated substitutional Cr d^4 has since been observed, and is found to have a zero-phonon line at 0.82 eV (see Section 2.3).

Ushakov et al. also noted that the 0.84 eV luminescence was not seen in strongly n=type material, nor if the Fermi level were too low. They interpreted these results as showing that if the Fermi level were too high or too low then the appropriate chromium charge state for luminescence was not occupied. Others have subsequently pursued the same line of thought. Caution is necessary in interpreting the effects of experimental variables on the intensity of luminescence, because excitation, energy transfer and transitions within the luminescent center may each be affected. For example, the absence of luminescence in strongly n-type material can be the effect of Auger quenching[89] rather than a change of charge state, and also the charge state occupancy under excitation may be quite different from that in the dark.

A big step forward was made by Lightowlers and his colleagues[90,91] who made measurements of cathodoluminescence and absorption at much higher resolution than before. They found that the sharp line at 0.84 eV in fact was not a single line but had considerable structure. A typical spectrum is shown in Figure 21. The strongest line, called G, is at 839.37 ± 0.05 meV with width (FWHM) ~0.15 meV at 2K. Measurements of the fine structure could be made up to 25K before thermal broadening washed out the details. From the temperature dependence of the intensities of the lines it could be deduced which lines were from the lowest state and which from thermalized states. A total of 13 lines were seen. An energy level scheme with five levels in the ground state and seven in the excited state was deduced. This is shown in Figure 22, where the transition leading to the strong line G is indicated. To get adequate absorption to give a good signal strength, Lightowlers et al. used samples which were 5 cm thick. They found that the 0.839 eV absorption is not seen in low-resistivity n-type material containing chromium: this absence of absorption is more significant than the absence of luminescence.

Because the ground-state level scheme found by Lightowlers et al. is inconsistent with the levels of isolated chromium in any charge state found by esr, White[92] suggested that the 0.84 eV luminescence comes from a small fraction of the chromium atoms in some unusual configuration. (This illustrates a general problem of luminescence studies, that a strong signal may result from a few impurities, giving a distorted view of their

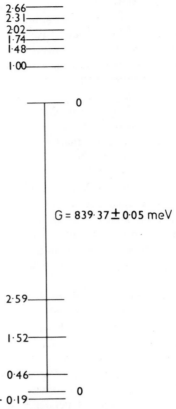

FIGURE 22 Energy level scheme for the 0.84 eV absorption and luminescence, after Lightowlers, Henry and Penchina[91].

relative significance.) If only a small fraction is involved, this accounts for the need of Lightowlers et al. to use exceptionally long path-lengths in absorption measurements even though the total chromium concentration was 10^{17} cm^{-3}. It also accounts for the fact that spin resonance of the luminescent center is not observed. White proposed a model for the center in which a chromium atom on a gallium site is associated with a donor impurity on a nearest-neighbor arsenic site in a [111] direction. Specifically, he proposed that the center is neutral over-all, with the ground state being $(Cr\,d^4)^- + D^+$. In the excited state, one electron is removed from the chromium d-shell to a loosely-bound orbit where it is attracted by the Coulomb potential of the resultant $(Cr\,d^3)^0 + D^+$. The simplicity of this scheme was obscured by a convoluted terminology in which the luminescence becomes "excitonic recombination at a deep iso-electronic trap." In the ground state of White's model, the $^5T_2$ levels of Cr d^4 are split by a field of $C_{3v}(3m)$ symmetry from the donor. In the excited state, the $^4T_1$ levels of Cr d^3 are split by the C_{3v} field but there is additional splitting from coupling of these levels with the spin of the loosely-bound electron. Zeeman effect studies of the luminescence by Eaves et al.[93] showed that the luminescent center appeared to have C_{3v} symmetry, in agreement with White's model of a chromium-donor complex. Killoran et al.[94] also made similar measurements. They showed that, although the symmetry of the center is approximately C_{3v}, it does not have exactly axial symmetry. Instead, the symmetry seems to be orthorhombic with principal axes in the $[111]$, $[1\bar{1}0]$ and $[\bar{1}\bar{1}2]$ directions. This is consistent with a Jahn-Teller distortion of a predominantly C_{3v} center. Only the two strongest luminescent lines were investigated.

An alternative to White's model was put forward by Picoli et al.[95,96] Again the ground state is that of the neutral center $(Cr\ d^4)^- + D^+$ but the excited state is within the same configuration, i.e. excitation is within the d-shell instead of involving the removal of an electron from it. The ^5D level of d^4 is split by the $C_{3v}(3m)$ field imposed by the donor into ^5E and $^5A_1$ levels, as in Figure 23. In order to explain the large number of levels in the ground state observed experimentally, ^5E must be lower than $^5A_1$ and this requires the nearest-neighbor to be positively charged, i.e. a donor rather than acceptor. A proper treatment of the level structure requires the crystal field, the spin-orbit coupling and the Jahn-Teller interaction to be taken into account simultaneously. This is a formidable task, and some approximations are necessary to make the problem tractable with reasonable effort. It is known that, for Cr d^4 on a tetrahedral site in II-VI compounds, the Jahn-Teller effect is small in the ^5E upper state but large in the $^5T_2$ ground state. Picoli et al. correspondingly neglect the Jahn-Teller effect in the excited state and consider only the second-order spin-orbit and C_{3v} crystal field interactions. In the lower ^5E state derived from $^5T_2$, they consider a dynamic Jahn-Teller effect and a spin-orbit coupling reduced by the Ham effect. Picoli et al. claim fairly good agreement between experiment and theory for the number and spacing of the levels in the ground and excited states. They also suggest that the transition from the upper ^5E to the $^5A_1$ level in Figure 23 is the origin of the slight peak at 0.75 eV which is always seen on the 0.8 eV band. This assignment must be regarded as tentative, for the evidence lies in the resolution of the luminescent band into two Gaussian components corresponding to these ^5E-$^5A_1$ and ^5E-^5E transitions: this may be an oversimplification. Because of the symmetry between electrons and holes in an $n\ell$ shell, a model of $(Cr\ d^6)^- + A^+$ would give the same energy spectrum. The chromium would then be

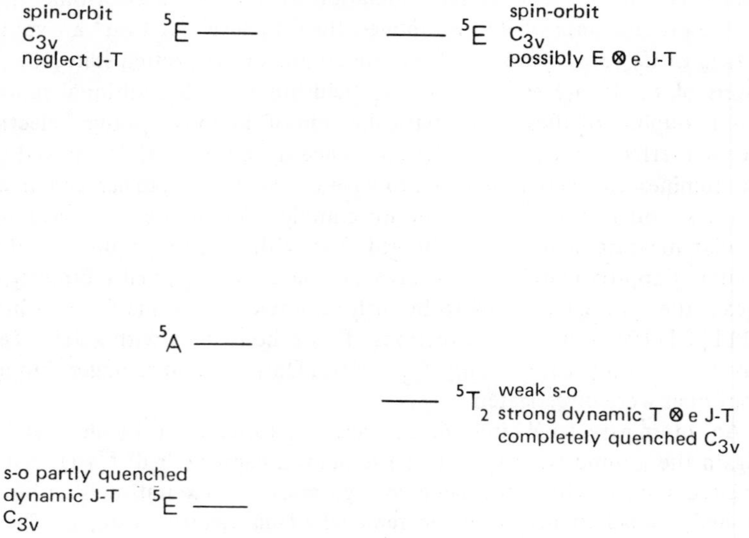

spin-orbit ^{5}E ——   —— ^{5}E   spin-orbit
C$_{3v}$ C$_{3v}$
neglect J-T   possibly E ⊗ e J-T

                 ^{5}A ——

                                            —— ^{5}T$_2$ weak s-o
                                                           strong dynamic T ⊗ e J-T
                                                           completely quenched C$_{3v}$

s-o partly quenched
dynamic J-T ^{5}E ——
C$_{3v}$

FIGURE 23 A comparison of the models of Picoli et al. and of Voillot et al. for the energy levels of the complex giving the 0.84 eV luminescence and absorption.

interstitial. An argument against this attribution is that the tetrahedral crystal-field splitting observed is not far from that of the isolated substitutional chromium (as witness the zero-phonon energies 0.84 eV and 0.82 eV) and this is unlikely if the chromium were interstitial. Eaves et al.[93] interpreted their Zeeman data above 2T with the model of Picoli et al.

Voillot et al.[97,98] advocated a modification of this model. Like Picoli et al. they considered the effect of the C$_{3v}$(3m) crystal field and second-order spin-orbit interactions in the ^{5}E state, with negligible or small E × e Jahn-Teller coupling. They treat the ground state rather differently, considering the dynamic Jahn-Teller effect to be large enough to quench completely the effect of the C$_{3v}$(3m) crystal field on the ^{5}T$_2$ state, leaving only the partly quenched spin-orbit interaction and the T × e Jahn-Teller interaction. They too claim good agreement between theory and experiment for levels and for the Zeeman data of Killoran et al.[94] (which are almost identical with those of Eaves et al.[93]). They also made measurements of the effect of uniaxial stress on the luminescence and found good agreement between theory and experiment. An interesting point is that the dynamic Jahn-Teller distortion deduced is so great that it is near the static limit, i.e. it is near the situation deduced from esr for isolated chromium.

The models of Picoli et al. and of Voillot et al. are broadly similar, differing only in details. At present, the latter has been more rigorously tested against experiment than the former. If the Jahn-Teller, spin-orbit and crystal-field effects could be treated simultaneously exactly, then the two models would constitute different approximations to the exact one. The difference between the models of White and of the French groups is more fundamental and lies in the nature of the excited state. Evidence definitely favoring an excited state within the d^4 configuration comes from the work of Kocot and Pearson[99] on luminescence in $Al_xGa_{1-x}As:Cr$ crystals. They found that the peak of the luminescence band does not change, within rather large error limits, with composition, whereas the band-gap does. It is improbable that the ground state of the center remains at the same depth below the conduction band as the composition changes. On White's model the excited state contains a loosely bound electron so its energy should follow, at least roughly, the conduction band. The luminescence energy should then change with composition, contrary to experiment. However, the interpretation is complicated by the presence of luminescence from isolated Cr d^4 at higher aluminium concentrations.

In summary, it seems fairly well established that the 0.84 eV luminescence arises from a small fraction of the chromium impurities, which have a donor as a nearest neighbor. The configuration has overall neutrality, being $(Cr\ d^4)^- + D^+$. Luminescence and absorption are produced by transitions within the d-shell, slightly perturbed by the donor. The perturbation produces a wealth of fine structure in the spectra and there is still work to be done on the details of this. Lightowlers noted that the absorption at 0.84 eV is absent in strongly n-type material, which suggests that another charge state exists with its ionization level just below the conduction band edge. Since both the donor level and the d^5-d^4 level of isolated chromium are close to the band edge, the charge state is likely to have strong configuration interaction between

$$(Cr\ d^5)^{2-} + D^+ \quad and \quad (Cr\ d^4) + D^o.$$

A major unresolved problem is the identity of the donor.

4.3 The 0.56 eV Luminescence

The luminescence band at 0.57 eV apparently associated with the presence of chromium in GaAs was reported by Peka et al.,[100] who used optical excitation from 77K to 300K, and by Aleksandrova et al.,[101] who measured cathodoluminescence at 80K and room temperature. It was noted that the band overlapped other bands not necessarily associated with chromium. Koschel et al.[86,87] made measurements at lower temperatures and reported that the luminescence was seen in semi-insulating GaAs:Cr, but not in moderately n=type material (10^4 Ω-cm).

A major advance was made by Lightowlers and Penchina.[90] They made high-resolution measurements of the cathodoluminescence spectrum (Figure 24). Three zero-phonon lines were found, the strongest at 573.49 $\pm$ 0.02 meV the two weaker at 574.96 meV and 576.1 meV. The intensities of the weaker bands increased between 7.7 K and 21.2 K which suggests that the initial state is split into three levels between which thermalization

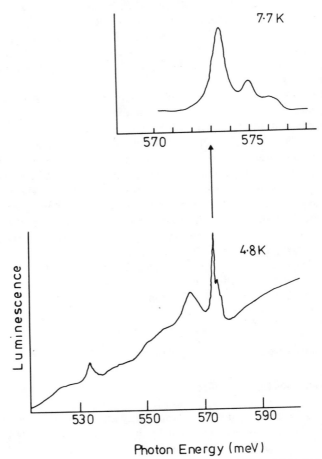

FIGURE 24   The 0.56 eV luminescence of GaAs:Cr and some of the fine structure associated with it, after Lightowlers et al.[90,91]

occurs. An interesting feature is that the lines are broad compared with those of the 0.84 eV spectrum. The width (FWHM) of the strongest is 1.0 meV at both 7.7K and 21K, and the two weaker lines have similar widths. No absorption corresponding with the luminescence has been found (Lightowlers, private communication). A local phonon mode at 40.7 meV from the strong line is seen in the luminescence.

The nature of the center responsible is not known. The lack of perceptible absorption suggests that only a small fraction of the chromium atoms is involved. If it is true that the luminescence is not seen in moderately n-type material then the center may require the Fermi level to be sufficiently low for it to be in the necessary charge state, but no quantitative statement is yet possible. Both Koschel et al.[86] and Leyral et al.[102] have found the excitation spectrum to be similar to that of σ_p^o for d^3-d^4, i.e. for the transition in which an electron in the valence band goes onto Cr d^3 to give Cr d^4 plus a free hole. They speculate that recapture of the hole may give an excited state derived from Cr d^3. The evidence is insubstantial, because the hole might be captured at some different center associated with chromium. (This illustrates a general problem in luminescence studies that excitation of one center can result from excitation of another center, followed by energy transfer or particle transfer). Picoli et al.[96] have attempted to accommodate the transition on a configuration coordinate diagram in which there is a large lattice relaxation in the d^4-d^3 transition of chromium associated with a donor, but they require the vertical transition to be at 0.57 eV, which is inconsistent with having the zero-phonon line at the same value, yet with strong phonon coupling. They base their model on the idea that the 0.84 and 0.57 eV bands are transitions within the same center, but this has yet to be shown.

In summary, it is probable that the 0.57 eV band comes from a small fraction of the chromium atoms. The nature of the center, e.g. whether it be a complex of substitutional chromium, or involves interstitial chromium, is not known.

5. Other Properties

This review is mainly about energy levels and transitions between them. No attempt is made to cover other properties, but a brief discussion of diffusion and solubility is necessary.

When chromium is diffused into GaAs, the resultant concentration profile is that which is usual for diffusion of transition metal impurities into compound semiconductors. There is a strong gradient near the surface and a very shallow gradient extending much further.[103] This behavior is readily explained if chromium diffuses rapidly on interstitial sites, slowly on substitutional sites, and if the interstitial concentration is much lower than the substitutional one. The diffusion data therefore suggest that some chromium is on interstitial sites at typical diffusion temperatures (e.g., 750–1150°C). There are insufficient data for one to predict the ratio of interstitial to substitutional chromium at room temperature.

Surprisingly, there has been no proper investigation of the solubility of chromium in GaAs. The equilibrium solubility of chromium in pure GaAs is, from the phase rule, a function of two thermodynamic variables. A convenient pair is the temperature and the ambient arsenic pressure. If the GaAs contains additional electrically active impurities such as shallow donors or acceptors the solubility will also be a function of their concentration. None of the variations has been studied quantitatively. Brozel et al.[76] found that, in material grown from the melt in horizontal boats, they could get chromium concentrations up to 5×10^{17} cm^{-3} when silicon was present predominantly as a donor, and observed qualitatively that, as the silicon concentration decreased, so did the chromium concentration. Their data are shown in Figure 25. Ippolitova et al.[43] measured the chromium concentration in Czochralski-grown GaAs, not necessarily at saturation. In material with no deliberately added donors and a background

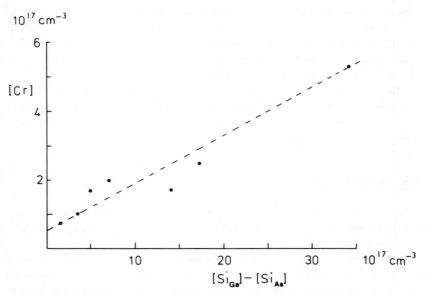

FIGURE 25 The chromium concentration in boat-grown GaAs as a function of the excess of donor silicon over acceptor silicon, after Brozel et al.[76]

donor concentration about 5×10^{15} cm^{-3}, they found chromium concentrations up to 3.3×10^{17} cm^{-3}. Tuck and Adegboyega[103] found surface concentrations in diffusion measurements up to 3×10^{17} cm^{-3} but, as they point out, the surface was not in equilibrium with other phases.

A combined study of diffusion and solubility can in principle provide data on the concentration of interstitial chromium as a function of total chromium concentration and of the thermodynamic variables. At present we have no sound knowledge of the properties of interstitial chromium.

6. Conclusions

Despite the considerable effort spent in the last twenty years, it is surprising how little is firmly established concerning the properties of GaAs:Cr. Some points have been labored over in great detail, such as fine structure in the 0.84 eV luminescence under different conditions. Elsewhere there are large gaps in our knowledge, such as the identity of the associated center in this luminescence. In this concluding section an attempt is made to sort what is known from what is not.

First we consider configurations. It is quite unsatisfactory that the evidence for Cr d^2 rests on a single broad esr line. In principle optical transitions to excited states should be seen in absorption in moderately p-type material, if the shallow acceptors do not complex with a high proportion of the chromium. The existence of Cr d^3 is more firmly established by its esr signal. Although the details are not settled, it is clear that there is an appreciable Jahn-Teller distortion. Again there are no optical data, although absorption to excited states should be visible in p-type semi-insulating material. (It is possible that the 0.57 eV luminescence is an internal transition within d^2 or d^3 and that no corresponding absorption has been seen because material with low enough Fermi level has not been investigated.) Chromium d^4 is seen in esr, optical absorption and luminescence. Again the Jahn-Teller distortion is established but the details are in dispute. Linewidths in optical absorption have been treated in an oversimplified configuration coordinate model and there is room for a more rigorous treatment. Until this has been done the strength of coupling to the lattice is in doubt. The second acceptor level of chromium has been associated here with the existence of Cr d^5 but the only evidence for this configuration is indirect, namely that it exists in II-VI compounds. Experiments seeking further confirmation would be welcome.

Next we consider energy levels. The position of the d^3-d^2 level is only known very roughly. Better data should be easy to obtain, so this gap in our knowledge should be closed soon. The d^4-d^3 level has been widely investigated because of its importance in producing GaAs close to its maximum resistivity. Thermal measurements have given the Gibbs energy to within 10 meV at and near room temperature. Because of the difficulty in deconvoluting electronc and vibrational contributions to the shape of photoionization spectra, there is uncertainty about the strength of vibrational coupling in transitions to and from the level. There is some divergence of opinion, with French groups maintaining that it is strong with $Sh\omega \simeq 0.15$ eV while other evidence suggests a lower value. Pressure experiments have shown that the d^5-d^4 level (if the configuration is indeed d^5) lies close to the bottom of the conduction and could well be above the band edge at liquid nitrogen and higher temperatures. These experiments need repeating with better quality material. If, as in the II-VI compounds, there is strong vibrational coupling in the transition this will need to be taken into account in the interpretation of experimental data.

Thermal transition rates for d^4-d^3 and d^3-d^4 at or near room temperature are known with good accuracy. Capture of an electron by the neutral d^3 center is two orders of magnitude less than capture of a hole by $(d^4)^-$. The difference seems too large to be explained only by the Coulomb potential. Photoionization cross-sections for transitions from d^4 to the conduction band as determined by different workers are in fair agreement, although there is dispute about the relative strengths of transitions directly into the band and those via the d^4 excited state. No good quantitative data are available concerning either thermal or optical rates for the d^3-d^2 and d^5-d^4 levels.

Because chromium has four accessible charge states in GaAs, it would be a good system for studying how the crystal field varies with the charge state of a given impurity in a particular host crystal. Similarly if the electron-electron interaction is parameterized by a Racah coefficient B, or by covalency coefficients, one would like to know how the coefficients vary with the charge state. It is therefore particularly disappointing that at present we have no knowledge of the excited states of Cr d^2, d^3 and d^5. The crystal field splitting Δ is about 4500 cm^{-1} for Cr d^4 in II-VI compounds and about 7000 cm^{-1} for Cr d^4 in GaAs, but without more information we cannot say if the difference is mainly because of the charge (Cr is $(d^4)^0$ in II-VI compounds and $(d^4)^-$ in III-V compounds) or covalency differences or a combination of the two effects.

There is at present no adequate theory with which to compare the experimental positions of the levels in the gap. Allen[104,11] used an interpolation scheme[10] and the experimental levels of GaAs:V and GaAs:Co to deduce that the Cr d^4-d^3 level is 0.73 eV below the conduction band. This is in good agreement with experiment but the method is only an interpolation scheme, not an ab initio one. Hemstreet and Dimmock[105] have made scattered-wave calculations for a five-atom cluster of chromium in GaAs, and Il'in and Masterov[106] used the Haydock-Heine-Kelly continued fraction method with a 29-atom cluster as a basis for extrapolation. Although these calculations are instructive, they are only semi-quantitative and cannot be compared directly with experiment. A surprising feature of the experimental data concerns the Mott-Hubbard U. (The value of U for d^4, for example, is the energy difference between the d^5-d^4 and d^4-d^3 levels.) If, in round numbers the d^5-d^4, d^4-d^3 and d^3-d^2 levels are 0, 0.7 and 1.1 eV below the conduction band, the values of U are 0.7 eV for d^4 and 0.4 eV for d^3. Because the half-filled d^5 configuration is especially stable, one expects the d^4 U-value to be the smaller, or in other words the d^5-d^4 level is expected to be lower than it is because of the d^5 stability. There is a vibrational contribution to the experimental value of U, but because the coupling in the d^5-d^4 transition is greater than that in the d^4-d^3 transition, this makes the discrepancy in the electronic contribution worse. An unresolved problem therefore exists.

Another intriguing result, and one which has caused confusion in the past, is that the excitation and ionization energies of Cr d^4 in GaAs are nearly equal so the excited level lies close to the conduction band edge. Also the d^5-d^4 level lies at approximately the same position. These near-coincidences are likely to be the result of chance rather than something more fundamental, as similar phenomena do not happen often with other combinations of host and impurity.

It may be thought that, after so much work, GaAs:Cr is well understood and only details remain to be settled. It has been the purpose of this review to show what is known and so show where large gaps remain in our knowledge. My conclusion is that although some areas are well founded, some others which are of fundamental importance have, as yet, hardly been touched.

Acknowledgements. I am deeply grateful to all those who helped me in the preparation of this review by sending me preprints and unpublished work. I am also grateful to those who gave me permission to use their data, especially Dr. E. C. Lightowlers and Dr. B. Deveaud for their spectra. I thank my wife, Dr. H. R. Szawelska, for her help in preparing and editing this work.

Note Added in Proof. J. J. Krebs and G. H. Stauss (Phys. Rev. B*26*,

2296 (1982)) have observed splitting under uniaxial stress of the electron spin resonance signal attributed to Cr d^2. This confirms the attribution to substitutional chromium in the d^2 configuration and is strong evidence against the signal being that of interstitial Cr d^5.

References

1. R. W. Haisty and G. R. Cronin, Proc. 7th Int. Conf. on Physics of Semiconductors, Paris 1964 (Dunod: Paris) p. 1161.
2. G. R. Cronin and R. W. Haisty, J. Electrochem. Soc. *111*, 874 (1964).
3. F. D. M. Haldane and P. W. Anderson, Phys. Rev. B*13*, 2553 (1976).
4. V. N. Fleurov and K. A. Kikoin, J. Phys. C: Solid State Phys. *9*, 1673 (1976).
5. K. A. Kikoin and V. N. Fleurov, J. Phys. C: Solid State Phys. *10*, 4295 (1977).
6. V. N. Fleurov and K. A. Kikoin, J. Phys. C: Solid State Phys. *12* 61 (1979).
7. G. W. Ludwig and H. H. Woodbury, Solid State Physics *13* 223 (1962).
8. M. G. Clark, J. Phys. C: Solid State Phys. *13*, 2311 (1980).
9. W. Shockley and J. T. Last, Phys. Rev. *107*, 392 (1957).
10. J. W. Allen, Proc 7th Int. Conf. on Physics of Semiconductors, Paris 1964 Dunod: Paris) p. 781.
11. J. W. Allen, Proc. Conf. on Semi-Insulating III-V Materials, Nottingham 1980. (Shiva; Orpington) p. 261.
12. R. I. Gloriozova and L. I. Kolesnik, Sov. Phys. Semicond. *12*, 66 (1978).
13. U. Kaufmann and J. Schneider, App. Phys. Lett. *36*, 747 (1980).
14. G. H. Stauss, J. J. Krebs, S. H. Lee, and E. M. Swiggard, Phys. Rev. B*22*, 3141 (1980)
15. U. Kaufmann and J. Schneider, Sol. St. Comm. *20*, 143 (1976).
16. G. H. Stauss and J. J. Krebs. Proc. Conf. Gallium Arsenide and Related Compounds, Edinburgh 1976 (IOP: Bristol) p. 84.
17. B. Frick and D. Siebert, Phys. Stat. Sol. (a) *41*, K185 (1977).
18. J. J. Krebs and G. H. Stauss, Phys. Rev. B*15*, 17 (1977).
19. G. H. Stauss and J. J. Krebs, Phys. Rev. B*22*, 2050 (1980).
20. M. Bacci, A. Ranfagni, M. Cetica, and G. Viliani, Phys. Rev. B*12*, 5907 (1975).
21. J. T. Vallin, G. A. Slack, and S. Roberts, Sol. St. Comm. *7*, 1211 (1969).
22. J. T. Vallin, G. A. Slack, S. Roberts, and A. E. Hughes, Phys. Rev. B*2*, 4313 (1970).
23. J. T. Vallin and G. D. Watkins, Phys. Lett. *37*A, 297 (1971).
24. J. T. Vallin and G. D. Watkins, Sol. St. Comm. *9*, 953 (1971).
25. J. T. Vallin and G. D. Watkins, Phys. Rev. B*9*, 2051 (1974).
26. B. Nygren and J. T. Vallin, Sol. St. Comm. *11*, 35 (1972).
27. G. Grebe and H.-J. Schulz, Phys. Stat. Sol. (b), K69 (1971).
28. G. Grebe and H.-J. Schulz, Z. Naturforsch. *29a*, 1803 (1974).
29. G. Grebe, G. Roussos, and H.-J. Schulz, J. Luminescence *12/13*, 701 (1976).
30. G. Grebe, G. Roussos, and H.-J. Schulz, J. Phys. C: Solid State Phys. *9*, 4511 (1976).
31. M. Kaminska, J. M. Baranowski, S. M. Uba, and J. T. Vallin, J. Phys. C: Solid State Phys. *12*, 2197 (1979).
32. J. J. Krebs and G. H. Stauss, Phys. Rev. B*16*, 971 (1977).

33. J. J. Krebs and G. H. Stauss, Phys. Rev. B20, 795 (1979).
34. H. Tokumoto and T. Ishiguro, J. Phys. Soc. Japan 46 84 (1979).
35. R. J. Wagner and A. M. White, Sol. St. Comm. 32, 399 (1979).
36. A. M. Hennel, W. Szuszkiewicz, M. Balkanski, G. Martinez and B. Clerjaud, Phys. Rev. B23, 3933 (1981).
37. V. Narayanamurti, M. A. Chin, and R. A. Logan, Appl. Phys. Lett. 33, 481 (1978).
38. E. M. Ganopol'skii, Sov. Phys. Sol. State 16, 1868 (1975).
39. P. Bury, L. J. Challis, P. J. King, D. J. Monk, A. Ramdane, V. W. Rampton, and P. Wiscombe, Proc. Conf. on Semi-Insulating III-V Materials, Nottingham 1980 (Shiva: Orpington) p. 214.
40. A. S. Abhvani, S. P. Austen, C. A. Bates, J. R. Fletcher, P. J. King, and L. W. Parker, J. Phys. C: Solid State Phys. 14 L199 (1981).
41. A. Goltzene, C. Schwab, and G. M. Martin. Proc. Conf. on Semi-Insulating III-V Materials, Nottingham 1980 (Shiva: Orpington) p. 221.
42. S. A. Abagyan, G. A. Ivanov, Yu. N. Kuznetsov, Yu. A. Okunev, and Yu. E. Shanurin. Sov. Phys. Semicond. 7, 989 (1974).
43. G. K. Ippolitova, E. M. Omel'yanovskii, and L. Ya. Pervova, Sov. Phys. Semicond. 9 864 (1976).
44. D. Bois and P. Pinard, Phys. Rev. B9, 4171 (1974).
45. B. Clerjaud, A. M. Hennel, and G. Martinez, Sol. St. Comm. 33, 983 (1980).
46. P. J. Williams, L. Eaves, P. E. Simmonds, M. O. Henry, E. C. Lightowlers, and Ch. Uihlein, J. Phys. C: Solid State Phys. 15, 1337 (1982).
47. J. M. Baranowski, J. M. Noras, and J. W. Allen, Proc. 12th Int. Conf. on Physics of Semiconductors, Stuttgart 1974 (Teubner: Stuttgart) p. 416.
48. B. Deveaud and G. Martinez, Third "Lund" Conf. on Deep Level Impurities in Semiconductors, Connecticut 1981, unpublished.
49. H. G. Drickamer, Solid State Physics 17, 1 (1965).
50. B. Deveaud, B. Lambert, H. L'Haridon, and G. Picoli, J. Luminescence 24/25, 273 (1981).
51. K. Kocot, R. A. Rao, and G. L. Pearson, Phys. Rev. B19, 2059 (1979).
52. G. A. Allen, Brit. J. App. Phys. Series 2, 1, 593 (1968).
53. D. C. Look, J. Phys. Chem. Solids 36, 1311 (1975).
54. D. C. Look, J. App. Phys. 48, 5141 (1977).
55. J. B. Mullin, D. J. Ashen, G. G. Roberts, and A. Ashby, Proc. Conf. on Gallium Arsenide and Related Compounds, Edinburgh 1976. (IOP: Bristol) p. 91.
56. J. Jimenez-Lopez, J. Bonnafe, and J. P. Fillard, J. App. Phys. 50, 1150 (1979).
57. C. E. Jones and A. R. Hilton, J. Electrochem. Soc. 113, 504 (1966).
58. H. J. Von Bardeleben, A. Goltzene, C. Schwab, and R. S. Feigelson, J. App. Phys. 52, 5037 (1981).
59. A. A. Gutkin, A. A. Lebedev, R. K. Radu, G. N. Talalakin, and T. A. Shoposhnikova, Sov. Phys. Semicond. 6, 1674 (1973).
60. A. M. Hennel, W. Szuszkiewicz, G. Martinez, and B. Clerjaud, Rev. Phys. Appl. 15, 697 (1980).
61. G. Martinez, A. M. Hennel, W. Szuszkiewicz, M. Balkanski, and B. Clerjaud, Phys. Rev. B23, 3920 (1981).
62. R. F. Broom, J. App. Phys. 38, 3483 (1967).
63. H. R. Szawelska and J. W. Allen, J. Phys. C: Solid State Phys. 12, 3359 (1979).
64. A. Chantre, G. Vincent, and D. Bois, Phys. Rev. B23, 5335 (1981).

65. D. V. Lang and R. A. Logan, J. Electronic Materials *4*, 1053 (1975).
66. A. Mitonneau, A. Mircea, G. M. Martin, and D. Bois, Revue Phys. Appl. *14*, 853 (1979).
67. G. M. Martin, A. Mitonneau, D. Pons, A. Mircea, and D. W. Woodard, J. Phys. C: Solid State Phys. *13*, 3855 (1980).
68. C. T. Sah, R. N. Noyce, and W. Shockley, Proc. IRE *45*, 1228 (1957).
69. J. S. Blakemore, J. App. Phys. *53*, 520 (1982).
70. C. D. Thurmond, J. Electrochem. Soc. *122*, 1133 (1975).
71. S. Makram-Ebeid, G. M. Martin, and D. W. Woodard, Proc. 15th Int. Conf. on Physics of Semiconductors, Kyoto 1980.
72. S. Braun and H. G. Grimmeiss, Sol. St. Comm. *11*, 1457 (1972).
73. D. Pons and S. Makram-Ebeid, J. de Physique *40*, 1161 (1979).
74. S. Makram-Ebeid, App. Phys. Lett. *37*, 464 (1980).
75. H. G. Grimmeiss, L.-A. Ledebo, and E. Meijer, App. Phys. Lett. *36*, 397 (1980).
76. M. R. Brozel, J. Butler, R. C. Newman, A. Ritson, D. J. Stirland, and C. White-head, J. Phys. C: Solid State Phys. *11*, 1857 (1978).
77. N. K. Goswami, R. C. Newman, and J. E. Whitehouse, Sol. St. Comm. *36* 897 (1980).
78. L. J. Challis and A. Ramdane, Proc. Int. Conf. on Phonon Scattering in Condensed Matter, Brown Univ. 1980 (Plenum: New York) p. 121.
79. L. J. Challis, M. Locatelli, A. Ramdane, and B. Salce, J. Phys. C: Solid State Phys. *15*, 1419 (1982).
80. A. M. Hennel, W. Szuszkiewicz, G. Martinez, B. Clerjaud, A. M. Huber, G. Morillot, and P. Merenda, Proc. Conf. on Semi-Insulating III-V Materials, Nottingham 1980 (Shiva: Orpington) p. 228.
81. A. M. Hennel and G. Martinez, Phys. Rev. B*25*, 1039 (1982).
82. H. R. Szawelska, unpublished.
83. M. Kaminska, J. M. Baranowski, and M. Godlewski, Proc. 14th Int. Conf. on Physics of Semiconductors, Edinburgh 1978 (IOP: Bristol) p. 303.
84. W. J. Turner and G. D. Pettit, Bull. Am. Phys. Soc. *9*, 269(A) (1964).
85. H. J. Stocker and M. Schmidt, Proc. 13th Int. Conf. on Physics of Semiconductors, Rome 1976, p. 611.
86. W. H. Koschel, S. G. Bishop, and B. D. McCombe, Proc. 13th Int. Conf. on Physics of Semiconductors, Rome 1976 p. 1065.
87. W. H. Koschel, S. G. Bishop, and B. D. McCombe, Sol. St. Comm. *19* 521 (1976).
88. V. V. Ushakov, A. A. Gippius, and B. V. Kornilov, Soc. Phys. Semicond. *12*, 207 (1977).
89. N. T. Gordon and J. W. Allen, Sol. St. Comm. *37*, 441 (1981).
90. E. C. Lightowlers and C. M. Penchina, J. Phys. C: Solid State Phys. *11*, L405 (1978).
91. E. C. Lightowlers, M. O. Henry, and C. M. Penchina, Proc. 14th Int. Conf. on Physics of Semiconductors, Edinburgh 1978 (IOP: Bristol) p. 307.
92. A. M. White, Sol. St. Comm. *32*, 205 (1979).
93. L. Eaves, Th. Englert, and Ch. Uihlein, "Physics in High Magnetic Fields." Springer Series in Solid State Sciences (1981).
94. N. Killoran, B. C. Cavenett, and W. E. Hagston, Sol. St. Comm. *35* 333 (1980).
95. G. Picoli, B. Deveaud, and D. Galland, Proc. Conf. on Semi-Insulating III-V Materials, Nottingham 1980 (Shiva: Orpington) p. 254.

96. G. Picoli, B. Deveaud, and D. Galland, J. de Physique *42*, 133 (1981).
97. F. Voillot, J. Barrau, M. Brousseau, and J. C. Brabant, J. Phys. C: Solid State Phys. *14*, 1855 (1981).
98. J. Barrau, F. Voillot, M. Brousseau, J. C. Brabant, and G. Poiblaud, J. Phys. C: Solid State Phys. *14*, 3447 (1981).
99. K. Kocot and G. L. Pearson, Sol. St. Comm. *25*, 113 (1978).
100. G. P. Peka and Yu. I. Karkhanin, Sov. Phys. Semicond. *6*, 261 (1972).
101. G. A. Aleksandrova, V. A. Vil'kotskii, D. S. Domanevskii, and V. D. Tkachev, Sov. Phys. Semicond. *6*, 266 (1972).
102. P. Leyral, F. Litty, S. Loualiche, A. Nouailhat, and G. Guillot, Sol. St. Comm. *38*, 333 (1981).
103. B. Tuck and G. A. Adegboyega, J. Phys. D: Appl. Phys. *12*, 1895 (1979).
104. J. W. Allen and G. L. Pearson, Stanford University Solid-State Electronics Laboratories Technical Report 5115-1 (1967).
105. L. A. Hemstreet and J. O. Dimmock, Phys. Rev. B*20*, 1527 (1979).
106. N. P. Il'in and V. F. Masterov, Sov. Phys. Semicond. *11*, 864 (1977).

CHAPTER 10

Chromium in II-VI Compounds

J. M. Baranowski

Institute of Experimental Physics
University of Warsaw
Warsaw, Poland

1. INTRODUCTION
2. ELECTRON SPIN RESONANCE
   2.1 $3d^4$ Configuration of Cr^{2+}
   2.2 $3d^5$ Configuration of Cr^{1+}
   2.3 $3d^3$ Configuration of Cr^{3+}
3. OPTICAL STUDY OF THE $Cr^{2+}(d^4)$ STATE
   3.1   The $^5\hat{B}$-$^5\hat{E}$ Absorption
   3.2   The $^5T_2$-^5E Absorption
   3.3   The ^5E-$^5T_2$ Luminescence
4. PHOTOIONIZATION PROCESSES
5. SUMMARY

1. Introduction

Chromium, an example of a transition-metal impurity, appears to be sub-stitutional on the cation site in II-VI compounds. A free $Cr(3d^54s^1)$ atom introduced into a II-VI compound provides two 4s electrons to bonding and the neutral chromium charge state is $Cr^{2+}(d^4)$. It has been found that the occupancy of the 3d-shell is preserved in going from the free atom to the impurity in the solid. Therefore chromium is an example of a highly

691

correlated many-electron system. An electron may be added to or removed from the d-shell giving $Cr^{1+}(d^5)$ or $Cr^{3+}(d^3)$. The electronic configuration may be established by electron spin resonance, which also gives the site symmetry. Optical absorption to excited states of the d-shell gives the configuration and site symmetry as well. The two methods are complementary; the $Cr^{2+}(d^4)$ is detected by the optical method and the $Cr^{1+}(d^5)$, $Cr^{2+}(d^4)$ and $Cr^{3+}(d^3)$ by electron spin resonance.

For transition metal impurities, neither the one electron approximation nor the expansion of the impurity wave function in terms of Bloch or Wannier functions of the pure crystal is valid. Instead of these approaches, experimentalists consider transition-metal impurities as atomic systems immersed in a modifying environment. Most of the experimental data on chromium impurities in II-VI compounds have been successfully interpreted within the framework of the above model.

The ESR investigations of the Cr^{2+}, Cr^{1+} and Cr^{3+} states will be described in Section 2. The present stage of optical investigation of the excited states of the $Cr^{2+}(d^4)$ will be presented in Section 3. The photoionization processes of the Cr impurity and the resulting positions of the acceptor and donor levels in II-VI compounds will be discussed in Section 4.

2. Electron Spin Resonance

2.1 $3d^4$ Configuration of Cr^{2+}

The ESR spectra of the cation substitutional $Cr^{2+}(d^4)$ charge state have been reported in almost all II-VI compounds: ZnS in Refs. 1 and 2; ZnSe in Refs 1, 2, 3, and 4; ZnTe in Refs. 1 and 2; CdS in Refs. 1, 5, 6, 7, 8, 9, 10, 11, 12, and 15; and CdTe in Refs. 1, 2, and 13. In all the II-VI compounds studied, a complex anisotropic spectrum at $\leqslant 4.2$ K was observed and identified with Cr^{2+}. The spectra have been analyzed with a spin Hamiltonian of the following form:

$$H_s = \mu_B \, g_{/\!/} \, S_z \, H_z + \mu_B \, g_\perp \, (S_x H_x + S_y H_y) + D \, S_z^2 + E(S_x^2 - S_y^2) +$$
$$\frac{F}{180} \left[35 \, S_z^4 - 30 \, S(S+1) \, S_z^2 + 25 \, S_z^2 \right] + \frac{1}{6} a \left[S_1^4 + S_2^4 + S_3^4 \right]$$

where $\mu_B = \frac{eh}{2mc}$ is the Bohr magneton, $\overline{H}$ is the magnetic field and $S = 2$. For ZnS, ZnSe, ZnTe and CdTe, $M_s = \pm 4, \pm 2, \pm 1$ transitions have been observed allowing complete analysis of the $S = 2$ spin Hamiltonian.[1] For cubic crystals, the x, y, z principal axes have been found to coincide with the cubic (1, 2, 3) axes, and the ESR spectrum could be resolved into three identical spectra, each given by the above spin Hamiltonian but with the z axis of each along a different cubic axis (1, 2, 3). For cubic crystals, E = 0. For hexagonal crystals, six identical spectra have been observed. The spin Hamiltonian parameters determined by Vallin and Watkins[1] are given in Table 1.

The dominant fine-structure term in the spin Hamiltonian is D which is mainly responsible for the zero field splitting and the anisotropy in the ESR spectrum. It was shown by Vallin and Watkins[1] that the anisotropy in the ESR spectrum was due to a tetragonal static Jahn-Teller distortion around the Cr^{2+} ion. The ground $^5T_2$ state of the $Cr^{2+}(d^4)$ is likely to suffer a Jahn-Teller distortion. Using uniaxial stress,[1] it was found that the relative intensities of the spectral components associated with differently-oriented defects change dramatically versus stress. Careful measurements revealed that the changes reflected redistribution among the three defect orientations. This easy reorientation at low temperatures ruled out the alternative possibility that the anisotropy results from the Cr^{2+} ion paired to a defect or an impurity. It gave unambiguous evidence for tetragonal static Jahn-Teller distortion at the ground $^5T_2$ state of the Cr^{2+} impurity. The local symmetry around the Cr^{2+} impurity is reduced from T_d to D_{2d}. The absence of an E term in the spin Hamiltonian in cubic II-VI compounds reveals a pure tetragonal distortion. Stress-alignment experiments allowed one also to estimate the Jahn-Teller coupling coefficients and the magnitude of the Jahn-Teller energies as about ~ 500 cm^{-1}.[1] This estimation, based on the assumption that the ligands move under the applied stress in the same way as they would in the perfect lattice, overestimates by about 50% the Jahn-Teller energies obtained in a direct way.[45,46]

The large variations in the spin-Hamiltonian parameters D and a and different sign of D for tellurium compounds than for sulphur and selenium

TABLE 1 Spin Hamiltonian parameters for $Cr^{2+}(d^4)$ in II-VI compounds

	CdTe	CdS	ZnTe	ZnSe	ZnS
$g_{//}$	1.980	1.934	1.97	1.961	1.94
$g_\perp$	1.980	1.970	1.99	1.98	1.98
D cm^{-1}	+0.260	−1.805	+2.30	−2.48	−1.86
E cm^{-1}		+0.0225			
a cm^{-1}	+0.05	+0.150	+0.140	±0.024	+0.193
F cm^{-1}	−0.05				−0.14

compounds (Table 1) have been explained by a contribution of the spin-orbit interaction at the ligand cores.[1] Because the ligand spin-orbit constant is much larger than that of the $Cr^{2+}(d^4)$ ion, even small admixtures of ligand wave functions may have significant effects. Including a ligand contribution which varies substantially as the ligand changes from S to Se and to Te to the spin-orbit interaction of the $Cr^{2+}(d^4)$ ion leads even to a change of sign of the spin-orbit constant in CdTe and ZnTe. This causes reversed order of the spin-orbit levels in the ground $^5T_2$ state of Cr^{2+} in CdTe and ZnTe. The free-ion Cr^{2+} spin-orbit constant $\lambda = +57$ cm^{-1} is changed from $+46$ cm^{-1} for sulphur compounds to -57 cm^{-1} and -74 cm^{-1} for CdTe and ZnTe respectively.[1] This has been also confirmed in the analysis of the $^5E-^5T_2$ luminescence within $Cr^{2+}(d^4)$.[46]

The characteristic feature of the Cr^{2+} ESR spectrum in II-VI compounds is its strong temperature dependence. As the temperature is raised above ~ 8 K, the ESR line widths broaden abruptly.[1] The line width of the Lorentzian lines increased exponentially with increase of temperature, characteristic of a thermally activated process. Transitions over the barrier spearating three equivalent Jahn-Teller minima are clearly responsible for this process. The obtained activation energies ranging from 0.0035–0.0085 eV for different II-VI compounds[1] indicate a very small barrier, considerably smaller than the corresponding Jahn-Teller energies.

However, it was shown that the presence of the spin-orbit coupling, although it is exponentially quenched by the Ham effect,[16] drastically reduces the barrier between three equivalent Jahn-Teller minima.[46]

2.2 3d⁵ Configuration of Cr¹⁺

The ESR Cr^{1+} charge state has been found in all zinc compounds: ZnS in Refs. 14, 17, 18, 19, 20, 21, 22, 23, 24, 25, 26, 27, 28, and 29; ZnSe in Refs. 30, 31, 32, 33, and 34; ZnTe in Refs. 30, 31, 32, and 35; and CdTe in Refs. 36, 37, and 13. The stable Cr^{1+} signal has been found in samples which have been annealed in zinc or cadmium vapors, thus in samples with a position of the Fermi level close to the conduction band. Therefore it is believed that Cr^{2+} acts as an electron trap and produces an acceptor level which after filling gives the Cr^{1+} ESR signal. The ground state of the cation substitutional Cr^{1+} is the $^6A_1$ and the ESR spectra in cubic crystals have been described by the spin Hamiltonian characteristic for the $3d^5$ configuration with the spin $S = 5/2$:

$$H_s = \mu_o \, g \, \overline{S} \, \overline{H} + \frac{a}{6} \left[S_x^4 + S_y^4 + S_z^4 \right] + \overline{A} \, \overline{S} \overline{J} + \sum_n \overline{S} \, A_n' \, \overline{J}_n \,,$$

where the sum over n is taken over all nuclei having an appreciable interaction with the impurity spin.

The second term of H_s is the cubic field term, the third is the hyperfine interaction with the impurity nuclens and the fourth is the superhyperfine interaction with lattice nuclei. Values of the isotropic g-factors, the fine structure constant a, the hyperfine structure constant A and the average value for superhyperfine constant A' are given in Table 2.

In several cases the ESR signal of Cr^{1+} has been found under influence of additional illumination of light. A typical ESR spectrum of the light-induced Cr^{1+} charge state in $CdTe^{13}$ is shown in Figure 1. The characteristic feature of the Cr^{1+} photo-induced signal is that it can be observed at much higher temperatures than the Cr^{2+} signal. At low temperatures, the ESR signal has a metastable character. It is believed that a trapping of free holes is responsible for this process.[13,38]

The spectral dependence of the excitation of the Cr^{1+} ESR signal allowed the identification of the character of photoionization absorption observed in some II-VI compounds as due to the transfer of an electron from the valence band into the Cr acceptor level.[13,38] Detection of the Cr^{1+} charge state in ZnS, ZnSe, ZnTe and CdTe gives a direct evidence of an acceptor-like character of the Cr impurity in these compounds. In CdS and CdSe, the Cr^{1+} charge state has never been found and there are reasons to believe that the $Cr^{2+}(d^4)$ in these compounds do not behave as an acceptor but as a donor.

2.3 $3d^3$ Configuration of Cr^{3+}

There is much less information about the Cr^{3+} charge state in II-VI compounds. The ESR $Cr^{3+}(d^3)$ signal has only been reported in $ZnSc^{39,40}$ and $CdS.^6$ The ESR signal was anisotropic which is in agreement with the symmetry of the ground $^4T_1$ state of the $Cr^{3+}(d^3)$ which can suffer Jahn-Teller distortion. The ESR signal of Cr^{3+} has been found also to be photo-

TABLE 2   Spin Hamiltonian parameters for $Cr^{1+}(d^5)$ in II-VI compounds

	ZnS	ZnSe	ZnTe	CdTe
g	1.9995	2.0016	2.0023	1.9997
$a \times 10^{-4}$ cm^{-1}	3.9	5.35	6.6	3.1
$A \times 10^{-4}$ cm^{-1}	13.4	13.3	12.4	12.78
$A' \times 10^{-4}$ cm^{-1}	1.23	1.81	3.59	10.6

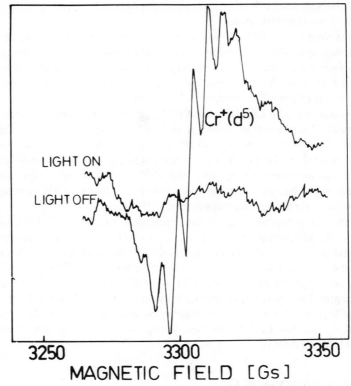

$Cr^+(d^5)$

LIGHT ON

LIGHT OFF

3250   3300   3350
MAGNETIC FIELD [Gs]

FIGURE 1   ESR spectrum of the light induced $Cr^{1+}(d^5)$ state in CdTe at T = 103 K.[13]

sensitive. The photosensitivity of the Cr^{3+} charge state indicates the possibility of the ionization process of the $Cr^{2+}(d^4)$ suggesting a donor-like character of the $Cr^{2+}(d^4)$ charge state in ZnSe and CdS.

3. Optical Study of the $Cr^{2+}(d^4)$ State

The optical study of the transition-metal ions in II-VI compounds has been used as a tool complementary to ESR for identification of the charge state of an impurity. It has been found that the crystal-field theory[41,42] is a good starting point for giving an account of the energy levels of the 3d-shell electrons of these ions. The $Cr^{2+}(d^4)$ state is a good example of such analysis. First the Jahn-Teller absorption within the split ground $^5T_2$ state will be described. The crystal field absorption $^5T_2 - ^5E$ and the corresponding $^5E-^5T_2$ luminescence are presented in sections 3.2 and 3.3 respectively.

3.1 The $^5\hat{B}-^5\hat{E}$ Absorption

It is known from EPR studies[1] that the $^5T_2$ state is strongly coupled to a Jahn-Teller active E-mode. This finding has been confirmed in optical study.[43,44,45,46] There are three energy minima corresponding to static distortions along each one of the three cubic axes. The local site symmetry of the impurity in each minimum is D_{2d} and the orbital $^5T_2$ splits into $^5\hat{B}$ and $^5\hat{E}$ states. The Hamiltonian for the $^5T_2$ state which includes E-mode Jahn-Teller coupling has the following form:[16]

$$\hat{H} = V_1 \, (Q_\Theta \, \hat{\epsilon}_\Theta + Q_\epsilon \, \hat{\epsilon}_\epsilon) + \frac{1}{2} m\omega^2 \, (Q_\Theta^2 + Q_\epsilon^2) \, \hat{I} + \lambda \, \hat{L} \, \hat{S} \,.$$

The first term describes the linear Jahn-Teller effect, V_1 is the coupling coefficient related to the Jahn-Teller energy $E_{JT}(^5T_2)$ by the expression $V_1^2 = E_{JT}(^5T_2)/2m\omega^2$, Q_Θ and Q_ϵ are the two E-mode distortions[47] and $\hat{\epsilon}_\Theta$ and $\hat{\epsilon}_\epsilon$ are the electronic operators defined by Ham.[48] The second term is the elastic energy, m is the ligand mass, ω is the frequency of the E-mode[48] and $\hat{I}$ is the electronic unit matrix. The third term corresponds to the spin-orbit coupling, and λ is the spin-orbit parameter. The above form of Hamiltonian corresponds to the static limit where the kinetic energy connected with momenta conjugate to distortion has been neglected.

The eigenvalues of $\hat{H}$ which gives the adiabatic potential surfaces for the $^5T_2$ state in the (Q_Θ, Q_ϵ) space have been calculated by Kaminska et al.[46]

The cross section through the adiabatic potential surfaces along one of the E-mode distortions is shown in Figure 2. The calculations presented in Figure 2 are appropriate for the ZnTe:Cr system and demonstrate the stable energy minimum of a static Jahn-Teller distortion. There are three such stable distortions corresponding to a compression along each one of the cubic axes. The broken curves in Figure 2 show the splitting of the orbital $^5T_2$ state without taking into account spin-orbit coupling. Near the stable energy minimum spin-orbit coupling is indeed quenched by the Ham effect[16] and has little influence on the separation between the lower $^5\hat{B}$ and upper $^5\hat{E}$ branch. This ordering of the $^5\hat{B}$ and $^5\hat{E}$ states is given by a negative value of the Jahn-Teller coupling coefficient V_1 and corresponds to a contraction of the tetrahedron. This separation is equal to $3E_{JT}(^5T_2)$ and thus absorption connected with $^5\hat{B}-^5\hat{E}$ transition provides a direct measurement of the Jahn-Teller energy.

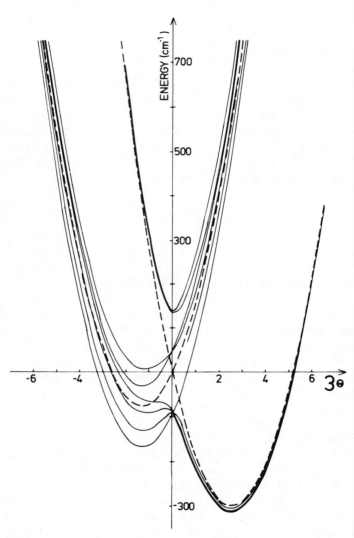

FIGURE 2 Cross section along the z_o axis for the ${}^5T_2$ state with coupling to an E-mode with $\hbar\omega = 50$ cm^{-1}, $\lambda = -74$ cm^{-1} and $E_{JT}({}^5T_2) = 320$ cm^{-1}, all parameters appropriate for ZnTe:Cr(d^4) system. The broken curves show the analogous cross section without spin orbit splitting but with unchanged values of $E_{JT}({}^5T_2)$ and $\hbar\omega$. The unit on the abscissa is $(\frac{\hbar}{m\omega})^{1/2}$.[46]

The absorption due to the $^5\hat{B}$–$^5\hat{E}$ transition has been found in ZnSe: Cr,[45,46] ZnTe:Cr,[46] and CdTe:Cr[46] in the 1000 cm^{-1} wavenumber region. The obtained Jahn-Teller energies and Jahn-Teller coupling coefficients are given in Table 3. The Jahn-Teller coupling coefficients have been obtained assuming that the Jahn-Teller active modes correspond to the TA phonons from the L point of the Brillouin zone.

The E_{JT} given in Table 3, with the exception of ZnS, have been obtained in the direct way with accuracy of a few percent. The E_{JT} obtained earlier from the analysis of the influence of uniaxial stress on the ESR spectra, thus not in a direct way, overestimated the E_{JT} by about 50%.[1] The E_{JT} collected in Table 3 do not show noticeable variation with the host lattice. It seems that they are characteristic for orbital triplet states also for other transition-metal impurities in II-VI compounds. For example, the established E_{JT} in the $^4T_1(F)$ term of Co(d^7) is 350 cm^{-1} and 425 cm^{-1} for ZnS[49] and ZnSe[50] respectively.

The Huang-Rhys parameter $S = E_{JT}/\hbar\omega$ (where $\hbar\omega$ is the energy of the Jahn-Teller active mode—after Ref. 46), given in Table 3 increases along the series ZnS, ZnSe, ZnTe and CdTe. This definite trend is connected with the energy of TA phonons of the host crystals, which are the Jahn-Teller active phonons, decreasing along the series.[1,44,46]

3.2 The $^5T_2$–^5E Absorption

Chromium occupies a cation site in the II-VI compounds and is, therefore, neutral with respect to the lattice where it is in the Cr^{2+}(3d^4) configuration. The ground ^5D state of the Cr^{2+}(3d^4) free ion is split by a crystal field of tetrahedral or predominantly tetrahedral symmetry. The result is a ground $^5T_2$ and an excited ^5E state. The separation between the $^5T_2$ and ^5E states is equal to Δ—the crystal field parameter. The higher excited states of the Cr^{2+}(d^4) ion, spin triplets and singlets, are also affected by the crystal field. However, their energies depend not only on the crystal

TABLE 3 Jahn-Teller energies E_{JT}, Jahn-Teller coupling coefficients V_1, and Huang-Rhys parameters $S = E_{JT}/\hbar\omega$ for the $^5T_2$ state of Cr^{2+}(d^4) in II-VI compounds

	$E_{JT}[cm^{-1}]$	$V_1[cm^{-1}\ \text{Å}^{-1}]$	S
ZnSe	340	2790	4.8
ZnTe	320	2460	6.4
CdTe	370	1850	10.6
ZnS	300	2150	3.3

field parameter but also on the two parameters describing interaction between d-electrons—B and C Racah parameters.[41,42] The energy diagram
for all states of the configuration has been calculated by Tanabe and
Sugano[51] and is shown in Figure 3. All optical and ESR measurements of
transition-metal ions in II-VI compounds indicate that the crystal field is
relatively weak. Therefore the ground state of divalent ions is always on
the side of high spin complex and the Cr^{2+} is no exception here.

The diagram given in Figure 3 allows one to understand the optical properties of Cr^{2+} ion in II-VI compounds. Taking into account selection rules
of optical transitions, it is evident that the strongest absorption should
result from the $^5T_2$–^5E transition. Due to the small value of the spin-orbit
coupling constant (free ion value $\lambda = 57$ cm^{-1}) spin stays a good quantum
number and the transition $^5T_2$–^5E is the only one spin-allowed within the

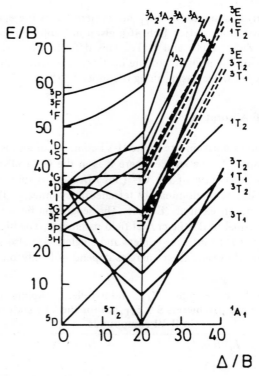

FIGURE 3   Energy levels of the $Cr^{2+}(d^4)$ state in T_d symmetry plotted against crystal field parameter Δ (both in units of Racah parameter B) as calculated by Tanabe
and Sugano.[51]

$3d^4$ configuration. The characteristic absorption has been found in all wide gap II–VI compounds: ZnS,[44,52,46] ZnSe,[53,44,46] ZnTe,[44,54,46] CdS,[55,44] CdSe,[56,57,58] and CdTe.[44,46,59]

The example of the absorption spectrum of the $Cr^{2+}(d^4)$ impurity in ZnTe[44] is shown in Figure 4. The broad absorption peak at about 5500 cm^{-1} wavenumbers with a very weak zero phonon line is due to the $^5T_2$–^5E transition. Forbidden optical transitions to upper spin triplet states are probably masked by strong photoionization absorption with a threshold at about 10000 cm^{-1} wavenumbers. This spectrum for ZnTe:Cr is representative for other II–VI compounds doped with Cr. Another example of the $Cr^{2+}(d^4)$ spectrum in CdS is shown in Figure 5. The main Cr peak position for various materials is given in Table 4. The presence of weak zero-phonon lines has been detected in ZnS, ZnSe, ZnTe and CdS. The energies of the zero-phonon lines (ZPL) are given in Table 4.

The intensity of the ZPL decreases along the series ZnS, ZnSe and ZnTe.[44] This is consistent with the trend of the Huang-Rhys parameter S given in Table 3. The intensity of the ZPL is roughly proportional to e^{-S}, and indeed the ZPL is strongest in ZnS, weaker in ZnSe, weakest in ZnTe and undetectable in CdTe.[44]

The variation of magnitude of the absorption coefficient of the $^5T_2$–^5E band versus the chemically established total Cr concentration for ZnSe is

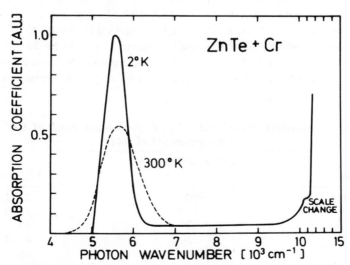

FIGURE 4 Optical absorption coefficient versus photon wavenumber for Cr-doped ZnTe.[44]

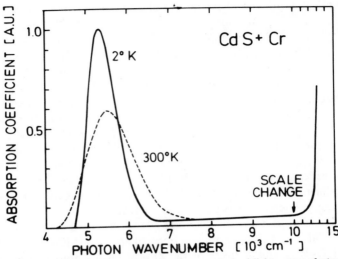

FIGURE 5   Optical absorption coefficient for unpolarized light versus photon wave-number for Cr-doped CdS.[44]

shown in Figure 6. The assumption that all of the Cr present in the crystal is Cr^{2+} gives a value $f = 5 \pm 1 \times 10^{-4}$ of the oscillator strength of the $^5T_2$-^5E transition.[44] The oscillator strength of this transition has also been determined for CdS—$4.5 \times 10^{-4} \leqslant f \leqslant 15 \times 10^{-4}$,[55] ZnS $f = 4 \times 10^{-3}$,[74] ZnTe—$f = 8.8 \times 10^{-4}$,[54] CdSe—$f = 5 \times 10^{-4}$[56] and CdTe—$f = 4.2 \times 10^{-4}$.[59] Variations of f observed from material to material seem to be due to experimental errors in establishing the Cr^{2+} concentration. Values of the oscillator strength for the d-d transitions of the order 5×10^{-4} indicate that

TABLE 4   Characteristics of the $^5T_2$-^5E absorption band of Cr^{2+} in II-VI compounds (Ref. 44)

Crystal	Peak position at 2 K cm^{-1}	ZPL cm^{-1}
ZnS cubic	5800	5224
ZnS hex	5780	5220
ZnSe	5525	4975
ZnTe	5530	4994
CdS	5300	4686
CdSe	5100	not observed
CdTe	5170	not observed

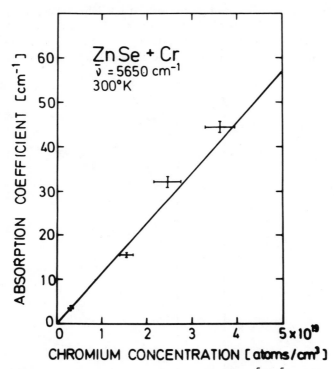

FIGURE 6   Optical absorption coefficient of ZnSe at the $^5T_2$-^5E absorption peak versus the chemically measured total Cr concentration. The temperature is 300 K.[44]

the odd parity wavefunction admixture within the core 3d region does not exceed a few percent.

To get a value of the crystal field parameter Δ responsible for the separation between the $^5T_2$ and ^5E state requires more detailed comparison between experiment and theory. It is important to treat the lattice as a dynamical system coupled to the electronic d-shell of the impurity ion. Besides the Jahn-Teller deformation, already discussed for the $^5T_2$ state, the presence of the Jahn-Teller effect in the ^5E state is important for evaluation of the crystal field parameter Δ. A schematic configuration-coordinate diagram for the $^5T_2$ and ^5E states both suffering Jahn-Teller effect when the spin-orbit interaction is neglected is shown in Figure 7. The symmetry-allowed optical transition $^5\hat{B}_2$-$^5\hat{A}_1$ is indicated with an arrow. Thus the center of gravity of the $^5T_2$-^5E absorption band is at an energy higher than the crystal field splitting Δ. In particular, the crystal field parameter Δ will depend on the sign of the Jahn-Teller coupling co-efficient V_2 in the ^5E state, which determines the order of the $^5\hat{A}_1$ and $^5\hat{B}_1$ levels. The sign of V_2 has been established in the combined luminescence-absorption experiments and will be discussed in the next section.

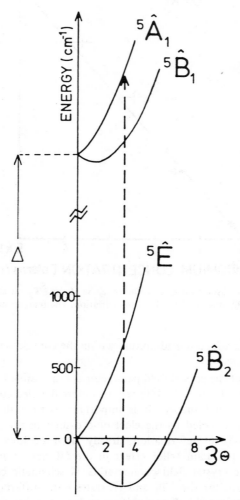

FIGURE 7 Schematic configuration coordinate diagram for the $^5T_2$ and ^5E states. The diagram is drawn for values of parameters corresponding to ZnSe:Cr ($V_1 < 0$, $V_2 > 0$). The allowed optical transition $^5\hat{B}$ $-^5\hat{A}_1$ is indicated with an arrow. The unit on the abscissa is $(\frac{\hbar}{m\omega})^{1/2}$.[46]

3.3 The $^5E-^5T_2$ Luminescence

In several II-VI compounds doped with Cr the luminescence corresponding to the $^5E-^5T_2$ transition has been found. This luminescence has been reported in ZnS:Cr,[60,61,62,46] ZnSe:Cr,[63,64,46] ZnTe:Cr,[46] and CdTe:Cr.[46] The luminescence consists of a broad triple-peak spectrum for ZnS, ZnTe and CdTe and a double peak for ZnSe.[46] A weak zero-phonon line has been found in ZnS:Cr[62] and ZnSe:Cr[63,64] which corresponds well to the absorption zero-phonon line of the $^5T_2-^5E$ transition.

It was shown that the shape of the luminescence spectrum of the Cr^{2+} impurity is determined mainly by the adiabatic potential surfaces of the $^5T_2$ state shown in Figure 2 and of the ^5E state. The presence of Jahn-Teller coupling in the ^5E state is a decisive factor giving the width of the observed band. For zero Jahn-Teller effect in the ^5E state one can expect the luminescence transitions in the center of the diagram shown in Figure 2. These transitions should give the shape of the resulting luminescence mainly determined by the spin-orbit structure of the $^5T_2$ state, which is known from ESR experiments.

It was found that it is necessary to include a small Jahn-Teller interaction in the ^5E state to be able to account for the broad shape of the observed spectra. The example of the $^5E-^5T_2$ luminescence and $^5T_2-^5E$ absorption in ZnSe:Cr is shown in Figure 8.[46] Monte-Carlo calculations within a semiclassical model are shown by histograms in Figure 8.[46] This comparison of theory and experiment allowed one to find the sign of the Jahn-Teller coupling coefficient in the ^5E state. The sign of this coupling coefficient decides the order of the $^5\hat{A}_1$ and $^5\hat{B}_1$ levels originating from the ^5E state shown in Figure 7. The agreement with the experiment is obtained only for positive sign of Jahn-Teller coupling coefficient V_2 which corresponds to the splitting of the ^5E state as shown in Figure 7. The determined Jahn-Teller energies and coupling coefficients for ZnS:Cr, ZnSe:Cr, ZnTe:Cr and CdTe:Cr are given in Table 5. The Jahn-Teller coupling coefficient has been obtained under the assumption that the dominant Jahn-Teller coupling is with TA phonons from the L point of the Brillouin zone, which are believed to be the Jahn-Teller active modes.[1,44,46]

The determination of Jahn-Teller energies in the ground $^5T_2$ and the excited ^5E states gave a more accurate determination of the crystal field parameter Δ in ZnS:Cr, ZnSe:Cr, ZnTe:Cr and CdTe:Cr systems. The obtained values of Δ[46] are given in Table 6.

Values of Δ given in Table 6 are about $1000\,cm^{-1}$ smaller than the energy of the $^5T_2-^5E$ absorption peak given in Table 4. Therefore the values of Δ for CdS and CdSe where the Jahn-Teller analysis has not been

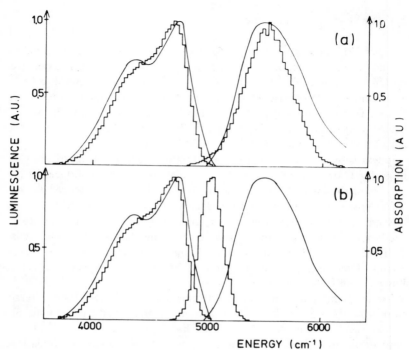

FIGURE 8 Calculated spectral histograms of luminescence (left-hand peaks) and absorption (right-hand peaks) for ZnSe:Cr for two sign of the Jahn-Teller coupling coefficient V_2: (a) $V_2 > 0$; (b) $V_2 < 0$. The experimental luminescence and absorption spectra are shown as smooth curves.[46]

done can be estimated at about 4300–4100 cm^{-1} respectively. Values of Δ for Cr(d^4) quoted throughout the literature are not far away from values given in Table 6. Very often they have been obtained without taking Jahn-Teller coupling into account or, in other cases, the values of Jahn-Teller energies have not been properly estimated.

TABLE 5 The Jahn Teller energies E_{JT} and coupling coefficient V_2 for the ^5E state of Cr^{2+}(d^4) in II-VI compounds[46]

	$E_{JT}[cm^{-1}]$	$V_2[cm^{-1}\,A^{-1}]$
ZnS	60	960
ZnSe	40	960
ZnTe	40	870
CdTe	40	610

TABLE 6   Crystal field parameter Δ for Cr(d^4) in II-VI compounds

	$\Delta \, [cm^{-1}]$
ZnS	4800
ZnSe	4600
ZnTe	4575
CdTe	4150

Excitation spectra of the ^5E-$^5T_2$ luminescence have been reported for ZnS[61] and ZnSe.[63] These spectra revealed several peaks up to the energy corresponding to the energy gap of ZnS and ZnSe. These peaks have been interpreted as due to transitions to higher triplet states of the $3d^4$ configuration. A fit to Tanabe-Sugano diagrams gave values of the Racah parameters $B = 500 \, cm^{-1}$ and $C = 2850 \, cm^{-1}$ for ZnS and $B = 510 \, cm^{-1}$ and $C = 3050 \, cm^{-1}$ for ZnSe. Obtained values of B are reduced compared with a free-ion value $B_o = 830 \, cm^{-1}$, which is in agreement with results for several other transitions-metal impurities in II-VI compounds. For example the reduction of B for Co impurity was about 50%–60%.[65,66] It is believed that a certain screening of d-d interactions by bonding electrons is responsible for this effect.[65] This can also be considered as a manifestation of some hybridization of the d-electrons. The obtained values of C for Cr(d^4) are rather high—the ratio of C/B is 5.7 and 6 for ZnS and ZnSe respectively. The ratio of C/B for all transition-metal free ions is close to 4–4.5 and it is generally preserved when ions are incorporated into the crystal lattice.

Besides the ^5E-$^5T_2$ luminescence, other bands have been reported at $7000 \, cm^{-1}$ and $10000 \, cm^{-1}$ for ZnSe:Cr.[63] However the origin of these bands is not yet identified.

4. Photoionization Processes

A characteristic feature of the absorption spectrum of Cr doped II-VI compounds is the presence of the strong photoionization absorption. The examples of this absorption for ZnTe:Cr and CdS:Cr are shown in Figures 4 and 5 respectively. The photoionization absorption has been found to be strongly temperature dependent in ZnS:Cr,[38] ZnSe:Cr,[38] ZnTe:Cr[67] and CdTe:Cr.[67] The temperature dependence of the normalized photoionization absorption spectrum in ZnSe:Cr is shown in Figure 9. The photoionization character of the transitions in this region has been unambiguously determined by photo-induced ESR. The photo-induced ESR signal charac-

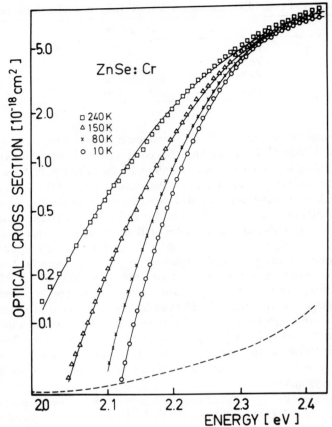

FIGURE 9 Photoionization absorption cross section spectra of Cr impurity in ZnSe for different temperatures. The excitation spectrum of ESR signal of $Cr^{1+}(d^5)$ at 175 K is shown by the dashed line.[38]

teristic of the $Cr^{1+}(d^5)$ charge state has been observed with the excitation spectrum corresponding to the absorption spectrum as shown in Figure 9. By this method it was confirmed that the photoionization absorption is due to an electron transition from the valence band to the deep Cr^{2+} level, as a result of which the Cr^{1+} charge state is created. The same situation as in ZnSe:Cr exists in ZnS:Cr,[38] ZnTe:Cr[67] and CdTe:Cr.[67]

The known oscillator strength of the crystal field $^5T_2-^5E$ transition in $Cr^{2+}(d^4)$ in ZnSe[44] allowed a determination of the Cr concentration N_{Cr} in the investigated sample and establishing the absorption cross section $\sigma(\hbar\omega) = \alpha(\hbar\omega)/N_{Cr}$, where $\alpha(\hbar\omega)$ is the photoionization absorption coefficient. The plot of the spectral dependence of the absorption cross section $\sigma(\hbar\omega)$ for ZnSe:Cr is given in Figure 9. The σ_{max} is of the order 7×10^{-18} cm^2 for ZnSe, 5×10^{-18} for ZnS[38] and 7×10^{-18} for ZnTe,[67] respectively.

The strong temperature dependence of $\sigma(\hbar\omega)$ indicates that the photo-ionization process has to be treated as an impurity-lattice relaxation process. The change of the number of the electrons in the 3d-shell may, due to electrostatic interaction with the nearest neighbors, change the average impurity-ligand distance. This leads to the configuration coordinate diagram for Cr^{2+} and Cr^{1+} as shown in Figure 10. The optical ionization energy E_I^{op} is different from the thermal ionization energy E_I^{th} by an amount equal to the lattice-relaxation energy. The thermal population of the vibronic levels in the ground state as the temperature is rising leads to an increase of the absorption below E_I^{op}. The analysis of the photoionization spectra shown in Figure 9 has been done for strong impurity-lattice coupling and for a semiclassical approximation.[68] The pure electronic

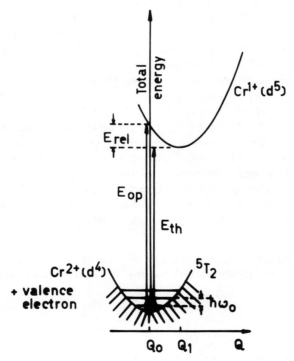

FIGURE 10   Configuration coordinate diagram for Cr^{2+} and Cr^{1+} charge states.

cross section $\sigma_{el} \sim (\hbar\omega - E_I)^{1/2} (\hbar\omega)^{-3}$ (Ref. 69) was used as for allowed optical transition from the p-type valence band to the d-type Cr acceptor level. The optical ionization energies E_I^{op} with respect to the top of the valence band and lattice relaxation energies of the $Cr^{1+}(d^5)$ obtained by this method are given in Table 7.

The spectral distribution of the photoionization cross section for transitions from the $Cr^{1+}(d^5)$ to the conduction band has been obtained in photocapacitance measurements of n-type ZnSe:Cr.[70] The optical ionization energy for this process was close to 1 eV. The sum of optical ionization energies for creation of the $Cr^{1+}(d^5)$ by the transfer of an electron from the valence band and annihilation of the $Cr^{1+}(d^5)$ by the transfer of an electron to the conduction band should be equal to the energy gap plus twice the lattice relaxation energy. The photocapacitance measurements of n-type ZnSe:Cr confirmed the large lattice-relaxation energy of the order 0.2–0.3 eV.

The obtained values of lattice-relaxation energy for Cr impurity in II-VI compounds are about one order of magnitude larger than the Jahn-Teller stabilization energy of the ground $^5T_2$-$Cr^{2+}(d^4)$ state. Because the ESR spectrum of the Cr^{1+} is isotropic—in agreement with the symmetry $^6A_1$ of the ground state, one can conclude that the lattice relaxation taking place in the change of Cr^{2+} to Cr^{1+} charge state is connected with totally symmetric distortion of the nearest neighbors.

The lattice relaxation energies have been obtained under the assumption of equal phonon energies in the ground $Cr^{2+}(d^4)$ and $Cr^{1+}(d^5)$ states which may be a crude approximation. The phonons which are most effectively coupled to the $^5T_2$ state of $Cr^{2+}(d^4)$ are the Jahn-Teller active ones, which bend the tetrahedron of the nearest neighbors. On the other hand, the $^6A_1$ state of $Cr^{1+}(d^5)$ can be coupled only to A_1 modes which stretch the tetrahedron. The local environment of the impurity may be softer for bending and harder for stretching. However, even if the lattice relaxation energies for II-VI compounds given in Table 7 are overestimated, they have been obtained in the same way and can be affected by some systematic

TABLE 7 Optical ionization energy E_{op}^I for the $Cr^{2+}(d^4) \rightarrow Cr^{1+}(d^5)$ process and lattice relaxation energy E_{rel} in II-VI compounds

	E_I^{op} eV	E_{rel} eV	Ref.
ZnS	2.78	0.37	38
ZnSe	2.26	0.33	38
ZnTe	1.46	0.16	71
CdTe	1.50	0.16	71

error. Therefore it is justifiable to compare them. They show some systematic trend: E_{rel} is the same for Te compounds, higher for Se, and highest for S. This suggests that only the nearest neighbors are important.

The study of the photoionization process of the Cr impurity in II-VI compounds allowed us for the first time to get a sufficient set of experimental data to look for a correlation between the ionization energy and the host crystal. The plot of the optical ionization energy of the Cr acceptor level against the top of the valence band calculated in the LCAO model is shown in Figure 11. It is known that the description of the valence band and the relative positions of the top of the valence band in different semiconductors are properly described by the LCAO model.[72] The trend obtained in Figure 11 is in full agreement with the theory based on the Anderson model which describes the positions of the transition metal impurity level in the II-VI compounds.[73] The positions of the top of the valence band for CdSe and CdS are indicated in Figure 11. The predicted position of the Cr acceptor levels in CdSe and CdS is at an energy higher than the corresponding energy gaps in these materials. Therefore, the Cr acceptor level is expected to be inside the conduction band of CdSe and CdS. This explains a puzzling fact that the ESR spec-

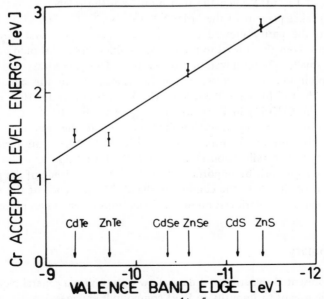

FIGURE 11   Experimental position of the $Cr^{1+}(d^5)$ in respect of the valence band in II-VI compounds versus the top of the valence band relative to the vacuum level calculated in LCAO model.

trum of $Cr^{1+}(d^5)$ has never been found in these materials. It is well established also that Cr in CdSe does not have an acceptor-like character.[58] Introducing Cr into CdSe leads to creation of a deep donor state only, connected with the $Cr^{2+}(d^5)$ charge state.[57,58] The ionization energy of the $Cr^{2+}(d^4)$ is 1.25 eV at 80 K.[58] Information about the presence of the $Cr^{2+}(d^5)$ donor level within the gap of other II-VI semiconductors is rather poor. The $Cr^{2+}(d^4)$ donor level at about 1.5 eV below the bottom of the conduction band of CdS has been reported.[55] This is consistent with ESR observation of the $Cr^{3+}(d^3)$ state.[6] The $Cr^{3+}(d^3)$ charge state has been also reported in ZnSe[39,40] indicating that it is possible to ionize the $Cr^{2+}(d^4)$. Unfortunately, the $Cr^{2+}(d^4)$ donor level in ZnSe is not known and, therefore, the energy which separates the $Cr^{1+}(d^5)$ and $Cr^{2+}(d^4)$ charge states is unknown. Nevertheless because three charge states Cr^{1+}, Cr^{2+} and Cr^{3+} have been observed in ZnSe that means that at least two levels, one acceptor and one donor, are present within the forbidden energy gap of ZnSe. Thus the energy which separates the $Cr^{1+}(d^5)$ and $Cr^{2+}(d^4)$ charge states is not higher than 1.5-2.0 eV.

There are also experimental results indicating that not all properties of transition-metal impurities in II-VI compounds are determined by the core 3d-shell region. The maximum of the photoionization cross section for creation of $Cr^{1+}(d^5)$ when the transition from the valence band into the $Cr^{2+}(d^4)$ takes place is of the order 5×10^{-18} cm^2. This transition corresponds to the parity-allowed one. Thus, one can expect that forbidden transitions from the d-level to the s-type conduction band should produce a much smaller photoionization cross section. This is contrary to the observed photoionization cross section for transitions to the conduction band for the $Cr^{2+}(d^4)$ in CdSe where $\sigma(h\omega)$ reaches the order 10^{-17} cm^2[58] and for the $Cr^{1+}(d^5)$ in ZnSe where σ_{max} is $6-8 \times 10^{-17}$ cm^2.[70] This high photoionization cross section for transitions to the conduction band indicates that the Cr impurity contains beside the 3d-like core region the odd parity p-like tail region coming from the host anion wave function. The tail region must be important at least in the magnitude of the photoionization of the Cr to the conduction band. The presence of the tail region indicates the hybridization of the d-electrons indeed takes place.

5. Summary

All the presented experimental data allow one to give a general picture in which the role of Cr impurity in II-VI compounds emerges.

The correlation of 3d electrons within the core region is important as the ESR and optical experiments indicate. Considering the chromium impurity as an atomic system immersed in a modifying environment seems to be valid. It is possible to add or remove an electron from the 3d shell which gives the acceptor and also donor level in several II-VI compounds. On the other hand, the photoionization cross sections indicate that in addition to the core region there is a tail region of the chromium impurity wave function. There is a definite need for additional experiments which could better clarify this basic problem.

Very little is also known about transport properties of II-VI's doped with chromium. The measurements of the negative magnetoresistance in CdSe:Cr[75] is perhaps the only one example of such investigation.

It is known that transition-metal impurities quench the band to band luminescence in II-VI compounds. The transfer of energy to the 3d shell takes place and luminescence within it often takes place. The mechanism of this energy transfer is not understood. The chromium impurity which can be in some compounds the electron trap, in others the hole trap, and finally, like in ZnSe, the trap for both carriers, could be an interesting system for an experimental study in which transfer of energy could be clarified. Unfortunately the quenching of the band to band luminescence by the chromium impurity in II-VI compounds has not been studied.

Also important and an open problem is whether incorporation of Cr impurity into the crystal lattice induces an additional lattice defect. There is lack in the literature of any information which could rule out or confirm this possibility.

Therefore, in spite of a progress in recent years of understanding of the role of transition-metal impurities in II-VI compounds, and chromium is a good representative of this class of impurities, there are still open and unsolved problems.

References

1. J. T. Vallin and G. D. Watkins, Phys. Rev. *B9*, 2051 (1974).
2. J. T. Vallin and G. D. Watkins, Sol. St. Comm. *9*, 953 (1971).
3. M. DeWit, A. R. Reinberg, W. C. Holton, and T. L. Estle, Bull. Am. Phys. Soc. *10*, 389 (1965).
4. K. Suto, M. Aoki, N. Nakada, and S. Ibuki, J. Phys. Soc. Japan *22*, 1121 (1967).
5. K. Morigaki, J. Phys. Soc. Japan *18*, 733 (1963).
6. K. Morigaki, J. Phys. Soc. Japan *19*, 187 (1964).
7. P. B. Dorain and D. Locker, Rev. Sci. Instr. *34*, 359 (1963).
8. J. T.Vallin and G. D. Watkins, Phys. Lett. *37A*, 297 (1971).
9. P. B. Dorain and D. Locker, Bull. Am. Phys. Soc. *1*, 306 (1962).
10. G. R. Wagner and J. G. Castle, Bull. Am. Phys. Soc. *11*, 906 (1966).
11. D. C. Look and D. R. Locker, Bull. Am. Phys. Soc. *14*, 835 (1969).

12. D. R. Locker and P. B. Dorain, Bull. Am. Phys. Soc. *15*, 249 (1970).
13. M. Godlewski and J. M. Baranowski, Phys. Stat. Sol. b *97*, 281 (1980).
14. A. Räuber, J. Schneider and F. Matosi, Z. Naturforsch. *17a*, 654 (1962).
15. T. L. Estle, G. K. Walters, and M. DeWit, Paramagnetic Resonance, edited by W. Low, Academic, New York, Vol. 1, p. 144, 1963.
16. F. S. Ham, Phys. Rev. *138*, A1727 (1965).
17. R. S. Title, Phys. Rev. *131*, 623 (1963).
18. J. Dieleman, R. S. Title, and W. V. Smith, Phys. Letters *1*, 334 (1962).
19. F. Matossi, A. Räuber, and F. W. Küpper, Z. Naturforsch. *18a*, 818 (1963).
20. T. Taki and H. Bo, J. Phys. Soc. Japan *25*, 1324 (1968).
21. H. D. Fair Jr., R. D. Ewing, and E. E. Williams, Phys. Rev. *144*, 298 (1966).
22. P. Jaszczyn Kopec, J. Gallagher, and B. Kramer, Phys. Rev. *A140*, 1309 (1965).
23. A. Raüber and J. Schneider, Z. Naturforsch *17a*, 266 (1962).
24. R. Parrot and C. Blanchard, Phys. Rev. *B5*, 819 (1972).
25. T. Buch, B. Clerjaud, and B. Lambert, Phys. Rev. *B7*, 184 (1973).
26. B. Lambert, C. Marti, and R. Parrot, J. Lumin. *3*, 21 (1970).
27. T. Ray, Sol. St. Commun. *9*, 911 (1970).
28. B. Lambert, T. Buch, and B. Clerjaud, Sol. St. Commun. *10*, 25 (1972).
29. H. Poppischer, W. Elssner, and H. Böttner, Phys. Stat. Sol. *27A*, 375 (1975).
30. R. S. Title, Phys. Rev. *133*, A1613 (1964).
31. T. L. Estle and W. C. Holton, Bull. Am. Phys. Soc. *10*, 57 (1965).
32. T. L. Estle and W. C. Holton, Phys. Rev. *150*, 159 (1966).
33. K. Suto, M. Aoki, M. Nakada, and S. Ibuki, J. Phys. Soc. Japan *22*, 1121 (1967).
34. H. Poppischer, H. Böttner, and W. Elssner, Phys. Stat. Sol. *B63*, K85 (1974).
35. K. Suto and M. Aoki, J. Phys. Soc. Japan *22*, 149 (1967).
36. G. Ludwig and M. R. Lorentz, Phys. Rev. *131*, 601 (1963).
37. M. Cieplak, M. Godlewski, and J. M. Baranowski, Phys. Stat. Sol. *70B*, 323 (1975).
38. M. Kaminska, J. M. Baranowski, and M. Godlewski, Proc. 14th Int. Conf. Phys. Semic. Edinburg 1978, Inst. Phys. Conf. Ser. No 43, 303 (1979).
39. R. Rai, J. Y. Savard, and B. Tousiguont, Phys. Lett. *25A*, 443 (1967).
40. R. Rai, J. Y. Savard, and B. Tousiguont, Can. J. Phys. *47*, 1147 (1969).
41. J. S. Griffith, The Theory of Transition-Metal Ions, Cambridge, At the University Press (1964).
42. S. Sugano, Y. Tanabe, and H. Kamimura, Multiplets of Transition Metal Ions in Crystals, Academic Press (1970).
43. J. T. Vallin, G. A. Slack, S. Roberts, and A. E. Hughes, Solid. St. Commun. *7*, 1211 (1969).
44. J. T. Vallin, G. A. Slack, S. Roberts, and A. E. Hughes, Phys. Rev. *B2*, 4313 (1970).
45. B. Nygren, J. T. Vallin, and G. A. Slack, Solid St. Commun. *11*, 35 (1972).
46. M. Kamińska, J. M. Baranowski, S. M. Uba, and J. T. Vallin, J. Phys. C: Solid St. Phys. *12*, 2197 (1979).
47. M. D. Sturge, Solid St. Phys. *20*, 91 (1967).
48. F. S. Ham, Phys. Rev. *166*, 307 (1968).
49. P. Koidl, O. F. Schirmer, and U. G. Kaufmann, Phys. Rev. *B8*, 4926 (1973).
50. S. M. Uba and J. M. Baranowski, Phys. Rev. *B17*, 69 (1978).
51. Y. Tanabe and S. Sugano, J. Phys. Soc. Japan *9*, 753 (1954).
52. C. S. Kelley and F. E. Williams, Phys. Rev. *B2*, 3 (1970).

53. E. M. Wray and J. W. Allen, J. Phys. C *4*, 512 (1971).
54. H. Komura and M. Sekinobu, J. Phys. Soc. Japan *29*, 1100 (1970).
55. R. Pappalardo and R. E. Dietz, Phys. Rev. *123*, 1188 (1961).
56. J. M. Langer and J. M. Baranowski, Phys. Stat. Sol. b *44*, 155 (1971).
57. J. M. Baranowski and J. M. Langer, Phys. Stat. Sol. b *48*, 863 (1971).
58. L. Jastrzebski and J. M. Baranowski, Phys. Stat. Sol. b *58*, 401 (1973).
59. P. A. Stodowy and J. M. Baranowski, Phys. Stat. Sol. b *49*, 499 (1972).
60. H. Nelkowski and G. Grebe, J. Lumin. *1/2*, 88 (1970).
61. G. Grebe and H. J. Schulz, Phys. Stat. Sol. b *54*, K69 (1972).
62. G. Grebe and H. J. Schulz, Z. Naturf. *29a*, 1803 (1974).
63. G. Grebe, G. Roussos, and H. J. Schulz, J. Lumin. *12/13*, 701 (1976).
64. G. Grebe, G. Roussos, and H. J. Schulz, J. Phys. C: Solid St. Phys. *9*, 4511 (1976).
65. J. M. Baranowski, J. W. Allen, and G. L. Pearson, Phys. Rev. *160*, 627 (1967).
66. A. M. Hennel, J. Phys. C: Sol. St. Phys. *11*, L389 (1978).
67. M. Kamińska, unpublished.
68. U. Piekara, J. M. Langer, and B. Krukowska-Fulde, Sol. St. Commun. *23*, 583 (1977).
69. A. A. Kopylov and A. N. Pikhtin, Fiz. Tverd. Tela *16*, 1837 (1974), Sov. Phys. Sol. St. *16*, 1200 (1975).
70. H. Szawelska and J. W. Allen, unpublished.
71. J. M. Baranowski, Postepy Fizyki *31*, 19 (1980) (in Polish).
72. W. A. Harrison, Electronic Structure and the Properties of Solids, W. H. Freeman and Co. San Francisco (1980).
72. P. Vogl and J. M. Baranowski, to be published.
74. C. S. Kelley and F. Williams, Bull. Am. Phys. Soc. *13*, 726 (1968).
75. D. M. Finlayson, J. Irvine, and L. S. Peterkin, Philosophical Magazine B *39*, 253 (1979).

CHAPTER 11

The Optoelectronic Properties of Copper in Zinc-cation II-VI Compound Semiconductors

D. J. Robbins, P. J. Dean[†], P. E. Simmonds[*‡] and H. Tews[*]

Royal Signals and Radar Establishment
St. Andrews Road
Malvern, Worcestershire, UK

1. INTRODUCTION
2. EXPERIMENTAL
3. Cu_{Zn} AS A DEEP LEVEL IMPURITY
   3.1 The electronic configuration of Cu_{Zn}
   3.2 General deep level models
   3.3 Effects of admixture of 3d and VB orbitals and of d-orbital occupancy
4. GENERAL TRENDS IN THE OPTICAL SPECTRA OF $ZnX:Cu_{Zn}$
   4.1 ZnO:Cu
   4.2 ZnS:Cu
   4.3 ZnSe:Cu
   4.4 ZnTe:Cu
5. THE I_1^{DEEP} BOUND EXCITON IN ZnSe
   5.1 Experimental properties
   5.2 The I_1^{DEEP} localization energy
   5.3 A model for I_1^{DEEP}

[*] Max Planck Institut fur Festkorperforschung, Stuttgart, Federal Republic of Germany
[‡] Present address: Department of Physics, University of Wollongong, NSW 2500 Australia
[†] Deceased

5.4 BE states with excitonic e-h interaction dominant
5.5 BE states with h-h interaction dominant
5.6 Relative energies of the two-hole states
5.7 A correlation approach for I_1^{DEEP}
5.8 The Zeeman spectrum of I_1^{DEEP}
6. BOUND EXCITON RECOMBINATION AT Cu_{Zn}^o IMPURITIES
7. APPENDIX: RELATIVE TRANSITION PROBABILITIES FOR THE
ZEEMAN TRANSITIONS FOR I_1^{DEEP}

1. Introduction

There are three motivations for a better understanding of the behavior of Cu in the II-VI compounds. First, it is the best known activator of efficient luminescence, particularly in ZnS where it introduces a series of broad luminescence bands throughout the visible and into the near infrared (IR) spectral region. IN order to produce some particular commercially important phosphor materials, the concentration of Cu impurity is deliberately controlled to enable one or more of these luminescence bands to predominate in electron-hole recombination. Conversely, the second reason for the study of Cu concerns its prominent role as a residual contaminant in purified crystals of the II-VI compounds. Cu is a major impurity in Zn (and in Cd), and is also frequently active in the environments from which single crystals of these II-VI compounds are prepared. The presence of inadvertent Cu has two general consequences. The first is an enhancement of the tendency of the wider-gap II-VI compounds towards high resistance, since Cu_{Zn} is a very deep acceptor in these materials. The traditional view of the physical chemistry of the II-VI compounds emphasizes the role of V_{Zn} in providing this strong compensation of n-type electrical behavior and strong limitation of free carrier density in p-type materials. However, recent developments based upon the careful optical and electrical studies of single crystals prepared under better conditions than were available in most of the early work have shown that Zn-site chemical substituents, particularly Cu and Li, play a much more important role in these aspects than was suspected hitherto.[1-3] The second effect of the inadvertent Cu is the promotion of a variety of optical phenomena which in turn form a sensitive indicator of the presence of Cu. These phenomena may be exploited in impurity analysis, once the involvement of Cu is firmly established.

The third reason for a study of Cu in II-VI compounds is to achieve a fuller understanding of its optoelectronic properties and of the rich variety of recombination mechanisms promoted by this single impurity. The most striking of these optical phenomena involve electron-hole recombination luminescence, including both bound exciton (BE) and distant donor-acceptor pair (DAP) recombination. These spectra contain much information about carrier localization and energy relaxation effects involving the Cu impurity.

The main aims of this article are to review the principal optical phenomena associated with Cu impurity in the chalcogenides ZnX (X = O, S, Se, Te), and from this to present a systematic account of the point defect substitutional Cu_{Zn} acceptor throughout this family of compounds. Particular emphasis will be given to a *microscopic* description of the Cu center in each compound, and to the nature of the associated electronic potential which controls carrier localization and the form of the BE recombination luminescence in each case. It will be seen that Cu_{Zn} in the compounds ZnX provides, by itself, excellent illustrations of the complete range of electronic behavior expected for an open-shell atomic species, such as a member of a transition series, in a semiconducting crystal.

Cu_{Zn} forms a deep impurity in all these compounds, in the sense that the acceptor binding energies E_A and the hole wavefunctions cannot be well-described within the confines of effective-mass (EM) theory (although $ZnTe:Cu_{Zn}$ comes close to this EM limit, as also does Cu_{Cd} in CdTe). The microscopic description of the Cu_{Zn} acceptor and the subsequent discussion of the trends on going from ZnO to ZnTe depend heavily on a knowledge of the symmetries of the various electronic states to which this center can give rise. These state symmetries for Cu_{Zn} have been derived largely from analysis of Zeeman data for both intracenter d-d and BE optical transitions, and to a lesser extent, from electron paramagnetic resonance (EPR) and optically-detected magnetic resonance (ODMR) measurements.

The binding energy of the deep Cu acceptor, derived from donor acceptor pair luminescence bands, decreases monotonically with anion atomic number in the series ZnX (X = O, S, Se, Te). This change produces dramatic differences in the form of the optical data, especially those related to acceptor bound exciton (ABE) states. It will be shown that these changes reflect changes in the ground state character of the deep acceptor, with a transition from a Γ_7 atomic-like ($3d^9$) form (t_2 orbital symmetry) appropriate for ZnO and ZnS to a Γ_8 host-like form (p-like symmetry) appropriate for ZnTe. The bound hole is formally described by a linear combination of d-like and p-like functions, and the transition to effective Γ_8 character occurs between ZnSe and ZnTe. However, analysis of new and existing ABE data shows that the transition in the behavior of the BE occurs between ZnS and ZnSe. In ZnO and ZnS the exciton is bound by electron core interactions. The much smaller anion electronegativities in

ZnSe and ZnTe increase the effective d-orbital occupancy for Cu_{Zn}, and thereby reduce the core binding of the excitonic electron. Hole-hole exchange interactions then become important in the exciton binding, and the localization energy and phonon couplings are both considerably reduced, producing a spectrum typical of an exciton bound by a conventional neutral acceptor (A^o, X). However, analysis of new Zeeman data for the Cu_{Zn} BE in ZnSe demonstrates that the two holes behave as unlike particles, and the electron-hole interaction predominates in the binding, a novel result for ABE states in semiconductors. The conventional (A^o, X) BE with equivalent holes cannot account for the small zero-field splitting, the number of magnetic subcomponents or their polarization properties. However, the (A^o, X) BE model is appropriate for Cu_{Zn} in ZnTe.

The plan of the article is therefore as follows. Section 2 gives a brief indication of the techniques used for the new experimental results reported in this paper. The general trends in electronic configuration of Cu_{Zn} as the anion X is changed in ZnX are discussed in Section 3.1 with experimental evidence from photoemission. Evidence from appropriate theoretical models of deep energy states is considered in Section 3.2, and Section 3.3 introduces a descriptive model for the interaction of TM 3d and host VB orbitals. The general optical properties of the deep Cu_{Zn} acceptor in ZnO, ZnS, ZnSe and ZnTe are compared and contrasted in Sections 4.1–4.4, respectively. Section 5 deals in detail with the properties of the Cu_{Zn}-related I_1^{DEEP} BE in ZnSe. The experimental properties are described in Section 5.1, the BE localization energy is discussed in Section 5.2, and a detailed model is introduced in Section 5.3. Sections 5.4 and 5.5 address different limits in the inter-particle interactions within the BE, while the relative energies of the resulting BE substates are compared in Section 5.6. Correlation effects are considered in Section 5.7. The resulting model for this (A^o, X) BE is applied to the Zeeman data in Section 5.8. The changes in character of the electronic states of Cu_{Zn} within ZnX deduced from all these analyses are reviewed in Section 6, again using the generalized descriptive model for configuration interaction in the ground state of the deep Cu_{Zn} acceptor. Transition probabilities for the magnetic subcomponents of the I_1^{DEEP} BE are given in an Appendix.

2. Experimental

Much of the experimental data reviewed and in some cases re-interpreted in this paper have been published already. The reader is referred to the original publications cited throughout the text for details of the many different experimental techniques which have been employed. Almost all the evidence comes from optical spectroscopy in direct form, photoluminescence or optical absorption, or in some variant such as optically detected magnetic resonance, although conventional electron paramagnetic resonance is also used. Measurements are usually made at the lowest convenient temperatures in order to obtain the maximum relief possible from the effects of strong phonon coupling present in many of the spectra and also to ensure the thermal stability of those BE states of low binding energy. Original experimental results, presented mainly in Sections 4.3 (ZnSe) and 4.4 (ZnTe), are discussed in Section 5. The new results in ZnTe are Zeeman measurements on the Li and $P(A^\circ, X)$ BE at large magnetic fields. Selective DAP luminescence involving P acceptors is also examined at high magnetic fields from a 12T superconducting magnet. The samples were directly immersed in liquid He pumped below the λ point in these measurements, with luminescence excited either by an Ar^+ laser or with light from a tuneable dye laser. Similar techniques were employed in the study of the magneto-optical properties of the I_1^{DEEP} (A°, X) BE in ZnSe. Spectra were examined in both Voigt and Faraday configurations with polarization analysis of the emitted light in both systems.

3. Cu_{Zn} As A Deep Level Impurity

3.1 The Electronic Configuration of Cu_{Zn}

The optoelectronic properties of a transition series impurity in a semiconductor differ from those of a main group impurity largely because of the possibility of generating an open-shell configuration in the former case. For Cu_{Zn} in ZnX, the neutral Zn atom has the configuration [Ar] $3d^{10}$ $4s^2$, and neutral $Cu[Ar]$ $3d^{10}$ $4s^1$, where [Ar] represents the core electrons with the configuration of argon. In its chemistry Cu shows two common oxidation states, Cu^{1+} (d^{10}) and Cu^{2+} (d^9). Which of these best describes the configuration of Cu_{Zn} in ZnX depends very much on the energy difference between the Cu 3d orbitals and the valence orbitals of the ZnX host lattice. The limiting cases for substitution of Cu in ZnX may be described as follows:

1. The Cu impurity gives up only one electron in bonding to the lattice, retaining the filled d-shell and generating the formal oxidation state Cu_{Zn}^{1+} (d^{10}). The impurity acts as a conventional acceptor, with an s-like configuration and effective lattice charge of -1, and intro-

duces a hole into the host valence band (VB). The neutral acceptor configuration would then be $\left\{ [Cu_{Zn}^{1+} (d^{10})]^-; h \right\}$. This configuration is favored if the Cu 3d orbitals are significantly more stable than the upper VB orbitals of the host. It appears a good approximation to the situation for ZnTe:Cu.

2. The Cu_{Zn} impurity gives up two electrons to satisfy the local bonding requirements, promoting one from the d-shell. This generates an open-shell isovalent (or isoelectronic) impurity, with formal oxidation state Cu_{Zn}^{2+} (d^9). The impurity has no lattice charge, but can accept an electron to fill the d-shell. In particular this center can compensate donors, and in ZnS exhibits the classical type of DAP recombination process. The neutral acceptor configuration would then be $[Cu_{Zn}^{2+} (d^9)]^\circ$. The bound hole has strong atomic-like d-orbital character, and its energy states can be quite well-described in terms of crystal field theory. Thus, the 5-fold degenerate d-electron orbitals split in a tetrahedral (T_d) field into an upper triplet (t_2) and a lower doublet (e) set, as represented on a one-electron energy level diagram with electron energy increasing positively on the ordinate. In its ground state the bound hole occupies the t_2 orbital, generating an acceptor state of T_2 symmetry; the hole can be promoted to the e orbital set, generating an excited acceptor state of E symmetry. This situation will be favored when the Cu 3d orbitals are significantly higher in energy than the VB orbitals of the host. The crystal field description is particularly applicable to Cu_{Zn} in ZnO, where both the t_2 (T_d) and e (T_d) orbital levels lie far above the VB edge. We must also recognize in ZnO further small splittings due to a reduction in point group symmetry from T_d to C_{3v} in the wurzite lattice.[4]

This change in the character of the acceptor ground state as a function of the energy difference between the d-orbitals and the VB orbitals has been discussed schematically by two of the present authors[5] in terms of an oxidation state correlation diagram, Figure 1. On passing from left to right across the diagram the bound hole changes from being atomic-like, bound by short-range central-cell forces, to being effective-mass-like with the character of the host VB and bound by the long-range Coulombic interaction. ZnO:Cu and ZnTe:Cu are examples falling towards the left hand and towards the right hand sides of this correlation diagram, respectively. Using this approach, the changes in the Cu_{Zn} acceptor ground state configuration and binding energy across the ZnX series would be described in terms of a differential electron affinity $\chi_e(Cu)$, defined by the reaction

$$[Cu_{Zn}^{2+} (d^9)]^\circ = [Cu_{Zn}^{1+} (d^{10})]^- + h_v + \chi_e(Cu) \qquad (1)$$

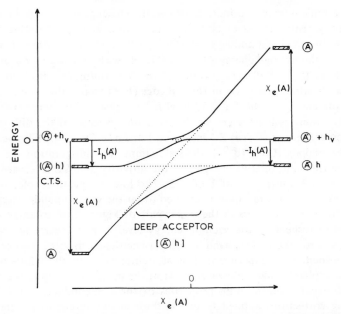

FIGURE 1 An oxidation state correlation diagram illustrating the possibility of configurational interaction between coulombic and core-hole states of a neutral open-shell impurity. The abscissa represents a differential electron affinity $\chi_e(A)$ defined w.r.t the VB edge by: $\chi_e(A) = I_e(A^-) - E_g$, where I_e is an electron ionization energy. The energy of the ionized acceptor Ⓐ + h_v is set equal to zero across the diagram. Only relative energy separations between configurations at any value of $\chi_e(A)$ are significant. Here Ⓐ = $[Cu_{Zn}^{2+}(d^9)]^0$, and Ⓐ⁻ = $[Cu_{Zn}^{1+}(d^{10})]^-$. CTS = charge transfer state. In the bracketed region the impurity shows the properties of a deep neutral acceptor with a hole wavefunction strongly perturbed by the configuration. ZnSe:Cu would fall in this region, with ZnO:Cu to the left and ZnTe:Cu to the right. [Adapted from reference 5].

where h_v represents a hole in the host VB. When $\chi_e(Cu)$ is negative it is energetically favorable to form the isovalent impurity configuration, i.e., the reaction (1) lies to the left. As discussed elsewhere,[6] a VB → impurity charge transfer transition can then be excited optically:

$$[Cu_{Zn}^{2+}(d^9)]^0 + h\nu \rightleftarrows [Cu_{Zn}^{1+}(d^{10})]^-; h_b \qquad (2)$$

where h_b represents a hole bound to the negatively-charged core by a combination of short-range and Coulombic interactions.

Formation of the isovalent configuration $[Cu_{Zn}^{2+}(d^9)]^0$ is favored for the strongly electronegative anion oxygen, i.e., $\chi_e(Cu)$ is negative. As the anion atomic number Z_X increases down the chalcogenide group the anion electronegativity decreases and the ionic radius increases, leading to an increase in covalent bonding in the ZnX lattice. Therefore, there is an increase in the average energy of the VB levels with increase in Z_X and an increase in the overall energy width of the VB continuum. Both of these factors contribute to a rise in the VB edge (E_v) towards the vacuum level. Since the energy of the Cu_{Zn} 3d level is less sensitive to changes in anion than E_v, there will be a general increase in the value of the differential electron affinity $\chi_e(Cu)$ as Z_X increases. For ZnTe:Cu, $\chi_e(Cu)$ has become positive, making $[Cu_{Zn}^{1+}(d^{10})]^-$ the stable impurity configuration.

These trends in relative energies of the impurity and VB levels are illustrated in the upper part of Figure 2. The low energy thresholds of UV photoemission spectra[7] have been used to set the valence band energies of the four semiconductors in the upper part of Figure 2 on to an absolute scale with respect to the vacuum energy level. It is necessary to apply a correction for the surface band bending, since the electron escape depth in the photoemission experiment is small compared with the width of the surface depletion layer. However, most of the optical techniques described in this paper refer to bulk properties of the semiconductor. The band bending corrections applied to obtain these energy levels for the interior region of the semiconductor are 0.01 eV, 1.2 eV, 0.58 eV and −0.18 eV for ZnO, ZnS, ZnSe and ZnTe, respectively.[7] Positive values indicate on enhancement of the electron binding at the top of the valence band in the semiconductor interior. In general, these data show the expected trend of a decrease in this energy with decrease in the electronegativity of the semiconductor anion. The value for ZnO deviates from this predicted trend. However we note that the VB spectrum for ZnO derived from X-ray photoemission[8] is also anomalous in showing a region of small but non-zero density of states extending ~1.4 eV above the steep onset of the main VB peak. It is not clear whether these anomalies arise from bulk defect states above the VB edge, from surface states or from the VB structure of ZnO. Because of these effects, it is possible that the VB energy of ZnO in Figure 2 is shown as ~1 eV too high relative to the other semiconductors. The energies and symmetries of the Cu_{Zn} impurity states shown in Figure 2 have been derived from a variety of optical and electrical measurements, mainly optical, which will be discussed later.

The idea that the Cu_{Zn} acceptor binding energy E_A (as well as the impurity configuration) can be related to the electron affinity difference between Cu and the host anion, as suggested by the correlation diagram in Figure 1, is supported by the trend illustrated in the lower part of Figure 2.

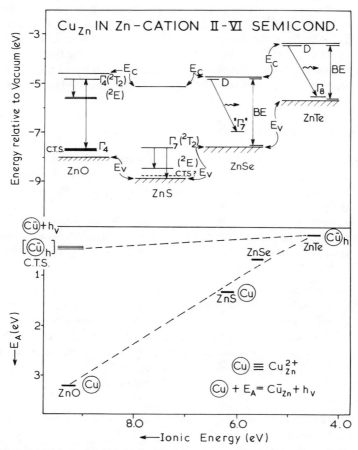

FIGURE 2 (a) The trend in level energies for Cu_{Zn} in the ZnX semiconductor series. The energies of the valence band maxima of the host crystal are referred to the energy of an electron in a vacuum as zero. There is some larger uncertainty in this position for ZnO. The photoelectron ionization data (Ref. 7) used to construct this figure are often quoted in deriving electron affinity values for the zinc chalcogenides (e.g., B. L. Sharma and R. K. Purohit, 'Semiconductor Heterojunctions,' (Pergamm Press, Oxford 1974)). However the energies of the CB minima indicated here differ from the quoted electron affinities, and from Ref. 6, through the inclusion of a correction for surface band bending (see text). (b) The energies of the Cu acceptor bound exciton (BE) across the ZnX hosts as a function of the ionic energy C from the Phillips electronegativity theory of the energy states in a semiconductor.

Here the estimated acceptor ionization energies of Cu_{Zn} are plotted against the ionic energy term in the binding of the host lattice, using Phillips' description of the effective single-oscillator energy gap for a zincblende (ZB) semiconductor.[9] To the extent that the trend in Cu-X ionicity reflects that in Zn-X, this ionic energy term should be some measure of the electron affinity difference between Cu and the anion. The trend in Figure 2 is remarkably smooth, despite the drastic change in the character of the lowest energy bound hole state of Cu_{Zn} near ZnSe discussed in Section 5.

3.2 General Deep-Level Models

The previous section has discussed the gap states introduced by Cu_{Zn} in terms of energy differences and interactions between Cu 3d orbitals and VB orbitals of the host lattices in which it substitutes. Recent calculations of deep level energies in ZB semiconductors, using both cluster[10] and Green's-function[11] methods, give useful insight into the physical nature of the VB orbitals involved. The picture which is emerging suggests that a good description of deep level formation is achieved by the following sequence of steps:

1. Remove a neutral host atom (here Zn) to form a lattice vacancy, leaving 4 dangling-bond orbitals which form a singlet (a_1) and a triplet (t_2) set. These vacancy orbitals contain (8-n) electrons, where n is the formal oxidation state in the lattice of the atom removed (n = 2 for Zn). As a general rule in ZB materials, the energy separation between the a_1 and t_2 vacancy orbitals appears to be $\sim$1-2 eV, the singlet lying lower. The t_2 vacancy orbital has the same symmetry as the top of the VB.

2. Introduce the impurity atom into the vacancy, allowing an interaction between its valence orbitals and the $a_1 + t_2$ vacancy orbitals.

Removing the host atom to create a vacancy produces a repulsive potential which pulls the $a_1 + t_2$ orbitals up from the VB, the magnitude of the potential depending on the atom removed; introducing the impurity atom tends to lower the energy of these vacancy orbitals. If the impurity is a first-row transition element, the orbitals most strongly interacting with the vacancy orbitals will be the 3d, 4s and 4p orbitals. When the impurity has t_2 (3d) orbitals higher in energy than t_2 (vac), the latter will tend to be pushed down into the VB and filled. The gap states of t_2 symmetry will then have considerable atomic character, although this may not be dominant, according to Hemstreet.[10] On the other hand, if the t_2 (3d) orbitals lie below t_2 (vac), the t_2 (3d) may be pushed down into the VB and appear as a resonance. The gap states of t_2 symmetry are then derived largely from t_2 (vac), with more-or-less VB character. If the impurity is an acceptor, a hole occupying the t_2 gap orbital may be more-or-less effective-mass-like, depending on the combination of short-range and Coulombic potentials introduced by the impurity. Once again, this is the trend described schematically in Figure 1, where the differential electron affinity $\chi_e(A)$ would be a measure of the energy difference between the impurity 3d orbitals and the vacancy orbitals. It should be noted that the e(3d) orbital doublet does not interact with the $a_1 + t_2$ vacancy orbitals by symmetry.

The most relevant calculation is that of Hemstreet,[10] who considered a particular host lattice (GaAs) with different impurity atoms drawn from the first-row transition series. The calculation used the $X\alpha$ scattered-wave method on a small "molecular" cluster to approximate the GaAs lattice. As the impurity atomic number Z_m increases across the transition series, the 3d electrons are stabilized by the increase in effective nuclear charge which results from imperfect screening of one d-electron for another. Therefore the energy difference between the 3d orbitals and the vacancy orbitals decreases as Z_m increases. The features of the calculation[10] are:

1. For the lighter transition metals (TM), t_2 and e orbitals appear in the gap. These have significant atomic 3d character, although the degree of hybridization with "lattice" orbitals is much greater for t_2 (3d) than e (3d).

2. For the heavier TM, the t_2 (3d) and e (3d) orbitals appear as resonances deep in the VB, and only t_2 gap orbitals appear. The transition between (a) and (b) occurs at the impurity Co.

3. As Z_m increases, the 3d character in the t_2 gap orbitals decreases, and this level moves steadily towards the VB edge as the impurity atom approaches Ga in the Periodic Table.

The case of Cu_{Zn} in the series of compounds ZnX considered here is complementary to that discussed by Hemstreet.[10] The study of substitution of Cu is easier in the more ionic II-VI's, as a general consequence of their larger energy gaps and in particular because the neutral form Cu^{2+} ($3d^9$) is a common valence state of Cu. The principal factors determining the relative energies of Zn-vacancy dangling-bond orbitals in ZnX should be the anion electronegativity, its size, and the covalence of the lattice. One might expect the vacancy repulsive potential to be larger for the more covalent lattices, where the valence electrons have a greater probability of being found at the Zn site. In general, therefore, the energy difference between the Cu 3d orbitals and the Zn-vacancy orbitals should decrease as Z_X increases. We suggest that ZnO:Cu is analogous to the situation of a light TM in GaAs (e.g., GaAs:Cr) as described by Hemstreet, with t_2 and e orbitals of largely 3d character lying relatively close to the conduction band (CB). In addition there are more extended orbitals near the VB edge accessible optically.[12-14] ZnTe:Cu, on the other hand, is analogous to a heavy TM in GaAs (e.g. GaAs:Ni or Cu), where the Cu 3d orbitals occur as a resonance below the VB edge and are filled. The hole introduced by the Cu impurity occupies a t_2 gap orbital of largely VB character. Thus, Cu_{Zn} acts as a conventional, but still intermediate deep,[15] acceptor in ZnTe.

3.3 Effects of Admixture of 3d and VB Orbitals and of d-Orbital Occupancy

The extent of admixture of Cu 3d and host VB orbitals and the effective d-orbital occupancy will prove to be important factors in trying to rationalize the optoelectronic properties of Cu_{Zn} in ZnX. The wavefunction of the bound hole in the neutral acceptor configuration can be written as a linear combination:

$$\Phi_h = (1 - \alpha)^{\frac{1}{2}} \Phi_{3d} + \alpha^{\frac{1}{2}} \Phi_{VB} \qquad (0 \leqslant \alpha \leqslant 1) \qquad (3)$$

As shown in previous sections, the hole will have t_2 orbital symmetry in the ground state, and the $Cu_{Zn} t_2$ (3d) and $V_{Zn} t_2$ (vac) orbitals will generally dominate the above expansion. Alternatively, the acceptor ground state could be described in terms of a configuration interaction between the two neutral configurations $[Cu_{Zn}^{2+} (d^9)]^0$ and $\{[Cu^{1+} (d^{10})]^-; h_b\}$. Whichever description is used, the interaction between the 3d orbitals and the lattice orbitals has important consequences for both the effective 3d-orbital occupancy and for the acceptor spin-orbit coupling constant, which are discussed in turn below.

As a simple heuristic device we shall express the ground state configuration of the neutral acceptor whose wavefunction is given by (3) in the following way:

$$Cu_{Zn}^o \equiv [Cu_{Zn} (d^{9+\alpha})]^{-\alpha} \cdot (\alpha h_b) \qquad (4)$$

Then, as $\alpha \to 0$, $Cu_{Zn}^o \equiv [Cu_{Zn} (d^9)]^0$ which is the limit applicable to ZnO:Cu. As $\alpha \to 1$, $Cu_{Zn}^o \equiv [Cu_{Zn} (d^{10})]^- \cdot h_b$, which is the limit corresponding to ZnTe:Cu. The effective 3d-orbital occupancy in this approximation is $(9 + \alpha)$. According to the calculation by Hemstreet,[10] the co-efficient α, different for orbitals of e and t_2 symmetry, will be a minimum when the 3d orbitals lie well above the VB edge. It will increase steadily as these d-orbitals become deeper in energy and move across the gap towards the VB edge, and will increase rapidly towards unity as the 3d orbitals fall below the VB edge. The magnitude of the coefficient α has an important bearing on the electron binding potential of the Cu_{Zn} impurity, and will be discussed in more detail in Section 6 in relation to the localized electron-hole recombination spectra.

A second consequence of the 3d-VB interaction is that the spin-orbit coupling constant and the g-value for the acceptor bound hole both be-come sensitive to the magnitude of the coefficient α. There is an isomor-phism between t_2 orbital sets derived from d-type and p-type orbitals in a cubic field, but the sign of the orbital angular momentum is reversed due to the different transformation properties of the basis functions.[16] That is, using a complex basis, $t_2(p) \pm 1 \equiv t_2(d) \mp 1$. As a result the spin-orbit coupling constants for $t_2(p)$ and $t_2(d)$ orbitals are opposite in sign. Thus, the p-like VB of a ZB semiconductor splits with the Γ_8 electron level above Γ_7, whereas the $t_2(3d)$ shell of a TM splits with the Γ_8 electron level below Γ_7, producing for example the Γ_7 ground state for the single atomic-like hole of Cu_{Zn}^{2+} (d^9) in ZnS. Two examples of situations in which admixture of d- and p-like orbitals affects the sign of spin-orbit coupling are as follows:

1. The spin-orbit splitting of the upper VB in ZnO is anomalous.[17] The O anion is light enough that even a small admixture of Zn 3d charac-ter into the VB functions is sufficient to reverse the splitting pre-dicted for purely p-like functions.[18] This inverted splitting due to 3d-orbital admixture becomes even more marked when VB-like holes are localized at Cu_{Zn} impurity, as in the excited states of the α, β, γ transitions of ZnO:Cu.[14]

2. Interpretation of the EPR spectra of Cr^{2+} in II-VI compounds[19] re-quires a description in terms of ligand field theory, involving inter-action between Cr 3d and anion p-orbitals. The spin-orbit coupling constant λ for the Cr^{2+} impurity has been observed to decrease from its free-ion value $\lambda_o = 57$ cm^{-1} as the anion atomic number Z_X in-creased, i.e. as covalency increases the orbital reduction factor. In fact, λ becomes negative for selenide and telluride compounds. The trend could be approximated by the expression:

$$\lambda = \lambda_o \ [1 - \beta(\zeta_X/\zeta_M)] \tag{5}$$

where ζ_X and ζ_M are the anion and metal one-electron spin-orbit constants. When $\zeta_X \gg \zeta_M$, as for the heavy anions, the observed impurity spin-orbit parameter λ can change sign through admixture of anion wavefunction into the 3d orbitals.

As one crosses the series of compounds ZnX:Cu from X = O to X = Te, both the coefficient α in the neutral acceptor wavefunction and the anion spin-orbit constant increase. For the lighter anions the spin-orbit splitting of the t_2 hole orbital will be determined by the Cu 3d interactions, giving an acceptor ground state of Γ_7 symmetry. Conversely, for the heavier anions the t_2 orbital splitting can be reversed, giving a Γ_8 acceptor ground state. As we shall see, this discontinuity in the symmetry of the Cu_{Zn}^o ground state occurs between ZnSe and ZnTe.

4. General Trends in the Optical Spectra of ZnX:Cu$_{Zn}$

The principal luminescence bands which appear to correlate with substitutional Cu_{Zn} impurity in the ZnX series of compounds are illustrated schematically in Figure 3. Also indicated are the approximate configurations of the excited and ground states involved in the transitions. These spectra and their interpretations are reviewed below for each semiconductor in turn.

4.1 ZnO:Cu

It is well established that Cu_{Zn} in ZnO exhibits the Cu_{Zn}^{2+} (d^9) configuration unless the Fermi level is very high due to doping by Zn_I, H or In donors. Circularly-polarized Zeeman measurements on the intra-d shell t_2 (T_d) $\rightarrow$ e (T_d) infrared absorption give a negative g-value [$g_\parallel \approx -0.74$ (± 0.3)] for the Γ_4 (C_{3v}) ground state,[4,14] consistent with largely d-orbital character. The position of the ground state of Cu_{Zn}^{2+} can be determined from observations of the quenching of the characteristic EPR signal as the Fermi level is raised by co-doping.[20] Our understanding of the origin of the no-phonon transitions observed in the blue spectral region both in optical absorption[12] and in luminescence[12,13,21] (top, Figure 3) in terms of a VB $\rightarrow$ impurity charge transfer process[14] [see Equation (2)] shows that the Γ_4 acceptor ground state must lie significantly higher than 2.86 eV above the VB edge. Combination of these experimental data suggests a threshold energy for the ionization (6) of $\sim$3.26 eV:

$$[Cu_{Zn}^{2+} (d^9)]^\circ + h\nu \rightarrow [Cu_{Zn}^{1+} (d^{10})]^- + h_v \tag{6}$$

Thus, Cu_{Zn} is an outstandingly deep acceptor in ZnO.

The α, β, γ no-phonon lines of the VB → impurity charge transfer transition represent excitation of the acceptor hole from the d-shell of the Cu_{Zn}^{2+} (d^9) impurity into more extended states, but with the hole remaining bound by a combination of short range and long-range forces.[14] The symmetries and g-values of the final states are determined by the symmetry properties of the VB functions, but the energies are determined largely by spin-orbit interaction with the Cu core.

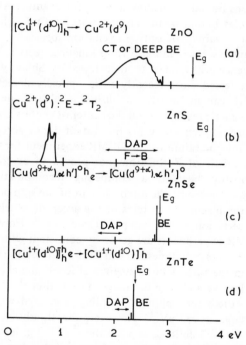

FIGURE 3 Schematic representations of the various luminescence transitions attributable to Cu_{Zn} in the ZnX semiconductor series. The donor-acceptor pair (DAP) bands involve straightforward changes of charge state, either in the $3d^9$ core (for ZnS and ZnSe) or on the surrounding ligands (for ZnTe). Electronic reactions involving the BE are more complex, changing with host anion as illustrated by the Equations in (a), (c) and (d).

4.2 ZnS:Cu

The situation is qualitatively similar in ZnS. The presence of Cu_{Zn}^{2+} (d^9) is confirmed by the Zeeman effect in the intra-d shell $t_2 \rightarrow e$ optical absorption.[22] The g-value of the Γ_7 (T_d) ground state is again negative $[g_{11} = -0.7]$, as expected for a hole with largely 3d orbital character.[14,23] Luminescence is observed from the inverse transitions within Cu_{Zn}^{2+} (d^9), (Figure 3). The presence of Cu_{Zn} is supported by the observation of EPR from a variety of associates, e.g. Ga_{Zn}–Cu_{Zn},[24] but no signals have ever been detected from isolated Cu_{Zn} on cubic lattice sites. The energy position of the Γ_7 $(^2T_2)$ ground state of Cu_{Zn}^{2+} has been determined through the recognition that the broad green luminescence characteristic of ZnS:Cu, whose position is also indicated in Figure 3, involves recombination between electrons bound to shallow donors, for example $A\ell_{Zn}$ with $E_D \sim 0.1$ eV, and holes bound in the ground state of Cu^{2+}. This model was established from a variety of optical excitation spectra for the EPR signal and visible luminescence, and also from kinetic effects in the EPR and visible luminescence which have been reviewed by Shionoya.[25] The attribution has been tested more recently from optically detected magnetic resonance which can frequently establish a reliable connection between a particular magnetic resonance and the center or centers responsible for a given luminescence, particularly for the distant donor acceptor pair (DAP) process. The interpretation of the ODMR experiment for the green luminescence characteristic of ZnS:Cu has not proved to be completely straightforward, possibly because a competing recombination path described in Section 4.3 prevents the establishment of the spin population difference between the magnetic substates of the acceptor which is a prerequisite to the observation of an acceptor resonance.[26] However, the results of these ODMR experiments are generally consistent with a distant DAP model for the Cu-green luminescence in ZnS. No-phonon transitions cannot be resolved in the various luminescence, absorption and excitation spectra[27,28] which involve a change of charge state within the Cu d shell in ZnS, because of the strong phonon coupling associated with a hole binding energy as large as ~ 1 eV.[29] However, the thresholds of these spectra are sufficiently well-defined to place the $\Gamma_7(^2T_2)$ ground state of Cu_{Zn}^{2+} (d^9) at ~ 1.25 eV above E_V. This implies a hole binding energy of ~ 0.4 eV for the ^2E (d^9) acceptor excited state, which can promote energy transfer by inter-site hopping processes even at low temperatures.[30]

4.3 ZnSe:Cu

The character of the ground state of Cu_{Zn} in ZnSe is much less certain. There is no clear evidence from intra-d shell IR absorption for a dominantly atomic-like $3d^9$ configuration of Cu_{Zn}^{2+}, and no EPR or satisfactory ODMR were observed in the early work.[31] This may result from the increased d-p admixture predicted to occur as Z_X increases, and is qualitatively consistent with the trends established for ZnO and ZnS in Figure 2, supported by evidence from optical spectra concerning the depth of the Cu acceptor. Present opinion[31,32] is that the broad red luminescence band apparently characteristic of ZnSe:Cu, and also indicated schematically in Figure 3, involves distant DAP transitions at a shallow donor with $E_D \sim 28$ meV and acceptor with $E_A \sim 0.65 - 0.7$ eV. This follows from inter-comparison of optical absorption and luminescence.[32] Direct support for this assignment is provided by the spectrum for optical quenching of the Cu-red luminescence. This quenching spectrum is identical to the hole photoionization spectrum obtained from transient photocapacitance measurements on Cu-doped ZnSe.[33] Both spectra show a threshold near ~ 0.68 eV, in good agreement with the thermal activation energy for hole emission, ~ 0.71 eV. The complementary electron ionization spectrum for this Cu-induced level in ZnSe has also been observed, with threshold as expected near ~ 2.15 eV.[33]

However, the energy of this dominant Cu-level in ZnSe is made somewhat uncertain because both the hole ionization and luminescence quenching spectra show a peak above threshold, with maximum ~ 0.85 eV, which is absent from the photoconductivity spectrum of the same level.[33] This difference may indicate that the final states of the transitions giving rise to the peak are localized, rather than extended Bloch-like states. Such transitions might produce a signal in photocapacitance measurements, due to the large junction field which can sweep carriers out of the bound states, but not in photoconductivity measurements where large fields are absent. The nature of the localized excited hole states in ZnSe:Cu is uncertain, but there exists a formal analogy with the localized final states of the VB → impurity charge transfer transition of ZnO:Cu discussed in Section 4.1. In that case an assignment of the excited bound hole states can be made because of the observation of the sharp, discrete α, β, γ ZPL's in the absorption and luminescence spectra, and the knowledge of the Cu_{Zn}^{2+} (d^9) impurity ground state in ZnO. It is interesting to note that the electron photoionization spectrum in ZnSe:Cu also shows a peak above threshold, similar to the complementary hole ionization. The final states contributing to this peak may be analogous to the excited bound electron states recently observed in Co-doped ZnSe.[33]

In addition to the DAP transitions, the other principal luminescence spectrum attributed to Cu_{Zn} in ZnSe in Figure 3 is a bound exciton (BE) transition with no-phonon line at 2.783 eV. This BE transition seems to have many of the characteristics of the conventional neutral acceptor BE (A°, X) very familiar for main group acceptor impurities in the less ionic semiconductors.[35] This attribution to Cu°_{Zn}, and the "Γ_7" labelling of the acceptor ground state in Figure 2, are central issues in the present paper and will be discussed in detail in Section 5.

It is interesting that, as for ZnS:Cu, EPR of the isolated Cu_{Zn} center in ZnSe has proved difficult to observe, and yet is much more straightforward for associates such as Cu_{Zn}-Cu_I. This complex is probably the neutral hole trap in the Cu-green luminescence of ZnSe, which also originates from a distant inter-impurity-tpe recombination process.[34] This striking difference may be due to the removal of orbital degeneracy in the orthorhombic field of the complex, producing a more free-spin-like hole wavefunction with larger g-value than for isolated Cu^{2+}_{Zn} (d^9) in a T_d crystal field. We shall suggest in Section 5.8 that the g-value for Cu°_{Zn} in ZnSe is very small, which would make EPR almost impossible to detect for that center.

Very weak ODMR signals have recently been reported from heavily Cu-doped ZnSe.[36] Godlewski et al.[36] attribute a resonance near g = 1.54 observed on the broad red DAP band, already implicated with the deep Cu acceptor, to the Γ_7 $(^2T_2)$ ground state expected from conventional crystal field theory for Cu^{2+} $(3d^9)$. Their assignment was based on the observation that this g value is substantially shifted below the free electron value of g = 2 and is isotropic and broadened to a degree which gives an orbital reduction (Ham) factor similar to Ni^+ $(3d^9)$ in ZnSe. It is assumed that the broadening is produced by Cu hyperfine interactions, unresolved due to further inhomogeneous broadenings from random strains in the cubic crystal. The isotropic g-shift might be expected for a hole relatively tightly bound to Cu_{Zn} if there is appreciable d character as discussed in Section 3. However, while we agree with the symmetry attributed to Cu°_{Zn}, we believe it unlikely that this ODMR signal is directly associated with the Cu-related acceptor responsible for the red luminescence, for the following reasons:

1. It has proved very difficult to observe this ODMR signal. Many crystals which show the strong red DAP band attributed to Cu_{Zn} and

which possess generally high quality optical spectra do not show any resonance, other than those attributed to the Cu associates which promote spectrally distinct resonances.[34]

2. The $g = 1.54$ ODMR resonance causes a *decrease* in the intensity of the red DAP luminescence. Godlewski et al.[36] point out that this is possible if recombinations through Cu_{Zn} are predominantly non-radiative, so that the non-radiative rate increases at resonance. This is a possible explanation, perhaps attributable to a strong multi-phonon non-radiative recombination channel. However, it is quite inconsistent with the observed ODMR behavior in similarly broadened spectra involving isolated Cu_{Zn} in other semiconductors, including ZnS,[26] and in spectra involving Cu-associate hole binding centers in several semiconductors including ZnSe.[34,36] Moreover, the intrinsic efficiency of the red Cu-related luminescence in ZnSe is certainly not significantly less than the efficiencies of these other Cu-related bands in both ZnSe and ZnS, as it would have to be if the non-radiative process was intimately linked to the acceptor responsible for the red DAP recombinations.

3. A qualitatively similar type of interpretation has already been offered for the appearance of magnetic resonances which decrease the intensity of green luminescence in hexagonal ZnS:Cu.[26] However, the detailed mechanism must be different for ZnS:Cu because the donor as well as a second acceptor resonance both produce negative signals, whereas the donor resonance produces a luminescence *increase* for the red DAP band in ZnSe:Cu.[36] There was no temptation to attribute the non-radiative center responsible for the negative signal in ZnS:Cu to Cu_{Zn}, since the isotropic g value observed is slightly greater than 2, in fact close to that of the F^+ center in ZnS.

Evidently the $g = 1.54$ response observed from the ODMR occurs on some independent center, perhaps associated with very high concentrations of Cu. Recombinations through this center apparently compete with those through the ~ 0.7 eV acceptor responsible for the Cu-red band in ZnSe. The possibility of observing a strong magnetic resonance on a luminescence band even when the electronic states coupling to the microwave signal are not those directly responsible for the detected light is becoming a well-recognized featuer of the ODMR technique. In some systems, it may be very difficult to specify the exact nature of the kinetic interactions in the pathways of energy transfer which are responsible for this phenomenon.

In particular, we do not feel that the evidence linking the isotropic $g =$ prove the model for this center derived from Zeeman data on the (A^o, X) BE luminescence, and presented in Section 5. This model predicts a very small g-value for Cu_{Zn}^o. Unfortunately, the strong phonon coupling associated with the ~ 0.7 eV acceptor precludes the method of demonstration of an intimate link between this (A^o, X) BE and the red DAP band used for the Cu_{Zn} acceptor in ZnTe (Section 4.4). This same problem is also encountered for the significantly shallower Ag acceptor in ZnSe as recently discussed by Dean et al.[39] However, it is certainly true that the intensities of both the red DAP band and the I_1^{DEEP} BE are increased by Cu doping[40] and that both are frequently present in undoped material, consistent with the fact that Cu is a known persistent residual impurity in ZnSe.

4.4 ZnTe:Cu

The behavior of Cu in ZnTe represents almost the opposite extreme from Cu in ZnS, and particularly in ZnO, according to the general ideas presented in Ref. 5 and Figure 1. Copper is now recognized as introducing one of the most persistent point defect acceptors in ZnTe. This acceptor frequently controls the Fermi level in undoped ZnTe and has a binding energy of 149 meV.[41,42] This is only about three times the effective mass value for this relatively covalently bonded II-VI semiconductor. Observation of 'two-hole' (A^o, X) satellite transitions for the (A^o, X) BE, shown at the bottom of Figure 3, confirms that this acceptor BE has conventional (A^o, X) character, containing two holes and an electron. In addition, correlation of the detailed form of the acceptor excited states, available from both the 'two-hole' (A^o, X) BE satellites and from the photo-excitation and selectively excited photo-luminescence spectra[43,41] for the DAP luminescence also indicated schematically at the bottom of Figure 3, provides definitive evidence that the same acceptor is involved in both types of spectra. We have already seen in Section 4.3 that this technique cannot be used for arbitrarily deep acceptors, in part because of the large spectral broadening from the strong phonon coupling inevitably associated with charge transfer reactions at very deep acceptors.

Three no-phonon components may result from exchange splitting between the three electronic particles in the conventional (A^o, X) BE in a zincblende semiconductor like ZnTe.[34] The isotropic Zeeman spectrum of the lowest energy of these,[39] discussed further below (Figure 4), indicates that the corresponding exciton state has pure electron spin character, as is often true for (A^o, X) BE involving other than the most effective-mass-like acceptors.[34] The interpretation of the character of the ground state of the Cu acceptor in ZnTe, near the effective mass-like limit of the

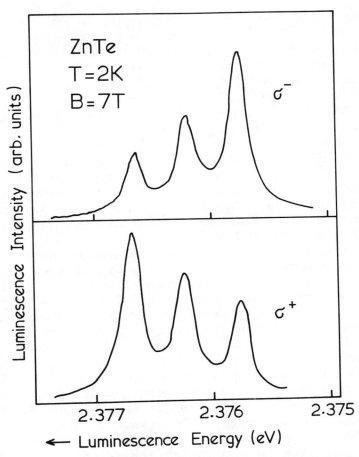

FIGURE 4 No-phonon luminescence of the $(A°, X)$ BE associated with Cu_{Zn} in ZnTe, in a magnetic field of 7T and for the Faraday polarization configuration, contrasted for the indicated circular polarizations ($\sigma^+ \equiv$ left circular polarized). The Voigt π-polarized components are shown in Ref. (1).

sequence in Figure 2, is intimately linked to the behavior of the electron. Evidence as to the symmetry of the acceptor state is best obtained from Zeeman analyses. These analyses always involve the electron as well as the hole, since the data are derived from the $(A°, X)$BE, from measurements of the free electron to neutral acceptor free-to-bound (AFB) recombinations,[44] from studies of the DAP luminescence using selective pair luminescence with a dye laser[45] or optically detected magnetic resonance,[46] or from measurements of spin-flip Raman scattering.[47] The problem has emerged that the results of some of these studies differ dramatically. We shall see that enough information is now available to achieve a resolution of this difficulty.

Zeeman analysis of the (A^o, X) BE in ZnTe was first applied to A_1^a, now firmly identified with Cu_{Zn}.[41,42] This BE contains a sharp, isolated line at zero field, well suited to this type of analysis. The observed Zeeman behavior has the form of a classic $J = \frac{1}{2} \to J = \frac{3}{2}$ recombination process. The σ-polarized luminescence spectrum, with only three resolved sub-components (Figure 4), can be readily understood only if the sign of the g value in the excited $[J = \frac{1}{2} (A^o, X) BE]$ state is opposite to those in the ground $J = \frac{3}{2}$ state (Figure 5a). The $J = \frac{3}{2}$ character is consistent with the Γ_8 acceptor ground state derived from a p-like valence band maximum at the zone center of a zincblende semiconductor such as ZnTe. The $J = \frac{1}{2}$ character of the (A^o, X) BE means that the $J = 0$ state formed by hole-hole exchange lies below the other $(J = 2)$ state possible from the Pauli exclusion principle. It must be well below to give a single, well-isolated zero-field line.[1] Such behavior is commonly observed for an (A^o, X) BE involving a relatively deep acceptor,[35] and $(E_A)_{Cu}$ is $\sim 3 \times E_A^{EMT}$ for ZnTe.[1,43] The only slightly unusual aspect is that the A_1^a BE is essentially isoenergetic with the more complicated zero-field structures associated with A^o, X for acceptors such as Li_{Zn} and P_{Te}, where $E_A \sim E_A^{EMT}$. This suggests that the (A^o, X) components $J = 2$ derived from the hole-hole state are repelled to significantly lower E_{BX} for the Cu acceptor. The original analysis of the $Cu(A^o, X)$ BE has been supported by several later studies, including those of the present authors and of Oka and Cardona.[41] However, there is a puzzle that the derived value of g_e, close to -0.40, is surprisingly far from the theoretical estimates $g_e = +0.47$ obtained with the 3 band model.[48] This estimate was obtained using a value of the optical matrix element $P^2 \sim 20$ eV which is both reasonable from other II-VI compounds and which is also consistent with the experimental value of $m_e^* = 0.12\, m_o$.[44] This puzzle was intensified when DAP magneto-optical studies gave values of g_e of essentially the same magnitude as the A_1^a BE Zeeman result but with *opposite* sign, ODMR providing the most accurate magnitude $|g_e| = 0.401 \pm 0.004$.[46] The agreement from *all* techniques that $|g_e|$ is close to 0.40 removes the credibility of an alternative possible interpretation of the Zeeman splitting of the A_1^a BE with $g_e = +1.65$ and $K = +0.55$. These assignments can reproduce the observed energy pattern of Figure 4 and the π polarized components quite well, although the agreement with the relative intensities of the magnetic sub-

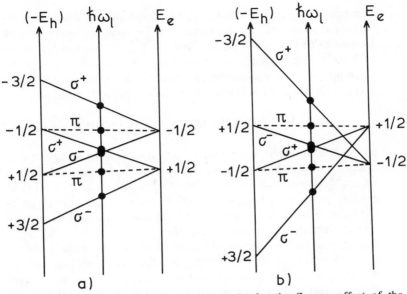

FIGURE 5 Two possible contrasting nomographs for the Zeeman effect of the (A°, X) BE in ZnTe (a) Is derived with $g_e \sim -0.4$ and an essentially isotropic hole g value as in Ref. 1. (b) Is derived with $g_e \sim +0.4$, which forces the very unusual hole g values shown, discussed in the text.

components is not as good as with the model in Figure 5a. With $g_e = +1.65$, the experimental g-shift of only -0.35 is now very much smaller than the theoretical estimate of -1.53. However, we reject this interpretation of Figure 4, mainly because of the disagreement with $|g_e| = 0.40$ obtained from so many independent measurements, since experience has shown that $(g_e)_{BE} \sim g_e$ for an electron bound at a repulsive center such as an acceptor.[35]

It is possible to find a transition nomograph which reproduces the observed (A°, X) BE Zeeman patterns, but with $g_e \sim +0.4$ (Figure 5b). This requires a very unusual magnetic splitting pattern of the hole in the ground state of the Cu acceptor, which perhaps might be reproduced from some appropriate admixture of ^2E character from the atomic states of the Cu TM. This assignment has the advantages of reconciling the DAP and (A°, X) BE magnetic data, including polarization effects, and also in giving g_e reasonably close to the theoretical estimate. Unfortumately, this assignment also has the six following drawbacks, which necessitate its rejection in favor of a straightforward analysis with $g_e \sim -0.40$.

1. It requires $(\Delta E)_{3/2} \gg 3 \times (\Delta E)_{1/2}$, where the ΔEs are magnetic splittings of the indicated m_J hole states, unlike the assignment of Figure 5a where $(\Delta E)_{3/2} \sim 3 \times (\Delta E)_{1/2}$ and the derived values of K are close to 0.60.[47] This equality can be violated so strongly only through the assumption of a strong contribution to the hole g value from the term L in the general form of the linear Zeeman effect of Γ_8 holes

$$E^h = E_o^h + \beta \, [K \, J.H + L(J_x^3 H_x + J_y^3 H_y + J_z^3 H_z)] \qquad (7)$$

where β is the Bohr magneton, E_o^h the hole binding energy in zero magnetic field and K and L are the isotropic and anisotropic hole g-values.[49] However, the assumption of large L/K in Equation (7) inevitably produces a large anisotropy in the linear Zeeman effect of the hole, which is not consistent with the essentially isotropic experimental Zeeman splittings of the A_1^a BE.

2. The unusual hole properties of Figure 5b could only occur from hybridization with d atomic states, and the exact splitting pattern should be a sensitive function of this interaction. However, we have observed very similar Zeeman behavior for (A^o, X) BE for the Cu, Ag and Au TM acceptors, and in addition also for AFB involving the main-group acceptors Li_{Zn} and P_{Te}.[44] This has been recently confirmed from the simple Zeeman behavior observed for the Li and $P(A^o, X)$BE at the large magnetic fields such that the zero field coupling is broken down (Figure 6). The relative transition probabilities can be derived quite simply assuming $L \sim 0$ in Equation (7). If the splitting at $H = 0$ is attributed to the two states $J = \frac{3}{2}$ and $J = \frac{5}{2}$ derived from electron coupling to a $J = 2$ hole-hole combination, the nomograph in Figure 6 shows how transitions to the $J = \frac{3}{2}$ final state produce a pattern of magnetic sub-components very similar to that for transitions from the $J = \frac{1}{2}$ BE state we believe to lie lowest for the Cu acceptor (Figure 5a), in the Paschen-Back limit of complete magnetic decoupling of electron and hole above $H \sim 8T$. The energy corrections to the BE magnetic substates arising from the zero-field splitting Δ are of order 0, $\Delta/4$ or $\Delta/2$, depending on the m_J of the substate. Therefore, they are at most ~ 0.1 meV. Together with other disregarded factors such as m_J-dependent diamagnetic shifts and contributions from a finite value of L in Equation 7, these energy corrections simply contribute to the width of the individual magnetic subcomponents. Disregarding thermalization, the sequence of intensities calculated for the high field subcomponents in Figure 6; (6.5, 10, 7, 10 and 6.5), do not differ greatly from the sequence for

$J = \frac{1}{2}$ lowest; (3, 4, 2, 4, 3). This simple behavior would not occur if $g_e \nsim -K$ and if K were not closely similar between the BE state and the single magnetic hole remaining after the (A^o, X) BE recombination, as shown in the nomograph of Figure 6. The important point here is that this simple isotropic magnetic behavior is qualitatively the *same* as for the Cu BE A_1^a.

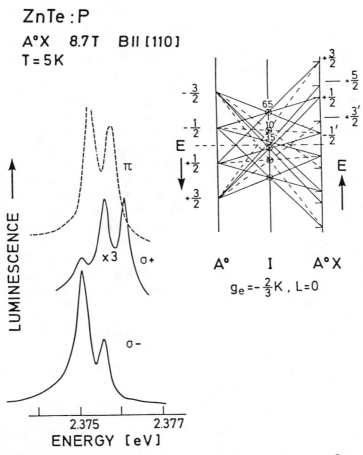

FIGURE 6 Zeeman effect of the no-phonon component of the $P_{Te}(A^o, X)$ BE in ZnTe, in a magnetic field of 8.7T, sufficiently large to reach the Paschen-Back limit for e-h exchange in the BE. This ensures overall splitting patterns very similar to Figure 4, although the (A^o, X) BE splittings, illustrated in the nomograph, are more complicated (compare Figure 5a).

3. The FE magneto-reflectance also indicates that $g_e < 0$ in ZnTe.[50]

4. Recent Zeeman measurements of the (A°, X)BE for a trigonal acceptor in ZnTe[51] confirm that $g_e = -0.50 \pm 0.05$, and is clearly negative from the form of the Zeeman spectra.

5. We have recently re-examined the magnetic properties of the selective DAP luminescence in ZnTe utilizing the LO replica of the exciting laser line and thus avoiding additional complications from magnetic splittings of the $n = 2$ acceptor state present in the earlier study of the 'two-hole' satellites.[45] These results will be reported in more detail elsewhere,[52] but we present here aspects relevant to the present discussion. No magnetic effects occur for $h\nu_{EXC} \sim h\nu_p$, where $h\nu_p$ is the energy of the distant DAP peak under non-selective excitation. Presumably, both electron and hole spins are conserved in these optical processes involving weakly interacting very distant DAP. Magnetic structure emerged when $h\nu_{EXC}$ was increased by a few meV. The form of this structure was similar to that reported by Tews,[45] though better resolved. The splittings of the outer σ-polarized subcomponents were appreciably smaller than in Figure 4, allowing for the influence of effects interpreted by the mixing of hole magnetic substates for B $\parallel$ <100>. This was the key observation leading to the derivation of a positive g_e by Tews.[45] However, it appears that the magnetic splittings can be described exclusively in terms of hole spin-flip processes in the new study,[52] using g-values similar to those from the (A°, X) BE Zeeman data. Thus, these spectra provide no information on the sign of g_e. Such conservation of electron-spin is connected with the weak spin-orbit coupling of electron states in the wide band gap II-VI semiconductors. This effect is also responsible for the long spin-lattice relaxation times which are a familiar feature of the magnetic behavior of BE[35] and of the medium-separation DAP which dominate in ODMR experiments.[31,46]

6. It seems probable that the evidence from ODMR is not as conclusive as appeared at first sight.[46] We recall that good quality ZnTe is weakly compensated p-type, and that long spin-lattice relaxation times T_e are possible for electrons on effective mass-like donors. Consider the following conditions, all perfectly feasible for high quality ZnTe at 2°K

$$T_h \ll \tau_{DAP} \ll T_e \qquad (6a)$$

where T_h is the hole spin-lattice relaxation time on shallow acceptors and τ_{DAP} is the donor acceptor pair recombination lifetime. Equation (6a) implies that the hole magnetic substates are well thermalized but the electron spins remain unthermalized. The DAP distribution can therefore be broken up into two non-communicating subgroups, (a) with electron spin up and total spin quantum numbers $m_J = 2$, 1, 0 and -1 and (b) with electron spin *down* and $m_J = 1$, 0, -1, -2. Thermal equilibrium exists within each of these two m_J subgroups, the lowest m_J level lying lowest in energy in each case, determined by the hole splitting. Transitions from $m_J = -2$ to the $J = 0$ ionized DAP state are forbidden. The population of system (b) should therefore significantly exceed that of (a) at the large carrier generation rates used in the ODMR experiment.[46] If we also postulate that $g_e < 0$, then the higher population spin down system (b) has the higher energy. Thus, the electron spin system becomes inverted. An increase in the σ_- luminescence intensity of 0.07% was induced by microwave radiation at the donor resonance energy, $|g| = 0.401 \pm 0.004$, but no acceptor resonances were observed.[46] This resonance was interpreted on the usual model of uncoupled electron-hole spins used for distant DAP ODMR effects. Assuming thermalization of the electron spin substates and +ve g_e, the microwave resonance must increase $m_J = +\frac{1}{2}$ at the expense of $m_J = -\frac{1}{2}$. The calculated changes in $I_{\sigma-}$ and $I_{\sigma+}$ assuming complete saturation of the donor state populations, a spin temperature of 2°K and the resonance field of 1.596T are, respectively, +7.0% and -3.8%.[46] These predicted values may be much larger than experiment because of incomplete microwave saturation and particularly because the temperature of the donor spin system is very unlikely to be as low as 2°K (Ref. 35) under 100–200 mW focused optical excitation employed. The conclusion that g_e must be positive is inescapable on this model of significant thermalization between the donor states before the microwaves are applied.[46] On the other hand, exactly the opposite conclusion is obtained on the assumption that the donor spin system is initially completely unthermalized, but is inverted under optical pumping as a consequence of the unequal DAP recombination rates in the magnetic field described above. The resonant microwave power then results in a transfer of intensity between subsystems (b) and (a), so that (a) is favored and the combined recombination lifetime correspondingly reduced. Consideration of Figure 3 of Killoran et al.[46] shows that the consequent increase in population of the lowest energy DAP substate with $|m_h, m_e > 1 - \frac{3}{2}, + \frac{1}{2}>$ also *increases* the σ_- luminescence intensity [σ_- luminescence corresponds to RCP light (Figure 6b) and to $\Delta m_J = +1$ in the transition (Figure 5)]. Thus, the initial conclusion that this increase of σ_--polarized luminescence establishes g_e as +ve cannot be regarded as conclusive. This misinterpretation of Killoran et al.[46] provides an important example of the potential lack of uniqueness in conclusions drawn from ODMR experiments, which all generally stem from unrecognized aspects of the kinetic pathways which control the intensities of the detected luminescence signals. These potential problems for the interpretation of ODMR effects have not yet received the attention they deserve.[37] In the present example, the effect is intimately linked with the possibility in p-type ZnTe that essentially all the detected luminescence arises from transiently photoneutralized donors. This explains why checking experiments on Cd Te, where g_e is −ve, appeared to confirm the conclusion obtained from the more usual ODMR analysis of Killoran et al.[46] CdTe is normally n-type or at least is not p-type with such weak compensation as occurs for the high quality ZnTe used in these experiments. We therefore conclude that these ODMR data do not conflict with the strong evidence from the many other experiments we discuss that g_e is negative in ZnTe, so that -0.401 ± 0.004 represents the most accurate estimate available.

According to the six arguments just presented, we must conclude that $g_e \sim -0.40$ in ZnTe, with corresponding simple behavior of g_h according to Figure 5a. The reason for the failure of the k.p theory to predict g_e to within nearly 1.0 unless P^2 is increased to the seemingly unreasonable value of ~ 35 eV must remain unexplained at present. It is possible to obtain a slightly larger theoretical negative g-shift from $g_e = +2.0$ than the value of -1.53 quoted above, by taking plausible contributions to the bare mass $m_e^* \sim 0.11 m_0$ from higher conduction bands to give $P^2 \sim 26$ eV, significantly larger than in CdTe.[53] This larger predicted negative g-shift yields $g_e \sim +0.2$. It certainly does not appear to be reasonable to look for any further increase in P^2 according to our knowledge of m_e^* in ZnTe, so we must take issue with some of the conclusions of the 5-band calculation of P^2 by Hermann and Weisbuch.[53] This limiting value of $g_e \sim +0.2$ still remains surprisingly far above the experimental consensus of -0.40. The only obvious further possibility[54] involves an interaction with the d-electron band from the Zn cation. Although this lies relatively high, it is still ~ 10 eV below the top of the valence band.[8] Such interactions are usually neglected, both because of the large energy denominators involved and the expected small magnitudes of the interaction matrix elements. However, it is perhaps just possible that this neglected interaction is responsible for the large residual under-estimate in the g-shift. It is not clear why this should be peculiar to ZnTe, since the host d-states are only ~0.7 eV deeper in CdTe than in ZnTe.[8]

The values of K derived for the Cu acceptor according to the analysis of Figure 5a (+0.60 ± 0.02) lie very close to those determined for the main group acceptors Li(+0.63), P(+0.64), Na(+0.65) and As(+0.8) by spin-flip Raman scattering[55] and for Li(+0.52 ± 0.10) from the Zeeman effect of the DAP luminescence.[45] We must therefore conclude that Cu_{Zn} produces an acceptor state of classical character in ZnTe, except that there is a relatively large central cell enhancement of E_A. Strong binding of the hole produces very little change in K, as is clear from our measurements of the Zeeman effect of the $Au(A^o, X)$ BE, since E_A is much larger for Au than for Cu.[56] We now contrast the complex behavior of Cu in ZnSe with this essentially classical story for ZnTe.

5. The I_1^{DEEP} Bound Exciton in ZnSe

The ZnSe spectrum illustrated in Figure 3, with origin at 2.783 eV, has been labelled the I_1^{DEEP} BE luminescence. It has already been pointed out that this has the general form of an (A^o, X) BE spectrum, but with many anomalous properties. In the following sections we review the salient experimental properties of this spectrum, its correlation with Cu impurity, and then develop a model which is consistent with the data and which can be used to fit ZnSe:Cu into the sequence of $ZnX:Cu_{Zn}$ systems.

5.1 Experimental Properties

1. The BE localization energy $E_{BX} \sim 19.7$ meV is anomalous, being about twice as large as for (A^o, X) BE involving the shallow, well-understood acceptors Li_{Zn} and Na_{Zn}[57] and N_P[58] in ZnSe. It is also much larger than for the almost comparably deep acceptors Ag_{Zn} and Au_{Zn}.[39]

2. It is a close doublet (not shown in Figure 3), with a very small zero-field splitting of ~0.21 meV.[57,59] This splitting can be clearly resolved only in crystals of the highest available quality, since the higher-energy component is always weak at the low temperatures necessary to minimize effects of thermal broadening,[32] and the splitting is smaller than the BE linewidth in the more heavily doped crystals.

3. This I_1^{DEEP} BE is a very persistent feature of photoluminescence in ZnSe. It must therefore derive from an intrinsic defect or a very persistent impurity species. The intensity ratio I_1^{DEEP}/I_2 measured in luminescence, where I_2 is due to recombination of (D°, X) BE from typical shallow donors, was observed to increase on anneal at 650-700°C for five hours in an initial vacuum of 5×10^{-6} mm Hg in a 0.5cc ampoule.[79] The average increase was ~5.3 fold for 4 crystals, with a spread between 2.7 and 8.0. The increase in optical absorption at I_1^{DEEP} for one crystal was the same as in luminescence, and increases in optical absorption at I_1^{DEEP} regularly occur under this treatment. Thus, there is no doubt that the concentration of centers responsible for I_1^{DEEP} increases under vacuum anneals. It has been traditional to argue that such evidence supports a relationship with cation vacancies.[60,61] However, comparison with recent advances in the understanding of the electronic properties of ZnTe[2,32] makes it likely that the center could be a freely available cation impurity substituent, whose solubility is enhanced under thermochemical conditions which promote vacancies. Copper is a very likely candidate, particularly since the other most common acceptor species in II-VI semiconductors, Li and Na, are known to produce other distinctive (A°, X) BE lines in ZnSe.[62]

4. Indications have been obtained that the relative strength of the I_1^{DEEP} luminescence line increases markedly upon diffusion of Cu.[40,63] This evidence, while direct, is not absolutely compelling because of the many complicated effects which are possible under heat treatments of II-VI compound semiconductors. However, the fact that only rather gentle heat treatments were necessary, for example only 250°C for two hours[63] and the striking spatial correlation which has been observed with the diffusion distance from a discretely-plated source of Cu metal on the semiconductor surface, lends strong support to the most straightforward interpretation of this result. Similar changes in I_1^{DEEP} do not occur for heat treatments at these low temperatures with no deliberate source of Cu.

5. Thermal quenching has been observed between the two sub-components of I_1^{DEEP} in luminescence but not in optical absorption. The ratio in absorption is 1:1.6 in favor of the lower energy sub-component, whereas it is 1:3.1 at 4.2°K and 1:5.0 at 1.6°K in luminescence. These results strongly suggest a small electronic splitting of 0.2_1 meV in the BE state, since the luminescence ratios predicted from the absorption result, assuming thermal equilibrium, are 1:2.9

at $4.2°K$ and $1:7.1$ at $1.6°K$. Alternative possibilities for the splitting, such as two slightly different environments for the center or a no-phonon isotope shift are not consistent with this thermal behavior. The intensity ratio predicted from the natural isotopic abundances of ^{65}Cu and ^{63}Cu is $1:2.2$, inconsistent with the experimental ratios.

6. The Zeeman data for I_1^{DEEP} show at least six lines, which all appear in both σ and π polarization in the Voigt configuration (see Figure 7). A least-squares fit assuming Lorentzian line-shape suggests that the

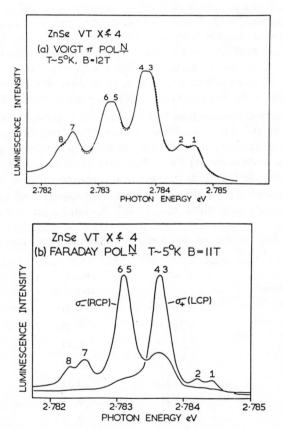

FIGURE 7 Zeeman effect of the no-phonon components of the $Cu_{Zn}(A^O, X)$ BE (I_1^{DEEP}) in ZnSe. (a) For Voigt π polarization. (b) For Faraday σ_+ and σ_- polarizations. The magnetic sub-components are labelled as in Table 1, which also gives an alternative labelling scheme when the central components (3,4) and (5,6) are treated as single lines.

two strongest lines in the center of the Zeeman pattern may be barely-resolved doublets, giving eight lines in all. (Table 1.) The circularly-polarized luminescence in Faraday configuration shows considerably larger differences between the intensities of the Zeeman sub-components; the three high-energy lines are predominantly left circularly-polarized (LCP) and the three low-energy lines predominantly right circularly-polarized (RCP). However the relative intensities of individual components remain similar in Voigt and Faraday configurations, and the effects of thermalization are quite small. The splitting pattern is relatively insensitive to crystal orientation in the magnetic field. This overall behavior of the Zeeman components is quite inconsistent with that expected for the $\Gamma_6 \rightarrow \Gamma_8$ transition of a conventional (A^o, X) BE, even for a deep acceptor like GaAs:Sn.[64] However, the large number of Zeeman subcomponents, with no evidence of orientational anisotropy, prove that this exciton in ZnSe must be bound to a point defect of lattice site symmetry T_d.

7. Sokolov and Konstantinov[65] have reported a relatively broad (on the scale of the PL linewidth in Figure 7) electro-absorption dip in ZnSe:Cu centered at 2.771 eV, which they attribute to a bound exciton state involving Cu^{2+}. This attribution is based upon a qualitative correlation of the strength of this feature with increase in [Cu] and in the infra-red absorption which they associate with the $T_2 \rightarrow E$ intra d-state excitations within Cu^{2+}. No luminescence was observed

TABLE 1   Decomposition of Voigt π Zeeman Spectrum at B = 12T (Figure 7a).

Sub-component		Energy (Ev)	Width[o] (meV)	Height (Relative)
1^+	1^*	2.78470	0.31	1.6
2	2	2.78445	0.31	1.2
3	3	2.78393	0.31	4.3
	4	2.78374	0.31	4.2
4	5	2.78329	0.31	2.6
	6	2.78312	0.31	2.8
5	7	2.78255	0.31	2.0
6	8	2.78232	0.31	1.0

* Notation in Figure 7 of Reference (32)
+ Notation in Figure 9
o Lorentzian sub-components.

from this BE. This fact and the substantial energy shift of the electro-absorption feature below the energy of I_1^{DEEP} (2.783 eV) may be associated with relatively high values of [Cu] in these crystals. In addition, the derivation of the exact value of an absorption peak from the form of the electro-absorption feature it produces is not completely straightforward. For example, Sokolov and Konstantinov also report a positive peak near 2.78 eV which they associate with the FE, although this has an energy of 2.802 eV. Therefore, it may be plausible to suggest that the 2.771 eV electro-absorption dip relates to the I_1^{DEEP} (A^o, X) BE.

5.2 The I_1^{DEEP} Localization Energy

Exciton localization at a neutral acceptor involves three particles in addition to the impurity core; these are the hole bound in the neutral acceptor, and the electron and hole which constitute the exciton. For a semiconductor like ZnSe with effective-mass ratio $m_h^*/m_e^* > 1$, exciton localization at a conventional neutral acceptor is usually described in terms of the hydrogen-molecule-like bound configuration, $A^-\,_h^h\,e$.[66] The case of ZnTe:Cu discussed in Section 4.4 provides a typical example where this model accounts for the optical properties. When the two holes are equivalent with symmetry Γ_8 ($J_h = \frac{3}{2}$) derived from the VB maximum in a ZB semiconductor, the exclusion principle allows only two hole-hole states, with total hole angular momenta $J_{hh} = 0$ and 2. The core A^- for a conventional acceptor has a closed-shell configuration with no net spin. The states of the (A^o, X) BE are therefore obtained by coupling the $s = \frac{1}{2}$ spin of the electron from the Γ_6 CB minimum to the hole-hole states, producing the familiar sequence of states $J_{AX} = \frac{5}{2}, \frac{3}{2}, \frac{1}{2}$, where J_{AX} is the total angular momentum of the (A^o, X) BE.[35]

The principal energy terms in the (A^o, X) localization are the core-hole interactions (including the short range central cell terms), the hole-hole correlation energy and the electron-hole interaction. Electron-core interactions, which are dominant in the charge transfer states of ZnO:Cu and ZnS:Cu (see Section 6), are negligible for localization at a conventional acceptor since the impurity core A^- is negatively charged and hence repulsive to the electron. The hole-hole correlation energy stabilizes the BE states of higher J_{AX}, and determines the ground state for shallow acceptors with $E_A \lesssim 2(E_A)_{EM}$. However, localization at deeper acceptors is commonly observed to invert this energy sequence, the $J_{AX} = \frac{1}{2}$ state lying lowest.[67] A possible explanation[68] involves a configuration interaction between this state and the higher-lying $J_{AX} = \frac{1}{2}$ state derived by coupling two Γ_7 ($J_h = \frac{1}{2}$) VB holes to give $J_{hh} = 0$, then coupling this hole-hole state to the electron spin. This $J_{hh} = 0$ state has an S-like orbital character which interacts strongly with the central cell potential of the impurity. This central cell term may be very large for a deep acceptor, and by configuration interaction can push the energy of the $J_{AX} = \frac{1}{2}$ state below the $\frac{5}{2}$ and $\frac{3}{2}$ states, as discussed above.

Table 2 compares the exciton binding energy E_X in the semiconductors GaAs, InP and ZnSe with the exciton localization energy E_{BX} at both shallow and deep acceptors. For the deep acceptors GaAs:Sn and InP:Ge, E_{BX} is significantly larger than E_X, showing that the core-hole and hole-hole energy terms dominate in the exciton binding. However for I_1^{DEEP} $E_{BX} \sim E_X$, even though E_{BX} is approximately twice as large as for the conventional shallow acceptors Na_{Zn}, Li_{Zn} and N_P. Since the exciton binding energy E_X is large in ZnSe, it is not clear that the electron-hole interaction can be neglected in comparison with the core-hole and hole-hole correlation terms when discussing the (A^o, X) BE states, despite the relatively large localization energy for I_1^{DEEP}.

5.3 A Model for I_1^{DEEP}

The evidence presented in Section 5.1 shows that I_1^{DEEP} is an (A^o, X) BE luminescence involving a T_d-symmetry (point defect) acceptor. In view of the correlation of this luminescence with the presence of Cu in ZnSe, and the evidence from Section 4.3 that Cu is a sufficiently deep acceptor compared with Li_{Zn} or Na_{Zn} to produce an anomalously large exciton localization energy, we investigate a model which postulates that Cu_{Zn}^o is the acceptor responsible for the I_1^{DEEP} luminescence.

Since the BE luminescence occurs at an energy only ~ 19 meV below the free exciton (FE) energy, it is clear that the recombination occurs between an electron and hole which have wavefunctions relatively slightly perturbed from those of the appropriate band extrema, i.e. the excitonic hole has Γ_8^- symmetry derived from the p-like VB and the electron has Γ_6^+ symmetry from the s-like CB. (The superscripts $\pm$ indicate the parity of the basis functions with respect to inversion, although this is not an operation of the T_d point group.[69]) There are three possibilities for the Cu_{Zn}^o acceptor hole:

1. A Γ_8^- ($J_h = \frac{3}{2}$) hole derived from the upper VB in ZnSe. This would produce a conventional (A^o, X) BE state with two equivalent holes, e.g. GaAs:Sn.[64] As pointed out previously, the zero-field splitting and unusual Zeeman behavior for I_1^{DEEP} are not consistent with such

TABLE 2 Comparison of exciton binding and A^o, X localization energies for both shallow and deep acceptors in ZB semiconductors

		GaAs	InP	ZnSe
E_g	(eV)	1.5188^a	1.4230^a	2.82
E_{gx}		1.5152	1.4182	2.802
Shallow (A^o, X)		~1.512	~1.4145	~2.792
Deep (A^o, X)		1.5063 (Sn)^b	1.3984 (Ge)^c	2.783 (I_1^{DEEP})
Exciton binding E_x	(meV)	3.8	4.8	~18
Exciton localization E_{BX} Shallow	(meV)	~3	~3.7	~10
Deep		~9	19.8	~19
Deep acceptor binding energy, E_A	(meV)	167	210	(~700) (Cu)

[a] A. M. White, P. J. Dean, L. L. Taylor, R. C. Clarke, D. J. Ashen, and J. B. Mullin, J. Phys. C 5 1727 (1972).
[b] Reference (64) in text.
[c] Reference (67) in text.

751

a model. This possible model must therefore be rejected. A conventional $(A°, X)$ BE model with inequivalent holes, involving a contribution from the spin-orbit (SO) split-off VB, can be rejected on energetic grounds, since the SO splitting of the valence band maximum in ZnSe is over 400 meV.

2. A Γ_8 $(J_h = \frac{3}{2})$ hole, but with a wavefunction different from that of the excitonic hole. Removing the restriction on the equivalence of the holes produces more allowed BE states, and offers some explanation for the zero-field splitting in terms of two $J_{AX} = \frac{1}{2}$ states stabilized by the central cell interaction. However, more detailed consideration of this model[70] reveals that it is inadequate to explain both the unequal energy spacing between the Zeeman sub-components and the strongly mixed σ/π polarization of all the lines in Voigt configuration. This attribution will not be considered further here.

3. The Cu^o_{Zn} acceptor hole is not of Γ_8 $(J_h = \frac{3}{2})$ symmetry. A model based on a Γ_7 acceptor ground state, typical of the Cu^{2+}_{Zn} (d^9) configuration, has been investigated and found to be consistent with most of the experimental data. This novel model for I^{DEEP}_1 is developed in some detail in the following sections.

Following Equation (3), the wavefunction for the Γ_7 ground state of the Cu^o_{Zn} acceptor will be written as:

$$\Phi_h (\Gamma_7) = (1 - \alpha)^{\frac{1}{2}} \Phi_{3d} (\Gamma_7) + \alpha^{\frac{1}{2}} \Phi_{VB} (\Gamma_7) \qquad (8)$$

Since the acceptor is deep (~ 0.7 eV) the function $\Phi_{VB} (\Gamma_7)$ will contain contributions from across the VB levels of the ZnSe host, the one-electron VB functions being multiplied by suitable envelope functions so that the final bound hole state transforms as Γ_7 in the impurity point group.

The principal interactions in the BE state considered in this model are electron-hole and hole-hole. The electron-core interaction, dominant for ZnO:Cu and ZnS:Cu (see Section 6), is neglected for the following reason. A strong electron-core interaction leads to capture of the electron into the Cu_{Zn} core, producing a closed-shell core configuration, and a strongly-phonon-coupled charge-transfer BE state typical of ZnO:Cu and ZnS:Cu. The characteristics of I^{DEEP}_1 are quite unlike this. In addition, for such a state the BE would have the symmetry of the excitonic hole bound to the negatively-charged core. In view of the relatively modest localization energy of I^{DEEP}_1, this hole would have Γ_8 symmetry, derived from the upper VB edge. Such a description of the BE state could not account for the observed zero-field splitting, nor for the number of component lines in the Zeeman spectrum. In the next two sections we determine the symmetry and wavefunctions of the BE states in the limit that, first the excitonic electron-hole interaction, and then the hole-hole interaction, is the dominant energy term.

5.4 BE States with Excitonic e-h Interaction Dominant

The excitonic states transform as[71]

$$\Gamma_8^- \text{ (hole)} \times \Gamma_6^+ \text{ (electron)} \rightarrow \overbrace{\Gamma_4^- + \Gamma_3^-}^{J_x = 2} + \overbrace{\Gamma_5^-}^{J_x = 1} \qquad (9)$$

The $J_x = 2$ exciton state lies lowest, and can be further split by the T_d crystal field, for example like the exciton bound to the deep isoelectronic center Bi_p in GaP^{35} where the Γ_4^- state lies ~ 0.3 meV below Γ_3^-. When an exciton is localized at an impurity which itself has a core spin, as for Cu_{Zn}^o (Γ_7), the exciton-core coupling gives rise to further BE states, as indicated in Table 3.

TABLE 3 Exciton-core interaction for Cu_{Zn}^o-BE

Exciton		Cu_{Zn}^o		BE	Wavefunction
$J_x = 2:$ $\begin{cases} \Gamma_4^- \\ \Gamma_3^- \end{cases}$	$\times$	Γ_7^+	$\rightarrow$	$\Gamma_7^- + \Gamma_8^-$	$\psi_7 + \psi_8$
	$\times$	Γ_7^+	$\rightarrow$	Γ_8^-	ψ_8''
$J_x = 1:$ Γ_5^-	$\times$	Γ_7^+	$\rightarrow$	$\Gamma_6^- + \Gamma_8^-$	$\psi_6 + \psi_8'$

The BE wavefunction can be conveniently written in the form:

$$| \psi_\Gamma, \gamma > = \sum_{m_1 m_2 m_s} | m_1 ; m_2 ; m_s > \cdot \alpha_{\Gamma\gamma} (m_1, m_2, m_s), \qquad (10)$$

where Γ, γ are the representation and the component of the localized exciton, $\alpha_{\Gamma\gamma}$ is an expansion coefficient, and m_1, m_2, m_s are the angular momentum components of the excitonic hole, the acceptor hole and the excitonic electron respectively. Representative wavefunctions which will be useful in the later discussion are as follows:

$$|\psi_7, -\frac{1}{2}> = -\frac{1}{\sqrt{6}}\left[|\frac{3}{2};\frac{1}{2};-\frac{1}{2}> - |-\frac{3}{2};-\frac{1}{2};-\frac{1}{2}> + |\frac{3}{2};-\frac{1}{2};\frac{1}{2}>\right.$$

$$\left.-\sqrt{3}\,|\frac{1}{2};\frac{1}{2};\frac{1}{2}>\right], \tag{11}$$

$$|\psi_8, -\frac{3}{2}> = \frac{\sqrt{3}}{2}\,|\frac{1}{2};-\frac{1}{2};\frac{1}{2}> - \frac{1}{2}\,|\frac{3}{2};-\frac{1}{2};-\frac{1}{2}>, \tag{12}$$

$$|\psi_8'', -\frac{3}{2}> = \frac{1}{\sqrt{2}}\left[|-\frac{1}{2};\frac{1}{2};\frac{1}{2}> + |\frac{1}{2};\frac{1}{2};-\frac{1}{2}>\right], \tag{13}$$

$$|\psi_8', -\frac{3}{2}> = -\frac{1}{\sqrt{12}}\,|\frac{1}{2};-\frac{1}{2};\frac{1}{2}> - \frac{1}{\sqrt{3}}\,|-\frac{1}{2};\frac{1}{2};\frac{1}{2}> - \frac{1}{2}\,|\frac{3}{2};-\frac{1}{2};$$

$$-\frac{1}{2}> + \frac{1}{\sqrt{3}}\,|\frac{1}{2};\frac{1}{2};-\frac{1}{2}>. \tag{14}$$

5.5 BE States with h-h Interaction Dominant

When the acceptor-bound hole has Γ_7^+ (predominantly d-like) symmetry and the excitonic hole Γ_8^- (p-like) symmetry, the two-hole states transform as:

$$\Gamma_7^+ \text{ (hole)} \times \Gamma_8^- \text{ (hole)} \rightarrow \underbrace{\overbrace{}^{J_{hh}=2}}_{\Gamma_4^- + \Gamma_3^-} + \underbrace{\overbrace{}^{J_{hh}=1}}_{\Gamma_5^-} \tag{15}$$

The BE states are obtained as the product of these h-h states with the Γ_6^+ representation of the excitonic electron, as shown in Table 4. Using the previous notation, representative wavefunctions are as follows:

$$|\phi_7, -\frac{1}{2}> = -\frac{1}{\sqrt{6}}\left[|\frac{3}{2};\frac{1}{2};-\frac{1}{2}> - |-\frac{3}{2};-\frac{1}{2};-\frac{1}{2}> + |\frac{3}{2};-\frac{1}{2};\frac{1}{2}>\right.$$

$$\left.-\sqrt{3}\,|\frac{1}{2};\frac{1}{2};\frac{1}{2}>\right], \tag{16}$$

$$|\phi_8, -\frac{3}{2}> = -\frac{1}{2}\,|\frac{3}{2};-\frac{1}{2};-\frac{1}{2}> + \frac{\sqrt{3}}{2}\,|\frac{1}{2};\frac{1}{2};-\frac{1}{2}>, \tag{17}$$

$$| \phi_8'', -\frac{3}{2} > \; = \; \frac{1}{\sqrt{2}} \left[| \frac{1}{2}; -\frac{1}{2}; \frac{1}{2} > \; + \; | -\frac{1}{2}; \frac{1}{2}; \frac{1}{2} > \right] \tag{18}$$

$$| \phi_8', -\frac{3}{2} > \; = \; \frac{1}{\sqrt{3}} \left[| \frac{1}{2}; -\frac{1}{2}; \frac{1}{2} > \; - \; | -\frac{1}{2}; \frac{1}{2}; \frac{1}{2} > \right] \; - \; \frac{1}{2} | \frac{3}{2}; -\frac{1}{2}; -\frac{1}{2} >$$
$$- \; \frac{1}{\sqrt{12}} | \frac{1}{2}; \frac{1}{2}; -\frac{1}{2} > . \tag{19}$$

The transformation linking the two sets of BE wavefunctions is:

$$| \phi_7 > \; = \; | \psi_7 > , \tag{20}$$

$$| \phi_8 > \; = \frac{1}{4} | \psi_8 > \; + \frac{3}{4} | \psi_8' > \; + \frac{\sqrt{3}}{2\sqrt{2}} | \psi_8'' > , \tag{21}$$

$$| \phi_8' > \; = \frac{3}{4} | \psi_8 > \; + \frac{1}{4} | \psi_8' > \; - \frac{\sqrt{3}}{2\sqrt{2}} | \psi_8'' > , \tag{22}$$

$$| \phi_8'' > \; = \frac{\sqrt{3}}{2\sqrt{2}} | \psi_8 > \; - \frac{\sqrt{3}}{2\sqrt{2}} | \psi_8' > \; + \frac{1}{2} | \psi_8'' > . \tag{23}$$

Thus, if one takes the functions $| \psi >$ as a basis and then allows the h-h interaction to increase in the BE state, there is a strong mixing of functions derived from the $J_x = 2$ and $J_x = 1$ exciton states. Since recombination from $J_x = 1$ only is dipole-allowed, the hole-hole interaction effectively transfers oscillator strength to the $J_x = 2$ level. This treatment is formally similar to that used by White[72] in discussing conventional (A^0, X) BE states, but differs in that we are considering inequivalent rather than equivalent holes in the BE configuration.

TABLE 4   Coupling between two-hole states and excitonic electron for Cu_{Zn}^0–BE

h-h state		Electron		BE	Wavefunction
$J_{hh} = 2$: $\begin{cases} \Gamma_4^- \\ \Gamma_3^- \end{cases}$	$\times$ $\times$	Γ_6^+ Γ_6^+	$\rightarrow$ $\rightarrow$	$\Gamma_6^- + \Gamma_8^-$ Γ_8^-	$\phi_6 + \phi_8'$ ϕ_8''
$J_{hh} = 1$: Γ_5^-	$\times$	Γ_6^+	$\rightarrow$	$\Gamma_7^- + \Gamma_8^-$	ϕ_7, ϕ_8

5.6 Relative Energies of the Two-hole States

The relative energies of the h-h states in the BE configuration can be estimated from a two-particle atomic model,[73] allowing for the inequivalence
of the two holes. This inequivalence stems from the difference in orbital
character; the excitonic hole has p-like symmetry derived from the VB,
but the acceptor hole in its ground state has t_2-like symmetry derived
from orbitals with $d(\ell = 2)$ character. We shall label the two holes p and
t_2 respectively to emphasize this difference, it being implicit in the remainder of this section that the hole designated 't_2' has d-like character.

The electrostatic interactions between the two holes are the same as
those for an atom with two holes in inequivalent p-shells, that is with the
configuration $(n'p^5.np^5)$. However because of the previously mentioned
difference in angular momentum properties of p- and $t_2(d)$-orbitals,[16]
the spin-orbit splitting of the t_2-hole is formally similar to that of a single
p-electron. The energies of the h-h interaction in the BE state can therefore be described in terms of an atomic configuration $(n'p^5.np^1)$, but with
a change of sign in the electro-static matrix elements. From Condon and
Shortley[73] it is straightforward to construct a schematic correlation diagram which illustrates the trends in energy of the h-h states on going from
the limit of strong spin-orbit coupling to that of strong electrostatic repulsion. This is shown in Figure 8.

The inverted spin-orbit splitting of the t_2-hole puts the $[p(\frac{3}{2}), t_2(\frac{1}{2})]$
two-hole configuration lower in energy than $[p(\frac{3}{2}), t_2(\frac{3}{2})]$, which would
have been the lowest energy configuration for two p-like holes. The elec-
trostatic repulsion splits the $[p(\frac{3}{2}), t_2(\frac{1}{2})]$ configuration, with the $J_{hh} = 1$
state lowest. In the limit of very strong electrostatic interaction this $J_{hh} =$
1 state connects to the lowest energy ^1P state of the $(n'p^5.np^5)$ atomic
configuration. This $J_{hh} = 1$ state is therefore the lowest energy two-hole
state in the Cu^0_{Zn} – BE, when spin-orbit and electrostatic energy terms are
included. The electrostatic splitting of the $[p(\frac{3}{2}), t_2(\frac{3}{2})]$ configuration is
also indicated in Figure 8, with stabilization of the $J_{hh} = 2$ state as previously discussed. The $J_{hh} = 0$ state is most sensitive to central cell interactions. However, as long as these are small compared with the spin-orbit
energy $\frac{3}{2}\zeta_d$, the $J_{hh} = 1$ state from the $[p(\frac{3}{2}), t_2(\frac{1}{2})]$ configuration will remain lowest in energy.

5.7 A Correlation Approach for I_1^{DEEP}

A simple spin-Hamiltonian for the Cu^0_{Zn} – BE configuration would be:

$$\mathcal{H} = -A\, j_1 \cdot s - a'j_2 \cdot s + B\, j_1 \cdot j_2\,, \tag{24}$$

where j_1, j_2 and s are the angular momenta of the excitonic hole, the Cu_{Zn}^o acceptor hole and the excitonic electron respectively. An estimate of the likely relative energies of the BE states can be made by considering the correlation between states for which the excitonic electron-hole interaction is dominant, and those for which the hole-hole interaction dominates. This can be done quite informatively by simplifying the spin-Hamiltonian, Equation (24), even further, so that in both limits it can be written as the sum of a zero-order and a perturbation term

1. Strong exciton electron-hole interaction

In this case the zero-order term will be the first term in (24), and the most significant perturbation will be the third term. The term $-a'\, j_2 \cdot s$, representing the interaction between the acceptor hole and the electron spin, will be neglected in this limit.

$$\mathcal{H}_o' + \mathcal{H}' = \mathcal{H} + a'\, j_1 \cdot s = -A\, j_1 \cdot s + B\, j_1 \cdot j_2 \tag{25}$$

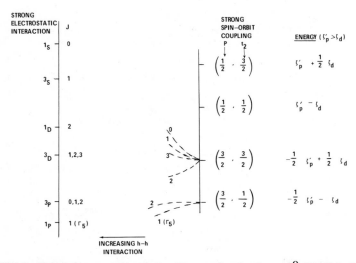

FIGURE 8 Hole-hole energy correlation diagram for the $Cu_{Zn}(A^o, X)$ BE in ZnSe. The two holes behave like holes in equivalent p shells of an atom, although the spin-orbit splitting of the t_2 hole is similar to a p electron. As the h-h interaction increases, the system evolves between the strong spin-orbit coupling limit shown at the right towards the limit for strong electrostatic interaction at the left.

2. Strong hole-hole interaction

In this case the zero-order term will be the third term in (24), and the electron-hole interaction is added as a perturbation. However, in the limit of strong h-h coupling it would seem reasonable to make the approximation that the electron couples to the *net* spin of the two-hole state. This is equivalent to assuming $A \sim a' = A'' \ll B$ in Equation (24). Therefore in this limit the simplified Hamiltonian becomes:

$$\mathcal{H}_o'' + \mathcal{H}'' = B\,j_1 \cdot j_2 - A''(j_1 + j_2) \cdot s$$

$$= B\,j_1 \cdot j_2 - A'' J_{hh} \cdot s \qquad (26)$$

The *schematic* correlation diagram linking these two limiting cases is illustrated in Figure 9. The right-hand side of this figure represents the Hamiltonian Equation (25) operating on the basis functions $|\psi\rangle$; the left-hand side represents Equation (26) operating on the basis set $|\phi\rangle$. On moving from one side of the diagram to the other the appropriate perturbation term ($\mathcal{H}'$ or $\mathcal{H}''$) increases in magnitude. The dotted line simply connect states of a given symmetry in order; the energies of these states

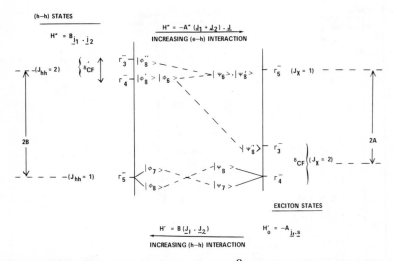

FIGURE 9 Correlation diagram for the Cu_{Zn} (A^o, X) BE in ZnSe. As the h-h interaction increases, and e-h interaction decreases, the system evolves from the limit of strong electron-hole coupling in the BE shown at the right towards the limit of dominant hole-hole coupling shown at the left. The experimental situation for the I_1^{DEEP} BE is believed to lie towards the right of this diagram.

will clearly not vary linearly in any realistic calculation. This correlation diagram is not truly symmetric in the sense that the two Hamiltonian Equations (25) and (26) are not equivalent in their treatment of the term $-a' j_1 \cdot s$ in Equation (24). However it is useful in allowing some qualitative insight into the nature of the I_1^{DEEP} states predicted by the model.

The most important overall conclusion is that the lowest energy states of the Cu_{Zn}^o-BE should have Γ_7 and Γ_8 symmetry. A simple physical picture of the binding is obtained by first imagining an exciton weakly localized at the Cu_{Zn}^o (Γ_7) center, corresponding to the right of Figure 9. As the strength of the hole-hole interaction increases, one state of Γ_8 symmetry derived from the $J_x = 2$ state increases rapidly in energy, leaving two low energy states of $\Gamma_7 + \Gamma_8$ symmetry. At the same time the h-h interaction mixes the $J_x = 2$ and $J_x = 1$ states, thereby allowing the possibility of dipolar exciton recombination from these lower energy states. Photoluminescence excitation (PLE) spectra of the I_1^{DEEP} line in high purity samples show positive excitation features only at energies corresponding to I_1^{Li} and above, for example Figure 10 of Dean et al.[57] This indicates that the combination of crystal field and h-h interaction energies in Figure 9 must produce a splitting $\gtrsim 10$ meV between the low energy $\Gamma_7 + \Gamma_8$ states and the three higher energy states of the BE.

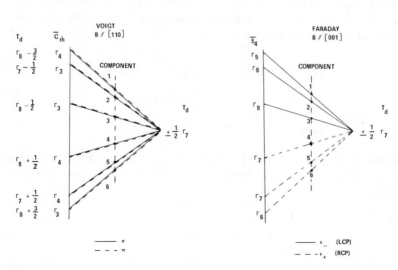

FIGURE 10 Theoretical fan diagrams for the Zeeman effect in the I_1^{DEEP} Cu_{Zn} (A^o, X) BE in ZnSe. The strongly mixed polarizations of the magnetic subcomponents in Voigt polarization geometry contasts with the situation in the Faraday configuration, in agreement with the experimental data in Figure 7.

We propose that transitions from the low energy, $\Gamma_7 + \Gamma_8$ BE states in Figure 9 are responsible for the pair of no-phonon transitions split by 0.21 meV contained within I_1^{DEEP}. Because the acceptor ground state has Γ_7 symmetry, the selection rules for the luminescence transitions are:[74]

$$\Gamma_8 \rightarrow \Gamma_7 \quad : \quad \text{dipole allowed}$$
$$\Gamma_7 \rightarrow \Gamma_7 \quad : \quad \text{dipole forbidden}$$

In order to explain the observation of the two zero-field lines there must be some symmetry-lowering perturbation which mixes the Γ_7 and Γ_8 excited states. This may be a local intra-center effect (e.g., a Jahn-Teller interaction involving the Cu_{Zn} center). However for such a small splitting (0.21 meV) the long-range perturbations always present in real crystals, involving other defects and impurities, will also be sufficient to give the necessary mixing. An external magnetic field automatically mixes the Γ_7 and Γ_8 excited states, so that transitions from all the excited Zeeman levels are allowed within the limit of the usual Δm_J selection rule.

5.8 The Zeeman Spectrum of I_1^{DEEP}

The g-values for the BE states can be estimated in the limits of strong e-h interaction and strong h-h interaction using the wavefunctions (Equations (11)–(14) and Equations (16)–(19)):

$$g(\psi_7) = g(\phi_7) = -\tfrac{1}{3}\, g_e + g_h^8 + \tfrac{1}{3}\, g_h^7 \, , \tag{27}$$

$$g(\psi_8) \qquad = -\tfrac{1}{6}\, g_e + \tfrac{1}{2}\, g_h^8 - \tfrac{1}{3}\, g_h^7 \, , \tag{28}$$

$$g(\phi_8) \qquad = -\left\{ -\tfrac{1}{3}\, g_e + \tfrac{1}{2}\, g_h^8 + \tfrac{1}{6}\, g_h^7 \right\} , \tag{29}$$

where g_e, g_h^8 and g_h^7 are the g-values of the excitonic electron and hole and of the $\text{Cu}_{\text{Zn}}^{\text{o}}$ (Γ_7) acceptor respectively. Quantitatively, for free particles:

$$g_e \sim +1.1 \quad (\text{Ref. 31}) \tag{30}$$

$$g_h^8 \sim -2\widetilde{\text{K}} \quad (\text{Ref. 75}), \tag{31}$$
$$\sim +0.42$$

where we use the convention that g describes the splitting of *electronic* functions. From comparison with the experimental data we suggest later that $g_h^7 \sim 0$. It was pointed out in Sections 2.3 and 4.1, 4.2 that the Cu_{Zn}^{2+} (d^9) configuration in the more ionic materials ZnO and ZnS has a negative g-value given by:

$$g_h^7 (Cu^{2+}, d^9) = -\tfrac{2}{3} [1 + 2k'] . \qquad (32)$$

Here, k' is a positive orbital reduction factor for d-like orbitals,[14] but is negative for p-like orbitals. Therefore, increased admixture of VB p-like character into the acceptor ground state of Cu_{Zn}^0 in ZnSe is the likely explanation of the accidentally-small g_h^7 value cited above. Using these g-values, and with the assumption that the functions $|\psi>$ provide the best zero-order basis for the BE states, it is predicted that both the Γ_7 and Γ_8 BE states will split with *negative* g-values.

The Zeeman selection rules appropriate for the crystal orientations used experimentally are given in Tables 5 (Voigt) and 6 (Faraday), derived from Bhattacharjee and Rodriguez.[74] Using these selection rules, simple nomographs have been constructed in Figure 10 to represent the Zeeman patterns and polarizations in Voigt and Faraday configurations. In these nomographs the relative energies of the Zeeman components are given by the intersections of the transition lines with the center line of each diagram.

TABLE 5 Zeeman Selection Rules for Voigt configuration ($B \parallel [110]$; $E_\perp \parallel [\bar{1}10]$)

BE State	Ground State (Γ_7)	Polarization	Relative Strength*
$\Gamma_8 + \tfrac{3}{2} \rightarrow$	$+ \tfrac{1}{2}$	$\parallel$	$\tfrac{1}{8} P$
$+ \tfrac{1}{2}$	$- \tfrac{1}{2}$	$\parallel$	$\tfrac{1}{8} Q$
$- \tfrac{1}{2}$	$+ \tfrac{1}{2}$	$\parallel$	$\tfrac{1}{8} Q$
$- \tfrac{3}{2}$	$- \tfrac{1}{2}$	$\parallel$	$\tfrac{1}{8} P$
$+ \tfrac{3}{2}$	$- \tfrac{1}{2}$	$\perp$	$\tfrac{1}{8} R$
$+ \tfrac{1}{2}$	$+ \tfrac{1}{2}$	$\perp$	$\tfrac{1}{8} S$
$- \tfrac{1}{2}$	$- \tfrac{1}{2}$	$\perp$	$\tfrac{1}{8} S$
$- \tfrac{3}{2}$	$+ \tfrac{1}{2}$	$\perp$	$\tfrac{1}{8} R$
$\Gamma_7 \rightarrow$	Γ_7	Forbidden in zero field (T_d symmetry)	

In magnetic field transitions from components $\Gamma_7 \pm \tfrac{1}{2}$ will show the same polarization behavior as corresponding components $\Gamma_8 \pm \tfrac{1}{2}$.

* See Appendix for definitions.

TABLE 6 **Zeeman Selection Rules for Faraday configuration (B ∥ [001])**

BE State	Ground State	Polarization	Relative Strength
$\Gamma_8 + \frac{3}{2} \rightarrow$	$+ \frac{1}{2}$	ϵ_+ (RCP)	$\frac{1}{4}$
$+ \frac{1}{2}$	$- \frac{1}{2}$	ϵ_+	$\frac{3}{4}$
$- \frac{1}{2}$	$+ \frac{1}{2}$	ϵ_- (LCP)	$\frac{3}{4}$
$- \frac{3}{2}$	$- \frac{1}{2}$	ϵ_-	$\frac{1}{4}$
$\Gamma_7 \rightarrow$	Γ_7	Forbidden in zero-field (T_d symmetry)	

In magnetic field transition from components $\Gamma_7 \pm \frac{1}{2}$ will show the same polarization behavior as corresponding components $\Gamma_8 \pm \frac{1}{2}$.

The nomographs were constructed assuming $g_h^7 \sim 0$, since this gives a pattern with 6 dominant lines comparable with the experimental data. For non-zero values of g_h^7 further splittings will occur, and this may be the cause of the 8-line pattern suggested by computer-fitting of the band-shape (Table 1). The Γ_7 and Γ_8 BE states are shown to split with negative g-values, as previously discussed. Since the magnetic field splittings are small compared with the sum of crystal field and hole-hole interactions (previously estimated as $\gtrsim 10$ meV), the Γ_7 and Γ_8 states will split as 2-fold and 4-fold degenerate levels. That is, the magnetic field will not de-couple the electron and hole spins. However there will be a field-induced interaction between the $\Gamma_7 \pm \frac{1}{2}$ and $\Gamma_8 \pm \frac{1}{2}$ components which will distort the first-order Zeeman pattern, making quantitative determination of g-values difficult. The excited state splittings shown in Figure 10 are chosen simply to reproduce approximately the experimentally-observed Zeeman patterns. The general form of the diagrams is based on the qualitative arguments presented above, but otherwise they are purely schematic.

The first significant property of this model is that it is consistent with the Zeeman polarization observed for I_1^{DEEP}. It predicts six major Zeeman components, all of which are allowed in both σ and π polarization in Voigt configuration, but with the three higher-energy lines being predominantly LCP and the three lower-energy lines RCP in Faraday configuration. From the Appendix, the line intensity parameters in Voigt configuration, listed in Table 5, are as follows:

$$P \sim R \sim 1 \tag{33}$$

$$Q \sim S \sim 3 \tag{34}$$

Using this approximation, the relative transition probabilities for each of the Zeeman components is presented in Table 7. The lines 2,5 obtain intensity (x) by field-induced mixing with lines 3, 4. From Table 7 it is clear that a similar pattern of relative intensities for the Zeeman components is predicted in all polarizations, as observed experimentally. A value x ~ 1 would be reasonably consistent with the experimental data, predicting (in the absence of thermalization effects) that the innermost lines should be approximately twice as intense as the outer lines. The small thermalization effects observed in the Zeeman spectra can be accounted for if it is assumed that relaxation is rapid only between components of the same symmetry in the magnetic field, as discussed in the Appendix.

In summary, the zero-field splitting and strongly anomalous Zeeman properties of the I_1^{DEEP} BE luminescence can be explained in terms of an (A^o, X) BE transition involving Cu_{Zn}^o, with inequivalence of the holes in the BE state. A model assuming Γ_7 symmetry of the Cu_{Zn}^o ground state, with $g_h^7 \sim 0$, provides a very good detailed description of the Zeeman spectra. Preliminary uniaxial stress measurements on I_1^{DEEP} have also been carried out, with P ‖ (112). After the measurements, the sample was found to be split, precluding quantitative analysis of energy shifts or of small changes in lineshape which were observed. However, with increasing stress, the I_1^{DEEP} line shifted to higher energy, and became preferentially polarized parallel to the stress direction. Similar uniaxial stress measurements on the FE transition in ZnSe[76] show that the barycenter of the stress-split components increases in energy with increasing stress, although the lowest component shifts slightly to low energy. However the *overall* upward shift of the I_1^{DEEP} line suggests that the ratio of hydrostatic stress to uniaxial stress components must be considerably larger for this BE transition. Such a difference in the stress behavior might also be explained by the perturbation of the excitonic hole wavefunction caused by interaction with the acceptor hole in the (A^o, X) BE state, reflected in the basis wavefunctions discussed in Section 5.4.

TABLE 7   **Relative transition probabilities for the Zeeman components***

		Component				
	1	*2*	*3*	*4*	*5*	*6*
Polarization						
Voigt σ	1 :	x :	3−x :	3−x :	x :	1
Voigt π	1 :	x :	3−x :	3−x :	x :	1
Faraday	1 :	x :	3−x :	3−x :	x :	1

* x is a magnetic field mixing parameter.

6. Bound Exciton Recombination at Cu_{Zn}^o Impurities

The present discussion of the optical and electronic properties of the neutral Cu_{Zn}^o acceptor in the Zn chalcogenides centers around the description of the bound hole wavefunction as a linear combination of d-like and p-like functions (Equation (3)). We have attempted to rationalize a number of trends by expressing the acceptor configuration symbolically in terms of a single occupancy parameter, α:

$$Cu_{Zn}^o \equiv [Cu_{Zn}(d^{9+\alpha})]^{-\alpha} \cdot (\alpha h_b).$$

This device indicates the relative disposition of charge as the orbital character changes. The hole wavefunction has t_2 orbital symmetry in a T_d lattice site. When $\alpha = 0$ (the ionic limit) the acceptor has the $[Cu_{Zn}^{2+}(d^9)]^o$ configuration and the bound hole is of pure d-character, with a ground state of Γ_7 symmetry determined by spin-orbit coupling. When $\alpha = 1$ (the effective mass limit) the acceptor has the configuration $\{[Cu_{Zn}^{1+}(d^{10})]^{-1}; h_b\}$ and the hole is of pure p-character with Γ_8 ground state. The magnitude of α has been related to the energy difference between Cu 3d-orbitals and V_{Zn} orbitals in Section 3.3. The picture then emerging from the previous discussion is that Cu_{Zn}^o in ZnO, ZnS and ZnSe has a Γ_7 ground state, and that the transition to a Γ_8 ground state with predominant p-character occurs for ZnTe.

When considering the form of localized electron-hole (or BE) states involving Cu_{Zn}^o, however, there appears to be a marked change in their character between ZnS and ZnSe. For ZnO:Cu the localized electron-hole state has the form of a VB → impurity charge transfer state,[6,14] giving the strongly-phonon coupled luminescence illustrated in Figure 3, with a no-phonon line ~0.6 eV below E_g. ZnS:Cu shows a transition of similar type in absorption, but this is not radiative. For ZnSe and ZnTe, however, the localized electron-hole states have the form of (A^o, X) BE states with much more modest phonon coupling and with much smaller localization energies. These states produce the BE luminescence illustrated in Figure 3. In order to understand these changes in the nature of the localized exciton states, and the reason that the major change should occur at an earlier point in the chalcogenide series than the transition in the acceptor symmetry, it is necessary to consider in more detail the forces binding the electron and hole in the localized state.

The most important of these forces are the electron-core and hole-core interactions, the hole-hole repulsion and correlation interactions, and the excitonic electron-hole interactions. The way in which these interactions change as the occupancy factor α increases is illustrated by considering three specific cases:

a) $\alpha \approxeq 0$ When α approaches zero, the Cu_{Zn} (d^9) impurity core is neutral with respect to the lattice. The e-core and h-core interactions are then of short range only, the former being most important.

The reason for the dominance of the e-core interaction in this limit is that the excitonic electron can occupy the empty d-orbital of the Cu_{Zn}^{2+} (d^9) core, where it experiences a high effective nuclear charge. Using an atomic analogy, if the electron were to occupy an s-like orbital outside the d-shell (e.g. the Cu 4s atomic orbital), it would move in a Coulombic field with an effective nuclear charge:

$$z_{eff}(s) = Z_{Cu} - (18 + 9) = +2 \tag{35}$$

since the 9 d electrons effectively screen the outer s-electron. If the electron occupies the empty d-orbital, however, the other d-orbitals are only partially effective in screening; using Slaters rules the inter-d electron screening factor is 0.35.[77] Therefore, the net effective nuclear charge experienced by the bound electron becomes:

$$z_{eff}(d) = Z_{Cu} - 18 - (9 \times 0.35) = +7.85 \tag{36}$$

This increased attractive force is partially offset by the e-e repulsions within the d-shell, but on balance it remains favorable for the electron to be bound in this shell. This is the major short-range force involved in exciton localization at Cu_{Zn}^o in the more ionic Zn chalcogenides, ZnO and ZnS. The hole is bound to the resulting negatively charged core by a combination of long and short-range forces.[14] Strong localization of the hole results in a marked weakening of the valence bonding around the impurity, leading to lattice relaxation and strong phonon coupling in the bound state.

b) $0 < \alpha < 1$ As the occupancy factor α in the acceptor ground state increases, the following changes in exciton binding forces occur:

1. The attractive short-range e-core interaction is reduced by a factor $\simeq (1 - \alpha)$ as the probability of the excitonic electron occupying the d-orbital is reduced.

2. There is a repulsive Coulombic (long-range) e-core interaction proportional to $\alpha e^2 / \epsilon r_e$, as the impurity core takes on an effective lattice charge of $-\alpha$.

3. For the same reason, a Coulombic attraction between the excitonic

hole and core is introduced, proportional to $\alpha e^2 / \epsilon r_h$. Both (2) and (3) will be partially screened by the acceptor hole.

4. Hole-hole repulsion and correlation terms become significant. The magnitude of the Coulomb repulsion for two holes (i,j) in s-like orbits around the impurity has the form[68]

$$V_{ij} = \frac{1}{2\epsilon^2} \frac{\sigma_i \sigma_j}{(\sigma_i + \sigma_j)^3} (3\sigma_i \sigma_j + \sigma_i^2 + \sigma_j^2) ,$$

where $\sigma_i = 2Z_i m_i^*$. There is therefore a tendency for the importance of h-h repulsion to be partially offset as α increases and the VB p-character of the acceptor hole increases, since σ for this hole becomes smaller.

c) $\alpha = 1$ In this effective mass limit the e-core interaction is purely repulsive, and the exciton localization is dominated by h-core and h-h interactions.

Experimentally, the exciton localization energy is largest when α is small and the character of the binding is determined by the e-core interaction. The changes in the form of the Cu_{Zn}^o localized exciton states along the series of Zn chalcogenides, and the associated spectral differences of Figure 3, can be understood in terms of a monotonic increase in the occupancy factor α from ZnO to ZnTe, with an associated weakening of e-core interactions. The major change, from a dominant e-core interaction to dominant h-core and h-h interactions, apparently occurs between ZnS and ZnSe. A significant change in α between these two compounds is indicated by the difference in Cu_{Zn}^o acceptor g-values deduced from the Zeeman spectra: $g_\parallel = -0.71$ for ZnS and $g_\parallel \sim 0$ for ZnSe. The trend in the nature of the e-core potential across the chalcogenides is illustrated schematically in Figure 11. For ZnO the potential is short-range and attractive; for ZnTe long-range and repulsive. For ZnSe we suggest the possibility of a barrier to capture of an electron into the core, which leads to a more convential (A^o, X) BE state, albeit with inequivalent hole symmetries and orbits.

It is interesting to note that the VB → impurity charge transfer states of ZnO:Cu and ZnS:Cu, which are the Cu_{Zn}^o–BE states in these compounds, have a configuration $[Cu_{Zn}^{1+} (d^{10})]^-; h_b$ which is entirely analogous to the Cu_{Zn}^o acceptor ground state in ZnTe. A detailed theoretical comparison of the bound hole states in the ZnO:Cu charge transfer state with the ground state properties of ZnTe:Cu may prove very informative about the properties of deep acceptors in general. There is however a difference between the properties of the charge transfer states of ZnO:Cu and ZnS:Cu,

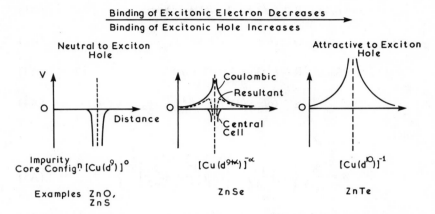

FIGURE 11 Schematic representation of the trend in the *e-core* component of potential within the (A^O, X) BE in Cu_{Zn}^O for the ZnX compound semiconductors. The potential ranges from short range attractive for ZnO to long-range repulsive for ZnTe, the latter being the more familiar situation as in main-group acceptors.

since the former is radiative while the latter is not. This difference is probably related to the energy difference between these charge transfer states and the Cu_{Zn}^{2+} (d^9) ligand field (LF) states in the two compounds.[6] The $^2E(T_d)$ LF state lies ~ 2.13 eV below the charge transfer state in ZnO:Cu, equivalent to ~ 30 LO phonons in ZnO. This energy difference inhibits non-radiative relaxation, so that for excitation above 2.86 eV the charge transfer state relaxes radiatively and the LF states are non-luminescent. For ZnS:Cu, on the other hand, the ^2E LF state probably lies less than 0.4 eV below the charge transfer state, corresponding to ~ 7 LO phonons in ZnS. Excitation of this charge transfer band leads to rapid non-radiative relaxation to the ^2E LF state, which then emits radiatively producing the luminescence shown in Figure 3.

Finally we note that the description of the Cu_{Zn}^O acceptor ground state and Cu_{Zn}^O-BE configurations for ZnSe presented in this paper differ from those given in Refs. 6 and 32. This reflects the conclusions emerging from the much more complete analysis of the I_1^{DEEP} Zeeman data presented in Section 5.

7. Appendix: Relative Transition Probabilities for the Zeeman Transitions for I_1^{DEEP}

From Bhattacharjee and Rodriguez[74] the intensity parameters listed in Table 5 are defined as follows:

$$P \simeq (\sqrt{3}\,\delta - 1)^2/(1 + \delta^2)\,, \tag{A1}$$

$$Q \simeq (\delta + \sqrt{3})^2/(1 + \delta^2)\,, \tag{A2}$$

$$R \simeq (\sqrt{3}\delta + 1)^2/(1 + \delta^2)\,, \tag{A3}$$

$$S \simeq (\delta - \sqrt{3})^2/(1 + \delta^2)\,, \tag{A4}$$

where

$$\delta = \frac{8}{3\sqrt{3}} \left\{ (\rho + \frac{17}{8}) - \left[(\rho + \frac{17}{8})^2 + \frac{27}{64} \right]^{\frac{1}{2}} \right\}, \tag{A5}$$

$$\rho = g_1'/g_2'\,, \tag{A6}$$

and the Zeeman operator for the Γ_8 state is:

$$H_Z^{(8)} = \mu_B\, g_1'\, (B.J) + \mu_B\, g_2'\, (B_x J_x^3 + B_y J_y^3 + B_z J_z^3)$$
$$+ \ldots . \text{ higher terms .} \tag{A7}$$

$$\text{If } \rho \gg 1, \text{ then } \delta \simeq 0 \text{ and } P \sim R \sim 1, Q \sim S \sim 3. \tag{A8}$$

The observed line strengths will follow the transition probabilities given in Table 7 as long as the excited state levels are equally populated, that is thermalization is negligible. The fact that the Zeeman pattern is approximately symmetrical about its center point shows that thermalization effects are indeed small.

At $B = 12T$, $T < 5K$ the experimentally-observed line intensities in Voigt configuration are:

$$\frac{I_5}{I_2} \sim 1.31\,, \frac{I_3}{I_4} \sim 1.41\,, \frac{I_1}{I_6} \sim 1\,.$$

As the temperature is increased, the ratio I_3/I_4 decreases. If the magnetic field components of the BE state are all excited at equal rates, and if thermal relaxation is faster than the luminescence lifetime only for those excited states with the same symmetry in the magnetic field (i.e., $\Gamma_8 \pm \frac{1}{2} \leftrightarrow \Gamma_7 \pm \frac{1}{2}$), then the relative line strengths will be as in Table A1. Such

TABLE A1 **Relative line strengths in the Zeeman spectrum, including thermalization**

Excited State	Component	Transition Probability	Population	Relative Strength*
$\Gamma_8 \quad -\frac{3}{2}$	1	1	1	1
$\Gamma_7, \Gamma_8 \begin{cases} -\frac{1}{2} \\ -\frac{1}{2} \end{cases}$	2	x	a	ax
	3	3−x	b	(3−x)b
$\Gamma_7, \Gamma_8 \begin{cases} +\frac{1}{2} \\ +\frac{1}{2} \end{cases}$	4	3−x	a	(3−x)a
	5	x	b	xb
$\Gamma_8 \quad +\frac{3}{2}$	6	1	1	1

*$a = e^{-\Delta/kT}/\Sigma$; $b = \frac{1}{\Sigma} = 2/(1 + e^{-\Delta/kT})$; $\Delta = |E(\Gamma_8 \pm \frac{1}{2}) - E(\Gamma_7 \pm \frac{1}{2})|$

a model predicts the following relative intensities, including thermalization, which are consistent with the low temperature data:

$$\frac{I_3}{I_4} = \frac{I_5}{I_2} = e^{\Delta/kT}$$

$$\frac{I_6}{I_1} = 1.$$

At $B = 12T$, $\Delta \sim 0.6$meV. The predicted and observed temperature variations of the principal lines for this field are:

$T(k)$	I_3/I_4 (5mW laser)	$e^{\Delta/kT}$
5	1.73	4.02
10	1.58	2.01
15	1.45	1.59

Agreement is reasonable at the higher temperatures. The discrepancy between the predicted and experimental ratios may simply reflect incomplete thermalization even between levels of the same symmetry. This is plausible since the overall lifetime of the I_1^{DEEP} BE luminescence at zero magnetic field is very short, $\tau_R \sim 1$ nsec,[78] and the condition for incomplete thermalization is $\tau_s \gg \tau_R$ where τ_s is the spin-lattice relaxation time.

References

1. P. J. Dean, H. Venghaus, J. C. Pfister, B. Schaub, and J. Marine, J. Luminescence, *16* 363 (1978).
2. J. C. Pfister, Rev. Phys. Appl. *15* 707 (1980).
3. G. F. Neumark, J. Appl. Phys. *51* 3383 (1980).
4. R. E. Dietz, H. Kamimura, M. D. Sturge, and A. Yariv, Phys. Rev. *132* 1559 (1963).
5. D. J. Robbins and P. J. Dean, Adv. in Phys. *27* 499 (1978).
6. D. J. Robbins, J. Luminescence *24/25* 137 (1981).
7. R. K. Swank, Phys. Rev. *153* 844 (1967).
8. L. Ley, R. A. Pollack, F. R. McFeely, S. P. Kowalczyk, and D. A. Shirley, Phys. Rev. B *9* 600 (1974).
9. J. C. Phillips "Bonds and Bands in Semiconductors" (Academic Press, London 1973).
10. L. A. Hemstreet, Phys. Rev. B *22* 4590 (1980).
11. G. A. Baraff, E. O. Kane, and M. Schluter, Phys. Rev. Lett. *47* 601 (1981); J. Bernholc, N. O. Lipari, S. T. Pantelides, and M. Scheffler, Phys. Rev. B *26* 5706 (1982).
12. I. J. Broser, R. K. F. Germer, H. J. E. Schulz, and K. P. Wisznewski, Solid State Electronics *21* 1597 (1978).
13. P. J. Dean, D. J. Robbins, S. G. Bishop, J. A. Savage, and P. Porteous, J. Phys. C. *14* 2847 (1981).
14. D. J. Robbins, D. C. Herbert, and P. J. Dean, ibid 2859 (1981).
15. A. M. White, P. J. Dean, and P. Porteous, J. Appl. Phys. *47* 3230 (1976).
16. J. S. Griffith, "The Theory of Transition Metal Ions," (Cambridge University Press, 1964).
17. D. G. Thomas, J. Phys. Chem. Solids *15* 86 (1960).
18. K. Shindo, A. Morita, and H. Kamimura, J. Phys. Soc. Japan *20* 2054 (1965).
19. J. T. Vallin and G. D. Watkins, Phys. Rev. B *9* 2051 (1974).
20. G. Muller, Phys. Stat. Sol. (b)*76* 525 (1976).
21. R. Dingle, Phys. Rev. Lett. *23* 579 (1969).
22. M. de Wit, Phys. Rev. *177* 441 (1969).
23. M. Wohlecke, J. Phys. C. *7* 2557 (1974).
24. H. D. Fair Jr., R. D. Ewing, and F. E. Williams, Phys. Rev. Lett. *15* 355 (1965).
25. S. Shionoya, J. Luminescence *1/2* 17 (1970).
26. J. L. Patel, J. J. Davies, J. E. Nicholls, and B. Lunn, J. Phys. C. *14* 4717 (1981).
27. I. Broser and K. H. Franke, J. Phys. Chem. Solids *26* 1013 (1965).
28. A. Suzuki and S. Shionoya, J. Phys. Soc. Japan *31* 1455, 1452 (1971).
29. J. J. Hopfield, J. Phys. Chem. Solids *10* 110 (1958).
30. M. Tabei and S. Shionoya, J. Luminescence *15* 201 (1977).
31. D. J. Dunstan, J. E. Nicholls, B. C. Cavenett, and J. J. Davies, J. Phys. C. *13* 6409 (1980).
32. P. J. Dean, Czech. J. Phys. *B.30* 272 (1980).
33. H. G. Grimmeiss, C. Ovren, W. Ludwig, and R. Mach, J. Appl. Phys. *48* 5122 (1977); H. G. Grimmeiss, C. Ovren, and R. Mach, *ibid 50* 6328 (1979). Photoconductivity spectra of ZnSe:Cu and final state effects in the excitation spectra of deep impurities are discussed by S. T. Pantelides and H. G. Grimmeiss, Solid State Comm. *35* 653 (1980). Bound states of holes and electrons excited from

the d-orbitals of transition metal impurities are discussed in References 6 and 14, and by D. J. Robbins, P. J. Dean, C. L. West, and W. Hayes, Phil. Trans. R. Soc. Lond. A *304* 499 (1982) for ZnSe:Co.

34. J. L. Patel, J. J. Davies, and J. E. Nicholls, J. Phys. C. *14* 5545 (1981).
35. P. J. Dean and D. C. Herbert "Excitons." ed. K. Cho, Chap. 3, (Bound excitons in semiconductors) (Springer, Berlin; 1979) p. 55.
36. M. Godlewski, W. E. Lamb, and B. C. Cavenett, Solid State Commun. *39* 595 (1981).
37. B. C. Cavenett, Adv. in Phys. *30* 475 (1981).
38. M. Gal, B. C. Cavenett, and P. J. Dean, J. Phys. C *14* 1507 (1981).
39. P. J. Dean, B. J. Fitzpatrick, and R. N. Bhargava, Phys. Rev. B *26* 2016 (1982).
40. R. N. Bhargava, Private communication (1982).
41. N. Magnea, D. Bensahel, J. L. Pautrat, and J. C. Pfister, Phys. Stat. Solidi *94* 627 (1979).
42. P. J. Dean, J. Luminescence *21* 75 (1979).
43. D. C. Herbert, P. J. Dean, H. Venghaus, and J. C. Pfister, J. Phys. C *11* 3641 (1978).
44. P. J. Dean, H. Venghaus, and P. E. Simmonds, Phys. Rev. *B18* 6813 (1978).
45. H. Tews, Phys. Rev. *B23* 587 (1981).
46. N. Killoran, B. C. Cavenett, and P. J. Dean, (unpublished work).
47. Y. Oka and M. Cardona, Phys. Rev. *B23* 4129 (1981).
48. M. Cardona, J. Phys. Chem. Solids *24* 1543 (1963).
49. J. M. Luttinger, Phys. Rev. *102* 1030 (1955); Y. Yafet and D. G. Thomas, Phys. Rev.
50. H. Venghaus and B. Jusserand (unpublished work).
51. J. L. Dessus, Le Si Dang, A. Nahami, and R. Romestain, Solid State Commun. *37* 689 (1981).
52. P. E. Simmonds, H. Venghaus, R. Sooryaleumar, and P. J. Dean, Solid State Comm. (to be published).
53. C. Hermann and C. Weisbuch, Phys. Rev. *B15* 823 (1977).
54. M. Cardona, private communication (1981).
55. D. J. Toms, J. F. Scott, and S. Nakashima, Phys. Rev. *B19* 928 (1979).
56. N. Magnea, J. L. Pautrat, K. Saminadayar, B. Pajot, P. Martin, and A. Bontemps, Rev. Phys. App. *15* 701 (1980).
57. P. J. Dean, D. C. Herbert, C. J. Werkhoven, B. J. Fitzpatrick, and R. N. Bhargava, Phys. Rev. *B,23* 4888 (1981).
58. W. Stutius, J. Cryst. Growth, *59* 1 (1982).
59. P. J. Dean and J. L. Merz, Phys. Rev. *178* 1310 (1969).
60. E. T. Handleman and D. G. Thomas, J. Phys. Chem. Solids *26* 1261 (1965).
61. F. A. Kroger, J. Phys. Chem. *69* 3367 (1965).
62. J. L. Merz, K. Nassau, and J. W. Shiever, Phys. Rev. *B8* 1444 (1972).
63. K. Kosai, H. G. Grimmeiss, B. J. Fitzpatrick, and R. N. Bhargava, Philips Labs. Tech. Rep. No 308 (1978) (unpublished).
64. A. M. White, I. Hinchliffe, P. J. Dean, and P. D. Greene, Solid State Comm. *10* 497 (1972); W. Schairer, D. Bimberg, W. Kottler, K. Cho, and M. Schmidt, Phys. Rev. *B,13* 3452 (1976).
65. V. I. Sokolov and V. L. Konstantinov, Solid State Commun, *33* 471 (1980).
66. J. J. Hopfield, Proc. Int. Conf. Semicond., Paris, 1964. (Dunod, Paris, 1964) p. 725.

67. A. M. White, P. J. Dean, and B. Day, Proc. Int. Conf. on Physics of Semicond., Rome 1976, p. 1057.
68. D. C. Herbert, J. Phys. C *10* 3327 (1977).
69. G. F. Koster, J. O. Dimmock, R. G. Wheeler. and H. Statz, "Properties of the Thirty-two Point Groups" (Cambridge, Mass: MIT Press, 1963).
70. D. J. Robbins and P. J. Dean (unpublished work).
71. Throughout Section 5 we use the character tables and coupling coefficients listed in Reference (69).
72. A. M. White, J. Phys. C *6* 1971 (1973).
73. E. U. Condon and G. H. Shortley, "The Theory of Atomic Spectra," (Cambridge University Press, 1957).
74. A. K. Bhattacharjee and S. Rodriguez, Phys. Rev. *B,6* 3836 (1972).
75. H. Venghaus, Phys. Rev. *B,19* 3071 (1979).
76. D. W. Langer, R. N. Euwema, K. Era, and T. Koda, Phys. Rev. *B2* 4005 (1970).
77. C. A. Coulson, "Valence," (Oxford University Press, 1952).
78. J. O. Williams (private communication).
79. P. J. Dean (unpublished work).

Index

A center 13, 46

acoustic paramagnetic resonance 549ff

antisite defect 45, 47, 383, 415, 435, 479

Arrhenius relation 18

athermal migration 13, 68

Auger transitions 62, 67, 223, 234, 279, 322, 352, 675

bound excitons 36, 63, 194ff, 719ff

Bourgoin-Corbett mechanism 18, 68

carrier capture 65, 98, 222, 265ff, 359, 514, 659

cascade capture 63, 98, 128, 373

CdS:Cr 691ff

CdTe:Cr 691ff

charge-state effects 65

cluster calculations 7, 14, 42
CNDO 39, 42
conductivity 3
configuration coordinate 59, 261, 273, 458ff, 463, 494ff, 709
continued-fraction method 9

dangling bonds 7, 13, 38, 48, 151, 310, 322, 727
deep level transient scpectroscopy (DLTS) 5, 44, 54, 103, 116, 126, 163, 242, 258ff, 362, 403,432, 507ff, 535, 573, 660
defect identification 43ff, 101, 151ff, 382ff, 391ff, 399ff
defect migration 13
defect molecule 7, 175
density functional theory 11
diffusion 17, 74, 180, 433
divacancy 12, 46, 165
donor-acceptor pairs 188, 730ff
DX centers 61, 489ff

effective-mass theory 6, 26, 88, 106
electron nuclear double resonance (ENDOR) 3, 142, 151, 558
electron paramagnetic resonance (EPR) 3ff, 12ff, 30, 46, 102, 111, 127, 141ff, 151, 382, 436, 545ff, 634ff, 692ff, 719
EL2 36, 45, 47, 333, 399ff
excited states 35ff, 94ff, 220, 301, 314, 722
extended x-ray absorption fine structure (EXAFS) 6
extended Huckel theory (EHT) 7, 14, 39, 50

formation energy 17, 72
formation entropy 17
Franck-Condon shift 59, 200, 455, 478

GaAs:Cr 49, 414, 627ff
GaAs:Fe 541ff

GaAs:O 45, 333, 401ff
GaAsP:O 41
GaP:Be 246
GaP:Cu 43, 253
GaP:Fe 541ff
GaP:Mg 246
GaP:Ni 391ff
GaP:O 16, 43, 44, 185ff
GaP:Zn 245
Green's function 6ff, 20, 25, 27, 31, 39, 42, 48, 78, 95, 176, 294

Hall effect 3, 125, 160, 648, 669
Huang-Rhys factor 495, 642
Hund's rule 14, 96, 154
hyperfine interactions 14, 102, 141, 734

InAs:O 335
InP:Fe 541ff
infrared absorption 3
internal transitions 35, 691ff
interstitialcy 17, 71, 75
irradiation 3, 422, 437

Jahn-Teller effect 13, 33, 48, 50, 73, 154, 178ff, 563, 634, 677, 696
junction techniques 4, 20, 107, 242, 505, 659

kick-out mechanism 75
killer centers 45, 48

lattice relaxation 49ff, 274, 492ff, 709
luminescence 3ff, 188ff, 243ff, 467, 543, 637, 672ff, 705, 730ff

many-body effects 13, 31, 176ff

metastable state 287, 311, 441, 461

migration energy 17, 167

migration entropy 17

MNDO 39, 42, 51

Mossbauer spectroscopy 593

multiphonon capture 63, 98, 279, 372, 493

multiplets 32, 38

muon spin resonance 35

negative U 53, 71, 161, 490

non-radiative transitions 7, 62, 352ff

nuclear spin 4

optical absorption 3, 108, 119, 123, 230, 255, 432, 465, 547ff, 640ff, 696ff

optical detection of magnetic resonance (ODMR) 5, 47, 320, 576, 719, 732

persistent photoconductivity 57, 491

phosphorus-vacancy pair 66, 164

photoconductivity 3, 57, 585, 652

photodetectors 4

pinning 25, 41, 77

process-induced defects 76

pseudopotentials 11

radiation-induced defects 12ff, 20

Raman scattering 3

recombination enhanced motion 65, 70, 74, 167, 211

self-diffusion 17, 74, 180

self-interaction correction 38

self-interstitials 13, 17ff, 28, 58, 68, 75, 149

self-trapping 57, 62, 491, 503ff

Si:Ag 16

Si:Al 13, 65, 73, 532
Si:Au 16, 44
Si:Fe 14, 73
Si:H 35
Si:Mn 14
Si:N 50
Si:O 13, 16, 50, 87ff
Si:S 16, 28, 87ff
Si:Se 87ff
Si:Te 87ff
Si:Zn 16, 29
spin polarization 31
supercell method 8, 12
surfaces 12

thermal activation energy 3
tight-binding method 9, 25, 40, 328
transition metal impurities 14ff, 29ff, 48ff, 326, 628

vacancy 7, 12, 14, 17, 22ff, 38, 42, 48, 147ff, 203, 384, 629, 726
vibrational modes 10, 42, 201

ZnO:Cu 717
ZnS:S 691ff
ZnS:Cu 717ff
ZnSe:Cr 691ff
ZnSe:Cu 36, 717ff
ZnTe:Cr 691ff
ZnTe:Cu 717ff

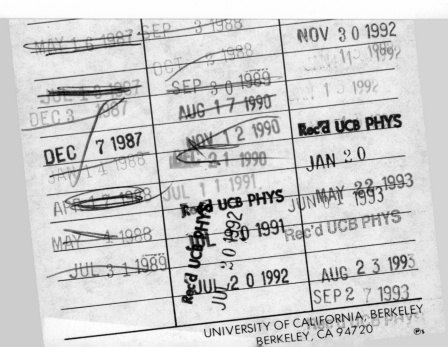